LACEWINGS IN THE CROP ENVIRONMENT

Lacewings are predatory insects which attack and kill large numbers of insect pests. *Lacewings in the Crop Environment* addresses both the theoretical and practical aspects of lacewing biology and their use in crop protection.

The book opens with a section on lacewing systematics and ecology. Next, the role of lacewings as predators in a wide variety of commercially important crops is reviewed and this is followed by a section on the principles of using lacewings in pest control. The possible impact of genetically modified crops on lacewing populations is also discussed. Finally, a fascinating array of case studies of lacewing use in many crops from around the world is presented, and future uses of lacewings speculated upon.

Lacewings in the Crop Environment is an essential reference work and practical handbook for students and researchers of biological control, integrated pest management and agricultural science, and for field workers using lacewings in pest management programmes worldwide.

PETER MCEWEN is Managing Director at Insect Investigations Ltd, a product-testing and development laboratory in Cardiff, Wales. He has written over 70 academic publications on the biology of lacewings, triatomine bugs, leafhoppers, and ecotoxicology.

TIM NEW is Reader and Associate Professor in Zoology at La Trobe University, Victoria, Australia. He is an entomologist with wide interests in systematics, ecology, and conservation of insects and has published extensively on these subjects.

ANDREW WHITTINGTON is Curator of Entomology at the National Museums of Scotland, Edinburgh. His major research interests include the taxonomy of Neuroptera and the Diptera families Syrphidae and Platystomatidae.

LACEWINGS
in the Crop Environment

Edited by:

P.K. McEwen
Cardiff University

T.R. New
La Trobe University

A.E. Whittington
National Museums of Scotland

CAMBRIDGE
UNIVERSITY PRESS

CAMBRIDGE UNIVERSITY PRESS
Cambridge, New York, Melbourne, Madrid, Cape Town, Singapore, São Paulo

Cambridge University Press
The Edinburgh Building, Cambridge CB2 8RU, UK

Published in the United States of America by Cambridge University Press, New York

www.cambridge.org
Information on this title: www.cambridge.org/9780521772174

First published 2001
This digitally printed version 2007

A catalogue record for this publication is available from the British Library

Library of Congress Cataloguing in Publication data

Lacewings in the crop environment / edited by Peter McEwen, Tim New.
  p. cm.
ISBN 0 521 77217 6 (hardback)
1. Lacewings. 2. Agricultural pests – Biological control. I. McEwen, Peter (Peter K.) II.
New, T. R.
SF562.L33 L23 2001
632′.96–dc21 00–045533

ISBN 978-0-521-77217-4 hardback
ISBN 978-0-521-03729-7 paperback

Contents

Contributors

Gilberto S Albuquerque
State University of North Fluminense,
Plant Protection Laboratory,
Avenida Alberto Lamego 2000,
28015–620 Campos, RJ,
Brazil

Alberto Alma
Entomologia e Zoologia applicate all'Ambiente 'Carlo
    Vidano',
via Leonardo da Vinci 44,
10095
Grugliasco, TO,
Italy

Franz Bigler
Eidgenössische Forschungsanstalt für Agrarökölogie
    und Landbau,
Reckenholzstrasse 191,
CH-8046 Zürich,
Switzerland

András Bozsik
Department of Plant Protection,
University of Agricultural Sciences,
Debrecen,
Boszormenyi utca 138,
H-4032,
Hungary

Stephen J. Brooks
Department of Entomology,
Natural History Museum,
Cromwell Road,
London SW7 5BD,
UK

Mercedes Campos
Consejo Superior de Investigaciones Cientificas,
Estacion Experimental del Zaidin,
Profesor Albareda 1,
18008 Granada,
Spain

Michel Canard
Laboratoire d'Entomologie,
Université Paul-Sabatier,
118 route de Narbonne,
F-31062 Toulouse,
France

Allen C. Cohen
Biological Control and Mass Rearing Research Unit,
PO Box 5367,
Mississippi State University,
MS 39762–5367,
USA

Kent M. Daane
Center for Biological Control,
Division of Insect Biology,
Department of Environmental Science, Policy and
    Management,
University of California,
Berkeley, CA 94720,
USA

Gavino Delrio
Dipartimento di Protezione delle Piante,
Sezione di Entomologia,
via Enrico De Nicola,
07100 Sassari, SS,
Italy

Luisa M. Díaz-Aranda
Departemento Biología Animal,
E-28871 Alcala de Henares,
Madrid,
Spain

Peter Duelli
Swiss Federal Institute for Forest, Snow and
    Landscape Research,
CH-8903 Birmensdorf,
Switzerland

The Late **Kenneth S. Hagen**
Center for Biological Control,
Division of Insect Biology,
Department of Environmental Science, Policy and
    Management,
University of California,
Berkeley, CA 94720,
USA

Charles S. Henry
Department of Ecology and Evolutionary Biology,
PO Box U-43, 75 North Eagleville Road,
University of Connecticut,
Storrs, CT 06269,
USA

Angelika Hilbeck
Swiss Federal Institute of Technology Zurich (ETH),
Geobotanical Institute,
Zuerichbergstrasse 38,
CH-8044 Zürich,
Switzerland

Paul A. Horne
IPM Technologies P/L,
PO Box 560,
Hurstbridge, VIC 3099
Australia

James B. (Ding) Johnson
Division of Entomology,
Department PSES,
University of Idaho,
Moscow, ID 83844–2339,
USA

Andrea Lentini
Dipartimento di Protezione delle Piante,
Sezione di Entomologia,
via Enrico De Nicola,
07100 Sassari, SS,
Italy

J. C. Maisonneuve
SRPV,
14 rue du Colonel Berthaud,
29283 Brest,
Cedex France

F. Marín
Departamento de Biología Animal,
Universidad de Alcala de Henares,
E-28871 Alcala de Henares,
Madrid,
Spain

Peter K. McEwen
Insect Investigations Ltd,
Units 10–12,
CBTC 2,
Off Parkway,
Capital Business Park,
Wentloog,
Cardiff, CF3 2PX,
Wales, UK

Víctor J. Monserrat
Departamento de Biología Animal I,
Catedra de Entomologia,
Universidad de Complutense,
E-28040 Madrid,
Spain

Marco Mosti
Bioplanet – Sistemi e tecnologie agroambientali
Via Masiera I, 1195
I-47020 Martorano-Cesena, FC,
Italy

T.R. New
Department of Zoology,
La Trobe University,
Bundoora, VIC 3083,
Australia

Donald A. Nordlund
Biological Control and Mass Rearing Research Unit,
PO Box 5367,
Mississippi State University,
MS 39762–5367
USA

John D. Oswald
Department of Entomology,
Texas A&M University,
College Station, TX 77843–2475,
USA

Roberto A. Pantaleoni
Istituto di Entomologia agraria,
Università degli Studi,
via Enrico De Nicola,
07100 Sassari, SS,
Italy

D. Papacek
Bugs for Bugs (Integrated
Pest Management P/L)
28 Orton Street
Mundubbera, QLD 4626
Australia

Mihaela Paulian
Research Institute for Plant Protection,
Bd Ion Ionescu de la Brad No. 8,
Sector 1,
RO – 71592,
Bucuresti,
Romania

Peter M. Ridland
Institute for Horticultural Development
Private Bag 15
South Eastern Mail Centre, VIC 3176,
Australia

Çetin Sengonca
Rheinische Friedrich-Wilhelms-Universität Bonn,
Institut für Pflanzenschutz,
Abteilung Entomologie und Pflanzenschutz,
Nussallee 9,
D-53115 Bonn,
Germany

Lara J. Senior
Insect Investigations Ltd,
Units 10–12,
CBTC 2,
Off Parkway,
Capital Business Park
Wentloog,
Cardiff, CF3 2PX,
Wales, UK

Rebecca A. Smith
Biological Control and Mass Rearing Research Unit,
PO Box 5367,
Mississippi State University,
MS 39762
USA

Ferenc Szentkirályi
Plant Protection Institute,
Hungarian Academy of Sciences,
Herman Otto utca 15
PO Box 102,
H-1525 Budapest,
Hungary

Catherine A. Tauber
Department of Entomology,
Comstock Hall,
Cornell University,
Ithaca, NY 14853–0901
USA

Maurice J. Tauber
Department of Entomology,
Comstock Hall,
Cornell University,
Ithaca, NY 14853–0901,
USA

Dominique Thierry
35 rue Eblé,
F49000 Angers,
France

M. Grazia Tommasini
CREA – Centre Research Environmental Agriculture,
Centrale Ortofrutticola,
Via Masiera I, 1191
I-47020 Martorano-Cesena, FC,
Italy

Elisa Viñuela
Unidad de Protección de Cultivos,
ETSI Agrónomos,
Universidad Politecnica de Madrid,
E-28040 Madrid,
Spain

Heidrun Vogt
Biologische Bundesanstalt für Land- und
    Forstwirtschaft,
Institut für Pflanzenschutz im Ostbau,
Schwabenheimer Strasse 101,
D-69221 Dossenheim,
Germany

Tatiana A. Volkovich
Biological Institute,
St Petersburg University,
Orianienbaumskoye Sh. 2,
RU-198904,
St Petersburg,
Russia

Andrew E. Whittington
National Museums of Scotland
Chambers Street,
Edinburgh EH1 IJF,
Scotland,
UK

I.G. Yazlovetsky
Institute of Biological Plant Protection,
blvd Dacia 58,
2060 MD, Kishinev,
Moldova

Preface

Lacewings in the Crop Environment is the first book to seriously address the lacewings since *Biology of the Chrysopidae* (Canard *et al.*, 1984). In the intervening period a number of important developments have been made in the field and this new book is thus greatly overdue.

The book addresses both the theoretical and practical aspects of lacewing biology and their use in crop protection. It provides a summary of many recent developments and information which has hitherto been scattered in the primary scientific literature, and some up-to-date examples of lacewings as biological control agents in a variety of crop contexts.

The book is divided into five parts. Part 1 deals with lacewing systematics and ecology and provides a background to the subject area. The taxonomic nomenclature used in this book was frozen at July 2000. Thus, any taxonomic changes or advances made after that point are not included in the text. For examples, because of the past widespread use of '*Chrysopa*' as a holding genus, it is almost inevitable that some names used in this book may reflect older practice. We have attempted to standardise these, by using the generic combinations included in the revision by Brooks & Barnard (1990), but some ambiguities and misidentifications may persist. We have not attempted to differentiate formally between use of *Mallada* and *Dichochrysa*, as a number of the species properly referable to the latter have not been critically appraised or transferred formally from *Mallada*. In such cases, we have followed the generic allocation adopted by individual authors. Likewise, use of the epithet '*Chrysoperla carnea*' should not be taken necessarily to exclude other members of this complex. Chapter 3 deals specifically with the sibling species question in the *Chrysoperla carnea* species complex. Progress in elucidating this complex has enormous implications for the successful use of lacewings in crop protection. Delimitation of sibling species is an area where further rapid developments can be expected, and Chapter 3 provides both a snapshot of the current state of thinking and a useful introduction to this important area of insect biology.

Part 2, 'Lacewings in crops', provides reviews of lacewings in a large number of crops and vegetation associations and suggests which species might be significant predators in these environments. It is intended as a useful reference chapter for students of biological control and integrated pest management who use lacewings, but also as a source for field workers wondering what is known about lacewings in their own particular working environment. Thus, biologists in crop environments as varied as tea plantations in India and apple orchards in Europe will find something of value.

Part 3 introduces the principles of using lacewings in pest control. Biological control, mass-rearing techniques, and sampling methodology exemplify areas of direct practical use. The importance of integrated pest management is emphasised, and interactions between lacewings and pesticides are covered in some detail. Traditional pesticides are covered as well as the so-called 'soft' pesticides such as insect growth regulators which have on occasion been claimed to be compatible with use of natural enemies. Possible consequences of genetic modification of crops in terms of their impact on natural enemy populations are also discussed.

Part 4 extends these principles to practice in presenting case studies on the use of lace-wings in crop protection from around the world. Cases range from using highly defined protocols, as in parts of Europe and North America, to more preliminary trials which open up some fascinating potentials for future development. Presentations vary from work on the interactions between green lacewings and leafhoppers in California, strawberry pest control in Italy, and control of artichoke pests in France. Combined with contributions from Australia, Spain, Romania and elsewhere, this section demonstrates just how important lace-wings are in crop protection in many parts of the world.

The final section, Part 5, considers the future of lacewings in the crop environment. Very few species of lacewings are actively used in crop protection at present and we can expect to see more work on additional species, to increase the spectrum of predators for various contexts. This in turn will require a more detailed knowledge of lacewing biology to underpin rational selection of optimal taxa for use against an increasing variety of pest arthropods. As the world increasingly moves towards agricultural systems that strive to reduce pesticide use, biological control and integrated pest management are likely to become increasingly important options. We can expect lacewings, as widespread predators, to play increasingly important crop protection roles as we enter the 21st century.

We have deliberately not deterred some level of overlap and repetition in different chap-ters. The different perspectives and contexts in which some basic biological knowledge is reiterated, in our opinion, are complementary. They help to emphasise how our limited understanding of lacewings has been applied in many extrapolations and interpretations for use in some very practical contexts. Likewise, many authors reiterate their confusion over identities of one notable taxon, *Chrysoperla carnea s. lat.*; although this is addressed in detail in Chapter 3, knowledge of the different practical applications of the name is necessary for readers to be able to assess and compare the numerous studies which involve members of this complex.

Any book such as this is a 'team effort'. We are deeply grateful to the many authors who have contributed so willingly to this collective enterprise, many of them meeting our limita-tions of time and length ably, and tolerating our editorial interference with remarkable for-bearance. We also thank Ward Cooper and his staff at Cambridge University Press, for their enthusiastic support for this venture.

The Editors
July 2000

REFERENCES

Brooks, S.J. & Barnard, P.C. (1990) The green lacewings of the world: a generic review (Neuroptera: Chrysopidae). *Bulletin of the British Museum (Natural History) Entomology* 59, 117–286.

Canard, M., Séméria, Y. & New, T.R. (eds.) (1984) *Biology of Chrysopidae*, Series Entomologica, vol. 27, ed. Spencer, K.A. Dr W. Junk, The Hague.

PART 1

Lacewing systematics and ecology

CHAPTER 1

Introduction to the Neuroptera: what are they and how do they operate?

T.R. New

1.1 INTRODUCTION

Neuroptera (also known as Planipennia), the lacewings, antlions, and their allies, are one of the smaller and more primitive orders of holometabolous insects. About 6000 species are known. They are commonly linked with the Megaloptera (alderflies, dobsonflies) and Raphidioptera (snakeflies) in the superorder Neuropteroidea (or Neuropterida), and are of world-wide occurrence, although many taxa are more restricted and some are characteristic faunal elements of particular regions. Neuroptera have a long evolutionary record, and there are many fossils from the Permian and Jurassic which are referable clearly to this order (Carpenter, 1992), mainly on details of the complex wing venation which gives lacewings their common name. Most are terrestrial, although a few (such as the specialised Sisyridae, the spongeflies, whose larvae are predators of freshwater sponges, and Nevrorthidae) frequent freshwater environments. Adults and larvae of most families are predatory, and the main interest of the order to many entomologists is that which forms the major theme of this book – that some lacewings can be of considerable value as manipulable predators for use in biological control programmes. This brief introduction is a broader survey of the Neuroptera, to facilitate a more general perspective of the order. Thus, whereas this book is about members of only three families (Coniopterygidae, Hemerobiidae, Chrysopidae with the last by far predominant), up to 18 Recent families are recognised, and a number of others are extinct. The phylogenetic relationships between some of these are still debated.

Adults and larvae are both diverse in appearance, but the larvae in particular suggest monophyly of the families by the presumed apomorphic states of (1) the mandible and maxilla of each side linked to form a suctorial tube for ingestion of liquid or semiliquid food, (2)

the junction between the mid and hindgut blocked, so that larval food remains can be egested only at the end of larval life, and (3) a single-segmented tarsus. Adult unifying apomorphies are elusive, though Meinander (1972) noted that the head is usually hypognathous, unlike the orthognathous condition of the putatively closely related Megaloptera.

1.2 DEFINITION

Neuroptera may be defined broadly as follows: Minute to large holometabolous insects, oviparous, bisexual, with an exarate decticous pupa. Adults: head orthognathous to hypognathous, mouthparts chewing, with mandibles usually strong; antennae usually filiform or moniliform, rarely pectinate or clubbed; compound eyes large; ocelli usually absent. Prothorax short, except in forms with raptorial fore legs (Mantispidae, some Berothidae), legs usually slender and cursorial, or short and sturdy; usually two pairs of large membranous wings, often patterned and with venation complex; major veins often with extensive 'twigging' near wing margins, trichosors often present. Abdomen ten-segmented (apparently nine-segmented in Chrysopidae due to fusion), female ovipositor usually not exserted (exceptions are Dilaridae and some Mantispidae); cerci absent, cercal callus usually evident as a flattened disc, with trichobothria. Larvae: usually active, campodeiform, occasionally scarabaeiform (Ithonidae) or vermiform (Mantispidae); prognathous, mouthparts strongly projecting with mandible and maxilla on each side strongly associated to form a 'sucking tube', maxillary palpi absent; tarsi one-segmented.

1.3 CLASSIFICATION

The phylogenetic classification reflected in Fig. 1.1 (Aspöck, 1992) differs from previously published ones

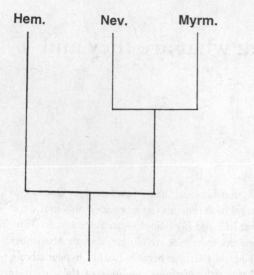

Fig. 1.1. Suggested phylogeny of major lineages of Neuroptera. Hem., Hemerobiiformia; Nev., Nevrorthidae; Myrm., Myrmeleontiformia. (After Aspöck, 1992.)

(for which Withycombe's 1925 scheme has been widely used as a basis) in suggesting the very isolated nature of the Nevrorthidae, based on the form of the larval gula. Nevertheless, several major associations of families (broadly, superfamilies) are evident. Following U. Aspöck, the order probably comprises three main monophyletic groups, Nevrorthidae, Hemerobiiformia, and Myrmeleontiformia.

Nevrorthidae constitutes a very distinct lineage, separated from all other families on the form of the larval gula. Hemerobiiformia (to which all families treated in this book belong) includes up to 12 families, and Myrmeleontiformia, the remaining five families. In earlier schemes, Nevrorthidae (as Neurorthidae) had generally been allied with Sisyridae and Osmylidae, and the adults resemble those of these families to a far greater extent than do the very unusual and characteristic larvae.

Hemerobiiformia contain a number of lineage groupings. Traditionally, Coniopterygidae has been treated as an early-evolved and isolated group, and Hemerobiidae and Chrysopidae as closely related families. However, U. Aspöck (1992) suggested that their close alliance had been assumed uncritically, on the basis of such characters as the curved larval jaws, and was unable to find any convincing synapomorphies for larvae of the two families: she regarded their curved

jaws as plesiomorphic. The other families in this series are Ithonidae, Rapismatidae (the only family for which early stages are entirely unknown, and sometimes included in Ithonidae), Polystoechotidae, Osmylidae, Sisyridae, Dilaridae, Mantispidae, and Berothidae, with the Rhachiberothidae sometimes separated from the last-named rather than included as the subfamily Rhachiberothinae.

The Myrmeleontiformia, antlions and their allies, include many of the largest and most spectacular Neuroptera. Myrmeleontidae (antlions) are one of the largest and most widely distributed families of Neuroptera, and the large strongly-flying Ascalaphidae (owlflies) are vigorous aerial predators, sometimes likened to Anisoptera but separable by their long clubbed antennae. The other members of this series are Nymphidae, Nemopteridae, and Psychopsidae.

The recent illustrated essay by Aspöck & Aspöck (1999) provides fuller background to systematic arrangements within the order.

1.4 BIOLOGY

There is considerable biological complexity and variety in the order. However, it must be emphasised that biological knowledge of most families is fragmentary, and many of the inferences in papers and texts are based on detailed knowledge of only very few species which, perhaps misleadingly, are presumed to be representative of entire larger groups. Thus, for example, relatively few antlion species have larvae which construct the conical pits for which the group is famous. However, the following generalities seem to apply (partly from New, 1986, 1989):

1. Eggs are laid singly or in batches, with the form of deposition and the oviposition site sometimes characteristic for particular taxa. Eggs of some families (Chrysopidae, Nymphidae, Mantispidae, Berothidae) are commonly stalked. Eggs are often laid generally on vegetation or other substrate, but are sometimes more closely associated with supplies of larval food.

2. Larvae of most families are free-living predators. Many are generalist feeders and take a wide variety of small arthropod prey, but others are more specialised. Exceptions to this are the triungulin first instars of Mantispidae, which seek out hosts (spiders, spider egg cases, social wasps) after hatch-

ing from very large batches, and Ithonidae, whose subterranean larvae may feed on decaying vegetation – a very unusual habit in the Neuroptera. Three larval instars are usual in most families, although five may occur in Ithonidae, and supernumerary instars are occasionally reported to occur under stressed rearing conditions.

3. Pupation takes place within a silken cocoon. Pupae are decticous and exarate, and both the form of the cocoon and the site of pupation can be characteristic in particular taxa.

4. Developmental time ranges from a few weeks to a few years, with some larger myrmeleontids taking at least three years to mature. Most species are at least univoltine, and diapause or some equivalent period of dormancy is common. Most lacewings in temperate regions overwinter as a prepupa within the cocoon, but a few diapause as adults or some other stage. Some members of the Coniopterygidae, Hemerobiidae, and Chrysopidae are the only lacewings which are known regularly to undergo two or more generations each year, with such rapid development enhancing their values as responsive predators in biological control.

5. Adults are mostly fully winged (brachyptery occurs in a few species) and are generally capable of flight. Some species are highly vagile, whereas others seem to disperse little. Many lacewings are characteristically crepuscular or nocturnal (with many attracted to lights) but some brightly coloured Nemopteridae and Myrmeleontidae are diurnal. Sex ratios are generally close to unity.

6. Most species are predators as adults. However, some Sisyridae and Osmylidae take animal and plant food, and adults of many Chrysopidae are specialised honeydew or pollen feeders. There are few detailed studies on feeding requirements of many families: recent studies by Stelzl (1990) have suggested previously unsuspected breadth of diet in some Hemerobiidae, for example, and the critical importance of understanding food needs for manipulating species is a recurring theme in this book.

Whereas many Neuroptera, in addition to Raphidioptera, have been inferred to have relevance in the biological control of a variety of arthropod pests, the spectrum of taxa proven to have such importance is very limited (New, 1999). Although, for example, the abundance of antlions in some arid/semiarid regions suggests strongly their importance as predators in natural communities, this capability is not equivalent to that of the more rapidly breeding, fecund and widespread taxa associated primarily with vegetation in more mesic environments. It is there that the needs for integrated pest management (IPM) on crops are greatest and most diverse, and where pre-adapted predators can be enhanced to increase their impacts in cropping systems of many kinds. The values of such lacewings in IPM are the central topic explored in this book

REFERENCES

Aspöck, U. (1992) Crucial points in the phylogeny of the Neuroptera (Insecta). In *Current Research in Neuropterology, Proceedings of the 4th International Symposium on Neuropterology*, ed. Canard, M. Aspöck, H. & Mansell, M.W., pp. 63–73. Sacco, Toulouse.

Aspöck, U. & Aspöck, H. (1999) Kamelhalse, Schlammfliegen, Ameisenlowen – wer sind sie? (Insecta: Neuropterida: Raphidioptera, Megaloptera, Neuroptera). *Stapfia* 60, 1–34.

Carpenter, F.M. (1992) *Superclass Hexapoda. Treatise on Invertebrate Palaeontology. Part R, Arthropoda 4*, vols. 3 and 4. Geological Society of America, Boulder CO and University of Kansas, Lawrence KA.

Meinander, M. (1972) A revision of the family Coniopterygidae (Planipennia). *Acta Zoologica Fennica* 136, 1–357.

New, T.R. (1986) A review of the biology of the Neuroptera Planipennia. *Neuroptera International* Supplement 1.

New, T.R. (1989) *Planipennia. Lacewings*. Handbuch der Zoologie, ed. Fischer, M., vol. 4, part 30. Walter de Gruyter, Berlin.

New, T.R. (1999) Neuroptera and biological control. *Stapfia* 60, 147–66.

Stelzl, M. (1990) Nahrungsanalytische Untersuchungen an Hemerobiiden–Imagines (Insecta: Planipennia). *Mitteilungen der deutschen Gesellschaft fur allgemeine und angewandte Entomologie* 7, 670–6.

Withycombe, C.L. (1925) [1924] Some aspects of the biology and morphology of the Neuroptera, with special reference to the immature stages and their possible phylogenetic significance. *Transactions of the Entomological Society of London* (1924), 303–411.

CHAPTER 2

Introduction to the systematics and distribution of Coniopterygidae, Hemerobiidae, and Chrysopidae used in pest management

T.R. New

2.1 INTRODUCTION

Not altogether coincidentally, the three families of Neuroptera discussed in this book are those on which most biological and taxonomic information and understanding is available. They are the most widespread, predominant, and diverse families in the northern temperate regions, where the foundations of early knowledge on the order were laid, and the use of lacewings in pest control was pioneered. All are more widespread, and all have recently been subject to comprehensive global review and reappraisal, so that much of the early scattered information has been brought together, re-evaluated and synthesised, keys to generic level produced, and some constructive comment on their diversity, evolution, and biogeography is feasible. However, biological information on most species, and many genera in other parts of the world, is not available or is fragmentary. Although some inferences can be made from the species already studied, the considerable biological variability within each family suggests that extrapolation from the few well-studied species in each should be cautious. Likewise, despite the impetus provided by recent overviews, many new species await discovery and description, and the minefields of over-simplified treatment of complex species groups are exemplified well in the following chapter on *Chrysoperla*, a single genus of Chrysopidae (Henry *et al.*, this volume).

This chapter is a broad introduction to the systematics and biogeography of the Coniopterygidae, Hemerobiidae, and Chrysopidae, to provide a perspective for the remainder of this book, and to help readers appreciate the complexity of the groups considered – as well as the opportunities they may furnish for study and manipulation in the future. Biological knowledge of the three families is highly uneven, with the predominant interest in Chrysopidae clearly evident. Hemerobiidae have attracted attention in applied contexts in several parts of the world. Despite early advocacy for the use of Coniopterygidae in biological control (Withycombe, 1924), their current interest is relatively minor. Nevertheless, they are included here because current attempts to foster the benefits from a wide variety of native predators in pest control in an increasing variety of contexts are likely to see their values increase. Penny *et al.* (1997) suggested that Coniopterygidae are indeed important predators but usually overlooked because of their small size, and Tauber & Adams (1990) noted that studies on this family are 'vital, given the importance of the group in the biological control of insect pests'. Some other recent commentators (such as Lo Verde & Monserrat, 1997) recapitulate sequences of prey records for particular species, and additional records continue to accumulate, mainly with little quantitative appraisal. However, similar inferential comments on the other families are also widespread, and are based sometimes on observations of increased abundance of particular species in disturbed or agricultural ecosystems compared to nearby native habitats. Most of the lacewing species which have been noted as useful in biological control are simply those which have been studied or found in association (however tentative) with pest arthropods. Many others may be equally useful once they are evaluated more effectively.

This chapter, however, is not a complete taxonomic account of the three families. Rather, I attempt to introduce their main features and diversity, and provide information to help in identifying the taxa that have been implicated or mentioned in biological control literature. Other species will undoubtedly be found in association with pests, but many of the genera most likely to be encountered are noted in the following account.

It must be emphasised that species-level identification of Neuroptera is often difficult, and that specialist advice should be sought in all contexts where precision in species identifications are needed, including all cases of predator evaluation in the field or laboratory. It can prove very misleading to rely on 'approximate' or unconfirmed names, because closely related taxa can differ greatly in their biology. Even in relatively well-known faunas, some species complexes are difficult to separate clearly and consistently, and it is easy to err even when using the most up-to-date literature available. Specimens submitted for identification should be mature (that is, if reared they should be allowed to 'harden' rather than killed whilst teneral) and in good condition. Coniopterygidae should be preserved in alcohol (although additional carded specimens of highly marked species are also valuable); Hemerobiidae and Chrysopidae may be preserved either in alcohol or as pinned/staged specimens with the wings of one side (at least) spread to facilitate examination of venation. Chrysopidae fade badly in alcohol, and good set specimens are preferred by many workers. As the following keys demonstrate, wing venation is very important in initial characterisation of many lacewings, and this must therefore be displayed clearly wherever possible. For examination of Coniopterygidae, it is often necessary to make a slide preparation of the wings of one side of the specimen, because veins are often pale and not easily differentiated from the rest of the wing at low-power observation.

The other key to succesful identification of lacewings is genitalic examination, which necessitates critical treatment and dissection of the abdominal apex. The structures present are often difficult to homologise between families and are noted under the individual family headings. General accounts of the genitalia of the order are provided by, *inter alios*, Acker (1960), Adams (1969), and Tjeder (1970). Examination procedures are largely common to all. After maceration with dilute potassium hydroxide or sodium hydroxide, the apex of the abdomen can be dissected, slide-mounted or stored in small vials associated with the donor specimen. For the small and delicate Coniopterygidae, Meinander (1972) recomended boiling the abdomen in 5% potassium hydroxide for 3 minutes, then passing it through water and alcohol to glycerol for examination before mounting it in euparal on a slide. Other workers recommend less rapid treatment, such as maceration in cold 10% potassium hydroxide overnight (Oswald, 1993, for Hemerobiidae) with subsequent storage, after washing and dehydration, in glycerine. For critical examination, such as discerning fine details of the female structures of Chrysopidae, staining for 1–2 minutes in saturated 70% ethyl alcohol with chlorazol black, then washing in 70% alcohol, is useful. To see such fine structures essentially *in situ*, Adams (1969) recommended removing one side of the abdominal segments carefully, so that the structures are lying in 'the half-shell'. Such dissection must be done carefully; as diagrams in this account show, both the form of the apical sclerites and details of the internal genitalia have diagnostic values, and may be an integral part of identifying genera and species.

Problems of interpretation are enhanced for Chrysopidae, in particular, by the complexity of the male genitalic structures and the variety of these which may or may not be present in different combinations. Because of the difficulties of examining some small structures, it is important not to over-clear these.

2.2 CONIOPTERYGIDAE BURMEISTER

An apparently isolated lineage within the Neuroptera, Coniopterygidae differ considerably from all other lacewings in appearance. They occur throughout the world. Their distinguishing features (Meinander, 1972) include their very small size (fore wing length up to about 5 mm, usually much less), the body and wings and (often) legs covered with white or greyish wax from specialised hypodermal glands (a feature which gives the insects their common name of 'dusty wings'), wing venation strongly reduced with costal area narrow and with two or fewer unforked costal crossveins; few other crossveins (except in the unusual Brucheiserinae), and – more technically – the unusual genitalic structures. Larvae have a protruding labium and straight, forward-projecting jaws (those of Brucheiserinae are unknown). Fossil forms occur from the Jurassic onward (Meinander, 1972; Carpenter, 1992), and the older ones can be allocated clearly to one major subfamily, Aleuropteryginae.

Meinander (1972) revised the world fauna of Coniopterygidae and later (Meinander, 1990) provided a revised checklist in which 423 recent species were noted: several additional species have been described since then, for example from the Yemen (Monserrat, 1996) and Tanzania (Sziraki, 1990). The current total of nearly 450 described species are divided among three recent subfamilies, two of which are diverse and widespread and the third (Brucheiserinae) comprises only the two known species of *Brucheiser* from South America, and was formerly regarded as a distinct family, Brucheiseridae (Riek, 1975). Nothing is known of the biology of Brucheiserinae, but the other subfamilies (Coniopteryginae, Aleuropteryginae) both contain a variety of genera ranging from widespread to highly restricted in distribution. Both these groups are characterised by several apomorphic features, and each is considered monophyletic (Meinander, 1972). As Meinander (1992) noted, a final decision over whether Brucheiserinae is really distinct from Aleuropteryginae must await discovery of the larva.

Key to subfamilies
1. Wing venation strongly reticulate (Fig. 2.2a)..............................Brucheiserinae
— Wing venation not reticulate, with few crossveins...2
2. Central region of fore wing with one radio-medial crossvein (Fig. 2.1b). Hind wing vein *Rs* not branching from *R* near the base of the wing. Galea one-segmented. Abdomen without plicaturae (paired membranous areas on the sternites of the central segments). Larvae with antennae about twice as long as labial palps; jaws short, not projecting from beneath labrum................................Coniopteryginae
—Central region of fore wing with two radio-medial crossveins (Fig. 2.1a). Hind wing vein *Rs* branching from *R* very close to base of the wing. Galea three-segmented. Abdomen with plicaturae on three to six segments. Larvae with antennae about the same length as labial palps; jaws projecting from beneath labrum................................Aleuropteryginae

Both the major subfamilies comprise groups of genera placed in separate tribes. However, the two tribes of Coniopteryginae can be separated mainly by characteristics of male genitalia, and are thus sometimes difficult to use. Thus, male genitalia of Coniopterygini have the gonarcus present or, if obliterated, the styli apparently arising from the ventral part of the ectoprocts, whereas

in Conwentziini the gonarcus is invariably obliterated but the styli (when evident) arise from the sclerotised ring of segment *ix*. Females, in particular, are best keyed directly to genera.

The Coniopterygini is dominated by the large, cosmopolitan genus *Coniopteryx*, which Meinander (1972) divided into six subgenera and later (Meinander, 1981) grouped the species within the subgenera. Several of the subgenera may not be monophyletic (Meinander, 1992). The other genera in this tribe are relatively small, of more limited distribution, and are reasonably distinctive. Within Conwentziini, the largest genus is *Semidalis* with nearly 60 described species and which occurs in most parts of the world except Australia. Some of these species are not clearly defined, and intraspecific variability suggests the need for more careful study. *Conwentzia* is widely distributed in the Holarctic, extending into Africa and southeast Asia, and has also been implicated in pest control.

The three tribes of Aleuropteryginae are separable more clearly, by features of wing venation, as follows:
1. Hind wing radial crossvein strikes *Rs* on branch *R2+3*..............................Fontenelleini
—Hind wing radial crossvein strikes *Rs* on stem of fork...2
2. Both wings with vein *M* forked. Fore wing: anterior branch of *M* coalescing with, or connected by a crossvein to *R4+5*................................Aleuropterygini
—Both wings with vein *M* simple or forked anterior branch of *M* in fore wing not connected to *R4+5*................................Coniocompsini

Further groupings are possible within these tribes, and members of many have been reported as valuable predators. Thus, three well-defined groups occur in Fontenelleini, representing distinct evolutionary lines. The first (including *Paraconis*, *Cryptoscenea*, *Neoconis*, *Pampoconis*) is basically Gondwanan, with *Cryptoscenea* known also from West Africa. The second group (containing only *Spiloconis*) is also ancient, and the third group (*Bidesmia*, *Helicoconis*, *Pseudoconis*, *Vartiana*) is predominantly northern hemisphere with a smaller African representation. The predominant genus, *Helicoconis*, is divided into four somewhat controversial subgenera (*Capoconis*, *Ohmopteryx*, *Fontenellea*, *Helicoconis*). In the Aleuropterygini, *Heteroconis* is predominantly from Australia and southeast Asia; many of the 50 or so species have heavily spotted wings, in

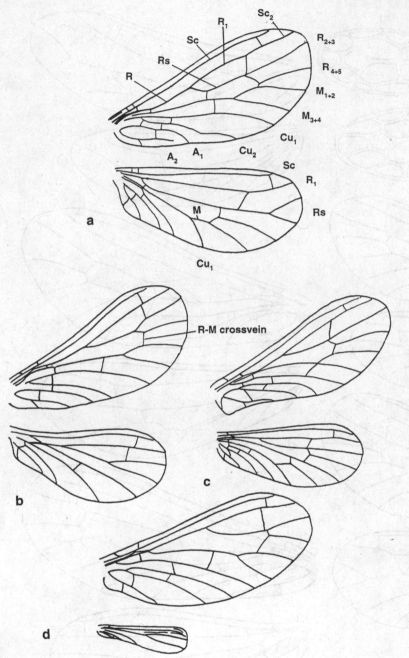

Fig. 2.1. Wings of representative Coniopterygidae, fore and hind wings shown: a, *Aleuropteryx*, with main veins labelled; b, *Coniopteryx*; c, *Semidalis*; d, *Conwentzia*. (Parts simplified after Meinander, 1972.)

contrast to most members of the family. The other genus in this group, *Aleuropteryx*, is divisible into three main species groups. Two of these are mainly Holarctic, and a third small group occurs in southern Africa. Coniocompsini includes only *Coniocompsa* from the Old World tropics with a 12th species from Hawaii.

It is difficult to estimate how complete current taxonomic knowledge of Coniopterygidae is. Females, in particular, are often difficult to identify to species level and, for North America, Tauber & Adams (1990) noted 'apparently, numerous, undescribed sibling species occur . . . and it is projected that there will be a 35%

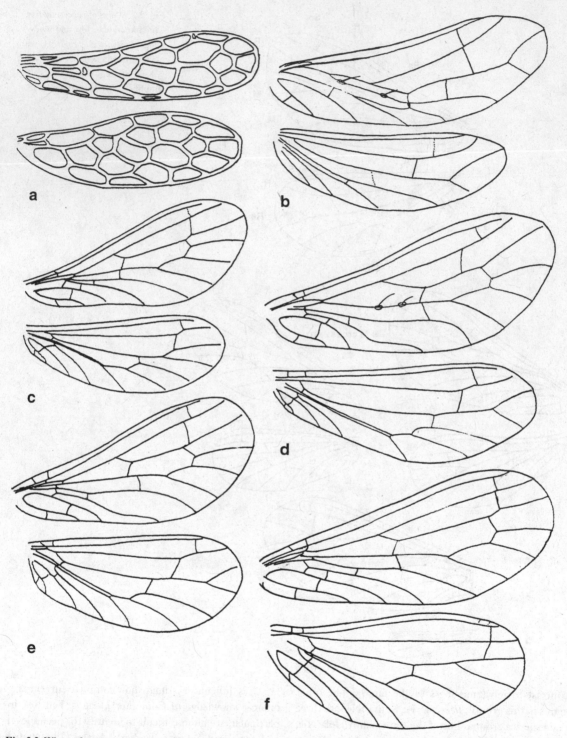

Fig. 2.2. Wings of representative Coniopterygidae, fore and hind wings shown: a, *Brucheiser*; b, *Coniocompsa*; c, *Cryptoscenea*; d, *Heteroconis*; e, *Aleuropteryx*; f, *Helicoconis* (key characters indicated in text). (Parts simplified after Meinander, 1972.)

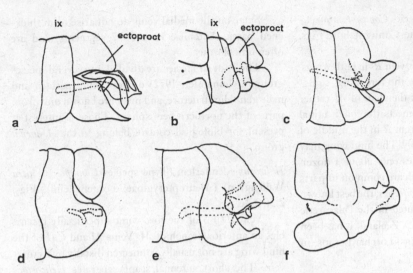

Fig. 2.3. Abdominal apex (posterior to right) of representative Coniopterygidae: a, b, *Coniopteryx*; c, d, *Conwentzia*; e, f, *Semidalis* (a, c, e, male; b, d, f, female) (Parts simplified after Meinander, 1972.)

increase in the number of species when the microhabitats and larval morphology of some genera are examined'. Similar sentiments may be valid for other parts of the world.

The abdomen of Coniopterygidae is weakly sclerotised, and usually shrivels considerably in dry specimens. The genitalia can be examined only after clearing the abdomen and careful orientation in liquid; female genitalia are very weakly sclerotised and, whereas male structures are often hard and complex, examination is facilitated greatly by staining either with chlorazol black (as noted earlier), or acid fuchsin. In Aleuropteryginae, some sternites have 'plicaturae', strongly folded (possibly glandular) structures on an ovoid membranous area. As Meinander (1972) noted, homologisation of genitalic structures even within the family is difficult, and the examples given here reflect a combination of morphologically satisfactory and more individualistic 'descriptive' terms, following Meinander's treatment. Unlike in other Neuroptera, there is no trace of cerci.

The female genitalia are particularly difficult to appraise, because they show few features of taxonomic value; consequently, as noted above, many female coniopterygids can be identified only from fortuitous association with males. The ectoproct is present, but a distinct sternite *ix* occurs only in Fontenelleini and Coniocompsini. The 'lateral gonapophyses' (gonocoxites *ix*) are short plates arising from the lower hind margin of segment *viii*, and may be fused; no styli are present. There is often a small sclerite (the 'subanale': Tjeder, 1970) (?sternite *x*) between the anal and genital openings, and all internal structures are weakly sclerotised. Male genitalia are elaborate, and it is necessary to examine these in lateral, caudal, dorsal, and ventral aspects to determine the relationships and possible homologies of the various structures. Those of the genera listed below exemplify some of the characteristic forms.

The genera noted as of possible value as biological control agents are as follows.

Coniopteryx Curtis (Type species *Coniopteryx tineiformis* Curtis) (Coniopteryginae: Coniopterygini). Figs. 2.1b, 2.3a, b.

The major key character for *Coniopteryx* is the unforked vein *M* in the hind wing, and this was used by Meinander (1972) to separate females from all other genera of the subfamily. Vein *Rs* in both wings is forked, but this character is shared with other genera of Coniopteryginae (except *Nimboa*).

This cosmopolitan genus contains many very similar species, divided among six subgenera (Meinander, 1972, 1990) on features of male genitalia. Some northern hemisphere species can be locally abundant, and seem to be the most common coniopterygids in regions where they occur. Sziraki (1992) provided keys to separate females of some subgenera and species occurring in Hungary on features of internal genitalia, providing useful leads for similar studies elsewhere.

Conwentzia Enderlein (Type species *Conwentzia pineticola* Enderlein) (Coniopteryginae: Conwentziini). Figs. 2.1d, 2.3c, d.

Hind wing vein *M* is forked, as is vein *Rs* in both wings. The *m-cu1* crossvein (always in the fore wing, sometimes in the hind wing) strikes the stem of *M* rather than its posterior branch, and meets the longitudinal veins at right angles. *Rs* forks from *R* in the middle of the wing rather than more basally. The hind wing may be reduced. *Conwentzia* includes only about a dozen described species, some of which are common in parts of the Holarctic. *Conwentzia psociformis*, the best known of these, is very widely distributed in the Palaearctic and has been introduced into New Zealand. It has been implicated as a valuable predator in orchard crops, in particular.

Semidalis Enderlein (Type species *Coniopteryx aleyrodiformis* Stephens) (Coniopteryginae: Conwentziini). Figs. 2.1c, 2.3e, f.

Hind wing vein *M* is forked, as is vein *Rs* in both wings. Unlike *Conwentzia*, the *m-cu1* crossvein in both wings is oblique, and usually strikes the posterior branch of *M*. *Semidalis* occurs worldwide except in the Australian region, and comprises five species groups defined on male genitalic structure (Meinander, 1972). Further work is needed to determine the limits of variation within many of the 60 or so recognised taxa. *Semidalis aleyrodiformis* is a valuable predator on various aphids and mites.

Heteroconis Enderlein (Type species *Heteroconis ornatus* Enderlein) (Aleuropteryginae: Aleuropterygini). Figs. 2.2d, 2.4 g, h.

Setae on fore wing medial veins arising from distinct basal thickenings. Antennae have 18 (rarely, 17) segments, often with some segments pale and others dark brown or black.

Closely related to *Aleuropteryx* (below), *Heteroconis* is the most diverse genus of Coniopterygidae in the Australian region and, perhaps, in southeast Asia. Nearly 60 species have been described, and others are known. Many species have the wings heavily patterned, and the pattern of wing and antennal coloration is often diagnostic.

Aleuropteryx Loew (Type species *Aleuropteryx loewii* Klapalek) (Aleuropteryginae: Aleuropterygini). Figs. 2.2e, 2.4a, b.

Setae on the medial veins do not arise from thickened bases. Antennae have 21–7 segments, and are often unicolorous.

The 30 or so species are divided into several species groups (Meinander, 1972) on fore wing pattern and male genitalic structure, and most are known only from parts of the northern hemisphere. Those of interest at present for biological control belong to the *A. loewii* group.

Helicoconis Enderlein (Type species *Coniopteryx lutea* Wallengren) (Aleuropteryginae: Fontenelleini). Figs. 2.2f, 2.4e, f.

Vein *R4 + 5* of the fore wing superficially resembles an anterior branch of *M*. Veins *M* and *Cu1* of the hind wing are not usually connected distally by a crossvein. The short antennal scape separates *Helicoconis* from *Spiloconis*, in which the scape is at least three times as long as broad, and the presence of plicaturae on abdominal sternite *vii* of the male abdomen enables differentiation from *Pampoconis* and *Neoconis*.

Helicoconis includes about 25 species from the Holarctic region and Africa. The species are very similar except in male genitalic features.

Cryptoscenea Enderlein (Type species *Cryptoscenea vesiculigera* Enderlein) (Aleuropteryginae: Fontenelleini). Figs. 2.2c, 2.4c, d.

Vein *R4 + 5* of the fore wing as in *Helicoconis*, above. Veins *M* and *Cu1* in hind wing running closely parallel for more than half the length of *Cu1*, with no membrane visible between them. This character combination is shared with *Paraconis*, from which *Cryptoscenea* is distinct by lacking plicaturae on the eighth sternite of the male abdomen.

2.3 HEMEROBIIDAE LATREILLE

The Hemerobiidae, brown lacewings, contains around 550 described species and is of world-wide occurrence. Hemerobiidae may be differentiated from all other Neuroptera by the following character combination (Oswald, 1993): anterior radial trace with at least two (up to 12) radial sector branches and nygmata absent. As their common name implies, they are most commonly brownish or greyish, more rarely green or other colours, and are generally fairly small, with fore wing length 3–18 mm. Historically, the family was considered as two distinct families by Comstock (1918), who

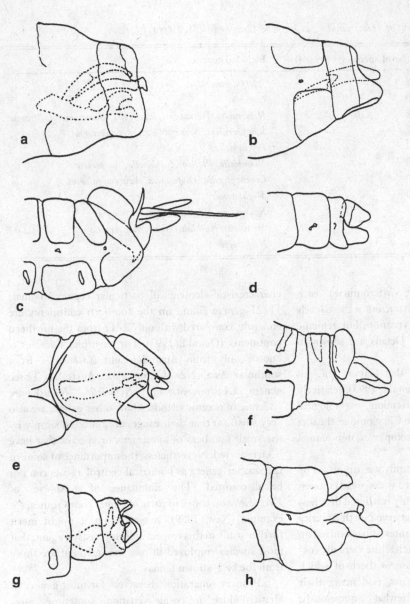

separated Hemerobiidae (all genera with three or more radial branches) from Sympherobiidae (the genera with only two radial branches), but Oswald (1993, 1994) convincingly showed the lack of need for this division. Other former subdivisions of the family [notably by Krüger (1922), into five subfamilies and by Nakahara (1960), into Notiobiellinae (*Notiobiella*) and Hemerobiinae (all other genera)] have not been adopted generally. Tjeder (1961) rejected all previously proposed subdivisions of the Hemerobiidae, and Oswald's carefully considered arrangement is likely to persist for

some time. Fossil Hemerobiidae are known from the Eocene to Holocene (Carpenter, 1992).

In an effort to settle previous confusions over subfamilies and genera, Oswald (1993, 1994) has recently reappraised all genus-level taxa in the family, so that the 25 genera recognised at present are now defined clearly. Oswald allocated these to ten subfamilies (Table 2.1), of varying complexity and distribution. Three subfamilies (Carobiinae, Psychobiellinae, Megalominae) each contain only a single genus (*Carobius*, *Psychobiella*, *Megalomus*, respectively), and the largest subfamilies

Table 2.1. *Taxonomic arrangement of Hemerobiidae, as given by Oswald (1993, 1994a, b)*

Subfamily	Number of species (at 1993/4)	Included genera
Adelphohemerobiinae	2	*Adelphohemerobius*
Carobiinae	9	*Carobius*
Hemerobiinae	198	*Hemerobius, Wesmaelius, Nesobiella, Hemerobiella, Biramus*
Sympherobiinae	*c.* 62	*Sympherobius, Nomerobius, Neosympherobius*
Psychobiellinae	2	*Psychobiella*
Notiobiellinae	*c.* 70	*Notiobiella, Psectra, Zachobiella, Anapsectra*
Drepanacrinae	10	*Conchopterella, Drepanacra, Austromegalomus*
Megalominae	*c.* 40	*Megalomus*
Drepanepteryginae	37	*Neuronema, Drepanepteryx, Gayomyia*
Microminae	*c.* 121	*Micromus, Nusalala, Noius, Megalomina*
(unplaced)	3	*Notherobius*

(Hemerobiinae, Notiobiellinae, Microminae) each include only four genera. Key characters at these levels rely heavily on details of wing venation, but genitalic characters are also important. Details are shown in Figs. 2.5–2.7.

Despite the likelihood that larvae of all Hemerobiidae are predatory (Oswald, 1993), relatively few species have attracted attention as biological control agents (New, 1988). As in Chrysopidae, the diet of adults of some species encompasses non-animal foods (Stelzl, 1990).

The zoogeography of the family was discussed by Oswald (1993). Hemerobiidae are widespread, though many of the subgroups have highly restricted distributions. Seven genera are shared between North America and Europe/Asia. Most applied interest has centred on species of widespread genera, such as the virtually cosmopolitan *Hemerobius* and *Micromus* (both of which have numerous generic synonyms, evidencing their considerable variability). Somewhat anomalously, *Hemerobius* is scarce in the Australian region, and the otherwise widespread *Wesmaelius* and *Sympherobius* appear to be absent from it. *Micromus* is the most widespread genus of Hemerobiidae, and has undergone dramatic radiation in the Hawaiian archipelago (Zimmerman, 1957). As with *Hemerobius*, many of the constituent species of *Micromus* have limited distributions, and all the above genera include taxa endemic to particular continents. However, as Oswald (1993) noted, *Micromus* is possibly absent from South America. In contrast, many of the smaller groups are

characteristic elements of particular regional faunas; 11/21 genera found on the southern continents are endemic, compared with only 2/12 from the northern continents (Oswald, 1993). For example, *Carobius* is known only from Australia, and *Zachobiella* from southeast Asia, New Guinea, and Australia. Three genera (*Austromegalomus, Nesobiella, Noius*) are endemic to oceanic islands. Many other genera are also very small, so that their ranges are reflected simply by the small numbers of specimens or species that have been reported. Nevertheless, the importance of some of the smaller genera as biological control agents can not be discounted. The abundance of a species of *Zachobiella* on coffee in parts of Papua New Guinea, for example (New, 1989), suggests that it might merit further study in this respect and emphasises, again, that most species employed in pest management are those from the best-known faunas.

Primary separation characters for most genera of Hemerobiidae are wing venation, sometimes augmented by head features and genitalic differences, and Oswald (1993) has provided a careful reinterpretation of the details of hemerobiid genitalia, critical in species separations. As in Chrysopidae (q.v.) the posterior segments of the abdomen are modified, and ectoprocts have a cercal callus (Fig. 2.6). Females have tergite *ix* extended ventrally and subtending the ectoproct (which, unlike the heavily ornamented condition in many male hemerobiids) is simple. Tergite *viii* is also deepened, and may be divided in the dorsal midline; sternite *viii* is absent. The paired gonocoxites *ix*

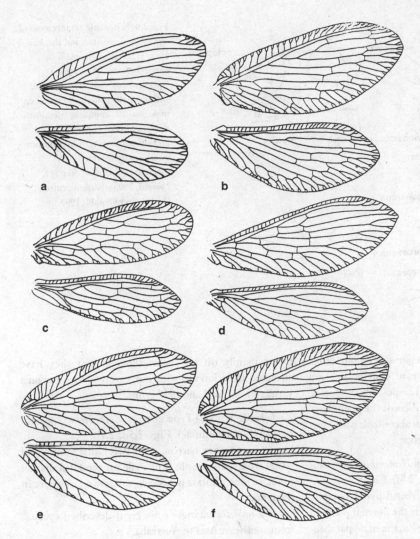

Fig. 2.5. Representative wing venation of (a) Sisyridae, 'spongeflies', the family taxon most likely to be confused with Hemerobiidae, to illustrate different form of *Rs* in fore wing and other differences, and (b–f) Hemerobiidae: b, *Hemerobius*; c, *Micromus*; d, *Zachobiella*; e, *Sympherobius*; f, *Wesmaelius* (key characters indicated in text).

('posterior gonapophyses') are commonly strongly rounded posteriorly, and a short stylus may or may not be present. A discrete spermatheca is absent, but Oswald noted that the term has been applied, imprecisely, to any sclerotised or expanded region of the main invagination–fertilisation canal in hemerobiids. Males have tergite *ix* less expanded than in the female but, like the ectoproct, it can be elaborated in various ways. Sternite *ix* is very variable, from a simple rounded or transverse plate to more extended and complex shapes; occasionally it is lost or fused with sternite *viii*. The genitalia comprise a very complex gonarcus, the parabaculum (Oswald's designation for the structure called 'parameres' by most earlier workers, because it is not a homologue of gonocoxites *ix*), and a small hypandrium

internum. The last is of little taxonomic value, whereas details of the gonarcus and parabaculum are of great value in specific diagnoses.

The genera noted as of possible value as biological control agents are as follows:

Hemerobius L. (Type species *Hemerobius humulinus* L.) (Hemerobiinae). Figs. 2.5b, 2.7a, b.
Wings as in Fig. 2.5b; differs from other Hemerobiinae by absence of crossvein *2r-m*, or it being proximal to crossvein *2m-cu*. Other diagnostic features (after Oswald, 1993) are the prominent sagittal seta on the clypeus and the parabaculum in the male genitalia completely divided.

Hemerobius is cosmopolitan, but appears to be most

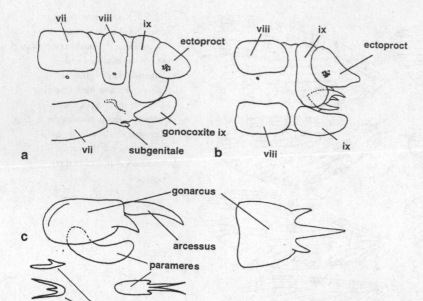

Fig. 2.6. Schematic arrangement of hemerobiid genitalia and abdominal apex: a, female, with small subgenitale and elongate 'spermatheca' indicated; b, male, with genitalic complex in place; c, male genitalic complex, lateral with dorsal aspect of gonarcus (right), and ventral aspects of hypandrium internum (bottom left) and 'parameres' (parabaculum of Oswald, 1993) (bottom centre). (Partly after Oswald, 1993.)

diverse in northern temperate regions. The numerous generic synonyms reflect this broad distribution and the high diversity, with about 125 described species. Most appear to be arboreal. Members of the genus have been implicated widely in biological control, with most data from North America or Europe.

Wesmaelius Kruger (Type species *Hemerobius concinnus* Stephens) (Hemerobiinae). Figs. 2.5f, 2.7c, d.

Rather similar to *Hemerobius*, but differs from it by fore wing crossvein *2r-m* being in the normal position near or distal to the crossvein *2m-cu*, the male parabaculum not completely divided, and the male ectoproct having a distinct 'pecten' (Fig. 2.7c). The clypeus also lacks the characteristic sagittal seta found in *Hemerobius*.

Micromus Rambur (Type species *Hemerobius variegatus* Fabricius) (Microminae). Figs. 2.5c, 2.7e, f.
Humeral area of fore wing very narrow, and its margin without trichosors. Fore wing lacking crossveins *2sc-r* and *2m-cu*, and with crossveins *1cua-cup* and *2cua-cup*.

About 100 species are known, and the genus is nearly cosmopolitan. However, *Micromus* is a complex genus, without clear synapomorphies (Oswald, 1993), and may eventually need to be subdivided to reflect the considerable variability amongst the included species. Various species may be predominantly arboreal or

found mainly on low vegetation, so that they have attracted attention in a wide range of pest contexts on a variety of orchard and field crops.

Zachobiella Banks (Type species *Zachobiella punctata* Banks) (Notiobiellinae). Figs. 2.5d, 2.7i.
Fore wings rather narrow, with two radial sectors; all costal crossveins simple. The male pregenital abdomen can have long digitate processes, unknown elsewhere in the family.

Zachobiella includes only eight described species, from southeast Asia to Australia.

Sympherobius Banks (Type species *Hemerobius amiculus* Fitch) (Sympherobiinae). Figs. 2.5e, 2.7g, h.

Wings as in Fig. 2.5e; usually only two radial sector veins; crossvein *2sc-r* absent (subfamily features). Fore wing vein *CuP* not forked proximal to crossvein *2cua-cup*. The male ectoproct has prominent processes.

Widely distributed, with about 55 species described, *Sympherobius* is easily the most diverse hemerobiid genus with only two fore wing radial sectors.

2.4 CHRYSOPIDAE SCHNEIDER

The Chrysopidae, green lacewings, is the most diverse family of the three considered here. Historically, the classification of the family has been difficult, with a

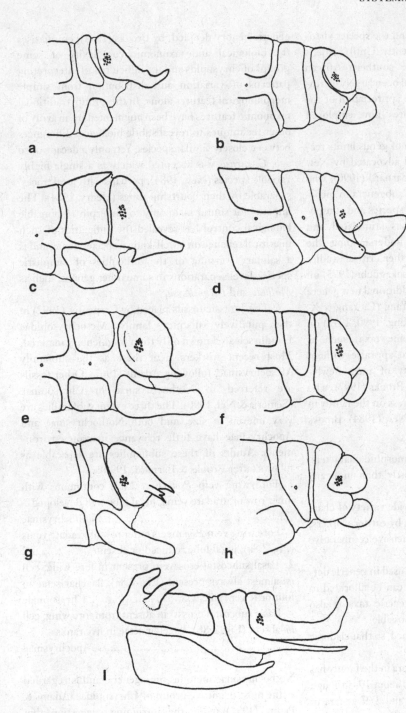

Fig. 2.7. Abdominal apex (posterior to right) of representative Hemerobiidae: a, b, *Hemerobius*; c, d, *Wesmaelius*; e, f, *Micromus*; g, h, *Sympherobius*; i, *Zachobiella* (a, c, e, g, i, male; b, d, f, h, female). (Partly after Oswald, 1993.)

great number of species from many parts of the world being referred uncritically to *Chrysopa* Leach, following the early adoption of this by Schneider in his classic monograph (Schneider, 1851). '*Chrysopa*' thereby became a 'holding genus' for a considerable variety of

taxa, and information on relationships and distribution of the various genera has been correspondingly difficult to interpret. Only the more distinctive other genera could be separated reliably from *Chrysopa* until recently, and many of the generic names earlier proposed in the family were poorly characterised and the species virtually impossible to allocate consistently. Until Tjeder's (1966) critical evaluation of the southern African Chrysopidae, generic diagnoses had been largely on features of wing venation, with little appreciation of the critical values of genitalic features now employed widely.

The need for a global revision of genus limits recognised by Killington (1937) and advocated by New (1984) was fulfilled by Brooks & Barnard (1990), who recognised 75 valid genera and 11 subgenera to incorporate the almost 1200 recognised species and subspecies. That landmark appraisal has stimulated and greatly facilitated progress in understanding the systematics of the family, and further critical studies have ensued – such as those by Tsukaguchi (1995) on the Japanese fauna (in which two additional new genera were recognised) and by Yang & Yang (C. Yang & X. Yang, 1989, 1990; X. Yang & C. Yang, 1990, 1992) on the complex Chinese fauna. Some taxa (such as *Tibetochrysa* Yang) are difficult to appraise without further examination. The validity of most newly-described genera was evaluated by Brooks (1997), who carefully reviewed taxonomic progress on the family in relation to the needs specified by New (1984). Brooks noted that:

1. Genera are now based on apomorphic characters, so that they defined more 'robustly' than under any previous arrangement.

2. Genera are now defined on a wide variety of characters, many of them not used by earlier workers; this has enabled more comprehensive comparative surveys throughout the family.

3. Male and female characters are used in generic definitions, so that many females can be allocated to genus. Recognition of new generic taxa is also much clearer than previously possible.

4. Tribes and subfamilies are defined, so that cladistic analyses can be attempted.

5. The number of monotypic genera in the family has been reduced from around 30 to about 10, although 6/13 genera of Apochrysinae and 3/9 genera of Nothochrysinae remain monotypic.

Many of the genera and species can thus now be recognised reasonably clearly, though the boundaries between many remain obscure. The next chapter, dealing with a sibling complex of species within *Chrysoperla* Steinmann [revised, with the complex

generic history detailed, by Brooks (1994)] emphasises the biological and taxonomic complexity of some groups of chrysopids and the difficulties of interpreting patterns of variation and distribution from simple morphological features alone. In this group, traditional taxonomic features have been augmented by an array of other techniques to reveal subtle biological differences between closely similar species. Yet, only a decade or so ago, *C. carnea* was accepted widely as a single highly variable species (New, 1984), perhaps with local 'races' separable by their courtship songs (Henry, 1984). The impetus for sound taxonomy to underpin responsible biological control has revealed the difficulties of such uncritical reliance on a well-known species name, and is a salutary warning of the difficulties of simplistic species-level separations in some other genera, such as *Mallada* and *Dichochrysa*.

The Jurassic fossils placed by Carpenter (1992) in the putatively distinct family Mesochrysopidae Handlirsch are known only from fragmentary material. Most recent workers treat these as the subfamily Mesochrysinae, following Adams (1967). Other fossils are referred to Nothochrysinae or Chrysopinae (Séméria & Nel, 1990). The three extant subfamilies are very uneven in size, and both Nothochrysinae and Apochrysinae have little relevance in crop environments. Adults of these subfamilies are separable as follows (after Brooks & Barnard, 1990):

1. Fore wing vein *Psm* (Fig. 2.8d) continuous with inner row of gradate veins; jugal lobe well developed ....Nothochrysinae
—Fore wing vein *Psm* meets outer row of gradate veins (Fig. 2.8a); jugal lobe reduced or absent 2
2. Basal subcostal crossvein present in fore wing; cell *im* almost always present (Fig. 2.8a); flagellar setae in four or fewer ranks ....Chrysopinae
—Basal subcostal crossvein absent from fore wing; cell *im* absent (Fig. 2.8b); flagellar setae in five ranks ....Apochrysinae

Nothochrysinae includes nine genera, and is regarded as the most primitive group of Chrysopidae. Adams & Penny (1992) made the intriguing suggestion that because Nothochrysinae lack the alar tympanic organ found in Chrysopinae they might be more subject to natural predation by bats, and this putative high mortality of adults might in part explain the paucity of living species in this group. Most Apochrysinae, with 13 genera, are tropical forest-dwelling species. Many

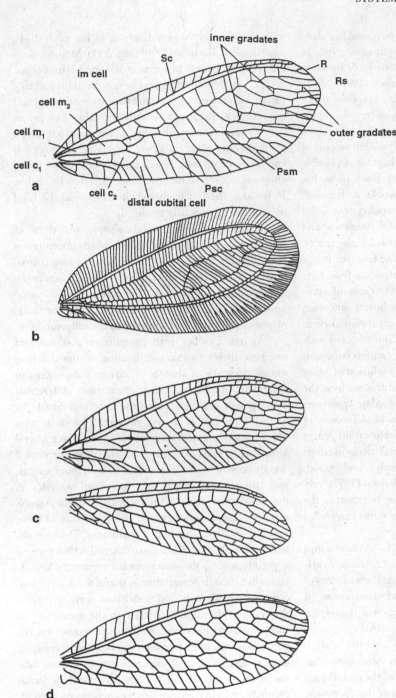

Fig. 2.8. Representative wing venation of Chrysopidae, with main diagnostic features indicated: a, *Chrysopa*, fore wing, with main veins labelled; b, Apochrysinae (*Nobilinus*), fore wing; c, *Anomalochrysa*, fore and hind wings; d, Nothochrysinae (*Triplochrysa*), fore wing. (Part simplified after Brooks & Barnard, 1990.)

are large lacewings with broad wings and very dense venation, on which most taxa are diagnosed. As Brooks (1997) noted, this plasticity of venational features hinders progress toward phylogenetic understanding in this subfamily. The inclusion of *Nothancyla* in this lineage is problematic, and it is possible that it should

constitute a distinct monotypic subfamily (Brooks, 1997). Chrysopinae includes more than 97% of known chrysopid species and comprises four tribes: Ankylopterygini (5 genera), Belonopterygini (15 genera), Chrysopini (more than 30 genera), and Leucochrysini (7 genera). However, the tribes are difficult to separate critically by conventional key characters but are linked by the subfamily autapomorphy of only four rings of flagellar setae (Brooks & Barnard, 1990). A cladogram by Brooks (1997) links Belonopterygini and Leucochrysini as one major lineage, and Chrysopini with Ankylopterygini as the other. The major group of relevance here is the Chrysopini, which includes virtually all chrysopids of any economic or applied interest. However, a possible species of *Nacaura* (Apochrysinae) from India has larvae predatory on mealybugs (Brooks & Barnard, 1990). The biology of most other Apochrysinae and a high proportion of Chrysopini is still unknown, and enhancement of biological knowledge is a high priority. Nevertheless, in order to recognise the taxa that might be of applied interest, or which simply occur from time to time in crop environments, a wide range of taxonomic features involving wings, mouthparts, antennae, and genitalia are used, with greatest confusion likely to occur in interpreting small but critical features of male genitalia. As Barnard (1984) noted, caution is needed not to rely on wing venation to the exclusion of other characteristics because, whereas features such as the shape of cell *im* and the number of gradate crossveins have been considered historically to be diagnostic of some genera, these can be more variable within genera and species than commonly supposed. Relationships within Chrysopini are still poorly understood. Tentative generic groupings by Brooks (1997) rely heavily on male genitalic complement, but most of the larger genera are in need of revision to refine knowledge of species limits and distributions.

Representative character states of chrysopid wings and genitalia are shown in Figs. 2.8–2.13. Generic variations in wing venation include the numbers of rows of gradate veins, and the numbers and arrangement of veins in each row; the shape and position of the *im* cell; the shape (straight or sinuous) of the radial crossveins; the relative lengths of cells *c1* and *c2*; and the setation of veins. The shape of the wing also varies somewhat between some genera, and the width of the costal space can be diagnostic. Male genitalia form the diagnostic basis for many genera but, as Brooks (1997) emphasised, these may not be entirely reliable because the presence of some structures in Chrysopini appears to be plesiomorphic, and they may have been lost on more than one occasion.

The arrangements of the abdominal apex and gen-

italia of Chrysopidae were described by Barnard (1984), and Brooks & Barnard (1990) (Fig. 2.11). Modifications are confined largely to segment *viii* and beyond (Figs. 2.12, 2.13). In both sexes, tergite *ix* is fused (or partially fused) with the ectoprocts, and the extent of fusion and the ectoproct shape can have generic value. Ectoprocts have a prominent cercal callus, and males often have a strengthening inner apodeme. In males, sternites *viii* and *ix* are commonly fused, though this fusion is not always complete, so that the suture line may be visible. In females, the last main sternite is *vii*, and the hind margin can be variously modified.

Female genitalia are quite simple, and those of many genera are rather similar. A bilobed subgenitale is present as is a discrete spermatheca with a long convoluted duct. The form of the spermatheca can vary in the size and shape of the vela, the extent of the ventral impression, and details of the duct. A 'praegenitale' may occur as a small lobe ventral to the subgenitale.

As noted earlier, male genitalia are complex and can be difficult to examine because of the differing mosaic presence or absence of various components of taxonomic importance and, even, their differential development over age within the same individual. An arched gonarcus is always present, though it may become more markedly transverse in some taxa; lateral lobes may be present on the gonarcus (these are termed 'entoprocessus' when they articulate with the gonarcus, and 'gonocornua' when they are fused with it). A median slender process, the arcessus, extends posteriorly from the dorsal region of the gonarcus and there may be a smaller, more dorsal mediuncus. The arcessus may be detached from the gonarcus, and is then termed a 'pseudopenis' – the structures are apparently homologous but their differentiation is useful for description. Some genera have an additional transverse structure, the tignum, dorsal and anterior to the gonarcus. All these structures lie on an eversible membranous sac, the gonosaccus, the apex of which can bear large, conspicuous gonosetae. A small gonapsis may be present near the apex of sternite *ix* (which may itself be ornamented), and a hypandrium internum is always present: it may be derived from sternite *x* (Acker, 1960) and is always associated with the gonopore.

All three subfamilies are widely distributed, but many of the genera are characteristic of only limited regions of the world. Thus, Apochrysinae occur widely but two genera occur only in Africa, four in the

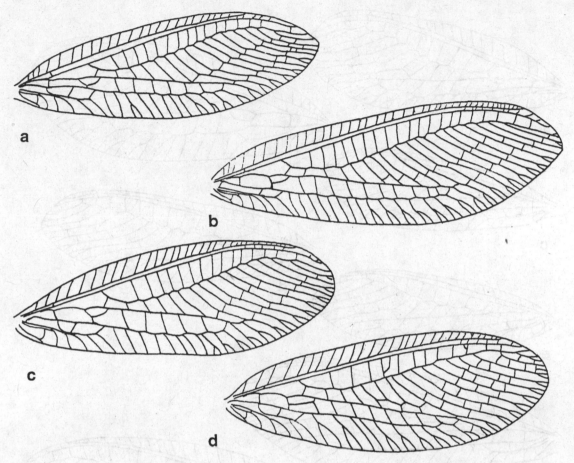

Fig. 2.9. Fore wings of representative genera of Chrysopidae: Chrysopinae: a, *Meleoma*; b, *Nineta*; c, *Plesiochrysa*; d, *Rexa*. (Parts simplified after Brooks & Barnard, 1972.)

Neotropics, six in the Oriental region or Australia, and one is found in the eastern Palaearctic. Likewise, each genus of Nothochrysinae is endemic to only a small part of the subfamily range. Distribution of the genera is summarised by Brooks & Barnard (1990). Within the Chrysopinae, some genera are relatively widespread, although past confusion over generic recognition has distorted interpretation of some of the ranges now recognised. Many others show very restricted distributions and narrow endemism in various continents.

The genera noted as of possible value as biological control agents are as follows:

Anomalochrysa Mclachlan (Type species *Anomalochrysa hepatica* McLachlan) (Chrysopinae: Chrysopini). Figs. 2.8c, 2.12h.

The venational features of *Anomalochrysa* include the presence of an ovate *im* cell in the fore wing and at least three series of gradate veins in the hind wing. The antennal scape is, at most, only slightly longer than broad, and the head is relatively small. The venation is often complex, with numerous small variations and irregular crossveins. The apex of the male abdomen is distinctive, being considerably flattened with tergite *vii* much reduced.

Anomalochrysa is endemic to Hawaii, and any superficially similar taxa are Oriental or from the eastern Palaearctic. It is apparently related to the *Mallada* group of genera (Brooks & Barnard, 1990). Unusually in Chrysopidae, eggs are not stalked and, whereas the typical habitat is native forest, larvae have been reported feeding on sugarcane aphids and

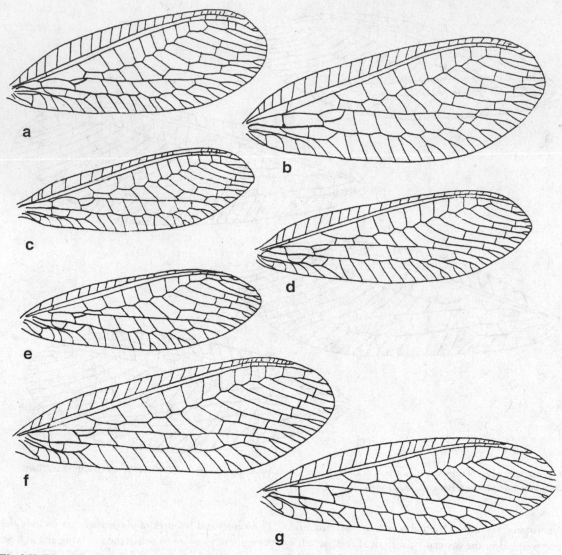

Fig. 2.10. Fore wings of representative genera of Chrysopidae: Chrysopinae: a, *Apertochrysa*; b, *Chrysopa*; c, *Brinckochrysa*; d, *Chrysoperla*; e, *Ceraeochrysa*; f, *Ceratochrysa*; g, *Mallada*. (Part simplified after Brooks & Barnard, 1990.)

sugarcane leafhopper (*Perkinsiella saccharicida*) in Hawaii (Zimmerman, 1957). Adults are also predatory.

Apertochrysa Tjeder (Type species *Chrysopa umbrosa* Navas) (Chrysopinae: Chrysopini). Figs. 2.10a, 2.12f. *Apertochrysa* is one of a number of genera closely related to *Mallada* and which are virtually indistinguishable on external features: fore wings have a narrow ovate *im* cell, and two parallel series of gradate veins. In the male genitalia, the tignum is absent, and a T-shaped

gonapsis is present, as are entoprocessus. The fore wing costal setae are longer than in most species of *Mallada* (Brooks & Barnard, 1990).

About 16 species of *Apertochrysa* occur in the Palaearctic and Old World tropics, and the genus occurs also in Australia. Adults are predatory.

Brinckochrysa Tjeder (Type species *Chrysopa peri* Tjeder) (Chrysopinae: Chrysopini). Figs. 2.10c, 2.11d, 2.12g, 2.13g.

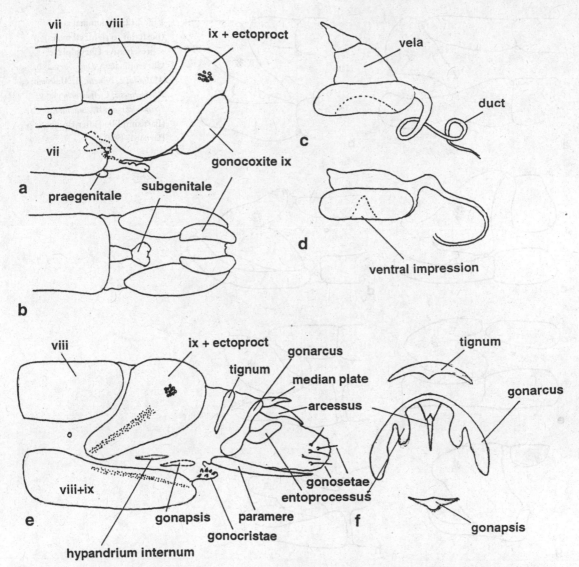

Fig. 2.11. Schematic arrangement of chrysopid genitalia and abdominal apex: a–d, female; e, f, male: a, b, abdominal apex in (a) lateral and (b) ventral aspects; c, d, spermatheca of (c) *Chrysoperla* and (d) *Brinckochrysa*, to indicate different states of vela and ventral impression; e, male, lateral aspect with genitalic complex extruded; f, genitalic complex, posterior aspect. (Part simplified after Brooks & Barnard, 1990.)

Generally rather small chrysopids with the fore wing of most species less than 11 mm long (rarely up to 15 mm). The fore wing is characteristically rather narrow, and the *im* cell is short and narrow; the gradate veins are in two parallel series. The male genitalia lack tignum, gonapsis and median plate, and pseudopenis; male ectoprocts are long and greatly extended ventrally.

Sternite vii of the female has a deep apical indentation. All species have a stridulatory structure, and the above abdominal features differentiate *Brinckochrysa* from all other chrysopids which have any similar mechanism.

Brinckochrysa is widespread in the Old World, and most of the 16 described species occur in the African tropics.

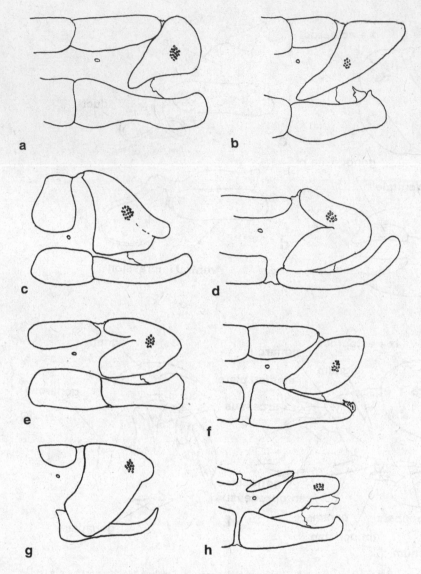

Fig. 2.12. Abdominal apex (posterior to right; all males) of representative Chrysopidae: Chrysopinae: a, *Dichochrysa*; b, *Mallada*; c, *Nineta*; d, *Meleoma*; e, *Plesiochrysa*; f, *Apertochrysa*; g, *Brinckochrysa*; h, *Anomalochrysa*. (Part simplified after Brooks & Barnard, 1990.)

Ceraeochrysa Adams (Type species *Chrysopa cincta* Schneider) (Chrysopinae: Chrysopini). Figs. 2.10e, 2.13a.

As Brooks & Barnard (1990) noted, 'Species of *Ceraeochrysa* have no outstandingly distinctive external characters, although members . . . can often be recognised by the presence of a red lateral stripe on the pronotum or scape and the dark radial crossveins'. Male genitalia have a straight elongate gonapsis and a 'horned' median plate, which are distinctive among chrysopid genera; the tignum and entoprocessus are

absent. In the female, the spermatheca is very small and narrow, and these features are also distinctive.

Ceraeochrysa, with more than 40 known species, is the predominant genus of Chrysopinae in the Neotropics, and occurs also throughout North America and in the Caribbean region. Gut contents of adults examined by Brooks & Barnard (1990) did not include insect remains.

Ceratochrysa Tjeder (Type species *Chrysopa ceratina* Navas) (Chrysopinae: Chrysopini). Figs. 2.10f, 2.13d.

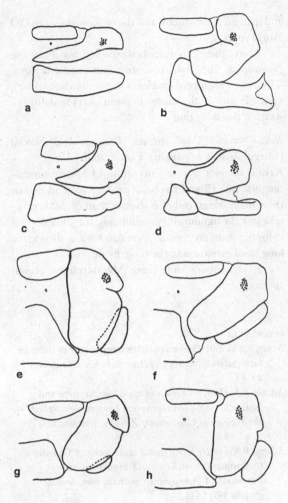

Fig. 2.13. Abdominal apex (posterior to right; a–d, male; e–h, female) of representative Chrysopidae: Chrysopinae: a, *Ceraeochrysa*; b, *Chrysopa*; c, h, *Chrysoperla*; d, *Ceratochrysa*; e, *Nineta*; f, *Rexa*; g, *Brinckochrysa*. (Part simplified after Brooks & Barnard, 1990.)

Diagnostic features of *Ceratochrysa* include the very short inner gradate crossvein series, with the outer gradates black and the number of inner gradates half or fewer than those of the outer row. Cell *im* is very narrow, and cell *c1* is shorter than *c2*. In the male genitalic complex the very long entoprocessus (usually protruding beyond the apex of the abdomen) are also diagnostic.

Ceratochrysa has few species, all from Africa, Madagascar, or Mauritius.

Chrysopa Leach (Type species *Hemerobius perla* L.) (Chrysopinae: Chrysopini). Figs. 2.10b, 2.13b.

Chrysopa is included here mainly as the former 'holding genus' to which a high proportion of described green lacewings were at some time allocated. In the restricted sense in which *Chrysopa* is now applied (Brooks & Barnard, 1990), it is characterised in part by broad, oval fore wings; black markings on the head and thorax; the ectoproct and tergite *ix* being only partially fused in both sexes; sternites *viii* and *ix* separate in the male; and dorsal horns on the male entoprocessus.

The genus is Holarctic, and includes about 50 described species, some with defined subspecies. Adults are predatory.

Chrysoperla Steinmann (Type species *Chrysopa carnea* Stephens) (Chrysopinae: Chrysopini). Figs, 2.10d, 2.13c, h.

In most species of *Chrysoperla*, the fore and hind wings are both narrow, with fore wing *im* cell short and ending before the first crossvein from *Rs*: this feature is shared with some *Mallada s. lat.* species. Males are distinctive by having a 'lip' at the apex of sternites *viii* + *ix*. Most species have a yellow median thoracic stripe, though this can occur also in some other genera.

Brooks (1994) recognised 36 valid species of *Chrysoperla*, the predominant genus used in biological control programmes. Sibling species are sometimes morphologically similar, and their separation in the least-derived species group (the *carnea*-group) is discussed by Duelli *et al.* (this volume). *Chrysoperla* is widely distributed.

Dichochrysa Yang (nom.nov. for *Navasius* Yang & Yang) (Type species *Navasius eumorphus* Yang & Yang) (Chrysopinae: Chrysopini). Fig. 2.12a.

Very similar to *Mallada* (below) and differentiated from it on features of male genitalia. Sternites *viii* + *ix* lack any projection at the apex of sternite *ix*; the ectoprocts are rounded and broad; and the genitalic complement is complete, with tignum, gonapsis, entoprocessus, and pseudopenis all present.

Dichochrysa is proving to be one of the largest genera in the Chrysopidae, and most of the species allocated to *Mallada* by Brooks & Barnard (1990) more properly belong here (Brooks, 1997). The genus is widespread and, at present, is difficult to enumerate.

Mallada Navás (Type species *Mallada stigmatus* Navás) (Chrysopinae: Chrysopini). Figs. 2.10g, 2.12b.

See notes under *Dichochrysa* (above). *Mallada* is now a reasonably small genus, but with several closely related species widespread in the Pacific region, in particular. It is distinct from *Dichochrysa* by males with a lobe or projection at the apex of sternites *viii + ix*; ectoprocts being narrowed and often angled apically, and the genitalic complex being reduced from the complete state.

The biological differences between *Mallada* and *Dichochrysa* are still largely unclear.

Meleoma Fitch (Type species *Meleoma signoretii* Fitch) (Chrysopinae: Chrysopini). Figs. 2.10d, 2.13a, h.

Despite considerable variability within *Meleoma* (Brooks & Barnard, 1990), many species have the head broad, antennae shorter than the fore wing and widely separated at the base; males often have intricate ornamentation on the head (including a pronounced frontal cavity), or stridulatory mechanism. The broad arcessus has an upturned pseudopenis.

This New World genus includes about 26 described species.

Nineta Navas (Type species *Hemerobius flavus* Scopoli) (Chrysopinae: Chrysopini). Figs. 2.9b, 2.12a, 2.13e.

Species of *Nineta* are large lacewings (much larger than most other Chrysopini), with fore wing length usually 16–22 mm (Brooks & Barnard, 1990). Antennae are shorter than the fore wing; cell *im* is narrow and ovate; most radial crossveins are sinuate; and cell *c1* is shorter than *c2*. The male genitalia are reduced, with tignum, gonapsis, pseudopenis, and median plate absent.

Nineta occurs only in the Holarctic, with most of the species Palaearctic. Brooks & Barnard (1990) believed it to be closely related to *Tumerochrysa* Needham, in which there are three series of gradate veins, rather than two as in *Nineta*.

Plesiochrysa Adams (Type species *Chrysopa brasiliensis* Schneider) (Chrysopinae: Chrysopini). Figs. 2.9c, 2.12e.

Plesiochrysa has long narrow wings with the fore wing cell *im* ovate. The prothorax is often elongate (especially in New World species). Ectoprocts of both sexes are broad and have an apical invagination. Male sternites *viii + ix* are not fused or incompletely fused. Male genitalia are reduced, with arcessus absent, gonapsis

and median plate absent, and the tignum absent in Old World species.

More than 20 described species occur in the Neotropics, Oriental, Australian and Pacific regions. The most widespread species is *P. ramburi* (Schneider), which is one of the more common chrysopids in the western Pacific region.

Rexa Navás (Type species *Rexa lordina* Navás) (Chrysopinae: Chrysopini). Figs. 2.9d, 2.13f.

Rexa is distinct among Chrysopini by having a quadrangular *im* cell and irregular gradate veins in two or three main series; cell *c1* is shorter than *c2*. Male genitalia lack the tignum and pseudopenis; the gonapsis and entoprocessus are broad. Females have a distinctive long basal extension to the subgenitale.

Rexa contains only three Mediterranean region species.

REFERENCES

Acker, T.S (1960) The comparative morphology of the male terminalia of Neuroptera (Insecta). *Microentomology* 24, 25–84.

Adams, P.A. (1967) A review of the Mesochrysinae and Nothochrysinae (Neuroptera: Chrysopidae). *Bulletin of the Museum of Comparative Zoology, Harvard* 135, 215–238.

Adams, P.A. (1969) A new genus and species of Osmylidae (Neuroptera) from Chile and Argentina, with a discussion of Planipennian genitalic homologies. *Postilla* 141, 1–11.

Adams, P.A. & Penny, N.D. (1992) New genera of Nothochrysinae from South America (Neuroptera: Chrysopidae). *Pan-Pacific Entomologist* 68, 216–221.

Barnard, P.C. (1984) Adult morphology related to classification. In *Biology of Chrysopidae*, ed. Canard, M., Séméria, Y. & New, T.R., pp. 19–29. Dr W. Junk, The Hague.

Brooks, S.J. (1994) A taxonomic review of the common green lacewing genus *Chrysoperla* (Neuroptera: Chrysopidae). *Bulletin of the British Museum (Natural History) Entomology* 63, 137–210.

Brooks, S.J. (1997) An overview of the current status of Chrysopidae (Neuroptera) systematics. *Deutsche entomologische Zeitschrift* 44, 267–275.

Brooks, S.J. & Barnard, P.C. (1990) The green lacewings of the world: a generic review (Neuroptera: Chrysopidae). *Bulletin of the British Museum (Natural History) Entomology* 59, 117–286.

Carpenter, F.M. (1992) *Superclass Hexapoda. Treatise on Invertebrate Palaeontology. Part R, Arthropoda 4*, vols. 3 and 4. Geological Society of America, Boulder CO and University of Kansas, Lawrence, KA.

Comstock, J.H. (1918) *The Wings of Insects*. Comstock Publishing, Ithaca, New York.

Henry, C.S. (1984) The sexual behavior of green lacewings. In *Biology of Chrysopidae*, ed. Canard, M., Séméria, Y. & New, T.R., pp. 101–110. Dr W. Junk, The Hague.

Killington, F.J. (1937) *A Monograph of the British Neuroptera*, vol. 2. Ray Society, London.

Krüger, L. (1922) Hemerobiidae. Beitrage zu einer Monographie der Neuropteren-Familie der Hemerobiiden. *Stettiner entomologisches Zeitung* 83, 138–172.

Lo Verde, G. & Monserrat, V.J. (1997) Nuovi dati sui Coniopterygidae siciliani. *Naturalista siciliana* 21, 57–66.

Meinander, M. (1972) A revision of the family Coniopterygidae (Planipennia). *Acta Zoologica Fennica* 136, 1–357.

Meinander, M. (1981) A review of the genus *Coniopteryx* (Neuroptera: Coniopterygidae). *Annales Entomologica Fennica* 47, 97–110.

Meinander, M. (1990). The Coniopterygidae (Neuroptera: Planipennia). A check-list of the species of the world, descriptions of new species and other new data. *Acta Zoologica Fennica* 189, 1–95.

Meinander, M. (1992) A review of the family Coniopterygidae (Insecta: Neuroptera). In *Current Research in Neuropterology, Proceedings of the 4th International Symposium on Neuropterology*, ed. Canard, M., Aspöck, H. & Mansell, M.W., pp. 255–260. Sacco, Toulouse.

Monserrat, V. (1996) Nuevos datos sobre los Coniopterygidos de Yemen (Neuroptera: Coniopterygidae). *Annales del Museo Civico Storia Naturale 'G.Doria'* 91, 1–26.

Nakahara, W. (1960) Systematic studies on the Hemerobiidae (Neuroptera). *Mushi* 34, 1–69.

New, T.R. (1984) The need for taxonomic revision in Chrysopidae. In *Biology of Chrysopidae*, ed. Canard, M., Séméria, Y. & New, T.R., pp. 37–41. Dr W.Junk, The Hague.

New, T.R. (1988) Neuroptera. In *Aphids, their Biology, Natural Enemies and Control*, vol. B, ed. Minks, A.K & Harrewijn, P., pp. 249–258. Elsevier, Amsterdam.

New, T.R. (1989) [1988] Hemerobiidae (Insecta: Neuroptera) from New Guinea. *Invertebrate Taxonomy* 2, 605–632.

Oswald, J.D. (1993) Revision and cladistic analysis of the world genera of the family Hemerobiidae (Insecta: Neuroptera). *Journal of the New York Entomological Society* 101, 143–299.

Oswald, J.D. (1994a) A new phylogenetically based subfamily of brown lacewings from Chile (Neuroptera: Hemerobiidae). *Entomologica Scandinavica* 25, 295–302.

Oswald, J.D. (1994b) A new genus and species of brown lacewing from Venezuela (Neuroptera: Hemerobiidae), with comments on the evolution of the hemerobiid forewing radial vein. *Systematic Entomology* 18, 363–70.

Penny, N.D., Adams, P.A. & Stange, L.A. (1997) Species catalog of the Neuroptera, Megaloptera and Raphidioptera of America north of Mexico. *Proceedings of the California Academy of Sciences* 50, 39–114.

Riek, E.F. (1975) On the phylogenetic position of *Brucheiser* Navas 1927 and description of a second species from Chile (Insecta: Neuroptera). *Studies on Neotropical Fauna and Environment* 10, 117–126.

Schneider, G.T. (1851) *Symbolae ad monographium generis Chrysopae*. Vratislava.

Séméria, Y. & Nel, A. (1990) *Paleochrysopa monteilsensis* gen. et sp. nov., a new fossil of Chrysopidae from the Upper Eocene formation of Monteils (France), with a review of known chrysopid fossils (Insecta: Neuroptera). In *Advances in Neuropterology, Proceedings of the 3rd International Symposium on Neuropterology*, ed. Mansell, M.W. & Aspöck, H., pp. 27–32. South African Department of Agricultural Development, Pretoria.

Stelzl, M. (1990) Nahrungsanalytische Untersuchungen an Hemerobiiden-Imagines (Insecta: Planipennia). *Mitteilungen der deutschen Gesellschaft fur allgemeine und angewandte Entomologie* 7, 670–676.

Sziraki, G. (1990) Two *Aleuropteryx* species from Tanzania (Planipennia: Coniopterygidae). *Folia Entomologica Hungarica* 51, 117–121.

Sziraki, G. (1992) Coniopterygidae of Hungary with a key to the identification of *Coniopteryx* Curtis females (Insecta: Neuroptera: Coniopterygidae). In *Current Research in Neuropterology, Proceedings of the 4th International Symposium on Neuropterology*, ed. Canard, M., Aspöck, H. & Mansell, M.W., pp. 359–366. Sacco, Toulouse.

Tauber, C.A. & Adams, P.A. (1990) Systematics of the Neuropteroidea: present status and future needs. *Virginia Polytechnic Institute and State University College of Agricultural and Life Sciences Information Series* 90, 151–64.

Tjeder, B. (1961) Neuroptera-Planipennia. The lacewings of southern Africa. 3. Family Hemerobiidae. *South African Animal Life* 8, 296–408.

Tjeder, B. (1966) Neuroptera-Planipennia. The lacewings of southern Africa. 5. Family Chrysopidae. *South African Animal Life* 12, 228–534.

Tjeder, B. (1970) Neuroptera. In *Taxonomist's Glossary of Genitalia in Insects*, 2nd edn, ed. Tuxen, S.L., pp. 89–99. Munksgaard, Copenhagen.

Tsukaguchi, S. (1995) *Chrysopidae of Japan (Insecta, Neuroptera)*. Privately published, Osaka.

Withycombe, C.L. (1924) Notes on the economic value of the Neuroptera, with special reference to the Coniopterygidae. *Annals of Applied Biology* 11, 112–125.

Yang, C.-K. & Yang, X.-K. (1989) Fourteen new species of green lacewings from Shaanxi Province (Neuroptera: Chrysopidae). *Entomotaxonomia* 11, 13–30.

Yang, C.-K. & Yang, X.-K. (1990) A review of the Chinese *Mallada* (Neuroptera: Chrysopidae). *Scientific Treatise on Systematics and Evolutionary Zoology* 1, 135–49.

Yang, X.-K. & Yang, C.-K. (1990) Examination and redescription of the type specimens of some Chinese Chrysopidae (Neuroptera) described by L. Navás. *Neuroptera International* 6, 75–83.

Yang, X.-K. & Yang, C.-K. (1992) Study on the genus *Chrysoperla* (Neuroptera: Chrysopidae). *Acta Entomologica Sinica* 35, 78–86.

Zimmerman, E.C. (1957) Order Neuroptera. In *Insects of Hawaii*, vol. 6, pp. 19–169. University of Hawaii Press, Honolulu.

CHAPTER 3

The common green lacewing (*Chrysoperla carnea s. lat.*) and the sibling species problem

C.S. Henry, S.J. Brooks, D. Thierry, P. Duelli, and J.B. Johnson

3.1 INTRODUCTION

Of the many genera of Chrysopidae, *Chrysoperla* Steinmann is the most important in agricultural programs of biological control (New, 1975; Karelin *et al.*, 1989; Brooks, 1994). Within *Chrysoperla*, the names of several of its 36 recognised species (Brooks, 1994) figure prominently in the literature of biological control, including *C. rufilabris* (Burmeister), *C. externa* (Hagen), *C. zastrowi* (Esben-Petersen), *C. pudica* (Navás), and *C. congrua* (Walker) (Barnes, 1975; Brettell, 1982; Albuquerque *et al.*, 1994; Legaspi *et al.*, 1996). The most frequently used control agent, however, is the common green lacewing, *Chrysoperla carnea* (Stephens). Long assumed to be a single widespread morphological species with an Holarctic distribution (Tjeder, 1960), 'C. carnea' has been mass-reared and released in croplands all over the world, including Europe (Alrouechdi, 1981; Bozsik, 1991; Sengonca *et al.*, 1995), western Asia (Gerling *et al.*, 1997), north Africa (Farag & Bleih, 1991), India (Thierry & Adams, 1992), North America (Ridgway & Jones, 1969; Tauber & Tauber, 1975; Chang *et al.*, 1995; Daane *et al.*, 1996), Central America (Tauber *et al.*, 1997), and even New Zealand (Dumbleton, 1936). However, evidence has been accumulating that 'C. carnea' is not a single species, but rather a complex of several to many remarkably similar, cryptic species (Tauber & Tauber, 1973a). This widespread mega-species is now more accurately called the *carnea*-group, one of four such species-groups into which *Chrysoperla* has recently been divided (Brooks, 1994).

Conclusions regarding species boundaries and diversity in the *carnea*-group depend upon species concept (Mayr, 1963; Templeton, 1989). Application of a morphological species concept placed the many, widespread populations of 'C. carnea' together in the first place (Tjeder, 1960), and it has been argued or assumed by some authors that lumping populations is the best way to answer systematic or ecophysiological questions (Tauber & Tauber, 1986*a*, *b*; Tauber *et al.*, 1997). On the other hand, a property of most valid species is reproductive isolation from all other populations, and formidable barriers to gene flow can exist between two populations without obvious morphological differences. Thus, the biological or reproductive species concept embraces any type of variation in ecological niche, seasonality, or reproductive behaviour that prevents two populations from exchanging genes under natural conditions (Templeton, 1989). Growing awareness of such biotic barriers to gene flow among populations has compelled current students of lacewing biology to recognise that there is much more to the *carnea*-group than just '*Chrysoperla carnea*'.

In this chapter, we critically examine the evidence for distinct species within the *carnea*-group, focusing on North America and Europe, where most of the recent systematic work has been done. First, we review gross morphology, fine structure, and ecophysiology, principally of adults but also of larvae. We then show how the confusing and conflicting results can be resolved by analysing traits contained within the unique courtship songs of these insects. Finally, results are described from recent molecular work enhancing and clarifying our understanding of lacewing systematics by testing hypotheses of phylogenetic relationship with independent data. We conclude by assessing the potential impact of these systematic findings on future programmes for biological control.

3.2 MORPHOLOGICAL CRITERIA

The use of morphology has a long history in the delineation of species boundaries in green lacewings. Many of the 1200 species in the family are delimited reasonably

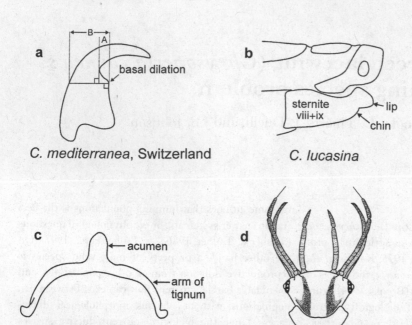

Fig. 3.1. Adult morphological traits of *carnea*-group green lacewings that have been used to distinguish cryptic species. a, Claw shape, quantified as the ratio B/A, where B is drawn perpendicular to the outer margin of the basal dilation; b, diagnostic lip and chin at apex of sternite *viii + ix* of male abdomen; c, tignum of male genitalia, showing acumen and arm from which a shape ratio is calculated; d, third instar larval head capsule, showing dorsal pigmentation patterns.

well by colour patterns and markings, wing venation, chaetotaxy, and especially male and female genitalia (Adams, 1957; Tjeder, 1966; Adams, 1969; Principi, 1977; Barnard, 1984; Brooks & Barnard, 1990; Henry *et al.*, 1992). However, the genus *Chrysoperla* is problematical. Although members of its *comans*-, *nyerina*-, and *pudica*-groups are rich in variable morphological characters, the *carnea*-group is exceptionally homogeneous (Brooks, 1994). Consequently, species diagnoses are ambiguous, which has led to the erection of many species of dubious validity. When Brooks (1994) undertook his revision of the genus, at least 81 different species or varietal names had been applied to all or parts of *C. carnea s. lat.*, based almost entirely on observed (or misinterpreted) morphological traits.

Lacewing systematists have gradually accepted the cryptic nature of species in the *carnea*-group and most now acknowledge the limited value of traditional morphology for delimiting these species. However, there are cogent practical reasons for identifying valid species morphologically, so the search has continued, particularly in Europe, for more reliable characters. In recent years, adult traits perceived to have that quality include ground colour and colour markings on the head, stipites, palps, abdomen, and wing veins; setal

length, colour, and distribution on the pronotum, costal and other wing veins, and abdominal sternites; wing proportions and shape at its apex; relative size and shape of the basal dilation of the pretarsal claw (Fig. 3.1a); shape of the lip on abdominal sternite *viii + ix* of the male (Fig. 3.1b); and shape of the components of the male genitalia (Fig. 3.1c; Brooks, 1994; Henry *et al.*, 1996, 1999a). In the larva (Fig. 3.1d), work has focused on patterns of head pigmentation (Tauber, 1974; Díaz-Aranda & Monserrat, 1990; Thierry *et al.*, 1992; Henry *et al.*, 1996).

3.2.1 Colour and markings of adults

Adult coloration and colour markings have the advantage of being obvious, but can change hue in specimens preserved in alcohol or become obscured by wrinkling or fading in dried specimens. Colours can also be locally variable across the range of species, which reduces their value as systematic characters.

Two examples of apparently valid species in the *carnea*-group that are characterised most obviously by body colour are North American *C. downesi* (Smith) and European *C. mediterranea* (Hölzel), both of which are darker green than other *Chrysoperla*. The colour is such that living individuals of the two species blend well with the foliage of coniferous trees and shrubs, and their strong natural association with conifers suggests that colour is adaptive (Hölzel, 1972; Tauber & Tauber, 1976). However, the total picture is more complicated. As discussed later in this chapter, the vibrational courtship songs of both *downesi* and *mediterranea* indicate that each taxon encompasses a wide range of morphology over a very broad geographical area, and includes variants that are lighter green or even yellowish. In *C. mediterranea*, for example, specimens from the eastern part of the range (Switzerland, Austria, and Slovakia) are a lighter shade of green, and are much more difficult to pick out of a collection than 'typical' individuals from the western shore of the Mediterranean Sea (Henry *et al.*, 1999*a*). The situation in *C. downesi* is even more extreme (Table 3.1). One population from the American southwest, unequivocally belonging to *C. downesi* by song, is so deviant from the type in colour and markings that it was considered a distinct species, *C. mohave* (Banks), diagnosed by yellow ground colour and black crossveins (Tauber & Tauber, 1973*b*; Henry, 1993*b*). That population is not associated with conifers, instead preferring desiccated deciduous shrubbery in parched chaparral (P.A. Adams, unpublished data). Thus, neither *mediterranea* nor *downesi* is a clear-cut example of a valid, colour-delimited species in the *carnea*-group. In fact, several cryptic species apparently occur within the morphological boundaries of *C. downesi*, suggesting the existence of a North American *C. downesi* complex (Wells & Henry, 1998) equivalent to the *C. carnea* complex in Europe (Thierry & Adams, 1992). On the other hand, the taxonomic unity of *C. mediterranea* is well supported by additional morphological, behavioural, and molecular data (Leraut, 1991; Cianchi & Bullini, 1992; Séméria, 1992; Brooks, 1994; Wells & Henry, 1998; Henry *et al.*, 1999*a*).

Chrysoperla lucasina (Lacroix) is another European species delimited at least partly by colour: adults have a dark pleural line on the first two abdominal segments. First synonymised with *carnea* by Aspöck *et al.* (1980), reinstated by Leraut (1991), but synonymised again by Brooks (1994), *lucasina* has been upheld as a valid species by colour markings (Leraut, 1991), male genitalia (Séméria, 1992), chaetotaxy and maxillary markings (Thierry *et al.*, 1992), pretarsal claw dimensions (Thierry *et al.*, 1998), and courtship songs combined with adult morphology (Henry *et al.*, 1996). As in the two species described above, however, inconsistent results are obtained outside of the central range of the species. For example, the dark pleural stripe, thought to be an autapomorphy of *lucasina*, is also found in acoustically distinct members of the *carnea*-group collected in Kyrgyzstan, central Asia (Wells & Henry, 1998; and S. J. Brooks, unpublished data). Nonetheless, the validity of *C. lucasina* is not seriously in doubt.

3.2.2 Chaetotaxy

In central France, Leraut (1991) distinguished between two forms of '*C. carnea*', one with light-coloured, long setae on the wing veins and the other bearing short, dark setae, which he called '*carnea* A' and '*carnea* B', respectively. Both types have relatively large basal dilations of the pretarsal claw, thus avoiding confusion with *C. renoni* (Lacroix) or *C. mediterranea*, which in France are diagnosed by a minute basal dilation (Leraut, 1991; and see later discussion in this chapter). Subsequently, Leraut (1992) assigned '*carnea* A' to *C. carnea s. str.* and '*carnea* B' to *C. kolthoffi* (Navás). Thierry and his colleagues then substantiated Leraut's interpretation of *kolthoffi* and *carnea s. str.* using additional approaches, including studies of ecophysiology (Thierry *et al.*, 1995) and larval colour (Thierry *et al.*, 1992) and two multivariate analyses of adult morphology (Thierry *et al.*, 1992, 1998). Thierry *et al.* (1998) furthermore specified the presence of dark rather than pale setae on the abdominal sternites to distinguish *kolthoffi* from *carnea s. str.*, and modified Leraut's original diagnosis of *kolthoffi* from short to long costal setae. Unfortunately, problems arise when one looks beyond central France, or incorporates additional characters in the analyses. Despite efforts to apply the concept of *C. kolthoffi* and *C. carnea s. str.* across the breadth of Europe (Thierry *et al.*, 1996), it has proved impossible to reconcile the two morphological phenotypes with the three song phenotypes found in populations of the *carnea*-group in southern Europe (Duelli, 1996; and C. S. Henry, S. J. Brooks, P. Duelli, and J. B. Johnson, unpublished data). Even if '*kolthoffi*' proves to have validity, it will need another name, because *C. kolthoffi* is presently a synonym of *C. nipponensis* (Okamoto) (Brooks, 1994). We will discuss the status of these two problematical species again later in this chapter.

A species erected largely because of its long costal and prothoracic setae is *C. ankylopteryformis* (Monserrat & Diaz-Aranda, 1989), described as a rare species from Spain occupying extremely hot, dry habitats. However,

Table 3.1. *Summary of existing information about the cryptic song species of the* Chrysoperla carnea *complex in North America and Europe*

Song species	Morphotypes	Song type	Geographic range	Habitats	Body colour	Pronotal stripe	Head marks	Gradate veins	Claw dilation[a]	Diapause colour	Setae[b]	Lip[c]	Fore wing apex	Other
North America														
C. plorabunda	(none)	plorabunda	N.A.	field/forest	green	yellow	br/blk	green	M	brown	S/P	??	rounded	
C. adamsi	(none)	adamsi	W. N.A.	mixed for.	green	yellow	br/blk	green	M	mixed	S/D	??	pointed	
C. johnsoni	(clinal variation)	johnsoni	W. N.A.	mixed for.	bright gr	yellow	br/blk	gr/bl	M	no △	S/D	??	rounded	
	mohave	johnsoni	coastal Cal	chaparral	yel-gr	yellow	red spot	black	B	yellow	S/P		rounded	
C. downesi	downesi-1	downesi	E. N.A.	conifers	dark gr	white	br/blk	green	N	no △	S/D	??	pointed	
	mohave	downesi	S. Cal.	chaparral	yel-gr	yellow	red spot	black	B	yellow	S/P		rounded	univoltine
C. 'downesi-2'	(not known)	slo-downesi	W. N.A.	mixed for.	bright gr	white	red spot	green	N	no △	S/D	??	pointed	
C. 'downesi-3'	(not known)	odd-downesi	S. Cal.	conifers	emerald	indistinct	pale gr	green	??	no △	S/D	??	pointed	
Europe														
C. lucasina	(none)	lucasina	C./S. Eur., N. Africa	field/forest	green	yellow	br/blk	green	M	no △	S/D	L/B/P	pointed	pleural stripe
C. mediterranea	(clinal variation)	mediterranea	C./S. Eur., N. Africa	conifers	dark to bright gr	white	red band	green	vN to M	no △	S/D	S/N/D	narrow, pointed	
C.c.2	carnea s. str. Leraut/Thierry	slow-motorboat	W. and C. Europe	decid. for.	green	yellow	br/blk	green	M to B	mixed	L/P	S/N/P	rounded	
C.c.3	(not known)	maltese	S. Europe	field/forest	green	yellow	br/blk	green	M	mixed	S/D	S/B/P	pointed	
C.c.4	kolthoffi/ affinis/ carnea (Stephens)	motorboat	C. and N. Europe	field/forest	green	yellow	br/blk	green	M to B	brown	S/D,[b] S/P	L/B/P	rounded	
C.c.5	C. sillemi	generator	W. Asia and N. Africa	semi-desert	light green	yellow	pale br	green	N	mixed	S/D, S/P	S/N/P	pointed	straight unquis
Undetermined														
C. renoni		unknown	C. Europe	decid. for.	green	yellow	br/blk	green	vN	??	L/P	??	rounded	long setae
ankylopteryformis		unknown	Spain	semi-desert	green	yellow	br/blk	green	vN		L/P		rounded	

Notes:

[a] Basal dilation of claw: B=broad, M=medium, N=narrow, vN=very narrow.

[b] Setae on costal margin of wing and abdominal sternites: S=short, L=long; P=pale, D=dark. Note: Thierry *et al.* (1998) now consider setal length in '*kolthoffi*' to be long, not short.

[c] Lip on abdominal sternite *viii*+*ix*: S=short, L=long; N=narrow, B=broad or bulbous; D=downturned, P=directed posteriorly.

both Leraut (1991) and Aspöck (1992) detected a strong resemblance between *ankylopteryformis* and a previously described southern European species associated with moist, deciduous habitats, *C. renoni*. The new species was formally synonymised with the latter by Brooks (1994) in his revision of *Chrysoperla*, although several workers continue to recognise both species, based on habitat differences (above) and relative breadth of the fore wing (broader in *ankylopteryformis*; Thierry *et al.*, 1998). At the present time, *renoni* (including *ankylopteryformis*) remains tenuously characterised by the simultaneous presence of long setae and a minute dilation of the pretarsal claw (Brooks, 1994). That combination of traits is also thought to differentiate *renoni* from *carnea s. str.*, discussed earlier. However, in view of how geographically variable setal length and basal dilation have proved to be in other cryptic lacewing species (e.g. Henry *et al.*, 1999*a*), the ultimate validity of *renoni* is uncertain.

Colour and length of setae on the costal margin and other wing veins were also used by Leraut (1991) to substantiate the identity of *lucasina* and *mediterranea*. Thierry *et al.*'s (1992) morphometric study supporting Leraut's groupings utilises several setal characters of the wing and prothorax as well. However, setal colour and length have generally proved to be unreliable for species identification as our knowledge of the *C. carnea* complex has grown more sophisticated (Thierry *et al.*, 1997).

3.2.3 Shape parameters

Shape is a slippery concept, and although it promises considerable power it must be used with caution in any systematic study. In the *carnea*-group, shape analysis has been applied to wings, male genitalic structures, sternites near the end of the male abdomen, and pretarsal claws.

The length/width ratio of the wing describes how narrow it is, and this measure has been used to separate several closely related species in the *carnea*-group. In North America, for example, the late Dr Ellis MacLeod was convinced (unpublished data) that the wings of *C. downesi* were consistently narrower than those of its sympatric sibling *C. plorabunda*. In Europe, a parallel situation exists in narrow-winged *C. mediterranea* versus broader-winged *C. renoni*, with which *mediterranea* shares a minute dilation of the pretarsal claw and might therefore be confused (Brooks, 1994). However,

later work on both continents (C. S. Henry, unpublished data; and Henry *et al.*, 1999*a*) has demonstrated that wing proportions are too variable to be truly diagnostic in any of the above taxa, especially over the entire geographic range of each species. As a case in point, the fore wing of *mediterranea* is narrower than that of other members of the *C. carnea* complex on average (3.32 vs. 3.02, $n = 65$ and 151), but its range of variation (2.83–3.72) overlaps that of other cryptic species. The shape of the apex of the fore wing has also been used to delineate some species within the *carnea*-group, e.g., *C. lucasina* from three other European song species (Henry *et al.*, 1996). Unfortunately, intraspecific variation also limits the value of this feature when comparing *lucasina* to cryptic species other than those three (Brooks, 1994).

Male genitalia (Fig. 3.1c) are a potentially rich source of taxonomic characters in lacewings, but unfortunately vary little among species of the *carnea*-group. Nonetheless, Séméria (1992) claims to have successfully separated six species (*C. renoni*, *lucasina*, *mediterranea*, *C. nanciensis* Séméria, *carnea s. str.*, and '*kolthoffi*') in France using the size and shape of the lateral plates of the gonarcus. Brooks (1994) was unable to replicate Séméria's results, probably because the gonarcal arms become increasingly sclerotised with age in living individuals, leading to apparent changes in shape when genitalia are compressed under a cover slip for microscopic examination (Adams & Penny, 1987).

In profile view, the sternites of the terminus of the male abdomen assume slightly different shapes that may be specific to some taxa in the *carnea*-group. The presence of a lip on sternite *viii + ix* (Fig. 3.1b) is autapomorphic for *Chrysoperla* (Brooks & Barnard, 1990). A survey of this character throughout the cryptic species so far segregated within the *C. carnea* complex of Europe has revealed that the shape of the lip and associated chin is usually diagnostic. For example, in *C. mediterranea*, the lip is shorter and narrower than in *C. lucasina* and, instead of being directed posteriorly, it is downturned (Henry *et al.*, 1999*a*). This feature is worthy of further investigation.

Shape of the pretarsal claw (Fig. 3.1a) has a recent history of use in the *carnea*-group and remains one of the most promising diagnostic features. It may be interpreted as the relative size of the basal expansion expressed as a binary character (large vs. minute; see Brooks & Barnard, 1990; Leraut, 1991), as a simple ratio (Brooks, 1994; Henry *et al.*, 1996), or as a multi-variate statistic encompassing several angles and distances (Thierry *et al.*, 1998). Using the simple ratio, Henry *et al.* (1999a) were unable to separate all specimens of *C. mediterranea* (initially identified by their unique courtship song) from other *carnea s. lat.* of Europe. On the other hand, application of the multivariate statistical approach successfully resolved *mediterranea*, *lucasina*, *carnea s. str.*, and '*kolthoffi*' from Portugal to Poland and Bulgaria (Thierry *et al.*, 1998). However, the challenge remains to ascertain whether taxa diagnosed principally by tarsal claw morphology are robust enough to withstand an independent test of taxon groupings based on other characters more closely associated with reproductive isolation. As yet there is no evidence to support the hypothesis that these morpho-taxa are reproductively isolated entities.

3.2.4 Colour and markings of larvae

Larvae have been little used as sources of systematic characters in the Chrysopidae. Tauber & Tauber (1976) used dorsal head markings (Fig. 3.1d) of all three larval stadia to identify and distinguish six common species of the *carnea*- and *pudica*-groups of North American *Chrysoperla*, but their descriptions and keys are compromised by geographical variation. Years later, when Thierry *et al.* (1992) applied similar criteria to first-instar *carnea*-group larvae from central France, they found that larval head colour co-varied with adult colour markings and chaetotaxy, allowing them reliably to segregate *C. lucasina*, '*kolthoffi*', and *carnea s. str.* using multivariate statistics. Most recently, however, two studies of *C. lucasina* and *C. mediterranea*, which included detailed analyses of larval morphology over most of Europe, failed to find any consistent, useful differences in the colour patterns of the larval head capsule of those species (Henry *et al.*, 1996, 1999a). Again, we are forced to conclude that the general value of larval morphology to the systematics of the *carnea*-group is questionable, although morphological differences of larvae can be useful for segregating species from one another in local areas of sympatry.

3.3 ECOPHYSIOLOGICAL CRITERIA

It is fitting that ecophysiology should remain a valuable source of systematic traits in the *carnea*-group, considering that the genus *Chrysoperla* was itself erected to accommodate taxa which overwintered as adults (Séméria, 1977). Active research has centred on seasonal cycles, diapause characteristics, and adult dietary preferences.

Much work has been published by the Taubers (1970, 1976, 1977, 1981, and papers cited therein) supporting genetically-based differences between North American *C. plorabunda* (called *C. carnea* by the authors) and *C. downesi* in diapause physiology, life cycles, and habitat choice. Those studies suggest that the two taxa are well-defined sibling species. Certainly in the eastern United States, *downesi* is univoltine, conifer-associated, and green throughout the winter, whereas *plorabunda* is a multivoltine meadow species with a reddish-brown winter phase. However, both species have wide ranges, and in the western part of the continent are much less distinct ecophysiologically. Also present in the west are aberrant populations like '*C. mohave*', which in Strawberry Canyon (Berkeley, California) has a facultative summer diapause, terminated by the presence of aphid prey (Tauber & Tauber, 1973b). The Taubers have attributed such increased variation in western North America to local adaptation mediated by phenotypic plasticity of life-history traits within '*carnea*' (= *plorabunda*) and *downesi* (Tauber & Tauber, 1986a, 1992, and papers cited therein). As discussed below, it is more likely that some of these 'plastic' responses are instead phenotypic expressions of distinct genotypes associated with cryptic song species hidden in the *plorabunda* and *downesi* complexes.

Ecophysiological studies in Europe have yielded important insights into the systematics of the *carnea*-group there. For example, both *C. mediterranea* and *C. lucasina* overwinter without changing colour, while most other European *carnea*-group species turn from green to reddish-brown (Thierry *et al.*, 1995). That fact has unanticipated systematic significance. In his original description of *C. carnea* from London, Stephens (1835) specifically noted the insect's reddish winter phase (hence the name '*carnea*'). To date, only *lucasina* and a song species resembling Leraut/Thierry's '*kolthoffi*' (*C.c.4* 'motorboat') have been documented as present in the British Isles. We know that Stephens description could not refer to *lucasina*, because that species remains green throughout the winter. By default, then, 'true *carnea*' becomes *C.c.4*/ '*kolthoffi*', which is the most common cryptic species of the

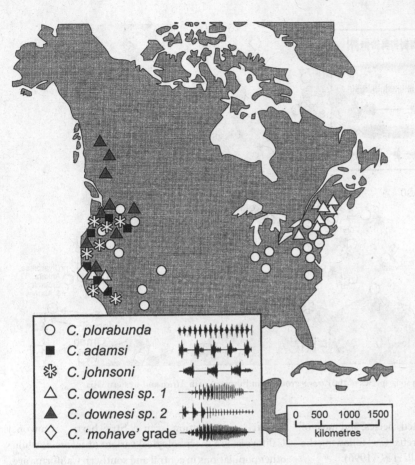

Fig. 3.2. Collecting sites of five known song species of the *carnea*-group from North America. Note that *C. mohave* is considered a grade, with populations that are acoustically either *C. johnsoni* or *C. downesi*.

Legend:
- ○ *C. plorabunda*
- ■ *C. adamsi*
- ✳ *C. johnsoni*
- △ *C. downesi sp. 1*
- ▲ *C. downesi sp. 2*
- ◇ *C. 'mohave' grade*

0   500   1000   1500
kilometres

carnea-group in northern Europe and turns reddish in winter. However, a third taxon has also been tentatively identified from a few British specimens based on certain, somewhat variable, morphological traits. That taxon apparently corresponds to a different song type (*C.c.2*), which also has a reddish winter phase, but as yet no populations have been found in Britain which produce songs of this type. The problem is discussed further below, but at present we cannot be certain whether it is *C.c.2* or *C.c.4* that corresponds to the 'true *carnea*' of Stephens (1835).

3.4 MATING SIGNALS

As just suggested, another approach to existing systematic confusion and conflicts lies in the study of mating signals. Sexually receptive green lacewings of the *carnea*-group produce low-frequency, substrate-borne signals by vibrating their abdomens, and signals thus produced are picked up by subgenual organs in the tibiae of the legs of potential mates (Devetak & Pabst, 1994; Devetak & Amon, 1997). Songs are quite elaborate and identically expressed in both sexes, and mating will not occur under natural conditions unless the participants engage in a prolonged and accurate duet (Wells & Henry, 1992). Mating signals such as these are part of specific mate recognition systems and therefore under strong stabilising selection in each species (Paterson, 1986; Butlin & Ritchie, 1994). Thus, plasticity within a species in its non-sexual morphological characters may be expected, but variation in courtship songs will be slight. In lacewings, as in other animals, mating signals will be reliable indicators of species identity over wide geographical ranges.

To date, we have described in more or less detail the songs of five cryptic song species in North America (Figs. 3.2, 3.4), six cryptic song species in Europe (Figs. 3.3, 3.4), and four more in central and eastern Asia. In some cases, the song species correspond roughly to existing taxa. However, others are undescribed, or

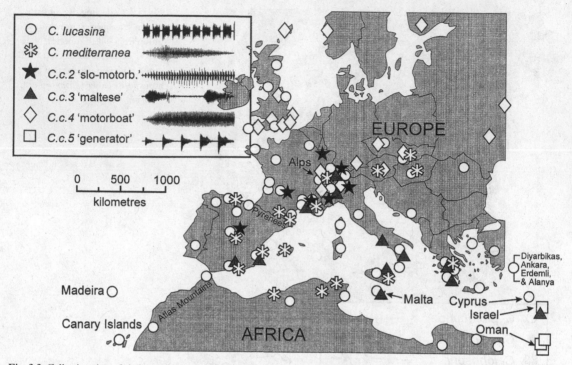

Fig. 3.3. Collecting sites of six known song species of the *carnea*-group from Europe, north Africa and western Asia.

uncertainly associated with named species, and most of those are referred to using the prefix '*C.c.*' following the nomenclature proposed by Duelli *et al.* (1996).

The importance of songs in the reproductive isolation of cryptic lacewing species was first recognised in North America (Wells & Henry, 1992). Initially, the distinct songs of *plorabunda* and *downesi* were described (Henry, 1979, 1980). Then, two new species, *C. adamsi* and *C. johnsoni*, were erected for song variants hidden in the *C. plorabunda* complex (Henry, 1991, 1993a; Henry *et al.*, 1993). These four species have broad, extensively overlapping geographic ranges (Fig. 3.2) yet do not hybridise in nature (Wells & Henry, 1994), and their songs show little intraspecific variation over thousands of kilometres (Henry & Wells, 1990). Additional cryptic species remain to be described in western North America, including at least two members, '*D2*' and '*D3*', of the *C. downesi* complex (Wells & Henry, 1998). The power of songs to resolve systematic problems is illustrated by the case of '*C. mohave*', discussed earlier as a confusing taxon. Song analysis shows clearly that '*mohave*' is really a morphological grade, attained independently by local populations of two different species in response to arid conditions (Henry, 1993b). One

'*mohave*' population, in Strawberry Canyon, California, is a morphological variant of *johnsoni*, while other populations in central and southern California are variants of *downesi* (Table 3.1).

In Europe (Figs 3.3, 3.4), the courtship songs of *lucasina* and *mediterranea* have been studied in detail (Henry *et al.*, 1996, 1999a). Both studies found much greater morphological variability within each song-defined species than had previously been suspected, leading to revised species diagnoses. Another song species, *C.c.5* 'generator' collected from Israel, Dubai, and Oman, is very likely part or all of *C. sillemi* (Esben-Petersen), based on range and similar, extreme pretarsal claw morphology (Brooks, 1994; Henry *et al.*, 1996). The three other (known) European song species, *C.c.2, 3,* and *4,* are less easily assigned to currently accepted morphological species. Each is widespread in Europe and, like *downesi, johnsoni, lucasina* and *mediterranea,* exhibits confusing and overlapping patterns of morphological variability (C. S. Henry, S. J. Brooks, J. B. Johnson, and P. Duelli, unpublished data). The southerly distributed *C.c.3* 'maltese' has no clear morphological counterpart. In contrast, *C.c.2* 'slow motorboat' is probably Leraut/Thierry's '*C. carnea s.*

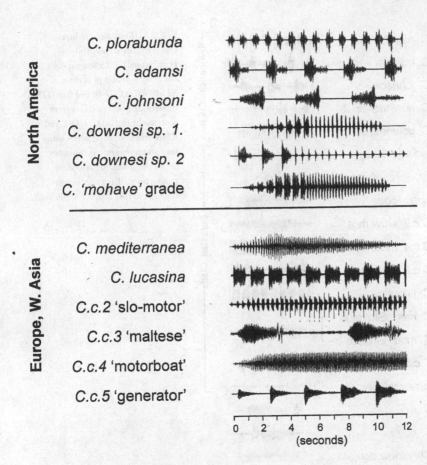

Fig. 3.4. Oscillographs of typical vibrational mating signals of the North American and European song species of the *carnea*-group shown in Figs. 3.2 and 3.3. Twelve seconds of each song is shown.

str.', at least in part, while *C.c.4* 'motorboat' corresponds most closely to Leraut/Thierry's '*kolthoffi*'. It remains unresolved, however, which song species corresponds to the true *carnea* of Stephens (1835). The most obvious candidate in Britain, *C.c.4*, differs from the lectotype of *C. carnea* (Stephens) in having darker-coloured setae and a narrower basal dilation of the claw, which make it more similar to the lectotype of *C. affinis* (Stephens) (synonymised with *C. carnea* by Schneider 1851). *C.c.2* has light-coloured sternal setae and a relatively large basal dilation of the claw, as in the lectotype of *C. carnea* (Stephens), but it has not been found in the British Isles; if it has merely been overlooked, *C.c.2* could prove to be the 'true *carnea*'. In any case, the names '*carnea*' and '*affinis*' can be correctly assigned to specific song types only after additional study of the degree of variation of the morphological characters in those taxa and further clarification of the distribution of the song species.

Known song species from both Europe and North America are matched as accurately as possible to described 'species', morphological characters, and eco-physiological features in Table 3.1.

3.5 MOLECULAR SYSTEMATICS

DNA sequence data have been used recently to explore the genetic integrity of song-defined species and infer phylogenetic relationships among them (Wells & Henry, 1998). Fifteen cryptic species of the *carnea*-group, plus three outgroup lacewing taxa, were compared with respect to 1110 nucleotide sites from the COII and ND2 mitochondrial genes (Henry *et al.*, 1999*b*). Results from analyses using maximum parsimony, minimum evolution, and maximum likelihood confirm that individuals within each cryptic species are more closely related to one another than to members of any other taxon (Fig. 3.5). The data also show that the song species form two monophyletic clades, one North American, the other Eurasian. Relationships among five North American species are resolved, but those among ten Eurasian species are ambiguous. This study

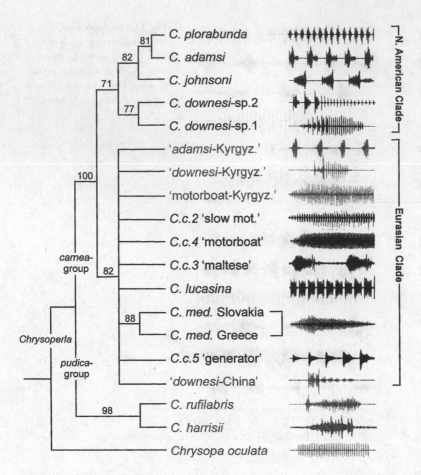

Fig. 3.5. Bootstrap phylogeny (50% majority rule, 100 replications) of 15 song species of the *carnea*-group plus three additional taxa, inferred from DNA nucleotide sequence data from sections of the ND2, COII, and *t*-RNA mitochondrial genes using maximum parsimony. Numbers above the nodes are bootstrap proportions. Taxa specifically relevant to this chapter are in bold.

strengthens the validity of song-based species in the *carnea*-group because, in every case, multiple individuals sharing a song type group together in the phylogenetic analysis.

In contrast, multilocus electrophoretic studies of morphological species and ecophysiological populations of the *carnea*-group in Europe show groupings of individuals that do not correspond well with traditional systematic or ecological ideas (Bullini & Cianchi, 1984; Bullini *et al.*, 1984; Cianchi & Bullini, 1992). That is probably due to the existence of cryptic song species, unrecognised by the authors when the studies were performed. It would be useful to redo these experiments using specimens whose song phenotypes are known.

3.6 CONCLUSION AND PROGNOSIS

Recognition of the existence of song species within the *carnea*-group is a two-edged sword. On the one hand, the new behavioural approach makes sense of the con-

fusing assemblage of incongruent morphological observations that has marked lacewing systematics for many years, by providing at last a robust set of stable characters with which to recognise biological species over broad geographical areas. However, this knowledge has also weakened existing diagnoses of what were considered to be good morphological species, including *downesi* and *plorabunda* in North America and *mediterranea*, *lucasina*, 'carnea s. str.', and 'kolthoffi' in Europe. Instead of being morphologically uniform, song species typically display disconcertingly high levels of intraspecific variation, often as great as levels previously thought to exist among different morphological species. Only in local areas of sympatry can species of the *carnea*-group be segregated morphologically with any confidence. Even then, for example in central and western Europe, we must rely upon suites of characters rather than single diagnostic traits to determine species identities.

Knowledge gained from courtship songs about lacewing systematics is also difficult to apply. Without possession of living, sexually receptive specimens, it is impossible to perform the necessary identifications. The conscientious observer can learn to recognise the different song types by eye without the need for recording equipment, but that still presumes access to living, non-diapausing individuals.

Regardless of such difficulties, it is very important that practitioners of biological control determine which cryptic species within the *carnea*-group they are rearing and releasing. It is clear that complex ecophysiological, as well as behavioural, differences exist within and between such species, which often occur sympatrically. Those differences undoubtedly make certain taxa or populations in the *carnea*-group more suitable as predators of plant lice than others, which may be characterised by inappropriate ecological or seasonal adaptations and incompatible mating signals.

Unfortunately, there are no solutions to the *C. carnea* sibling species problem that will be acceptable to all biologists. Systematists working with preserved specimens of these lacewings will always be faced with ambiguous suites of morphological and ecophysiological features, which will often be divergent at different sites. Applied entomologists will be forced to adopt a conservative approach to biological control, rearing and releasing only local *carnea*-group populations at any given site. Evolutionary biologists will be happy, however, because this swarm of rapidly evolving cryptic species provides a rare window into a complex, ongoing process of differentiation and speciation.

ACKNOWLEDGMENTS

This work was supported in part by NSF Award DEB-9220579 to Charles S. Henry and by the Research Foundation of University of Connecticut to C. S. Henry and M. L. M. Wells. All molecular studies were carried out in the laboratory of Dr Chris Simon, to whom we express deepest gratitude. Special thanks go to Dr Marta Lucía Martínez Wells (University of Connecticut), who was largely responsible for the DNA sequence work. Drs M.L.M. Wells and Cynthia S. Jones (University of Connecticut) reviewed an earlier version of the manuscript and provided constructive comments. We dedicate the chapter to the late Phil Adams (1929–1998), mentor and dear friend of the first author. He was a biologist and lacewing systematist without peer, whose life's work met the very highest standards of scholarship.

REFERENCES

Adams, P.A. (1957) A new genus and new species of Chrysopidae from the western United States, with remarks on the wing venation of the family (Neuroptera). *Psyche* 63, 67–74.

Adams, P.A. (1969) New species and synonymy in the genus *Meleoma* (Neuroptera, Chrysopidae), with a discussion of genitalic homology. *Postilla* 136, 1–18.

Adams, P.A. & Penny, N.D. (1987) Neuroptera of the Amazon Basin. Part 11a. Introduction and Chrysopini. *Acta Amazonica* 15, 413–479.

Albuquerque, G.S., Tauber, C.A. & Tauber, M.J. (1994) *Chrysoperla externa* (Neuroptera, Chrysopidae): life history and potential for biological control in Central and South America. *Biological Control* 4, 8–13.

Alrouechdi, K. (1981) Relations comportementales et trophiques entre *Chrysoperla carnea* (Stephens) (Neuroptera; Chrysopidae) et trois principaux ravageurs de l'olivier. *Neuroptera International* 1, 122–134.

Aspöck, H. (1992) The Neuropteroidea of Europe: a review of the present knowledge. In *Current Research in Neuropterology, Proceedings of the 4th International Symposium on Neuropterology*, ed. Canard, M., Aspöck, H. & Mansell, M.W., pp. 43–56. Sacco, Toulouse.

Aspöck, H., Aspöck, U. & Hölzel, H. (1980) *Die Neuropteren Europas. Eine zusammenfassende Darstellung der Systematik, Ökologie und Chorologie der Neuropteroidea (Megaloptera, Raphidioptera, Planipennia)*. Goecke and Evers, Krefeld.

Barnard, P.C. (1984) Adult morphology related to classification. In *Biology of Chrysopidae*, ed. Canard, M., Séméria, Y. & New, T.R. pp. 19–29. Dr W. Junk, The Hague.

Barnes, B.N. (1975) The susceptibility of *Chrysopa zastrowi* Esb.-Pet. (Neuroptera: Chrysopidae) to two insecticides in the laboratory. *Phytophylactica* 7, 131–132.

Bozsik, A. (1991) Effect of chemicals on aphidophagous insects – response of adults of common green lacewing *Chrysoperla carnea* to pesticides. In *Behaviour and Impact of Aphidophaga*, ed. Polgár, L., Chambers, R.J., Dixon, A.F.G. & Hodek, I., pp. 297–304. SPB Academic Publishing, The Hague.

Brettell, J.H. (1982) Green lacewings (Neuroptera: Chrysopidae) of cotton fields in central Zimbabwe. 2. Biology of *Chrysopa congrua* and *C. pudica* and toxicity of certain insecticides to their larvae. *Zimbabwe Journal of Agricultural Research* 20, 77–84.

Brooks, S.J. (1994) A taxonomic review of the common green lacewing genus *Chrysoperla* (Neuroptera: Chrysopidae). *Bulletin of the British Museum (Natural History) Entomology* 63, 137–210.

Brooks, S.J. & Barnard, P.C. (1990) The green lacewings of

the world: a generic review (Neuroptera: Chrysopidae). *Bulletin of the British Museum (Natural History) Entomology*, 59, 117–286.

Bullini, L. & Cianchi, R. (1984) Electrophoretic studies on gene-enzyme systems in chrysopid lacewings. In *Biology of Chrysopidae*, ed. Canard, M., Séméria, Y. & New, T.R., pp. 48–56. Dr W. Junk, The Hague.

Bullini, L., Principi, M.M. & Cianchi, R. (1984) Electrophoretic studies in the genus *Chrysopa* (s.l.), evolutionary and phylogenetic inferences. In *Progress in World's Neuropterology, Proceedings of the 1st International Symposium on Neuropterology*, ed. Gepp, J., Aspöck, H. & Hölzel, H., pp. 57–59. Thalerhof, Graz.

Butlin, R.K. & Ritchie, M.G. (1994) Mating behaviour and speciation. In *Behaviour and Evolution*, ed. Slater, P.J. & Halliday, T.R., pp. 43–79. Cambridge University Press, Cambridge.

Chang, Y.F., Tauber, M.J. & Tauber, C.A. (1995) Storage of the mass-produced predator *Chrysoperla carnea* (Neuroptera: Chrysopidae): influence of photoperiod, temperature, and diet. *Environmental Entomology* 24, 1365–1374.

Cianchi, R. & Bullini, L. (1992) New data on sibling species in chrysopid lacewings: the *Chrysoperla carnea* (Stephens) and *Mallada prasinus* (Burmeister) complexes (Insecta: Neuroptera: Chrysopidae). In *Current Research in Neuropterology, Proceedings of the 4th International Symposium on Neuropterology*, ed. Canard, M., Aspöck, H. & Mansell, M.W., pp. 99–104. Sacco, Toulouse.

Daane, K.M., Yokota, G.Y., Zheng, Y. & Hagen, K.S. (1996) Inundative release of common green lacewings (Neuroptera: Chrysopidae) to suppress *Erythroneura variabilis* and *E. elegantula* (Homoptera: Cicadellidae) in vineyards. *Environmental Entomology* 25, 1224–1234.

Devetak, D. & Amon, T. (1997) Substrate vibration sensitivity of the leg scolopidial organs in the green lacewing, *Chrysoperla carnea*. *Journal of Insect Physiology* 43, 433–437.

Devetak, D. & Pabst, M.A. (1994) Structure of the subgenual organ in the green lacewing, *Chrysoperla carnea*. *Tissue and Cell* 26, 249–257.

Díaz-Aranda, L.M. & Monserrat, V.J. (1990) Estadíos larvarios de los Neurópteros Ibéricos. VI: *Chrysoperla carnea* (Stephens, 1836), *Chrysoperla mediterranea* (Hölzel, 1972) y *Chrysoperla ankylopteryformis* Monserrat y Díaz-Aranda, 1989 (Insecta, Neuroptera: Chrysopidae). *Boletín de Sanidad Vegetal Plagas* 16, 675–689.

Duelli, P. (1996) The working group '*carnea*-complex:' Report on activities, results and cooperative projects. In *Pure and Applied Research in Neuropterology, Proceedings*

of the 5th International Symposium on Neuropterology, ed. Canard, M., Aspöck, H. & Mansell, M.W., pp. 307–311. Sacco, Toulouse.

Duelli, P., Henry, C.S. & Johnson, J.B. (1996) Kryptische Arten am Beispiel der Florfliegen: Eine Herausforderung für die Systematik, die angewandte Entomologie und den Naturschutz. In *Verhandl 14 International Symposium über Entomofaunistik in Mitteleuropa, SIEEC,* Munich, 1994, ed. Gerstmeier, R., pp. 383–387. SIEEC, Munich.

Dumbleton, L.J. (1936) The biological control of fruit pests in New Zealand. *New Zealand Journal of Science and Technology* 18, 588–592.

Farag, A.I. & Bleih, S.B. (1991) Effect of sequential insecticide treatments on developmental stages of *Chrysoperla carnea* in Egyptian cotton fields. In *Behaviour and Impact of Aphidophaga*, ed. Polgár, L., Chambers, R.J., Dixon, A.F. G. & Hodek, I., pp. 313–317. SPB Academic Publishing, The Hague.

Gerling, D., Kravchenko, V. & Lazare, M. (1997) Dynamics of common green lacewing (Neuroptera: Chrysopidae) in Israeli cotton fields in relation to whitefly (Homoptera: Aleyrodidae) populations. *Environmental Entomology* 26, 815–827.

Henry, C.S. (1979) Acoustical communication during courtship and mating in the green lacewing *Chrysopa carnea* (Neuroptera: Chrysopidae). *Annals of the Entomological Society of America* 72, 68–79.

Henry, C.S. (1980) The courtship call of *Chrysopa downesi* Banks [*sic*] (Neuroptera: Chrysopidae): its evolutionary significance. *Psyche* 86, 291–297.

Henry, C.S. (1991) The status of the P2 song morph, a North American green lacewing of the *Chrysoperla carnea* species-group (Neuroptera: Chrysopidae). *Canadian Journal of Zoology* 69, 1805–1813.

Henry, C.S. (1993a) *Chrysoperla johnsoni* Henry (Neuroptera: Chrysopidae): acoustic evidence for full species status. *Annals of the Entomological Society of America* 86, 14–25.

Henry, C.S. (1993b) *Chrysoperla mohave* (Banks) (Neuroptera: Chrysopidae): two familiar species in an unexpected disguise. *Psyche* 99, 291–308.

Henry, C.S. & Wells, M.M. (1990) Geographical variation in the song of *Chrysoperla plorabunda* in North America (Neuroptera: Chrysopidae). *Annals of the Entomological Society of America* 83, 317–325.

Henry, C.S., Penny, N.D. & Adams, P.A. (1992) Chapter 28. The neuropteroid orders of Central America: Neuroptera and Megaloptera. In *Insects of Panama and Mesoamerica: Selected Studies*, ed. Quintero, D. & Aiello, A., pp. 432–458. Oxford University Press, Oxford.

Henry, C.S., Wells, M.M. & Pupedis, R.J. (1993) Hidden

taxonomic diversity within *Chrysoperla plorabunda* (Neuroptera: Chrysopidae): two new species based on courtship songs. *Annals of the Entomological Society of America* 86, 1–13.

Henry, C.S., Brooks, S.J., Johnson, J.B. & Duelli, P. (1996) *Chrysoperla lucasina* (Lacroix): a distinct species of green lacewing, confirmed by acoustical analysis (Neuroptera: Chrysopidae). *Systematic Entomology* 21, 205–218.

Henry, C.S., Brooks, S.J., Duelli, P. & Johnson, J.B. (1999*a*) Revised concept of *Chrysoperla mediterranea* (Hölzel), a green lacewing associated with conifers: Courtship songs across 2800 kilometers of Europe (Neuroptera: Chrysopidae). *Systematic Entomology* 24, 335–350.

Henry, C.S., Wells, M.M. & Simon, C. (1999*b*) Convergent evolution of courtship songs among cryptic species of the *carnea*-group of green lacewings (Neuroptera: Chrysopidae: *Chrysoperla*). *Evolution* 53, 1165–1179.

Hölzel, H. (1972) *Anisochrysa (Chrysoperla) mediterranea* n. sp. eine neue europäische Chrysopiden-Spezies (Planipennia, Chrysopidae). *Nachrichtenblatt bayerischen Entomologen* 21, 81–83.

Karelin, V.D., Yakovchuk, T.N. & Danu, V.P. (1989) Development of techniques for commercial production of the common green lacewing, *Chrysopa carnea* (Neuroptera: Chrysopidae). *Acta Entomologica Fennica* 53, 31–35.

Legaspi, J.C., Correa, J.A., Canuthers, R.I., Legaspi, B.C. & Nordlund, D.A. (1996) Effect of short-term releases of *Chrysoperla rufilabris* (Neuroptera: Chrysopidae) against silverleaf whitefly (Homoptera: Aleyrodidae) in field cages. *Journal of Entomological Science* 31, 102–111.

Leraut, P. (1991) Les *Chrysoperla* de la faune de France (Neuroptera Chrysopidae). *Entomologie Gallica* 2, 75–81.

Leraut, P. (1992) Névroptères des Alpes centrales françaises (Neuroptera). *Entomologie Gallica* 3, 59–65.

Mayr, E. (1963) *Animal Species and Evolution*. Belknap Press of Harvard University Press, Cambridge MA.

Monserrat, V.J. & Diaz-Aranda, L.M. (1989) Nuevos datos sobre los crisópidos ibéricos (Neuroptera, Planipennia, Chrysopidae). *Boletín del Asociación Española Entomológica* 13, 251–267.

New, T.R. (1975) The biology of Chrysopidae and Hemerobiidae (Neuroptera) with reference to their usage as biocontrol agents: a review. *Transactions of the Royal Entomological Society of London* 127, 115–140.

Paterson, H.E.H. (1986) The recognition concept of species. In *Species and Speciation,* Transvaal Museum Monograph no. 4, ed. Vrba, E.S., pp. 21–29. Transvaal Museum, Pretoria.

Principi, M.M. (1977) Contributi allo studio dei Neurotteri Italiani. XXI. La morfologia addominale ed il suo valore per la discriminazione generica nell'ambito delle Chrysopinae. *Bollettino dell'Istituto di Entomologia dell'Università di Bologna* 31, 325–360.

Ridgway, R.L. & Jones, S.L. (1969) Inundative releases of *Chrysopa carnea* for control of the bollworm and the tobacco budworm on cotton. *Journal of Economic Entomology* 62, 177–180.

Schneider, W.G. (1851) *Symbolae ad monographiam generis Chrysopae, Leach.* Vrotislavia.

Séméria, Y. (1977) Discussion de la validité taxonomique du sous-genre *Chrysoperla* Steinmann (Planipennia, Chrysopidae). *Nouvelle Revue d'Entomologie* 7, 235–238.

Séméria, Y. (1992) Données numériques relatives aux génitalia mâles des *Chrysoperla* Steinmann et leur valeur dans la discrimination spécifique (Insecta: Neuroptera: Chrysopidae). In *Current Research in Neuropterology, Proceedings of the 4th International Symposium on Neuropterology,* ed. Canard, M., Aspöck, H. & Mansell, M.W, pp. 333–339. Sacco, Toulouse.

Sengonca, Ç., Griesbach, M. & Lochte, C. (1995) Suitable predator–prey ratios for the use of *Chrysoperla carnea* (Stephens) eggs against aphids on sugar beet under laboratory and field conditions. *Zeitschrift für Pflanzenkrankheiten und Pflanzenschutz* 102, 113–120.

Stephens, J.F. (1835) *Illustrations of British Entomology; or, A Synopsis of Indigenous Insects: Containing their Generic and Specific Distinctions, with an Account of their Metamorphoses, Times of Appearance, Localities, Food, Economy.* Mandibulata, vol. 6. Balwin & Cradock, London.

Tauber, C.A. (1974) Systematics of North American chrysopid larvae: *Chrysopa carnea* group (Neuroptera). *Canadian Entomologist* 106, 1133–1153.

Tauber, C.A. & Tauber, M.J. (1973*a*) Diversification and secondary intergradation of two *Chrysopa carnea* strains (Neuroptera: Chrysopidae). *Canadian Entomologist* 105, 1153–1167.

Tauber, C.A. & Tauber, M.J. (1977) A genetic model for sympatric speciation through habitat diversification and seasonal isolation. *Nature* 268, 702–705.

Tauber, C.A. & Tauber, M.J. (1986*a*) Ecophysiological responses in life-history evolution: evidence for their importance in a geographically widespread insect species complex. *Canadian Journal of Zoology* 64, 875–884.

Tauber, C.A. & Tauber, M.J. (1986*b*) Genetic variation in all-or-none life-history traits of the lacewing *Chrysoperla carnea*. *Canadian Journal of Zoology* 64, 1542–1544.

Tauber, C.A. & Tauber, M.J. (1992) Phenotypic plasticity in *Chrysoperla*: genetic variation in the sensory mechanism and correlated reproductive traits. *Evolution* 46, 1754–1773.

Tauber, M.J. & Tauber, C.A. (1973*b*) Nutritional and photoperiodic control of the seasonal reproductive cycle in *Chrysopa mohave*. *Journal of Insect Physiology* 19, 729–736.

Tauber, M.J. & Tauber, C.A. (1975) Criteria for selecting *Chrysopa carnea* biotypes for biological control: adult dietary requirements. *Canadian Entomologist* 107, 589–595.

Tauber, M.J. & Tauber, C.A. (1976) Environmental control of univoltinism and its evolution in unicyclic insect species. *Canadian Journal of Zoology* 54, 260–265.

Tauber, M.J. & Tauber, C.A. (1981) Seasonal responses and their geographic variation in *Chrysopa downesi*: ecophysiological and evolutionary considerations. *Canadian Journal of Zoology* 59, 370–376.

Tauber, M.J., Tauber, C.A. & Denys, C.J. (1970) Adult diapause in *Chrysopa carnea*: photoperiod control of duration and colour. *Journal of Insect Physiology* 16, 949–955.

Tauber, M.J., Tauber, C.A. & Lopez-Arroyo, J.I. (1997) Life-history variation in *Chrysoperla carnea*: implications for rearing and storing a Mexican population. *Biological Control* 8, 185–190.

Templeton, A.R. (1989) The meaning of species and speciation: a genetic perspective. In *Speciation and Its Consequences*, ed. Otte, D &. Endler, J.A., pp. 3–27. Sinauer Associates, Sunderland MA.

Thierry, D. & Adams, P.A. (1992) Round table discussion on the *Chrysoperla carnea* complex (Insecta: Neuroptera: Chrysopidae). In *Current Research in Neuropterology, Proceedings of the 4th International Symposium on Neuropterology*, ed. Canard, M., Aspöck, H. & Mansell, M.W,. pp. 367–372. Sacco, Toulouse.

Thierry, D., Cloupeau, R. & Jarry, M. (1992) La Chrysope commune *Chrysoperla carnea* (Stephens) *sensu lato* dans le centre de la France: mise en évidence d'un complexe d'espèces (Insecta: Neuroptera: Chrysopidae). In *Current Research in Neuropterology. Proceedings of the 4th International Symposium on Neuropterology*, ed.

Canard, M., Aspöck, H. & Mansell, M.W., pp. 379–392. Sacco, Toulouse.

Thierry, D., Cloupeau, R. & Jarry, M. (1995) Variation in the overwintering ecophysiological traits in the common green lacewing West-Palaearctic complex (Neuroptera: Chrysopidae). *Acta Oecologica* 15, 593–606.

Thierry, D., Cloupeau, R. & Jarry, M. (1996) Distribution of the sibling species of the common green lacewing *Chrysoperla carnea* (Stephens) in Europe (Insecta: Neuroptera: Chrysopidae). In *Pure and Applied Research in Neuropterology, Proceedings of the 5th International Symposium on Neuropterology*, ed. Canard, M., Aspöck, H. & Mansell, M.W., pp. 233–240. Sacco, Toulouse.

Thierry, D., Ribodeau, M., Foussard, F. & Jarry, M. (1997) Allozyme polymorphism in a natural population of *Chrysoperla carnea sensu lato* (Neuroptera: Chrysopidae): a contribution to the status of constitutive taxons in western Europe. *European Journal of Entomology* 94, 311–316.

Thierry, D., Cloupeau, R., Jarry, M. & Canard, M. (1998) Discrimination of the West-Palaearctic *Chrysoperla* Steinmann species of the *carnea* Stephens group by means of claw morphology (Neuroptera, Chrysopidae). *Acta Zoologica Fennica* 209, 255–262.

Tjeder, B. (1960) Neuroptera from Newfoundland, Miquelon, and Labrador. *Opuscula Entomologia* 25, 146–149.

Tjeder, B. (1966) Neuroptera, Planipennia. The lacewings of Southern Africa. 5. The family Chrysopidae. *South African Animal Life* 12, 228–534.

Wells, M.M. & Henry, C.S. (1992) The role of courtship songs in reproductive isolation among populations of green lacewings of the genus *Chrysoperla* (Neuroptera: Chrysopidae). *Evolution* 46, 31–42.

Wells, M.M. & Henry, C.S. (1994) Behavioral responses of hybrid lacewings (Neuroptera: Chrysopidae) to courtship songs. *Journal of Insect Behaviour* 7, 649–662.

Wells, M.M. & Henry, C.S. (1998) Songs, reproductive isolation and speciation in cryptic species of insects: a case study using green lacewings. In *Endless Forms: Species and Speciation*, ed. Howard, D. & Berlocher, S., pp. 217–233. Oxford University Press, New York.

CHAPTER 4
Recognition of larval Neuroptera

V.J. Monserrat, J.D. Oswald, C.A. Tauber, and L.M. Díaz-Aranda

4.1 CONIOPTERYGIDAE
V.J. Monserrat

4.1.1 Introduction

Three subfamilies of Coniopterygidae are usually recognised: Aleuropteryginae, Brucheiserinae, and Coniopteryginae, with 12, 1, and 10 described genera respectively. Of the approximately 450 recent described species, no more than 23 species belonging to 9 genera have preimaginal stages more or less known, sometimes with imprecise descriptions. There are no preimaginal stages known for Brucheiserinae and many other genera of the other two subfamilies (see list below).

4.1.2 General biology and morphology of coniopterygid preimaginal stages

Their small size (2–3 mm) accounts for the scarcity of knowledge of the preimaginal stages. Eggs are laid in abundance, 2–5 single eggs per day for about a month (up to 226 eggs per female have been recorded) attached to bark or on the surface or margins of leaves and needles, sometimes in close proximity to future food colonies (Withycombe, 1923b, 1925b; Killington, 1936; Narayanan, 1942; Collyer, 1951; Muma, 1967, 1971; Henry, 1976). Eggs are oval, white, yellow, orange, purple or pink coloured, with a conical micropylar projection and rather flattened chorion and with polygonal impressions of the follicular cells (Figs. 4.1; 4.2). It takes from six days to some weeks for eggs to hatch, depending on temperature. An egg burster (Fig. 4.2a) is present in the embryo (Withycombe, 1923b, 1925b; Killington, 1936; Badgley et al., 1955; Kimmins & Wise, 1962; Muma, 1967; Meinander, 1972; Castellari, 1980; Gepp, 1992). Revised knowledge on coniopterygid eggs is provided in Gepp (1990).

Larval stages are active predators, free-living on vegetation, and are valuable potential biological control agents of eggs, nymphs, and imagos of soft insects such as aphids or coccids, but especially mites, and are capable

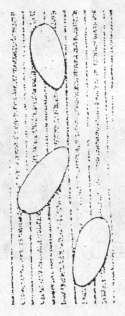

Fig. 4.1. Eggs of *Conwentzia sinica*. (After Yang, 1974)

of surviving for some time without feeding (Withycombe, 1923b, 1924b; Killington, 1936; Kimmins & Wise, 1962; Putman & Herne, 1966; Muma, 1967; Stimmel, 1979; Castellari, 1980; New, 1989). Adults (and presumably larvae) have been recorded linked to specific plant substrates (see Chapter 22, this volume). Three larval instars are usual, but records of four instars in *Heteroconis* (Badgley et al., 1955) and in *Semidalis* (Muma, 1967, 1971) need to be confirmed. Developmental times between 41 and 75 days were recorded by Gepp & Stürzer (1986) in *Semidalis aleyrodiformis*.

Larvae are soft, campodeiform (Fig. 4.3), somewhat fusiform, with head prognathous (Fig. 4.4), labrum covering the base or the entire jaws, mandibles straight, with acute apex (Fig. 4.4c), maxillae with cardo and stipes present but without maxillary palp,

43

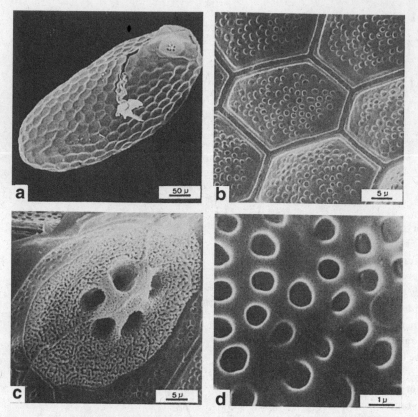

Fig. 4.2. *Conwentzia* eggs: a, egg-shell of *Conwentzia psociformis* (after Gepp, 1990); b, follicular impressions on egg chorion of *Conwentzia psociformis* (after Gepp, 1990); c, micropyle of egg of *Conwentzia psociformis* (after Gepp, 1990); d, chorionic pattern on egg of *Conwentzia psociformis* (after Gepp, 1990.)

labium small and labial palp short, two-segmented, with the apical segment expanded; eyes with five or six ommatidia; antennae two-segmented bearing long setae (Figs. 4.4d, f, 4.5); thorax wide, length of legs variable, tibiae and one-segmented tarsus not freely articulated; two tarsal claws and empodium are present; no lateral processes on thorax or abdomen (Withycombe 1923b, 1925a, b; Killington, 1936; Narayanan, 1942; Badgley *et al.*, 1955; Kimmins & Wise, 1962; Meinander, 1972).

Relative length of antennae segments and antennae, legs and cephalic setation is characteristic in each known genus, and differences in setal position and morphology (see key) as well as body colour pattern seem to show specific differences (Fig. 4.6) (Killington, 1936; Badgley *et al.*, 1955; Rousset, 1960, 1966; Ward, 1970; Meinander, 1972, 1974; Monserrat, 1978). In contrast Gepp (1992) and Gepp & Stürzer (1986) found intraspecific variation in the setation and coloration, variably dependent on prey in central European larvae.

Pupation occurs in a flattened circular cocoon of silk with double or triple layers (Fig. 4.7a), in leaf litter, under bark, etc. (Emerton, 1906; Withycombe, 1925b;

Killington, 1936; Badgley *et al.*, 1955; Kimmins & Wise, 1962; Meinander, 1972). Pupae are decticous and exarate, as is usual in Neuroptera (Fig. 4.7c–d).

4.1.3 Key to known subfamilies and genera of larvae

From the first tentative larval key given by Löw (1885) to the most recent one given by Meinander (1974), modified by Monserrat & Hölzel (1987), it is clear that it is not easy to provide an accurate key to genera of this family. This is largely because of a deficiency of correct and precise information in many genera. A general key is provided here, based on the available published information.

1. Antennae truncated at apex (Fig. 4.5a). Labrum covering only the basal part of the jaws (Fig. 4.4c). Integument with frequent squamiform setae (Figs. 4.4c; 4.7e)....................................Aleuropteryginae (2)
— Antennae tapered at apex (Fig. 4.5b). Labrum covering the jaws completely (Fig. 4.4d–f). Integument with usual (not squamiform) setae (Fig. 4.7f) ....Coniopteryginae (5)

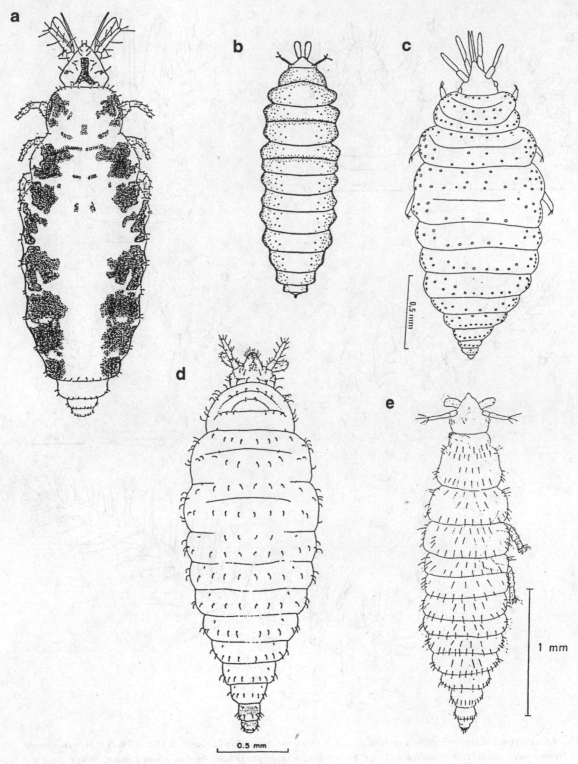

Fig. 4.3. Larvae of: a, *Heteroconis picticornis* (after Badgley *et al.*, 1955); b, *Cryptoscenea australiensis* (after Kimmins & Wise, 1962); c, *Aleuropteryx loewii* (after Rousset, 1966); d, *Hemisemidalis pallida* (after Monserrat & Hölzel, 1987); e, *Helicoconis lutea* (after Greve, 1974.)

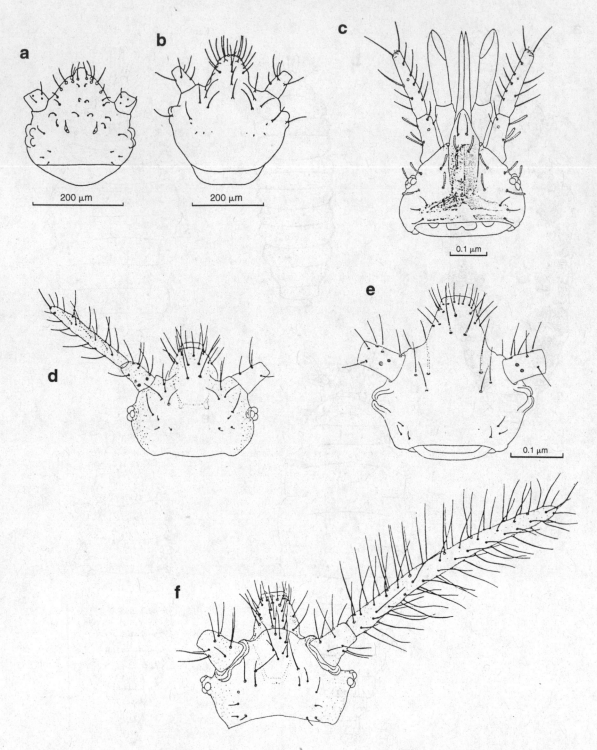

Fig. 4.4. Head parts of coniopterygids: a, cephalic capsule of *Helicoconis lutea* (after Monserrat & Hölzel, 1987); b, cephalic capsule of *Hemisemidalis pallida* (after Monserrat & Hölzel, 1987); c, head of *Aleuropteryx loewii* (modified from Rousset, 1966); d, head of *Semidalis vicina* (after Meinander, 1974); e, cephalic capsule of *Coniopteryx pygmaea* (modified from Rousset, 1966); f, head of *Conwentzia barretti* (after Meinander, 1974.)

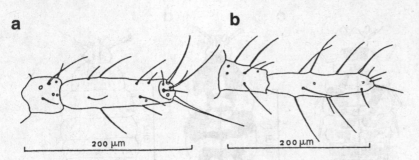

a

b

200 μm

200 μm

Fig. 4.5. Antennae of: a, *Helicoconis lutea* (after Monserrat & Hölzel, 1987); b, *Hemisemidalis pallida* (after Monserrat & Hölzel, 1987.)

2. Clypeolabral setae short and curved (Fig. 4.4a) ....*Helicoconis.*
—Clypeolabral setae long and scarcely curved (Fig. 4.4c)..3
3. Body subcylindrical, thoracic segments as wide as the first abdominal segment (Fig. 4.3b) ....*Cryptoscenea*
—Body fusiform, thoracic segments clearly wider than first abdominal segment (Fig. 4.3a, c) 4
4. Four pairs of clypeolabral setae, flagellum less than twice as long as scape (Fig. 4.4c) ....*Aleuropteryx*
—Six pairs of clypeolabral setae, flagellum three times as long as scape (Fig. 4.3a) ....*Heteroconis*
5. Legs much longer than width of thorax (Fig. 4.6g–i). Flagellum six times as long as scape (Fig. 4.4f). More than ten irregularly placed medial setae on clypeolabrum (Fig. 4.4f). ....*Conwentzia*
—Legs as long as or shorter than width of thorax (Fig. 4.6a–f). Flagellum three times as long or less than scape (Fig. 4.4d). Fewer than five regularly placed medial setae on clypeolabrum (Fig. 4.4b, d–f).......................6
6. Flagellum almost twice as long as scape (Fig. 4.5b). Clypeolabral setae in a single medial row (Fig. 4.4b) ....*Hemisemidalis*
—Flagellum more than twice as long as scape (Fig. 4.4d). Most of clypeolabral setae on two rows in both sides of the midline. (Fig. 4.4d)................................7
7. Two pairs of setae on clypeolabrum posterior to the single median seta (Fig. 4.4d).........................*Semidalis*
— One pair of setae on clypeolabrum posterior to the single median seta (Fig. 4.4e) ....*Coniopteryx*

4.1.4 List of species for which preimaginal stages have been partially or fully described

A previous summary of the knowledge of coniopterygid larvae was provided by Gepp (1984) and brought up to date for central European species by Gepp (1986). This list of taxa is modernised and amplified using the correct valid names.

ALEUROPTERYGINAE

Aleuropteryx loewii Klapalek, 1894. Lacroix, 1924: 82; Rousset, 1966: 71, 1969: 108.

Aleuropteryx juniperi Ohm, 1968. Ward, 1970: 75; Henry, 1976: 199.

Heteroconis picticornis (Banks, 1937). Badgley *et al.*, 1955: 77.

Cryptoscenea australiensis (Enderlein, 1906). Kimmins & Wise, 1962: 39.

Helicoconis lutea (Wallengren, 1871). Löw, 1885: 74; Dziedzielewicza, 1905: 382; Tullgren, 1906: 7; Mjöberg, 1909: 152; Withycombe, 1923*b*: 588; Greve, 1974: 20; Monserrat & Hölzel, 1987: 137.

CONIOPTERYGINAE

Coniopteryx abdominalis Okamoto, 1906. Yang, 1951: 344.

Coniopteryx esbenpeterseni Tjeder, 1930. Castellari, 1980: 168.

Coniopteryx exigua Withycombe, 1925*a*. Narayanan, 1942: 2.

Coniopteryx pygmaea Enderlein, 1906. Withycombe, 1923*b*: 582; Lacroix, 1924: 74; Killington, 1936: 203; Genay, 1953: 23; Rousset, 1956*a*: 1, 1956*b*: 1, 1958: 1; Ghilarov, 1962: 412; Rousset, 1966: 8, 1969: 108; Meinander, 1972: 31.

Coniopteryx tineiformis Curtis, 1834. Vine, 1895: 265; Withycombe, 1923*b*: 584; Lacroix, 1924: 73; Killington, 1936: 198.

Coniopteryx westwoodi (Fitch, 1856). Muma, 1971: 287.

Semidalis aleyrodiformis (Stephens, 1836). Withycombe, 1923*b*: 585; Lacroix, 1924: 78; Withycombe, 1925*a*: 20; Killington, 1936: 208; Collyer, 1951: 563;. Yang, 1951: 342; Hoffmann, 1962: 326; Rousset, 1966: 34, 1969: 108; Lauterbach, 1972: 143; Agekyan, 1978: 509; Gepp & Stürzer, 1986: 241.

Semidalis candida Navás, 1916. Monserrat, 1978: 372.

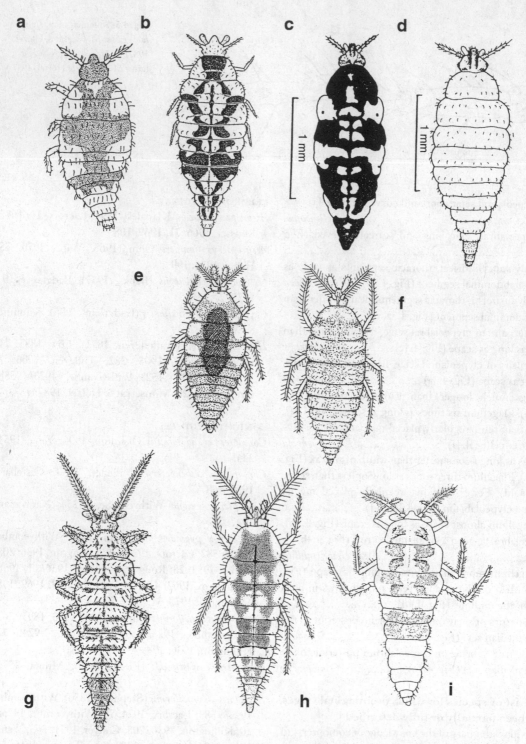

Fig. 4.6. General aspect of the larvae of: a, *Semidalis vicina* (after Meinander, 1974); b, *Semidalis aleyrodiformis* (after Collyer, 1951); c, *Semidalis pseudouncinata* (after Monserrat *et al.*, 1991); d, *Semidalis pluriramosa* (after Monserrat *et al.*, 1991); e, *Coniopteryx pygmaea* (after Killington, 1936); f, *Coniopteryx tineiformis* (after Killington, 1936); g, *Conwentzia sinica* (after Yang, 1974); h, *Conwentzia psociformis* (after Killington, 1936); i, *Conwentzia barretti* (after Meinander, 1974.)

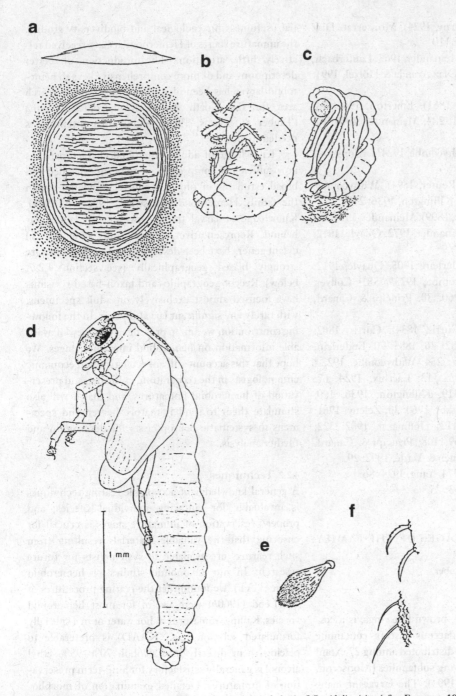

Fig. 4.7. Cocoons and pupae: a, coccon (section and dorsal view) of *Semidalis vicina* (after Emerton, 1906); b, prepupal larva of *Semidalis vicina* (after Emerton, 1906); c, pupa of *Semidalis vicina* (after Emerton, 1906); d, pupa of *Helicoconis lutea* (after Greve, 1974); e, squamiform seta of *Heteroconis picticornis* (after Badgley *et al.*, 1955); f, non-squamiform setae of *Hemisemidalis pallida* (after Monserrat & Hölzel, 1987.)

Semidalis pluriramosa (Karny, 1924). Monserrat, Diaz Aranda & Hölzel, 1991: 110.

Semidalis pseudouncinata Meinander, 1963. Lauterbach, 1972: 142; Monserrat, Diaz Aranda & Hölzel, 1991: 110.

Semidalis vicina (Hagen, 1861). Emerton, 1906: 74; Muma, 1967: 286, 1971: 284; Meinander, 1974: 14; Stange, 1981: 1.

Hemisemidalis pallida (Withycombe, 1924*b*). Monserrat & Hölzel, 1987: 139.

Parasemidalis fuscipennis (Reuter, 1894). Withycombe, 1922*b*: 171, 1923*b*: 588; Killington, 1936: 213.

Conwentzia barretti (Banks, 1899). Meinander, 1974: 14.

Conwentzia californica Meinander, 1972. Quayle, 1912: 506.

Conwentzia pineticola Enderlein, 1905. Quayle, 1912: 506, 1913: 87; Withycombe, 1923*b*: 581; Collyer, 1951: 557; Rousset, 1960: 30; Principi & Canard, 1974: 161.

Conwentzia psociformis (Curtis, 1834). Curtis, 1862: 528; Schlechtendal, 1882: 26, 1883: 70; Enderlein, 1906: 185; Arrow, 1917: 254; Withycombe, 1922*c*: 224, 1923*b*: 578, 1924*b*: 115; Lacroix, 1924: 67; Withycombe, 1925*b*: 319; Killington, 1936: 190; Collyer, 1951: 557; Rousset, 1960: 30; Zelený, 1961: 170; Ghilarov, 1962: 412; Hoffmann, 1962: 322; Rousset, 1966: 53, 1969: 108; Principi & Canard, 1974: 161; Sinacori, Mineo & Verde, 1992: 29.

Conwentzia sinica Yang, 1974. Yang, 1974: 86.

Conwentzia sp. Genay, 1953: 25.

4.2 PREIMAGINAL STAGES OF THE FAMILY HEMEROBIIDAE

J.D. Oswald and C.A. Tauber

4.2.1 Introduction

The family Hemerobiidae, brown lacewings, is a cosmopolitan group of predaceous insects containing approximately 550 species distributed among 27 extant genera and placed in 10 living subfamilies (Monserrat, 1990; Oswald, 1993*a*, *b*, 1994). The larvae of many hemerobiid species, especially of the genera *Hemerobius* and *Micromus*, commonly prey upon economically important pest insects in agricultural, horticultural, and forest environments. However, despite their significance as predators in these situations and their poten-

tial usefulness for ecological and biodiversity studies, the immature stages of Hemerobiidae have received relatively little attention. The production of better descriptions and of more comprehensive keys to hemerobiid larvae has been identified as a priority research area for the North American neuropteran fauna (Tauber & Adams, 1990), and similar attention is needed worldwide.

Knowledge of adult hemerobiid characteristics is now sufficiently comprehensive to permit detailed and broad estimates of phylogenetic relationships within the family Hemerobiidae (Oswald, 1993*a*, *b*, 1994). Knowledge of larval traits, however, has lagged far behind. Representative larvae of only nine (of 27) extant genera have been described, and these studies are strongly biased geographically (see section 4.2.7, below). Recent geographic- and taxon-based revisions have focused almost exclusively on adult specimens, with rarely any significant larval content. In the following contribution we aim to present an overview of available information on hemerobiid immature stages. We hope that this account will aid ecologists and economic entomologists in the recognition, culturing, and preservation of hemerobiid immatures, and that it will also stimulate them to send immature hemerobiid specimens to systematic neuropterists for description and further analysis.

4.2.2 Techniques

A general knowledge of acceptable rearing techniques is invaluable for culturing individual species, and proper preservation of immature stages is crucial for ensuring that the valuable materials resulting from such rearings are available to systematists for future research. In our taxonomic studies of hemerobiid larvae (CAT), we found that the rearing procedures of MacLeod (1960*b*) were useful for most hemerobiid species. Killing immatures in hot water or in a specially formulated solution (e.g., KAAD) is preferable to placing them directly into alcohol. 70%–95% ethyl alcohol is generally satisfactory for long-term preservation of immatures. Detailed examination of morphological structures may require maceration in KOH or lactic acid, which may be followed by staining and/or slide mounting. Additional information on rearing and preservation of immature stages is available in Stehr (1987).

4.2.3 Morphology of immature stages

General

Hemerobiids possess the following preimaginal stages: egg/prelarva, three instars, and a pupa. The 'prepupal' stage frequently referred to in the hemerobiid literature refers only to the quiescent mature third instar, which typically assumes an immobile, C-shaped state following cocoon spinning and prior to pupation.

Egg/Prelarva

Hemerobiids lay sessile (not stalked) eggs that are deposited singly or in small groups (Figs. 4.8, 4.9) Newly-laid eggs are white or cream. Prior to eclosion, however, the egg assumes the colour of the developing first instar, which is visible through the translucent chorion. The egg chorion always bears a small polar micropylar process and typically possesses some degree of fine surface microsculpture. The first instar escapes from the egg by using a saw-like oviruptor to pierce and create a longitudinal, subpolar, rent in the egg chorion. The oviruptor is not attached to the cuticle of the first instar; rather, it is part of a separate 'prelarval' cuticle. Eclosion of the first instar thus involves escape from the prelarval cuticle and the egg chorion.

First instar

The pre-engorged first instar is typically pale in coloration. Body setation is sparse and simple. First instars can be distinguished from later instars by the presence of trumpet-shaped pretarsal empodia (Fig. 4.10) and a relatively large head.

Second instar

Second instars are typically very similar to third instars except for their smaller size and fewer secondary setae.

Third instar

Third instars achieve full size and maximum numbers of secondary setae. The mature third instar spins a cocoon using silk produced in the Malpighian tubules, and which leaves the body through the anus. The cocoon is typically double-walled, with a loose outer mesh that encloses and supports a denser (sometimes opaque) inner envelope. After spinning the cocoon, the third instar assumes the typically C-shaped 'prepupal' form, becomes quiescent, and subsequently pupates

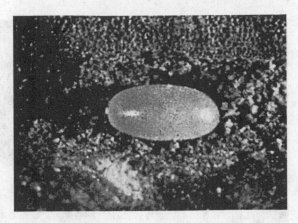

Fig. 4.8. *Wesmaelius betulinus* (Strøm) eggs. (From Killington, 1936, as *Boriomyia betulina*.)

within the cocoon. The fully grown third instar (prior to spinning the cocoon) is the stage that generally exhibits the highest degree of interspecific differentiation. Consequently, it has been heavily emphasised for the purposes of description and identification. However, informative characters may occur in all three instars (e.g., see MacLeod, 1960b).

Hemerobiid larvae appear to be divisible into two general body forms, each with a distinctive set of morphological characteristics. Type 1 larvae possess a slender body, an elongate, tapered abdomen, a head that is not distinctly retracted into the prothorax, and prominent pro-, meso-, and metathoracic sclerites (Fig. 4.11). Type 2 larvae have the body swollen (often somewhat physogastric), the abdomen more distinctly lobed, a head that retracts (to a greater or lesser degree) into the prothorax, and the meso- and metathoracic sclerites reduced to absent (Fig. 4.12). Type 1 larvae appear to be active, fast-moving predators, whereas type 2 larvae tend to be less active, slow-moving or 'sit-and-wait' predators. The distinction between these two larval types is not always clear, but the categories are broadly useful.

Pupa

The hemerobiid pupa is characterised by legs and antennae that are free from (not fused to) the body wall (exarate) and fully functional mandibles (decticous) (Fig. 4.13). Prior to adult eclosion, the pupal mandibles are used to create an irregular opening in the cocoon,

Fig. 4.9. *Drepanepteryx phalaenoides* L. eggs. (From Fulmek, 1941.)

through which the pharate adult emerges. The pharate adult may move away from the cocoon before eclosion. Hemerobiid pupae also appear to be characterised by the presence of two pairs of oppositely-directed 'hooks' on the terga of the third and fourth abdominal segments (Fig. 4.13b). The function, if any, of these structures is unknown.

Hemerobiidae or Chrysopidae?

Larval hemerobiids share many features with larvae of the apparently closely related family Chrysopidae, including inwardly-curving, opposable jaws that lack prominent teeth, two pretarsal claws and relatively well-developed antennae and labial palpi (see Chapter 4, this volume, and Tauber, 1991). Hemerobiid larvae differ from chrysopids in the following principal features: (1) their heads are smaller, with shorter, stouter jaws, (2) their bodies generally bear fewer, and less modified, setae, (3) they lack setiferous tubercles on the thorax and abdomen, (4) their first three abdominal segments are similar in size, and (5) second and third

instars lack trumpet-shaped empodia between the pretarsal claws (Fig. 4.10).

4.2.4 General biology

Most known hemerobiid larvae are relatively active, plant-frequenting predators of soft-bodied insects or their eggs. Unlike many chrysopid larvae, and contrary to some early reports in the literature, no hemerobiid larvae are known to 'carry trash'. Very little is known about the prey specificity of hemerobiid larvae or adults. Although most species are thought to be generalist predators, many species appear to exhibit considerable fidelity to specific habitats or plant species (Killington, 1936, 1937; Monserrat & Marín, 1996), which may be a reflection of restricted prey diets.

Preimaginal development can usually be completed in one to two months. The number of generations per year varies as a function of species characteristics, temperature, prey availability, and other factors. Several hemerobiid species appear to develop and reproduce under relatively low temperatures

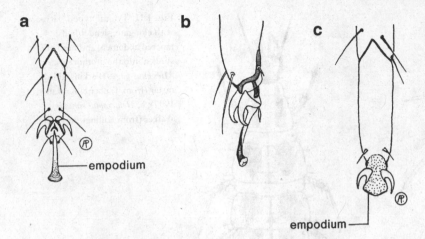

Fig. 4.10. Larval tarsi of: a, third-instar chrysopid, with trumpet-shaped empodium (from Tauber, 1990); b, first-instar hemerobiid, with trumpet-shaped empodium, *Drepanepteryx phalaenoides* L. (from Fulmek, 1941); c, third-instar hemerobiid, with pad-shaped empodium, *Hemerobius* (from Tauber, 1991.)

(Neuenschwander, 1975, 1976; Syrett & Penman, 1981) and thus may function effectively as biological control agents early in the season before other natural enemies are active. Unfortunately, very little is known about the developmental and overwintering biology of hemerobiids, and this characteristic has not been exploited in biological control or integrated pest management programmes.

4.2.5 Key

Veenstra *et al.* (1990) presented a working key for identifying second and third instars of European hemerobiids to genus or subgenus. Their key was based on descriptions and reared material of all European hemerobiid genera and provides the best existing framework for the construction of more comprehensive generic keys to hemerobiid larvae. To facilitate generic comparisons of hemerobiid larvae on a worldwide basis, we present below an updated version of the Veenstra *et al.* key that includes known, non-European genera (i.e. *Drepanacra* and *Nusalala*) and reflects recent changes in hemerobiid classification (i.e. the elimination of subgeneric taxa in the genera *Micromus* and *Wesmaelius*).

Although the key presented below is generally consistent with published descriptions of larvae from North America, Hawaii, Europe, Asia, and Australia, some genera (e.g. *Psectra*, *Drepanepteryx*) are known only from specimens and/or descriptions of a very limited number of species. The generality of the diagnostic characters attributed to these genera will require additional confirmation. For this reason, our key should be regarded as a 'preliminary' key, that is, one subject to further testing and revision through the comparison of additional species and genera.

4.2.6 Key to late instars of known hemerobiid genera (after Veenstra *et al.*, 1990)

1. Antenna length $\geq$ head capsule width ....3[1]
1'. Antenna length < head capsule width 2
2. (1') Labial palp projecting only slightly beyond mandible ....*Sympherobius*
2'. Labial palp projecting well beyond mandible
....*Psectra*
3. (1) Antenna length < 2.0 times head capsule width OR antenna length < 2.0 times mandible length 5
3'. Antenna length > 2.0 times head capsule width AND antenna length > 2.0 times mandible length 4
4. (3') Labial palp fusiform, at most slightly longer than mandible ....*Micromus*
4'. Labial palp swollen, distinctly longer than mandible ....*Nusalala*
5. (3) Mandible length < head capsule width.............6
5'. Mandible length *c.* 1.5 times head capsule width ....*Wesmaelius* (part)
6. (5) Labial palp length < mandible length; labial palp width < mandible width 7
6'. Labial palp length > mandible length; labial palp width > mandible width ....*Megalomus*
7. (6) Maxilla width < 1.8 times mandible width for half their length...8

---

[1] The illustration of *S. matsucocciphagus* depicts antennae longer than the head capsule width (Yang, 1980*a*); neither of the present authors has seen specimens of this species

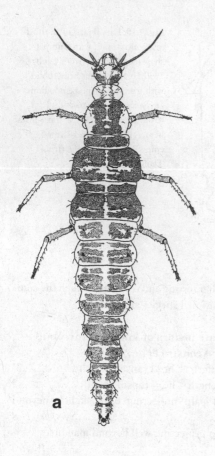

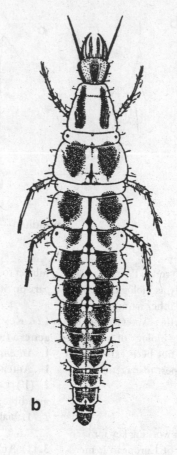

Fig. 4.11. Typical 'type 1' larvae with elongate, slender body, tapered abdomen, and head not sunken into the prothorax: a, *Micromus vagus* (Perkins) third instar (from Tauber & Krakauer, 1997); b, *Hemerobius simulans* Walker (from Killington, 1932*b*).

7′. Maxilla width > 2.0 times mandible width for half their length ....*Wesmaelius* (part)[2]
8. (7) Proleg length 1 to 1.5 times combined head capsule and mandible lengths 9
8′. Proleg length nearly 2.0 times combined head capsule and mandible lengths ....*Drepanepteryx*
9. (8) Antenna with a distinct subdivision between pedicel and flagellum ....*Hemerobius*
9′. Antenna without a distinct subdivision between pedicel and flagellum ....*Drepanacra*

4.2.7 Larval diagnoses by genus

Below we provide a tentative list of larval characteristics that distinguish each hemerobiid genus for which one or more species have been described. References to the primary literature containing descriptive information about hemerobiid immature stages are summarised

in Table 4.1. This literature is strongly biased toward taxa from the northern hemisphere, especially species in the genera *Hemerobius*, *Wesmaelius*, *Sympherobius*, and *Micromus*, which are well represented in northern temperate areas. This bias partly stems from the historical distribution of workers who have published on hemerobiid immature stages and biology, which has been dominated by workers in Europe and the United States, and partly from the numerical dominance of these four genera in terms of numbers of described species [these genera together account for approximately 63% (345 of 550) of known world hemerobiid species]. Only three hemerobiid species endemic to the southern hemisphere have their immature stages described (*Drepanacra binocula*, *Micromus tasmaniae*, *Nusalala uruguaya*). For the northern hemisphere, 47 species in seven genera have been described; however, the descriptions for many of these species are very brief and lack the level of detail needed to make meaningful comparisons. Together only ~50 species (9% of the

---

[2] Some *Wesmaelius* species may not key to either couple 5′ or 7′, e.g. *W. navasi* (see Monserrat, 1983).

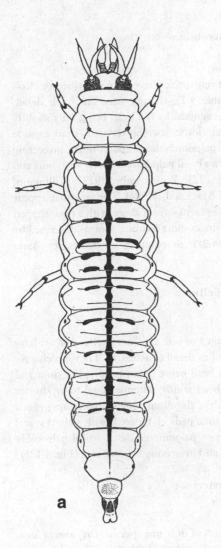

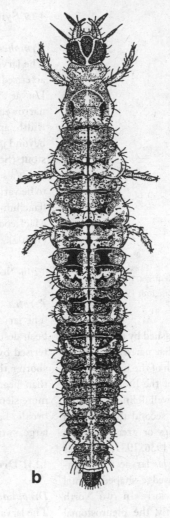

Fig. 4.12. Typical 'type 2' larvae with physogastric body, lobed abdomen, and head retracted into prothorax: a, *Megalomus fidelis* (Banks) third instar (from MacLeod, 1960*b*, as *Boriomyia fidelis*); b, *Psectra diptera* (Burm.) third instar (from Killington, 1946.)

a

b

world fauna) have been described. Clearly, substantial additional work is needed in this area.

From the standpoint of phylogenetics and evolutionary studies, the immature stages of many southern genera are of particular interest. No immature stages have been described from three (Adelphohemerobiinae, Carobiinae, Psychobiellinae) of the ten hemerobiid subfamilies recognised by Oswald (1993*a*, *b*, 1994). All three of these monogeneric subfamilies are restricted to the southern hemisphere and immatures from their nominotypical genera (*Adelphohemerobius, Carobius, Psychobiella*) would be of special interest. Perhaps the most anomalous absence from the list of genera for which immatures are completely unknown is the genus *Notiobiella*, which is very broadly pantropical in distri-

bution and contains approximately 40 valid species. Another notable absence from this list is the eastern Asian genus *Neuronema* with at least 25 known species.

4.2.8 Hemerobiinae

Hemerobius
The larvae of 14 species have been described or characterised (Table 4.1). *Hemerobius* is one of two hemerobiid genera that is commonly found in agricultural situations, especially in temperate regions; it is also frequently encountered in forests. Some species are active under relatively low temperatures and thus have potential as biological control agents when other natural enemies are inactive (e.g. Neuenschwander, 1975).

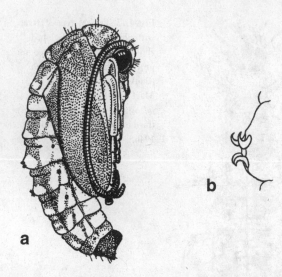

Fig. 4.13. *Wesmaelius quadrifasciatus* (Reut.): a, pupa; b, dorsal hooks on abdomen of pupa. (From Killington, 1934.)

Hemerobius larvae can be distinguished by the following traits (Fig. 4.11b): head somewhat narrowed behind; jaw slightly shorter than the width of the head; antenna slender, longer than the width of the head; flagellum much longer than scape; pedicel well delineated; apical palpomere longer than first and second together, but not projecting beyond apex of jaw or greatly swollen (Withycombe, 1923*b*; Killington, 1936, 1937; Krakauer & Tauber, 1996). Most *Hemerobius* larvae are distinguished by pale heads with a wedge-shaped frontal marking that is broadened anteriorly. In two North American species studied recently, the pleurostomal suture is bifurcated (cf. *Micromus*; Krakauer & Tauber, 1996).

Wesmaelius

The larvae of eight species have been described or characterised (Table 4.1). *Wesmaelius* larvae resemble those of *Hemerobius* as follows: antenna longer than jaw; flagellum much longer than scape; apical palpomere longer than first and second together, but not projecting beyond apex of jaw nor greatly swollen. However, in most described species of *Wesmaelius*, unlike *Hemerobius*, the jaws are particularly long, approximately 1.5 times as long as the width of the head capsule. This characteristic is not definitive of the genus and unfortunately other distinguishing characters are not known.

4.2.9 Sympherobiinae

Sympherobius

The larvae of ten species have been described or characterised (Table 4.1), though none in much detail. Larvae are recognised by the head being only slightly narrowed posteriorly; jaw shorter than head capsule width; apical palpomere large and swollen, projecting beyond jaw; two basal palpomeres short; legs short and stout (Smith, 1923; Withycombe, 1923*b*; Killington, 1937; also see MacLeod, 1960*b*). The antennae appear to be variable in form – from short, with a flask-shaped flagellum bearing a short terminal seta, as described by MacLeod (1960*b*), to elongate as figured by Yang (1980*a*).

4.2.10 Notiobiellinae

Psectra

The larval stages of one species (*Psectra diptera*) have been described in detail (Table 4.1). They are characterised by the head being broad basally; jaw stout and shorter than head width; antenna equal to or shorter than head width; flagellum longer than scape, much more slender than pedicel, tapered, with a short apical bristle; basal two palpomeres short, apical palpomere large, swollen and projecting beyond jaws (Fig. 4.12*b*).

4.2.11 Drepanacrinae

Drepanacra

The larval stages of only one species (*Drepanacra binocula*) have been described (Table 4.1). These larvae are characterised by the head being somewhat retracted into the prothorax (especially in second and third instars); meso- and metathoracic sclerites reduced; body slightly physogastric; antenna long, flagellum relatively long, with a long apical seta; labial palp inflated (as in *Drepanepteryx*, *Sympherobius* and *Psectra;* but not as inflated as *Megalomus*) and about two-thirds the length of jaw; jaw shorter than head width.

4.2.12 Megalominae

Megalomus

The larvae of two species have been described (Table 4.1), one with particularly good detail (MacLeod, 1960*b*). They have broad heads posteriorly, antenna

Table 4.1. *Synopsis of descriptive and comparative morphological information on the immature stages of the brown lacewings of the world.*

Hemerobiidae

Adelphohemerobiinae
Adelphohemerobius [2 spp.; Argentina, Chile] – immature stages unknown.

Carobiinae
Carobius [9 spp.; Australia] – immature stages unknown.

Hemerobiinae
Biramus [1 sp.; Venezuela] – immature stages unknown.
Hemerobius [*c.* 125 spp.; cosmopolitan, but poorly represented in the Australian region].
    atrifrons McLachlan: Killington, 1932*a* (E, L1, L2, L3*), 1937 (E, L1, L2, L3, C).
    conjunctus Fitch: Krakauer & Tauber, 1996 (L1, L2, L3).
    discretus Navás: Mitchell, 1962 (E*, L1, L2, L3*, P*, C, as *Hemerobius neadelphus*).
    humuli [*sic*] of Authors: see *humulinus* Linnaeus.
    humulinus Linnaeus: Brauer, 1867 (L, as *Hemerobius humuli* [*sic*]); Smith, 1923 (O*, E*, L1, L2*, L3*, P*, C*, as *Hemerobius humuli* [*sic*]); Withycombe, 1923*b* (E, L1, L2, L3*, P, C, as *Hemerobius humuli* [*sic*]); Killington, 1936 (E*, L3*, as *Hemerobius humuli* [*sic*]), 1937 (E, L1, L2, L3, C); Nakahara, 1954 (L3*, P*, as *Hemerobius obtusus*); MacLeod, 1964 (L3*); Agekyan, 1973 (E*, L3*, P*); Krakauer & Tauber, 1996 (L1, L2, L3*).
    japonicus Nakahara: Nakahara, 1954 (E*, L1*, L3*).
    lutescens Fabricius: Withycombe, 1923*b* (O*, E, L1, L2, L3), 1925*b* (L3*); Killington, 1937 (E, L1, L2, L3, C); Genay, 1953 (L*).
    marginatus Stephens: Killington, 1937 (E).
    micans Olivier: Withycombe, 1923*b* (E, L1, L2, L3*, P, C); Killington, 1936 (L3*), 1937 (E, L1, L2, L3, C).
    neadelphus Gurney: see *discretus* Navás.
    nitidulus Fabricius: Withycombe, 1923*b* (E, L1, L2, L3*, P, C); Killington, 1936 (L3*), 1937 (E, L1, L2, L3, C); Genay, 1953 (L3*); Bänsch, 1964 (L*).
    obtusus Nakahara: see *humulinus* Linnaeus.
    pacificus Banks: Quayle, 1912 (E, L*, C); Moznette, 1915*a* (E*, L1, L3*, P*, C*), 1915*b* (E*, L1, L3*, P*, C).
    perelegans Stephens: Killington, 1934 (E, L1*, L3*, C), 1936 (O*, L1*, L3*), 1937 (E, L1, L2, L3, C).
    pini Stephens: Killington, 1932*c* (E, L1, L2, L3*, C), 1936 (L3*), 1937 (E, L1, L2, L3, C).
    simulans Walker: Killington, 1932*b* (E, L1, L3*, C), 1936 (L3*), 1937 (E, L1, L2, L3, C).
    stigma Stephens: Withycombe, 1922*a* (E*, L3*, P*, C), 1923*b* (E, L1, L2, L3*, P, C); Smith, 1923 (O*, E*, L1*, L2*, L3*, C*, as *Hemerobius stigmaterus*); Laidlaw, 1936 (E*, L1*, L3*, C); Killington, 1936 (L3*), 1937 (E, L1, L2, L3, C); Miller & Lambdin, 1982 (E*), 1984 (L1*, L2*, L3*).
    stigmaterus Fitch: see *stigma* Stephens.
Nesobiella [1 sp.; Hawaiian Islands] – immature stages unknown.
Wesmaelius [*c.* 65 spp.; widely distributed in North America, Africa, Europe and Asia].
    bihamitus (Yang): Yang, 1980*b* (E*, L3*, P*, as *Kimminsia bihamita*).
    betulinus (Strøm) [=? *Wesmaelius nervosus*]: Strøm, 1788 (L3?*, C*, as *Hemerobius betulinus*); Killington, 1934 (E*, L1, L2, L3*, C, as *Boriomyia betulina*); Morton, 1935 (L3, as *Boriomyia nervosa*); Killington, 1935 (L, as *Boriomyia betulina*), 1936 (E*, L3*, as *Boriomyia betulina*), 1937 (E, L1, L2, L3, C, as *Boriomyia betulina*).
    concinnus (Stephens): Withycombe, 1923*b* (E*, L2, L3*, P, C, as *Boriomyia concinna*); Killington, 1937 (E, L2, L3*, C).
    navasi (Andréu): Monserrat, 1983 (O*, E*, L1*, L2, L3*).
    nervosus (Fabricius) [see also *betulinus* (Strøm)]: Withycombe, 1923*b* (L, P, C, as *Boriomyia nervosa*); Miles, 1924 (E, L, P, C, as *Boriomyia nervosa*).

Table 4.1 (*cont.*)

quadrifasciatus (Reuter): Killington, 1934 (E*, L2, L3*, P*, C*), 1936 (E*, L3*, P*, C*), 1937 (E, L2, L3*, C).

ravus (Withycombe): Withycombe, 1923*a* (L); Killington, 1937 (L, as *Boriomyia rava*).

subnebulosus (Stephens): Withycombe, 1923*b* (E, L1, L2, L3*, C, as *Boriomyia subnebulosa*); Killington, 1935 (L, as *Boriomyia subnebulosa*), 1936 (L3*, as *Boriomyia subnebulosa*), 1937 (E, L1, L2, L3, C, as *Boriomyia subnebulosa*); Agekyan, 1973 (E, L3, P, as *Boriomyia subnebulosa*).

Hemerobiella [2 spp.; Panama (new record), Venezuela, Ecuador] – immature stages unknown.

Sympherobiinae

Nomerobius [6 spp.; southern South America] – immature stages unknown.

Neosympherobius [1 sp.; Argentina] – immature stages unknown.

Sympherobius [*c.* 55 spp.; widespread in North and South America, Africa, Europe and temperate Asia, apparently absent from Australia, India, and most or all of tropical southeast Asia].

   amiculus (Fitch): Smith, 1923 (E*, L1, L2*, L3*, P*, C*).

   amicus Navás: see *fallax* Navás.

   angustus (Banks): see *californicus* Banks.

   barberi (Banks): Smith, 1934 (E, L1, L3).

   californicus Banks: Essig, 1910 (L3?*, P*, C, as *Sympherobius angustus*, misidentification).

   domesticus Nakahara: Nakahara, 1954 (E*, L3*, P*).

   elegans (Stephens): Withycombe, 1923*b* (E, L); Killington, 1937 (E, L).

   fallax Navás: Bodenheimer, 1930 (L, as *Sympherobius (Nefasitus) amicus*).

   fuscescens (Wallengren): Withycombe, 1923*b* (E, L1, as *Sympherobius inconspicuus*), 1925*b* (L1*); Killington, 1931 (E, L1, L3*, P, C), 1936 (L3*), 1937 (E, L1, L2, L3*, C).

   inconspicuus McLachlan: see *fuscescens* (Wallengren).

   matsucocciphagus Yang: Yang, 1980*a* (L*).

   pellucidus (Walker): New, 1967*a* (E, L3*, C).

   pygmaeus (Rambur): Withycombe, 1923*b* (O*, E, L1, L3*); Killington, 1937 (E, L3*); Silvestri, 1942 (L3*); New, 1967*b* (P*, C).

   tessellatus Nakahara: Nakahara, 1954 (E).

Psychobiellinae

Psychobiella [2 spp.; Australia] – immature stages unknown.

Notiobiellinae

Notiobiella [*c.* 40 spp.; pantropical] – immature stages unknown.

Psectra [*c.* 25 spp.; widespread in the Old World, 1 sp. in North America (possibly introduced)].

   diptera (Burmeister): Killington, 1946 (O*, E*, L1*, L2*, L3*, P, C).

Anapsectra [2 spp.; Nigeria, Zaire] – immature stages unknown.

Zachobiella [8 spp.; southeast Asia and Australia] – immature stages unknown.

Drepanacrinae

Conchopterella [4 spp.; southern South America, Juan Fernandez Islands] – immature stages unknown.

Austromegalomus [2 spp.; French Polynesia (Tahiti and Rapa Islands)] – immature stages unknown.

Drepanacra [4 spp.; southeast Asia, Australia, New Zealand].

   binocula (Newman): New, 1975 (O*, E, L1, L2, L3*, P*, C).

Megalominae

Megalomus [*c.* 40 spp.; widely distributed in North and South America, Europe, northern Africa and parts of Asia].

   fidelis (Banks): MacLeod, 1960 (L1*, L3*, as *Boriomyia fidelis*), [1961] (O*, E, L1*, L2, L3*, C, as *Boriomyia fidelis*).

   hirtus (Linnaeus): Killington, 1934 (E*, L1*, L2*, L3*, P, C), 1936 (O*, E*, L1*), 1937 (E, L1, L2, L3*, C).

Table 4.1 (*cont.*)

Drepanepteryginae

Neuronema [*c.* 25 spp.; eastern Asia] – immature stages unknown.

Gayomyia [3(?) spp.; South America] – immature stages unknown.

Drepanepteryx [6 spp.; Europe and Asia].

   phalaenoides (Linnaeus): Réaumur, 1737 (L*); Gleichen, 1770 (L*); Brauer, 1867 (L, in *Drepanopteryx* [*sic*]); Standfuss, 1906(P), 1910 (L*, figures from Gleichen, 1770, P*, C*); Morton, 1910 (E*, L*, figures from Gleichen, 1770, P*, C*); Killington, 1937 (L, C); Fulmek, 1941 (E*, L1*, L3, P); Christensen, 1943 (L*, figures from Gleichen, 1770).

Microminae

Noius [2 spp.; islands of the southwestern Pacific Ocean] – immature stages unknown.

Nusalala [*c.* 20 spp.; Central and South America, Antilles]

   uruguaya (Navás): Souza, 1988 (E, L1*, L2*, L3*, P*, C).

Megalomina [3 spp.; Australia, New Guinea] – immature stages unknown.

Micromus [*c.* 100 spp.; cosmopolitan, but species recorded from South America may belong to *Nusalala*].

   aphidivorus (Schrank): see *Micromus angulatus* (Stephens).

   angulatus (Stephens): Aubrook, 1935 (E, L1, L3, C, as *Micromus aphidivorus*); Killington, 1936 (E, L1, L3, C, as *Eumicromus angulatus*); Stroyan, 1949 (C).

   australis Froggatt: see *tasmaniae* (Walker).

   numerosus Navás: Nakahara, 1954 (E*, L1*, L3*, P*, as *Eumicromus numerosus*); Kawashima, 1958 (L3*, P*, C*, as *Eumicromus numerosus*).

   paganus (Linnaeus): Withycombe, 1923b (O*, E*, L1, L2, L3*, C); Killington, 1936 (E*, L1, L2, L3*, C, as *Eumicromus paganus*).

   posticus (Walker): Smith, 1922*b* (O*, L1), 1923 (E*, L1, L2, L3*, P*, C*); Selhime & Kanavel, 1968 (E, L1, L2, L3, P); Miller & Cave, 1987 (E*, L1, L2, L3*, C); Krakauer & Tauber, 1996 (L1, L2, L3).

   rubrinervis (Perkins): Tauber & Krakauer, 1997 (L1, L2, L3).

   subanticus (Walker): Selhime & Kanavel, 1968 (E, L1, L2, L3, P); Krakauer & Tauber, 1996 (L1, L2, L3*).

   tasmaniae (Walker): Froggatt, 1904 (L3?*, C, as *Micromus australis*); New & Boros, 1983 (O*, E, L1, L2, L3*, P*, C).

   timidus Gerstaecker: Williams, 1927 (E, L, P, as *Micromus vinaceus*).

   vagus (Perkins): Terry, 1908 (E, L3, C, as *Nesomicromus vagus*); Tauber & Krakauer, 1997 (L1, L2, L3*).

   variegatus (Fabricius): Brauer, 1871 (L3*); Withycombe, 1924a (E, L); Killington, 1936 (E, L1, L3 from Brauer, 1871); Kimmins, 1939*a* (L3*, P*); Dunn, 1954 (L3*, C); Agekyan, 1973 (E, L3, P).

   variolosus Hagen: Smith, 1934 (E, L1, L2, L3).

   vinaceus Gerstaecker: see *timidus*.

Genera unplaced to subfamily

Notherobius [3 spp.; Australia] – immature stages unknown.

Notes:

Data are organised within the framework of a comprehensive classification of world hemerobiid genera, for which synoptic information on species diversity and distribution are also presented. Species are listed alphabetically within each genus. Papers that discuss hemerobiid biology but lack descriptive information on immature stages have been omitted. The data on hemerobiid immature stages presented here are corrected, updated and expanded from the earlier listings of Gepp (1984) and Oswald (1993a).

   Abbreviations: C, cocoon; E, egg; L#, larva (# indicates instar, where known); O, oviruptor; P, pupa. An asterisk (*) indicates a figure of the item that it follows, e.g. P* = figure of all or part of the pupa.

slightly longer than head width, with flagellum slender, scarcely longer than scape, and bearing an apical bristle as long as flagellum; apical palpomere very long and swollen, projecting beyond tip of jaw; jaw stout and slightly shorter than head width (Fig. 4.12a).

4.2.13 Drepanepteryginae

Drepanepteryx
The larval stages of only one species (*Drepanepteryx phalaenoides*) have been described (Table 4.1). The head of the larva is broad basally with a U-shaped frontal suture and a distinct coronal suture. The antennae are similar to those of *Micromus*, i.e. much longer than the jaws and with the second and third segments not distinctly separated (cf. Krakauer & Tauber, 1996). The apical palpomere is swollen and longer than the two basal palpomeres together, but projects only slightly, if at all, beyond the jaws. The legs are relatively long. The eggs are laid in clusters and have a characteristic shape and prominent micropylar processes (Fulmek, 1941). There is significant contradiction in the literature regarding the body form of *Drepanepteryx* larvae. The very clear figures presented by Réaumur (1737) and Fulmek (1941) depict a type 1 larvae with an elongate body, head not retracted into the prothorax, and long legs. In contrast, MacLeod (1960b) described the mature *Drepanepteryx* larva as having a swollen, physogastric appearance, with the associated characteristics of the head deeply retracted within the prothorax and strongly reduced meso- and metathoracic sclerites.

4.2.14 Microminae

Micromus
The larvae of 11 species have been described or characterised (Table 4.1). Several *Micromus* species commonly occur in agricultural situations where they feed on economically important pests. Larvae are fusiform, elongate, and slender, with the following characteristics (Fig. 4.11a): head narrowed posteriorly; antennae at least twice as long as the jaws, with the flagellum and pedicel not distinctly separated; jaws slightly shorter than head width; apical labial palpomere fusiform, not swollen (Smith, 1923; Killington, 1936; Miller & Cave, 1987; Krakauer & Tauber, 1996; Tauber & Krakauer, 1997). In four species from North America and Hawaii, the pleurostomal suture is simple, not bifurcated (cf. *Hemerobius*).

Nusalala
The larvae of one species (*Nusalala uruguana*) have been described (Table 4.1). They strongly resemble those of *Micromus* in that they are fusiform, elongate, and slender; the antenna is very long, with the flagellum and pedicel not distinctly separated; and the jaw is slightly shorter than the head width. The labial palpi appear to be longer and slightly more swollen than those of known *Micromus* species.

Acknowledgements
We thank M.J. Tauber and A.H. Krakauer for their help; this work was supported in part by NRI/USDA Competitive Grant 9802447 (CAT, MJT), Regional Project W-185, CALS, and Cornell University's administration.

4.3 RECOGNITION OF EARLY STAGES OF CHRYSOPIDAE
L.M. Díaz-Aranda, V.J. Monserrat, and C.A. Tauber

4.3.1 Introduction
The larvae of the Chrysopidae are beginning to receive the systematic attention that they deserve given their significance as major predators in agricultural and horticultural situations and their attractiveness for ecological and biodiversity studies.

The European fauna has been studied especially well and immature stages of most species are included in modern, detailed keys and descriptions (see European fauna in this chapter). Keys and descriptions of the Japanese fauna have recently become available (Tsukaguchi, 1978, 1995). Larvae of some of the major North American taxa have been described (Tauber, 1969, 1974, 1975) but, unfortunately, the extensive faunae of North and South America, Africa, Australia, and most of Asia are in great need of further work (Tauber & Adams, 1990).

4.3.2 General morphological characteristics of chrysopid larvae
All neuropteran larvae, including the chrysopids, are characterised by suctorial mouthparts consisting of sickle-shaped, grooved mandibles and maxillae (laciniae) that form feeding tubes. The mouth is reduced to a small, narrow cleft and is closed off by the modified labrum and labium. Other neuropteran characteristics include the absence of maxillary palpi, the presence of one-segmented tarsi, and closure between the mid and

hindgut. Undigested food is stored during the entire larval period in the caudal part of the ventriculus and is egested immediately after the adult emerges. Most neuropterans (except the Coniopterygidae) have eight Malpighian tubules, six of which are attached to the hindgut and modified as silk-producing organs in the last instar.

Chrysopid larvae are distinguished from other neuropteran larvae by their large heads, usually long, toothless, sickle-shaped jaws, relatively long antennae, somewhat reduced first abdominal segment, usually well-developed setigerous tubercles on the thorax and abdomen, and trumpet-shaped empodia between the tarsal claws (all instars) (cf. hemerobiid larvae). Keys for identifying insect larvae to order and family are available in Stehr (1987, 1991).

Like most neuropterans, the green lacewings have three instars. First instars can be distinguished by their small size, pale body and simple setation, which usually consists of a small number of primary setae (Figs. 4.14, 4.15). Except in *Italochrysa* (Fig. 4.14f) the thoracic lateral tubercles bear, at most, two setae on the prothorax and three setae on the meso- and metathoracic segments. In all genera, each lateral tubercle on abdominal segments *ii-vii* bears two setae. Second and third instars have numerous primary and secondary setae (Figs. 4.16, 4.17), and the lateral tubercles bear several setae (usually more than two to three).

Within the Chrysopidae, there are two basic larval forms, each of which has a distinct set of morphological and behavioural characteristics. The first type, the debris-carrying larvae, assemble and carry packets of exogenous material on their dorsal surfaces. The packets may include plant parts, exuviae, waxy secretions or remains of prey, and general debris that protect the larvae from ants and other natural enemies (Eisner *et al.*, 1978; Milbrath *et al.*, 1994). Among the morphological characteristics of debris-carrying larvae are a humped (gibbous) abdomen having rows of hooked setae and greatly elongated thoracic tubercles bearing numerous, long setae (Fig. 4.16). The second type of chrysopid larvae, naked larvae (Fig. 4.17), have abdomens that are more or less flat, and short, straight setae; their thoracic tubercles are reduced in size and also bear short setae. In the most extreme cases, the lateral tubercles are absent (Fig. 4.17a).

In our taxonomic studies of chrysopid larvae, we made descriptions from specimens preserved in 70%–95% ethyl alcohol; to examine structures in detail, we also cleared specimens of each instar in hot lactic acid or KOH. The cephalic setae are indicated by Rousset's (1966) numerical system (Fig. 4.18). The designation of thoracic and abdominal tubercles and setae follows that of Tsukaguchi (1978) and, for convenience, other morphological characters that are useful in distinguishing the chrysopid larvae are named as in Fig. 4.14a. The generic classification of the Chrysopidae follows Brooks & Barnard (1990).

4.3.3 Eggs

Except for the endemic Hawaiian *Anomalochrysa*, all chrysopid genera are typified by laying stalked eggs. Oviposition occurs either on the underside or tips of leaves and twigs or on tree trunks (Fig. 4.19); eggs can occur as (1) single eggs (Fig. 4.19a, b), when they are deposited singly or in small groups of two to six eggs; (2) batches (Fig. 4.19c, d), when the eggs are laid in groups of many eggs (usually 20–40); (3) clusters (Fig. 4.19e, f), when the eggs are laid in bundles and the stalks become loosely (Fig. 4.19e) or tightly intertwined (Fig. 4.71f). Newly-laid eggs can be white, cream, bluish or green; egg colour and the pattern of egg deposition is characteristic for most species. Studies with mutants in two species indicate that egg colour may reflect the maternal genotype (Tauber & Tauber, 1971; Tauber *et al.*, 1976).

4.3.4 European genera of chrysopid larvae

In Europe, the family Chrysopidae currently is represented by about 63 described species, arranged in 13 genera and two subfamilies (Aspöck, 1992; Monserrat & Rodrigo, 1992). The taxonomic status of several species is confusing (particularly in *Chrysoperla* and *Cunctochrysa*), and it is possible that the number of species will change when the taxonomic problems are clarified. During the last ten years the larval stages of several European species have been described (e.g. Canard & Labrique, 1989; Gepp, 1989; Labrique & Canard 1989; Labrique, 1990; Diaz-Aranda, 1992; Diaz-Aranda & Monserrat, 1988, 1990a, 1990b, 1991, 1992, 1994, 1995, 1996), and, at present we know the preimaginal stages of approximately 80% of the European species (the current state of knowledge is summarised in Table 4.2).

In his recent study of the Japanese Chrysopidae, Tsukaguchi (1995) divided the cosmopolitan genus *Mallada* into two genera (*Mallada* and *Pseudomallada*), based on male and female genitalia, wing venation, and

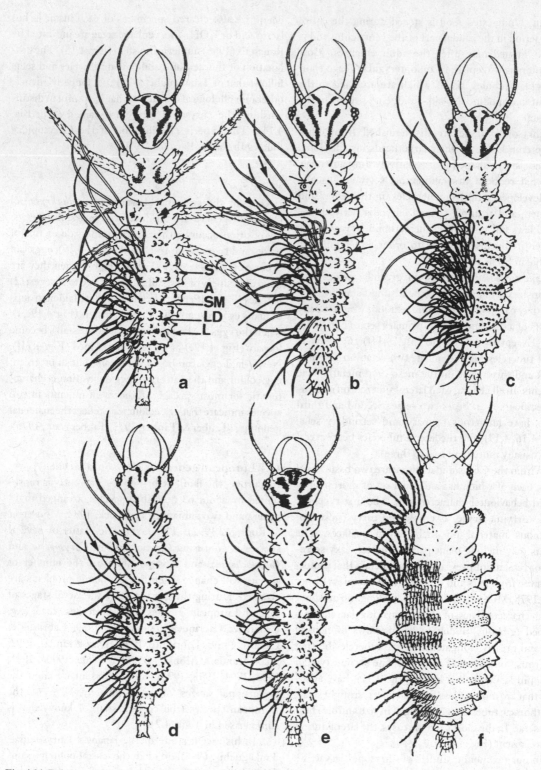

Fig. 4.14. Debris-carrying first instars a, *Mallada (M. venosus)*; b, *Cunctochrysa (C. baetica)*; c, *Suarius (S. walsinghami)*; d, *Chrysopidia (C. ciliata)*; e, *Rexa (R. lordina)*; f, *Italochrysa (I. stigmatica)*.

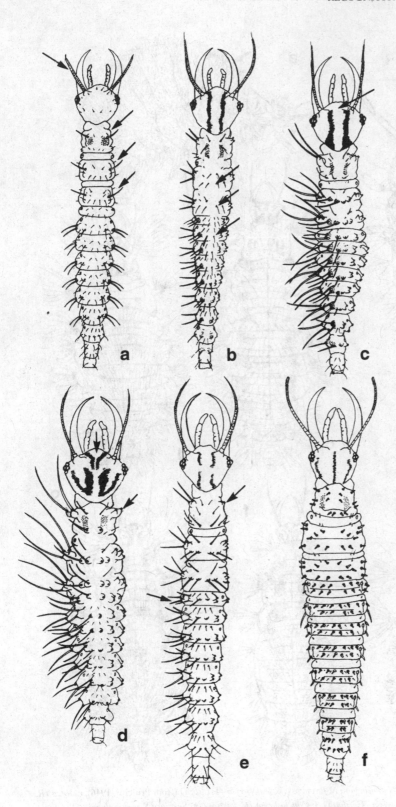

Fig. 4.15. Naked first instars. a, *Brinckochrysa (B. nachoi)*; b, *Peyerimhoffina (P. gracilis)*; c, *Chrysoperla (C. carnea)*; d, *Chrysopa (C. regalis)*; e, *Nineta (N. guadarramensis)*; f, *Hypochrysa (H. elegans)*.

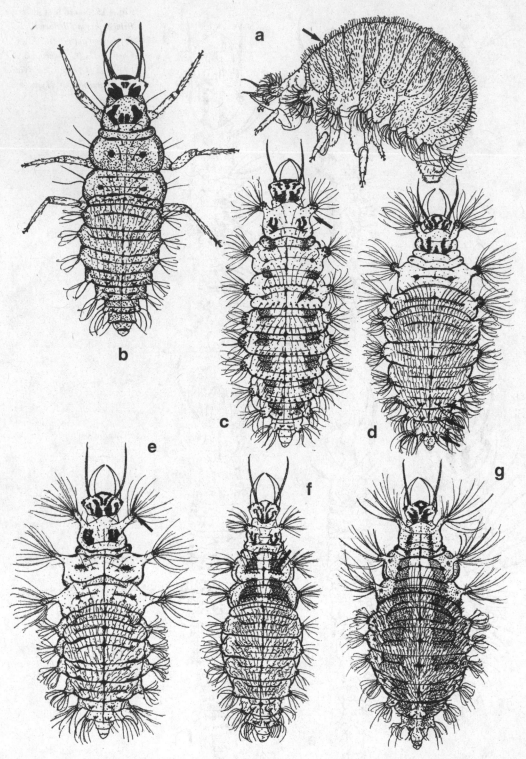

Fig. 4.16. Debris-carrying third instars. a, *Nothochrysa (N. fulviceps)*; b, *Italochrysa (I. italica)* (from Principi, 1946); c, *Rexa (R. lordina)*; d, *Mallada (M. venosus)*; e, *Chrysopidia (C. ciliata)*; f, *Cunctochrysa (C. baetica)*; g, *Suarius (S. walsinghami)*.

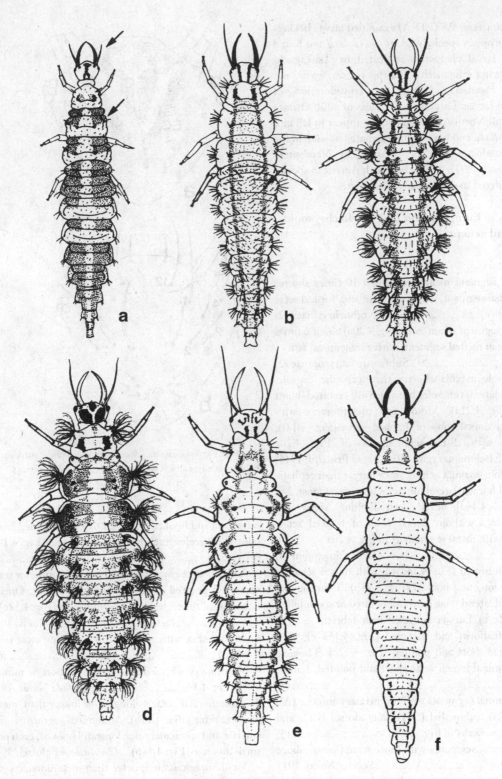

Fig. 4.17. Naked third instars. a, *Brinckochrysa (B. nachoi)*; b, *Peyerimhoffina (P. gracilis)*; c, *Chrysoperla (C. carnea)*; d, *Chrysopa (C. regalis)*; e, *Nineta (N. guadarramensis)*; f, *Hypochrysa (H. elegans)*.

larval characters. We (LD-A) examined larvae belonging to European species of *Mallada* (*s. lat.*) and found that the larval characters indicated by Tsukaguchi (1995) are not informative. For this reason, we did not include *Pseudomallada* in our consideration of European larvae. Largely on the basis of adult characteristics, all North American species appear to fall into *Pseudomallada* and have been designated as such in the recent catalogue of North American Neuroptera (Penny *et al.*, 1997). Thus, the North American species are considered under *Pseudomallada* here.

4.3.5 Key to European and Old-World chrysopid tribes and genera

First instars

1. Distal segment of antenna about 10 times shorter than medial segment, with a group of small apical setae (Fig. 4.20b)................Subfamily Nothochrysinae (12)
—Distal segment of antenna (Fig. 4.20a) about 6 times shorter than medial segment, with a long apical seta ....Subfamily Chrysopinae (2)
2. Mandible–maxilla shorter than cephalic capsule. Thoracic lateral tubercles bearing 8–10 serrated–blunt setae (Fig. 4.21a). Meso- and metathoraces with numerous dorsal rows of hooked setae (Fig. 4.14f). Antenna with distal segment as in Fig. 4.20c ......Tribe Belonopterygini (*Italochrysa* Principi, 1946)
—Mandible–maxilla as long as, or longer than cephalic capsule. Thoracic lateral tubercles bearing, at most, 2–3 setae (Fig. 4.14a), never serrated–blunt. Meso- and metathoraces without dorsal rows of hooked setae. Antenna with distal segment as in Fig. 4.20a ....Tribe Chrysopini (3)
3. Body humped (Fig. 4.14a–e), with dorsal setae on abdomen long and hooked (Fig. 4.21b). Thoracic and abdominal lateral setae long and pointed or spoonbilled (Fig. 4.22c, e). Larva coats itself with debris...............4
—Body fusiform and flattened (Fig. 4.15a–e), with dorsal setae short and pointed (Fig. 4.22c). Thoracic and abdominal lateral setae short and pointed. Larvae naked ....(8)
4. Abdominal segments with few primary anterior (A), spiracle (S), submedial (SM), laterodorsal (LD) and lateral (L) setae as in Fig. 4.14a..............................(5)
—Abdominal segments with more setae than as above (Fig. 4.14c) ....*Suarius* Navás, 1911
5. Abdominal segment (at least *i–iv*) with anterior-

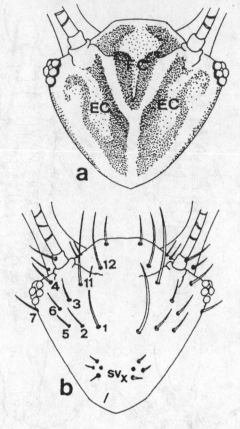

Fig. 4.18. Cephalic capsule (dorsal view). a, cephalic marking: FC = frontoclypeal, EC = epicranial; b, cephalic chaetotaxy.

setae long and hooked, on small tubercles, with 2 laterodorsal tubercles, each bearing 1 long seta (Fig. 4.14a, b, d) ....(6)
—Abdominal segments (at least *i–iv*) with anterior setae short and pointed. Anterior tubercles absent. Only 1 laterodorsal tubercle, bearing 1 long seta (Fig. 4.14e) ....*Rexa* Navás, 1920
6. Mesothorax with a dorsal row of 4 long setae (Fig. 4.14b, d)..(7)
—Mesothorax with a dorsal row of 4 short or minute setae (Fig. 4.14a)........................*Mallada* Navás, 1925
7. Mesothoracic setae as long as, or longer than, metathoracic setae (Fig. 4.14b). Abdominal segment *i* with anterior and spiracular setae long and hooked, each on a small tubercle (Fig. 4.14b) ...*Cunctochrysa* Hölzel, 1974
—Mesothoracic setae shorter than metathoracic setae (Fig. 4.14d). Abdominal segment *i* with anterior and

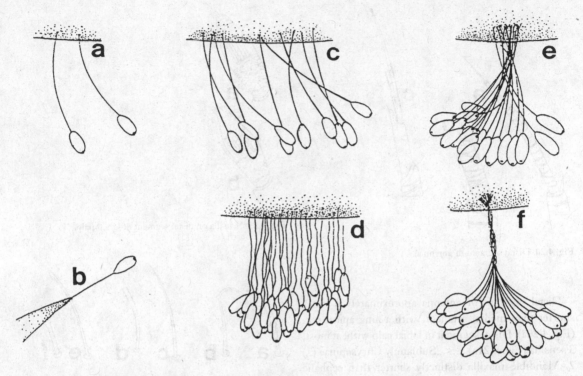

Fig. 4.19. Pattern of egg deposition. a, b, Single; c, d, batches; e, f, clusters.

spiracular setae short and unhooked, without tubercles (Fig. 4.14d)..............................*Chrysopidia* Navás, 1910
8. Antenna shorter than mandible–maxilla (Fig. 4.15a). Thoracic lateral tubercles absent or small, bearing 1 short seta (Fig. 4.15a). Thoracic and abdominal setae multipointed apically (Fig. 4.22d)
....*Brinckochrysa* Tjeder, 1966
—Antenna longer than mandible–maxilla. Thoracic lateral tubercles large, bearing 2 setae on prothorax and 3 on meso- and metathoraces (Fig. 4.15b–e). Thoracic and abdominal setae with single point apically (Fig. 4.22c) 9
9. Thorax and abdomen with lateral and laterodorsal setae short (shorter than femur), (Fig. 4.15b, e) .....(10)
—Thorax and abdomen with lateral and laterodorsal setae long (as long as, or longer than femur) (Fig. 4.15c, d)...(11)
10. Body (including mouthparts) longer than 3 mm. Thorax and abdomen with lateral and laterodorsal setae longer than setae on legs (Fig. 4.15e)
....*Nineta* Navás, 1912
—Body (including mouthparts) shorter than 3 mm.

Thorax and abdomen with lateral and laterodorsal setae about as long as setae on legs (Fig. 4.15b)
....*Peyerimhoffina* Lacroix, 1920
11. Cephalic seta S12 (Fig. 4.18b) present. Frontoclypeal marking (Fig. 4.18a) absent or only slightly developed (Fig. 4.15c)
....*Chrysoperla* Steinmann, 1964
—Cephalic seta S12 (Fig. 4.18b) absent (except *Chrysopa viridana*). Frontoclypeal marking (Fig. 4.18a) present (Fig. 4.15d)....................*Chrysopa* Leach, 1815
12. Thoracic and abdominal dorsal setae hooked and long ....*Nothochrysa* McLachlan, 1868
—Thoracic and abdominal dorsal setae knobbed apically (Fig. 4.22f) and very short (Fig. 4.15f)
....*Hypochrysa* Hagen, 1866

Third instars
1. Distal segment of antenna approximately 10 times shorter than medial segment, with a group of small apical setae (Fig. 4.20b). Distal segment of labial palp with 4 or more sensilla on its outer edge (Fig. 4.21b)
....................................Subfamily *Nothochrysinae* (12)

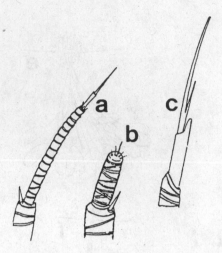

Fig. 4.21. Sensilla on distal segment of labial palp.

Fig. 4.20. Distal segment of antennae.

—Distal segment of antenna approximately 6 times shorter than medial segment, with a long apical seta (Fig. 4.20a). Distal segment of labial palp with, at most, 3 sensilla (Fig. 4.21a) ....Subfamily Chrysopinae (2)
2. Mandible–maxilla distinctly shorter than cephalic capsule. Distal segment of antenna as in Fig. 4.20c. Cephalic capsule and thoracic and abdominal lateral tubercles with serrated–blunt setae (Fig. 4.22a). Thoracic and abdominal segments with numerous dorsal rows of hooked setae (Fig. 4.16). Larvae associated with ant nests.
......Tribe Belonopterygini (*Italochrysa* Principi, 1946)
—Mandible–maxilla as long as, or longer than cephalic capsule. ·Distal segment of antenna as in Fig. 4.20a. Cephalic setae never serrated-blunt. Thoracic and abdominal lateral setae usually unserrated. Usually free living...Tribe Chrysopini (3)
3. Abdomen humped (Fig. 4.16b–g). Thoracic lateral tubercles elongate, bearing long setae. Abdomen with spherical lateral tubercles bearing long setae. Abdominal segments with several dorsal rows of hooked setae. Larva coats itself with debris 4
—Abdomen not humped, more or less flattened (Fig. 4.17). Thoracic lateral tubercles not elongate, bearing comparatively short setae (Fig. 4.17). Abdomen without several dorsal rows of hooked setae. Larva usually naked...8
4. Mandible–maxilla about as long as cephalic capsule. Thoracic lateral tubercles robust, with thick stalks. All thoracic and abdominal lateral setae with black bases

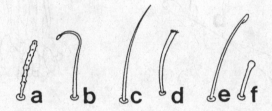

Fig. 4.22. Types of setae. a, Serrated–blunt; b, hooked–blunt; c, pointed; d, multipointed; e, spoonbilled; f, knobbed; g, serrated–pointed.

(Fig. 4.16c). Abdominal segments with 3–4 dorsal rows of setae, of which the posterior row is more conspicuous, with setae arising from small black protuberances (Fig. 4.16c)........................*Rexa* Navás, 1920
—Mandible–maxilla longer than cephalic capsule. Thoracic lateral tubercles with thin stalk. Not all thoracic and abdominal lateral setae with black bases. Abdominal segments with setae in all dorsal rows of similar size (Fig. 4.16d–g)...5
5. Cephalic seta S12 (Fig. 4.18b) minute or absent. Abdominal segments *v,vi*, and *vii* each with laterodorsal tubercles bearing 2 long setae (Fig. 4.16d)
....*Mallada* Navás, 1925
—Cephalic seta S12 (Fig. 4.18b) present. Abdominal segments *v* without laterodorsal tubercles (present on abdominal segments *vi* and *vii*) (Fig. 4.16e–g) 6
6. Thoracic lateral tubercles very long and thin, not or only slightly enlarged apically. Larva white, with conspicuous dark laterodorsal sclerites on pronotum (Fig. 4.16e) ....*Chrysopidia* Navás, 1910

Table 4.2. *European species of* Chrysopidae *with preimaginal stages described*

Subfamily Chrysopinae

Tribe Chrysopini
Brinckochrysa
  nachoi   Diaz-Aranda & Monserrat (1992)

Chrysopa
  abbreviata   Killington (1937), Gepp (1983)
  commata   Gepp (1983)
  dorsalis   Alderson (1911a), Killington (1937), Fraser (1945), Gepp (1983, 1989)
  formosa   Principi (1947), Tsukaguchi (1978, 1995), Gepp (1983), Pantaleoni (1983)
  hungarica   Gepp (1983)
  nigricostata   Brauer (1850), Gepp (1983, 1989)
  pallens   Withycombe (1923b), Killington (1937), Principi (1940), Tsukaguchi (1978, 1995), Pantaleoni (1983),
                  Gepp (1983, 1989)
  perla   Brauer (1850), Lacroix (1921), Withycombe (1923b), Killington (1937), Gepp (1983)
  phyllochroma   Withycombe (1923b), Killington (1937), Tsukaguchi (1978), Gepp (1983, 1989)
  regalis   Diaz-Aranda (1992), Diaz-Aranda & Monserrat (1995)
  viridana   Principi (1954), Gepp (1983), Pantaleoni (1983)
  walkeri   Gepp (1983, 1989)

Chrysoperla
  ankylopteryformis Diaz-Aranda & Monserrat (1990a)
  carnea   see Diaz-Aranda & Monserrat (1990a)
  mediterranea   Diaz-Aranda & Monserrat (1990a)

Chrysopidia
  ciliata   Withycombe (1923b), Killington (1937), Gepp (1983, 1989), Diaz-Aranda (1992), Diaz-Aranda &
                  Monserrat (1995), Tsukaguchi (1995)

Cunctochrysa
  albolineata   Withycombe (1923b), Killington (1937), Agekjan (1973), Tsukaguchi (1977), Gepp (1983, 1989), Diaz-
                  Aranda & Monserrat (1994)
  baetica   Diaz-Aranda & Monserrat (1994)

Mallada
  clathratus   Principi (1956), Pantaleoni (1983)
  flavifrons   Lacroix (1925), Withycombe (1923b), Killington (1937), Principi (1940, 1956), Gepp (1983)
  genei   Monserrat (1984)
  granadensis   Diaz-Aranda & Monserrat (1988)
  iberica   Labrique & Canard (1989)
  inornata   Lacroix (1925), Gepp (1983)
  picteti   Labrique (1990), Diaz-Aranda & Monserrat (1991)
  prasinus   Withycombe (1923b), Principi (1956), Gepp (1983)
  subcubitalis   Monserrat (1989)
  venosus   Diaz-Aranda & Monserrat (1990b)
  ventralis   Withycombe (1923b), Killington (1937), Gepp (1983, 1989)

Nineta
  carinthiaca   Gepp (1983)
  flava   Alderson (1911b), Withycombe (1923b), Killington (1937), Gepp (1983, 1989), Diaz-Aranda (1992)

Table 4.2 (*cont.*)

guadarramensis	Gepp (1983), Diaz-Aranda (1992)
inpunctata	Gepp (1983)
pallida	Brauer (1867), Gepp (1983), Diaz-Aranda (1992)
vittata	Killington (1937), Gepp (1983, 1989), Tsukaguchi (1995)
Peyerimhoffina	
gracilis	Gepp (1983), Diaz-Aranda (1992), Diaz-Aranda & Monserrat (1995)
Rexa	
lordina	Canard & Labrique (1989), Diaz-Aranda (1992), Diaz-Aranda & Monserrat (1995)
Suarius	
iberiensis	Diaz-Aranda & Monserrat (1996)
tigridis	Diaz-Aranda & Monserrat (1996)
walsinghmai	Diaz-Aranda & Monserrat (1996)
Tribe Belonopterygini	
Italochrysa	
italica	Principi (1946)
stigmatica	Diaz-Aranda & Monserrat (1995)
Subfamily Nothochrysinae	
Hypochrysa	
elegans	Brauer (1867), Principi (1956), Gepp (1983, 1989), Diaz-Aranda & Monserrat (1995)
Nothochrysa	
capitata	Withycombe (1923b), Killington (1937), Kimmins (1939b), Gepp (1983, 1989), Diaz-Aranda (1992)
fulviceps	Killington (1937), Gepp (1983), Diaz-Aranda & Monserrat (1995)

—Thoracic lateral tubercles enlarged apically, longer on prothorax than on meso- or metathorax. Body white, with brown or reddish markings on thoracic and abdominal segments. Pronotum with laterodorsal sclerites not very conspicuous 7

7. Mesothorax with a dorsal row of long setae (Fig. 4.16f). Abdomen slightly humped. Thoracic lateral tubercles slightly elongated, nearly spheroidal on meso- and metathorax. Thorax white with a pair of oblique brown or reddish markings on meso- and metathorax (Fig. 4.16f). Larva with little debris ....*Cunctochrysa* Hölzel, 1970

—Mesothorax without a dorsal row of long setae. Abdomen strongly humped. Thoracic lateral tubercles elongate. Thorax white with a pair of laterodorsal brown stripes (Fig. 4.16g). Larva completely covered with large amount of debris..........*Suarius* Navás, 1914

8. Antennae shorter than mandible–maxilla (Fig. 4.17a). Thoracic and abdominal setae multipointed api-

cally (Fig. 4.22d). Thoracic and abdominal segments without lateral, laterodorsal or submedial tubercles, with 1 lateral seta on thorax and 2 lateral setae on abdomen (Fig. 4.17a) ....*Brinckochrysa* Tjeder, 1966

—Antennae longer than mandible–maxilla. Thoracic and abdominal setae with single point apically (Fig. 4.22c). Thoracic and abdominal lateral tubercles large, bearing numerous setae...9

9. Thoracic and abdominal lateral tubercles small (Fig. 4.17b). Abdomen without submedial and laterodorsal tubercles. Abdominal segments dorsally with numerous secondary setae spatulate (Fig. 4.22e) and oriented to dorsomedian line (Fig. 4.17b) ....*Peyerimhoffina* Lacroix, 1920

—Thoracic and abdominal lateral tubercles large. Thorax and abdomen with submedial and laterodorsal tubercles (Fig. 4.17c–e). Secondary setae pointed apically..10

10. Larva (including mouthparts) about 12 mm long.

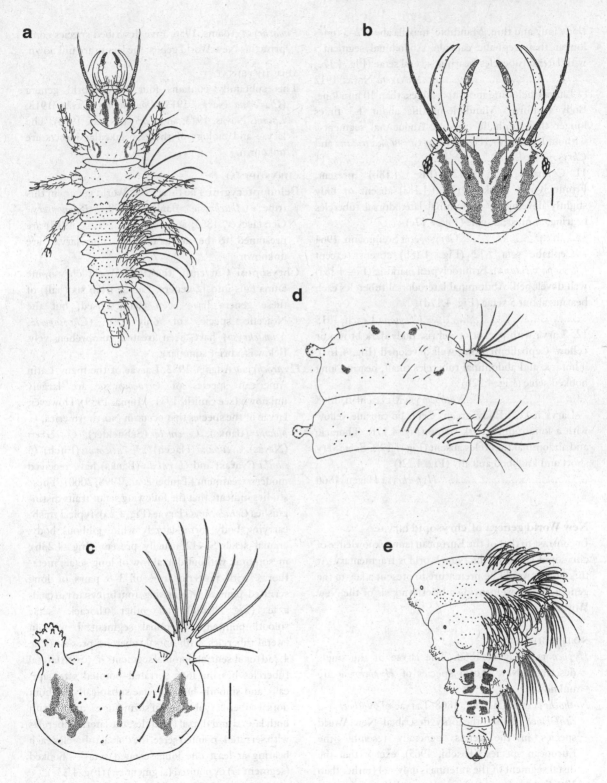

Fig. 4.23. *Ceraeochrysa (C. placita)*. a, First instar; b, detail of head; c, third instar prothorax; d, third instar meso–and metathorax; e, third instar abdominal segments v–x.

Body long and thin. Mandible–maxilla about 2–3 times longer than cephalic capsule. Abdominal segment *i* with lateral tubercles bearing several setae (Fig. 4.17e) ....*Nineta* Navás, 1912
—Larva (including mouthparts) less than 10 mm long. Body fusiform. Mandible-maxilla about 1.5 times longer than cephalic capsule. Abdominal segment *i* without lateral tubercles (except *Chrysopa viridana* and *Chrysopa pallens*)...11

11. Cephalic seta S12 (Fig. 4.18b) present. Frontoclypeal marking (Fig. 4.18a) absent or only slightly developed. Abdominal laterodorsal tubercles bearing, at most, 2 setae (Fig. 4.17c) ....*Chrysoperla* Steinmann, 1964
—Cephalic seta S12 (Fig. 4.18b) absent (except *Chrysopa viridana*). Frontoclypeal marking (Fig. 4.18a) well developed. Abdominal laterodorsal tubercles each bearing about 5 setae (Fig. 4.17d) ....*Chrysopa* Leach, 1815

12. Larva with packet of debris. Body dark brown or yellow. Cephalic marking well developed (Fig. 4.16a). Thoracic and abdominal tubercles small, bearing long hooked setae (Fig. 4.22b) ....*Nothochrysa* McLachlan, 1868
—Larva naked. Body green. Cephalic capsule yellow with a longitudinal dark stripe (Fig. 4.17f). Thoracic and abdominal tubercles absent (Fig. 4.17f). Setae very short and knobbed apically (Fig. 4.22f) ....*Hypochrysa* Hagen, 1866

New World genera of chrysopid larvae

In contrast to that of the European fauna, knowledge of chrysopid larvae of the New World is fragmentary. At this time it would be premature to present a key to the genera. Thus, we provide a brief synopsis of the New World genera.

NOTHOCHRYSINAE

Hypochrysa Hagen, 1866. The larvae of the single described New World species of *Hypochrysa* are unknown.

Nothochrysa McLachlan, 1868. Larvae of *Nothochrysa californica* Banks, the only described New World species in the genus, generally resemble the European species (Toschi, 1965), except that the distal segment of the antenna is only ~4 (rather than 10) times shorter than the medial segment as described for the European species.

Pimachrysa Adams, 1956. Five described species comprise this New World genus; the larvae are unknown.

APOCHRYSINAE

This subfamily contains four New World genera (*Claverina* Navás, 1913; *Domenechus* Navás, 1913; *Lainius* Navás, 1913; and *Loyola* Navás, 1913). The larvae and biology of these spectacular insects are unknown.

CHRYSOPINAE

Belonopterygini The three New World genera in this tribe (*Abachrysa* Banks, 1938; *Belonopteryx* Gerstaecker, 1863; and *Nacarina* Navás, 1915) are presumed to be ant-associated. The larvae are unknown.

Chrysopini Currently, the New World chrysopine fauna falls into 12 genera. Larvae from six (half) of these genera have been characterised, but the Nearctic species of only two (*Chrysoperla*, *Yumachrysa*) have been treated comprehensively. Below is a brief summary.

Ceraeochrysa Adams, 1982. Larvae of the many Latin American species of *Ceraeochrysa* are largely unknown (see Smith, 1931; Muma, 1959). However, larvae of the species that occur in North America, *C. placita* (Banks), *C. cincta* (Schneider), *C. claveri* (Navás), *C. cubana* (Hagen) *C. lineaticornis* (Fitch), *C. smithi* (Navás) and *C. valida* (Banks) have received modern treatment (Tauber *et al.*, 1998, 2000). These studies indicate that the following set of traits distinguishes *Ceraeochrysa* larvae (Fig. 4.23): typical trash-carrying body form; largely white gibbous body; cranial setae S1–12 usually present (Fig. 4.23b); mesothorax without dorsal row of long setae; metathorax with posterior row of 3–8 pairs of long serrated-pointed setae arising from brown to tan chalazae (Fig. 4.23c); all other thoracic setae smooth–pointed; abdominal segment I without lateral tubercles, with dorsal tubercles present (Fig. 4.23d) or absent; abdominal segments *ii–v* with lateral tubercles bearing long serrated–pointed setae apically and smooth–hooked setae subapically, without dorsal tubercles; abdominal segments *vi* and *vii* with both lateral and dorsal tubercles; the lateral tubercles with serrated–pointed setae, the dorsal tubercles each bearing at least one long serrated seta — hooked (segment *vi*) or pointed (segment *vii*) (Fig. 4.23e).

Chrysopa Leach, 1815. In the New World, this genus is restricted to North America; the larvae appear to

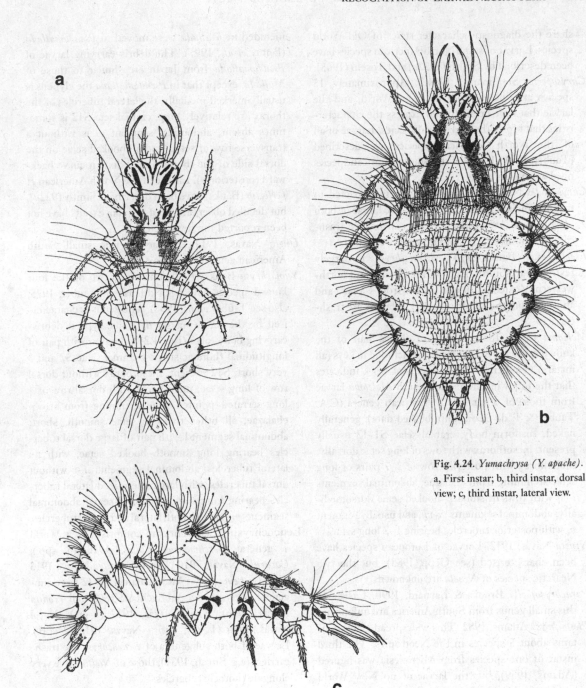

Fig. 4.24. *Yumachrysa (Y. apache).*
a, First instar; b, third instar, dorsal view; c, third instar, lateral view.

share the diagnostic character states of Old World species. Larvae of some North American species have been described by Smith (1922a) and Toschi (1965).

Chrysoperla Steinmann, 1964.3. Approximately 15 species are known from the New World, and the larvae that have been studied express the characteristics that typify the Old World species. Larvae of all of the North American species are described (Tauber, 1974); larvae of the South American species need attention.

Chrysopodes Navás, 1913. This genus is restricted to tropical and subtropical America. The larvae of both subgenera (*Chrysopodes* and *Neosuarius*) are trash-carriers (Smith, 1931), but none are described.

Eremochrysa Banks, 1903. *Eremochrysa* is largely restricted to southwestern United States and northwestern Mexico (with one species from Canada and another from the West Indies). The larvae are trash-carriers.

Meleoma Fitch, 1855. Larvae of almost half of the known species are described and included in keys (all instars) (Tauber, 1969). Subsequent work indicates that the following traits distinguish *Meleoma* larvae from those of other North American genera (C.A. Tauber & T. de Leon, unpublished data): generally naked, fusiform body, cranial setae S1–12 usually present; mesothorax with rows of long setae dorsally; metathorax with posterior row of 2–7 pairs of long setae stemming from chalazae; abdominal segments *i–vi* with rows of smooth–hooked setae dorsomedially; abdominal segments *vi–vii*, and usually segment *v*, with posterior tubercles bearing 1–2 long setae.

Nineta Navás, 1912. Larvae of European species have been characterised (see Gepp, 1983), but the two Nearctic species of *Nineta* are unknown.

Parachrysopiella Brooks & Barnard, 1990. Larvae of this small genus from South America are unknown.

Plesiochrysa Adams, 1982. This widespread genus contains about 5 species in the Neotropics. The third instar of one species from Micronesia was figured (Adams, 1959), but the larvae of no New World species are described.

Pseudomallada Tsukaguchi, 1995. This genus was recently separated from *Mallada* on the basis of differences in adult and larval characteristics among species in the Japanese fauna (Tsukaguchi, 1995). Subsequently, on the basis of adult characteristics, all four described North American species previously included in *Mallada* were moved to *Pseudomallada* (Penny *et al.*, 1997). The debris-carrying larvae of *Pseudomallada* from Japan are similar to those of *Mallada*, except that in *Pseudomallada* the clypeus is usually marked medially, the lateral tubercles of the thorax are relatively long, cranial seta S12 is sometimes absent, abdominal segment *vi* is without a transverse row of setae, and all hooked setae on the dorsal side of the thorax and abdomen curve backward (posteriorly). Larvae of the North American *P. perfectus* (Banks) have been described (Smith 1922a), but detailed observations of the above traits have not been reported for any species.

Ungla Navás, 1914. Larvae of this small South American genus are unknown.

Yumachrysa Banks, 1950. Larvae of two of the four known species have been described (Tauber, 1975; also see Tauber *et al.*, 1998). The genus is characterised by the following set of traits: typical debris-carrying body form (Fig. 4.24); dorsum with pair of longitudinal thoracic stripes; cranial setae S2 and 3 very short, S12 present; mesothorax without dorsal row of long setae; metathorax with dorsal row of 4 long serrated–pointed setae stemming from brown chalazae; all other thoracic setae smooth, short; abdominal segment I with pair of large dorsal tubercles bearing long smooth–hooked setae, with no lateral tubercles; abdominal segments *ii–v* without dorsal tubercles, with large sclerotized lateral tubercles bearing long smooth–hooked setae; abdominal segments *vi–vii* with both lateral and dorsal tubercles.

Leucochrysini Leucochrysini comprises New World 7 genera: *Berchmansus* Navás, 1913 (3 spp.); *Cacarulla* Navás, 1910 (1 sp.); *Gonzaga* Navás, 1913 (7 spp.); *Leucochrysa* McLachlan, 1868 (2 large subgenera, *Leucochrysa* McLachlan, 1868 and *Nodita* Navás, 1916); *Neula* Navás, 1917 (1 sp.); *Nuvol* Navás, 1916 (1 sp.); *Vieira* Navás, 1913 (2 spp.). Larvae of both subgenera of *Leucochrysa* are trash-carriers (e.g. Smith, 1931); those of *Nodita* have very elongate thoracic tubercles.

Acknowledgements

C.A. Tauber acknowledges the help of M.J. Tauber, T. de Leon, P.A. Adams (deceased), and N. Penny; this work was supported in part by NRI/USDA Competitive Grant 9802447 (CAT, MJT), Regional Project W-185, and Cornell University, CALS.

REFERENCES

Adams, P.A. (1959) Neuroptera: Myrmeleontidae and Chrysopidae. *Insects of Micronesia* 8, 13–33.

Agekyan, N.G. (1973) Neuroptera feeding on bamboo aphids in Adzharia and their parasites *Entomologicheskoe Obozreni* 52, 549–564. (in Russian) English translation: *Entomological Review* 52, 362–371.

Agekyan, N.G. (1978) A little-known entomophagous insect *Semidalis aleyrodiformis* (Neuroptera, Coniopterygidae) in Adzharia, USSR. *Entomologicheskoe Obozreni* 57, 509–512. (in Russian) English translation: *Entomological Review* 57, 348–350.

Alderson, E.M. (1911a) Notes on *Chrysopa dorsalis* Burm. *Entomologist's Monthly Magazine* 22, 49–57.

Alderson, E.M. (1911b) Notes on the life-history of *Chrysopa flava* Scopoli. *Entomologist* 4, 126–130.

Arrow, G.J. (1917) The life-history of *Conwentzia psociformis* Curt. *Entomologist's Monthly Magazine* 53, 254–257.

Aspöck, H. (1992) The Neuropteroidea of Europe: a review of present knowledge (Insecta). In *Current Research in Neuropterology, Proceedings of the 4th International Symposium on Neuropterology*, ed. Canard, M. Aspöck H. & Mansell, M.H. pp. 43–56. Sacco, Toulouse.

Aubrook, E.W. (1935) Notes on the biology of *Micromus aphidivorus* Schrank (Planipennia, Hemerobiidae). *North Western Naturalist* 10, 34–36.

Badgley, M.E., Fleschner, C.A. & Hall, J.C. (1955) The Biology of *Spiloconis picticornis* Banks (Neuroptera:Coniopterygidae). *Psyche* 62, 75–81.

Bänsch, R. (1964) Das Beutefangverhalten der aphidivoren Hemerobiidenlarven. *Zoologischer Anzeiger* 173, 278–281.

Bodenheimer, F.S. (1930) *Sympherobius (Nefasitus) amicus* Nav. in Die Schädlingsfauna Palästinas. *Monographien zur angewandten Entomologie [= Beihefte zur Zeitschrift für angewandte Entomologie]* 10, 137–140.

Brauer, F. (1850) Beschreibung und Beobachtung der österreichischen Arten der Gattung *Chrysopa*. *Naturwissenschaftliche Abhandlungen Wien* 4, 1–14.

Brauer, F. (1867) Larve von *Hypochrysa nobilis* Heyd. *Verhandlungen der kaiserlich-königlichen zoologische-botanischen Gesellschaft in Wien* 17, 27–30.

Brauer, F. (1871) Beiträge zur Kenntniss der Lebensweise und Verwandlung der Neuropteren (*Micromus variegatus* Fabr., *Panorpa communis* L., *Bittacus italicus* Klg. und *Bittacus Hagenii* Brau.). *Verhandlungen der kaiserlich-königlichen zoologische-botanischen Gesellschaft in Wien* 21, 107–116.

Brooks, S.J. & Barnard, P.C. (1990) The green lacewings of the world: a generic review (Neuroptera: Chrysopidae). *Bulletin of the British Museum (Natural History) Entomology* 59, 117–286.

Canard, M. & Labrique, H. (1989) Bioécologie de la chrysope Méditerranéenne *Rexa lordina* Navás (Neuroptera: Chrysopidae) et description de ses stades larvaires. *Neuroptera International* 5, 151–158.

Castellari, P.L. (1980) Indagini biologische su *Coniopteryx* (*Metaconiopteryx*) *esbenpeterseni* Tjeder (Neur.Coniopterygidae), predatore di acari tetranichidi sul pesco. *Bollettino dell'Istituto di Entomologia dell'Università di Bologna* 35, 157–180.

Christensen, P.J.H. (1943) Om morphologien og biologien hos *Drepanepteryx phalaenoides* L. *Flora og Fauna* 49, 1–13.

Collyer, E. (1951) The separation of *Conwentzia pineticola* End. from *Conwentzia psociformis* (Curt.) and notes on their biology. *Bulletin of Entomological Research* 42, 555–564.

Curtis, J. (1862) *British Entomology*, vol. 4 *Hymenoptera*, part 2, *Neuroptera: Trichoptera*. Lowell Reeve, London.

Diaz-Aranda, L.M. (1992). Estadíos preimaginales de los Crisûpidos ibéricos (Insecta, Neuroptera: Chrysopidae). PhD thesis, Universidad de Alcala, Madrid.

Diaz-Aranda, L.M. & Monserrat, V.J. (1988) Estadíos larvarios de los Neurópteros ibéricos. IV: *Mallada granadensis* (Pictet, 1865) (Planipennia: Chrysopidae). *Neuroptera International* 5, 111–119.

Diaz-Aranda, L.M & Monserrat, V.J. (1990a). Estadíos larvarios de los Neurópteros ibéricos. VI: *Chrysoperla carnea* (Stephens, 1836), *C. mediterranea* (Hölzel, 1972) y *C. ankylopteryformis* Monserrat & Diaz Aranda, 1989 (Planipennia: Chrysopidae). *Boletín de Sanidad Vegetal Plagas* 16, 675–689.

Diaz-Aranda, L.M. & Monserrat, V.J. (1990b) Estadíos larvarios de los Neurópteros ibéricos. VIII: *Mallada venosus* (Rambur, 1842) (Planipennia: Chrysopidae). *Neuroptera International* 6, 95–105.

Diaz-Aranda, L.M. & Monserrat, V.J. (1991) Estadíos larvarios de los Neurópteros ibéricos. VII. *Mallada picteti* (McLachlan, 1880) (Planipennia: Chrysopidae). *Neuroptera International* 6, 141–147.

Diaz-Aranda, L.M. & Monserrat, V.J. (1992) Descripción de los estadíos larvarios de *Brinckochrysa nachoi* Monserrat, 1977 (Neuroptera: Chrysopidae). *Nouvelle Revue d'Entomologie* 9, 207–214.

Diaz-Aranda, L.M. & Monserrat, V.J (1994) The larval stages of genus *Cunctochrysa* Hölzel, 1970 in Europe (Neuroptera: Chrysopidae). *Deutsche entomologische Zeitschrift, Berlin* N.F. 41, 163–171.

Diaz-Aranda, L.M. & Monserrat, V.J. (1995) Aphidophagous predator diagnosis: key to genera of European chrysopid larvae (Neur.: Chrysopidae). *Entomophaga* 40, 169–181.

Diaz-Aranda, L.M. & Monserrat, V.J. (1996) On the larval stages of genus *Suarius* (Neuroptera: Chrysopidae) in Europe. *Deutsche entomologische Zeitschrift, Berlin* N.F. 43, 89–97.

Dunn, J.A. (1954) *Micromus variegatus* Fabricius (Neuroptera) as a predator of the pea aphid. *Proceedings of the Royal Entomological Society of London (A)* 29, 76–80.

Dziedzielewicza, J. (1905) Bielotki Galicyi Slaska (Coniopterygidae Haliciae et Silesiae). *Kosmos* 30, 377–385.

Eisner, T., Hicks, K., Eisner, M. & Robson, D.S. (1978) 'Wolf in-sheep's-clothing' strategy of a predaceous insect larva. *Science* 199, 790–794.

Emerton, J.H. (1906) Cocoons and young of *Coniopteryx vicina*. *Psyche* 13, 64–65.

Enderlein, G. (1906) Monographie der Coniopterygiden. *Zoologische Jahrbücher (Systematik, Geographie und Biologie)* 23, 173–242.

Essig, E.O. (1910) The natural enemies of the citrus mealybug. *Pomona College Journal of Entomology and Zoology* 2, 143–146.

Fraser, F.C. (1945) Biological notes on *Chrysopa dorsalis* Burm. (Neuroptera). *Proceedings of the Royal Entomological Society of London* 20, 116–121.

Froggatt, W.W. (1904) Experimental work with the peach aphis (*Aphis persicae-niger*, Sm.). *Agricultural Gazette of New South Wales* 15, 603–612.

Fulmek, L. (1941) Über die Aufzucht von *Drepanepteryx phalaenoides* L. ex ovo. (Neuroptera: Planipennia, Hemerobiidae). *Arbeiten über morphologische und taxonomische Entomologie aus Berlin-Dahlem* 8, 127–130.

Genay, A. (1953) Contribution à l'étude des Névroptères de Bourgogne. *Travaux du Laboratoire de Zoologie et de la Station Aquicole Grimaldi de la Faculté des Sciences de Dijon* (N.S.) 3, 1–30.

Gepp, J. (1983) Schlüssel zur Freilanddiagnose mitteleuropischer Chrysopidenlarven (Neuroptera, Chrysopidae). *Mitteilungen des naturwissenschaftlichen Vereins für Steiermark* 113, 101–132.

Gepp, J. (1984) Erforschungsstand der Neuropteren-Larven der Erde (mit einem Schlüssel zur Larvaldiagnose der Familien, einer Übersicht von 340 beschreibenen Larven und 600 Literaturzitaten). In *Progress in World's Neuropterology, Proceedings of the 1st International Symposium on Neuropterology*, ed. Gepp, J., Aspöck, H. & Hölzel, H., pp. 183–239. Thalerhof, Graz.

Gepp, J. (1986) Biology and larval diagnosis of central European Neuroptera (A review of present knowledge). In: *Recent Research in Neuropterology, Proceedings of the 2nd International Symposium on Neuropterology*, ed.

Gepp, J., Aspöck, H. & Hölzel, H., pp. 137–144. Thalerhof, Graz.

Gepp, J. (1989) Zurkologischen Differenzierung der primaginaln Stadien baumbewohnender Chrysopiden im Alperaum (Planipennia: Chrysopidae). *Sitzungsberichte der österreicheschen Akademie der Wissenschaften, mathematische-naturwissenschaftliche Klasse I* 197, 1–73.

Gepp, J. (1990) An illustrated review of egg morphology in the families of Neuroptera (Insecta: Neuropteroidea). In *Advances in Neuropterology, Proceedings of the 3rd International Symposium on Neuropterology*, ed. Mansell, M.W. & Aspöck, H., pp. 131–149. South African Department of Agricultural Development, Pretoria.

Gepp, J. (1992) Contribution to the knowledge of coniopterygid larvae of Central Europe (Insecta: Neuroptera: Coniopterygidae). In *Current Research in Neuropterology, Proceedings of the 4th International Symposium on Neuropterology*, ed. Canard. M., Aspöck, H. & Mansell, M.W., pp. 151–152. Sacco, Toulouse.

Gepp, J. & Stürzer, C. (1986) *Semidalis aleyrodiformis* (Steph., 1836) Biologie, Ökologie und Larvenstadien (Planipennia, Coniopterygidae). *Mitteilungen des naturwissenschaftlichen Vereins für Steiermark* 116, 241–262.

Ghilarov, M.S. (1962) The larva of *Dilar turcicus* Hag. and the position of the Family Dilaridae in the order Planipennia. *Entomologicheskoe Obozreni* 41, 402–416.

Gleichen, W.F.F. von (also named Russworm) (1770) *Versuch einer Geschichte der Blatläuse und Blatlausfresser des Ulmenbaums*. Nuremberg.

Greve, L. (1974) The larvae and pupa of *Helicoconis lutea* (Wallengren, 1871) (Neuroptera, Coniopterygidae). *Norsk Entomologisk Tidsskrift* 21, 19–23.

Henry, T.J. (1976) *Aleuropteryx juniperi*: a European scale predator established in North America (Neuroptera: Coniopterygidae). *Proceedings of the Entomological Society of Washington* 78, 195–201.

Hoffmann, J. (1962) Faune des Névroptèroïdes du Grand-Duché de Luxembourg. *Archives. Institut Grand-ducal de Luxembourg* (N.S.) 28, 249–332.

Kawashima, K. (1958) Bionomics and earlier stages of some Japanese Neuroptera (II) *Eumicromus numerosus* (Navás) (Hemerobiidae). *Mushi* 32, 43–46.

Killington, F.J. (1931) Notes on the life-history of *Sympherobius fuscescens* Wall. (= *inconspicuus* McL.) [Neuroptera]. *Entomologist* 64, 217–223.

Killington, F.J. (1932a) The life-history of *Hemerobius atrifrons* McLach. [Neuroptera]. *Entomologist* 65, 201–203.

Killington, F.J. (1932b) The life history of *Hemerobius simulans* Walker (= *orotypus* Wall.) (Neuroptera, Hemerobiidae). *Entomologist's Monthly Magazine* 68, 176–180.

Killington, F.J. (1932c) Notes on the life-history of *Hemerobius pini* Steph. (Neuroptera). *Transactions of the Entomological Society of the South of England* 8, 41–44.

Killington, F.J. (1934) On the life-histories of some British Hemerobiidae (Neur.). *Transactions of the Society for British Entomology* 1, 119–134.

Killington, F.J. (1935) Remarks on the name of *Boriomyia betulina* (Strom) (= *Hemerobius nervosus* Fabr.) (Neur.): a reply to Mr. K. J. Morton. *Entomologist's Monthly Magazine* 71, 188–194.

Killington, F.J. (1936) *A Monograph of the British Neuroptera*, vol. 1. Ray Society, London.

Killington, F.J. (1937) *A Monograph of the British Neuroptera*, vol. 2. Ray Society, London.

Killington, F.J. (1946) On *Psectra diptera* (Burm.) (Neur., Hemerobiidae), including an account of its life-history. *Entomologist's Monthly Magazine* 82, 161–176.

Kimmins, D.E. (1939a) Supplementary notes on the life-history of *Micromus variegatus* (Fab.) (Neur.). *Journal of the Society for British Entomology* 1, 239–240.

Kimmins, D.E. (1939b) The first instar larva of *Nathanica capitata* (Fabr.). *Journal of the Society for British Entomology* 1, 240–241.

Kimmins, D.E. & Wise, K.A.J. (1962) A record of *Cryptoscenea australiensis* (Enderlein) (Neuroptera: Coniopterygidae) in New Zealand, with a redescription of the species. *Transactions of the Royal Society of New Zealand* 2, 35–39.

Krakauer, A.H. & Tauber C.A. (1996) Larvae of *Micromus*: generic characteristics and a description of *Micromus subanticus* (Neuroptera: Hemerobiidae). *Annals of the Entomological Society of America* 89, 203–211.

Labrique, H. (1990) Description de la larve de la chrysope méditerranéenne *Mallada picteti* (McLachlan, 1880) (Neuroptera: Chrysopidae). *Nouvelle Revue d'Entomologie* (N.S.) 7, 427–434.

Labrique, H. & Canard, M. (1989) Description de la larve de *Mallada ibericus* (Navás) (Neur. Chrysopidae). *Bulletin de la Société Entomologique de France* 94, 59–68.

Lacroix, J.L. (1921) Études sur les Chrysopides. Premier mèmoire. *Annales de la Société Linnéenne de Lyon* 68, 51–104.

Lacroix, J.L. (1924) Faune de Planipennes. Coniopterygidae. *Bulletin de la Société d'Étude des Sciences Naturelles d'Elboeuf* 42, 53–84.

Lacroix, J.L. (1925) Études sur les Chrysopides. Époque du coconnage chez larves du groupe *Chrysopa prasina* Burm. *Bulletin de la Société d'Étude des Sciences Naturelles d'Elboeuf* 43, 87–91.

Laidlaw, W.B.R. (1936) The brown lacewing flies (Hemerobiidae): their importance as controls of *Adelges cooleyi* Gillette. *Entomologist's Monthly Magazine* 72, 164–174.

Lauterbach, K.E. (1972) Die Planipennier oder echten Netzflügler der Umgebung von Tübingen (Insecta–Neuroptera). *Veröffentlichung landesstelle Naturschutz Landschaftpflege Baden-Württemberg* 40, 141–144.

Löw, F. (1885) Beitrag zur Kenntnis der Coniopterygiden. *Sitzungsberichte der Akademie der Wissenschaften in Wien, mathematische-naturwissenschaftliche Klasse I* 91, 73–89.

MacLeod, E.G. (1960a) Morphological studies on the head capsule of the larval stages of *Boriomyia fidelis* (Banks) (Neuroptera: Hemerobiidae). MSc thesis, University of Maryland, USA.

MacLeod, E.G. (1960b) [1961] The immature stages of *Boriomyia fidelis* (Banks) with taxonomic notes on the affinities of the genus *Boriomyia* (Neuroptera: Hemerobiidae). *Psyche* 67, 26–40.

MacLeod, E.G. (1964) A comparative morphological study of the head capsule and cervix of larval Neuroptera (Insecta). PhD thesis, Harvard University, USA.

Meinander, M. (1972) A revision of the family Coniopterygidae (Planipennia). *Acta Zoologica Fennica* 136, 1–357.

Meinander, M. (1974) The larvae of two North American species of Coniopterygidae (Neuroptera). *Notulae Entomologicae* 54, 12–16.

Meinander, M. (1990) The Coniopterygidae (Neuroptera, Planipennia). A check-list of the species of the world, descriptions of new species and other new data. *Acta Zoologica Fennica* 189, 1–95.

Milbrath, L.R., Tauber, M.J. & Tauber, C.A. (1994) Larval behavior of predacious sister-species: orientation, molting site, and survival in *Chrysopa*. *Behaviorial Ecology and Sociobiology* 35, 85–90.

Miles, H.W. (1924) On the life history of *Boriomyia (Hemerobius) nervosa*, Fab. (Planipennia, Hemerobiidae). *Bulletin of Entomological Research* 14, 249–250.

Miller, G.L. & Cave, R.D. (1987) Bionomics of *Micromus posticus* (Walker) (Neuroptera: Hemerobiidae) with descriptions of the immature stages. *Proceedings of the Entomological Society of Washington* 89, 776–789.

Miller, G.L. & Lambdin, P.L. (1982) *Hemerobius stigma* Stephens (Neuroptera: Hemerobiidae): external morphology of the egg. *Proceedings of the Entomological Society of Washington* 84, 204–207.

Miller, G.L. & Lambdin, P.L. (1984) Redescriptions of the larval stages of *Hemerobius stigma* Stephens (Neuroptera: Hemerobiidae). *Florida Entomologist* 67, 377–382.

Mitchell, R.G. (1962) Balsam wooly aphid predators native to Oregon and Washington. *Oregon State University, Agricultural Experiment Station, Technical Bulletin* 62, 1–63.

Mjöberg, E. (1909) Nätvingar: Neuroptera 1 Planipennia. Svensk Insektfauna 8 (1). *Entomologisk Tidskrift Arg.* 30 H 3–4,130–161.

Monserrat, V.J. (1978) Sobre los Neurópteros de las Islas Canarias II. *Semidalis candida* Navás, 1916 (Neur., Coniopterygidae). *Nouvelle Revue d'Entomologie* 8, 369–376.

Monserrat, V.J. (1983) Sobre los Neurópteros de las Islas Canarias IV. *Wesmaelius (Kimminsia) navasi* (Andreu, 1911) (Neur., Plan., Hemerobiidae). *Boletín de la Asociación española de Entomologia* 6, 209–224.

Monserrat, V.J. (1984) Estadíos larvarios de los Neurópteros ibéricos. III: *Anisochrysa genei* (Neuroptera, Planipennia, Chrysopidae). *Neuroptera International* 3 (1), 13–21.

Monserrat, V.J. (1989) Estadíos larvarios de los Neurópteros ibèricos. II: *Mallada subcubitalis* (Planipennia: Chysopidae). *Neuroptera International* 5, 125–132.

Monserrat, V.J. (1990) A systematic checklist of the Hemerobiidae of the world (Insecta: Neuroptera). In *Advances in Neuropterology, Proceedings of the 3rd International Symposium on Neuropterology,* ed. Mansell, M.W. & Aspöck, H., pp. 215–262. South African Department of Agricultural Development, Pretoria.

Monserrat, V.J. & Hölzel, H. (1987) Contribución al conocimiento de los neurópteros de Anatolia (Neuropteroidea, Planipennia). *Eos* 63, 133–142.

Monserrat, V.J. & Marín, F. (1996) Plant substrate specificity of Iberian Hemerobiidae (Insecta: Neuroptera). *Journal of Natural History* 30, 775–787.

Monserrat, V.J. & Rodrigo, F. (1992) Nuevas citas sobre los crisópidos ibéricos (Insecta, Neuroptera: Chrysopidae). *Zoologica Baetica,* 3, 123–138.

Monserrat, V.J., Diaz-Aranda, L.M. & Hölzel, H. (1991) Contribución al conocimiento de los neurópteros de Marruecos (Insecta, Neuropteroidea). *Eos* 66, 101–115.

Morton, K.J. (1910) Life-history of *Drepanepteryx phalaenoides,* Linn. *Entomologist's Monthly Magazine* 46, 54–62.

Morton, K.J. (1935) Remarks on the name of *Boriomyia nervosa* Fabr. *Entomologist's Monthly Magazine* 71, 95–100.

Moznette, G.F. (1915a) Notes on the brown lace-wing

(*Hemerobius pacificus,* Bks.). *Journal of Economic Entomology* 8, 350–354.

Moznette, G.F. (1915b) The brown lacewing *Hemerobius pacificus* Banks. In *Second Biennial Crop Pest and Horticultural Report, 1913–1914,* pp. 181–183. Oregon Agricultural College Experiment Station, Corvallis OR.

Muma, M.H. (1959) Chrysopidae associated with citrus in Florida. *Florida Entomologist* 42, 21–29.

Muma, M.H. (1967) Biological notes on *Coniopteryx vicina* (Neuroptera: Coniopterygidae). *Florida Entomologist* 50, 285–293.

Muma, M.H. (1971) Coniopterygidae (Neuroptera) on Florida citrus trees. *Florida Entomologist* 54, 283–288.

Nakahara, W. (1954) Early stages of some Japanese Hemerobiidae including two new species. *Kontyû* 21, 41–46.

Narayanan, E.S. (1942) On the bionomics and life-history of *Coniopteryx pusana* Withycombe. Coniopterygidae (Neuroptera). *Indian Journal of Entomology* 4, 1–4.

Neuenschwander, P. (1975) Influence of temperature and humidity on the immature stages of *Hemerobius pacificus. Environmental Entomology* 4, 215–220.

Neuenschwander, P. (1976) Biology of the adult *Hemerobius pacificus. Environmental Entomology* 5, 96–100.

New, T.R. (1967a) Biological notes on *Sympherobius pellucidus* (Walk.), with a description of the larva (Neuroptera: Hemerobiidae). *Entomologist's Gazette* 18, 50–52.

New, T.R. (1967b) *Sympherobius pygmaeus* (Ramb.) (Neur., Hemerobiidae) in Berkshire, with a description of the cocoon and pupa. *Entomologist's Monthly Magazine* 103, 171–172.

New, T.R. (1975) The immature stages of *Drepanacra binocula* (Neuroptera: Hemerobiidae), with notes on the relationships of the genus. *Journal of the Australian Entomological Society* 14, 247–250.

New, T.R. (1989) *Planipennia. Lacewings.* Handbuch der Zoologie, ed. Fischer, M, vol. 4, part 30. Walter de Gruyter, Berlin.

New, T.R. & Boros, C. (1983) The early stages of *Micromus tasmaniae* (Neuroptera: Hemerobiidae). *Neuroptera International* 2, 213–217.

Oswald, J.D. (1993a) Revision and cladistic analysis of the world genera of the family Hemerobiidae (Insecta: Neuroptera). *Journal of the New York Entomological Society* 101, 143–299.

Oswald, J.D. (1993b) A new genus and species of brown lacewing from Venezuela (Neuroptera: Hemerobiidae), with comments on the evolution of the hemerobiid forewing radial vein. *Systematic Entomology* 18, 363–370.

Oswald, J.D. (1994) A new phylogenetically basal subfamily of brown lacewings from Chile (Neuroptera: Hemerobiidae). *Entomologica Scandinavica* 25, 295–302.

Pantaleoni, R.A. (1983) Riconoscimento in campo delle larve di crisopidi. *Informatore Fitopatologico* 7/8, 31–36.

Penny, N.D., Adams, P.A. & Stange, L.A. (1997) Species catalog of the Neuroptera, Megaloptera, and Raphidioptera of America north of Mexico. *Proceedings of the California Academy of Sciences* 50, 39–114.

Penny, N.D., Tauber, C.A. & DeLeón, T. (2000) A new species of Chrysopa from western North America with a key to North American species (Neuroptera: Chrysopidae). *Annals of the Entomological Society of America* 93, 776–84.

Principi, M.M. (1940) Contributi allo studio dei neurotteri italiani. 1. *Chrysopa septempunctata* Wesm. e *Chrysopa flavifrons* Brauer. *Bollettino dell'Istituto di Entomologia dell'Università di Bologna*, 12, 63–144.

Principi, M.M. (1946) Contributi allo studio dei 'Neurotteri' italiani. 4. *Nothochrysa italica* Rossi., *Bollettino dell'Istituto di Entomologia dell'Università di Bologna* 1943–1946, 85–102.

Principi, M.M. (1947) Contributi allo studio dei neurotteri italiani. 5. Ricerche su *Chrysopa formosa* Brauer e su alcuni suoi parassiti. *Bollettino dell'Istituto di Entomologia dell'Università di Bologna* 16, 134–175.

Principi, M.M. (1954) Contributi allo studio dei neurotteri italiani. 11. *Chrysopa viridana* Schn. *Bollettino dell'Istituto di Entomologia dell'Università di Bologna* 20, 359–376.

Principi, M.M. (1956) Contributi allo studio dei neurotteri italiani. 13. Studio morfologico, etologico e sistematico di un gruppo omogeneo di specie del gen. *Chrysopa* Leach (*C. flavifrons* Brauer, *prasina* Burm. e *clathrata* Schn.). *Bollettino dell'Istituto di Entomologia dell'Università di Bologna* 21, 213–410.

Principi, M.M. & Canard, M. (1974) Les Névroptères. In *Les Organismes Auxiliaires en Verger des Pommiers*, pp. 151–162. OILB/SROP, Wegeningen, The Netherlands.

Putman, W.L. & Herne, D.H. (1966) The role of predators and other biotic agents in regulating the population density of phytophagous mites in Ontario peach orchards. *Canadian Entomologist* 98, 808–820.

Quayle, H.J. (1912) Red spiders and mites of citrus trees. *University of California Agricultural Experiment Station Bulletin* 234, 483–530.

Quayle, H.J. (1913) Some natural enemies of spiders and mites. *Journal of Economic Entomology* 6, 85–88.

Réaumur, R.A.F. (1737) *Mémoires pour servir à l'histoire des insectes*, vol. 3. Paris. [Note: Neuropterida parts contained in the 'Onzième Mémoire: Histoire des vers mangeurs de pucerons', pp. 363–412.]

Rousset, A. (1956a) Sur l'anatomie céphalique des larves de Coniopterygidae (Névroptères Planipennes). *Comptes Rendus Hebdomadaires des Séances de l'Académie des Sciences, Paris* 242, 933–936.

Rousset, A. (1956b) Sur l'anatomie des larves de Coniopterygidae (Névroptères Planipennes). Les stylets et leur musculature. *Comptes Rendus Hebdomadaires des Séances de l'Académie des Sciences, Paris* 243, 869–872.

Rousset, A. (1958) Sur le système nerveaux central céphalique de la larve de *Coniopteryx* (Névroptères Planipennes). *Comptes Rendus Hebdomadaires des Séances de l'Académie des Sciences, Paris*. 246, 842–845.

Rousset, A. (1960) Contribution à la faune de France des Névroptères. *Travaux du Laboratoire de Zoologie et de la Station Aquicole Grimaldi de la Faculté des Sciences de Dijon* 35, 23–33.

Rousset, A. (1966) Morphologie céphalique des larves de Planipennes (Insectes Névroptèroïdes). *Mémoires du Muséum National d'Histoire Naturelle, Paris* (A) 42, 1–199.

Rousset A. (1969) Morphologie thoracique des larves de Planipennes (Insectes, Névroptèroïdes). I. Squelette et musculature des régions antérieures du thorax chez les larves de *Chrysopa* et Conioptérygidés. *Annales des Sciences Naturelles, Zoologie et Biologie Animale* series 2, 12, 97–138.

Schlechtendal, D.H.R. von (1882) *Coniopteryx psociformis* Curtis, als Schmarozer in Spinneneiern. *Jahresbericht des Vereins für Naturkunde zu Zwickau* 1881, 26–31.

Schlechtendal, D.H.R. von (1883) Nachträgliche Berichtigung über *Coniopteryx psociformis* Curtis. *Jahresbericht des Vereins für Naturkunde zu Zwickau* 1882, 45–47.

Selhime, A.G. & Kanavel, R.F. (1968) Life cycle and parasitism of *Micromus posticus* and *M. subanticus* in Florida. *Annals of the Entomological Society of America* 61, 1212–1214.

Silvestri, F. (1942) Ordo Neuroptera. In *Compendio di Entomologia Applicata (Agraria – Forestale – Medica – Veterinaria)*, vol. 2, pp. 1–33.

Sinacori, A., Mineo, G. & Verde, G. Lo (1992) Osservazioni su *Aphanogmus steinitzi* Priesner (Hym.Ceraphronidae) parassitoide di *Conwentzia psociformis* (Curtis) (Neur. Coniopterygidae). *Phytophaga* 4, 29–48.

Smith, R.C. (1922a) The biology of the Chrysopidae. *Memoirs of the Cornell University Agricultural Experiment Station* 58, 1287–1372.

Smith, R.C. (1922b) Hatching in three species of

Neuroptera. *Annals of the Entomological Society of America* 15, 169–176.

Smith, R.C. (1923) The life histories and stages of some hemerobiids and allied species (Neuroptera). *Annals of the Entomological Society of America* 16, 129–151.

Smith, R.C. (1931) The Neuroptera of Haiti, West Indies. *Annals of the Entomological Society of America* 24, 798–821.

Smith, R.C. (1934) Notes on the Neuroptera and Mecoptera of Kansas, with keys for the identification of species. *Journal of the Kansas Entomological Society* 7, 120–145.

Souza, B. de (1988) Aspectos morfológicos e biológicos de *Nusulala uruguaya* (Navás, 1923; Neuroptera, Hemerobiidae) em condições de laboratório. MSc thesis, Escola Superior de Agricultura de Lavras, Lavras, Minas Gerais, Brazil.

Standfuss, M. (1906) Bewegliche Puppen bei Insekten mit vollkommener Verwandlung. *Mitteilungen der schweizerischen entomologischen Gesellschaft* 11, 154.

Standfuss, M. (1910) Notes on the biology of *Drepanepteryx phalaenoides* L. *Entomologist's Monthly Magazine* 46, 60–62. [Note: This work consists of an English translation of notes by Standfuss that are contained wholly within Morton (1910), cited above.]

Stange, L.A. (1981) The Dustywings of Florida. Part 1. Genera (Neuroptera: Coniopterygidae). *Florida Department of Agriculture and Consumer Services, Division of Plant Industry, Entomology Circular* 233, 1–2.

Stehr, F.W. (1987) *Immature Insects*, vol. 1. Kendall/Hunt, Dubuque IA.

Stehr, F.W. (1991) *Immature Insects*, vol. 2. Kendall/Hunt, Dubuque IA.

Stimmel, J.F. (1979) Seasonal history and distribution of *Carulaspis minima* (Targ.-Tozz) in Pennsylvania. *Proceedings of the Entomological Society of Washington* 81, 222–229.

Strom, H. (1788) Nogle insect larver med deres forvandlinger. *Ny Samling af det Kongelige Norske Videnskabers Selskabs Skrifter* 2, 375–400c.

Stroyan, H.L.G. (1949) A note on the biology of *Eumicromus angulatus* (Stephens) (Neuroptera, Hemerobiidae). *Entomologist* 82, 273–274.

Syrett, P. & Penman, D.R. (1981) Developmental threshold temperatures for the brown lacewing, *Micromus tasmaniae* (Neuroptera: Hemerobiidae). *New Zealand Journal of Zoology* 8, 281–283.

Tauber, C.A. (1969) *Taxonomy and Biology of the Lacewing Genus* Meleoma *(Neuroptera: Chrysopidae)*. University of California Press, Berkeley.

Tauber, C.A. (1974) Systematics of North American

chrysopid larvae: *Chrysopa carnea* group (Neuroptera). *Canadian Entomologist* 106, 1133–1153.

Tauber, C.A. (1975) Larval characteristics and taxonomic position of the lacewing genus *Suarius*. *Annals of the Entomological Society of America* 68, 696–700.

Tauber, C.A. (1991) Order Neuroptera. In *Immature Insects*, vol. 2, ed. Stehr, F.W., pp. 126–143. Kendall/Hunt, Dubuque IA.

Tauber, C.A. & Adams, P.A. (1990) Systematics of the Neuropteroidea: present status and future needs. In *Systematics of the North American Insects and Arachnids: Status and Needs*, ed. Kosztarab, M. & Schaefer, C.W., Virginia Agricultural Experiment Station Information Series 90–1, pp. 151–164. Virginia Polytechnic Institute and State University, Blacksburg.

Tauber, C.A. & Krakauer, A.H. (1997) Larval characteristics and generic placement of endemic Hawaiian hemerobiids (Neuroptera). *Pacific Science* 51, 413–423.

Tauber, C.A., Tauber, M.J. & Nechols, J.R. (1976) Yellow-body, a sex-linked recessive mutant in *Chrysopa*. *Journal of Heredity* 67, 119–120.

Tauber, C.A., De León, T., Lopez Arroyo, J.I. & Tauber, M.J. (1998) *Ceraeochrysa placita* (Neuroptera: Chrysopidae): generic characteristics of larvae, larval descriptions, and life cycle. *Annals of the Entomological Society of America* 91, 608–618.

Tauber, C.A., DeLeón, T., Penny, N.D. & Tauber, M.J. (2000) The genus Ceraeochrysa (Neuroptera: Chrysopidae) of America north of Mexico: larvae, adults, and comparative biology. *Annals of the Entomological Society of America* 93, 1195–221.

Tauber, M.J. & Tauber, C.A. (1971) An autosomal recessive (Neuroptera: Chrysopidae) mutant in a neuropteran. *Canadian Entomologist* 103, 906–907.

Terry, F.W. (1908) Notes on the life-history of an endemic hemerobiid (*Nesomicromus vagus* Perk.). *Proceedings of the Hawaiian Entomological Society* 1, 174–175.

Toschi, C.A. (1965) The taxonomy, life histories, and mating behavior of the green lacewings of Strawberry Canyon (Neuroptera: Chrysopidae). *Hilgardia* 36, 391–431.

Tsukaguchi, S. (1977) Biology and rearing of green lacewings. *Insectarium* 14, 180–184.

Tsukaguchi, S. (1978) Descriptions of the larvae of *Chrysopa* Leach (Neuroptera, Chrysopidae) of Japan. *Kontyû* 46, 99–122.

Tsukaguchi, S. (1995) *Chrysopidae of Japan (Insecta, Neuroptera)*. Osaka, Japan.

Tullgren, A. (1906) Zur Kenntnis schwedischer Coniopterygiden. *Arkiv för Zoologi* 3, 1–15.

Veenstra, C., Feichter, F. & Gepp, J. (1990) Larval diagnosis of the European genera of Hemerobiidae (Insecta:

Neuroptera). In *Advances in Neuropterology, Proceedings of the 3rd International Symposium on Neuropterology*, ed. Mansell, M.W. & Aspöck, H., pp. 211–213. South African Department of Agricultural Development, Pretoria.

Vine, H.C.A. (1895) Predacious and parasitic enemies of aphides (including a study of hyper-parasites). *International Journal of Microscopy* 5, 254–267 and 395–408.

Ward, L.K. (1970) *Aleuropteryx juniperi* Ohm (Neur. Coniopterygidae) new to Britain feeding on *Carulaspis juniperis* (Hem. Diaspididae). *Entomologist's Monthly Magazine* 106, 74–78.

Williams, F.X. (1927) The brown Australian lacewing (*Micromus vinaceus*). *Hawaiian Planters' Record* 31, 246–249.

Withycombe, C.L. (1922a) The life-history of *Hemerobius stigma*, Steph. *Entomologist* 55, 97–99.

Withycombe, C.L. (1922b) *Parasemidalis annae* End., a coniopterygid new to Britain, with notes on other British Coniopterygidae. *Entomologist* 55, 169–172.

Withycombe, C.L. (1922c) The wing venation of the Coniopterygidae. *Entomologist* 55, 224–225.

Withycombe, C.L. (1923a) A new British hemerobiid (Order Neuroptera). *Entomologist* 56, 202–204.

Withycombe, C.L. (1923b) Notes on the biology of some British Neuroptera. *Transactions of the Entomological Society of London* 1922, 501–594.

Withycombe, C.L. (1924a) Further notes on the biology of some British Neuroptera. *Entomologist* 57, 145–152.

Withycombe, C.L. (1924b) Notes on the economic value of the Neuroptera with special reference to the Coniopterygidae. *Annals of Applied Biology* 11, 112–125.

Withycombe, C.L. (1925a) A contribution towards a monograph of the Indian Coniopterygidae (Neuroptera). *Memoirs of the Department of Agriculture of India, Entomological Series* 9, 1–20, pls.1–4.

Withycombe, C.L. (1925b) Some aspects of the biology and morphology of the Neuroptera. With special reference to the immature stages and possible phylogenetic significance. *Transactions of the Entomological Society of London* 1924, 303–411.

Yang, C. (1951) New records of neuropterous insects from China: Coniopterygidae Enderlein, 1905. *Annales Entomologici Sinici* 1, 341–347.

Yang, C.-K. (1974) Notes on Coniopterygidae (Neuroptera). *Annales Entomologici Sinici* 17, 83–91.

Yang, C.-K. (1980a) Three new species of *Sympherobius* from China (Neuroptera: Hemerobiidae). *Peiching Nung Yeh Ta Hsueh Pao* [= *Acta Agriculturae Universitatis Pekinensis*] 6, 87–92.

Yang, C.-K. (1980b) Some new species of the genera *Wesmaelius* and *Kimminsia* (Neuroptera: Hemerobiidae). *K'un Ch'ung Hsueh Pao* [= *Acta Entomologica Sinica*] 23, 54–65.

Zelený J. (1961) A contribution to the identification of the family Coniopterygidae (Neuroptera) in Bohemia. *Casopis Ceskoslovenské Spolecnosti Entomologické* [= *Acta Societatis Entomologicae Cechosloveniae*] 58, 169–179.

CHAPTER 5
Ecology and habitat relationships

F. Szentkirályi

5.1 INTRODUCTION

During the 15 years since the publication of the last comprehensive book on lacewings (Canard *et al.*, 1984), some excellent reviews have been published on the ecology of lacewings (New, 1986, 1989; Bay *et al.*, 1993). Certain chapters of the present book cover ecological issues (such as food selection and utilisation, life-cycle characteristics, diapause, assessment of the impact of lacewings, natural enemies of lacewings, biodiversity conservation, enhancing chrysopids in field crops) and for this reason this chapter mainly focuses on issues relating to field ecology, in particular on lacewing patterns that are characteristic of different temporal and spatial scale levels, organisation of lacewing guilds, and chemical relationships between lacewings and the environment (chemical ecology).

The study of various spatial and temporal scale levels has great importance (Wiens *et al.*, 1986) in ecological issues. Knowledge of the dynamic processes of differing scale levels may be used in conservation of natural enemies and enhancement of their impact in agricultural or forest habitats.

5.2 TEMPORAL PATTERNS OF LACEWINGS

5.2.1 Long-term fluctuation patterns and population dynamics

Few studies have been carried out on year-to-year population changes of lacewings. In order to recognise real population dynamics, changes in the abundance of species and environmental factors should be observed over numerous generations in the same localities.

Ressl (1971) and Gepp (1973) published long-term data on individual numbers of chrysopids without any evaluation. Over ten (1961–1970) and nine (1964–1972) years' study in Austria, Ressl and Gepp respectively reported the yearly numbers of lacewings. The yearly abundance level of lacewings was highest in 1963, 1966, and 1968 in the 1961–1970 collecting period, while the peak was in 1966/7 and in 1972 in the 1964–1972 period. The longest period (1959–1978) for the fluctuation pattern of *Chrysoperla carnea* was published from Germany (Heil *et al.*, 1980). The difference between the extreme values of abundance was 15 times over the 20-year period, based on light-trap catches. The population peaks were highest in 1959 and 1964 and lowest in 1969, 1972, and 1976. Elliott & Kieckhefer (1990) collected *Chrysoperla plorabunda* by sweep-netting in alfalfa fields over 13 years (1973–1985) in America (South Dakota). The annual abundance of this chrysopid varied by about one order of magnitude over the studied years. Two low peaks (1974, 1982) and a high peak (1985) were detected.

Ressl (1974) published data on yearly changes of abundance of hemerobiid species in Austria. Over 12 years (1961–1972) the maximum catches of most of 32 brown lacewing species were recorded in 1962, 1966, and 1968. Long-term monitoring with light traps provided a time series of yearly changes of species number and abundance of hemerobiids in Hungary (Szentkirályi, 1992, 1998, 1999; Szentkirályi *et al.*, 1998). According to these studies fluctuation patterns of lacewing assemblages between 1981 and 1992 showed differences in each locality (Szentkirályi, 1998, 1999), but the numerical peaks in highland and lowland areas at regional scales overlapped. Both peaks of annual species number and abundance were detected every third year (1983, 1986, and 1989) by time-series analysis. The periodical fluctuation of hemerobiids every 3–4 years was supported by results of light-trap collection in apple orchards and maize fields in Hungary in 1976–1989. Collection peaks were registered in 1979, 1983, 1986, and 1989 (F. Szentkirályi, unpublished data).

Reviewing available data on long-term fluctuation patterns of population dynamic changes of lacewings, it can be stated that maxima are detected every 3–5 years in both abundance of species and number of species in assemblages. The existence of regular periodicity of fluctuations should be studied over a longer time series. According to data from Hungary and Austria the long-term fluctuation patterns of lacewings at the regional level are more or less synchronised. Factors such as climatic fluctuations and outbreaks of prey populations (as occurred with aphids in 1986 in Hungary) could be responsible to a large extent for the synchronicity of changes of lacewing population dynamics over a large area. Time-series analysis of lacewing data in Hungary suggests that mean winter air temperature (Szentkirályi, 1992) or yearly drought level (Szentkirályi et al., 1998) may play a role in the occurrence of fluctuations of hemerobiid abundance level. According to regression analysis it seems that the increase in yearly abundance and number of species of brown lacewings correlates with increasing winter temperature, while they decrease with increasing summer precipitation. Applying various drought indices, both the regional number of individuals and species of hemerobiids increased with the advance of drought. However, periods of strong drought (for instance, 1990–1993) were unfavourable for brown lacewing populations and a decrease of abundance and number of species was recorded (Szentkirályi et al., 1998). The possible impact of increasing aridity was demonstrated by Paulian (1991) who found reduced numbers of eggs laid per plant by lacewings in maize fields during dry seasons. In order to recognise the environmental factors governing fluctuations in the population dynamics of lacewings, further long-term monitoring is needed. Analysis of possible effects of climate change on lacewing assemblages will be relevant in this respect in different agricultural areas of Europe in future.

5.2.2 Seasonal patterns of lacewings

Recognition of the characteristics of lacewing seasonality is necessary to enhance the efficiency of biological protection, and in particular to assess whether the synchrony between the given predator and prey is adequate. There are numerous examples of patterns representing the seasonal dynamics of developmental stages of various species of lacewings, especially in the agricultural areas of Europe and North America (see Chapters 9 and 10, this volume).

By studying the flight activity pattern of lacewing adults through regular trapping, certain categories may

be established on the basis of the length of season. According to long-term light-trapping data from Hungary we can differentiate among brown lacewing species that have short (*Psectra diptera*), medium (e.g. *Hemerobius lutescens*, *Sympherobius elegans*) or long seasons of flight activity (Szentkirályi, 1997). Zelený (1965) also recorded shorter (*Chrysopa phyllochroma*) and longer (*Chrysoperla carnea*) seasonal activity of chrysopids on herbaceous plant stands from agricultural areas. Lacewings with shorter seasonality showed a unimodal pattern of activity (for instance, *P. diptera*, *Micromus variegatus*, *Chrysopa phyllochroma* and univoltine chrysopids), whilst the pattern of species that are active for more months during the season was polymodal (e.g. *Hemerobius humulinus*, *Wesmaelius subnebulosus*, *Sympherobius pygmaeus*, *Chrysopa formosa*) in central Europe (Szentkirályi, 1997, 1998; Zelený, 1965). Taking into account the intraspecific similarity of the flight pattern of hemerobiids, Szentkirályi (1997) detected that activities were similar (>70%) across regions. In addition, the maximum activity in hilly regions lagged on average one week behind peaks in the lowlands. This shift of patterns may be explained by the cooler climate of highland areas. The mass flight period of chrysopids covers June–August under the temperate climate of Europe, while for hemerobiids it occurs later, from mid-July to late September (Szentkirályi, 1984, 1986; see further examples in Chapters 9 and 10, this volume). Accordingly, synchronies of green and brown lacewings with their aphid prey can be different. Two growing periods of aphid abundance may usually be detected during the season in temperate climate regimes. The first occurs between May and mid-July, and the second from early September to late October. This double wave of aphid abundance can be recorded in the case of the major annual arable crops [for instance, on maize: Szentkirályi, 1986; Coderre & Tourneur, 1988; on lucerne (alfalfa), etc.], as well as on perennial bushes and trees (e.g. on apple trees: Szabó & Szentkirályi, 1981). Increases in aphid abundance can be seen on weed species from year to year (Szentkirályi, 1991; F. Szentkirályi, unpublished data). According to the above-described general activity pattern of adults, chrysopids are mainly associated with the first wave of aphid abundance, while the hemerobiids are associated with the second wave. Naturally, numerous local variations are possible in synchrony between lacewings and aphids. On the basis of 10 years' data from maize fields and apple orchards in Hungary, the seasonal pattern of egg-laying and larval activity corresponds with this relationship (Szabó & Szentkirályi, 1981; Szentkirályi, 1986, 1991; Rácz et al., 1986). For instance, the majority of eggs were deposited onto corn plants and weedy plants in maize fields by chrysopids in July, the activity of larvae was detected in late July – early August, and is associated with the first abundance period of aphids. Chrysopids reacted to the second wave of abundance of aphids only in some years and the amount of deposited eggs was also limited to the first third of September. On the other hand, the period of egg-laying and larval activity of hemerobiids on the corn plants took from early August to late October with a peak in September. Brown lacewings deposited eggs in parallel with the two waves of aphids, mainly from May to mid-July and in September–October in herbaceous weedy field margin of maize stands (Szentkirályi, 1986, 1991). Overlap analysis made between seasonal patterns of oviposition of lacewing predators and potential aphid prey showed that chrysopids were less synchronised with aphids than were hemerobiids. For instance Micromus species from hemerobiids appeared on the first aphid colonies, early in April, infesting the herbaceous plants (Urtica, Arctium, Carduus spp.) in maize field margins.

5.2.3. Diel activity patterns of lacewings

Lacewings, mainly hemerobiids, are known as nocturnally active insects. Distribution of flight activity has been studied by a number of authors (Banks, 1952; Lewis & Taylor, 1965; Duelli, 1984a, 1986a; Ábrahám & Vas, 1999; Vas et al., 1999). These studies were made in field conditions through suction trapping and sticky trapping, and also in the laboratory. The nocturnal flight activity of five chrysopid species (Chrysopa formosa, C. pallens, C. perla, Dichochrysa prasina, Chrysoperla carnea) and two hemerobiid species (Hemerobius humulinus, H. lutescens) were clearly demonstrated through suction trapping by Vas et al. (1999). Banks (1952), applying suction traps, and Williams & Killington (1935), sampling hourly with light traps, showed that flight of hemerobiids started during twilight (18.00–20.00 hours) and finished around dawn (03.00–04.00). Mass flight of hemerobiids occurred between 21.00 and 01.00 with a peak at 21.00–22.00 hours. A few individual chrysopids in Banks's study flew before midnight, while Williams & Killington

found that chrysopids were active after midnight. Using sticky traps in potato fields, Mack & Smilowitz (1979) surveyed the diel activity of C. carnea s. lat. According to their data the highest activity period of C. carnea was between 20.30 and 09.00 hours. Jones et al. (1977) analysed the diel periodicity of feeding, mating, and oviposition of C. carnea. The adults were actively feeding mainly at 18.00–22.00 hours as well as at 02.00–09.00 hours, with a maximum at 07.00–08.00 hours. The majority of mating activity was recorded between 20.00 and 22.00 hours. The most intensive egg-laying period was at 20.00–01.00 hours, and the peak of activity was registered at 21.00–22.00 hours. Schotzko & O'Keeffe (1989) studied the diel activity of larvae and adults of Chrysopa spp. by sweep-netting in a lentil field. They demonstrated a higher peak of activity in the period 21.00–02.00 hours, and a lower one between 10.00 and 13.00 hours.

It was Duelli (1984a, 1986a) who examined the various types of daily flight activity of green lacewings in detail. He recognised four types of activity. Most species of lacewing belong to the carnea-type that starts flight after sunset and finishes before sunrise. Its peak of activity can be observed mainly in the first part of the night. According to recent experiments H. humulinus, H. micans, H. atrifrons, Nineta flava, Chrysotropia ciliata, and Dichochrysa prasina belong to this type (Ábrahám & Vas, 1999). Maximum activity of these species was detected during the period of 19.00–21.00 hours, with the exception of N. flava that peaked between 20.00 and 01.00 hours. Moreover, its lower daytime peak was also demonstrated between 13.00 and 15.00 hours. The second type of flight activity is covered by Mallada basalis. Its diel flight pattern is characterised by two twilight peaks. The third type is represented by Chrysopa perla. Its flight starts during the afternoon and terminates in total darkness. Ábrahám & Vas (1999) classified Chrysopa dorsalis to this type that had a high activity level between 12.00 and 22.00 hours. Although Micromus variegatus is considered to belong to this type by these authors, its flight pattern is rather similar to that of N. flava: it has a lower but characteristic peak of activity between 13.00 and 16.00 hours. Nothochrysa fulviceps is closer to the 'flava'-type. The fourth type is made up of lacewings belonging to 'hypochrysodes' lacewing species, such as H. elegans, that is active in the daytime.

5.3 SPATIAL PATTERNS OF LACEWINGS

5.3.1 Changes of lacewings at the geographical scale

There are significant changes in environmental factors, such as day-length, climate, vegetation type, number of potential prey species, etc. that relate to spatial dynamics at a geographical scale, in particular in latitudinal zones in the north–south direction. Lacewings have had to adapt to both spatial and temporal changes at the geographical scale. For this reason, numerous questions of evolutionary ecology connected with lacewings may be answered only through studies made at this spatial scale, as has already been emphasised by Tauber & Tauber (1993). They demonstrated (Tauber & Tauber, 1986, 1993) the physiologically as well as ecologically close correlation between diapause and migration of lacewings. They studied the geographical variations in photoperiodic responses of three biotypes of the *Chrysoperla carnea* species complex which were in reproductive isolation. They also demonstrated that life-history traits and behavioural elements of *Chrysopa quadripunctata*, such as feeding and larval camouflaging, varied significantly at a geographic scale. They concluded that disparate prey resources could lead to the differentiation of locally adapted behaviour and the evolution of specialisation (Tauber *et al.*, 1995).

The number of species of chrysopids and hemerobiids may also change on a geographical scale. Aspöck *et al.* (1980) demonstrated an increase with latitudinal lacewing species richness from the region of north Europe toward south Europe. The maximum number of species of both lacewing families can be found in the Mediterranean region, south of 45° N. Due to the lack of faunistic research this pattern of species diversity of lacewings cannot be verified in other continents, but it is very probable that there is a similar trend in North America. It is worthy of note that aphids and scale insects (being the potential major prey groups) show characteristic latitudinal changes in species diversity across Europe. Dixon *et al.* (1987) detected that the number of species of aphids decreased from the temperate region towards the tropical region. According to these authors, this could be explained by such specialism of aphids as their high degree of host specificity, their low efficiency in host-plant location, and additionally, the fact that they can survive only short periods without food. In parallel, as Kozár (1995) demonstrated, species richness of scale insects increases from north to south in Europe, and its maximum is reached in the Mediterranean region. Species richness of scale insects then decreases toward tropical areas, although 80% of these homopteran insects were mono- or oligophagous. It has been unclear how the trend in diversity of these two prey groups contributes to the creation of latitudinal patterns of lacewing species richness in Europe. For all cases, the presence of the coccidophagous *Dichochrysa* genus (great numbers of which are characteristic of the Mediterranean region) corresponds well with the maximum diversity of scale insects.

5.3.2 Lacewings at the landscape scale

Agricultural and forest landscapes consist of numerous mosaic units of various types including arable fields, orchards, gardens, vineyards, meadows, and forest stands. Suitable habitat types for lacewings may be scattered a long distance from one to another in a spatially heterogeneous landscape. For this reason, it is important to raise the question as to how the lacewings reach habitat patches.

Long-distance dispersal of lacewings

Greve (1969) produced data on the airborne transport of lacewings. She demonstrated that 13 hemerobiids and one chrysopid species passively drifted in the wind from a fjord to a mountain plateau at 1200 m a.s.l. Of these hemerobiid species, 94% were associated with pine, for instance, *Wesmaelius concinnus*, *H. pini*, *H. nitidulus*, *H. atrifrons* or *H. stigma*. In Hungary *H. nitidulus* and *H. atrifrons* were also recorded in a monocultural maize field, deriving from a probable source (*Pinus silvestris* plantation) 2–3 km away (Szentkirályi, 1989). Sugg *et al.* (1994) monitored the recovery of eradicated lacewing populations from barren, devastated habitats of Mount St Helens over a seven-year period following the 1980 eruption. They collected 227 immigrating individuals of 13 hemerobiids and seven chrysopids over the period. From the location of the survey, the next undisturbed area was 10 km away; thus long-distance dispersal of lacewings had occurred. According to the authors not only active flight but also passive transport by wind currents played an important role in immigration. Larvae of hemerobiids frequently appeared in pitfall traps close to small patches of feeble vegetation reflecting the start of the process of lacewing recolonisation in the eradicated field. Air currents would strongly contribute to long-distance habitat colonisation of lacewings because their flying speed is low. The mean flying speed of newly hatched individuals of *C. carnea s. lat.* was 0.51 m/sec during their first night while it was 0.74–0.79 m/sec during their second night, according to flight-mill tests carried out by Duelli (1984*b*). Taking into account that wind velocity increases with growing elevation, the estimated mean groundspeed of a preovipository female *C. carnea* flying at 6–12 metres height would be 4.2–7.3 m/sec. On this basis, Duelli calculated the flight distance of migrating females as 40 km per night. During the two days of obligate flight they can cover up to 80 km. At higher altitudes, wind transports lacewings further. Duelli (1984*a*) recorded the vertical distribution of flying individuals of *C. carnea* on a 115-metre high tower in Switzerland. He found that the number of individuals was reduced at higher altitude, and that the majority of migrating female lacewings was flying below 30 metres.

Changes of lacewing characteristics with altitude

As one moves up a mountain, environmental conditions alter dramatically (e.g. shorter vegetation period, cooler and more extreme climate, decreasing species diversity and complexity of vegetation, limited number of prey). In parallel, the adaptive life-history traits of lacewings vary. Gepp (1980, 1988/9) collected data from the Alps in 1962–82 on how the life history characteristics of lacewings changed according to altitude zones.

He ascertained, as was expected, that with increasing altitude the yearly number of generations of lacewings is reduced. Thus *Chrysopa perla* provides two complete and a partial third generation in a warm belt at 300–600 m a.s.l., while in a cold belt of the same altitude only two generations occur. *C. perla* has only one generation yearly at 700–1000 m a.s.l. in the Alps.

Gepp's studies showed that the length of seasonal flight activity gradually decreased with higher altitude by one week per 100 metres for *C. perla*. Seasonal activity of adults of *Chrysotropia ciliata* and *Nineta flava* was also reduced, by one month per 500 metres approximately, so that activity occurred from late May to September at 0–500 m a.s.l., while above 1000 m a.s.l. it was detected typically only in July–August.

Where the altitude is higher, the abundance and species richness of lacewing assemblages is lower in the Alps, according to Gepp (1988/9). Only a few second and third instar larvae of lacewing species were found above 1200 m a.s.l. including *Chrysoperla carnea* and *Chrysopa abbreviata*. The majority of chrysopid species deposited their eggs only in areas below 1000 m: *N. flava* up to 700–800 m, *Chrysoperla carnea* up to 300–700 m, *H. elegans* up to 500–600 m, and *Chrysopa pallens* up to 300–500 m a.s.l. Greve *et al.* (1987) also demonstrated an altitudinal decrease of number of lacewing species in the hilly part of Norway. They trapped lacewings between 900 and 1450 m a.s.l. There were no catches in the middle alpine zone with open, patchy vegetation at 1400 m. No lacewing individuals appeared in the low alpine zone (1100–1400 m) over 1200 m. Only in the lower part of the low alpine zone with a closed plant-cover were six hemerobiid species encountered. In addition, seven hemerobiid and one chrysopid species (*Chrysoperla carnea*) were recorded in the sub-alpine zone (900–1000 m) covered by the vegetation of the birch belt. On the basis of their data they concluded that the flight period of numerous species was significantly shorter in mountains than in lowlands.

Interhabitat patterns of lacewings

Hierarchical mosaic patterns of vegetation patches of various habitat types characterise the spatial structure of landscapes. Agricultural landscapes predominantly consist of cultivated fields (for instance, stands of arable field crops, orchards, gardens, meadows) with more or less natural or semi-natural habitat patches, such as wood patches, grasslands, or abandoned orchards scattered among them. These undisturbed habitat mosaic units can ensure the colonisation and high biodiversity level of aphidophagous insects in agricultural fields (Duelli, 1988, 1997). According to this concept lacewings living in the patchy agricultural environment have a set of interacting local populations. The metapopulation dynamics of lacewings (i.e. rate of habitat patch occupancy and extinction) may be defined by the distribution of habitat patch sizes and interpatch distances (isolation) (Hanski & Gilpin, 1991). Evidence supporting this concept has been found for numerous insect species, where the rate of occupancy or colonisation probability increased with growing patch size and with decreasing isolation, while the extinction probability reduced with increasing patch area and level of population size (Hanski, 1994). Not only the metapopulation structure of lacewings but also their species diversity at the regional landscape scale depends significantly on the type, size, and density of uncultivated habitat patches. Through long-term light-trap monitoring in agricultural sites it has been shown in Hungary (Szentkirályi, 1992) that species richness of brown lacewing assemblages was much greater in the highland region, which has a higher average level of natural vegetation diversity due to many forest patches (hedgerows) than has the lowland region. According to the resource–diversity hypothesis the frequent wood patches and uncultivated bushy field margins between the agricultural habitats of the highland region can maintain more hemerobiid species, in contrast to the lowland region where natural–semi-natural field borders were rare (Szentkirályi & Kozár, 1991; Szentkirályi, 1992). Forest stands between cultivated habitats can contribute in great part to maintenance of species richness of lacewings at the landscape level. This is confirmed by light-trapping observations in Hungary where the mean local number of species, and species diversity, of chrysopids and hemerobiids associated with forests was significantly higher than that of lacewing assemblages living in agricultural areas (in the case of hemerobiids the difference was a factor of two) (Szentkirályi, 1984).

There is an important role for natural–seminatural vegetation patches in the maintenance and concentration of agricultural lacewing populations because they provide essential and alternative food sources, and suitable sites and microhabitats for mating, oviposition, sheltering, cocooning, or overwintering. The most common uncultivated vegetation type in agricultural landscape can be found in field margins frequently forming ecotones. These are narrow transitional zones in the interface area between two adjacent, well-defined habitat units. Due to the special vegetation structure of habitat interfaces the ecotones are characterised by abrupt, sharp changes of environmental conditions (Jeník, 1992). There are numerous types of field margins according to their floral composition and structure, such as hedgerow, windbreak, shelterbelt, fence, ditch, or weedy strip. Depending on habitat quality, the zones of boundary vegetation may serve as a reservoir, ecological corridor, or even a barrier for lacewings. In order to demonstrate the lacewing species living in field margins and to recognise the reservoir

role of windbreaks, hedgerows, shelterbelts, several investigations were carried out. It was Galecka & Zelený (1969) who made one of the earliest studies on lacewing assemblages in wooded field margins. They found that *C. carnea* was represented in the greatest number (over 90%) while some individuals of *Chrysopa commata*, *C. phyllochroma*, *D. prasina* and *C. perla* were found in the wooded shelterbelt containing *Robinia* and *Sambucus*. Bowden & Dean (1977) collected larvae of *Chrysoperla carnea*, *Wesmaelius subnebulosus* and *Dichochrysa* species from a high hedgerow that was species rich (ash, oak, elder, maple, dogwood, and dense shrub layer). *C. carnea* was the most abundant and common species, although the majority of individuals were not resident in the hedgerow. Tsibulskaya *et al.* (1977) found ten chrysopid species in wooded vegetation of windbreaks (oak, elm, ash, acacia, and maple) in the Ukraine. The recorded species were as follows in order of dominance: *C. carnea* (73.3%), *D. prasina* (8%), *Chrysopa formosa* (6.3%), *C. pallens* (4.5%), *C. perla* (3.3%), *Nineta flava*, *D. ventralis*, *D. flavifrons*, *C. abbreviata*, and *C. phyllochroma*. Of the brown lacewings only *Hemerobius humulinus* was observed. Pantaleoni & Sproccati (1987) collected a similar chrysopid assemblage from wooded hedges (*Robinia*, *Salix*, *Populus*, and *Ulmus*) in Italy. The most abundant lacewing species were *Chrysoperla carnea*, *Chrysopa formosa*, and *D. prasina*. Some individuals of *D. flavifrons*, *C. perla*, *C. pallens*, and *D. picteti* were also present. Only two hemerobiid species, *H. humulinus* and *Micromus angulatus*, were recorded. From a hedgerow (*Ulmus* and *Acer*), also in Italy, Paoletti *et al.* (1997) collected only *C. carnea* from the green lacewings, and three species of brown lacewings, namely *H. humulinus*, *H. simulans*, and *M. angulatus*.

The existence of tree and bush species in field margins can contribute to the maintenance of lacewing assemblages with higher species diversity inside the ecotone zone. It is probable, however, that only green and brown lacewings associated with herbaceous plants immigrate to stands of arable field crops and colonise them. Most investigations of lacewings living on herbaceous or weedy plants of field margins have been made in Europe. Honěk (1981) sampled aphidophagous complexes from three weedy species infested by aphids, and he found five chrysopid species among them: *Chrysoperla carnea*, *Chrysopa commata*, *C. perla*, *C. formosa*, and *C. pallens*. Each species of lacewing was recorded on *Artemisia vulgaris*. Only *Chrysoperla carnea*, *Chrysopa commata*, and *C. perla* were present on *Urtica dioica*, and only *Chrysoperla carnea*, *Chrysopa commata*, and *C. pallens* were present on *Matricaria maritima*. Völkl (1988) collected *Chrysoperla carnea* and *Chrysopa perla* from aphidophagous assemblages on the weedy plant *Cirsium arvense* infested by four aphid species along the field edges. Szentkirályi (1991) studied lacewing subguilds associated with aphids living on *Carduus acanthoides* in weedy strips that bordered maize fields in Hungary. Eggs of *Chrysoperla carnea*, *Chrysopa phyllochroma*, and *C. formosa* were recorded on aphid colonies of *Brachycaudus cardui* infesting taller, flowering individuals of common thistle. The hemerobiids *Micromus angulatus* and *M. variegatus* laid their eggs both on the leaf-rosette and flowering stages of thistles. The leaf-rosette plants infested with the *Brachycaudus helichrysi* aphids were preferred by these brown lacewings. Observing the oviposition of green lacewings in olive groves, McEwen & Ruiz (1993) found eggs of *Chrysoperla* and *Dichochrysa* species on many weedy plants. They detected that *Chrysoperla* spp. expressed no preference among weeds and that their eggs were deposited on *Carduus*, *Malva*, *Reseda*, or *Cichorium* plants, while *Dichochrysa* lacewings laid their eggs only on *Sonchus* and *Psolarea* species.

In North America, Altieri & Whitcomb (1979) recognised the role of weedy patches bordering agricultural fields in the colonisation of predatory insects. A hemerobiid species, *Micromus posticus*, was recorded in stands of *Chenopodium ambrosioides* near maize fields as well as on maize plants. According to their observations this lacewing consumed *Uroleucon* aphid species infesting weeds.

The role of weed cover in the maintenance of lacewing populations can be deduced through experiments made in uncultivated areas. Harris & Phillips (1986) reported the detrimental effects of the mowing of spring weeds growing along roadsides, field-edge ditches, and other uncultivated areas. In samples from 17 dicotyledonous weed species it was demonstrated that larval density of lacewings was reduced by 45%–50% as a result of regular mowing as compared with the unmown weedy vegetation. The detrimental effect of mowing on lacewings was highest in the case of crimson clover.

Although field margins provide habitats for lacewings during their whole life cycle in undisturbed sites, it is questionable whether they serve as real colonisation sources in all cases for stands of cultivated plants. It may occur that eggs deposited onto non-crop plants actually reduce the lacewing population in stands of the crop plant (Szentkirályi, 1991; McEwen & Ruiz, 1993). Plants at field margins may attract and arrest egg-laying females, if the increase of prey populations in crop and non-crop vegetation is synchronised. Szentkirályi (1991) detected this phenomenon, observing the growth of aphid abundance in parallel on maize plants and individuals of common thistle. Chrysopid females laid hundreds of eggs on foliage of this weedy plant in the margin of the maize fields. The transferring impact of field margin herbaceous vegetation on colonisation of lacewings can be observed in other cases. Aphid colonies developing on certain weedy plants (*Urtica*, *Matricaria*, and *Arctium*) in late May and early June attracted a great number of adults of *Chrysopa phyllochroma* and *C. formosa* close to maize fields.

The predatory adults consumed aphids or pollen of *Urtica*. Until mid-June the number of individuals of lacewings strongly decreased in field margins while in the same period adults of both chrysopid species appeared inside the maize field and commenced oviposition on maize plants. This immigration was promoted by the growth of the first aphid colonies on maize as compared with weeds, which started to develop later (first third of June). The same concentration of *Chrysoperla carnea* adults in field margins has also been detected. In the daytime hundreds of newly emerged adults were observed consuming pollen and nectar on *Conium maculatum* that was flowering from mid-June as well as on other Umbelliferae species (*Daucus*, *Pastinaca*). During the last third of June *C. carnea* adults shifted from the field border to the inside of maize stands (F. Szentkirályi, unpublished data). The accumulation of *C. carnea* adults close to the barriers of hedgerows or windbreaks has also been demonstrated (Lewis, 1968; Bowden & Dean, 1977). According to Bowden & Dean (1977) this distribution of lacewings may be explained by the combined impacts (hypsotaxis, deceleration, filter effect, arresting by food sources) of the structural complexity of hedges. These authors recognised that the great majority of *C. carnea* adults did not develop in the hedgerow studied but had come from elsewhere, being airborne visitors. Duelli (1988) estimated the immigration and emigration rate of *C. plor-*

abunda in crop fields of America, and of *C. carnea* in Europe. He concluded that the importance of immigration of these lacewing species was much more relevant than production of adults inside the crop habitat. His results indicated that immigration and emigration (extinction) of these lacewings were largely independent of area and distance of other crop fields. Monitoring between-habitat movements of arthropods by trapping systems in agricultural landscapes, Duelli & Obrist (1995) could establish various dispersal types. They demonstrated that the dispersal movement of *C. carnea* as a typical nomadic species depended on overwintering or hibernation places outside the crop fields. For this reason the maximal abundance of adults was in early springtime (April) and in autumn (October–November) adjacent to or inside natural habitats for instance, forest edge). In late spring–summer (June–July) the peak abundance of *C. carnea* adults was within crop fields where reproduction occurred. This interhabitat dispersal pattern of common green lacewings occurs twice yearly by migratory flights between sites of hibernation and reproduction (Duelli & Obrist, 1995).

Comparing the lists of lacewing species living in field margins in Europe (see above) with ones in agricultural habitats (see tables in Chapter 9, this volume) and in forests (see tables in Chapter 10) we may draw some conclusions: (1) each lacewing species living in agricultural areas – including the field margins – can find suitable living conditions in forests, too, and that is why these habitats are the most important reservoirs for chrysopids and hemerobiids; (2) the most important and dominant chrysopid and hemerobiid species of agricultural habitats live in lacewing assemblages of field margins, such that the non-crop vegetation is a potential colonisation source of lacewing guilds associated with cultivated fields; (3) depending on floristic composition, the European field margins can contribute to crop field lacewing assemblages with the following species: *Chrysoperla carnea, Chrysopa perla, C. pallens, C. formosa, Dichochrysa prasina, D. flavifrons, Nineta flava, Hemerobius humulinus, Wesmaelius subnebulosus* coming from wooded field margins and colonising mainly orchard habitats; and *Chrysoperla carnea, Chrysopa formosa, C. phyllochroma, C. commata, Micromus angulatus, M. variegatus,* and *H. humulinus* coming from the herbaceous type of field margins and colonising arable field crops.

5.3.3 Lacewings at the habitat scale: intrafield patterns

Horizontal distribution of lacewings within cultivated habitats

Several environmental factors within agricultural fields may influence the spatial distribution of lacewings. The structure and quality of vegetation inside the given habitat and in adjacent surroundings have the most relevant effect on the activity of lacewings. Wooded field margins have an edge effect on agricultural fields, so that certain lacewing species occur only at the edge of crop plant stands. Paoletti *et al.* (1997) observed such an exclusive marginal occurrence of *Hemerobius humulinus* in a wheat field in Italy. At the same time, distribution of the widely spread *Chrysoperla carnea* in an alfalfa field was not influenced by the hedgerow in the field margin. According to their samples, other species, such as *H. simulans*, remained exclusively within the hedgerow. Mayse & Price (1978) monitored the distribution of eggs of *C. carnea s. lat.* from the edge to the middle of stands of soya-bean fields weekly. No consistent distribution was found in the absence of a hedgerow, and seasonal changes in the number of eggs was similar in the whole field. If a hedgerow (pine trees) was situated at the margin of the field, the trend of oviposition increased towards the centre of the soya-bean field.

The non-crop vegetation (weeds, groundcover plants) serving as alternative resources inside the field can greatly influence the distribution of lacewings. Bosch (1987) studied the impact of some dominant weed species (*Matricaria, Lamium*) in a sugarbeet field on predatory arthropods in weedy and weed-free plots. He counted the larvae of lacewings on plants heavily infested by aphids in experimental plots. Bosch recorded chrysopid larvae only on sugarbeet from weedy plots. Through similar experiments in weedy and weedless collard stands Horn (1981) sampled eggs deposited by the green lacewing *Chrysopa oculata*. In one year significantly more chrysopid eggs were found in weedy plots because aphids on *Chenopodium album* served as an alternative food source for lacewings. In another year, in mid-summer, there were significantly more chrysopid eggs laid on collard plants of the weedless plot. This distribution was caused by a stronger infestation of green peach aphid in weedless plots compared with the previous experimental year.

The cultivation methods and management techniques in crop fields may also modify the spatial distribution of lacewings. Neuenschwander & Michelakis (1980) observed the significant impact of irrigation on patterns of chrysopids in olive orchards. Adults and larvae of *Dichochrysa* species (*flavifrons, zelleri*) and of *Suarius nanus* were found in higher numbers on non-irrigated olive trees. Despite this, adults and larvae of *Chrysoperla carnea* could be found in greater numbers on irrigated trees. Until recently there have been few experiments conducted to characterise the intrafield distribution of types of lacewings. Pantaleoni & Ticchiati (1990) studied the distribution of larvae of *C. carnea*, while Lapchin (1991) investigated the distribution of eggs laid by *C. carnea* in wheat fields. Both distributions were of the aggregative type of spatial distribution.

Vertical intrahabitat distribution of lacewings
The preference of European lacewings for different vertical strata of vegetation is well known (Aspöck *et al.*, 1980). Despite this, there has been little experimental work to verify it. In general it is the vertical distribution of adults in airspace that has been studied. These studies are summarised as follows:

• New (1967) in a park in the UK used suction trapping at heights of 1.2 m and 9 m;
• Nielsen (1977) in old-growth beech forest in Denmark used light trapping at heights of 0.6 m (herb layer), 10 m (trunk layer), 21 m (mid-crown layer), and 30 m (above canopy);
• Sziráki (1996) in oak forests in Hungary applied sticky traps at heights of 2 m (shrub layer) and 10 m (crown layer);
• Szentkirályi and Markó (unpublished data) in an oak forest in Hungary used Malaise-trapping for three years at heights of 1.5 m (shrub layer), 12.5 m (mid-crown layer), and 25 m (above canopy).

Based on these experiments the preference of various lacewing species for vertical strata is known. Summing up the results, the herb and shrub layers (0.6–2 m) were preferred by *Chrysopa phyllochroma*, *C. pallens*, *Dichochrysa prasina*, *D. ventralis*, *Micromus angulatus*, *M. variegatus*, *Hemerobius humulinus*, and *H. micans*, and the following species showed preference for the higher tree-crown layer (9–21 m): *Chrysoperla carnea*, *Nineta flava*, *Hypochrysa elegans*, *Chrysotropia ciliata*, *Hemerobius gilvus*, *H. lutescens*, *H. stigma*, *H. pini*, *Wesmaelius nervosus*, *W. quadrifasciatus*, *Sympherobius pygmaeus*, and *S. elegans*. Generally the vertical distribution of adults of these lacewings corresponds well with the preference of given species to vegetation strata. The vertical activity peaks of some arboreal species, such as *H. lutescens*, *H. stigma*, or *W. nervosus*, followed the changes of height of the tree-crown layers in various forest types (in oak forest: 9 m, in beech forest: 21 m).

Intrahabitat species diversity of local lacewing assemblages
Reviewing the number of recorded lacewings in different agricultural or horticultural fields (see tables in Chapter 9, this volume) it can be said that species richness of local chrysopid and hemerobiid assemblages may vary strongly relating even to the same crop plant. The number of lacewing species within cultivated fields is determined by diversity and quality of surrounding habitats (colonisation sources); in addition it is partly influenced by intrafield impacts, such as the structure of crop stands and management practice (agricultural techniques and pesticide treatments).

According to results from long-term Hungarian studies in apple orchards and maize fields, the intrahabitat species richness of entomophaga is fundamentally determined by the diversity of the adjacent vegetation that is responsible for available resources ('resource diversity hypothesis') (Szentkirályi & Kozár, 1991). This hypothesis was proven in relation to assemblages of lacewings both in apple orchards and maize fields, namely the more diverse the extrafield vegetation, the higher the recorded intrafield species richness of lacewings (Szentkirályi, 1986, 1989, 1992). Where there was a higher number of plant species in field margin vegetation (shrubs and trees) of several maize stands, such that the number of aphid prey sources was also higher, consequently the amount of eggs laid by chrysopids and hemerobiids was greater (Szentkirályi, 1986, 1989). Although the number of lacewing species immigrating to apple orchards initially depends on the diversity of surrounding vegetation, the pest-management practice over the season determines the success of colonisation. As intensity of pesticide treatments increases, the number of recorded lacewing species in the tree crown decreases significantly (Szentkirályi, 1992, and unpublished data).

Impacts of cropping systems on lacewing abundance
Modern cropping systems are expected to reduce the use of pesticides and to promote pest controls, in part through conservation and augmentation of natural enemies. It is possible to increase the number of entomophagous insects by crop diversification, where more crop plant species are planted together inside a single stand. Depending on the number of cultivated crop plant species, mono-, di-, tri- or polyculture may be defined. There are numerous cultivation systems depending on intrafield spatial arrangements of plant components, such as intercropping, strip cropping, cover cropping, etc. Reduction of soil cultivation (no tillage) or lack of herbicide treatments increases the species number of weedy plants in the crop stand. Increased floral diversity and decreased number and intensity of disturbances inside the crop field generally contribute to an increase in the population of predatory insects, including lacewings. The additional plant species of polycultural cropping systems provide the aphidophaga with alternative sources of prey and nectar, as well as egg-laying and sheltering sites, increasing in this way their abundance.

Most data on the positive impacts of polycultural cropping on lacewings have been documented from intercropping experiments made on the most common arable field plants. Nazarov & Prygunkov (1986) reported that stands of maize mixed with sunflower or mallow (*Malva*) attracted aphidophagous groups, including chrysopids, and that the number of individuals was four or five times higher than in maize monocultural fields. The rate of increase of *Chrysoperla carnea s. lat.* was higher (57.9%) in soya-bean interplanted with maize diculture in comparison to a stand of soya-bean monoculture (Wang & Yue, 1998). The majority of assessments of the impact of intercropping have been made in relation to natural enemies of pests (aphids, noctuids) of cotton. Wu *et al.* (1991) found that cotton interplanted with maize in differing density (1790–2240 plants/ha) resulted in the growth of the population of aphidophaga, including chrysopids, that increased by 63%–116% compared with a monoculture. The impact of dicultural stands of cotton with a *Brassica* oil crop (Wu, 1986) or safflower (Li, 1987) on the augmentation of chrysopids was studied in China. Safflower plants were infested with the *Myzus persicae* aphid, which acted as alterative prey and attracted great numbers of *Chrysopa formosa*, *C. pallens*, and

Chrysoperla carnea s. lat. lacewings to the intercropping field. The growing population of lacewings significantly decreased the infestation caused by the aphid (*Aphis gossypii*) on cotton in the diculture field as compared with the monoculture. Besides the abovementioned three chrysopid species, *Chrysopa intima* was also among the dominant predators attracted by the cotton–*Brassica* mixed cropping stands. The aphidophagous insects controlled the population of *A. gossypii* effectively in this cotton diculture as compared with the monocultural cotton field. At the same time there is a counter-example, indicating that the abundance of lacewings may decrease under the influence of diculture in cotton intercropping fields. Schultz (1988) studied the number of deposited eggs of *Chrysoperla* species on plants in experimental plots of cotton cropped together with bean, maize, and weeds. He found that the mean number of eggs per plant in diculture was significantly lower (with beans: 1.4, with maize: 1.1, and with weeds: 0.7) than in cotton monoculture (2.2–2.7). It is most probable that these differences between egg numbers were caused by varying aphid abundance levels in mono- and dicultures. The greatest number of aphids was recorded in the control monocultural plots and the lowest in weedy plots, whilst an intermediate aphid infestation was detected in bean and maize treatments.

Various alternative tillage systems may influence phytophagous insects and their lacewing predators. Andow (1992) studied three conservation tillage systems in maize and their impact on egg-predators of *Ostrinia nubilalis*. The three conservation tillage systems (> 30% residual cover) were as follows: spring chisel plough, ridge till, and no tillage. He demonstrated that predation rate of chrysopid larvae on cornborer eggs was highest in the no-tillage system whilst it was lowest in the chisel-plough system. Due to an inverse relationship between chewing predators and *Chrysopa* spp. predation, Andow explained the different lacewing-consuming rate by its speciesspecific preferences to various tillage environments.

5.3.4 **Lacewing patterns at individual plant scale**
From the viewpoint of the ecology of lacewings this spatial scale is very important because the whole life history, from egg-laying to pupation, is associated with plants. The architectural complexity of plants influences lacewings in a number of their activities,

such as egg-laying, movement of larvae, prey-searching behaviour, sheltering, cocooning. Depending on plant species the architectural complexity varies from simple (for instance, herbaceous species) to complicated (for instance, forest trees). In accordance with the hierarchical structure of plants, patterns of lacewings have until now been investigated at two spatial levels: plant organs and microhabitats of plant surfaces.

Intraplant allocation of lacewing patterns

Most investigations on within-plant distributions of lacewings relate to herbaceous plants. The distribution of lacewing eggs on plants relates to their oviposition site preference and may be species-specific. Wilson & Gutierrez (1980) observed distribution of eggs of *Chrysoperla carnea* (probably *C. plorabunda*) on cotton while Patel & Vyas (1985) did the same for *Brinckochrysa scelestes*. The overwhelming majority of deposited eggs of *C. carnea* (93.7%) were located on leaves, with only a small percentage on fruits (5.9%), and a negligible percentage on branches (0.4%). Although the largest percentage of eggs of *B. scelestes* females were also laid on leaves (63.8%) the cotton branches were more attractive (21.4%), unlike with *C. carnea* females. Both species deposited more eggs on the bottom surface of leaves. The distribution of eggs of *B. scelestes* within leaves was as follows: lower surface: 38.4%, upper surface: 19.05%, petioles: 6.35%. The remaining eggs were recorded on flowers (6.35%), on stalks (5.6%), on bolls (2.4%), and on branches.

Distribution of the same lacewing species on various crop plants can be different. Patel & Vyas (1985) also studied the preference of oviposition sites of *B. scelestes* on green gram (*Vigna radiata*). They found that the rate of eggs deposited on green gram was significantly greater on branches (40.5%) and on pods (14.3%) while it was lower on leaves (35.7%) compared with cotton. On maize the majority (65%) of eggs of *Chrysopa oculata* were deposited (with exception of ear and tassel) on the lower six leaf-levels under the ear (Coderre *et al.*, 1987). The distribution of the potential prey aphid, *Rhopalosiphum padi*, was similar to the lacewing egg pattern (70% of infestation was found under the ear). At the same time 87.5% of chrysopid eggs were deposited on maize plants with no aphid infestation, so the oviposition preference for the lower plant parts was due to *C. oculata* being a typical inhabitant on herbaceous vegetation (see Chapter 9, this volume).

Marín (1987) reported a similar vertical distribution relating to eggs of *Chrysoperla carnea*, 77% of which were deposited on the lower half of maize plants. Szentkirályi (1986) studied the vertical distribution of eggs of chrysopids (predominantly *Chrysoperla carnea*, *Chrysopa phyllochroma*, and *C. formosa*) as well as hemerobiids (mainly *Micromus* spp. and *Hemerobius humulinus*) on weedy plants in margins of maize fields. He found that the vertical distribution pattern of eggs deposited by these two lacewing groups differed in that 80% of hemerobiid eggs were placed below the height of 60 cm with a maximum at the soil surface, while the majority of chrysopid eggs (85%) were detected between the heights of 65 and 130 cm. The distribution of hemerobiid eggs on maize plants differed from that found on weeds in the same experimental site. On maize plants the females laid the most eggs at a height of 30–130 cm with a peak on ears at 90–110 cm. The vertical distribution pattern of chrysopid eggs on maize differed only slightly from that found on non-crop plants, the majority of eggs being laid between 60 and 180 cm above the ground.

The distribution of moving larvae on crop plants may diverge significantly from that of eggs. In the experiments of Wilson & Gutierrez (1980) 53.2% of *Chrysoperla carnea* larvae were located on cotton bolls, while 45.7% of them were on leaves. The vertical distribution of *H. humulinus* larvae on maize plants overlapped with its aphid prey, *Rhopalosiphum maidis*. Both predator and prey equally preferred the ears and only the larvae of *H. humulinus* (from aphidophagous species) could follow the aphids hidden under the husk leaves (Coderre & Tourneur, 1986a). In Hungarian maize fields vertical activities of larval chrysopids and hemerobiids on plants were separated as it was found that the majority of lacewing larvae (82%) belonged to green lacewings, and only a minority (18%) to brown lacewings. At the same time 91% of recorded cocoons from soil samples of maize fields belonged to hemerobiids and only 9% to chrysopids. The proportion of larvae collected from the soil surface through pitfall trappings was similar to that in soil samples of the two lacewing families: 95% of them were hemerobiids and 5% were chrysopids (Szentkirályi, 1986).

To date there have been few clear experiments on the distribution of lacewings on trees. Gepp (1988/9) found that *P. gracilis* deposited eggs mainly in the mid-crown level and its larvae lived on upper, sunny, and warmer parts of coniferous trees until pupating, at ground level and to a lesser extent in the foliage. Oviposition and larval development of *H. elegans* occurred in the upper crown level while its cocooning was 100% among fallen leaves. According to Gepp, the vertical movements of *Dichochrysa prasina* were the most variable. Oviposition and early larval development occurs in the mid-crown region of coniferous trees, larvae (second to third instar stage) overwinter at ground level, and in springtime third-instar larvae climb up to the mid-crown belt to feed, before pupating in the soil. Hemerobiids also prefer certain parts or organs of trees. Deyrup & Deyrup (1978) recovered from Douglas-fir trees great numbers of mature hemerobiid larvae that were spinning cocoons inside the fallen cones.

Herbaceous architecture can greatly influence tritrophic interactions (host plant – aphid – aphidophaga) by means of reduced prey availability due to availability of microhabitats and protected locations. Clark & Messina (1998) studied prey searching behaviour of larvae of the green lacewing *C. plorabunda* on perennial grasses having two different architectures in laboratory experiments. Searching time was longer on crested wheat-grass plants (*Agropyron desertorum*) than on Indian rice-grass plants (*Oryzopsis hymenoides*). The chrysopid larvae could capture more Russian wheat aphids (*Diuraphis noxia*) living on Indian rice-grass than ones living on crested wheat-grass plants. The authors explained that crested wheat-grass has flat, broad leaves tightly rolled inward with blade–sheath junctions that provided protected places for aphid colonies. These microsites were too narrow for larvae to enter them. Despite the fact that Indian rice-grass had linear, narrow, tight leaf rolls or blade–sheath junctions these could not provide spatial refuges for aphids, and for this reason they were better exposed to predatory activity.

According to field observations, hemerobiid larvae, being narrow, elongated, and bobbin-shaped, predate much more successfully in protected microhabitats formed by plant organs than chrysopids do. Coderre & Tourneur (1986a, b) found that only larvae of *H. humulinus* (from the aphidophagous complex) preyed upon *Rhopalosiphum maidis* aphids that were available under corn ear-husk leaves. Szentkirályi (unpublished data) found a similar phenomenon in maize fields in Hungary. Larvae of *Micromus angulatus*

and *M. variegatus* were frequently located under the husk leaves and inside the narrow holes between the rolled sheets and stalks, where abundant colonies of *R. padi* were preyed upon by them in these microsites in autumn. No chrysopid larvae were found in these protected plant sites during these investigations.

Within-plant distribution of lacewings in microhabitats

Plant microhabitats used by chrysopids and hemerobiids may be considered microspaces that are created by specific plant surface architectures (e.g. crevices, trichomes, bristles, sheaths, veins, and other microsculptures) being within the size range of the developmental stages of lacewings. These microhabitats have a special morphology and microclimate and, in several cases, secrete chemical substances (such as waxes) that can assist or block various activities of lacewings (for instances, oviposition, searching movement, sheltering, cocooning, prey handling and availability). They are the direct sites of tritrophic interactions on plant surfaces.

One of the most frequent impacts of microhabitats on oviposition of lacewings results from the hairiness of plant surfaces. According to experiments made in maize fields (Duelli, 1986b; F. Szentkirályi, unpublished data), on weeds of field margins (F. Szentkirályi, unpublished data), and in cotton fields (Miller & Cave, 1987), brown lacewings deposit eggs at a great rate on hairs of the plant surface, such that the sides of oval eggs are glued on the hairs. This preference towards hairs during oviposition was significant in the *Micromus* species *M. angulatus* and *M. variegatus* in Europe, as well as *M. posticus* in North America. Other hemerobiids (for instance, *Hemerobius humulinus*, *Wesmaelius subnebulosus*, and *Sympherobius pygmaeus*) preferred fibrous material (such as loose balls of cotton strings) during egg-laying, as indicated by laboratory experiments (Miermont & Canard, 1975; Miller & Cave 1987; F. Szentkirályi, unpublished data). Eggs of *Micromus spp.* and *H. humulinus* were observed in webs of the spider mite *Tetranychus urticae* living on maize leaves of plants in fields (F. Szentkirályi, unpublished data), moreover eggs of *M. posticus* were also recorded in webs of the same spider mite species living on cotton (Miller & Cave, 1987). Several species of chrysopids, such as *C. carnea*, also prefer the top of hairs or bristles for oviposition, where they stick the base of the egg pedicel to top of the hairs. (Duelli, 1986b; F. Szentkirályi, unpublished data). Observing numerous eggs of lacewings on weeds in Hungary, their spatial distribution was as follows: (1) 94.8% of hemerobiid eggs (*Micromus angulatus*, *M. variegatus*, *H. humulinus*, *W. subnebulosus*) were laid on hairs while 5.2% were laid on the plant surface; (2) 8.6% of chrysopid eggs (mainly *Chrysoperla carnea*, *Chrysopa phyllochroma*, *C. perla*, *C. abbreviata*, *C. formosa*) were laid on top of hairs, 68.5% on plant tips (bristles, leaf-tops, scale-leaf tips of inflorescence), 17.1% on leaf-edges and 5.8% on plant surfaces (F. Szentkirályi, unpublished data).

According to samples from maize fields, also in Hungary, 72.6% of eggs of *Micromus* spp. and *H. humulinus* and more than 90% of chrysopid eggs (mainly of *Chrysoperla carnea*) were deposited on leaf hairs. The mean length of maize hairs that were preferred by hemerobiids was 1.86 mm, and the average density of hairs surrounding deposited eggs was 103.4 hairs/cm^2 (F. Szentkirályi, unpublished data). Duelli (1986b) and Miller & Cave (1987) considered this use of hairs for oviposition by hemerobiids as an intermediate step in the evolution of lacewing egg pedicels, providing a selective advantage for protection against oophagous insects. But not every species of lacewing prefers plant surfaces covered by hairs for oviposition. Obrycki & Tauber (1984) studied oviposition of *Chrysopa oculata* on potato clones, and they noted that the number of eggs laid on smooth-leaf cultivars was double that on glandular pubescent potato leaves.

A tendency can be observed among hemerobiid species living at the tree-crown level that they prefer the lower surface of pubescent leaves and vein nooks and junctions with hairy bunches for allocation of eggs, for instance, *H. humulinus* laying eggs on *Tilia* leaves (F. Szentkirályi, unpublished data). It was recorded that coniferous chrysopids (such as *Chrysopa dorsalis*, *Chrysoperla mediterranea*, *Cunctochrysa*, *Eremochrysa*, and *Chrysopiella* spp.) frequently laid eggs on to the tip of needles (Duelli, 1986b, 1987; Gepp, 1988/9).

Quality and microstructures of plant surfaces, such as waxiness and hairiness, influence not only oviposition but also movement, searching efficiency, and functional responses of lacewing larvae in great measure. Through laboratory experiments Arzet (1973) detected that the hard wax layer produced by various cultivars of cabbage inhibited the searching movement of *Chrysoperla carnea* larvae. Mobility of lacewing larvae on plant surfaces was also strongly determined by hair density and type. For instance, the strong felt-like hairiness of *Pelargonium*, *Phaseolus*, and *Nicotiana* plants inhibited movement of *C. carnea* larvae and reduced their efficiency (Arzet, 1973). According to experiments made by Elsey (1974) first- and second-instar larvae of *C. carnea* could move at higher mean speeds (7.5.–18.5 cm/min) on less hairy cotton foliage than on more hairy tobacco foliage (0.6–1.1 cm/min). The prey-searching speed of larvae was also affected by glandular trichomes that covered tobacco, compared with non-glandular trichomes on cotton. He demonstrated that the speed of *C. carnea* larvae on tobacco proportionally reduced as trichome density increased. Treacy *et al.* (1987) observed how non-glandular trichomes of cotton plants influenced the functional response of *C. rufilabris* larvae. Based on the density of hairiness there were three types of cotton cultivars: smooth-leaf cotton (0.9 hairs/cm^2), hirsute cotton (61.2 hairs/cm^2), and pilose cotton (481 hairs/cm^2). According to greenhouse tests trichomes were mechanical barriers that inhibited the mobility of larvae, so that their predatory ability was reduced. In all three larval stages, as density of hairiness increased, the searching distance as well as number of predated bollworm eggs decreased. Thus the average distance which the newly hatched chrysopid larvae walked away from the egg sites within 48 hours was over 100 mm on smooth potato leaves, 25–30 mm on hirsute leaves, and about 5 mm on pilose leaves.

5.4 LACEWINGS IN THE APHIDOPHAGOUS GUILD ORGANISATION

5.4.1 Lacewings and food webs

The feeding relations of arthropods in cultivated plant stands form food chains integrating into food webs within agroecosystems. Food webs are among the most important dynamic ecological structures of communities. Schoenly (1990) analysed prey–predator links within a great number of known terrestrial food webs and found that Neuroptera take part in predatory links at a low rate (below 1%). The participation of lacewings in food chains has been neglected in research. Eglin (1980) was among the first experts who recognised its importance, via studies over a 40-year period on food web connections of lacewing families living in mountainous forest natural parks in

Switzerland. His generalised scheme included food plants, homopteran and other prey groups, and natural enemies of lacewings. Canard (1984) described the relation of eggs deposited by *Nineta flava* within food chains. The first link within the chain was the presence of *Acer pseudoplatanus* trees infested by the second link, namely the aphid *Drepanosiphum platanoides*, which in turn connected to eggs of *N. flava* as the third link. A fourth, parasitoid, link within the trophic chain, *Telenomus acrobates*, attacked the lacewing eggs. Further predatory elements also participated in forming the food web, such as the chrysopid *C. pallens* and an anthocorid bug, *Anthocoris nemorum*. These all preyed on both aphids and on eggs of *N. flava* (including individuals of *T. acrobates* that were developing inside the eggs). In parallel, eggs of *Chrysopa pallens* were parasitised by *T. acrobates*.

5.4.2 Chrysopid and hemerobiid subguilds in aphidophagous guilds

Because of their high abundance and easy availability, aphids are frequently found as obligate or temporal food sources for numerous insect species. By focusing on the totality of parasitoids and predators consuming an aphid species inside a food web, one may recognise aphidophagous guilds belonging to trophic guilds. According to the original concept a trophic guild is a group of species regardless of their taxonomic relations that exploit the same class of food sources in a similar way (Simberloff & Dayan, 1991). Aphidophagous guilds can be recruited from various taxonomic groups. Frazer (1988) listed 22 families from six insect orders where predatory members consumed aphids. Moreover, other generalist predators, such as Araneae, Acari, Opiliones, frogs, birds, or bats may be temporal facultative elements of aphidophagous guilds. Obligate predators consuming aphids belong to the constant members of the aphidophagous guilds and they derive from the Coccinellidae, Syrphidae, Cecidomyiidae, Chrysopidae, and Hemerobiidae. A guild exploiting single aphid species can be formed by a great number of predatory and parasitoid insect species. Bouchard *et al.* (1986) detected 11 families and 64 predators or parasitoids from the entomophagous complex associated with *Aphis pomi*, the green apple aphid. What conditions of coexistence are required inside an aphidophagous complex of a huge number of species? Resource partitioning among guild members spatially and temporally

is expected, and a certain level of intraguild niche separation will be characteristic. Accordingly, those guild members that utilise food resources at the same time and location, in a similar way, and are separated from other members while they are close to each other taxonomically, can create aphidophagous subguilds. Bouchard *et al.* (1986) refer to this intraguild organisation; by sampling the aphidophagous assemblage of *A. pomi*, and using multivariate statistical methods (PCA, PCoA) they detected several characteristic clusters. These were well-defined taxonomic groups that represented prey–predator relationships. For instance, a cluster representing the Miridae family that was formed at low aphid density in early summer was one of these groups. The growing chrysopid and hemerobiid populations belonged to another cluster that followed (temporally) the declining mirids. Increasing populations of parasitoids and Cecidomyiidae in close relation to *A. pomi* created a further cluster.

Spatio-temporal separations and organisation at subguild level

There are data on the participation of lacewings in guild organisation from long-term investigations of maize fields in Hungary and Canada. According to results from studies on aphid–aphidophaga relations between field margins and maize fields in Hungary (1976–1985), chrysopids and hemerobiids could be divided into two subguilds (Szentkirályi, 1986, 1989, 1991). Temporal and spatial patterns of lacewings that related to subguilds repeated year to year in the same way both in maize fields and their field margins. This pattern-repeatability refers to the organisation inside aphidophagous guilds which results in subguild structures. The two lacewing subguilds were separated spatially and temporally. The spatial separation both in maize and weedy plants was expressed in a difference in vertical distribution between green and brown lacewings (see pp. 92–3 above) and by differing microhabitat use (see pp. 93–4). Temporal separation of chrysopid and hemerobiid subguilds was formed by their various seasonality patterns. The seasonal activity patterns of the green lacewing subguild (oviposition, larval activity) with a time delay were predominantly associated with aphid populations increasing in June–July, and to a limited degree with the second aphid seasonal increase both in stands of maize and herbaceous plants. Brown lacewings in maize stands were more associated with the second mass increase of aphid abundance in late summer and early autumn. The activity of members of the hemerobiid subguild on weedy plants of field margins was synchronised with both aphid population waves during the seasons (see section 5.2.2 above). There was a difference between the numerical responses of members of the two lacewing subguilds on the changing abundance of aphid prey populations. Hemerobiids made a close reproductive numerical response, while chrysopids were characterised by delayed reproductive response (Szentkirályi, 1991; F. Szentkirályi, unpublished data). By these characteristics the members of the two lacewing families in a given habitat can establish not only taxonomic but also real trophic subguilds.

Although the same predatory species were associated with the first and second mass increases of aphids, there were two different aphidophagous guilds that developed in maize stands. This is explained by the observation that more than 95% of colonies were formed by two different species during the two aphid infestation periods. Throughout the first aphid wave in June–July the dominant species was *Metopolophium dirhodum* whilst in the second wave of September–October *Rhopalosiphum padi* was dominant. There were 32 potential members of aphidophagous guild (5 syrphid, 8 coccinellid, 10 chrysopid, 4 hemerobiid, 2 anthocorid, and 3 nabid species) recruiting to exploit populations of *M. dirhodum*. During the experimental years (1980–1985) only 18 species were constant and only 10 species were dominant within this guild. From among the lacewing subguilds *Chrysoperla carnea*, *Chrysopa phyllochroma*, *C. formosa*, *Micromus angulatus*, and *M. variegatus* were dominant predators in this guild preying on *Metopolophium dirhodum*. Besides these dominant species, *C. perla*, *C. pallens*, and *H. humulinus* as well as *W. subnebulosus* could be regarded as constant lacewings. In order to consume *R. padi* a new guild consisting of 33 species was created by reorganisation from the prior aphidophagous guild. This newly formed second aphidophagous guild associated with *R. padi* involved 16 constant and only 8 dominant members. Among the latter were the lacewings *Chrysoperla carnea*, *Micromus angulatus*, *M. variegatus*, and *H. humulinus*. Of the lacewings *Chrysopa formosa*, *C. phyllochroma*, and *W. subnebulosus* were also constant guild members. Diversity levels of surrounding vegetation strongly influenced potential number of lacewing members participating in aphidophagous guilds within maize stands (Szentkirályi, 1989; and see also section 5.3.3).

According to seasonal changes in the abundance of *Metopolophium dirhodum* a temporal succession in the numerical response of certain taxonomic groups could be observed in the aphidophagous guilds (Rácz *et al.*, 1986; F. Szentkirályi, unpublished data). Thus syrphids were present for a short time in the early developmental period of aphid populations. Coccinellids and green lacewings followed later. Lastly bugs (anthocorids and nabids) preyed upon colonies of *M. dirhodum*. The activity peaks of chrysopids and bugs were found in declining periods of aphid abundance. The seasonal activity patterns of guild members associated with the second wave of *R. padi* were much more synchronised with each other and with the aphid species. For this reason only two groups following each other could be detected temporally. During the period of increasing aphid abundance the overlapping populations of hemerobiids, anthocorids, and nabids formed a characteristic group that was followed by another group of syrphids, coccinellids, and chrysopids at the peak of aphid abundance. It seems that each of these differing aphidophagous taxonomic groups can be considered as a subguild. This can be supported not only by their spatial and temporal separations, but also by their feeding mode (e.g. lacewings feed by piercing and sucking while coccinellids are chewing predators) or differing searching mobility of larvae (syrphids have low mobility, coccinellids medium, and chrysopids high).

Spatio-temporal separations and organisation at species level

Coderre & Tourneur (1986*a*, *b*) and Coderre *et al.* (1987) indicate that this structure of subguilds within aphidophagous guilds associated with maize may be characteristic not only in central Europe but also in Canada. During studies conducted in Canadian maize fields they analysed the spatial and temporal distribution of the six most important species from the aphidophagous complex and their aphid prey as well as relationships between predators and maize aphids. There were two aphids, *Rhopalosiphum padi* and *R. maidis*, as co-dominant species recorded by Coderre *et al.* (1987) inside maize stands. Seasonal changes in the abundance of maize aphids were synchronised with each other, and, similarly to conditions in Hungary, the aphids showed two growing periods, in late July and in mid-August–late September. Predator species associated with the two waves of aphid were as follows: *Hemerobius humulinus* and *Chrysopa oculata* (lacewings), *Hippodamia tredecimpunctata*, *Coccinella septempunctata*, *Coleomegilla maculata* (coccinellids), and *Sphaerophoria philanthus* (syrphids). A temporal separation of the activities of these species could be observed during the season, similar to results from Hungary. *Sphaerophoria philanthus* appeared earliest, at the beginning of the first wave of aphids. Populations of syrphids were followed by coccinellids in late July. During August *Hemerobius humulinus* was most abundant although it continued to be present in maize stands until late September. *Chrysopa oculata* was active in July–August with a peak in late August. Predators were also generally separated from one another spatially. This spatial separation was expressed in horizontal distribution among certain species inside maize stands: *Coleomegilla maculata* preferred the middle of the stand, while *Hippodamia tredecimpunctata* was found more at field edges. However syrphid and chrysopid species were evenly distributed within maize stands (Coderre *et al.*, 1987). The most characteristic spatial distribution of members of the aphidophagous complex occurred in their vertical distributions within plants (Coderre & Tourneur, 1986*a*). These vertical distributions followed the allocation of aphids on the plants, changing in accordance with phenophase of maize. Populations of both aphid species appeared in the lower level of the leaves in the vegetative period. Later in the period of tasselling, *R. padi* remained on the six lower leaf levels but *R. maidis* mainly infested the tassels. During maturation, 80%–100% of the population of both aphid species were present on ears. As plants dried, aphids were most abundant on the lower leaves and their infestation level showed a gradually decreasing trend on the upper plant parts.

The oviposition and larval activity patterns of predators also followed, but in a different way, changes in the vertical distribution of aphids on the maize plant (Codere & Tourneur, 1986*a*). *Sphaerophoria philanthus* and *C. maculata* laid their eggs exclusively on the six leaves under the ear. Most eggs of the coccinellids *Coccinella septempunctata* and *H. tredecimpunctata* and of the lacewing *Chrysopa oculata* (range: 62%–68%) were also located on lower leaf levels while some

(range: 32%–35%) were deposited on the upper leaf levels and on the tassels. The vertical distribution of aphidophagous larvae on maize plants was more specific to species. Activity of larvae of *Hemerobius humulinus* showed the best fit to the seasonal changes of aphid distribution. Hemerobiid larvae appeared seasonally during the tasselling, and they were present in equal amounts on the lower leaves, infested mainly by *R. padi*, and upper leaf levels and on tassels infested strongly by *R. maidis*. At maturation about 75% of brown lacewing larvae stayed on ears closely following the infestation maximum of both aphids on ears (80%–100%). In the period of drying the vertical distribution of brown lacewing larvae overlapped the distribution of aphids, their abundance gradually decreasing from the lower leaves towards upper ones. The immobile larvae of syrphids remained exclusively on lower leaves from the vegetative phase to maturation. The larvae of *Hippodamia tredecimpunctata* were active mainly on lower leaves (80%–90%) during the vegetative period and tasselling while later in the phase of maturation they relocated up to higher leaves (30%) and in drying period to ears (50%). Larvae of *Coccinella septempunctata* and *Coleomegilla maculata* stayed on lower leaves (75%–80%) during tasselling. At maturation, larvae of *Coccinella septempunctata* were evenly distributed on the plant while larvae of *Coleomegilla maculata* were more active on lower leaves (60%) and to a lesser extent on ears (30%). In drying maize stands the majority of larvae of *Coccinella septempunctata* appeared on ears (80%), and larvae of *Coleomegilla maculata* preferred the upper leaf levels (about 45%) rather than the lower leaves (about 30%) or ears (25%). Coderre & Tourneur (1986*a*) published no data on vertical distribution of larvae of *Chrysopa oculata*. Comparing these distributions of aphidophagous species with one another it can be said that the major portion of the population of *Hemerobius humulinus* larvae changed its vertical position such that they did not (or at least only partly) overlap with the other aphidophaga during the season avoiding or decreasing the interspecific competition with other intraguild members. Applying multivariate statistical techniques Coderre & Tourneur (1986*b*) and Coderre *et al.* (1987) analysed the synchrony and numerical response of each aphidophagous component relating to changing abundance of aphids. They showed that *Sphaerophoria philanthus* and *Hippodamia tredecimpunctata* were closely synchronised and associated with populations of *R. padi*. Individuals of *Hemerobius humulinus* and *Coccinella septempunctata* responded rather to changes in abundance of *R. maidis* but the hemerobiid species also showed a reaction to peak density of *R. padi*. *Coleomegilla maculata* gave a weak positive response while the green lacewing *Chrysopa oculata* produced a rather negative reproductive numerical response to the occurrence of aphids.

As Lorenzetti *et al.* (1997) demonstrated, members of the aphidophagous guild were separated not only horizontally within maize fields or vertically on plants, but also among individual plants. A series of coloured maize plants from yellowish to green were produced by changes of nutrient supply. Adult coccinellids and chrysopids were attracted to the coloured plants at different rates. Coccinellids selected the yellowish plants and were more abundant on them. Adult lacewings preferred green maize plants for oviposition. Thus at the time of colonisation the plant colour was used as a distinctive cue in order to find the proper location for egg-laying and feeding sites that could result in a spatial separation among aphidophagous guild groups, minimising detrimental intraguild interactions (Lorenzetti *et al.*, 1997).

Lacewing niche segregation and guild organisation
The above-described spatial and temporal separation of aphidophagous guild members and subguild coalitions in maize stands, as well as their differing numerical responses, may allow the partitioning of shared prey resources. This resource utilisation contributes to the ability of obligatory members to coexist within the guild. According to results of investigations in maize fields conducted in Hungary and Canada there are spatial and temporal separations both at the levels of subguild and of species. Moreover, green and brown lacewings show characteristic differences from each other and from other intraguild components in this respect. It is probable that niche segregation is the basis of this resource partitioning within aphidophagous guilds of maize, as has been emphasised before (Szentkirályi, 1986; Coderre et al., 1987; Lorenzetti *et al.*, 1997). This niche segregation is necessary for spatio–temporal organisation within aphidophagous guilds, which in turn enables the obligatory and facultative guild members to coexist in the same way.

5.4.3 Intraguild interactions of lacewings
In order to exploit the same aphid prey source the obligatory and facultative members of aphidophagous guilds are often recruited from different species and other taxa in great number, as has been described above. The probability of competition or predation risk may be high among the members of the guild when the food supply is limited (for instance, due to low aphid density in the early period of infestation or the late period of aphid population decline). The spatial and temporal separations referred to above (section 5.4.2.) may result not only in resource partitioning but also in the reduction of these detrimental interspecific interrelations. Additionally, further mechanisms, like feeding specialisation, switching to other prey by generalist lacewing predators, or use of oviposition-deterring synomone (see section 5.5.2) may contribute to this reduction.

Lacewings' performance in intraguild predation
Individuals of different predatory species of an aphidophagous guild encountering one another may attack or kill each other. This is termed intraguild predation (IGP). The level of IGP grows if the availability of extraguild aphid prey is limited, its abundance is low, or colony-size distribution is disadvantaged. IGP can be classified as: symmetric (mutual predation) or asymmetric (regularly one of the two interacting species is the predator and the other is the prey) (Polis *et al.*, 1989). In IGP the outcome of an encounter between two guild partners depends on their following general characteristics: relative body size, mobility, vigour, aggressiveness, defensive strategies, and degree of feeding specificity.

Despite earlier opinion to the contrary, IGP among arthropods seems to be widespread under field conditions (Polis *et al.*, 1989). For some years the number of experiments on the role of lacewings in IGP, in particular in the laboratory, has been increasing. As far as lacewings are concerned most of these studies have looked at larvae of *Chrysoperla carnea s. lat.* and *C. rufilabris*. Both obligatory and facultative generalist predators were included among their partners in an aphidophagous guild, such as coccinellid, cecidomyiid, reduviid, carabid, ant, and spider species. IGP was shown to be stage-specific by Sengonca & Frings (1985) since the same-aged or older and stronger larvae of *C. carnea* killed the larvae of *Coccinella septempunctata* in the absence of other aphid prey. Although the IGP was more or less symmetric, only the older coccinellid larvae could prey upon lacewing larvae. Consumption of eggs was mutual but lacewings preyed on coccinellid egg batches at a higher rate (>95%). In the presence of aphid prey the rate of IGP was reduced. However, adult coccinellids were very aggressive and even in the presence of aphids attacked all larval stages of *Chrysoperla carnea*. Lucas *et al.* (1997) also found stage-specific IGP between larvae of *C. rufilabris* and the coccinellid *Coleomegilla maculata*. Older lacewing larvae were superior to coccinellid larvae although the proportion of lethal contact caused by them was reduced as the prey–predator size ratio grew. The recorded mortality of coccinellid larvae in first instar was 35% while in third and fourth instars it was only 2%. The vulnerability of coccinellid larvae decreased with age because defensive mechanisms or escaping reactions (e.g. wriggling, biting) of older larvae was more effective against lacewing attack. Lucas et al. (1998) studied IGP characteristics relating to various developmental stages of *Chrysoperla rufilabris*, *Coleomegilla maculata*, and *Aphidoletes aphidimyza*. Larvae first to third instars of *Chrysoperla rufilabris* consumed each stage of the gall midge (eggs, young and old larvae) due to obvious differences between the two species in mobility and feeding specificity. The high rate of mortality caused by lacewing larvae remained after addition of aphid prey, consequently 70% of eggs and 100% of old larvae of gall midges became victims of IGP. The IGP proved symmetrical between *C. rufilabris* and *Coleomegilla maculata* depending on body-size ratio and degree of aggressiveness of these two generalist, mobile predators. Consumption of eggs and larvae was mutual but lacewing larvae of the same age preyed upon larvae of coccinellids. Introduction of aphids stopped IGP between larvae in first instar, while the IGP rate of third instar lacewings on third instar *C. maculata* remained at 100%. Lucas *et al.* (1998) explained this behaviour as being due to the changing level of encountering risk, namely, in the absence of risk IGP was maintained by lacewings.

Two members of the aphidophagous guild on the cotton aphid (*Aphis gossypii*) *Chrysoperla carnea* (eggs and its 1–4-day-old larvae), and a reduviid bug *Zelus renardii* (larvae and adults) were studied in the context of IGP by Cisneros & Rosenheim (1997). Larvae (first

to fifth instars) and adult bugs killed the lacewings equally. Survival of lacewings decreased in the presence of older bug larvae. It was undoubtedly proved through experiments that mean prey size increased as body size of the predatory bug increased; thus older stages of *Z. renardii* preferred the larger larvae of *C. carnea*. However the younger bug larvae (first and second instars) were unable to prey upon the old (third-instar) lacewing larvae.

Dinter (1998*a*, *b*) studied guild members consuming *Sitobion avenae*, including larvae (second instar) of *C. carnea*, two erigonid spider species (*Erigone atra* and *Oedothorax apicatus*), and a carabid beetle, *Pterostichus melanarius*. Without extraguild prey both spiders and the carabid beetle caused severe mortality of lacewing larvae (75%–100%). The assymetric IGP of spiders on lacewings was sex-specific and species-specific. The predation rate of *O. apicatus* and also of the female spiders was higher than was the case with *E. atra* and the male spiders. When the supply of alternative aphid prey grew, the mortality rate of lacewing larvae decreased significantly. The carabid beetle behaved as top predator and preyed mainly upon female *E. atra* besides the lacewing larvae. Taking the experimental combination of an individual female spider and a living lacewing larva, the number of individuals of *S. avenae* became significantly lower than when it was recorded only in the presence of any single predator. This interspecific additive impact on aphids was observed in connection with both (spider–lacewing species) combinations. Dinter demonstrated through these microcosm experiments that larvae of lacewings could be involved in complicated interaction systems due to IGP within aphidophagous guilds.

Ant species may be classified as true members of aphidophagous guilds where they utilise honeydew as food resource but protect aphid colonies against natural enemies. Ants as bodyguards for aphids or scale insects often participate in IGP as attackers or predators of aphidophages. The imported red fire ant, *Solenopsis invicta*, is known as an aggressive predator that strongly associates with *Aphis gossypii* in USA cotton fields. According to Vinson & Scarborough (1989) workers of *S. invicta* kill third-instar larvae of *C. carnea* on cotton plants reducing the predatory effectiveness of lacewings significantly. Through laboratory and field experiments, Morris *et al.* (1998) studied the impacts of six ant species on exposed eggs of *C. carnea* living in olive orchards of Spain. They recorded a species-specific oophagy that produced the highest predation rates by *Crematogaster scutellaris* (19%–93%), *Tapionoma nigerrimum* (up to 100%), and *Formica subrufa* (36%–57%). These results demonstrate the important role that ants play as antagonists of lacewings.

Indirect effects of IGP and predator-mediated interactions

Interactions within aphidophagous guilds can affect guild members and the population of common prey not only directly but also indirectly. When a superior predator in IGP reduces the abundance of a relevant, effective aphidophagous partner, then the control of the common extraguild prey can be disrupted resulting in a population release effect. Such an effect relating to lacewings was observed by Vinson & Scarborough (1989): the predatory ability of old larvae of *Chrysoperla carnea* was greatly reduced by *Solenopsis invicta*. Thus *Aphis gossypii* populations were able to double in comparison to levels seen when the lacewings were present on ant-free cotton plants. It was enough to interrupt the control of aphids by lacewing larvae if at least two ants were present per leaf. In the experiments of Cisneros & Rosenheim (1997) the bug *Zelus ranardii* preyed upon *A. gossypii* but was unable to control aphid abundance. At the same time larvae of *C. carnea* strongly suppressed the growth of cotton aphid populations. By adding *Z. renardii* to this prey–predator system the control effect of lacewings was frequently disrupted as bugs consumed mainly the most voracious third-instar larvae of *C. carnea*. The older fourth and fifth instar larvae and adults of the reduviid bug played a more effective role in disruption of the control, so that the growth rate of the cotton aphid population increased.

For members of an aphidophagous guild there is frequently more than one aphid prey species available. In this case an extraguild prey species can influence not only the consumption of intraguild predators directly, but also, indirectly through aphidophagy, the population dynamics of other aphid prey species. An example on this predator-mediated interaction was represented in experiments relating to cereal aphids made by Bergeson & Messina (1997, 1998). They found that larvae of *C. carnea* preyed equally, without any preference, upon *Rhopalosiphum padi* and *Diuraphis noxia* aphids on the same host plant, if prey species infested the individual plants alone. However, if colonies of both aphid species were present on the same plant, the lacewing predator could not suppress the population of *D. noxia* because the *C. carnea* larvae preferentially consumed the more available individuals of *R. padi*. Thus the presence of *R. padi* influenced the common predator producing a significantly higher rate of increase of *D. noxia* population. The differing rates of predation by *C. carnea* larvae were caused by the fact that they encountered individuals of *R. padi* more frequently than colonies of *D. noxia* living in sheltered, protected parts of the host plant (Bergeson & Messina, 1998). These results support the hypothesis that a non-target, more available, alternative aphid prey can inhibit the short-term efficiency of lacewing larvae as biological control agents against the target aphid pest.

5.5 CHEMICAL ECOLOGY OF LACEWINGS

Chemical ecology relating to aphidophagous insects is a new, dynamically developing branch of study. Numerous chemical compounds play an important role in ecological relationships between various developmental stages of lacewings and their environment. There are some major research issues belonging to the chemical ecology of lacewings: (1) ecological impacts of chemical composition of food (such as effect on development, reproduction, fitness) (Hagen, 1986); (2) information-carrying role of infochemicals (Dicke & Sabelis, 1988; Vet & Dicke, 1992) in multitrophic systems formed by host plants, homopteran prey, lacewings, and natural enemies of lacewings; (3) assessment of the impact of pesticides (Darvas & Polgár, 1998) and other chemical compounds (e.g. fertilisers) applied in crop and sylvicultural systems. No discussion is made here on chemo-ecological effects of foods and pesticides on lacewings as this is dealt with elsewhere in this book. For this reason, the following pages cover intra- and interspecific functions of infochemicals produced by lacewings (pheromones, defensive allomones, and synomones), as well as the effects of allelochemicals released by prey and their host plants (kairomones, synomones) in modifying lacewing behaviour, and mechanisms of how these compounds assist in the location of suitable habitats and prey. Insecticidal impacts of trans-genic crop plants and their risks for tritrophic interactions are also covered.

5.5.1 Sex pheromones

Intraspecific individuals of lacewings belonging to opposite sexes can find each other before mating through acoustic or chemical (pheromonal) signals. Although there is evidence for the existence of sex pheromones in certain lacewing species, this area has been neglected. It has been known for a long time that males of *Chrysopa perla* secrete a volatile substance to attract females. Wattebled *et al.* (1978) made detailed studies of the anatomy and histology of the pheromone-producing organ of male *C. perla*. During courtship the males extrude two eversible vesicles repeatedly from the genital aperture. The vesicles bear tubercules with an apical hair. It is suggested that an aphrodisiac pheromone is secreted from these vesicles to stimulate the female. Unfortunately, the chemical nature of this secretion was not determined.

5.5.2 Infochemicals in defensive interactions of lacewings

Insects can obtain their defensive substances by synthesis or by sequestration from external sources. For example, lacewings follow the first method in order to protect themselves from their natural enemies, while several of their prey (such as aphids) apply the second method, consuming toxic secondary plant substances in food and storing these substances in their body or transferring them into honeydew for defensive purposes.

Oviposition-deterring pheromones and synomones
Rearing a Nearctic (*Chrysopa oculata*) and three Palaearctic lacewing species (*C. perla*, *C. commata*, and *Chrysoperla carnea*) under laboratory conditions, Růžička (1994, 1997a, 1998) discovered that lacewing larvae secreted an egg-laying-deterrent substance which was effective on females of the same species. This defensive substance against intraspecific egg-laying was named by Růžička (1994) an oviposition-deterring pheromone (ODP). The presence of active ODP could be recorded equally on tracks and in anal secretions of larvae. It is possible that ODP is associated with adhesive/defensive secretion. According to analysis the active ODP extracted from larval contaminated

substrates is a volatile and water-soluble substance. The repellent effect of ODP lasts in the open air for 3–4 weeks, while in a closed place (e.g. in a petri dish) for 20 weeks. The effectiveness of material produced by young larvae (first instar) was greater than by older larvae (third instar). The degree of conspecific response to ODP was more or less species-specific although both predatory and non-predatory adult females were equally repelled. The egg-laying rate of *C. oculata* in choice tests was lower on larval contaminated substrates than that of any of *Chrysoperla carnea*, *Chrysopa perla*, or *C. commata*.

Further choice tests revealed that ODP substances produced by larvae of each lacewing species had also a significant deterrent effect on oviposition of other heterospecific chrysopids (Růžička, 1998). ODP secretions produced by *C. oculata* and *C. perla* had a stronger repellent effect on oviposition of other chrysopids. The difference was explained by Růžička by the fact that the amount of ODP was larger on substrate due to bigger body size and more frequent abdominal marking of larvae (first instar) of these species.

According to results of further experiments (Růžička, 1997b; Růžička & Havelka, 1998) oviposition-deterring substances can produce mutual interspecific responses not only among the members of Chrysopidae but also of other aphidophagous taxonomic groups. Such heterospecific repellent effects have been recorded between *Chrysopa oculata* and *Coccinella septempunctata*. The response of *C. septempunctata* females was significant on oviposition-deterring secretions of chrysopid larvae if the substrate was immediately tested after contamination. The effect of reducing egg-laying was not significant after 24 hours. The response of *Chrysopa oculata* to ODP produced by larvae of *Coccinella septempunctata* was negligible. The oviposition rate of another aphidophagous species, the cecidomyiid *Aphidoletes aphidimyza*, decreased significantly on plants contaminated with ODP from larvae of *Chrysopa oculata*, *C. perla*, or *Coccinella septempunctata* compared with control plants. Among lacewing secretions *Chrysopa oculata* larvae produced the most deterrent heterospecific effect, while the ODP of *Chrysoperla carnea* had no significant effect on females of *A. aphidimyza*.

The oviposition-deterring substances mentioned above in interspecific relations are called allomones by

Růžička (1997*b*, 1998; Růžička & Havelka, 1998). This name in this context may be disputed because the response of females to heterospecific larval repellent markings promotes a more equal spatial distribution of predatory larvae, decreasing the probability of detrimental encounters among them, a result that is beneficial for both species. Therefore according to recent allelochemical terminology modified by Dicke & Sabelis (1988) it is more acceptable to use the term of oviposition-deterring synomone (ODS). The existence of ODP and ODS is relevant for the organisation of aphidophagous guilds. If conspecific or heterospecific larvae have previously been present on a plant surface, the decreased oviposition rate in the same place reduces the chance not only for cannibalism (by ODP) but also for interspecific intraguild predation (by ODS) as discussed above in this chapter. In order to identify the chemical composition of these substances that are responsible for the egg-laying repellent effect, further analyses are required.

Allomones released by lacewings against natural enemies

Lacewings use various defensive methods for protection against natural enemies. Their numerous effective defence mechanisms are based on the application of chemical substances. All developmental stages of lacewings from eggs to adults have chemical weapons for combating attackers.

It is a well-known phenomenon in the majority of lacewing species from the *Chrysopa* genus that adults exude some strong, stinking odour when attacked. This offensive scent is released by the prothoracic stink glands. These paired tubular glands are located on the anterior part of the prothorax, and in the case of attack they discharge a liquid secretion. Blum *et al.* (1973) identified the chemical composition of the stinking substance: tridecene (90%) with the remaining part composed of skatole. This scent may be considered as an allomone that gives a defensive chemical signal to attacking insects. It also has a repellent effect on predatory small mammals or lizards. Odorous chrysopid adults offered to mice (Blum *et al.*, 1973) or *Chrysopa formosa* and *C. pallens* offered to a lizard (*Lacerta muralis*) were rejected by the predators, while only non-odorous *Dichochrysa* species were accepted as food (Séméria, 1979).

Deposited lacewing eggs may also have chemical protection against predators. Eisner *et al.* (1996) recorded that during oviposition females of the green lacewing *Ceraeochrysa smithi* (USA) coated the egg pedicel with oily liquid droplets. They found that this oily substance consisted of fatty acids, an ester, and numerous straight-carbon-chain aldehydes. This defensive secretion protected the eggs from attack by the ant *Monomorium destructor*. The components of the repellent allomone covering on egg stalks were proved in a laboratory experiment to be an effective chemical irritant for a cockroach species. Eisner and his co-workers observed that neonate chrysopid larvae just after hatching sucked out these droplet-coated egg pedicels. It can be supposed that this substance is serving not only as a food resource but also for further chemical protection of young lacewing larvae against predators.

Since the major food sources of lacewings belong to homopteran groups, their larvae frequently encounter honeydew-collecting ants. During attack by ants, and presumably other predators, the lacewing larvae apply chemicals as allomones in defence. Generally, naked chrysopid larvae alone synthesise defensive chemical substances while trash-carrier larvae camouflage themselves with substances produced by prey aphids or coccids.

It has been observed that naked chrysopid larvae exude a droplet from their anus, which they coat onto the head or antennae of attacking ants. Due to this action ants immediately interrupt the attack and clean off the repellent substance (Kennett, 1948). The older larvae with bigger body size apply this defensive method against ants more successfully than do young larvae (first instar). LaMunyon & Adams (1987) made tests on larvae of five lacewing species (*Chrysoperla plorabunda, C. comanche, C. mohave, C. downesi,* and *Eremochrysa punctinervis*), and observed how they applied and produced the defensive substance against *Iridomyrmex humilis* ants and what was its effect. They recognised that secretions probably blocked sensorial receptors on the head and antennae. Using scanning electron micrographs it could be seen that the secreted substance glued setae on the antennae of the ant together, such that it had to stop combat and clean itself. Implementation of anal defensive secretion against predators seems to be widespread among

lacewings, such as in the genera of *Chrysoperla,*
Eremochrysa, Chrysopa, and Ceraeochrysa (Spiegler,
1962; LaMunyon & Adams, 1987). According to analysis the anal defensive substance was the same as the previously known adhesive substance that helped to stick
the abdominal tip to the substrate (Spiegler, 1962). The
anal secretion has further functions: it participates in
the excretion of uric acid and in the production of prepupal silk (Spiegler, 1962; LaMunyon & Adams, 1987).

Among the debris-carrying chrysopid larvae the
food-specialist species that are strongly associated with
certain wax-secreting homopteran insects (aphids, scale
insects) partly or totally coat their own dorsal body
surface with waxy substances in order to camouflage
themselves against bodyguard ants. For instance, larvae
of *Ceraeochrysa cincta* used the waxy packets produced
by the pseudococcid *Plotococcus eugeniae* (Eisner &
Silberglied, 1988). The woolly alder aphid *Prociphilus
tesselatus* serves as an exclusive food resource for larvae
of the American lacewing *Chrysopa slossonae.* The lacewing larvae make a shield on the dorsal surface of their
body from thin, dense wax filaments secreted by aphids
which provide effective protection against the aphid-
attending ants, like *Camponotus* and *Formica* species
(Eisner *et al.,* 1978; Milbrath *et al.,* 1993).

Due to waxy compounds the chrysopid larvae pos-
sessing a wax covering were camouflaged not only vis-
ually but also chemically, thus ants could detect only
homopteran-specific volatile or contact-cue chemicals
at encounters. The waxy substances secreted by aphids
or coccids are probably multifunctional, having excre-
tory or defensive roles equally. It is supposed that they
can consist also of aphid alarm pheromone
(Whittington & Brothers, 1991). Their chemical composition may be simple or more complicated. Wax
secreted by integumental glandular cells of *P. tesselatus*
consists of a long-chained ketoester (Eisner et al.,
1978). The chemical composition of waxy substances
that are produced by scale insects can vary (Tamaki,
1997). From the four larger groups of compounds the
true waxes consist of mixture of long-chained lipids,
esters of higher fatty acids, and fatty alcohols. Resinous
materials or terpenoids (e.g. cyclic sesterterpenoids)
form another relevant group of components. The waxy
substances of scale insects consist of hydrocarbons and
different kinds of pigment. It is very probable that ants
are able to perceive the cues of these chemical groups to

identify scale insects, so lacewing larvae covered by wax
makes them cryptic for ants.

Aphid allelochemical defence against lacewings
There are various direct and indirect pathways in tritrophic interactions among host plant, aphid, and predator to transfer the allelochemicals derived from plants
towards aphidophaga (van Emden, 1995). Certain
aphids use them to obtain defensive allomones through
sequestration of toxic allelochemicals from host plants
against lacewings. Predators exposed to detrimental,
toxic chemicals that are mediated by prey may react in
different ways. According to Malcolm's classification
(Malcolm, 1992) the reactions are as follows: (1) the
predator is 'included', if it is able to avoid detrimental
effect due to fast detoxification; (2) the predator is
'excluded', if it is unable to tolerate the allelochemical
toxic effect; (3) the predator is 'peripheral', if it is able
to consume the toxic prey but at some cost to its performance. The toxic allelochemicals that are mediated by
aphid prey toward lacewings may have several negative
impacts on predators, such as reduction of the feeding
rate, prolongation of developmental periods and additional mortality.

Philippe (1972) reported such an effect of a toxic
compound on lacewings that was derived from a plant
and mediated by aphid prey. He fed *Chrysoperla carnea*
larvae with *Aphis sambuci* aphids living on elder. It was
observed that a sambunigrin compound taken up from
the host plant by the prey was toxic for lacewing larvae.
However, the adults and larvae of the brown lacewing
Drepanepteryx phalaenoides take *A. sambuci* as prey
without detriment ('included predator') (F.
Szentkirályi, unpublished data). Experiments by
Mendel *et al.* (1992) on legumes containing alkaloid
substances provide further examples of toxic effects of
prey-mediated allomones. They studied the relations
between scale insects (*Icerya spp., Lepidosaphes ulmi,
Planococcus citri*) and aphids (*Aphis craccivora*) living on
two toxic plants (*Spartium junceum, Erythrina corallodendrum*), and their predators. It was proved that
feeding efficiency of the lacewings *C. carnea* and
Sympherobius fallax was reduced by sequestration of
prey-mediated plant alkaloids as well as their transfer
into honeydew. However, it may also be that homopteran prey excrete their own defensive allomones against
lacewings. Whittington & Brothers (1991) found that a

substance was excreted from siphuncula onto the head of larvae of chrysopid *D. handschini* by rose aphids (*A. gossypii*, *Macrosiphum rosae*, and *Rhodobium porosum*) which reduced the feeding rate and prolonged the developmental periods of lacewings.

5.5.3 Allelochemicals in tritrophic interactions of lacewings

In tritrophic interactions, information from numerous chemical mediators guides the predatory activity and oviposition of lacewings, from finding the proper habitat, through location of prey patches, to encounter with prey items. These allelochemicals which are regularly volatile and beneficial for lacewings are derived directly from prey and its activity or they are released from the host plants. The cues from volatile and contact allelochemicals evoke appropriate behavioural responses from lacewing adults and larvae in searching for prey and selecting oviposition sites.

Kairomones in prey-foraging of lacewings
Kairomones as infochemicals derive from prey (emitter) and after perception by lacewings (receiver) attract the latter to the prey. Behavioural response to kairomones is beneficial for lacewings but disadvantageous for potential prey. There are several sources of kairomones as chemical cues, such as faeces, cuticle, exuviae, glandular secretions, honeydew, pheromones, scales from the body, or haemolymph (Vet & Dicke, 1992).

Lewis *et al.* (1977) and Nordlund *et al.* (1977) first showed the utilisation of kairomones by lacewings through experiments. They demonstrated that from scales of the moth *Heliothis zea* kairomones were released that attracted *C. carnea* larvae. The presence of kairomones (scales or hexane extract from scales) increased the predation rate of *C. carnea* larvae on eggs of the noctuid moth. The experiments confirmed that kairomones were present in materials that related to egg-laying. The kairomone played a role in finding and acceptance of prey (eggs) by lacewings.

Recent studies have proved the prior hypothesis that lacewings can use the sex pheromone of aphids and scale insects as a kairomone. Boo *et al.* (1998) tested responses given by a chrysopid species, *Chrysopa cognata* (= *C. pallens*) from Korea to a common sex pheromone of many aphid species. Through olfactometer studies, electroantennogram (EAG) responses,

and field trappings it was recorded that both components of the pheromone (4aS, 7S, 7aR)-nepetalactone and (1R, 4aS, 7S, 7aR)-nepetalactol attracted the lacewing adults. Their mixture in ratios of 4:1 or 1:4 provided the highest rate of capture in field trapping. Through experiments Boo *et al.* (1998) showed that the sex pheromone of prey was utilised as a kairomone by a generalist predator, adults of *C. cognata*. Boo *et al.* (1999) tested sticky traps baited with aphid sex pheromones in field conditions. The traps with nepetalactol attracted 13 times more chrysopid individuals than traps with nepetalactone. Besides *C. cognata* a few adult individuals of *C. formosa* and *C. phyllochroma* were also captured by baited traps, which reflects the possibility that perception of aphid sex pheromones is widespread among lacewings.

Brown lacewings can also use sex pheromones as kairomones produced by their prey. Exposing sex-pheromone traps of three *Matsucoccus* scale insect species (*M. josephi*, *M. feytaudi*, and *M. matsumurae*) in stands of *Pinus pinaster* in Portugal, Mendel *et al.* (1997) recorded that they attracted significant amounts of *Hemerobius stigma*. This hemerobiid species was closely associated with *Matsucoccus* species as prey (Covassi *et al.*, 1991). Adults of *H. stigma* were strongly attracted by the sex pheromone of all three scale insect species.

Aphids release other important volatile infochemicals besides sex pheromones, such as alarm pheromones. It is a cornicle secretion that is exuded by aphids when they are attacked by predators. The major component of the alarm pheromone of numerous aphids is the sesquiterpene hydrocarbon (E)-β-farnesene. The alarm pheromone of the aphid is also supposed to act as a kairomone attracting aphidophagous insects to their prey source (Pickett *et al.*, 1992). This presumption has not been proved. Boo *et al.* (1998) also studied the reaction of adults of *Chrysopa cognata* lacewings to aphid alarm pheromone. There was neither an EAG response nor a significant difference in olfactometer choice tests. The absence of a behavioural response of *C. cognata* was an unexpected outcome in comparison with the result of a prior experiment in that the EAG response of *Chrysoperla carnea* to β-farnesene was 30% higher than to aphid sex pheromone (Boo *et al.*, 1998). This latter finding with β-farnesene related to *C. carnea* is also supported by Dodds & McEwen (1998).

Honeydew excreted by homopteran potential prey may serve not only as a food source for lacewings but also, due to its volatile components, as a kairomone for finding prey habitats (Hagen, 1986). It consists of several nutritional chemicals such as various kinds of sugar (fructose, glucose, trehalose, trisaccharide), vitamins (such as ascorbic acid, B-vitamins), nitrogen compounds as free amino acids although the presence of complete essential amino acid composition is rare in honeydew. The necessary essential amino acids for adult lacewings are provided by production of yeast symbiotes in the gut crop. In field tests, applying artificial honeydew that contained protein hydrolysates, it was found that a volatile kairomone source in natural honeydew could be tryptophan. According to olfactometer analysis adults of *C. carnea* are attracted by acid-hydrolysed tryptophan, isomers of tryptophan, and hydrogen peroxide tryptophan. Results of the tests showed that breakdown products of tryptophan such as indol-acetaldehyde attracted the adults of *C. carnea* (Hagen, 1986). Attempts to develop field manipulation of *C. carnea* through the spraying of artificial honeydew, incorporating acid-hydrolysed L-tryptophan, have been made (McEwen *et al.*, 1993, 1994). However, new research makes the role of L-tryptophan questionable. Harrison & McEwen (1998) found that there was very little evidence for the presence of L-tryptophan in insect honeydew. Moreover, there was no evidence that L-tryptophan could be broken down by acid hydrolysis. Furthermore, they showed (1998) that L-tryptophan had no further reaction and no breakdown products in acid solution (hydrochloric acid). They suggest that the absence of hydrolysis and lack of a volatile character to L-tryptophan or tryptophan-hydrochloride reactions indicates that other mechanisms must be attracting lacewing adults during field-sprays of L-tryptophan. One possible mechanism would be damage to the plant surface by the acidic spray that releases the plant's attractive volatile chemicals. Another possible pathway would be a breakdown of L-tryptophan by microfauna transforming it to attractive compounds (Harrison & McEwen, 1998).

There is evidence on the role of honeydew not only in long-range habitat location but also in utilisation of short-range kairomones by lacewings. The result of antennal contact by female adult *C. carnea* with honeydew of the olive black scale (*Saissetia oleae*) is reduced walking speed, increased frequency of turning, and also increased turning angle (McEwen *et al.*, 1993). The same behavioural responses could be also observed before the period of antennal contact in the presence of honeydew. The arrestant effect of honeydew on *C. carnea* adults may contribute to more effective searching for optimal oviposition sites, around prey colonies, in particular if deposition of honeydew is really close to potential prey of lacewing larvae. Further studies are needed to recognise whether the same chemical compounds are responsible for the effect of long-range attraction and short-distance arrestment (McEwen *et al.*, 1993).

According to observation, honeydew has an arrestant effect on searching by chrysopid larvae (Downes, 1974) and may operate also as a short-range kairomone. Kawecki (1932) published data on the latter case where larvae of *C. carnea* could perceive the volatile released from scale-insect honeydew from 5 cm distance.

Synomones in habitat and prey location by lacewings

In tritrophic interactions several plant-derived, long-range attractants may guide lacewings to find suitable habitats and within them, patches of food sources. These allelochemicals detected via olfaction are beneficial not only for the receiving lacewings but also for the host plant due to the reduction of plant damage by attracted predators consuming prey. The presence of homopteran herbivores can trigger the emission of attractive volatiles from host plants, consequently they can contribute to the earlier appearance of aphidophaga. Following the terminology of Dicke & Sabelis (1988) the plant volatile compounds that influence the searching behaviour of lacewings would be considered as synomones.

One known plant volatile synomone is caryophyllene. It is a sesquiterpene hydrocarbon compound ($C_{15}H_{24}$) being present in essential oils of several plants. Flint *et al.* (1979) recorded that both caryophyllene and β-caryophyllene had an attractive effect on adults of *C. carnea*. According to experiments this synomone is a volatile cue giving information on plant habitat differing from the signal from honeydew that informs lacewings of intrahabitat food sources. Simultaneous perception of two volatile substances is needed for lacewings, *C. carnea*, to locate successfully the habitat to land on and to find the prey (Hagen, 1986). The occurrence of caryophyllene in fields (e.g. in cotton stands) changes seasonally depending on plant phenology, thus, for example, the greatest activity was recorded in the early period of the season (Flint *et al.*, 1979). Biosynthesis of β–caryophyllene relates to β–farnesene aphid alarm pheromone. The β–caryophyllene inhibits the activity of the aphid alarm pheromone. A wide circle of plants produce both the alarm pheromone and its inhibitor. Reducing the dispersive effect of alarm pheromone by the simultaneous production of the inhibitor may assist also to enable predation as well as attracting *C. carnea* (Pickett *et al.*, 1992). Recently, Dodds & McEwen (1998) noticed that *C. carnea* gave a relatively low antennal response on α and β isomers of caryophyllene in EAG experiments. A high level of EAG activity of both sexes of *C. carnea* was detected in tests of β–farnesene.

Other plant habitat synomones were also observed as attractants for certain lacewing species: eugenol for *Chrysoperla carnea*, methyl eugenol for *Mallada basalis*, terpenyl acetate for *C. carnea* and *Chrysopa nigricornis* (Hagen, 1986). Molleman *et al.* (1997) tested volatile methyl salicylate released by pear trees that were infested by *Psylla pyri*. They showed that *Chrysoperla carnea* adults were significantly attracted by this synomone from June to late September. However the baited traps could not attract hemerobiids. Release of synomone was triggered from trees because production of methyl salicylate was started due to feeding of psyllids.

Dodds & McEwen (1998) reported EAG tests on several further plant compounds as attractants for lacewings. It was detected that electrophysiological responses of *C. carnea* at the age of four days were higher than at one day. This might be explained by a transition from migratory phase to appetitive phase over this time. From 14 plant-derived volatile chemicals that were tested and mentioned by the authors some were found promising on the basis of stronger EAG responses (e.g. camphor).

5.5.4 Lacewings and transgenic crop plants: non-target impacts and ecological risks

Due to rapid developments in biotechnology, applying molecular methods by genetic engineering, novel solutions have been implemented to handle pest management problems during the last decade. Thus genetically modified organisms (GMOs) were born and among them also the so-called transgenic plants. Inserting genes from other organisms (microorganism, plant, animal) changes the features of host plants or gives them new features. From the perspective of plant protection a positive feature would be resistance against insect pests. The transgenic insect-resistant (or insecticidal) crop plants contain new genes that govern the production of compounds having toxic effects on certain target pest or taxonomic groups of insects. In this way the entomopathogenic chemicals (mainly proteins) produced in cells may express in each part or several parts of the gene-manipulated plant individuals. There have already been transgenic cultivars of about a dozen important plant species from arable and industrial crops that are resistant against insect pests. These include maize, cotton, tobacco, potato, soyabean, rice, oilseed rape, tomato, sugarcane, and apples.

Through genetic manipulation there are two approaches to create plants that are insect-resistant and produce entomopathogenic proteins. The first route is that gene sequences coding for delta-endotoxins of *Bacillus thuringiensis* (*Bt*) bacterium are introduced to plant genomes. The other approach is based on implementation of further plant-derived genes, such as proteinase inhibitors or lectin producers, for genetic manipulation.

Effects of Bt-expressing transgenic crop plants

Bacillus thuringiensis (*Bt*) is a spore-forming bacterium appearing commonly in nature. During the period of sporulation proteins of *Bt* consist of parasporal crystalline toxins (Cry) that are also named delta-endotoxins (Darvas & Polgár, 1998). Actually there have been more than 40 Cry genes recorded. The natural originated delta-endotoxins have differing structures; thus they are toxic for various insect orders. According to specificity of their activity the natural *Bt* endotoxins are divided into four groups: Cry1 of Lepidoptera; Cry2 of Lepidoptera and Diptera; Cry3 of Coleoptera; Cry4 of Diptera. Moreover, the four groups consist of 27 serotypes or 34 serovars. Transgenic plants are usually modified to produce variants of Cry1 and Cry3.

Solubilising the proteinaceous crystals (parasporal body) in the mid gut of susceptible insects releases protoxins decomposing into smaller active toxic polypeptides. These activated toxin proteins bind to receptors in the mid gut epithelium. Toxins generate pores in cell membranes. The gut will be paralysed and the insects will stop feeding. The dissolved membrane causes lesions that are followed by a lethal septicaemia.

Bt genes coding active insecticide toxins have been

introduced into different crop plants. For this reason toxins expressed in transgenic plants require no activation in the gut of insects contrary to the inactive parasporal protoxins. Consequently, the specificity of active toxins is smaller than protoxins, so they can be toxic for other non-target natural enemies, too. Insecticide preparations from *Bt* have been applied for a long time. In 1970 *Bt* serovar *kurstaki* was described and showed 200 times higher toxicity level for certain insect species than for other serovars (Dulmage, 1970). The *Bt* serovar *kurstaki* is one of the *Bt* variants used most frequently in pest management practice. Adylov *et al.* (1990) studied the non-target effect of this *Bt* serovar on cotton fields. The *Bt* preparation sprayed in 0.4–1.0 kg/ha concentrations had no adverse effect on adults or larvae of *Chrysoperla carnea*. However *Bt* ssp. *thuringiensis* in 2.0 kg/ha concentration reduced slightly the abundance of *C. carnea* although the decrease ceased within two weeks. Salama *et al.* (1982) recorded detrimental effects on larvae of *C. carnea* that were fed with *Spodoptera littoralis* caterpillars and *Aphis durantae* aphids treated with *Bt*. They found that lacewing larval duration was significantly increased, while the predatory rate was decreased compared with controls. According to these previous results certain detrimental effects of *Bt* crops on lacewings could be expected, too.

The transgenic *Bt*–maize hybrids were developed against the target pest, European corn-borer (ECB), *Ostrinia nubilalis*. Hilbeck *et al.* (1998a, b, 1999) studied the direct and indirect effects of *Bt* toxin (Cry1A$_b$) expressed in a transgenic maize hybrid in lacewing feeding experiments. They used the chrysopid *C. carnea s. lat.* as a known predator of eggs and young larvae of *O. nubilalis*. The lacewing was reared on larvae of *Bt*–maize-fed ECB and a non-target noctuid pest, *Spodoptera littoralis*. The mean total mortality rate of lacewings was 66% on *Bt*-fed ECB prey, and 59% on *Bt*-fed *S. littoralis*, respectively. These rates were significantly greater than the 37% mortality of lacewings that consumed *Bt*-free (control) prey. The toxic *Bt*-fed ECB prey prolonged the developmental time of *C. carnea* larvae. The reduced fitness of chrysopid larvae found by Hilbeck *et al.* (1998a, b) was due to the combined direct effect of *Bt* proteins and the indirect effect of lower nutritional quality of sick prey. Hilbeck *et al.* (1999) reared *C. carnea* larvae on meridic-diet-fed caterpillars of *S. littoralis*. The artificial food of *S. littoralis* contained activated toxin (Cry1A$_b$), and two

protoxins (Cry1A$_b$, Cry2A) of *Bt* in 25, 50, and 100 µg/g diet concentrations, respectively. The total (from first-instar larvae to adults) mortality rates of lacewings were dose-responsive and ranged on toxin-fed preys between 55% and 78%, and on protoxin-fed preys between 46% and 62%. The level of lacewing mortality was significantly higher than in the control (26%). In the case of prey fed on diet with the highest concentration of *Bt* toxin the duration of lacewing development was greater. The lethal effect of *Bt* toxin to lacewings was higher in indirect (prey-mediated) feeding tests than in direct (treated artificial diet without prey) feeding ones.

These bioassays carried out by Hilbeck *et al.* (1998a, b, 1999) revealed the following. (1) The effect of *Bt* toxin/protoxin was sublethal for non-target *S. littoralis* larvae, because, it seems, they were able to change (detoxify) the toxic proteins to less toxic ones. However these latter substances derived from *Bt* toxins proved highly lethal to *C. carnea* which consumed the treated prey. (2) The combined effects of toxic *Bt* proteins × herbivore prey (*S. littoralis*) interaction significantly increased the prey toxicity for the predator (*C. carnea*). (3) Along the tritrophic chains (*Bt*–maize – target and non-target herbivore prey – lacewings) the *Bt* toxins can be released from the transgenic crop plant and cause adverse prey-mediated effects for lacewings (*C. carnea*).

Cry toxins from *Bt* plants can go not only into food chains associated with phytophagous (target or non-target) insects but can also be expressed in pollen of transgenic crops. The toxic pollen can be spread in the environment through the wind or by insects consuming it. There are several pollini- or glyciniphagous species among the aphidophagous insects, such as lacewings. For this reason the making of ecological impact assessment of consumption of pollen derived from *Bt* plants on lacewings is particularly important.

Cry1A$_b$ is expressed in the pollen of *Bt*–maize hybrids. Pilcher *et al.* (1997) used the pollen produced by the *Bt*-transgenic maize for food source of three major predators in North America maize fields. The effect of *Bt* pollen on *C. carnea* larvae was investigated besides the effects on a coccinellid and an anthocorid species. They found no adverse effect of *Bt*–maize pollen on larvae of *C. carnea*, because there was no difference in the developmental time and mortality (49%) between lacewing larvae fed *Bt* pollen and *Bt*-free pollen. Also they did not find a significant difference between mean number of eggs or larvae/adults per plant of *Bt*–maize and non-*Bt*–maize stands with samples taken before, during and after pollen shed. However, there is no information about the real effect of *Bt* pollen on lacewing adults because of a lack of such an investigation. Sims (1995) studied the impacts of *Bt*–cotton (Cry1A$_c$) on several predatory insects. The transgenic cotton had no adverse effect on *C. carnea*. The aphids sucking plant juice from phloem where the toxins are generally absent seem to not mediate harmful effects for aphidophaga, as was found e.g. for a coccinellid fed on aphids living on maize (Hillbeck *et al.*, 1998c).

Possible impacts of transgenic crops expressing anti-aphid lectins

Transgenic crops manipulated by plant-derived genes presumably have detrimental effects on aphidophagous lacewings if their pest-resistance is based on the expression of lectins. Lectins are toxic for a wide range of insects, because they bind to gut of pests (including aphids) after feeding from transgenic plants. Birch *et al.* (1999) studied the effect of transgenic potato on the ladybird *Adalia bipunctata*. The applied potato line was genetically manipulated for resistance against aphids. The aphid resistance was provided by expression of snowdrop lectin in potato plants. The target aphid pest was *Myzus persicae* as prey of the ladybird. Birch and his team demonstrated that anti-aphid snowdrop lectin mediated by prey was toxic for adults of *A. bipunctata*. The adverse impacts were shown by decreased rate of egg production, egg fertility, and hatch from eggs. The most obvious effect was the reduction of lifespan of coccinellid females by 51% after a 12-day feeding period on *M. persicae*-infested transgenic potatoes. Although the impact on lacewings of transgenic crop plants expressing anti-aphid lectins has not been studied yet, detection of their toxic effect can be strongly predicted.

Ecological risks for lacewings by introduction of transgenic crops

During the 21st century one of the greatest challenges for entomologists will be the study of the ecological consequences of GMOs introduced into field cultivation. There are some expected non-target, hazardous impacts of transgenic crop plants, which can be predicted for lacewing populations in field conditions:

- Via tritrophic interactions the adverse or lethal effects of prey-mediated toxins (the originals or the toxins modified by pests) can decrease the population size of lacewings.
- Extended growing of transgenic plants can suppress regionally the target pests as potential prey; consequently the lacewing population level should decline even across larger agricultural areas, where large-sized monocultural arable fields are cultivated in various crop belts.
- The longer persistence of toxic substances over the seasonal field presence of a transgenic crop can not permit recovery of decreased lacewing populations such as occurs frequently after conventional pesticide treatments.
- If land use is characterised by continuous monocultural large-sized fields of the same transgenic crop, the suppression of target pests and the presence of sick non-target herbivores can cause area-wide synchronous shortages in sources of prey for lacewings. Consequently, in this situation it is very important to provide lacewings with alternative sources by natural–semi-natural vegetation patches, field margins scattered over the landscape.
- The prey shortage caused by transgenic crops grown wide-range at landscape level may change the lacewing interspecific interactions within the aphidophagous guilds (increasing competition or IGP).
- After feeding on transgenic pollen the reproductive output and fitness of lacewing adults can be decreased (decline of number of eggs laid, fertility and hatching rate, or shorter longevity of females).
- After some years *Bt*-resistant selected strains of target and non-target herbivores can reappear in transgenic crop stands, so the non-resistant lacewings attracted and trapped by the greater populations of these returning potential prey may be suppressed via an increased level of toxic food sources within such fields.
- Genetic manipulation of crop plants can change the emitted infochemicals (e.g. concentrations, mode of release, chemical composition, etc.) causing unpredictable shifts in behavioural responses of lacewings.
- By hybridisation via pollination the genes harmful for aphidophaga may be spread and expressed by other species, like weeds, wild plants, and other

crops, resulting in a higher encounter rate of lacewings with toxic prey items both within crop fields and outside over uncultivated lands.

In the near future it is necessary to study these above-mentioned possible impacts of GMOs in order to increase our knowledge about their ecological risks for ecosystems involving the associated lacewing assemblages.

5.6 ACKNOWLEDGEMENTS

I am very grateful to Peter McEwen (Insect Investigations Ltd, Cardiff School of Biosciences, Cardiff University, UK) and to Andy Whittington (National Museums of Scotland, Edinburgh, UK) for their editorial work and language corrections on the text. Many thanks to Ferenc Kádár (Plant Protection Institute of the Hungarian Academy of Sciences, Budapest, Hungary) for assistance in typing the bibliography. This work was partly supported by the Hungarian National Science Foundation (research grant number: OTKA T 023284).

REFERENCES

Ábrahám, L. & Vas, J. (1999) Preliminary report on study of the daily activity pattern of Neuroptera in Hungary. *Acta Phytopathologica et Entomologica Hungarica* 34, 153–164.

Adylov, Z.K., Khakmov, A., Babebekov, K. & Agzamova, K.K. (1990) Influence of microbiological preparations on the entomophages of the cotton agrocoenosis. *Zashchita Rastenii* 7, 34. (in Russian)

Altieri, M.A. & Whitcomb, W.H. (1979) Predaceous arthropods associated with Mexican tea in north Florida. *Florida Entomologist* 62, 175–182.

Andow, D.A. (1992) Fate of eggs of first generation *Ostrinia nubilalis* (Lepidoptera: Pyralidae) in three conservation tillage systems. *Environmental Entomology* 21, 388–393.

Arzet, H.R. (1973) Suchverhalten der Larven von *Chrysopa carnea* Steph. (Neuroptera: Chrysopidae). *Zeitschrift für angewandte Entomologie* 74, 64–79.

Aspöck, H., Aspöck, U. & Hölzel, H. (1980) *Die Neuropteren Europas. Eine zusammenfassende Darstellung der Systematik, Ökologie und Chorologie der Neuropteroidea (Megaloptera, Raphidioptera, Planipennia) Europas*, vol. 1. Göcke & Evers, Krefeld.

Banks, C.J. (1952) An analysis of captures of Hemerobiidae and Chrysopidae in suction traps at Rothamsted, July, 1949. *Proceedings of Royal Entomological Society, London* 27, 45–53.

Bay, T., Hommes, M. & Plate, H.P. (1993) Die Florfliege *Chrysoperla carnea* (Stephen). *Mitteilungen aus der biologischen Bundesanstalt für Land- und Forstwirtschaft* 288, 3–175.

Bergeson, E. & Messina, F.J. (1997) Resource- versus enemy-mediated interactions between cereal aphids (Homoptera: Aphididae) on a common host plant. *Annals of the Entomological Society of America* 90, 425–432.

Bergeson, E. & Messina, F.J. (1998) Effect of a co-occurring aphid on the susceptibility of the Russian wheat aphid to lacewing predators. *Entomologia Experimentalis et Applicata* 87, 103–108.

Birch, A.N.E., Geoghegan, I.E., Majerus, M.E.N., McNicol, J.W., Hackett, C., Gatehouse, A.M.R. & Gatehouse, J.A. (1999) Tritrophic interactions involving pest aphids, predatory 2-spot ladybirds and transgenic potatoes expressing snowdrop lectin for aphid resistance. *Molecular Breeding* 5, 75–85.

Blum, M.S., Wallace, J.D. & Fales, H.M. (1973) Skatole and tridecene: identification and possible role in a chrysopid secretion. *Insect Biochemistry* 3, 353–357.

Boo, K.S., Chung, I.B., Han, K.S., Pickett, J.A. & Wadhams, L.J. (1998) Response of the lacewing *Chrysopa cognata* to pheromones of its aphid prey. *Journal of Chemical Ecology* 24, 631–643.

Boo, K.S., Kang, S.S., Park, J.H., Pickett, J.A. & Wadhams, L.J. (1999) Field trapping of lacewings with aphid pheromones. *International Organisation for Biological and Integrated Control of Noxious Animals and Plants (IOBC), West Palaearctic Regional Section (WPRS) Bulletin* 22, 35–36.

Bosch, J. (1987) Der Einfluss einiger dominanter Ackerunkräuter auf Nutz- und Schadarthropoden in einem Zuckerrübenfeld. *Zeitschrift für Pflanzenkrankheiten und Pflanzenschutz* 94, 398–408.

Bouchard, D., Pilon, J.G. & Tourneur, J.C. (1986) Role of entomophagous insects in controlling the apple aphid, *Aphis pomi*, in southwestern Quebec. In *Ecology of Aphidophaga*, ed. Hodek, I., pp. 369–374. Academia, Prague and Dr W. Junk Dordrecht.

Bowden, J. & Dean, G.J.W. (1977) The distribution of flying insects in and near a tall hedgerow. *Journal of Applied Ecology* 14, 343–354.

Canard, M. (1984) Écologie des pontes de *Nineta flava* (Scopoli) (Neuroptera, Chrysopidae): disposition et facteurs biotiques antagonistes. In *Progress in World's Neuropterology, Proceedings of the 1st International*

Symposium on Neuropterology, ed. Gepp, J., Aspöck, H. & Hölzel, H., pp. 253–260. Thalerhof, Graz.

Canard, M., Séméria, Y. & New, T.R. (eds.) (1984) *Biology of Chrysopidae*, Series Entomologica, vol. 27, ed. Spencer, K.A. Dr W. Junk, The Hague.

Cisneros, J.J. & Rosenheim, J.A. (1997) Ontogenetic change of prey preference in the generalist predator *Zelus renardii* and its influence on predator–predator interactions. *Ecological Entomology* 22, 399–407.

Clark, T.L. & Messina, F.J. (1998) Foraging behaviour of lacewing larvae (Neuroptera: Chrysopidae) on plants with divergent architectures. *Journal of Insect Behaviour* 11, 303–317.

Coderre, D. & Tourneur, J.C. (1986a) Vertical distribution of aphids and aphidophagous insects of maize. In *Ecology of Aphidophaga*, ed. Hodek, I., pp. 291–296. Academia, Prague and Dr W. Junk, Dordrecht.

Coderre, D. & Tourneur, J.C. (1986b) Synchronization and voracity of aphidophagous insects on maize in Quebec, Canada. In *Ecology of Aphidophaga*, ed. Hodek, I., pp. 363–368. Academia, Prague and Dr W. Junk, Dordrecht.

Coderre, D. & Tourneur, J.C. (1988) Summer decline in aphid populations on maize. *Revue d'Entomologie du Quebec* 33, 16–24.

Coderre, D., Provencher, L. & Tourneur, J.C. (1987) Oviposition and niche partitioning in aphidophagous insects on maize. *Canadian Entomologist* 119, 195–203.

Covassi, M., Binazzi, A. & Toccafondi, P. (1991) Studies on the entomophagous predators of a scale of the genus *Matsucoccus Cock.* in Italy. I. Faunistical–ecological notes on species observed in pine forests in Liguria and Tuscany. *Redia* 74, 575–597.

Darvas, B. & Polgár, L. (1998) Novel-type insecticides: specificity and effects on non-target organisms. In *Insecticides with Novel Modes of Action: Mechanisms and Application*, ed. Ishaaya, I. & Degheele, D., pp. 188–259. Springer-Verlag, Berlin.

Deyrup, M. & Deyrup, N. (1978) Pupation of *Hemerobius* in Douglas-fir cones. *Pan-Pacific Entomologist* 54, 143–146.

Dicke, M. & Sabelis, M.W. (1988) Infochemical terminology: based on cost–benefit analysis rather than origin of compounds? *Functional Ecology* 2, 131–139.

Dinter, A. (1998a) Intraguild predation between erigonid spiders, lacewing larvae and carabids. *Journal of Applied Entomology* 122, 163–167.

Dinter, A (1998b) Interactions between spider and lacewing predators (Araneae: Erigonidae and Neuroptera: Chrysopidae) and their effects on the grain aphid *Sitobion avenae* Fab. (Homoptera: Aphididae). *IOBC Bulletin* 21, 91–101.

Dixon, A.F.G., Kindlmann, P., Lepš, J. & Holman, J. (1987) Why there are so few species of aphids, especially in the tropics. *American Naturalist* 129, 580–592.

Dodds, C. & McEwen P.K. (1998) Electroantennogram responses of green lacewings (*Chrysoperla carnea*) to plant volatiles: preliminary results. *Acta Zoologica Fennica* 209, 99–102.

Downes, J.A. (1974) Sugar feeding by the larva of *Chrysopa* (Neuroptera). *Canadian Entomologist* 106, 121–125.

Duelli, P. (1984a) Flight, dispersal, migration. In *Biology of Chrysopidae*, ed. Canard, M., Séméria, Y. & New, T.R., pp. 110–116. Dr W. Junk, The Hague.

Duelli, P. (1984b) Dispersal and oviposition strategies in *Chrysoperla carnea*. In *Progress in World's Neuropterology, Proceedings of the 1st International Symposium on Neuropterology*, ed. Gepp, J., Aspöck, H. & Hölzel, H., pp. 133–145. Thalerhof, Graz.

Duelli, P. (1986a) Flight activity patterns in green lacewings (Planipennia: Chrysopidae). In *Recent Research in Neuropterology, Proceedings of the 2nd International Symposium on Neuropterology*, ed. Gepp, J., Aspöck, H. & Hölzel, H., pp. 165–170. Thalerhof, Graz.

Duelli, P. (1986b) A 'missing link' in the evolution of the egg pedicel in lacewings? *Experientia* 42, 624.

Duelli, P. (1987) Eine isolierte Reliktpopulation von *Chrysoperla mediterranea* (Planipennia: Chrysopidae) in der Schweiz. *Mitteilungen der schweizerischen entomologischen Gesellschaft* 60, 301–306.

Duelli, P. (1988) Aphidophaga and the concepts of island biogeography in agricultural areas. In *Ecology and Effectiveness of Aphidophaga*, ed. Niemczyk, E. & Dixon, A.F.G., pp. 89–93. SPB Academic Publishing, The Hague.

Duelli, P. (1997) Biodiversity evaluation in agricultural landscapes: an approach at two different scales. *Agriculture, Ecosystems and Environment* 62, 81–91.

Duelli, P. & M. Obrist (1995) Comparing surface activity and flight of predatory arthropods in a 5 km transect. *Acta Jutlandica* 70, 283–293.

Dulmage, H. (1970) Insecticidal activity of HD-1, a new isolate of *Bacillus thuringiensis* subsp. *alesti*. *Journal of Invertebrate Pathology* 15, 232–239.

Eglin, W. (1980) Die Netzflügler des schweizerischen Nationalparks und seiner Umgebung (Insecta: Neuropteroidea). *Ergebnisse der wissenschaftlichen Untersuchungen im schweizerischen Nationalpark* 15, 281–351.

Eisner, T. & Silberglied, R.E. (1988) A chrysopid larva that cloaks itself in mealybug wax. *Psyche*, 95, 15–19.

Eisner, T., Hicks, K., Eisner, M. & Robson, D.S. (1978) 'Wolf-in-sheep's-clothing strategy of a predaceous insect larvae. *Science* 199, 790–794.

Eisner, T., Attygalle, A.B., Conner, W.E., Eisner, M., MacLeod, E. & Meinwald, J. (1996) Chemical egg defense in a green lacewing (*Ceraeochrysa smithi*). *Proceedings of the National Academy of Sciences*, 93, 3280–3283.

Elliott, N.C. & Kieckhefer, R.W. (1990) A thirteen-year survey of the aphidophagous insects of alfalfa. *Prairie Naturalist* 22, 87–96.

Elsey, K.D. (1974) Influence of plant host on searching speed of two predators. *Entomophaga* 19, 3–6.

van Emden, H.F. (1995) Host plant–aphidophaga interactions. *Agriculture, Ecosystems and Environment* 52, 3–11.

Flint, H.M., Slater, S.S. & Walters, S. (1979) Caryophyllene: an attractant for the green lacewing. *Environmental Entomology* 8, 1123–1125.

Frazer, B.D. (1988) Predators. In *Aphids: Their Biology, Natural Enemies and Control*, vol. B, ed. Minks, A.K. & Harrewijn, P., pp. 217–230. Elsevier, Amsterdam.

Galecka, B. & Zelený, J. (1969) The occurrence of predators of aphids of the genus *Chrysopa* spp. on crops growing on a four-crop field and in the neighbouring shelterbelts. *Ekologia Polska Seria A* 17, 351–360.

Gepp, J. (1973) Vergleichend-quantitative Untersuchungen der Dichten von Neuropterenimagines in den Jahren 1964 bis 1972 im Kaiserwald südwestlich von Graz. *Berichte der Arbeitsgemeinschaft für ökologische Entomologie in Graz* 1, 29–41.

Gepp, J. (1980) Hinweise zur Höhenverbreitung und Phänologie waldbewohnender Neuropteren Mitteleuropas. *Acta Musei Reginaehradecensis (Supplementum)*, 42–45.

Gepp, J. (1988/9) Zur ökologischen Differenzierung der präimaginalen Stadien baumbewohnender Chrysopiden im Alpenraum. *Sitzungsberichten der österreichische Akademie der Wissenschaften Mathematische-Naturwissenschaften Klasse 1*, 197, 1–73.

Greve, L. (1969) An aerial-drift of Neuroptera from Hardangervidda, western Norway. *Acta Universitatis Bergensis, Series Mathematica Rerumque Naturalium* 2, 3–15.

Greve, L., Solem, J.O. & Bretten, S. (1987) Distribution, abundance and phenology of adult Neuropteroidea (Orders Plannipennia, Raphidioptera and Megaloptera) and Mecoptera in the Dovrefjell mountains, South Norway. *Fauna Norvegica, Series B* 34, 57–62.

Hagen, K.S. (1986) Ecosystem analysis: plant cultivars (HPR), entomophagous species and food supplements. In *Interactions of Plant Resistance and Parasitoids and Predators of Insects*, ed. Boethel, D.J. & Eikenbary, R.D., pp. 151–197. Wiley, New York.

Hanski, I. (1994) Patch-occupancy dynamics in fragmented landscapes. *TREE* 9, 131–135.

Hanski, I. & Gilpin, M. (1991) Metapopulation dynamics: brief history and conceptual domain. *Biological Journal of the Linnean Society* 42, 3–16.

Harris, V.E. & Phillips, J.R. (1986) The effect of mowing spring weed hosts of *Heliothis* spp. on predatory arthropods. *Journal of Agricultural Entomology* 3, 77–86.

Harrison, S.J. & McEwen, P.K., (1998) Acid hydrolysed L-tryptophan and its role in the attraction of the green lacewing *Chrysoperla carnea* (Stephens) (Neuropt., Chrysopidae). *Journal of Applied Entomology* 122, 343–344.

Heil, M., Krämer, K. & Zotzmann, K.F., (1980) Das Auftreten von Florfliegen (*Chrysopa* spp.) in einer hessischen Obstanlage. *Gesunde Pflanzen* 32, 284–291.

Hilbeck, A., Baumgartner, M., Fried, P.M. & Bigler, F. (1998a) Effects of transgenic *Bacillus thuringiensis* corn-fed prey on mortality and development time of immature *Chrysoperla carnea* (Neuroptera: Chrysopidae). *Environmental Entomology* 27, 480–487.

Hilbeck, A., Moar, W.J., Pusztai-Carey, M., Filippini, A. & Bigler, F. (1998b) Toxicity of the *Bacillus thuringiensis* Cry1Ab toxin on the predator *Chrysoperla carnea* (Neuroptera: Chrysopidae) using diet incorporated bioassays. *Environmental Entomology* 27, 1255–1263.

Hilbeck, A., Moar, W.J., Frick, C., Zwahlen, C., Filippino, A. & Bigler, F. (1998c) Tritrophic interactions of transgenic *Bt*-corn, herbivores and natural enemies. In *Global Working Group on Transgenic Organisms in Integrated Pest Management and Biological Control*, Newsletter no. 1, ed. Hilbeck, A. & Raps, A., pp. 10–11. IOBC, Reckenholz, Zürich.

Hilbeck, A., Moar, W.J., Pusztai-Carey, M., Filippini, A. & Bigler, F. (1999) Prey-mediated effects of Cry1Ab toxin and protoxin and Cry2A protoxin on the predator *Chrysoperla carnea*. *Entomologia Experimentalis et Applicata* 91, 305–316.

Honěk, A. (1981) Aphidophagous Coccinellidae (Coleoptera) and Chrysopidae (Neuroptera) on three weeds: factors determining the composition of populations. *Acta Entomologica Bohemoslovaka* 78, 303–310.

Horn, D. (1981) Effect of weedy backgrounds on colonization of collards by green peach aphid, *Myzus persicae*, and its major predators. *Environmental Entomology* 10, 285–289.

Jeník, J. (1992) Ecotone and ecocline: two questionable concepts in ecology. *Ekológia (CSFR)* 11, 243–250.

Jones, S.L., Lingren, P.D. & Bee, M.J. (1977) Diel periodicity of feeding, mating, and oviposition of adult *Chrysopa carnea*. *Annals of the Entomological Society of America* 70, 40–47.

Kawecki, Z. (1932) Beobachtungen über das Verhalten und die Sinnesorientierung der Florfliegenlarven. *Bulletin int. Acad. pol. Sci. Lett.* 2, 91–106.

Kennett, C.E. (1948) Defense mechanism exhibited by larvae of *Chrysopa californica* Coq. *Pan-Pacific Entomologist* 24, 209–211.

Kozár, F. (1995) Species richness of scale insects (Homoptera: Coccoidea) in different regions of the world. In *Abstracts of the 7the European Ecological Congress*, ed. Demeter, A. & Peregovits, L., pp. 53. Hungarian Biological Society, Budapest.

LaMunyon, C.W. & Adams, P.A. (1987) Use and effect of an anal defensive secretion in larval Chrysopidae (Neuroptera). *Annals of the Entomological Society of America* 80, 804–808.

Lapchin, L. (1991) Spatial heterogeneity and density dependence in wheat aphids and their natural enemies. *Redia* 74, 223–230.

Lewis, T. (1968) Windbreaks, shelter and insect distribution. *Span* 11, 3–6.

Lewis, T. & Taylor, L.R. (1965) Diurnal periodicity of flight by insects. *Transactions of the Royal Entomological Society of London* 116, 393–479.

Lewis, W.J., Nordlund, D.A., Gross, H.J. Jr, Jones, L.R. & Jones, S.L. (1977) Kairomones and their use for management of entomophagous insects. V. Moth scales as a stimulus for predation of *Heliothis zea* (Boddie) eggs by *Chrysopa carnea* Stephens larvae. *Journal of Chemical Ecology* 3, 483–487.

Li, H.C. (1987) Augmentation of *Chrysopa* spp. to control cotton aphids by intercropping cotton and safflower. *Chinese Journal of Biological Control* 3, 109–111. (in Chinese).

Lorenzetti, F., Arnason, J.T., Philogène, B.J.R. & Hamilton, R.I. (1997) Evidence for spatial niche partitioning in predaceous aphidophaga: use of plant colour as a cue. *Entomophaga* 42, 49–56.

Lucas, E., Coderre, D. & Brodeur, J. (1997) Instar-specific defense of *Coleomegilla maculata lengi* (Col.: Coccinellidae): influence on attack success of the intraguild predator *Chrysoperla rufilabris* (Neur.: Chrysopidae). *Entomophaga* 42, 3–12.

Lucas, E., Coderre, D. & Brodeur, J. (1998) Intraguild predation among aphid predators: characterization and influence of extraguild prey density. *Ecology* 79, 1084–1092.

Mack, T.P. & Smilowitz, Z. (1979) Diel activity of green peach aphid predators as indexed by sticky traps. *Environmental Entomology* 8, 799–801.

Malcolm, S. (1992) Prey defense and predator foraging. In *Natural Enemies: The Population Biology of Predators, Parasites and Diseases*, ed. Crawley, M.J., pp. 458–475. Blackwell Scientific Publications, Oxford.

Marín, M. (1987) Occurrence of *Chrysopa carnea* Steph. (Neuroptera–Chrysopidae) in maize crops. *Analele Institutului de Cercetari pentru Protectia Plantelor* 21, 77–83. (in Romanian).

Mayse, M.A. & Price, P.W. (1978) Seasonal development of soybean arthropod communities in east central Illinois. *Agro-Ecosystems* 4, 387–405.

McEwen, P.K. & Ruiz, J. (1993) Relationship between non-olive vegetation and green lacewing eggs in Spanish olive orchard. *Antenna* 18, 148–150.

McEwen, P.K., Clow, S., Jervis, M.A. & Kidd, N.A.C. (1993) Alteration in searching behaviour of adult female green lacewings *Chrysoperla carnea* (Neur.: Chrysopidae) following contact with honeydew of the black scale *Saissetia oleae* (Hom.: Coccidae) and solutions containing acid-hydrolysed L-tryptophan. *Entomophaga* 38, 347–354.

McEwen, P.K., Jervis, M.A. & Kidd, N.A.C. (1994) Use of a sprayed L-tryptophan solution to concentrate numbers of the green lacewing *Chrysoperla carnea* in olive tree canopy. *Entomologia Experimentalis et Applicata* 70, 97–99.

Mendel, Z., Blumberg, D., Zehavi, A. & Weissenberg, M. (1992) Some polyphagous Homoptera gain protection from their natural enemies by feeding on the toxic plants *Spartium junceum* and *Erythrina corallodendrum* (Leguminosae). *Chemoecology* 3, 118–124.

Mendel, Z., Adar, K., Nestel, D. & Dunkelblum, E. (1997) Sex pheromone traps as a tool for the study of population trends of the predator of a scale insect and for the identification of potential predators for biological control. International Organisation for Biological and Integrated Control of Noxious Animals and Plants, West Palaearctic Regional Section *Bulletin* 20, 231–240.

Miermont, Y. & Canard, M. (1975) Biologie du prédateur aphidiphage *Eumicromus angulatus* (Neur.: Hemerobiidae): études au laboratoire et observations dans le sud-ouest de la France. *Entomophaga* 20, 179–191.

Milbrath, L.R., Tauber, M.J. & Tauber, C.A. (1993) Prey specificity in Chrysopa: an interspecific comparison of larval feeding and defensive behaviour. *Ecology* 74, 1384–1393.

Miller, G.L. & Cave, R.D. (1987) Bionomics of *Micromus posticus* (Walker) (Neuroptera: Hemerobiidae) with descriptions of the immature stages. *Proceedings of the Entomological Society of Washington* 89, 776–789.

Molleman, F., Drukker, B. & Blommers, L. (1997) A trap for monitoring pear psylla predators using dispensers with the synomone methylsalicylate. *Proceedings of the Section Experimental and Applied Entomology of the Netherlands Entomological Society* 8, 177–182.

Morris, T.I., Campos, M., Jervis, M.A., McEwen, P.K. & Kidd, N.A.C. (1998) Potential effects of various ant species on green lacewing, *Chrysoperla carnea* (Stephens) (Neuropt., Chrysopidae) egg numbers. *Journal of Applied Entomology* 122, 401–403.

Nazarov, S.S. & Prygunkov, V.A. (1986) Maize agrotechnics and entomophages. *Zaschita Rastenii* 11, 29.

Neuenschwander, P. & Michelakis, S. (1980) The seasonal and spatial distribution of adult and larval chrysopids on olive-trees in Crete. *Acta Oecologica* 1, 93–102.

New, T.R. (1967) The flight activity of some British Hemerobiidae and Chrysopidae, as indicated by suction-trap catches. *Proceedings of the Royal Entomological Society of London* 42, 93–100.

New, T.R. (1986) A review of the biology of Neuroptera Planipennia. *Neuroptera International* supplement 1, 1–47.

New, T.R. (1989) *Planipennia (Lacewings)*, Handbuch der Zoologie, ed. Fischer, M., vol. 4, Arthropoda: Insecta 30. Walter de Gruyter, Berlin.

Nielsen, E.S. (1977) Studies on lacewings (Neuroptera *s. str.*) in a Danish beech stand. *Entomologiske Meddelelser* 45, 45–64.

Nordlund, D.A., Lewis, W.J., Jones, R.L., Gross, H.R. Jr. & Hagen, K.S. (1977) Kairomones and their use for management of entomophagous insects. VI. An examination of the kairomones for the predator *Chrysopa carnea* Stephens at the oviposition sites of *Heliothis zea* (Boddie). *Journal of Chemical Ecology* 3, 507–511.

Obrycki, J.J. & Tauber, M.J. (1984) Natural enemy activity on glandular pubescent potato plants in the greenhouse: an unreliable predictor of effects in the field. *Environmental Entomology* 13, 679–683.

Pantaleoni, R.A. & Sproccati, M. (1987) I Neurotteri delle colture agrarie: studi preliminari circa l'influenza di siepi ed altre aree non coltivate sulle popolazioni di Crisopidi. *Bollettino dell'Istituto di Entomologia dell'Università di Bologna* 42, 193–203.

Pantaleoni, R.A. & Ticchiati, V. (1990) Neurotteri delle colture agrarie: esperienze sul metodo di campionamento per abbattimento chimico. *Bollettino dell'Istituto di Entomologia dell'Università di Bologna* 45, 143–154.

Paoletti, M.G., Boscolo, P. & Sommaggio, D. (1997) Beneficial insects in fields surrounded by hedgerows in North Eastern Italy. *Entomological Research in Organic Agriculture*, 311–323.

Patel, K.G. & Vyas, H.N. (1985) Ovipositional site preference by green lacewing, *Chrysopa (Chrysoperla) scelestes* Banks on cotton and green gram. *Gujarat Agricultural University Research Journal* 10, 79–80.

Paulian, M. (1991) The influence of climate and soil factors on the presence of species of *Chrysopa* in maize crops. *Analele Institutului de Cercetari pentru Protectia Plantelor, Academia de Stiinte Agricole si Silvice* 24, 145–151.

Phillipe, R. (1972) Biologie de la reproduction de *Chrysopa perla* (L.) (Neuroptera, Chrysopidae) en fonction d'alimentation imaginale. *Annales de Zoologie–Écologie Animale* 4, 213–227.

Pickett, J.A., Wadhams, L.J., Woodcock, C.M. & Hardie, J. (1992) The chemical ecology of aphids. *Annual Review of Entomology* 37, 67–90.

Pilcher, C.D., Obrycki, J.J., Rice, M.E. & Lewis, L.C. (1997) Preimaginal development, survival, and field abundance of insect predators on transgenic *Bacillus thuringiensis* corn. *Environmental Entomology* 26, 446–454.

Polis, G.A., Myers, C.A. & Holt, R.D. (1989) The ecology and evolution of intraguild predation: potential competitors that eat each other. *Annual Review of Ecology and Systematics* 20, 297–330.

Rácz, V., Szentkirályi, F. & Visnyovszky, É. (1986) Study of aphid–aphidophage connections in maize stands. In *Ecology of Aphidophaga*, ed. Hodek, I., pp. 317–322. Academia, Prague and Dr W. Junk, Dordrecht.

Ressl, F. (1971) Untersuchungen über die Chrysopiden des Bezirkes Scheibbs (Niederösterreich). Beitrag zur Kenntnis der Ökologie, Phänologie und Verbreitung der Chrysopiden Mitteleuropas. *Beiträge zur Entomologie* 21, 597–607.

Ressl, F. (1974) Untersuchungen über die Hemerobiiden (Neuroptera, Planipennia) des Bezirkes Scheibbs (NOE). Ein Beitrag zur Kenntnis der Verbreitung, Ökologie und Phänologie der Hemerobiiden Mitteleuropas. *Entomologische Gesellschaft Basel* 24, 10–28.

Růžička, Z. (1994) Oviposition-deterring pheromone in *Chrysopa oculata* (Neuroptera: Chrysopidae). *European Journal of Entomology* 91, 361–370.

Růžička, Z. (1997a) Persistence of the oviposition-deterring pheromone in *Chrysopa oculata* (Neur.: Chrysopidae). *Entomophaga* 42, 107–112.

Růžička, Z. (1997b) Recognition of oviposition-deterring allomones by aphidophagous predators (Neuroptera: Chrysopidae, Coleoptera: Coccinellidae). *European Journal of Entomology* 94, 431–434.

Růžička, Z. (1998) Further evidence of oviposition-deterring allomone in chrysopids (Neuroptera: Chrysopidae). *European Journal of Entomology* 95, 35–39.

Růžička, Z. & Havelka, J. (1998) Effects of oviposition-deterring pheromone and allomones on *Aphidoletes aphidimyza* (Diptera: Cecidomyiidae). *European Journal of Entomology* 95, 211–216.

Salama, H.S., Zaki, F.N. & Sharaby, A.F. (1982) Effect of *Bacillus thuringiensis* Berl. on parasites and predators of the cotton leafworm *Spodoptera littoralis* (Boisd.). *Zeitschrift für angewandte Entomologie* 94, 498–504.

Schoenly, K. (1990) The predators of insects. *Ecological Entomology* 15, 333–345.

Schotzko, D.J. & O'Keeffe, L.E. (1989) Comparison of sweep net, D-Vac, and absolute sampling, and diel variation of sweep net sampling estimates in lentils for pea aphid (Homoptera: Aphididae), nabids (Hemiptera: Nabidae), lady beetles (Coleoptera: Coccinellidae), and lacewings (Neuroptera: Chrysopidae). *Journal of Economic Entomology* 82, 491–506.

Schultz, B.B. (1988) Reduced oviposition by green lacewings (Neuroptera: Chrysopidae) on cotton intercropped with corn, beans, or weeds in Nicaragua. *Environmental Entomology* 17, 229–232.

Séméria, Y. (1979) Quelques données sur les convergences chromatiques chez les Chrysopinae (Neuroptera, Planipennia, Chrysopidae) en relation avec la nature de leurs sécrétions prothoraciques. *Bulletin Mensuel de la Société Linnéenne de Lyon* 48, 267–305.

Sengonca, Ç. & Frings, B. (1985) Interference and competitive behaviour of the aphid predators, *Chrysoperla carnea* and *Coccinella septempunctata* in the laboratory. *Entomophaga* 30, 245–251.

Simberloff, D. & Dayan, T. (1991) The guild concept and the structure of ecological communities. *Annual Review of Ecology and Systematics* 22, 115–143.

Sims, S.R. (1995) *Bacillus thuringiensis* var. *kurstaki* (Cry1Ac) protein expressed in transgenic cotton: effects on beneficial and other nontarget insects. *Southwestern Entomologist* 20, 493–500.

Spiegler, P.E. (1962) The origin and nature of the adhesive substance in larvae of the genus *Chrysopa* (Neuroptera: Chrysopidae). *Annals of the Entomological Society of America* 55, 69–77.

Sugg, P.M., Greve, L. & Edwards J.S. (1994) Neuropteroidea from Mount St. Helens and Mount Rainier: dispersal and immigration in volcanic landscapes. *Pan-Pacific Entomologist* 70, 212–221.

Szabó, S. & Szentkirályi, F. (1981) Communities of Chrysopidae and Hemerobiidae (Neuroptera) in some apple orchards. *Acta Phytopathologica Academiae Scientiarum Hungaricae* 16, 157–169.

Szentkirályi, F. (1984) Analysis of light trap catches of green and brown lacewings (Neuropteroidea: Planipennia, Chrysopidae, Hemerobiidae) in Hungary. Verhandlungen des X Internationalen Symposiums über Entomofaunistik Mitteleuropas (SIEEC) (ed. Kaszab, Z.) 10, 177–180.

Szentkirályi, F. (1986) Niche segregation between chrysopid and hemerobiid subguilds. In *Ecology of Aphidophaga*, ed. Hodek, I., pp. 297–302. Academia, Prague and Dr W. Junk, Dordrecht.

Szentkirályi, F. (1989) Aphidophagous chrysopid and hemerobiid (Neuropteroidea) subguilds in different maize fields: influence of vegetational diversity on subguild structure. *Acta Phytopathologica et Entomologica Hungarica* 24, 207–211.

Szentkirályi, F. (1991) Reproductive numerical response of chrysopids and hemerobiids (Neuropteroidea) to aphids on the common thistle. In *Behaviour and Impact of Aphidophaga*, ed. Polgár, L., Chambers, R.J., Dixon, A.F.G. & Hodek, I., pp. 273–280. SPB Academic Publishing, The Hague.

Szentkirályi, F. (1992) Spatio-temporal patterns of brown lacewings based on the Hungarian light trap network (Insects: Neuroptera: Hemerobiidae). In *Current Research in Neuropterology, Proceedings of the 4th International Symposium on Neuropterology*, ed. Canard, M., Aspöck, H. & Mansell, M.W., pp. 349–357. Sacco, Toulouse.

Szentkirályi, F. (1997) Seasonal flight patterns of some common brown lacewing species (Neuroptera, Hemerobiidae) in Hungarian agricultural regions. *Biologia (Bratislava)* 52, 291–302.

Szentkirályi, F. (1998) Monitoring of lacewing assemblages (Neuroptera: Chrysopidae, Hemerobiidae) by light trapping in region of Körös-Maros National Park. *Crisicum* 1, 151–167. (in Hungarian)

Szentkirályi, F. (1999) Long-term Insect Monitoring System (LIMSYS) based on light trap network. In *Long-Term Ecological Research in the Kiskunság, Hungary*, ed. Kovács-Láng, E., Molnár, E., Kröel-Dulay, G. & Barabás, S., pp. 22–24. Kiskun LTER, Institute of Ecology and Botany of the Hungarian Academy of Sciences, Vácrátót.

Szentkirályi, F. & Kozár, F. (1991) How many species are there in apple insect communities?: testing the resource diversity and intermediate disturbance hypotheses. *Ecological Entomology* 16, 491–503.

Szentkirályi, F., Leskó, K. & Kádár, F. (1998) Effects of

droughty years on the long-term fluctuation pattern of insect populations. In *Forest and Climate*, ed. Tar, K. & Szilágyi, K., pp. 94–98. (in Hungarian)

Sziráki, G. (1996) Ecological investigations of the Neuropteroidea of oak forests in Hungary (Insecta: Raphidioptera, Neuroptera). In *Pure and Applied Research in Neuropterology, Proceedings of the 5th International Symposium on Neuropterology*, ed. Canard, M., Aspöck, H. & Mansell, M., pp. 229–232. Sacco, Toulouse.

Tamaki, Y. (1997) Chemistry of the test cover. In *Soft Scale Insects: Their Biology, Natural Enemies and Control*, ed. Ben-Dov, Y. & Hodgson, C.J., pp. 55–72. Elsevier Science, Amsterdam.

Tauber, C.A. & Tauber, M.J. (1986) Ecophysiological responses in life-history evolution: evidence for their importance in a geographically widespread insect species complex. *Canadian Journal of Zoology* 64, 875–884.

Tauber, C.A., Tauber, M.J. & Milbrath, L.R. (1995) Individual repeatability and geographical variation in the larval behaviour of the generalist predator, *Chrysopa quadripunctata. Animal Behaviour* 50, 1391–1403.

Tauber, M.J. & Tauber, C.A. (1993) Adaptations to temporal variation in habitats: categorizing, predicting, and influencing their evolution in agroecosystems. In *Evolution of Insect Pests: Patterns of Variation*, ed. Kim K.C. & McPheron, B.A., pp. 103–127. Wiley, New York.

Treacy, M.F., Benedict, J.H., Lopez, J.D. & Morrison, R.K. (1987) Functional response of a predator (Neuroptera: Chrysopidae) to bollworm (Lepidoptera: Noctuidae) eggs and on smoothleaf, hirsute, and pilose cottons. *Journal of Economic Entomology* 80, 376–379.

Tsibulskaya, G.N., Kryzhanovskaja, T.V. & Lam, P.V. (1977) Neuropteroidea inhabiting windbreaks in Kiev region. *Revue Entomologie URSS* 56, 758–761. (in Russian)

Vas, J., Ábrahám, L. & Markó, V. (1990) Study of nocturnal and diurnal activities of lacewings (Neuropteroidea: Raphidioptera, Neuroptera) by suction trap. *Acta Phytopathologica et Entomologica Hungarica* 34, 149–152.

Vet, L.E.M. & Dicke, M. (1992) Ecology of infochemicals use by natural enemies in a tritrophic context. *Annual Review of Entomology* 37, 141–172.

Vinson, S.B. & Scarborough, T.A. (1989) Impact of the imported fire ant on laboratory populations of cotton aphid (*Aphis gossypii*) predators. *Florida Entomologist* 72, 107–111.

Völkl, W. (1988) Food relations in the Cirsium–aphid systems: the role of the thistles as a base for aphidophagous insects. In *Ecology of Effectiveness of Aphidophaga*, ed. Niemczyk, E. & Dixon, A.F.G., pp. 145–148. SPB Academic Publishing, The Hague.

Wang, Y. & Yue, Y. (1998) Efficacy of interplant and mixture sowing of maize and soybean on pest disease management in soybean. *Plant Protection* 24, 13–15. (in Chinese)

Wattebled, S., Bitsche, J. & Rousset, A. (1978) Ultrastructure of pheromone-producing eversible vesicles in males of *Chrysopa perla* L. (Insecta, Neuroptera). *Cell and Tissue Research* 194, 481–496.

Whittington, A.E. & Brothers, D.J. (1991) Notes on the biology of *Mallada handschini* (Navás) and comparisons with other southern African species (Neuroptera: Chrysopidae). *Annals of Natal Museum* 32, 215–220.

Wiens, J.A., Addicott, J.F., Case, T.J. & Diamond, J. (1986) Overview: the importance of spatial and temporal scale in ecological investigations. In *Community Ecology*, ed. Diamond, J. & Case, T.J., pp. 145–153. Harper & Row, New York.

Williams, C.B. & Killington, F.J. (1935) Hemerobiidae and Chrysopidae (Neur.) in a light-trap at Rothamsted Experimental Station. *Transactions of the Society for British Entomology*, 2, 145–150.

Wilson, L.T. & Gutierrez, A.P. (1980) Within-plant distribution of predators on cotton: comments on sampling and predator efficiencies. *Hilgardia* 48, 3–11.

Wu, G., Chen, Z., Ji, M., Dong, S., Li, H., An, J. & Shi, J. (1991) Influence of interplanting corn in cotton fields on natural enemy populations and its effects on pest control in southern Shananxi. *Chinese Journal of Biological Control* 7, 101–104. (in Chinese)

Wu, Q. (1986) Investigation on the fluctuations of dominant natural enemy populations in different cotton habitats and integrated application with biological agents to control cotton pests. *Natural Enemies of Insects* 8, 29–34. (in Chinese)

Zelený, J. (1965) Lacewings (Neuroptera) in cultural steppe and the population dynamics in the species *Chrysopa carnea* Steph. and *Chrysopa phyllochroma* Wesm. *Acta Entomologica Bohemoslovaka* 62, 177–194.

CHAPTER 6

Natural food and feeding habits of lacewings

M. Canard

6.1 INTRODUCTION

Feeding habits of lacewings are relatively well known in chrysopids but understudied in other groups. The review presented here concerns the ecological and behavioural aspects of feeding; the physiological aspects, the nutritional requirements as well as the artificial trophic relationships induced by mass-production and agricultural use of lacewings will be given elsewhere (see Chapter 11, this volume).

Feeding in Neuroptera larvae shows peculiarities derived from the structure of mouthparts and gut of which fine and complete descriptions are available (e.g. Killington, 1936; Rousset, 1966). Mandibles and maxillae are elongate, interlocked together to form strong jaws acting as sucking tubes. The jaws are longer than or approximately as long as the head and inwardly curved in green and brown lacewings (Fig. 6.1a), shorter and straight in dustywings (Fig. 6.1b). The buccal opening never functions; it is mechanically closed by the integument of the head welded just after hatching and each ecdysis, so that the sole communication to the pharynx crosses within the grooves of the jaws, and the food must be entirely fluid to be ingested. To do so efficiently, venom glands at the base of maxillae (Gaumont, 1965) (Fig. 6.1c) inject diastasic salivary secretions to predigest the tissues of the prey which are torn by backward-and-forward movements of the maxillae on the mandibles, acutely pointed and serrated near the apex of the mandibles (chrysopids, Fig. 6.1a) or of the maxillae (coniopterygids, Fig. 6.1b). After a preliminary preoral digestion, the juices of the prey are then sucked up. The digestive tube is closed behind the midgut, so that the hindgut is not functional. All insoluble remnants of food are stored during the whole pre-imaginal life in the distal part of the midgut and ejected by the newly emerged adult after imaginal ecdysis and full completion. Soluble remains of digestion are carried forward by haemolymph and are drained off by the Malpighian tubules.

6.2 FEEDING BY GREEN LACEWING LARVAE

The different phases in prey consumption by chrysopid larvae have often been described in detail and were reviewed by Principi & Canard (1984) and Canard & Duelli (1984). Since that time, no more major knowledge has appeared. The following sequence is the standard of various predators:

1. Active search leading to contact with the prey. Most green lacewing larvae are frequently active, scrambling up and down, mainly by night, either slowly as in *Dichochrysa* spp., or very quickly as *Nineta flava* does. When the larva is hungry, its activity increases and it seeks to find prey, this activity increasing in step-wise function with the duration of starvation (Sengonca *et al.*, 1995), but decreasing later until a comatose state is reached if no prey is encountered. The discovery of potential prey happens at random, independent of light, slightly stimulated by honeydew of sap-sucking insects or lepidopteran scales, always within a very short distance.

2. Prey identification. Actual discovery of food is conditioned by contact by the palpi and/or the antennae, followed by probing by the jaws, achieving chemosensory positive recognition.

3. Prey catching. Taking the prey is the outcome of fixed phases performed by the larva: slow approach to the target; stopping; opening the jaw; sudden attack by throwing the head forward and simultaneous jaw nipping, most often induced by moving of the prey; the catch is sometimes followed by lifting the prey.

4. Prey sucking. As noted above, salivary secretions are injected into the prey body while the internal tissues are lacerated in order to pound and to liquefy the content and to make available the nutritive juice which is then drawn up.

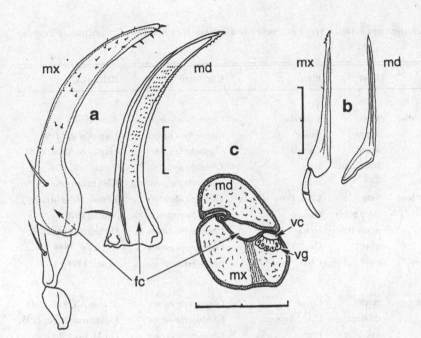

Fig. 6.1. Jaws of lacewings. a, In chrysopid, ventral view; b, in coniopterygid, ventral view; c, in chrysopid, section at basis. fc = food canal; md = mandible; mx = maxilla; vc = venom canal; vg = venom gland. Scales = 200μm.

5   Cleaning and post-prandial rest. The meal ends when a sufficient quantity of food is ingested, leading to satiety. The larva then cleans its mouthparts by rubbing them together and on the substrate. Resting posture is maintained until activity is resumed.

Behavioural factors sometimes influence the inability to feed. The prey, although estimated to be edible, may be too big; the hunting larva then fails to catch it, or sometimes climbs on and/or is itself lifted on the 'prey' so that it feeds as a temporary ectoparasite. The prey also may exhibit defence reactions, such as sudden movements with legs, or in Aphididae, release of defensive secretions by cornicles gluing the mouthparts if not the whole head of the predator.

The range of possible prey for the larva is broad in most species, as is often emphasised in the literature (e.g. Pariser, 1919; Killington, 1936; Principi & Canard, 1984; Hérard, 1986). A superficial look at such polyphagy would be confusing because it is not easy to separate occasional (if not wrong) observations from actual trophic relationships. In defined ecological conditions, the predators probably are more specific in their host relationships than was formerly and commonly believed; however, nutritional requirements may be better ensured by several prey than by only one: for

instance, the combination of two prey species achieves a better result than either prey does when given separately in Ceraeochrysa cubana fed on the aleyrodid Bemisia argentifolii together with the aphid Macrosiphum euphorbiae (Dean & Schuster, 1995).

There is no point in repeating extensive lists of possible prey here. Table 6.1 only gives some examples of the diversity of prey recorded in natural conditions of crop fields or urban environment, showing that the green lacewings are predominantly homopterophilic predators. But in confined experimental conditions in the laboratory or in mass-production insectaries, numerous foods are easily taken, allowing various kinds of rearing techniques (see Chapters 11 and 12, this volume). The prey are sometimes accepted even if unsuitable for growth and survival: e.g. Tetranychus gloveri, a spider-mite possibly encountered in the field, was avidly eaten by larvae of Chrysoperla rufilabris in the laboratory, but could not ensure its full development (Hydorn & Whitcomb, 1979). Conversely, an ecologically unexpected food such as cadavers of adult moth Anagasta kuehniella offered valuable nutrition and provided a means of cheap breeding and possible mass production (El Arnaouty, 1995).

However, some chrysopids are stenophagous. This is the case in (? all) Belonopterygini which have trash-carrier, myrmecophilous, and ant-feeder larvae. These

Table 6.1. *Examples of trophic relationships between green lacewings and various prey in natural conditions of crops or urban environment.*

	Prey	Host	Place	Chrysopid[a]	Reference
Homoptera					
Psyllidae	*Psylla acaciaebaileyanae*	acacia	Australia	*Chrysopa edwardsi*	New, 1982
Aleyrodidae	*Dialeurodes citri*	citrus	Turkey	*Chrysoperla carnea*	Uygun et al., 1990
Aphididae	*Aphis pomi*	apple	Canada	*Chrysoperla carnea*	Hagley & Allen, 1990
Lachnidae	*Cinara schmitscheki*	black pine	Italy	*Chrysopa dorsalis*	Pantaleoni, 1982
Pemphigidae	*Pemphigus mordvikoi*	—	India	*Cunctochrysa jubigensis*	Chakrabarti et al., 1991
Phylloxeridae	*Daktulosphaira vitifoliae*	vine	USA, Pennsylvania	*Chrysopa oculata*	Jubb & Masteller, 1977
Margarodidae	*Marchalina hellenica*	Alep pine	Greece	*Chrysopa pallens*	Santas, 1979
Pseudococcidae	*Planococcus citri*	grape fruit	Israel	*Chrysoperla carnea*	Berlinger et al., 1979
Coccidae	*Saissetia oleae*	olive	Greece	*Chrysoperla carnea*	Argyriou, 1984
Diaspididae	*Hemiberlesia lataniae*	olive	Chile	*Chrysoperla carnea*	Matta, 1979
Lepidoptera					
Noctuidae	*Spodoptera littoralis*	cotton	Egypt	*Chrysoperla carnea*	Maher-Ali et al., 1983
Plutellidae	*Plutella xylostella*	cabbage	USA, Arizona	*Chrysoperla carnea*	Eigenbrode et al., 1995
Tortricidae	*Tortrix viridana*	oak	France	*Chrysoperla carnea*	Du Merle, 1983
Yponomeutidae	*Prays oleae*	olive	Spain	*Chrysoperla carnea*	Campos & Ramos, 1985
Hymenoptera					
Tenthredinidae	*Hylotoma rosae*	rose	France	*Chrysopa perla*	Lucas, 1881
Coleoptera					
Chrysomelidae	*Leptinotarsa decemlineata*	potato	Moldavia	*Chrysoperla carnea*	Filippov, 1982
Diptera					
Anthomyiidae	*Pegomyia hyoscyami*	beet	Germany	*Chrysoperla carnea*	Groh & Tanke, 1980
Drosophilidae	*Drosophila melanogaster*	vine	France	*Chrysoperla carnea*	Baudry, 1996
Thysanoptera					
Thripidae	*Thrips tabaci*	cotton	Egypt	*Chrysoperla carnea*	Habib et al., 1980
Acarina					
Tetranychidae	*Tetranychus urticae*	hop	Czech Republic	*Chrysoperla carnea*	Zelený et al., 1981

Note:

[a] *Chrysoperla carnea* is here understood *sensu lato* as a complex of sibling species (see Chapter 3).

are worldwide-spread lacewings of the genus *Italochrysa* (Principi, 1946; New, 1983), the Neotropical *Nacarina*, and the Australian *Calochrysa* spp. (New, 1986). Other rare lacewing Chrysopini also have specialised diets, such as the American *Chrysopa slossonae* associated with the woolly alder aphid *Prociphilus tesselatus* (Tauber & Tauber, 1987; Brislow, 1988; Tauber et al., 1993), *Ceraeochrysa cincta* with the

mealybug *Plotococcus eugeniae* (Eisner & Silberglied, 1988), and the Mediterranean *Rexa lordina* with the psyllids *Euphyllura* spp. occurring on Oleaceae (Canard & Labrique, 1989); they all have slow-moving, trash-carrier larvae, covered (by crypsis) with wax produced by their prey (Eisner et al., 1978; Milbrath et al., 1993).

The amount of food that green lacewing larvae must consume to fulfil their growth is high. It depends of course on both the predator and prey species, but not or only weakly on the quality of food: a bad or poor-quality nutrition slows down growth in time and weight unless there is significant overconsumption. The quantitative nutritional requirements and the proportions eaten by the three larval stages have been variously estimated by different authors and have been previously reviewed by Killington (1936) and Principi & Canard (1984). For instance, during its larval life, laboratory-reared *Chrysoperla carnea* killed more than 12 500 eggs of *Tetranychus urticae* (Sengonca & Coeppicus, 1985); it needs an average of 312 eggs or 232 first-instar larvae of *Mamestra brassicae* to develop until cocoon spinning (Klingen *et al.*, 1996), 140 second-instar nymphs of *Myzus persicae* or 946 eggs of *Anagasta kuehniella* (El Arnaouty *et al.*, 1996), 932 eggs or 443 first-instar larvae of *Pectinophora gossypiella* (Shalaby *et al.*, 1999). The third stage always consumed the main part of the total intake, namely from 72% to 80% in the previous cited examples.

Little information is available on food consumption estimated by weight. All known data achieve in high consumption with respect to the conversion yield. *Chrysopa perla* and *Chrysoperla carnea* larvae fed on a high-quality aphid, *Myzus persicae*, swallowed during total weight gain time about 90 mg and 64 mg of fresh weight, with conversion rates of 22% and 17%, respectively; the latter eating eggs of *Mamestra brassicae* has a conversion rate of about 34% but only 15% when fed with first-instar caterpillar larvae (Klingen *et al.*, 1996). *Chrysoperla nipponensis* ate 43.3 mg of *Myzus persicae* and 23.1 mg of eggs of *Anagasta kuehniella*, with respective food efficiencies of 22.7% and 47.2% (El Arnaouty *et al.*, 1996).

Cannibalism is a natural tendency in all polyphagous green lacewing larvae as in many Neuroptera. Its occurrence is not clear in the field, but in experiments as well as in mass-rearing conditions, cannibalistic behaviour of lacewings is consistent and high, as reviewed by Canard & Duelli (1984). The main traits are the following.

1. Larvae eating eggs. Larvae, especially newly hatched larvae, may climb up the stalks of eggs which are easily pierced and emptied. When this happens, the profit for the cannibal is obvious, providing better growth and survival than an aphid. In natural situations where scarcity is the rule, this oophagy is probably rare. The pedicel in chrysopids may be a protection against oophagy (see Chapter 7).

2. Larvae eating larvae. This kind of cannibalism is a part of intraguild predation (see below), induced by starvation and scarcity of usual (? and better) prey, generally for the benefit of the larger and the more active of the two opponents. The trash-carrying habit may be a good protection against cannibalism.

3. Significance of larval cannibalism. As a function of prey availability, cannibalism may enhance or handicap the predation potential. It is estimated to be disadvantageous in optimal food conditions; but in conditions of severe scarcity of prey, it is the only possibility to prevent local extinction; it must be considered a delayed predation. In chrysopids with palyno-glycophagous habits in adults, such as *Chrysoperla carnea* (see below), oophagy may provide the possibility of surviving independently of any other food, because larvae were able to develop to maturity from ten conspecific eggs per day (Bar & Gerling, 1985).

6.3 FEEDING BY ADULT GREEN LACEWINGS

Based on some old isolated observations probably on *Chrysopa* spp., all adult green lacewings were for a long time considered predacious, essentially aphidophagous. Further studies showed that most of them have, at least partially, non-live food, mainly honeydew and other sweet juices. The mouthparts of the non-carnivorous green lacewings show adaptations to this diet: they have symmetrical mandibles without any incisor, and spoon-like lacinia for the consumption of honeydew and pollens. Their gut also is fitted for such feeding: they have at the distal part of the oesophagus a strongly muscular diverticulum (called the crop), ventilated by means of more or less large tracheal trunks providing air. In the diverticulum, symbiotic yeasts participate in first breakdown of the ingested saccharides, enabling them to be assimilated (Hagen & Tassan, 1966, 1972). The yeasts are picked up after emergence with food and/or by trophallaxis. Hagen *et al.* (1970) identified the yeasts as *Torulopsis* sp. and measured the diameter of the tracheal trunks supplying air to the diverticulum in some American predacious green lacewing species. Later, Neuenschwander *et al.* (1981) also measured the tracheal trunks in nine eastern Mediterranean green lacewing species, and Canard *et al.* (1990) established in 26 west European species an index giving an estimation of 'diverticulum air supply' independent of the size:

$$C = \frac{\text{diverticulum tracheal trunk diameter in } \mu\text{m}}{\text{forewing length in mm}}$$

characterising the availability of oxygen and correlatively, the importance of yeast participation in palyno-glycophagy by the species.

Most adult green lacewings have a glycophagous diet. This regime, sometimes mistakenly classified as phytophagous (Stelzl, 1991, 1992), is composed of both products of vegetable origin (nectar, various exudates) and insect honeydew. Pollen grains are commonly found in the gut contents, sometimes in such quantity that the occurrence did not seem to be by chance, like in *Chrysoperla carnea* and *Dichochrysa prasina* (Bozsik, 1992). The regime may thus be qualified palyno–glycophagous. But in many other cases, the diversity and the small quantity of pollen leads to the belief that they are swallowed accidentally being glued in the sticky juices and not systematically collected on flowers. In addition, one can always find other various particles like conidia and spores, unexpected fragments of arthropods, lepidopteran scales, and mineral fragments that lacewings must have taken when scraping honeydew from leaves and twigs (Bozsik, 1992). All these green lacewings have a strong ventilation in the diverticulum. The relevant air supply indices are high: for instance in the European *Nothochrysa* spp., $C = 7$ to 10 in both males and females, or in the females of the European *Dichochrysa* spp., $C = 6$ to 7. In some palyno-glycophagous species, there are large sexual differences in air supply to the diverticulum between males and females manifest by considerable differences between their nutritional requirements and digestive ability; for instance, the C value of *Chrysoperla mediterranea* is 2.3 in males and 6.0 in females, and *Dichochrysa ventralis* 1.7 in males and 7.6 in females (Canard *et al.*, 1990).

In glyco- and palyno-glycophagous adult green lacewings, food detection is achieved by an anemo-chemotactic approach induced by the scent plume issued from the plants; the kairomone may be produced either by the plant itself in cases of high level of substrate specificity such as in *Nineta pallida* and *Peyerhimoffina gracilis* closely associated with conifers (Monserrat & Marín, 1994), or more acutely by honeydew for eurytokous species such as *Chrysoperla carnea* (Duelli, 1987). The main volatile agent is L-tryptophan, a breakdown product of honeydew (van Emden & Hagen, 1976; Hagen *et al.*, 1976) which induces modifications in searching behaviour (McEwen *et al.*, 1993, 1994). Consequent field manipulations of natural and/or artificial volatile cues gave rise to agricultural developments in attracting (Brettell & Burgess, 1973; Harrison & McEwen, 1998) and managing natural populations of green lacewings.

However, some adult green lacewings have a tendency to be carnivorous: among 75 genera reviewed by Brooks & Barnard (1990), only three Chrysopini genera are qualified 'insectivorous', namely *Anomalochrysa* spp. occurring in the Hawaiian Islands, *Atlantochrysa* spp. occurring in the Atlantic Islands, and the widespread *Chrysopa* spp. The last genus shows adaptations to predation in relatively larger (Canard *et al.*, 1990) and asymmetrical mandibles with a strong incisor on the left side used to capture and to hold prey (Stelzl, 1992), and in a low-ventilated diverticulum ($C = 0.7$–2). Though *Chrysopa* green lacewings are usually considered aphidophagous, analysing the gut content leads to quite different conclusions: (1) the prey may be diverse including aphids as basic food, but also includes soft scales, armoured scales, and mites, manifesting an oligophagy with respect to the possible soft-bodied arthropod prey; (2) other items are regularly found in the gut, such as pollens of various plants, yeasts, and spores of fungi, appearing in such quantities and such frequency that it cannot be accidental; (3) for these reasons, these green lacewings are classified as omnivorous (Stelzl, 1991, 1992). However, not all *Chrysopa* spp. have the same feeding habits: for instance the common *Chrysopa pallens* eats more food of plant origin than the truly carnivorous *C. perla* (Bozsik, 1992), as well as *C. viridana* which shows tracheal trunks much more enlarged relative to the diverticulum ($C = 1.7$ in males and 3 in females).

The carnivorous females of *Chrysopa* may show typical cannibalistic behaviour by eating their own eggs encountered when hungry. However, they sometimes eat their own eggs at the tip of the genital duct before depositing them on the pedicel. This peculiar auto-oophagy, analysed in *Chrysopa perla* by Philippe (1972), does not result from hunger, but is a consequence of trouble in reproduction (lack of adequate secretion to produce normal stalks) due to an unusual delay in or a lack of copulation.

Some other species of green lacewings are fairly phytophagous, eating exclusively pollens (Brooks & Barnard, 1990). They all belong to the four nothochrysine genera, *Hypochrysa*, *Kimochrysa*, and *Pamochrysa* occuring in southern Africa (Tjeder, 1966), and *Pimachrysa* which are brown or black chrysopids from southwestern North America. The Palaearctic *Hypochrysa elegans*, whose foregut conformation and gut content were studied by Canard *et al.* (1990), has a peculiar morphological structure with no special tracheal trunks connecting the diverticulum in males and only very small ones in females; its air supply index to diverticulum is $C=0$ in males and $C=1.5$ in females reflecting the quite different process of digestion of pollens.

6.4 FEEDING BY BROWN LACEWINGS AND DUSTYWINGS

It seems that larvae of hemerobiids, as far is known, do not differ from the general predation scheme of chrysopids. Coniopterygid larvae are poorly investigated; nevertheless, their tiny size, the straightness of the jaws hindering the holding of prey, and the weak motility of the head do not allow the dustywing larvae to lift the prey. Mechanisms for the discovery of prey are unknown and may be supposed as in chrysopids to be randomly conditioned.

Hemerobiids and coniopterygids are always predacious, adults as well as larvae, feeding on slow-moving soft-bodied arthropods (Tjeder, 1961; Stelzl, 1990; Penny *et al.*, 1997). Differentiating larval and imaginal diets would be illusory, as also would be a separation of their predatory activity in the biological control of pests. As they occur on the same substrate, they both participate in reducing herbivore populations. Prey that is toxic but not rejected is also known in hemerobiids: *Nusulala uruguaya* cannot achieve larval development when fed on the aphid *Toxoptera citricidus* and does not spin cocoon when fed with *Brevicoryne brassicae* (Souza *et al.*, 1989).

In many brown lacewings, the true feeding requirements remain poorly known. The gut content of all species that have been analysed included aphid remains. However, the adults of brown lacewings do not only prey on aphids, but on various other arthropods. For example, some observations of relationships between noxious arthropods and natural populations of lacewings are given in Table 6.2. Most of them are Homoptera, but one also may find spider-mites as has been ascertained with *Brauerobius marginatus*, *Hemerobius humulinus*, *H. lutescens*, *H. micans*, *H. perelegans*, *H. pini*, *H. stigma*, *Micromus variegatus*, and *Sympherobius pygmaeus*. Occasionally the gut contains more surprising particles, such as fragments of Diptera (nematocerous antenna segments) and lepidopteran scales, evidently swallowed by accident. Nevertheless, some adult brown lacewings are fairly omnivorous. *Drepanepteryx phalaenoides*, *Hemerobius lutescens*, *H. nitidulus*, and *Micromus lanosus* are able to feed partially but regularly on pollen and honeydew (Stelzl, 1990, 1991). Diversity in possible prey occurs in several situations in which hemerobiids and coniopterygids are valuable biological control agents (see Parts 2 and 3 of this volume).

Substrate specificity commonly registered in the members of the family Hemerobiidae (e.g. Killington, 1936, 1937; Zelený, 1978; Monserrat & Marín, 1996) indicates that most species occur as larvae and adults in a well-defined microhabitat. They consequently depend on the same prey, often with a relatively straight food choice (stenophagy), especially on conifers. Other species even refused to feed on aphids and only accepted coccids, such as *Sympherobius amiculus* which seems specific of the *Dactylopius* of cactus (Mann, 1969), *S. angustus* of *Planococcus citri* (Colle, 1933), and *S. domesticus* of various mealybugs in Japan (Murakami, 1963).

The amount of prey eaten is always high. Related data are scattered in several papers and feature very different aspects of voracity. As examples, when the prey are small, and without estimating the overconsumption, the larva of *Hemerobius stigma* killed (? and consumed) an average of about 3000 eggs and first-instar larvae of various adelgids, and the adult male and female of the same brown lacewing, more than 13000 and 17000 respectively during two months of their adult life spans (Laidlaw, 1936). *Hemerobius pacificus* larva needed 350 various aphids to pupate (Moznette, 1915), while *Nusalala uruguaya* larva needed 135

Table 6.2. *Examples of prey eaten in natual conditions by larvae and/or adults of some hemerobiids in various countries*

	Prey	Host	Place	Hemerobiid	Reference
Homoptera					
Psyllidae	*Psylla pyricola*	pear	Canada	*Hemerobius pacificus*	McMullen & Jong, 1967
	Psylla acaciaebaileyanae	acacia	Australia	*Drepanacra binocula*	New, 1984
Triozidae	*Trioza vitreoradiata*	pittosporum	New Zealand	*Drepanacra binocula*	Wise, 1996
Aleyrodidae	*Aleurothrixus floccosus*	citrus	Brazil	*Megalomus* sp.	Chagas et al., 1982
Aphididae	Alfalfa aphids	lucerne	USA, California	*Hemerobius ovalis*	
				Hemerobius pacificus	Neuenschwander et al., 1975
	Acyrthosiphon pisum	lucerne	Czech Republic	*Micromus variegatus*	Zelený, 1965
		lucerne	New Zealand	*Micromus tasmaniae*	Syrett & Penman, 1980
	Myzus persicae	artichoke	USA, California	*Hemerobius pacificus*	Neuenschwander & Hagen, 1980
Lachnidae	Various	pine	Italy	*Wesmaelius submebulosus*	Pantaleoni, 1982
Pemphigidae	*Pemphigus spirothecae*	poplar	Italy	*Sympherobius pygmaeus*	Pantaleoni, 1982
	Eriosoma lanigerum	apple	USA, Ohio	*Hemerobius humulinus*	Holdsworth, 1970a
Phylloxeridae	*Phylloxera punctata*	oak	UK	*Sympherobius pygmaeus*	Withycombe, 1924
Adelgidae	*Adelges piceae*	pine	Canada	*Hemerobius stigma*	Garland, 1981
Margarodidae	*Matsucoccus matsumurae*	pine	China	*Sympherobius* sp.	
			Palestine	*Hemerobius lacuneris*	Weijun et al, 1983
Pseudococcidae	*Planococcus citri*	citrus		*Sympherobius amicus*	Rivnay, 1943
	Pseudococcus comstocki	—	Japan	*Sympherobius domesticus*	Murakami, 1963
Coccidae	*Saissetia nigra*	—	USA, California	*Hemerobius pacificus*	Flanders, 1959
Psocoptera	Various psocids	—	UK	*Hemerobius lutescens*	Withycombe, 1923
				Hemerobius stigma	
Lepidoptera					
Noctuidae	*Heliothis* spp.	cotton	Australia	*Micromus tasmaniae*	Samson & Blood, 1970
	Heliothis virescens	cotton	Peru	*Sympherobius* sp.	Aguilar & Lamas, 1980
Olethreutidae	*Laspeyresia pomonella*	apple	USA, Ohio	*Hemerobius pacificus*	
				Micromus posticus	Holdsworth, 1970b
Pterophoridae	*Platyptilia carduifolia*	artichoke	USA, California	*Hemerobius pacificus*	Neuenschwander & Hagen, 1980
Acarina					
Tetranychidae	*Paratetranychus urticae*	apple	USA	*Hemerobius pacificus*	Moznette, 1915
	Paratetranychus telarius	hop	Hungary	*Hemerobius humulinus*	Szabó & Szentkirályi, 1981

Dactynotus sp. to grow and an average of 4.7 aphids per day for adults to reproduce (Souza, 1988). *Wesmaelius subnebulosus* larvae consumed 160 and 190 *Myzus persicae*, or 57 and 72 *Aphis fabae* depending on the future sex of the individual, male or female respectively; the male and female adults consumed 9.5 or 17.2 *A. fabae* per day, respectively (Laffranque, 1973; Laffranque & Canard, 1975). *Hemerobius nitidulus* required about 80 full-grown *A. rumicis* (Withycombe, 1923) whereas *Micromus posticus* was satisfied with about 40 aphids during its larval life and an average of 10 aphids per day as an adult (Cutright, 1923). Adult males and females of *Sympherobius maculipennis* consumed about 5 and 15 nymphs of the mealybug *Phenacoccus manihoti* per day, respectively (B. Löhr & A.M. Varela, unpublished data). Conversion rate in hemerobiids probably is close to that of green lacewings: the growth efficiency expressed as the ratio of the dry weights of weight increase and consumed prey weight is estimated 41% to 43% in *Drepanacra binocula* and 34% to 42% in *Micromus tasmaniae*, as a function of the kind of prey (New, 1984).

The adult dustywings as other neuropteran insects were considered predaceous, as far as is known. Little information on the bionomics of these beneficial insects is available, mainly because of their size and their activity near dark and dusk which render them often overlooked in the field. Many species are closely associated with a specific species of tree or bush in the Iberian Peninsula (Monserrat & Marín, 1992) indicating a permanent relation with one or a few types of prey. The larvae and adults together feed on sessile food or tiny arthropods (Table 6.3) which are caught by adults in grooves on the maxillary palps acting as a comb (Stelzl, 1992). Most species have slow-moving larvae, the jaws of which are straight (Fig. 6. 1b) and not able to catch and lift the prey. Prey is most often motionless such as eggs, post-embryonic stages of spider-mites and eriophyids (Stelzl, 1991), or phylloxerids. Adults also seem able to feed on honey and/or honeydew (Lacroix, 1924; Stelzl, 1991).

6.5 INFLUENCE OF FOOD ON REPRODUCTIVE POTENTIAL, TRITROPHIC RELATIONS

The amount but also the intrinsic quality of the food eaten by larvae and adults of lacewings strongly influ-ence all development parameters, survival, reproduction ability, fecundity, and fertility, as reviewed for chrysopids by Principi & Canard (1984) and Rousset (1984). Little is known for hemerobiids and coniopterygids. Some aspects of these interactions are investigated in Chapter 7 (this volume).

The kind of plant host on which the aphid prey are feeding may also induce variations in the trophic relations with the predator. Indeed, the deleterious effects were more commonly registered in development time, survival rate or other factors. However, an example of an all-or-none ability to feed (rejection) is known in *Chrysoperla carnea* whose larvae ate *Aphis sambuci* living on *Sambucus nigra*, but refused it when growing on *Philadelphus coronarius* (Bänsch, 1964).

6.6 LACEWINGS IN THE APHIDOPHAGOUS GUILD

Lacewings are predators which may kill and consume other natural enemies belonging to the same guild, i.e. exploiting the same prey/host (Polis *et al.*, 1989; Polis & Holt, 1992; Rosenheim *et al.*, 1993). Their polyphagy predisposes them to eat guild co-members (Phoofolo & Obrycki, 1998). Induced effects in aphid population dynamics are ambiguous, either estimated as a negative factor in biological control (Rosenheim & Wilholt, 1995), or an advantage in surviving to drastic trophic conditions by feeding on other aphid antagonists. *Chrysoperla carnea* (Sengonca & Frings, 1985) and *C. rufilabris* (Lucas *et al.*, 1997) turn against coccinellids. Other green lacewings such as *Chrysopa pallens* sometimes pierce and feed from the eggs of *Nineta flava*, indiscriminately either healthy or parasitised by the scelionid *Telenomus chrysopae*, acting in this last case as a positive factor within the complex of the aphidophagous guild (Canard, 1984). Cannibalistic behaviour (see above) is also a part of intraguild predation. Lastly, lacewings may feed on lepidopterous eggs already parasitised by embryonic parasitoids, as for example *Chrysoperla carnea* (Alrouechdi & Vogelé, 1981) or *Brinckochrysa scelestes* (Krishnamoorthy & Mani, 1985) which ate eggs harbouring *Trichogramma* spp.; by so doing they reduced the natural or artificial pressure exerted by the antagonist on some pest populations.

Table 6.3. *Examples of prey eaten in natural conditions by larvae and/or adults of some coniopterygids in various countries*

Prey	Host	Place	Coniopterygid	Reference
Homoptera				
Aleyrodidae				
Dialeurodes citri	citrus	Turkey	*Coniopteryx hageni*	Soylu, 1980
			Coniopteryx psociformis	
			Semidalis aleyrodiformis	Pantaleoni, 1982
		Italy	*Coniopteryx esbenpetersemi*	
Phylloxeridae				
Phylloxera punctata	oak	UK	*Coniopteryx psociformis*	Killington, 1936
			Semidalis aleyrodiformis	Killington, 1936
Coccidae				
Lichtensia viburni	ivy	UK	*Coniopteryx psociformis*	Wise, 1996
Pseudococcidae				
Tryonimus wisei	apple	New Zealand	*Cryptoscenea australiensis*	
Diaspididae				
Aspidiotus abietis				
Leucaspis pini	pine	Germany	*Aleuropteryx loewi*	Lacroix, 1924
		UK	*Aleuropteryx lutea*	Killington, 1936
Carulaspis minima	juniper	USA, Pennsylvania	*Aleuropteryx juniperi*	Henry, 1976
Carulaspis juniperi	juniper	USA, Pennsylvania	*Aleuropteryx similima*	Wheeler, 1980
	juniper	Italy	*Aleuropteryx juniperi*	
	cypress	Italy	*Semidalis pseudouncinata*	De Marzo & Pantaleoni, 1994
Acarina				
Tetranychidae				
Panonychus ulmi	peach	Italy	*Coniopteryx esbenpetersemi*	Castellari, 1980
	apple	—	*Coniopteryx pineticola*	
Bryobia praetiosa	oak	UK	*Coniopteryx psociformis*	Principi & Canard, 1974
Metatetranychus citri	citrus	Georgia	*Semidalis aleyrodiformis*	Killington, 1936
Paratetranychus telarius	—	USA, California	*Coniopteryx psociformis*	Agekyan, 1965
			Coniopteryx hageni	Quayle, 1913

REFERENCES

Agekyan, N.G. (1965) On a little known entomophagous insect of the Georgian Soviet Republic – the neuropteran *Conwentzia psociformis* Curt. *Entomologicheskoye Obozrenie* 44, 84–88. (in Russian)

Aguilar, F. & Lamas, C. (1980) Apuntes sobre el control biológico y el control integrado de las plagas agrícolas en el Peru. 2. El cultivo del algodonero. *Revista Peruana de Entomología* 23, 91–97.

Alrouechdi, K. & Vogelé, J. (1981) Prédation des trichogrammes par les chrysopides. *Agronomie* 1, 187–189.

Argyriou, L. (1984) The soft scales of olive tree in Greece. In *Integrated Pest Control in Olive-Groves*, ed. Cavalloro, R. & Crovetti, A., pp. 147–151. Balkema, Rotterdam.

Bänsch, R. (1964) Vergleichende Untersuchungen zur Biologie und zum Beutefangverhalten aphidivorer Coccinelliden, Chrysopiden und Syrphiden. *Zoologische Jarbücher, Systematik, Ökologie und Geographie der Tiere* 91, 271–340.

Bar, D. & Gerling, D. (1985) Cannibalism in *Chrysoperla carnea* (Stephens) (Neuroptera: Chrysopidae). *Israel Journal of Entomology* 19, 13–22.

Baudry, O. (1996) *Reconnaître les Auxiliaires: Vergers et Vignes / Recognizing Natural Enemies: Orchards and Vineyards*. CTIFL, Paris.

Berlinger, M.J., Tzahor, J. & Gol'berg, A.M. (1979) Contribution to the phenology of *Chilochorus bipustulatus* L. (Coccinellidae) in citrus groves and the control of *Planococcus citri* (Risso) (Pseudococcidae) in Israel. In *31st International Symposium on Crop Protection, Medelingen van de Faculteit Landbouwwetenschappen Rijksuniversiteit Gent* 44, 49–54.

Bozsik, A. (1992) Natural adult food of some important *Chrysopa* species (Planipennia: Chrysopidae). *Acta Phytopathologica et Entomologica Hungarica* 27, 141–146.

Brettell, J.H. & Burgess, M.W. (1973) A preliminary assessment of the effect of some insecticides on predators of cotton pests. *Rhodesia Agricultural Journal* 70, 103–104.

Brislow, C.M. (1988) What makes a predator specialize? *Trends in Ecology and Evolution* 3, 1–2.

Brooks, S.J. & Barnard, P.C. (1990) The green lacewings of the world: a generic review (Neuroptera: Chrysopidae). *Bulletin of the British Museum (Natural History) Entomology* 59, 117–286.

Campos, M. & Ramos, P. (1985) Some relationships between the numbers of *Prays oleae* eggs laid on olive fruits and their predation by *Chrysoperla carnea*. In *Integrated Pest Control in Olive-Groves*, ed. Cavalloro, R. & Crovetti, A., pp. 237–241. Balkema, Rotterdam.

Canard, M. (1984) Écologie des pontes de *Nineta flava* (Scopoli) (Neuroptera: Chrysopidae): disposition et facteurs biotiques antagonistes. In *Progress in World's Neuropterology, Proceedings of the 1st International Symposium on Neuropterology*, ed. Gepp, J., Aspöck, H. & Hölzel, H., pp. 253–260. Thalerhof, Graz.

Canard, M. & Duelli, P. (1984) Predatory behavior of larvae and cannibalism. In *Biology of Chrysopidae*, Series Entomologica 27, ed. Canard, M., Séméria, Y. & New, T.R., pp. 92–100. Dr W. Junk, The Hague.

Canard, M. & Labrique, H. (1989) Bioécologie de la chrysope méditerranéenne *Rexa lordina* Navás (Neuroptera: Chrysopidae) et description de ses stades larvaires. *Neuroptera International* 5, 151–158.

Canard, M., Kokubu, H. & Duelli, P. (1990) Tracheal trunks supplying air to the foregut and feeding habits in adults of European green lacewing species (Insecta: Neuroptera: Chrysopidae). In *Advances in Neuropterology, Proceedings of the 3rd International Symposium on Neuropterology*, ed. Mansell, M.W. & Aspöck, H., pp. 227–286. South African Department of Agricultural Development, Pretoria.

Castellari, P.L. (1980) Indagini biologiche su *Coniopteryx (Metaconiopteryx) esbenpeterseni* Tjeder (Neur.: Coniopterygidae), predatore di Acari Tetranychidi sul pesco. *Bollettino dell'Istituto di Entomologia dell'Università di Bologna* 35, 157–180.

Chagas, E.F. das, Silveira Neto, S., Braz, A.J.B.P., Mateus, C.P.B. & Coelho, I.P. (1982) Fluctuação populacional de pragas e predadores em citros. *Pesquisa Agropecuária Brasileira* 17, 817–824.

Chakrabarti, S., Debnath, N. & Ghosh, D. (1991) Developmental rate, larval voracity and oviposition of *Cunctochrysa jubigensis* (Neuroptera: Chrysopidae), an aphidophagous predator in the western Himalaya. In *Behaviour and Impact of Aphidophaga*, ed. Polgár, L., Chambers, R.J., Dixon, A.F.G. & Hodek, I., pp. 103–113. SPB Academic Publishing, The Hague.

Cole, F.R. (1933) Natural control of the citrus mealybug. *Journal of Economic Entomology* 26, 855–864.

Cutright, C.R. (1923) Life history of *Micromus posticus* Walker. *Journal of Economic Entomology* 16, 448–456.

Dean, D.E. & Schuster, D.J. (1995) *Bemisia argentifolii* (Hom.: Aleyrodidae) and *Macrosiphum euphorbiae* (Hom.: Aphididae) as prey for two species of Chrysopidae. *Environmental Entomology* 24, 1562–1568.

De Marzo, L. & Pantaleoni, R.A. (1994) Sulla presenza in Italia di *Aleuropteryx juniperi* Ohm e *Semidalis pseudouncinata* Meinander (Neuroptera:

Coniopterygidae), predatori oligofagi di coccidei delle Cupressaceae. In *Atti del Convegno Lotta Biologica*, ed. Viggiani, G., pp. 107–111. Ministero dell'Agricoltura e delle Foreste, Rome.

Duelli, P. (1987) The influence of food on the oviposition-site selection in a predatory and a honeydew-feeding lacewing species (Planipennia: Chrysopidae). *Neuroptera International* 4, 205–210.

Du Merle, P. (1983) Les facteurs de mortalité des œufs de *Tortrix viridana* L. (Lep.: Tortricidae). 1. Le complexe des prédateurs (Hym.: Formicidae; Orth.: Phaneropteridae; Neur.: Chrysopidae). *Agronomie* 3, 239–246.

Eigenbrode, S.D., Moodie, S. & Castagnola, T. (1995) Predators mediate host plant resistance to a phytophagous pest in cabbage with glossy leaf wax. *Entomologia Experimentalis et Applicata* 77, 335–342.

Eisner, T. & Silberglied, R.E. (1988) A chrysopid larva that cloaks itself in mealybug wax. *Psyche* 95, 15–19.

Eisner, T., Hicks, K., Eisner, M. & Robson, D.S. (1978) 'Wolf-in-sheep's clothing' strategy of a predaceous insect larva. *Science* 199, 790–794.

El Arnaouty, S.A. (1995) Adult cadavers of *Ephestia kuehniella* as a source of food for *Chrysoperla carnea* 3rd instar larvae. *Egyptian Journal of Biological Control* 5, 11–13.

El Arnaouty, S.A., Ferran, A. & Beyssat-Arnaouty, V. (1996) Food consumption by *Chrysoperla carnea* (Stephens) and *Chrysoperla sinica* (Tjeder) of natural and substitute prey: determination of feeding efficiency (Insecta: Neuroptera: Chrysopidae). In *Pure and Applied Research in Neuropterology, Proceedings of the 5th International Symposium on Neuropterology*, ed. Canard, M., Aspöck, H. & Mansell, M.W., pp. 109–117. Sacco, Toulouse.

van Emden, H.F. & Hagen, K.S. (1976) Olfactory reactions of the green lacewing *Chrysopa carnea* to tryptophan and certain breakdown products. *Environmental Entomology* 5, 469–473.

Filippov, N. (1982) Integrated control of pests of vegetable crop grown in the open in Moldavia. *Acta Entomologica Fennica* 40, 6–9.

Flanders, S.E. (1959) Biological control of *Saissetia nigra* (Nietn.) in California. *Journal of Economic Entomology* 52, 596–600.

Garland, J.A. (1981) [1978] Reinterpretation of information on exotic brown lacewings (Neuroptera: Hemerobiidae) used in biocontrol programme in Canada. *Manitoba Entomologist* 12, 25–28.

Gaumont, J. (1965) Observations sur quelques Chrysopidae (Insectes: Planipennes) prédateurs d'aphides. *Annales de l'Université et de l'APERS*, Reims 3, 24–32.

Groh, K. & Tanke, N. (1980) Untersuchungen zum Einfluss von Herbizidapplikationen und die Fauna Zuckerrüben-Anbauflächen. In: *Proceedings of the International Symposium of OILB/WPRS on Integrated Control in Agriculture and Forestry*, ed. Russ, K. & Berger, H., pp. 439–497. Vienna.

Habib, A., Rezk, G.N., Farghaly, H.T. & Ragab, Z.A. (1980) Seasonal abundance of some predators in cotton fields and its relation to certain pests. *Bulletin de la Société Entomologique d'Égypte* 60, 191–196.

Hagen, K.S. & Tassan, R.L. (1966) The influence of protein hydrolysates of yeasts and chemically defined (diets) upon the fecundity of *Chrysopa carnea* Stephens (Neuroptera). *Věstník Československé Společnosti Zoologické* 30, 219–227.

Hagen, K.S. & Tassan, R.L. (1972) Exploring nutritional roles of extracellular symbiotes on the reproduction of honeydew feeding adult chrysopids and tephritids. In *Insect and Mites Nutrition*, ed. Rodriguez, J.G., pp. 323–351. North-Holland, Amsterdam.

Hagen, K.S., Tassan, R.L. & Sawall, E.F. Jr (1970) Some ecophysiological relationships between certain *Chrysopa*, honeydews and yeasts. *Bollettino del Laboratorio di Entomologia Agraria Filippo Silvestri* 28, 113–134.

Hagen, K.S., Greany, P., Sawall, E.F. & Tassan, R.L. (1976) Tryptophan in artificial honeydew as a source of an attractant for adult *Chrysopa carnea*. *Environmental Entomology* 5, 458–468.

Hagley, E.A.C. & Allen, W.R. (1990) The green apple aphid, *Aphis pomi* Degeer (Homoptera: Aphididae), as prey of polyphagous arthropod predators. *Canadian Entomologist* 122, 1221–1228.

Harrison, S.J. & McEwen, P.K. (1998) Acid hydrolysed L-tryptophan and its role in the attraction of the green lacewing *Chrysoperla carnea* (Stephens) (Neuroptera: Chrysopidae). *Journal of Applied Entomology* 122, 334–344.

Henry, T.J. (1976) *Aleuropteryx juniperi*: a European scale predator established in North America (Neuroptera: Coniopterygidae). *Proceedings of the Entomological Society of Washington* 78, 195–201.

Hérard, F. (1986) Annotated list of the entomophagous complex associated with pear psylla *Psylla pyri* (L.) (Hom.: Psyllidae) in France. *Agronomie* 6, 1–34.

Holdsworth, R.P. Jr (1970a) Aphids and aphid enemies: effects of integrated control in an Ohio apple orchard. *Journal of Economic Entomology* 63, 530–535.

Holdsworth, R.P. Jr (1970b) Codling moth control as part of an integrated programme in Ohio. *Journal of Economic Entomology*, 63, 894–897.

Hydorn, S.B. & Whitcomb, W.H. (1979) Effects of larval

diet on *Chrysopa rufilabris. Florida Entomologist* 62, 293–298.

Jubb, G.L. & Masteller, E.C. (1977) Survey of arthropods in grape vineyards of Erie county, Pennsylvania. *Environmental Entomology* 6, 419–428.

Killington, F.J. (1936) *A Monograph of the British Neuroptera,* vol. 1. Ray Society, London.

Killington, F.J. (1937) *A Monograph of the British Neuroptera,* vol. 2. Ray Society, London.

Klingen, I., Johansen, N.S. & Hofsvang, T. (1996) The predation of *Chrysoperla carnea* (Neur.: Chrysopidae) on eggs and larvae of *Mamestra brassicae* (Lep.: Noctuidae). *Journal of Applied Entomology* 120, 363–367.

Krishnamoorthy, A. & Mani, E.C. (1985) Feeding behaviour of *Chrysopa scelestes* Banks on parasitized eggs of some lepidopterous species. *Entomon* 10, 17–19.

Laidlaw, W.B.R. (1936) The brown lacewing flies (Hemerobiidae): their importance as controls of *Adelges cooleyi* Gillette. *Entomologist's Monthly Magazine* 72, 164–174.

Lacroix, J.-L. (1924) Faune des Planipennes de France. Coniopterygidae. *Société d'Étude des Sciences Naturelles d'Elboeuf* 42, 53–84.

Laffranque, J.-P. (1973) Biologie du prédateur aphidiphage *Boriomyia subnebulosa* (Stephens) (Neuroptera: Hemerobiidae) et influence de l'alimentation sur son potentiel de multiplication. Doctoral thesis, Université Paul-Sabatier, Toulouse.

Laffranque, J.-P. & Canard, M. (1975) Biologie du prédateur aphidiphage *Boriomyia subnebulosa* (Stephens) (Neuroptera: Hemerobiidae): études au laboratoire et dans les conditions hivernales du Sud-Ouest de la France. *Annales de Zoologie–Écologie Animale* 7, 331–343.

Löhr, B. & Varela, A.M. Estudio de la biología de *Sympherobius maculipennis* Kimmins (Neuroptera: Hemerobiidae), predator de la cochinilla harinosa de la mandioca *Phenacoccus manihoti* Matile-Ferrero (Homoptera: Pseudococcidae) a tres temperaturas. 15 pp. (Unpublished document, cited *in* Souza, 1988.)

Lucas, E., Coderre, D. & Brodeur, J. (1997) Instar-specific defence of *Coleomegilla maculata lengi* (Coleoptera: Coccinellidae): influence on the attack success of the intraguild predator *Chrysoperla rufilabris* (Neuroptera: Chrysopidae). *Entomophaga* 42, 3–12.

Lucas, H. (1881) *Hemerobius se nourrit de Hylotoma rosae. Bulletin de la Société Entomologique de France* 6, 30. (Cited *in* Pariser, 1919.)

Maher-Ali, A., Moftah, S.A. & Rizk, G.A. (1983) Evaluation of the impact of certain predators on the population density of egg-masses of the cotton leafworm, *Spodoptera littoralis* (Boisd.) in cotton fields. *Bulletin de la Société Entomologique d'Égypte* 62, 111–116.

Mann, J. (1969) *Cactus-feeding Insects and Mites.* Bulletin of the US National Museum, no. 256.

Matta, V.A. (1979) Enemigos naturales de las conchuelas blancas del olivo en el valle de Apaze, Arica-Chile. *Idesia* 5, 231–242.

McEwen, P.K., Clow, S., Jervis, M.A. & Kidd, N.A.C. (1993) Alteration in searching behaviour of adult female green lacewing *Chrysoperla carnea* (Neur.: Chrysopidae) following contact with honeydew of the black scale *Saissetia oleae* (Hom.: Coccidae) and solutions containing acid-hydrolysed L-tryptophan. *Entomophaga* 38, 347–354.

McEwen, P.K., Jervis, M.A. & Kidd, N.A.C. (1994) The influence of an artificial food on larval and adult performances in the green lacewing *Chrysoperla carnea* (Stephens). *International Journal of Pest Management* 42, 25–27.

McMullen, R.D. & Jong, C. (1967) New records and discussion of predators of the pear psylla, *Psylla pyricola* Forster, in British Columbia. *Journal of the Entomological Society of British Columbia* 64, 35–40.

Milbrath, L.R., Tauber, M.J. & Tauber, C.A. (1993) Prey specificity in *Chrysopa*: an interspecific comparison of larval feeding and defensive behavior. *Ecology* 74, 1384–1389.

Monserrat, V.J. & Marín, F. (1992) Plant substrate specificity of Iberian Coniopterygidae (Insecta: Neuroptera). In *Current Research in Neuropterology, Proceedings of the 4th International Symposium on Neuropterology,* ed. Canard, M., Aspöck, H. & Mansell, M.W., pp. 279–290. Sacco, Toulouse.

Monserrat, V.J. & Marín, F. (1994) Plant substrate specificity of Iberian Chrysopidae (Insecta: Neuroptera). *Acta Oecologica* 15, 119–131.

Monserrat, V.J. & Marín, F. (1996) Plant substrate specificity of Iberian Hemerobiidae (Insecta: Neuroptera). *Journal of Natural History* 30, 775–787.

Moznette, F.G. (1915) Notes on the brown lace-wing (*Hemerobius pacificus* Banks). *Journal of Economic Entomology* 8, 350–354.

Murakami, Y. (1963) *Sympherobius domesticus* Nakahara (Neuroptera: Hemerobiidae) predaceous on Comstock mealybug, *Pseudococcus comstocki* (Kuwana) (Homoptera: Coccoidea). *Japanese Journal of Applied Entomology and Zoology* 7, 3. (in Japanese)

Neuenschwander, P. & Hagen, K.S. (1980) Role of the predator *Hemerobius pacificus* in a non-insecticide treated artichoke field. *Environmental Entomology* 9, 492–495.

Neuenschwander, P., Canard, M. & Michelakis, S. (1981)

The attractivity of protein hydrolysate baited traps to different chrysopid and hemerobiid (Neuroptera) species in a Cretan olive orchard. *Annales de la Société Entomologique de France* (N.S.) 17, 213–220.

Neuenschwander, P., Hagen, K.S. & Smith, R.F. (1975) Predation on aphids in California's alfalfa fields. *Hilgardia* 43, 53–78.

New, T.R. (1982) Aspects of the biology of *Chrysopa edwardsi* Banks (Neuroptera: Chrysopidae) near Melbourne. *Neuroptera International* 1, 165–174.

New, T.R. (1983) The egg and first instar larva of *Italochrysa insignis* (Neuroptera: Chrysopidae). *Australian Entomological Magazine* 10, 29–32.

New, T.R. (1984) Comparative biology of some Australian Hemerobiidae. In *Progress in World's Neuropterology, Proceedings of the 1st International Symposium on Neuropterology*, ed. Gepp, J., Aspöck, H. & Hölzel, H., pp. 153–166. Thalerhof, Graz.

New, T.R. (1986) Some early stages of *Calochrysa* Banks (Neuroptera: Chrysopidae). *Australian Entomological Magazine* 13, 11–14.

Pantaleoni, R.A. (1982) Neuroptera Planipennia del comprensorio delle Valli di Comacchio: indagine ecologica. *Bollettino dell'Istituto di Entomologia dell'Università di Bologna* 37, 1–73.

Pariser, K. (1919) Beiträge zur Biologie und Morphologie der einheimischen Chrysopidae. *Archiv für Naturgeschichte* 83A, 1–57 + 2 pl.

Penny, N.D., Adams, P. A. & Stange, L.A. (1997) Species catalog of the Neuroptera, Megaloptera, and Raphidioptera of America north of Mexico. *Proceedings of the California Academy of Sciences* 50, 39–114.

Philippe, R. (1972) Biologie de la reproduction de *Chrysopa perla* (L.) (Neuroptera: Chrysopidae) en fonction de l'alimentation imaginale. *Annales de Zoologie–Écologie Animale* 4, 213–227.

Phoofolo, M.W. & Obrycki, J.J. (1998) Potential for intraguild predation and competition among predatory Coccinellidae and Chrysopidae. *Entomologia Experimentalis et Applicata* 89, 47–55.

Polis, G.A. & Holt, R.D. (1992) Intraguild predation: the dynamics of complex trophic interactions. *Trends in Ecology and Evolution* 7, 151–154.

Polis, G.A., Myers, C.A. & Holt, R.D. (1989) The ecology and evolution of intraguild predation: potential competitors that eat each other. *Annual Review of Ecology and Systematics* 20, 297–330.

Principi, M.M. (1946) Contributi allo studio dei neurotteri italiani. 4. *Nothochrysa italica* Rossi. *Bollettino dell'Istituto di Entomologia dell'Università di Bologna* 15, 85–102.

Principi, M.M. & Canard, M. (1974) Les Névroptères. In:

Les Organismes Auxiliaires en Vergers de Pommiers, pp. 152–162. OILB-SROP, Wageningen, The Netherlands.

Principi, M.M. & Canard, M. (1984) Feedings habits. In *Biology of Chrysopidae*, Series Entomologica 27, ed. Canard, M., Séméria, Y. & New, T.R., pp. 76–92. Dr W. Junk, The Hague.

Quayle, H.J. (1913) Some natural enemies of spiders and mites. *Journal of Economic Entomology* 6, 85–88.

Rivnay, E. (1943) A study on the efficiency of *Sympherobius amicus* Navás in controlling *Pseudococcus citri* Risso on citrus in Palestine. *Bulletin de la Société Fouad 1er d'Entomologie* 27, 57–77.

Rosenheim, J.A. & Wilholt, L.R. (1995) Predators that eat other predators disrupt cotton aphid control. *California Agriculture* 47 (5), 7–9.

Rosenheim, J.A., Wilholt, L.R. & Armer, C.A. (1993) Influence of intraguild predation among insect predators on the suppression of an herbivore population. *Oecologia* 96, 439–449.

Rousset, A. (1966) Morphologie céphalique des larves de Planipennes (Insectes: Névroptèroïdes). *Mémoires du Muséum National d'Histoire Naturelle* (N.S.) 42.

Rousset, A. (1984) Reproductive physiology and fecundity. In *Biology of Chrysopidae*, Series Entomologica 27, ed. Canard, M., Séméria, Y. & New, T.R., pp. 116–129. Dr W. Junk, The Hague.

Samson, P.R. & Blood, P.R.B. (1980) Voracity and searching ability of *Chrysopa signata* (Neuroptera: Chrysopidae), *Micromus tasmaniae* (Neuroptera: Hemerobiidae) and *Tropiconabis capsiformis* (Hemiptera: Nabidae). *Australian Journal of Zoology* 28, 575–580.

Santas, L.A. (1979) *Marchalina hellenica* (Gennadius) an important insect for apiculture of Greece. In *Proceedings of the 27th International Congress of Apiculture of Apimondia*, Athens, pp. 419–422.

Sengonca, Ç. & Coeppicus, S. (1985). Fraßaktivität von *Chrysoperla carnea* (Stephens) gegenüber *Tetranychus urticae* Koch. *Zeitschrift für angewandte Zoologie* 72, 335–342.

Sengonca, Ç. & Frings, B. (1985) Interference and competitive behaviour of the aphid predators, *Chrysoperla carnea* and *Coccinella septempunctata* in the laboratory. *Entomophaga* 30, 245–251.

Sengonca, Ç., Kotikal, Y.K. & Schade, M. (1995) Olfactory reactions of *Cryptolaemus montrouzieri* Mulsant (Col.: Coccinellidae) and *Chrysopa perla* Stephens (Neur.: Chrysopidae) in relation to period of starvation. *Anzeiger für Schlädlingskunde Pflanzenschutz Umweltschutz* 68, 9–12.

Shalaby, F.F., Nada, A.M., Hafez, A.A. & Hassan, K.A. (1999) Larvae of the common green lacewing *Chrysoperla carnea* (Stephens) *sensu lato* (Neuroptera:

Chrysopidae) preying on *Pectinophora gossypiella* (Saunders) (Lepidoptera: Gelechiidae. In *Proceedings of the 1st Regional Symposium for Applied Biological Control in Mediterranean Countries*, Cairo, Egypt, 1998, ed. Canard, M. & Beyssat-Arnaouty, V., pp. 217–222.

Souza, B. (1988) Aspectos morfológicos e biológicos de *Nusalala uruguaya* (Navás, 1923) (Neuroptera: Hemerobiidae) em condiçoes de laboratório. MSc thesis, Escola Superior de Agricultura de Lavras, Lavras, Minas Gerais, Brazil.

Souza, B., Ciociola, A.I. & Matioli, J.C. (1989) Biologia comparada de *Nusalala uruguaya* (Navás, 1923) (Neuroptera: Hemerobiidae) alimentada com diferentes espécies de afídeos. II. Fase de pré-pupa, pupa e adulta. *Anais da Sociedade Entomológica do Brasil* 18, 43–51.

Soylu, O.Z. (1980) Investigations on the biology and control of citrus whitefly *Dialeurodes citri* Ashmead injurious in citrus orchards in the Mediterranean region of Turkey. *Bitki Koruma Bülteni* 20, 36–53. (in Turkish)

Stelzl, M. (1990) Nahrungsanalytische Untersuchungen an Hemerobiiden-Imagines (Insecta: Planipennia). *Mitteilungen der deutschen Gesellschaft für allgemeine und angewandte Entomologie* 7, 670–676.

Stelzl, M. (1991) Untersuchungen zu Nahrungsspektren mitteleuropäischer Neuropteren-Imagines (Neuropteroidea, Insecta). Mit einer diskussion über deren Nützlichkeit als Opponenten von Pflanzenschädlingen. *Journal of Applied Entomology* 111, 469–477.

Stelzl, M. (1992) Comparative studies on mouthparts and feeding habits of adult Raphidioptera and Neuroptera (Insecta: Neuropteroidea). In *Current Research in Neuropterology, Proceedings of the 4th International Symposium on Neuropterology*, ed. Canard, M., Aspöck, H. & Mansell, M.W., pp. 341–347. Sacco, Toulouse.

Syrett, P. & Penman, D.R. (1980) Studies of insecticide toxicity to lucerne aphids and their predators. *New Zealand Journal of Agricultural Research* 23, 575–580.

Szabó, S. & Szentkirályi, F. (1981) Communities of Chrysopidae and Hemerobiidae (Neuroptera) in some apple-orchards. *Acta Phytopathologica Academiae Scientiarum Hungarica* 16, 157–169.

Tauber, C.A. & Tauber, M.J. (1987) Food specificity in predacious insects: a comparative ecophysiological and genetic study. *Evolutionary Ecology* 1, 175–186.

Tauber, M.J., Tauber, C.A., Ruberson, J.R., Milbrath, L.R. & Albuquerque, G.S. (1993) Evolution of prey specificity via three steps. *Experientia* 49, 1113–1117.

Tjeder, B. (1961) Neuroptera-Planipennia. The Lace-wings of Southern Africa. 4. Family Hemerobiidae. In *South African Animal Life: Results of the Lund University Expeditions in 1950–1951*, vol. 8, ed. Hanström, B., Brinck, P. & Rudebeck, G., pp. 296–408. Almqvist & Wiksells, Uppsala.

Tjeder, B. (1966) Neuroptera-Planipennia. The Lace-wings of Southern Africa. 5. Family Chrysopidae. In *South African Animal Life: Results of the Lund University Expeditions in 1950–1951*, vol. 12 ed. Hanström, B., Brinck, P. & Rudebeck, G., pp. 228–534. Swedish Natural Science Research Council, Stockholm.

Uygun, N., Ohnesorge, B. & Ulusay, R. (1990) Two species of whiteflies on citrus in Eastern Mediterranean: *Parabemisia myricae* (Kuwana) and *Dialeurodes citri* (Ashmed). Morphology, biology, host plants and control in southern Turkey. *Journal of Applied Entomology* 110, 471–482.

Weijun, M., Qingjie, G. & Hanye, Z. (1983) Studies of some major predaceous natural enemies of *Matsucoccus matsumurae* (Kuwana). *Journal of Nanjing Technological College of Forest Products* 1983, 19–29.

Wheeler, A.C. Jr (1980) First United States record of *Aleuropteryx simillima*, a predator of scale insects on ornamental juniper (Neuroptera: Coniopterygidae). *Southern Entomologist* 5, 51–52.

Wise, K.A.J. (1996) Records concerning biological control of insect pests by Neuropteroidea (Insecta) in New Zealand. *Records of the Auckland Institute and Museum* 32, 101–117.

Withycombe, C.L. (1923) Notes on the biology of some British Neuroptera (Planipennia). *Transactions of the Entomological Society of London* 1922, 501–594.

Withycombe, C.L. (1924) Further notes on the biology on some British Neuroptera. *Entomologist* 57, 145–152.

Zelený, J. (1965) Lacewings (Neuroptera) in cultural steppe and the population dynamics in the species *Chrysopa carnea* Steph. and *Chrysopa phyllochroma* Wesm. *Acta Entomologica Bohemoslovaca* 62, 177–194.

Zelený, J. (1978) Les fluctuations spatio-temporelles des populations de Névroptères aphidiphages (Planipennia) comme élément indicateur de leur spécificité. *Annales de Zoologie–Écologie Animale* 10, 359–366.

Zelený, J., Hrdy, I. & Kalushkov, P.K. (1981) Population dynamics of aphid and mite predators in hops: Bohemian hop-growing area. *Bulletin OILB/SROP* 1981, 87–96.

CHAPTER 7
Outlines of lacewing development

M. Canard and T.A. Volkovich

7.1 PATTERN OF INDIVIDUAL DEVELOPMENT

Green lacewings, brown lacewings, and dustywings exhibit three larval instars. This is the standard developmental pattern of Neuroptera in which only Ithonidae differ having five stages. When the third-instar larva is full grown, i.e. has reached its maximal weight, it looks for a place to pupate. It then spins a cocoon by moving its anal spinneret, unwinding a whitish silk-like substance developed from modified Malpighian tubules. The motionless third-instar rests inside and is called prepupa. Pupal ecdysis is performed within the cocoon giving rise to an exarate pupa. The mature pupa (= pharate adult) emerges from the cocoon through a small round cap broken with its strong mandibles. It then urgently climbs up a vertical substrate to moult into an adult. After full completion, the new adult drains off the frass accumulated during the preimaginal life (see Chapter 6, this volume). Some days later, conditional on adequate feeding, sexual maturity is attained, pairing occurs and the female begins to oviposit.

7.2 EMBRYONIC DEVELOPMENT

Embryonic development is not unusual in lacewings. Its duration varies for different species (Tables 7.1 and 7.2); for instance, incubation lasts 12 and 6 days at 21 °C in *Micromus paganus* and *Hemerobius nitidulus*, respectively. It also varies with temperature, from 13 to 2.5 days in *Chrysoperla carnea* between 15 °C and 35 °C. As far as is known, embryos develop more slowly in coniopterygids than in green and brown lacewings, commonly requiring more than two weeks to hatch.

The weight of eggs during embryogenesis regularly decreases as a linear function of total duration. In *Chrysoperla mediterranea*, 5% of the initial weight (69 μg) is lost during the 7 days of embryogenesis, and in *Nineta pallida*, 18% of the initial weight (237 μg) is lost during 12 days (Canard *et al.*, 1996) under the same rearing conditions, namely 20 °C and 60%–80% of relative humidity.

7.3 HATCHING

All lacewing embryos have a saw-like egg burster on the labrum and clypeus to break the eggshell along a longitudinal ventral slit. Hatching begins with an embryonic moult; the successive phases in green lacewings have been meticulously described and illustrated, e.g. in *Nothochrysa capitata* by Withycombe (1923) and in *Chrysopa pallens* by Principi (1940). No basic difference appears in brown lacewings and dustywings, whose eggs are unstalked, laid and attached on the substrate by the dorsal side.

Hatching – and ecdysis – is a highly hazardous period in lacewing development. The neonate larva resting a few hours on the empty chorion is motionless, totally defenceless and it needs quietness to complete the closing of its mouth (see Chapter 6, this volume). Any mechanical disturbance during this critical phase would have a catastrophic consequence, preventing later efficient food intake by means of well-arranged larval jaw grooves.

7.4 LARVAL GROWTH

Larval growth out of any arrest of development is undergone rapidly in most green and brown lacewings, in which a full generation often lasts less than one month. However, some *Dichochrysa* and *Nothochrysa* spp. exhibit longer early stages. The few data related to Palaearctic coniopterygids show much more slower development. Information related to the duration of preimaginal development of chrysopids was given by Canard & Principi (1984); these data are not in Table 7.1 in which only additional data are given.

Table 7.1. *Further data (see Canard & Principi, 1984) on the duration of preimaginal stages in some green lacewings reared in the laboratory under controlled temperature and long-day light regimes*

		Duration in days								
	°C	Embryo	1st	2nd	3rd	Prepupa	Pupa	Larval prey	Place	Reference
Chrysopidae										
Anomalochrysa frater	25	4.3	4.2	3.5	3.9		—— 12.0 ——	eggs of *Anagasta kuehniella* + *Phthorimea operculella* larvae	USA, Hawaii	Tauber et al., 1992
Anomalochrysa maclachlani	23	3.3	3.8	3.1	4.1	3.9	8.0	eggs of *Sitotroga cerealella* + *Myzus persicae*	USA, Hawaii	Tauber et al., 1990
Brinckochrysa scelestes	27	—	2.3	2.1	2.3		—— 6.1 ——	eggs of *Corcyra cephalonica*	India	Patel & Vyas, 1985
Ceraeochrysa placita	24	10	9	7	9		—— 24 ——	eggs of *Sitotroga cerealella* + *Myzus persicae*	USA, New York	Tauber et al., 1998
Chrysopa pallens	20	5.9	4.1	3.6	4.3	8.1	12.7	eggs of *Anagasta kuehniella* + *Acyrthosiphon pisum*	France	Grimal & Canard, 1991
Chrysopa regalis	20	8.1	6.9	4.9	8.1	[diapause]		*Myzus persicae* + *Acyrthosiphon pisum*	Spain	Canard, 1986a
Chrysoperla externa	24	5.9	4.0	3.0	4.0	3.0	7.1	eggs of *Alabama argillacea*	Brazil	Carvalho et al., 1998
Chrysoperla nipponensis	22	—	—	12.7			11.1	eggs of *Anagasta kuehniella*	France	El Arnaouty et al., 1996
	22	—	—	17.1			15.4	*Myzus persicae*	Italy	Principi & Sgobba, 1993
Dichochrysa clathrata	20	—	9.1	10.3	10.5		26.4	eggs of *Galleria mellonella*	France	Canard et al., 1990
Dichochrysa picteti	21	9.0	8.1	7.5	8.1		—— 22.1	eggs of *Anagasta kuehniella* + *Acyrthosiphon pisum*	France	Grimal & Canard, 1996
Hypochrysa elegans	20	10.6	7.3	7.3	12.5	[diapause?]	[diapause]	eggs of *Anagasta kuehniella*	China	Lee & Shih, 1982
Mallada desjardinsi	25	4	4.5	4.8	5.9		11.7	*Paurocephala psylloptera*	South Africa	Whittington & Brothers, 1991
Mallada handschini	—	8.1	8.5	14.7	16.9		31.6	various aphids	USA, California	Toschi, 1965
Meleoma emuncta	24	7.7	7.5	4.9	9.1	[diapause]	13.2	various aphids	USA, California	Tauber, 1969
Meleoma kennethi	21.1	5.8	6.3	6.6	12.5	[diapause?]	11.4	*Acyrthosiphon pisum* + *Myzus persicae*		
Meleoma schwarzi	21.1	8.2	9.6	10.8	11.9		—— 28–30	eggs of *Anagasta kuehniella* + various aphids		
Nineta pallida	21	12.2	9.0	7.0	9.6	23	15.2	eggs of *Anagasta kuehniella* + various aphids	France	Canard, 1988, 1989
Nothochrysa capitata	17.2	11	12	10	15	[diapause]		—	UK	Withycombe, 1923
Peyerimhoffina gracilis	20	7.5		14.2			16.0	eggs of *Anagasta kuehniella* + various aphids	France	Grimal, 1988
Rexa lordina	21	7.5	6.4	6.5	16.0	[diapause]		eggs of *Anagasta kuehniella* + various aphids	France	Canard & Labrique, 1989

Table 7.2. *Duration of preimaginal stages in some brown lacewings and dustywings reared in the laboratory under controlled temperature and long-day light regimes*

		Duration in days								
	°C	Embryo	1st	2nd	3rd	Prepupa	Pupa	Larval and adult prey	Place	Reference
Hemerobiidae										
Boriomyia fidelis	—	7.6	3.4	3.5	4.5		10.7	*Macrosiphum liriodendri*	USA, Maryland	MacLeod, 1960
Drepanacra binocula	25	—		9–12		—	—	*Psylla acaciaebaileyanae*	Australia	New, 1984
Hemerobius humulinus	18.3	8	2	2.5	6	6	11	various aphids		
Hemerobius nitidulus	18.3	6	6	8	11	15	17	*Lachnus pini*	UK	Withycombe, 1923
Hemerobius pacificus	18.3	6.7		11.7			16.4	*Acyrthosiphon pisum* + *Hyadaphis apii*	USA, California	Neuenschwander *et al.*, 1975
Micromus angulatus	25	4.5	2.0	1.8	2.0	2.4	5.2	*Myzus persicae*	France	Miermont, 1973
Micromus angulatus	20	7.5	3.1	2.5	2.6	2.9	7.2	*Myzus persicae*	France	Miermont & Canard, 1975
Micromus tasmaniae	25	5.8		5.8			10	lucerne aphids	Australia	Syrett & Penman, 1980
Nusalala uruguaya	25	5.0	2.1	2.2	2.4	4.0	7.9	*Dactynotus* sp.	Brazil	Souza, 1988
Sympherobius pygmaeus	17.8	12	8	12	[diapause]		—	*Phylloxera punctata*	UK	Withycombe, 1923
Wesmaelius subnebulosus	20	7.0	3.5	3.0	3.5	6.0	10.0	*Myzus persicae* + *Acyrthosiphon pisum*	France	Laffranque & Canard, 1975
Coniopterygidae										
Coniopteryx esbenpeterseni	25	9	—	—	—	—	—	*Panonychus ulmi*	Italy	Castellari, 1980
Conwentzia psociformis	18.3	13	3	7	10	4	10			
Semidalis aleyrodiformis	16.7	8–21	5	7	19	4	8	*Phylloxera punctata*	UK	Withycombe, 1923

Table 7.3. *Weights in mg and growth estimates in some green and brown lacewings*

| | Weight of | | | | | |
	egg	cocoon	Weight ratio	Growth index	Place	Reference
Chrysopidae						
Chrysopa pallens	0.112	22.3	199	186	France	Grimal & Canard, 1991
Chrysopa perla	0.123	14.0	126	109	France	
Chrysoperla lucasina	0.071	10.9	154	74	France	
Chrysoperla mediterranea	0.069	9.7	130	73	France	Canard *et al.*, 1996
Dichochrysa picteti	0.082	7.1	87	30	France	Canard & Grimal, 1993
Hypochrysa elegans	0.124	7.0	56	26	France	Grimal & Canard, 1996
Nineta pallida	0.237	35.7	151	121	France	Canard *et al.*, 1996
Peyerimhoffina gracilis	0.091	5.7	63	40	France	Grimal, 1988
Rexa lordina	0.089	9.9	111	35	Spain	Canard *et al.*, 1996
Hemerobiidae						
Micromus angulatus	0.094	5.6	60	96	France	Miermont, 1973
Nusalala uruguaya	0.101	8.5	84	127	Brazil	Souza, 1988

Growth in body weight begins when the first prey item is eaten by the newly hatched larva. It continues until the third instar ceases to move and to feed, before spinning the cocoon. Information on growth is given by the ratio of weights:

$$WR = \frac{\text{weight of recently spun cocoon}}{\text{weight of freshly laid egg}}$$

showing a realistic estimation of the voracity of a species (Table 7.3). Weight ratios vary widely within a range of about 4 times between the less and the more voracious of the studied species. But predatory efficiency also depends on the rapidity of destroying the prey population: the quicker the development, the more active the predator. In this way, a larval growth index

$$GI = \frac{\text{cocoon weight in mg}}{\text{free-living larval duration in days}} \times 100$$

gives a good idea of the level of predation exerted by a species. Growth index also varies widely and shows, for instance, that a slow-growing green lacewing such as *Dichochrysa picteti* puts pressure on a prey population much more moderately than a fast-growing one such as the south American hemerobiid *Nusalala uruguaya*, although the former has a slightly higher weight ratio.

7.5 COCOON SPINNING AND DEVELOPMENT INSIDE THE COCOON

The cocoon is spun by the third–instar larva which has finished its total weight gain growth, at the end of its active predatory life. Descriptions of spinning behaviour are well established for chrysopids and various data were summarised by Canard & Principi (1984). The cocoons are usually whitish, sometimes lightly coloured, yellow greenish in *Chrysoperla harrisii*, pink in *Rexa lordina*, or yellow in some coniopterygids. The shape is sub-spherical in chrysopids, elliptical in hemerobiids, flat and lenticular in coniopterygids. It always has a more or less marked double structure: an outer thin web being used for holding the true cocoon on the substrate, an inner layer enclosing the prepupa. The true cocoon is closely woven and parchment-like, transparent or opaque in chrysopids; loosely woven so that the (pre)pupa is visible through rough meshes in hemerobiids; silky and tight in coniopterygids. The weight of the empty true cocoon recorded in some west Palaearctic chrysopids represents from 6% to 12% of the total initial weight, in the non-diapausing *D. picteti* and the diapausing *Nineta flava*, respectively (Canard *et al.*, 1996).

Spinning the cocoon needs contact with anchoring points to allow good fixation of the external web and successful building of the proper walls. Every species probably has well-defined requirements in choosing a good site to do this. Surprisingly, our knowledge of actual pupation places in the field is poor. Occasional observations related to chrysopids and hemerobiids show that the correct place in many species is found thigmotactically, by moving towards a dark and dry place. Some lacewings regularly exhibit a tendency to move downwards and to spin the cocoon under the host plant, in the litter among fallen leaves; this has been reported in *Chrysopa abbreviata* (Babrikova, 1981) and *Hypochrysa elegans* (Principi, 1956a). Alternatively some species spin the cocoon in the ground, at a shallow depth, if the soil is light and sandy enough, such as *C. formosa* (Principi, 1947) and *C. oculata* (Burke & Martin, 1956). Curved leaves containing aphid colonies harbour cocoons of other lacewings with more aerial habits such as *C. pallens* (Principi, 1940); various galls on oak trap those of *Sympherobius pygmaeus* (Aspöck *et al.*, 1980); dried rolled leaves or bark crevices of trees, those of *Hemerobius stigma* and *Wesmaelius concinnus* (Withycombe, 1923). However, most (? all) coniopterygids as well some chrysopids and hemerobiids do not need dark shelter and spin their cocoons in open and sunny locations, on the under or even the upper surfaces of green leaves, as do *Dichochrysa prasina* (Lacroix, 1925) and *D. flavifrons* (Principi, 1956b); dustywings frequently take up the wrinkle of a leaf rib for their cocoons, as do *Conwentzia pineticola* and *C. psociformis* (Principi & Canard, 1974).

In multivoltine species which enter diapause as a prepupa within the cocoon, differences with respect to the fate of the relevant brood may occur: overwintering cocoons may be heavier and located deeper in the substrate, as in *Chrysopa perla* (Canard & Prudent, 1978), or even spun in quite different places, as in *Hemerobius micans*, whose pupation occurs in curved leaves or in bark crevices for summer broods, but in moss at the foot of the host tree in the overwintering generation (Withycombe, 1923).

The weight of the cocoon decreases from spinning to adult emergence. This change has only been investigated in some chrysopids (Canard *et al.*, 1996). During the first step, water evaporates from liquid secretion, inducing hardening which results in a small weight loss (estimated 4% of the initial weight in *Chrysoperla mediterranea*). Later, decrease varies as a function of time, dryness, and fate of the prepupa within the cocoon;

continuous development in *C. mediterranea* is characterised by 5% and 9% of weight losses at pupal ecdysis and adult emergence, respectively; in delayed development of the prepupa, they are higher, for instance 11% and 15% in *Nineta flava*, the main part occurring during aestivation of this univoltine species.

7.6 ADULT EMERGENCE

Only a few hours pass between cocoon breaking by the mature pupa and the first flight undertaken by the new adult in search of food. This short time constitutes the second critical period in development of lacewings. The insect is then slow-moving and vulnerable, even in protected conditions without any predator. First, the pupa must quickly find a nearby and adequate vertical substrate and firmly grasp it before moulting. Just after ecdysis when the body is fully spread out and inflated, the meconium must be expelled by the new imago. Without complete rejection, the digestive tract would remain blocked by the solid remnants of stored larval food and later feeding would be impossible. The size of the meconium is relatively large and is a function of the species and of the adequateness of larval feeding; for instance, the meconium averages 118 µg in a small green lacewing such as *Peyerimhoffina gracilis* and 1357 µg in the large-sized *Nineta pallida*, both fed on *Anagasta kuehniella* eggs and aphids (*Myzus persicae* and *Acyrthosiphon pisum*), namely 2% and 4% of the initial weights of the cocoon, respectively (Canard *et al.*, 1996).

7.7 PRE-PAIRING TIME, SEXUAL BEHAVIOUR, AND PREOVIPOSITION TIME

When emerging, lacewings as a rule have immature gonads. They cannot mate and females cannot oviposit. The pre-pairing time corresponds to the duration necessary for the gonads to mature and for partners to encounter each other, and the preoviposition time, to the deposit of the first egg by the (inseminated) female. In normal conditions, these two periods are confounded and referred to in the literature as preoviposition. In *Chrysoperla carnea*, during this life phase, the lacewings undertake a true migration; the timing (nocturnal), orientation (upwind), duration (2–3 nights), (high) elevation and the extent of this obligatory movement were summarised by Duelli (1984a).

As far is known in green lacewings, mating is only

possible when the oocytes are mature in the genital tract of the virgin female, and oviposition of the inseminated female is initiated and stimulated by copulation. Spermatogenesis, oogenesis and maturation of accessory glands in green lacewings were discussed by Rousset (1984) and the relative durations were found to be variable, especially with imaginal feeding. Only complementary data on chrysopids are given here in Table 7.4, together with the few available data relating to some hemerobiids. Only the green lacewings *Anomalochrysa frater* and *A. maclachlani* show a conspicuously long period for oviposition of about 21 days, surprisingly independent of temperature (Tauber *et al.*, 1990, 1992) and then simulating a short and spontaneously broken reproductive diapause.

Sexual behaviour has been deeply investigated in chrysopids, but little is known for other lacewings. The present state of knowledge has not greatly changed since the reviews of Henry (1984) and Principi (1986) on the structure and role of the spermatophore in chrysopids. New developments concern analysis of courtship songs and their use as basement criteria in the taxonomy of sibling species complexes (see Chapter 3).

7.8 OVIPOSITION

In chrysopids (except in the endemic Hawaiian genus *Anomalochrysa*) the eggs are deposited on the top of a more or less long pedicel constituted by wiredrawing a droplet of mucus secreted from the female accessory glands. A similar gelatinous substance is produced by hemerobiid and coniopterygid females, but not stretched up by abdominal movement, so that it serves to glue the egg directly onto the substrate. Oviposition as far as is known takes place at night with a marked peak in the two hours after sunset in *Chrysoperla carnea* (Duelli, 1984*a*, *b*).

Fecundity is evidently strongly influenced by imaginal feeding as well by environmental conditions. Several data concerning fecundity and the factors affecting it in chrysopids have already been given (Rousset, 1984) and additional data are presented in Table 7.4. The total fecundity is of course correlated to female longevity, so that the recorded results often are not a realistic estimation of the potential oocyte production. To gain a practical estimate of population reproduction rate, it is better to compare the average egg production per day.

7.9 SEDENTARISM AND DISPERSAL

Apart from some data relative to the common green lacewing *Chrysoperla carnea* and a few scattered observations, almost nothing is known about the actual degree of adult mobility, the possible dispersal of the females together with offspring, and the exchanges between natural biotopes and agroecosystems. Whatever the nature of dispersive flights, another kind of apparent sedentarism comes from substrate specificity (see Chapter 19, this volume). Besides, some hemerobiids have evolved flightlessness which compels true sedentarism, either by facultative brachyptery as in the Palaearctic *Psectra diptera*, or by wing broadening and/or hardening as in *Conchopterella* from the Juan Fernández archipelago off the coast of Chile.

7.10 ADULT LONGEVITY

As far as is known, adult chrysopids and hemerobiids are potentially rather long-lived insects. Longevity of course is a function of climatic environment (temperature, relative humidity) and of quantitative and qualitative trophic conditions, but it does not show significant difference between the sexes. Several and various data are found in the literature giving more pertinent information on the adequateness of rearing than on intrinsic durability of lacewings. The following are examples of long imaginal life spans arising from a strategy of reproductive diapause: the females of *Chrysoperla mediterranea*, fed as larvae on eggs of *A. kuehniella* and as adults on pollen and honey, survive in the laboratory an average of 153 days at 20 °C (Carvalho *et al.*, 1996); *Nusalala uruguaya* fed as larvae and adults on *Dactynotus* sp. survive 93 in males and 55 days in females at 25 °C (Souza, 1988); *Micromus angulatus* fed as larvae and adults on *Myzus persicae* survive 90 to 175 days at 20 °C and *Wesmaelius subnebulosus*, in the same rearing conditions, 48 and 53 days in males and females, respectively; *Meleoma dolicharthra* fed as larvae on various aphids and as adults on the classical yeast and levulose mixture survive 73 in males and 76 days in females (Tauber, 1969).

In natural conditions, unpublished preliminary observations on *Chrysoperla carnea* carried out by scrutinising the developmental state of the ovaries in mobile females caught by McPhail traps indicated a very much shorter average life expectancy. However, there are always some individuals that escape the hazards of the

Table 7.4. *Reproductive characteristics in some lacewings reared in the laboratory under controlled temperature and long–day light regimes*

	°C	Larval prey	Adult prey (food)	Preoviposition (days)	No. eggs per day	Total fecundity	Place	Reference
Chrysopidae								
Chrysopa quadripunctata	24	*Acyrthosiphon pisum* + *Myzus persicae*	idem	3.6	—	—	USA, California	Albuquerque et al., 1997
Chrysopa slossonae	24	*Prociphilus tesselatus*	idem	5.0	—	—	Brazil	Carvalho et al., 1996
Chrysoperla externa	25	eggs of *Alabama argillacea*	yeast + honey	3.0	28.8	2304	France	Lee & Shih, 1982
Chrysoperla mediterranea	20	eggs of *Anagasta kuehniella*	yeast + honey	5.8	22	2160	India	
Mallada desjardinsi	25	*Paurocephala psylloptera*	yeast + honey	7.3	12.7	512		
Meleoma emuncta	21.1	*Acyrthosiphon pisum* + *Myzus persicae*	yeast + sugar	9.2	—	321	USA, California	Tauber, 1969
Hemerobiidae								
Drepanacra binocula	25	*Psylla acaciaebaileyanae*	idem	4.5	10.7	206	Australia	New, 1984
Drepanacra binocula	25	*Brevicoryne brassicae*	idem	6.0	6.3	154		
Hemerobius pacificus	18.3	*Acyrthosiphon pisum* + *Hyadaphis apii*	+ Wheast ®	10.1	10.8	—	USA, California	Neuenschwander, 1976
Micromus angulatus	20	*Myzus persicae*	idem	6.1	23–41	1500–2300	France	Miermont & Canard, 1975
Micromus tasmaniae	25	*Psylla acaciaebaileyanae*	idem	4.0	17.4	252		
Micromus tasmaniae	25	*Brevicoryne brassicae*	idem	3.8	17.5	280	Australia	New, 1984
Nusalala uruguaya	25	*Dactynotus* sp. nom.	+ yeast	6.5	9.5	457	Brazil	Souza, 1988
Wesmaelius subnebulosus	20	*Myzus persicae*	+ pollen + honey	4.7	21.6	1045	France	Laffranque & Canard, 1975
Coniopterygidae								
Coniopteryx esbenpeterseni	25	natural prey	*Panonychus ulmi*	4.5	2–8	—	Italy	Castellari, 1980

field, namely imaginal parasitisation (Pantaleoni, 1984) and predation by birds or bats (Miller, 1984). It induces an unavoidable overlapping in generations during summer and renders contingent the estimation of the actual rank of specimens from multivoltine species cohorts.

Adult longevity in the species which undergo a reproductive diapause (see below) is of course long. The arrest of reproduction in *C. carnea* may last in the field in mild-temperate climates from August to May and induces a longevity of more than 9 months (Thierry *et al.*, 1994); likewise, the summer diapause in *Nineta flava* induces a longevity reaching up to 7 months (Canard, 1982).

7.11 PROTECTIVE DEVICES

Protective devices are various in lacewings, mainly known and studied in chrysopids. During embryonic development, the eggs of only chrysopids are, at least partially, protected by the hair-like pedicel which is not considered a way to potential food by some crawling predators (Duelli & Johnson, 1992; Růžička, 1997a). Consumption by predators also is discouraged by hardening of the embryonic serous membranes so that oophagy may only take place during early or late incubation (Bar & Gerling, 1985).

The larvae of some slow-moving green lacewing species more or less totally cover their thoracic and abdominal tergites with various debris caught by hooked setae. The packet includes skin and remains of prey, but also various particles encountered like lichens. Some larvae even actively coat themselves with wax of the nearby prey, as does *Ceraeochrysa cincta* which collects the cottony production of the mealybug *Plotococcus eugeniae* (Eisner & Silberglied, 1988) or *Chrysopa slossonae*, which collects the wax of the woolly alder aphid *Pociphilus tesselatus* (Milbrath *et al.*, 1993). These trash-carrying green lacewing larvae are then protected by the debris mass which offers camouflage and a physical shield against predators, as demonstrated for instance in *Chrysotropia* by New (1969) and *Meleoma* by Eisner *et al.* (1978). At cocoon spinning, the debris-carriers incorporate this trash as a cover over the cocoon. The trash-carrying habit is almost universal in the green lacewings of the Neotropics (Adams & Penny, 1987), but unknown in hemerobiids and coniopterygids.

Some fast-moving green lacewing larvae show the capability to use anal secretions in two ways. It is possibly a repellent emitted in the presence of a predator: this protective behaviour was observed in some American *Chrysoperla* spp. and in *Eremochrysa punctinervis;* when the adversary was located, the larva faced it and exuded from the anus a droplet of defensive fluid which is probably the same substance used to adhere to the substrate; swinging its abdomen over its head, it contacted the enemy which soon stopped attacking (Kennett, 1948; LaMunyon & Adams, 1987). The adhesive–defensive substance is also used in certain species of chrysopid such as *Chrysopa oculata* as an oviposition deterrent inducing a larger dispersion of the offspring, together with expulsion of the other intraguild predators (Růžička, 1997b). As far is known, no similar protective devices occur in brown lacewings or dustywings.

Some adult chrysopids such as *Chrysopa walkeri* are able to emit from prothoracic glands a stink fluid, whose repellent effect on predators has been demonstrated (Blum *et al.*, 1973; Güsten & Dettner, 1992). The common green lacewing *Chrysoperla carnea* by means of a sophisticated auditory receptor is able to escape the attacks of hunting echolocating insectivorous bats (Miller, 1984). Adult hemerobiids and coniopterygids frequently have cryptic coloration rending them difficult to see. Some hemerobiids such as *Boriomyia fidelis* (MacLeod, 1960) or *Wesmaelius subnebulosus*, resting and/or falling down when disturbed, carry out a feigning-death position with head and appendages tucked in and wings folded.

7.12 TRITROPHIC RELATIONS

Independent of the prey rejection induced by the aphid substrate (see Chapter 6), some factor of growth may be altered by the sole host plant. In the laboratory, *Chrysopa perla* larvae ate *Aphis nerii* living on *Nerium oleander;* when the rose-laurel was in vegetative rest, chrysopid survival rate was about 78%; as soon as the plant was in active growth, survival fell quickly and decreased to zero; the inability to develop was due to the cardenolides safely ingested by the aphid and stored in their tissues, but toxic for the chrysopid larvae (Canard, 1977). Developmental time and survival rate were modified in *Chrysoperla* spp. preying on the aleyrodid *Bemisia argentifolii* as a function of the host plant:

living on cucumber and cantaloup gave better performances than on lima beans and poinsettia (Legaspi *et
al.*, 1996). Physical characters of the host plants may
interact with the behaviour of chrysopid larvae: for
instance, the pubescence of grass induced variations in
Diuraphis noxia control by *Chrysoperla carnea* which
reduced aphid populations to zero on smooth slender-
leafed grasses such as Indian rice grass but failed on
pubescent cultivars (Messina *et al.*, 1995); the mobility
of *C. carnea* larvae was greater on cabbages with glossy
leaves than on those with a wax bloom and thus controlled eggs and caterpillars of the diamondback moth
Plutella xylostella better (Eigenbrode *et al.*, 1995, 1996).

7.13 VOLTINISM AND SEASONAL ADAPTATIONS

The annual number of broods undergone by lacewings
may vary and is in many instances difficult to appraise.
The longevity of some females can be considerable, and
then their oviposition occurs over a long time.
Conclusions about voltinism are therefore tentative and
they are drawn from field catches together with rearings
undertaken under subnatural conditions. The voltinism relates to several traits: genetic characteristics of
species, adaptive patterns of populations, ability to
arrest development as a function of external cues (see
below), and climatic conditions.

Many species exhibit a continuous life cycle
without any true diapause; all developmental stages can
be found throughout the year in the field. This holody-
namous apausic development is the rule in equatorial,
wet tropical, and subtropical regions, for instances
among chrysopids in the southern African *Chrysoperla
zastrowi* (Brettell, 1982), the American *C. externa*
which underwent 45 successive generations at natural
light in Buenos Aires (Crouzel & Botto, 1977), the
Asian *Mallada desjardinsi* which bred 10–11 overlapping generations a year in Guangzhou (China) (Wei *et
al.*, 1986), and the Australian *Chrysopa edwardsi* which
showed no true developmental arrest in winter (New,
1982). Probably most (? all) tropical hemerobiids and
coniopterygids yield to the same pattern. Besides, some
hemerobiids occurring in temperate zones have no diapause and are only regulated by low winter temperature
and food availability. It is the case, for instance, in the
Holarctic *Hemerobius stigma* (Killington, 1937), in the
three North American *H. pacificus* (Neuenschwander,

1976), *H. neadelphus* and *H. ovalis* (Kevan &
Klimaszewski, 1987), which all can be found in winter
as prepupae or pupae within the cocoon, reproductive
adults, even eggs or active larvae.

Multivoltinism is certainly the most usual condition of lacewing development. The cycle is then regulated by means of a diapause, most often dependent on
photoperiod. Several studies have been undertaken on
this topic and a detailed review on chrysopids was
recently published (Canard, 1998). Some species have
either multivoltine or univoltine populations depending on to the place of origin, such as *Chrysopa abbreviata* in Italy (Pantaleoni, 1982), or *Wesmaelius mortoni* in
the United Kingdom (Killington, 1937).

The univoltine strategy for development regulated
by complex cue actions is not very frequent in lacewings, and has been reviewed by Canard (1998). In
hemerobiids, only *W. concinnus* and *W. quadrifasciatus*
are known to be strictly univoltine. Some chrysopids,
univoltine in the field, were possibly demonstrated
bivoltine in rearing for a small part of the population,
such as *Cunctochrysa baetica* in Italy (Pantaleoni, 1982).
Others showed a tendency to extend their life cycle by
prolonged diapause (see below). Nothing is known
about this in eremic green lacewing species or in other
lacewings.

7.14 DIAPAUSE AND ITS HORMONAL CONTROL

All known types of diapause, except the embryonic one
in Chrysopidae (Principi, 1991), have been recorded in
lacewings, but imaginal and larval arrests of development were most commonly found. There is little information on the topic for Hemerobiidae and
Coniopterygidae, while it is known that in the temperate zones brown lacewings overwinter at different
stages from eggs to adults (Zelený, 1963), whereas dustywings mainly do so as prepupae (Killington, 1937).

Various ways are involved in diapause control
depending on ontogenetic stages on which morphogenesis is delayed. The neurosecretory systems of
Neuroptera have never been examined unlike those in
other insect orders, with the exception of neurosecretory brain cell morphology (Panov & Davydova, 1976).
Therefore we may here consider the hormonal control
of diapause taking into account that similar standard
mechanisms are involved in arthropods in the inhibition

and stimulation of neurosecretory system activity. Whatever the type of diapause, external cues (e.g. photoperiod) perceived by the central nervous system, mainly the brain, and translated into neurohormonal stimulus, cause changes in hormonal balance leading to an arrest of development.

Only embryonic diapause is known to be controlled by the specific hormone of diapause which penetrates from the female haemolymph into the eggs and produces an arrest in the embryo's development at early stages (Fukuda, 1952). Embryonic diapause is uncommon in lacewings, with only *Wesmaelius concinnus* and *W. quadrifasciatus* overwintering as diapausing embryos in the chorion (Killington, 1937), but we do not know what embryogenetic phase is concerned.

Ecdysone(s) and juvenile hormone (JH) are the major hormones governing insect development. Ecdysones produced by the prothoracic glands (PTG) are responsible for metabolic and morphogenetic processes in tissues. Corpora allata are the source of JH which prevents metamorphosis as long as its threshold is high enough. It also regulates the synthesis of vitellogenins and their transfer from haemolymph to the ovarioles. According to the view currently accepted, ecdysones manage the processes leading to a moult whereas the kind of the moult depends on the threshold of JH.

The arrest of growth and morphogenesis in diapausing larvae and pupae is mainly induced by low ecdysone threshold in haemolymph. Inhibition of ecdysone release from PTG in turn is conditioned by the incapability of brain cells to discharge the activation hormone (AH) which controls activity of PTG. Such a regulatory pathway is observed in a wide spectrum of species (Tyshchenko & Kind, 1983; Denlinger, 1985), so that it is logical to assume that it also takes place in Neuroptera with early larval or pupal diapause.

In arthropods entering diapause as pronymphae, JH may be involved in the control as well. In several species of Lepidoptera, JH plays a primary role in induction and sometimes in maintenance of this diapause (Chippendale, 1977). A high level of JH in haemolymph seems to depress AH production by neurosecretory brain cells, and we cannot exclude such a pathway in neuropteran diapause regulation.

Reproductive adult diapause is more distinct in females because it is only manifested as an absence of egg-laying due to delayed vitellogenesis or occasionally previtellogenesis. In addition, diapause in some *Chrysoperla* of the *carnea* complex is accompanied by a change in body coloration (Honěk, 1973; Duelli, 1992) that may be considered as a collateral effect of change in the hormonal balance. Mainly JH is involved in this diapause type; low threshold or lack of JH inhibits ovarian development and diverts metabolites to fat body reserves. Inactivity of the corpora allata in turn may be controlled by the brain through both interruption of nervous impulses and decrease of AH in haemolymph (Hodkova, 1979; Tyshchenko & Kind, 1983).

7.15 SENSITIVE AND RESPONSIVE STAGES

The period of sensitivity to signals occurs either in the same stage as the resulting diapause or in preceding ones, but never extends to the whole life cycle. Concerning adult diapause, the most sensitive stages to diapause-averting or -inducing stimuli were manifested in: (1) the third-instar free-living larvae, the prepupae within the cocoon, and the pupae in *Chrysoperla downesi* (Tauber & Tauber, 1976a), (2) only adults in *C. harrisii* (Tauber & Tauber, 1974) and *Micromus angulatus* (Miermont & Canard, 1975), and (3) any stages after spinning the cocoon including adults in *C. plorabunda* (Tauber & Tauber, 1970a) and probably in *Peyerimhoffina gracilis* (Grimal, 1988).

In species diapausing as prepupae within the cocoon – namely numerous genera of chrysopids such as *Chrysopa* spp., some hemerobiids, and most (? all) coniopterygids – the growing larvae perceive the signals. All larval instars were found sensitive to photoperiod in *Chrysopa nigricornis* (Tauber & Tauber, 1972a); sensitivity was highest (1) either in second or third instars as in *C. oculata* (Propp *et al.*, 1969), *C. pallens* (Grimal & Canard, 1991; Orlova, 1998), *C. perla* (Volkovich, 1996), and *C. abbreviata* (N. Orlova, personal communication), or (2) in the third instar as in *C. phyllochroma* (Volkovich, 1997). All the prepupae of the above-mentioned species became insensitive to diapause-inducing cues after the cocoon was spun.

A derivation from the common pattern has been found in a southern strain of *Nineta flava*. Although all prepupae enter winter diapause regardless of daylength, nevertheless the photoperiod the larvae experienced determined the time of pupal ecdysis the next spring. Resumption of morphogenesis and consequently adult emergence occurred earlier in spring if the larvae were previously reared under short day length conditions. Adults were also photosensitive, because they were able to enter a (second) summer reproductive diapause, only in long photoperiods (Canard, 1983, 1986*b*).

Diapause may occur in lacewings during free-living larval instars. In *Dichochrysa flavifrons* (Principi & Sgobba, 1987), *D. picteti* (Canard *et al.*, 1990), and *D. clathrata* (Principi & Sgobba, 1993), the diapause is manifested as a lengthening of the third (sometimes also the second) larval instar(s) in winter. The sensitivity to diapause-inducing cues was present throughout the larval stages while the second one played the most important role in determining the developmental pathway.

In *Nineta pallida*, photoperiodic signals are perceived both by the embryo and the newly hatched larva; photoperiodic responses (PhPR) were exhibited as successive slowing down of developmental speed of larvae, both in first instar under short daylength conditions (winter diapause) and in the [second + third instar(s)] under long daylength conditions (spring diapause) (Canard, 1988, 1990).

A rare situation for lacewings has been recorded in *Hypochrysa elegans* which overwinters as pupae. Its eggs and two early larval instars were shown to be sensitive to photoperiod, but the deep (winter) diapause only occurred in the pupae. So, there was a photorefractory period between sensitive and responsive stages. In addition, under natural conditions, this nothochrysine green lacewing exhibits a long prepupal instar which may be considered a short summer diapause (Grimal & Canard, 1996).

The duration of diapause induction termed as the required day number (RDN) varies markedly in chrysopids as well. For instance, two or three short-day stimuli before spinning the cocoon were enough for diapause to be induced in 50% of *Chrysopa phyllochroma* prepupae at 24–28 °C (Volkovich, 1997), whereas six or seven short-day stimuli were needed in *C. pallens* (Orlova, 1998). The interspecific variations in the RDN are likely to reflect the tendency to enter diapause in different species even in the same climatic zone. On the other hand, *C. perla* from both southern France and central Russia required equal RDN for diapause induction (Canard, 1976; Volkovich, 1996) indicating that intraspecific variations may be absent, even in far-off populations.

7.16 PHYSIOLOGICAL AND BIOCHEMICAL ASPECTS OF DIAPAUSE

Resistance to the action of any unfavourable factors is increased in diapausing insects as a result of morphogenetic arrest and lowered level of metabolism. Canard & Queinnec (1971) demonstrated that in diapausing *Chrysopa perla* prepupae, the rate of heartbeat is about eight times lower than in non-diapausing ones. Other aspects of energy metabolism were studied in *C. pallens* prepupae using nuclear magnetic resonance spectroscopy; the appropriate fluctuation of ATP and sugar phosphate levels gave evidence that an anaerobic breaking up of carbohydrates is the main means of energy production in diapausing prepupae (Grimal *et al.*, 1992). Such a metabolic pathway leads to accumulation of polyols and other cryoprotective agents increasing cold hardiness. The seasonal fluctuations of carbohydrates chiefly by trehalose and glycogen during the winter diapause were also studied in *C. walkeri* prepupae (Sagné *et al.*, 1986). Some physiological traits of diapause such as water content, fat content, and supercooling point (SCP) were investigated in adults of *Chrysoperla carnea* (Vannier, 1986, 1988) and the larvae of *Nineta pallida* (Vannier & Canard, 1989; Canard & Vannier, 1992). The species mentioned above and others are freeze-susceptible in overwintering stages and show high supercooling capacity: for instance, the ex-ovo and unfed first-instar larvae of *N. pallida* showed the SCP about −25 °C, when fed and free-living, on the average −17.9 °C. Vannier (1986) supposed that the depression of SPC recorded in field populations of *C. carnea* from −12.6 °C in September to -17.3 °C in November was due to the synthesis of glycerol, trehalose, and specific proteins of haemolymph acting as antifreeze. It may be assumed that the proteins produce a thermal hysteresis described in some insects (Duman *et al.*, 1982). However this problem remains open to study.

7.17 EXTERNAL FACTORS GOVERNING THE INDUCTION OF DIAPAUSE

The importance of daylength as a primary cue for diapause induction has now been featured in more than 25 Holarctic multivoltine green lacewings overwintering in diapause. Photoperiods shorter than a critical threshold are able to trigger hormonal processes leading to

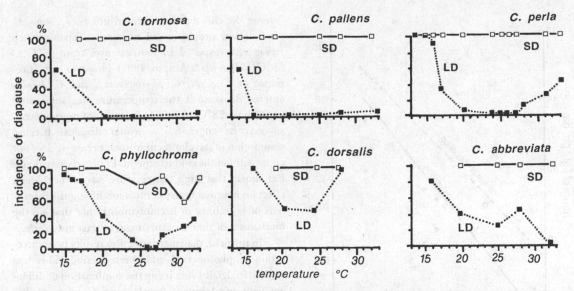

Fig. 7.1. Diapause induction (%) in prepupae of *Chrysopa* species originating from the forest–steppe zone of Russia and reared in the laboratory under short- (SD) or long daylength (LD) regimes at different temperatures.

diapause, whereas longer photoperiods prevent its induction. However, all constant photoperiods inhibited oviposition in females of *Chrysoperla downesi* and *Peyerimhoffina gracilis*; these two species are univoltine and spent the greater part of their life cycles in aestivo-hibernal reproductive diapause. Diapause may be avoided only by increasing daylength during either the last preimaginal or adult phases; the minimal gaps between the two regulating regimes were 4 and 2 hours in the two species respectively (Tauber & Tauber, 1976a; Grimal, 1988).

Few species of hemerobiids have been studied to clarify the role of environmental factors in their diapause control. As shown in experiments, adult reproductive diapause in *Micromus angulatus* was induced with both short daylengths (light:dark=8:16) and natural daylengths of November and December whereas it was prevented by long photophases (16:8 and 18:6) (Miermont & Canard, 1975). Similar PhPR was found in populations inhabiting the Far East: diapause was induced by 13:11 or 12:12 (light:dark) regimes, but prevented by 16:8 or 18:6 ones. In addition, gradual temperature decrease from 22°C to 15°C was favourable to diapause induction (Potjomkina, 1987, 1990). It may be expected that the diapause of many species of hemerobiids manifesting two or three generations per year is induced by short daylengths as well.

In some North American and European *Chrysoperla* species, decreasing daylengths promote diapause induction (Tauber & Tauber, 1970b; Sheldon & MacLeod, 1974; Volkovich, 1987). On the contrary, in other species such as *Chrysopa perla*, the induction of diapause only depended on the absolute daylength and not on its gradual change (Volkovich, 1996).

The mechanism of time measurement is temperature-dependent. Consequently, photoperiodic cues may be effective in the induction or in the prevention of diapause only above or below threshold temperatures which are species-specific (Danilevsky, 1961; Danks, 1987). Two species groups within *Chrysopa* spp. may be distinguished according to their temperature reaction. One includes *Chrysopa pallens*, *C. perla* (Fig. 7.1) and probably *C. oculata* as consistent with Tauber *et al.*(1987): they showed typical long PhPR over a wide range of temperature, for instance from 15°C to 33°C in *C. pallens*. On the other hand, the non-diapause development of the whole population of other species was possible in narrow ranges of either relatively high temperatures (*C. phyllochroma* and *C. abbreviata*) or moderate ones (*C. dorsalis*) (Fig. 7.1) (Volkovich, 1998).

In widely distributed species, such as *Chrysoperla plorabunda* and *Chrysopa oculata*, the critical photoperiodic thresholds for diapause induction are positively correlated to the latitudes (Fig. 7.2) (Tauber & Tauber,

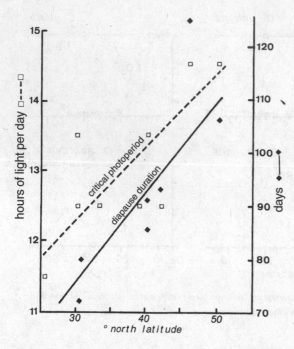

Fig. 7.2. Relationship between latitude and diapause responses in various geographical populations of *Chrysopa oculata:* critical photoperiods in open symbols and dotted line, diapause durations in solid symbols and solid line. (After Nechols *et al.*, 1987, by courtesy of Pergamon Journals Ltd.)

1972*b*, 1982; Nechols *et al.*, 1987). In the north, the growing seasons are shorter and daylengths in the late summer are longer than in the south; then long critical photoperiods in northern populations promote the earlier onset of diapause. Rather small variations in the critical photoperiods were found in other green lacewings such as in *C. perla* populations (Hinke, 1975; Canard, 1976; Volkovich & Arapov, 1994).

The critical photoperiod thresholds for diapause induction decreased more or less steadily as the temperature rose over the ranges of temperatures from 16 °C to 28 °C (*C. perla* and *C. pallens*) and from 20 °C to 28 °C (*C. phyllochroma* and *C. abbreviata*) (Volkovich & Arapov, 1994; Volkovich, 1997; Orlova, 1998). Only *Dichochrysa ventralis* showed a temperature-independent threshold of PhPR within the range from 20 °C to 26 °C. Diapause was induced at both higher (28 °C) and lower (16 °C) temperatures as well as in *C. dorsalis*.

Usually in insects with long day PhPR, the combination of high temperature + short photophase appears to prevent diapause (Danilevsky, 1961; Danks, 1987).

Among the chrysopids presented in Fig. 7.1, only *C. phyllochroma* prepupae did not enter diapause when larvae were exposed to temperatures from 25 °C to 33 °C (Volkovich & Arapov, 1993). However, some prepupae of *C. perla, C. phyllochroma,* and *C. dorsalis* entered diapause if the temperature was sufficiently high (above 28 °C or 30 °C). An arrest of morphogenesis may be appraised as winter diapause because resumption of development in such prepupae does not occur without the cold treatment. Preliminary evidence has shown that high temperature has an immediate effect on neurosecretory centres modifying the thresholds of hormones in haemolymph rather than on the mechanism of time measurement, at least in *C. perla.*

In the field, diapause induction results from interaction of photoperiod and thermoperiod. This was studied in detail when using the combination of different light and temperature rhythms. *Chrysopa phyllochroma* showed a slight increase (about 30 min) of the PhPR thresholds under thermoperiodic conditions as compared to those at constant temperatures. In addition, the diapause rate was inversely correlated to the cryophase temperatures, and thus restricts bivoltinism in northern regions. On the contrary, *C. perla* reacted similarly whatever the cryophase and thermophase, and only extremes – namely 5.5 °C and 33 °C – provoked diapause under long day conditions. Thus, in this species as well in *C. pallens, C. formosa,* and *Chrysoperla carnea*, natural temperature fluctuations were shown to be less favourable for diapause induction than constant temperatures, allowing these species to be diapause-free at low ambient temperature unlike the above-mentioned *Chrysopa phyllochroma* and *C. abbreviata* (Volkovich, 1996, 1997; Volkovich & Blumental, 1997; Orlova, 1998).

The effect of daily temperature cycles alone – the true thermoperiodic diapause induction – was recently revealed in *Chrysopa pallens* and *Chrysoperla carnea:* in continuous darkness, their diapause was induced by long cryophases and prevented by short ones, although the responses obtained are less clear than PhPRs (I.V. Sokolova & G. Nasier, personal communication).

There are few studies dealing with the effects of diet on lacewing diapause. The photoperiodic signals may be modified or even replaced by trophic ones as has long been known in *Chrysoperla mohave* which showed two types of diapause. A primary one which allows the adult to overwinter is a classical diapause induced by short photoperiods, whereas a secondary one is a food-mediated aestivation depending on abundance of larval prey: the lack of aphids causes a high incidence of later adult diapause even under diapause-preventing long-day length conditions (Tauber & Tauber, 1973a, 1982). The proportion of *Chrysopa perla* diapausing prepupae under long daylength increased from about a third of the larvae fed on aphids up to almost the whole population when fed on *Sitotroga* eggs (Ushtchjekov, 1989). Such an effect should be taken in account because often moth eggs are used as diet for chrysopid larvae in mass-rearing production (see Chapter 12, this volume).

7.18 DIAPAUSE INTENSITY

Species and geographical populations differ considerably in diapause intensity, both by duration and by depth. There are lacewings, usually diapausing as prepupae, with long and deep diapause, even in southern populations, such as *Nineta flava* or *Chrysopa perla*. Their morphogenesis cannot be resumed without any special external cues. On the other hand, reproductive diapause was expressed as slight lengthening of pre-oviposition time in southern populations of *Micromus angulatus* (Miermont & Canard, 1975). Larval diapause of *N. pallida* also is shallow and expressed only as a slackening of growth so that the larvae are able to move and feed in winter. Between intense and weak diapauses, numerous intermediates result from genetic and environmental factors.

The duration of diapause may vary with latitudinal origin: in *Chrysoperla plorabunda* and *Chrysopa oculata*, it is positively correlated to latitude (Fig. 7.2). Long arrests in development in northern populations are of adaptive value because these insects may thus enter diapause earlier in the season than in the south.

Photoperiod and thermoperiod, which are the basic inducing factors, can during the course of induction also modify the depth and the duration of diapause. In *Chrysoperla carnea*, the diapause duration was about 90 days when sensitive stages were exposed to constant photoperiods of 12 or 14 hours of light per day, and in contrast, it was 120 to 130 days when they were reared under photoperiods decreasing by 6 min per day from 18 or 17 to 12 hours per day (Volkovich, 1987). Similar lengthenings (1.3 and 2.4 times) of diapause were found in *C. mediterranea* as a result of transferring insects from a long photoperiod (16:8) to a medium

(12:12) or short (8:16) one (Canard *et al.*, 1994). However, the most marked difference was found in *Dichochrysa* spp. whose diapause is sometimes very brief. For instance, the duration of diapause in the third instar was extended by 4 to 6 times in *Dichochrysa picteti* if the photoperiod decreased from medium to short during the larval life (Canard & Grimal, 1993). Low temperature appeared to provoke a similar effect (Canard *et al.*, 1992). Working on *D. flavifrons*, Principi & Sgobba (1987) showed that the number of short-day stimuli perceived by the larvae had an effect on diapause duration: the longer the induction, the higher the diapause intensity. Indeed, the larvae remained in diapause for about 117 days at 25°C and 188 days at 15°C when sensitive second instars were reared at 15°C, while they completed their diapause faster in about 46 and 93 days, respectively, if the induction occurred at 25°C (Principi *et al.*, 1990). Therefore, shortening daylength together with falling temperature in the late growing season can maintain diapause, but also amplifies it.

7.19 DIAPAUSE DEVELOPMENT AND DIAPAUSE COMPLETION

Diapause completion is achieved by means of deblocking neurosecretory centres as the result of spontaneous reactivation and/or environmental factors. Two pathways lead to diapause ending, i.e. the resumption of morphogenesis: horotelic and tachytelic processes in Hodek's (1983) terminology. Under experimental conditions, many species complete diapause without any special cue. Then diapause development may be prolonged because only horotelic processes are evolved. Moreover there is evidence that the speed of proceeding processes depends on temperature (Principi *et al.*, 1990) and shows quantitative response to constant photoperiods (Tauber & Tauber, 1970b; Canard, 1990). Probably only southern species inhabiting mild winter climates are able to end diapause spontaneously in the field.

The time required for morphogenesis to resume spontaneously and the numbers of individuals concerned vary between species and conditions. For example, pupal ecdysis under short-day conditions occurred after three to five months in all prepupae of *Chrysopa nigricornis* (Tauber & Tauber, 1972a) and *C. oculata* (Nechols *et al.*, 1987), and the greater part (77%) of those of *C. pallens* (Grimal & Canard, 1991).

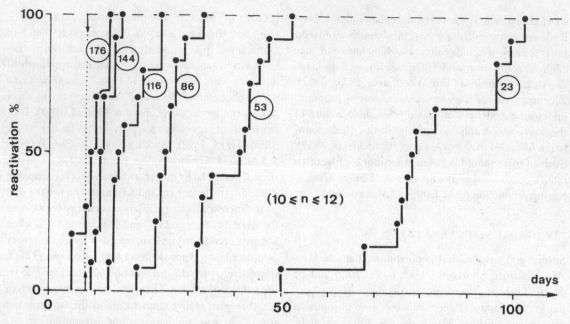

Fig. 7.3. Morphogenesis resumption (= pupal ecdysis) (%) in diapausing prepupae of *Chrysopa pallens* originating from southern France, reared in the laboratory at 20 °C under short daylengths, and maintained as cocoons during various durations (number of days circled) at 9 °C. Arrows and dotted line indicates the average minimal time off diapause necessary to perform pupal ecdysis (from Grimal & Canard, 1991.)

However, spontaneous reactivation seemed to be more typical for imaginal diapause such as in *Chrysoperla plorabunda, C. mediterranea, Micromus angulatus*, and the free-living diapausing larvae of *Dichochrysa* spp. whose diapause completion is appraised by food intake and re-increase of weight. That occurred only in a part of the population of *Dichochrysa picteti*, depending on the photoperiodic conditions the larvae were exposed to; the other diapausing larvae needed long days for reactivation, failing which they die after a long interval (Canard & Grimal, 1993).

In many temperate-inhabiting chrysopids, initiating morphogenesis depends on a more or less long exposure to low temperatures, usually from 10°C to 0 °C. Honěk & Hodek (1973) studied in detail the diapause development of *Chrysoperla carnea* in the field. The intensity of diapause was quantified by the duration of the preoviposition period in females when exposed to long photoperiods after transfer from outdoors to the laboratory. Preoviposition was the longest in the samples from September to late November, on the average from 17 to 22 days. It decreased to about half after the end of December and remained at the

same level up to late spring; at the same time the deviation from the average rate was reduced.

Low-temperature horotelic processes leading to the termination of development arrest at the brain level is likely to need a certain time. If transfer to a high temperature and long photoperiod occurs early in the course of diapause, the tachytelic activation lasts longer. Moreover, the response appears to depend on the kind of stimulus: long photoperiods are more efficient than high temperatures in *C. carnea*, because under short-daylength conditions all females collected in late October remained with unripe ovaries. Development resumed after a shorter delay if lacewings were activated while low-temperature processes were further advanced; all females laid eggs without any delay after horotelic processes were completed. In studying diapause termination in *C. plorabunda, C. harrisii*, and *Chrysopa nigricornis*, Tauber & Tauber (1972a, 1973b, 1974) obtained similar results. The reactivation time of *Chrysopa pallens* diapausing prepupae also was markedly shortened when the duration of chill exposure was prolonged (Fig. 7.3). All prepupae underwent pupal ecdysis as quickly as non-diapausing ones after chilling for about 25 weeks. The decrease in the delay necessary for resumption of development can be considered as a certain indication of the progress of horotelic processes with time.

While long days often hasten resumption of development in experiments on the above-mentioned green lacewings, few species exhibit photoperiodic reactivation in the field. There was no photorefractory period in *Meleoma signoreti* and the sensitivity to daylength persists throughout prepupal diapause. The critical photoperiod for diapause induction was shown to be two hours longer than for its ending, so that in the field, resumption of morphogenesis is avoided in autumn and diapause complete when vernal days exceed a critical length (Tauber & Tauber, 1975; Tauber *et al.*, 1986).

A sequence of short days followed by long days is a necessary prerequisite for oogenesis in *Chrysoperla downesi* and *Peyerimhoffina gracilis* (Tauber & Tauber, 1976a; Grimal, 1988). This requirement excludes reproduction by the newly emerged females during summer. It may be assumed that a two-step hormonal control of the reproduction takes place here, as has been described in carabid beetles which also lay eggs in spring. Previtellogenesis under short days is controlled by JH, whereas vitellogenesis under long days requires direct neurohumoral stimulation (Ferenz, 1977). In *C. downesi* the rate of diapause development directly correlated with the duration of photophase during winter, but after the vernal equinox adults became photorefractory (Tauber & Tauber, 1976b).

The fact that sensitivity to photoperiod gradually disappears during cold periods allows us to conclude that the ecological importance of photosensitivity should be sought rather in the maintenance of diapause than in its termination. Natural short days during autumn may retain the diapause state until low temperatures happen.

In some green lacewings diapausing within the cocoon, only some of the prepupae resume morphogenesis in the following spring whereas the rest have prolonged diapause. About 88% of *Chrysopa regalis* prepupae in the population inhabiting a montane environment remain in diapause for two years, so that their morphogenesis may be delayed up to a second summer (Canard, 1986a). The closely related species *C. dorsalis* is largely univoltine, though 37% of diapausing prepupae can overwinter twice and some individuals (about 10%) even three times (Volkovich, 1998). Such a long

arrest in development is possible on condition that the water and fat reserves accumulated during larval growth are used sparingly. In *C. regalis*, for instance, the weight losses due to dehydration and respiration were very small, less than 7% of the initial weight at the end of diapause. Some prepupae of *C. formosa*, *C. perla*, and *Nineta flava* may also extend diapause for up to two years (Canard & Principi, 1984). However, we still do not know the environmental factors responsible for maintenance and termination of long-term diapause. It may only be assumed that individual variation in the rate of horotelic processes is involved in diapause timing.

7.20 ENVIRONMENTAL CONTROL OF SUMMER DIAPAUSE

Aestivation as secondary diapause has been reported in four chrysopids inhabiting temperate climates. The trigger for induction may be either (1) long days in *Hypochrysa elegans* prepupae (Grimal & Canard, 1996), in second and third instars of *Nineta pallida* (Canard, 1988), and in adults of *N. flava* (Canard, 1983), or (2) as mentioned above in *Chrysoperla mohave*, the lack of larval food (aphids). The food-mediated aestivation was maintained until prey became available (Tauber & Tauber, 1973a). Whether reactivating factors are necessary for aestivation completion or whether it might be achieved spontaneously in the two *Nineta* species and in *Hypochrysa elegans* is not yet known. It is supposed that natural short days hasten tachytelic processes leading to the reduction of both the reproductive delay in *N. flava* and the metamorphosis delay in *N. pallida* larvae (Canard & Grimal, 1988).

7.21 TEMPERATURE REQUIREMENTS FOR LACEWING DEVELOPMENT

Once winter diapause ends, the subsequent phases of the seasonal cycle are primarily regulated by temperature. It acts rather on the speed of post-diapause development than as cues (Danilevsky, 1961; Tauber *et al.*, 1987). Most chrysopids have a lower development threshold (LDT) included in a common range in insects (Canard & Principi, 1984; Honěk & Kocourek, 1987; and further data in Table 7.5) depending on species and stages. The interspecies variation in threshold temperatures is reflected in the time of adult flight

Table 7.5. *Lower development thresholds in some green and brown lacewings*

	Embryo	Larvae	Prepupa and pupa	Species	Place	Reference
Chrysopidae						
Anomalochrysa frater	8.9	6.0–7.5	7.1	—	USA, Hawaii	Tauber *et al.*, 1992
Anomalochrysa maclachlani	9.5	8.7–10.3	9.9–10.3	9.8	USA, Hawaii	Tauber *et al.*, 1990
Ceraeochrysa cubana	13.1	12.0–13.4	11.9	12.4	Brazil	Silva, 1991
Chrysopa abbreviata	—	8.3	—	—	Russia	Volkovich, 1998
Chrysopa edwardsi	—	9.6	10.3	—	Australia	New, 1982
Chrysopa formosa	—	9.2	—	—	Russia	Volkovich, 1998
Chrysopa oculata	—	—	9.8	—	USA, Washington	
Chrysopa oculata	—	—	12.2	—	USA, New York	Nechols *et al.*, 1987
Chrysopa pallens		9.1	10.0	—	Russia	Volkovich & Orlova, 1998
Chrysopa perla	8.0	8.6	8.1	—	Russia	Volkovich, 1998
Chysopa phyllochroma	—	12.0	12.6	—	Russia	Volkovich & Orlova, 1998
Chrysopa signata	8.5	9.8	11.5	10.5	Australia	Samson & Blood, 1979
Chrysoperla carnea	—	—		≈ 10	Russia	Volkovich, 1998
Chrysoperla downesi	11.4	10.3–10.8	11.1/10.3	10.8	USA, New York	Tauber & Tauber, 1981
Chrysoperla harrisii	11.9	10.3–13.2	11.9/13.6	—	USA, New York	Tauber & Tauber, 1974
Dichochrysa ventralis	—	6.5–7.0	—	—	Russia	Volkovich, 1998
Hemerobiidae						
Drepanacra binocula	—	6.9	3.8	—	Australia	New, 1984
Hemerobius pacificus	0.4	4.1	0.6	—	USA, California	Neuenschwander, 1975
Hemerobius stigma	—	—	—	<5	Canada	Garland, 1981
Micromus tasmaniae	−0.1	2.6	1.4	−2.9	Australia	Samson & Blood, 1979
Micromus tasmaniae	4.8	5.7	6.0	5.8	New Zealand	Syrett & Penman, 1981
Micromus tasmaniae	—	3.2	2.7	—	Australia	New, 1984
Micromus variegatus	—	0.9	—	—	UK	Dunn, 1954
Wesmaelius subnebulosus	—	—	—	6	France	Laffranque, 1973

and oviposition in spring: for instance, among the Russian forest–steppe chrysopid fauna, *Chrysopa phyllochroma* emerged about two weeks later than *C. pallens* even if their prepupae spent winter under the same conditions (Volkovich & Orlova, 1998). The thermal adaptations unlike the seasonal ones (PhPR) seem to be relatively stable in geographical populations. For instance, the LDTs varied by only 1.2 °C in *C. oculata* (Tauber *et al.*, 1987) and 0.8 °C in *Chrysoperla downesi* (Tauber & Tauber, 1976*b*). By contrast, hemerobiids often have LDTs considerably lower than those of green lacewings. Geographical variations are sometimes important, as in the Australian *Micromus tasmaniae* (Table 7.5). It is very likely that intraspecies

variation occurs here, while differences are possibly rather more often due to the methods not being comparable.

Good tolerance to low temperature provides the advantage of storing insects to be used in biological control. Active females of *Hemerobius stigma* retain fecundity after storage of one month at 5 °C (Garland, 1981), as well as diapausing females of *Chrysoperla carnea* at the same temperature up to more than seven months (Tauber *et al.*, 1993). About three-quarters of one-day-old eggs and second-instar larvae of *M. angulatus* are able to survive after one week of storage at 5–6 °C and 70% relative humidity (Potjomkina, 1990). Larvae of *Wesmaelius subnebulosus* occur during autumn and winter and can complete their development at temperatures above 5°C (Laffranque & Canard, 1975). Finally, a North American species, *H. pacificus*, even exhibits reproductive activity from 4.4°C (Neuenschwander, 1975, 1976). Consequently, in contrast to some other aphidophagous predators (Honěk & Kocourek, 1987), most brown lacewings have low or very low LDTs. It allows them to maintain feeding activity at low temperature in early spring or even during winter in regions with a mild climate. Some green lacewing larvae exhibiting weak diapause also are able to catch and eat prey at temperature near to or even lower than 0°C (Canard, 1997).

Upper temperature thresholds for development are not well defined. There is evidence that high temperature retards development and often induces increasing mortality such as in *Micromus angulatus* (Potjomkina, 1987), *M. tasmaniae* (Syrett & Penman, 1981), and other species (Canard & Principi, 1984). For instance, at 28°C, mortality of *M. angulatus* pupae was markedly increased although the larvae could feed and develop even up to 35°C; so the larvae and pupae manifest different temperature requirements, 26–28°C and 22–26°C respectively (Potjomkina, 1987, 1990). However some chrysopids endure short-term exposure to higher temperatures. For instance, *Chrysoperla nipponensis* larvae subjected to 42–43°C survived 7–8 hours provided that the humidity was about 90% (Shuvakhina, 1987). Active larvae of *Nineta pallida* originating from a south European montane environment manifested conspicuously high thermostupor points in ex-ovo and second-instar larvae, 45°C and 52°C respectively (Canard & Vannier, 1992).

REFERENCES

Adams, P.A. & Penny, N.D. (1987) Neuroptera of the Amazon Basin. Part 11a. Introduction and Chrysopini. *Acta Amazonica* 15, 413–479.

Albuquerque, G.S., Tauber, M.J. & Tauber, C.A. (1997) Life-history adaptations and reproductive costs associated with specialization in predacious insects. *Journal of Animal Ecology* 66, 307–317.

Aspöck, H., Aspöck, U. & Hölzel, H. (unter Mitarbeit von H. Rausch) (1980) *Die Neuropteren Europas. Eine zusammenfassende Darstellung der Systematik, Ökologie und Chorologie der Neuropteroidea (Megaloptera, Raphidioptera, Planipennia) Europas*, 2 vols. Göcke & Evers, Krefeld.

Babrikova, T. (1981) Some morphological–bioecological characteristics of *Chrysopa abbreviata* Curtis (Chrysopidae: Neuroptera). *Gradinarska Lozarska Nauka* 18, 28–34. (in Bulgarian)

Bar, D. & Gerling, D. (1985) Cannibalism in *Chrysoperla carnea* (Stephens) (Neuroptera: Chrysopidae). *Israel Journal of Entomology* 19, 13–22.

Blum, M.S., Wallace, J.B. & Fales, H.M. (1973) Skatole and tridecene: identification and possible role in a chrysopid secretion. *Insect Biochemistry* 3, 353–357.

Brettell, J.H. (1982) Biology of *Chrysopa congrua* Walker and *Chrysopa pudica* Navás and toxicity of certain insecticides to their larvae. *Zimbabwe Journal of Agricultural Research* 20, 77–84.

Burke, H.R. & Martin, D.F. (1956) The biology of three chrysopid predators of the cotton aphid. *Journal of Economic Entomology* 49, 698–700.

Canard, M. (1976) La diapause chez *Chrysopa perla* (Linnaeus) (Neuroptera: Chrysopidae). *Annales de Zoologie–Écologie Animale* 8, 393–404.

Canard, M. (1977) Diminution du taux de survie du prédateur *Chrysopa perla* (L.) (Neuroptera: Chrysopidae) en relation avec le comportement du puceron *Aphis nerii* B. de F. (Homoptera: Aphididae). In *Mécanismes Éthologiques de l'Évolution*, ed. Médioni, J. & Bœsiger, E., pp. 49–51. Masson, Paris.

Canard, M. (1982) Diapause reproductrice photopériodique chez les adultes de *Nineta flava* (Scopoli) (Neuroptera: Chrysopidae). *Neuroptera International* 2, 59–68.

Canard, M. (1983) La sensibilité photopériodiques des larves de *Nineta flava* (Scopoli) (Neuroptera: Chrysopidae). *Entomologia Experimentalis et Applicata* 34, 111–118.

Canard, M. (1986a) Is the Iberian lacewing *Chrysopa regalis* a semivoltine species? *Ecological Entomology* 11, 27–30.

Canard, M. (1986b). A cautious univoltine strategy in the lacewing *Nineta flava* (Scopoli) (Neuroptera: Chrysopidae). In *Recent Research in Neuropterology*, *Proceedings of the 2nd International Symposium on Neuropterology*, ed. Gepp, J., Aspöck, H. & Hölzel, H., pp. 145–150. Thalerhof, Graz.

Canard, M. (1988) Seasonal change in photoperiodic response of the larvae of the lacewing *Nineta pallida*. *Entomologia Experimentalis et Applicata* 47, 153–159.

Canard, M. (1989) L'influence de la photopériode sur le développement à l'intérieur du cocon chez la chrysope *Nineta pallida* (Schneider) (Neuroptera: Chrysopidae): une diapause prénymphale relicte? *Annales de la Société Entomologique de France* (N.S.) 25, 25–32.

Canard, M. (1990) Effect of photoperiod on the first-instar development in the lacewing *Nineta pallida*. *Physiological Entomology* 15, 137–140.

Canard, M. (1997) Can lacewings feed in winter? (Neur.: Chrysopidae and Hemerobiidae). *Entomophaga* 42, 113–117.

Canard, M. (1998) Life history strategies of green lacewings in temperate climates: a review (Neuroptera: Chrysopidae). *Acta Zoologica Fennica* 209, 65–74.

Canard, M. & Grimal, A. (1988) Insect photoperiodism: various way of regulating univoltinism in lacewings (Planipennia: Chrysopidae). *Experientia* 22, 523–525.

Canard, M. & Grimal, A. (1993) Multiple action of photoperiod on diapause in the green lacewing *Mallada picteti* (McLachlan) (Neuroptera: Chrysopidae). *Bollettino dell'Istituto di Entomologia dell'Università di Bologna* 47, 233–245.

Canard, M. & Labrique, H. (1989) Bioécologie de la chrysope méditerranéenne *Rexa lordina* Navás (Neuroptera: Chrysopidae) et description de ses stades larvaires. *Neuroptera International* 5, 151–158.

Canard, M. & Principi, M.M. (1984) Development of Chrysopidae. In *Biology of Chrysopidae*, Series Entomologica 27, ed. Canard, M., Séméria, Y. & New, T.R., pp. 57–75. Dr W. Junk, The Hague.

Canard, M. & Prudent, P. (1978) Étude au laboratoire de la recherche du site de tissage du cocon par les larves de *Chrysopa perla* (Linné) (Neuroptera: Chrysopidae). *Entomologia Experimentalis et Applicata* 24, 11–21.

Canard, M. & Queinnec, Y. (1971) Modifications du rythme cardiaque au cours de la diapause de *Chrysopa perla* (Linnaeus) (Insectes, Névroptères). *Comptes Rendus Hebdomadaires des Séances de l'Académie des Sciences, Série D* 273, 1960–1963.

Canard, M. & Vannier, G. (1992) Adaptations of preimaginal stages of *Nineta pallida* (Schneider) (Insecta: Neuroptera: Chrysopidae). In *Current Research in Neuropterology, Proceedings of the 4th International Symposium on Neuropterology*, ed. Canard, M., Aspöck, H. & Mansell, M.W., pp. 75–95. Sacco, Toulouse.

Canard, M., Carvalho, C.F. & Sissoko, F. (1994) La diapause chez *Chrysoperla mediterranea* (Hölzel, 1972): influence de la photopériode sur la durée de préoviposition (Neuroptera: Chrysopidae). *Bulletin de la Société Entomologique de France* 99, 455–461.

Canard, M., Grimal, A. & Carvalho, C.F. (1996) Weight changes during preimaginal development in green lacewings (Insecta: Neuroptera: Chrysopidae). In *Pure and Applied Research in Neuropterology, Proceedings of the 5th International Symposium on Neuropterology*, ed. Canard, M., Aspöck, H. & Mansell, M.W., pp. 87–101. Sacco, Toulouse.

Canard, M., Grimal, A. & Hatté, M. (1990) Larval diapause in the Mediterranean green lacewing *Mallada picteti* (McLachlan) (Neuroptera: Chrysopidae): induction by photoperiod, sensitive and responsive stages. *Bollettino dell'Istituto di Entomologia dell'Università di Bologna* 44, 65–74.

Canard, M., Grimal, A. & Hatté, M. (1992) How does the green lacewing *Mallada picteti* (McLachlan) overwinter? (Insecta: Neuroptera: Chrysopidae). In *Current Research in Neuropterology, Proceedings of the 4th International Symposium on Neuropterology*, ed. Canard, M., Aspöck, H. & Mansell, M.W., pp. 87–93. Sacco, Toulouse.

Carvalho, C.F., Canard, M. & Alauzet, C. (1996) Comparison on the fecundity of the Neotropical *Chrysoperla externa* (Hagen) and the West-Palaearctic *Chrysoperla mediterranea* (Hölzel) (Insecta: Neuroptera: Chrysopidae). In *Pure and Applied Research in Neuropterology, Proceedings of the 5th International Symposium on Neuropterology*, ed. Canard, M., Aspöck, H. & Mansell, M.W., pp. 103–107. Sacco, Toulouse.

Carvalho, C.F., Souza, B. & Santos, T.M. (1998) Predation capacity and reproduction potential of *Chrysoperla externa* (Hagen) (Neuroptera: Chrysopidae) fed on *Alabama argillacea* (Hübner) eggs. *Acta Zoologica Fennica* 209, 83–86.

Castellari, P.L. (1980) Indagini biologiche su *Coniopteryx (Metaconiopteryx) esbenpeterseni* Tjeder (Neuroptera: Coniopterygidae), predatore di Acari Tetranychidi sul pesco. *Bollettino dell'Istituto di Entomologia dell'Università di Bologna* 35, 157–180.

Chippendale, G.M. (1977) Hormonal regulation of larval diapause. *Annual Review of Entomology* 22, 121–138.

Crouzel, I.S. de & Botto, E.N. (1977) Ciclo de vida de *Chrysopa lanata lanata* (Banks) y algunas observaciones biológicas en conditiones de laboratorio. *Revista de Investigaciones Agropecuarias INTA, Serie 5, Patología Vegetal* 13, 1–14.

Danilevsky, A.S. (1961) *Photoperiodism and Seasonal Development of Insects*. Leningrad State University, Leningrad, USSR. (in Russian) English edn 1965, Oliver & Boyd, London.

Danks, H.V. (1987) *Insect Dormancy: An Ecological Perspective*. Biological Survey of Canada, Ottawa.

Denlinger, D.L. (1985) Hormonal control of diapause. In *Comprehensive Insect Physiology, Biochemistry and Pharmacology*, vol. 8, ed. Kerkut, G.A. & Gilbert, R.I., pp. 353–412. Pergamon Press, New York.

Duelli, P. (1984a) Flight, dispersal, migration. In *Biology of Chrysopidae*, Series Entomologica 27, ed. Canard, M., Séméria, Y. & New, T.R., pp. 110–116. Dr W. Junk, The Hague.

Duelli, P. (1984b) Oviposition. In *Biology of Chrysopidae*, Series Entomologica 27, ed. Canard, M., Séméria, Y. & New, T.R., pp. 129–133. Dr W. Junk, The Hague.

Duelli, P. (1992) Body coloration and colour change in green lacewings (Insecta: Neuroptera: Chrysopidae). In *Current Research in Neuropterology, Proceedings of the 4th International Symposium on Neuropterology*, ed. Canard, M., Aspöck, H. & Mansell, M.W., pp. 119–123. Sacco, Toulouse.

Duelli, P. & Johnson, J.B. (1992) Adaptive significance of the egg pedicel in green lacewings (Insecta: Neuroptera: Chrysopidae). In *Current Research in Neuropterology, Proceedings of the 4th International Symposium on Neuropterology*, ed. Canard, M., Aspöck, H. & Mansell, M.W., pp. 125–134. Sacco, Toulouse.

Duman, J.G., Horwarth, K., Tomchaney, A. & Patterson, J.L. (1982) Antifreeze agents of terrestrial arthropods. *Comprehensive Biochemistry and Physiology* (A) 73, 545–555.

Dunn, J.A. (1954) *Micromus variegatus* Fabricius (Neuroptera) as a predator of the pea aphid. *Proceedings of the Royal Entomological Society of London* (A) 29, 76–80.

Eigenbrode, S.D., Moodie, S. & Castagnola, T. (1995) Predators mediate host plant resistance to a phytophagous pest in cabbage with glossy leaf wax. *Entomologia Experimentalis et Applicata* 77, 335–342.

Eigenbrode, S.D., Castagnola, T., Roux, M.-B. & Steljes, L. (1996) Mobility of three generalist predators is greater on cabbage with glossy leaf wax than on cabbage with a wax bloom. *Entomologia Experimentalis et Applicata* 81, 335–343.

Eisner, T. & Silbergleid, R.E. (1988) A chrysopid larva that cloaks itself in mealybug wax. *Psyche* 95, 15–19.

Eisner, T., Hicks, K., Eisner, M. & Robson, D.S. (1978) 'Wolf-in-sheep's clothing' strategy of a predaceous insect larva. *Science* 199, 790–794.

El Arnaouty, S.A., Ferran, A. & Beyssat-Arnaouty, V. (1996) Food consumption by *Chrysoperla carnea* (Stephens) and *Chrysoperla sinica* (Tjeder) of natural and substitute prey: determination of feeding efficiency (Insecta: Neuroptera: Chrysopidae). In *Pure and Applied Research in Neuropterology, Proceedings of the 5th International Symposium on Neuropterology*, ed. Canard, M., Aspöck, H. & Mansell, M.W., pp. 109–117. Sacco, Toulouse.

Ferenz, H.J. (1977) Two-step photoperiodic and hormonal control of reproduction in the female beetle, *Pterostichus nigrita. Journal of Insect Physiology* 23, 671–676.

Fukuda, S. (1952) Function of the pupal brain and suboesophageal ganglion in the production of non-diapause and diapause eggs in the silkworm. *Annotationes Zoologicae Japonenses* 25, 149–155.

Garland, J.A. (1981) Effect of low-temperature storage on oviposition in *Hemerobius stigma* Stephens (Neuroptera: Hemerobiidae). *Entomologist's Monthly Magazine* 116, 149–150.

Grimal, A. (1988) Exigences photopériodiques du cycle de développement de la chrysope *Tjederina gracilis. Entomologia Experimentalis et Applicata* 47, 189–194.

Grimal, A. & Canard, M. (1991) Modalités du développement de *Chrysopa pallens* (Rambur) (Neuroptera: Chrysopidae) au laboratoire. *Neuroptera International* 6, 107–115.

Grimal, A. & Canard, M. (1996) Preliminary observations on the effects of photoperiod on the life cycle of the green lacewing *Hypochrysa elegans* (Burmeister) (Insecta: Neuroptera: Chrysopidae: Nothochrysinae). In *Pure and Applied Research in Neuropterology, Proceedings of the 5th International Symposium on Neuropterology*, ed. Canard, M., Aspöck, H. & Mansell, M.W. pp. 119–127. Sacco, Toulouse.

Grimal, A., Palévody, C. & Canard, M. (1992) Monitoring energy metabolism during diapause in neuropteran prepupae by 31–Phosphorous nuclear magnetic resonance spectroscopy. In *Current Research in Neuropterology, Proceedings of the 4th International Symposium on Neuropterology*, ed. Canard, M., Aspöck, H. & Mansell, M.W., pp. 153–157. Sacco, Toulouse.

Güsten, R. & Dettner, K. (1992) The prothoracic gland of the Chrysopidae (Neuropteroidea: Planipennia). In *Proceedings of the 4th European Congress of Entomology*, ed. Zombori, L. & Peregovits, L., pp. 60–65. Hungarian Natural History Museum, Budapest.

Henry, C.S. (1984) The sexual behavior of green lacewings. In *Biology of Chrysopidae*, Series Entomologica 27, ed. Canard, M., Séméria, Y. & New, T.R., pp. 101–110. Dr W. Junk, The Hague.

Hinke, F. (1975) Autökologische Untersuchungen an mitteleuropäischen Neuropteren. *Zoologische Jahrbücher, Abteilung für Systematik* 102, 303–330.

Hodek, I. (1983) Role of environmental factors and endogenous mechanisms in the seasonality of reproduction in insect diapause as adults. In *Diapause and Life Cycle Strategy in Insects*, ed. Brown, V.K. & Hodek, I., pp. 9–33. Dr W. Junk, The Hague.

Hodkova, M. (1979) Hormonal and nervous inhibition of reproduction by brain in diapausing females of *Pyrrhocoris apterus* L. (Hemiptera). *Zoologische Jahrbücher, Abteilung für Anatomie und Ontogenie der Tiere* 83, 126–136.

Honěk, A. (1973) Relationship of colour changes and diapause in natural populations of *Chrysopa carnea* Stephens (Neuroptera: Chrysopidae). *Acta Entomologica Bohemoslovaca* 70, 254–258.

Honěk, A. & Hodek, I. (1973) Diapause of *Chrysopa carnea*

(Chrysopidae, Neuroptera) females in the field. *Věstník Československé Společnosti Zoologické* 37, 95–100.

Honěk, A. & Kocourek, F. (1987) Thermal requirements for development of aphidophagous Coccinellidae (Coleoptera), Chrysopidae, Hemerobiidae (Neuroptera), and Syrphidae (Diptera): some general trends. *Oecologia* 76, 455–460.

Kennett, C.E. (1948) Defense mechanism exhibited by larvae of *Chrysopa californica* Coq. *Pan-Pacific Entomologist* 24, 209–211.

Kevan, D.K.McE. & Klimaszewski, J. (1987) The Hemerobiidae of Canada and Alaska. Genus *Hemerobius* L. *Giornale Italiano di Entomologia* 3, 305–369.

Killington, F.J. (1937) *A Monograph of the British Neuroptera*, vol. 2. Ray Society, London.

Lacroix, J.-L. (1925) Étude sur les chrysopides. Époque du coconnage chez les larves du groupe *Chrysopa prasina* Burm. *Société d'Étude des Sciences Naturelles d'Elboeuf* 43, 87–91.

Laffranque, J.-P. (1973) Biologie du prédateur aphidiphage *Boriomyia subnebulosa* (Stephens) (Neuroptera: Hemerobiidae) et influence de l'alimentation sur son potentiel de multiplication. Doctoral Thesis, Université de Toulouse.

Laffranque, J.-P. & Canard, M. (1975) Biologie du prédateur aphidiphage *Boriomyia subnebulosa* (Stephens) (Neuroptera: Hemerobiidae): études au laboratoire et dans les conditions hivernales du Sud-Ouest de la France. *Annales de Zoologie–Écologie Animale* 7, 331–343.

LaMunyon, C.W. & Adams, P.A. (1987) Use and effect of an anal defensive secretion in larval Chrysopidae (Neuroptera). *Annals of the Entomological Society of America* 80, 804–808.

Lee, S.J. & Shih, C.I.T. (1982) Biology, predation, and field-cage release of *Chrysopa boninensis* Okamoto on *Paurocephala psylloptera* Crawford and *Corcyra cephalonica* Stainton. *Journal of Agriculture and Forestry* 31, 129–144.

Legaspi, J.C., Nordlund, D.A. & Legaspi, B.C. Jr (1996) Tritrophic interactions and predation rates in *Chrysoperla* spp. attacking the silverleaf whitefly. *Southwestern Entomologist* 21, 33–42.

MacLeod, E.G. (1960) The immature stages of *Boriomyia fidelis* (Banks) with taxonomic notes on the affinities of the genus *Boriomyia* (Neuroptera: Hemerobiidae). *Psyche* 67, 26–40.

Messina, F.J., Jones, T.A. & Nielson, D.C. (1995) Host plant affects the interaction between the Russian wheat aphid and a generalist predator, *Chrysoperla carnea*. *Journal of the Kansas Entomological Society* 68, 313–319.

Miermont, Y. (1973) Études au laboratoire de l'alimentation d'un prédateur aphidiphage: *Eumicromus angulatus* (Stephens) (Hemerobiidae). Doctoral Thesis, Université de Toulouse.

Miermont, Y. & Canard, M. (1975) Biologie du prédateur aphidiphage *Eumicromus angulatus* (Neuroptera: Hemerobiidae): études au laboratoire et observations dans le Sud-Ouest de la France. *Entomophaga* 20, 179–191.

Milbrath, L.R., Tauber, M.J. & Tauber, C.A. (1993) Prey specificity in *Chrysopa:* an interspecific comparison of larval feeding and defensive behavior. *Ecology* 74, 1384–1393.

Miller, L.A. (1984) Hearing in green lacewings and their responses to the cries of bats. In *Biology of Chrysopidae*, Series Entomologica 27, ed. Canard, M., Séméria, Y. & New, T.R., pp. 134–149. Dr W. Junk, The Hague.

Nechols, J.R., Tauber, M.J. & Tauber, C.A. (1987) Geographical variability in ecophysiological traits controlling dormancy in *Chrysopa oculata*. *Journal of Insect Physiology* 33, 627–633.

Neuenschwander, P. (1975) Influence of temperature and humidity on the immature stages of *Hemerobius pacificus*. *Environmental Entomology* 4, 215–220.

Neuenschwander, P. (1976) Biology of the adult *Hemerobius pacificus*. *Environmental Entomology* 5, 96–100.

Neuenschwander, P., Hagen, K.S. & Smith, R.F. (1975) Predation on aphids in California's alfalfa fields. *Hilgardia* 43, 53–78.

New, T.R. (1969) Notes on the debris-carrying habit in larvae of British Chrysopidae (Neuroptera). *Entomologist's Gazette* 20, 119–124.

New, T.R. (1982) Aspects of the biology of *Chrysopa edwardsi* Banks (Neuroptera: Chrysopidae) near Melbourne, Australia. *Neuroptera International* 1, 165–174.

New, T.R. (1984) Comparative biology of some Australian Hemerobiidae. In *Progress in World's Neuropterology*, *Proceedings of the 1st International Symposium on Neuropterology*, ed. Gepp, J., Aspöck, H. & Hölzel, H., pp. 153–166. Thalerhof, Graz.

Orlova, N. (1998) Effects of photoperiod and temperature on diapause induction in *Chrysopa pallens* (Rambur) (Neuroptera: Chrysopidae). *Acta Zoologica Fennica* 209, 195–202.

Panov, A.A. & Davydova, E.D. (1976) Neurosecretory cells of the medial protocerebrum and retrocerebral complex in the Neuropteroidea (Insecta). *Zoologicheskii Zhurnal* 55, 1824–1837. (in Russian with English summary)

Pantaleoni, R.A. (1982) Neuroptera Planipennia del comprensorio della Valli di Comacchio: indagine ecologica. *Bollettino dell'Istituto di Entomologia dell'Università di Bologna* 37, 1–73.

Pantaleoni, R.A. (1984) Note su alcuni parassiti (s.l.) di neurotteri planipenni con segnalazione del ritrovamento di acari foretici su di un crisopide. *Bollettino dell'Istituto di Entomologia dell'Università di Bologna* 38, 193–203.

Patel, K.G. & Vyas, H.N. (1985) Biology of the green lacewing *Chrysopa (Chrysoperla) scelestes* Banks (Neuroptera: Chrysopidae), an important predator in Gujarat. *Gujarat Agricultural University Research Journal* 11, 18–23.

Potjomkina, V.I. (1987) Influence of some ecological factors on *Micromus angulatus* Stephens (Neuroptera: Hemerobiidae). *Bulletin of All-Union Scientific Research Institute for Plant Protection* 68, 55–59. (in Russian with English summary)

Potjomkina, V.I. (1990) *Instructions, Rearing and Application of* Micromus angulatus *Stephens in the Aphid Control in Greenhouses.* All-Union Scientific Research Institute for Plant Protection, Leningrad. (in Russian)

Principi, M.M. (1940) Contributi allo studio dei neurotteri italiani. 1. *Chrysopa septempunctata* Wesm. e *Chrysopa flavifrons* Brauer. *Bollettino dell'Istituto di Entomologia dell'Università di Bologna* 12, 63–144.

Principi, M.M. (1947) Contributi allo studio dei neurotteri italiani. 5. Ricerche su *Chrysopa formosa* Brauer e su alcuni suoi parassiti. *Bollettino dell'Istituto di Entomologia dell'Università di Bologna* 16, 134–175.

Principi, M.M. (1956a) Reperti etologici su di un raro neurottero italiano, l'*Hypochrysa nobilis* Schneider. *Atti dell'Accademia delle Scienze dell'Istituto di Bologna Rendiconti Serie II*, 3, 152–154.

Principi, M.M. (1956b) Contributi allo studio dei neurotteri italiani. 13. Studio morfologico, etologico e sistematico di un gruppo omogeneo di specie del gen. *Chrysopa* Leach (*Chrysopa flavifrons* Brauer, *prasina* Burm. e *clathrata* Schn.) *Bollettino dell'Istituto di Entomologia dell'Università di Bologna* 21, 319–410.

Principi, M.M. (1986) Lo spermatoforo nei neurotteri crisopidi. *Frustula Entomologica* Nuova Serie 7–8 (20–21), 143–159.

Principi, M.M. (1991) Lo stado di diapausa negli insetti ed il sue manifestarsi in alcune specie di crisopidi (Insecta: Neuroptera) in dipendenza dell'azione fotoperiodica. *Bollettino dell'Istituto di Entomologia dell'Università di Bologna* 46, 1–30.

Principi, M.M. & Canard, M. (1974) Les Névroptères. In *Les Organismes Auxiliaires en Vergers de Pommiers*, pp. 151–162. OILB/SROP.

Principi, M.M. & Sgobba, D. (1987) La diapausa larvale in *Mallada (=Anisochrysa) flavifrons* (Brauer) (Neuroptera: Chrysopidae): cicli fotoperiodici responsabili dell'induzione, sviluppo di diapausa e attivazione, accrescimento ponderate dello stadio con diapausa. *Bollettino dell'Istituto di Entomologia dell'Università di Bologna* 41, 209–231.

Principi, M.M. & Sgobba, D. (1993) La diapausa larvale in *Mallada clathratus* (Schneider) (Neuroptera: Chrysopidae). *Bollettino dell'Istituto di Entomologia dell'Università di Bologna* 48, 75–91.

Principi, M.M., Memmi, M. & Sgobba, D. (1990) Influenza della temperatura sulla diapausa larvale di *Mallada flavifrons* (Brauer) (Neuroptera: Chrysopiade). *Bollettino dell'Istituto di Entomologia dell'Università di Bologna* 44, 37–55.

Propp, G.D., Tauber, M.J. & Tauber, C.A. (1969) Diapause in the neuropteran *Chrysopa oculata*. *Journal of Insect Physiology* 15, 1749–1757.

Rousset, A. (1984) Reproductive physiology and fecundity. In *Biology of Chrysopidae*, Series Entomologica 27, ed. Canard, M., Séméria, Y. & New, T.R., pp. 116–129. Dr W. Junk, The Hague.

Růžička, Z. (1997a) Protective role of the egg stalk in Chrysopidae (Neuroptera: Chrysopidae). *European Journal of Entomology* 94, 111–114.

Růžička, Z. (1997b) Recognition of oviposition-deterring allomones by aphidophagous predators (Neuroptera: Chrysopidae, Coleoptera: Coccinellidae). *European Journal of Entomology* 94, 431–434.

Sagné, J.-C., Moreau, R., Canard, M. & Bitsch, J. (1986) Glucidic variations in the lacewing *Chrysopa walkeri* during the prepupal diapause. *Entomologia Experimentalis et Applicata* 41, 101–103.

Samson, P.R. & Blood, P.R.B. (1979) Biology and temperature relationships of *Chrysopa* sp., *Micromus tasmaniae* and *Nabis capsiformis*. *Entomologia Experimentalis et Applicata* 25, 253–259.

Sheldon, J.K. & MacLeod, E.G. (1974) Studies on the biology of the Chrysopidae. 4. A field and laboratory study of the seasonal cycle of *Chrysopa carnea* Stephens in central Illinois (Neuroptera: Chrysopidae). *Transactions of the American Entomological Society* 100, 437–512.

Shuvakhina, Y.Y. (1987) Biological peculiarities of Aphis-lions. In *Introduction, Acclimatation and Selection of Entomophagous. Proceedings of the All-Union Scientific Research Institute for Plant Protection* 53, 69–77. (in Russian with English summary)

Silva, R.L.X. (1991) Aspectos bioecológicos e determinação das exigências térmicas de *Ceraeochrysa cubana* (Hagen, 1861) (Neuroptera: Chrysopidae) em laboratório. MSc thesis, Escola Superior de Agricultura de Lavras, Lavras, Minas Gerais, Brazil.

Souza, B. (1988) Aspectos morfológicos e biológicos de *Nusalala uruguaya* (Navás, 1923) (Neuroptera: Hemerobiidae) em condições de laboratório. MSc Thesis, Escola Superior de Agricultura de Lavras, Lavras, Minas Gerais, Brazil.

Syrett, P. & Penman, D.R. (1981) Developmental threshold temperature for the brown lacewing, *Micromus tasmaniae* (Neuroptera: Hemerobiidae). *New Zealand Journal of Zoology* 8, 281–283.

Tauber, C.A. (1969) Taxonomy and biology of the lacewing genus *Meleoma* (Neuroptera: Chrysopidae). *University of California Publications in Entomology* 58.

Tauber, C.A. & Tauber, M.J. (1982) Evolution of seasonal adaptation and life history traits in *Chrysopa:* response to diverse selective pressures. In *Evolution and Genetics of Life Histories*, ed. Dingle, H. & Hegmann, J.P., pp. 51–72. Springer-Verlag, New York.

Tauber, C.A., Tauber, M.J. & Nechols, J.R. (1987) Thermal requirements for development in *Chrysopa oculata:* a geographical stable trait. *Ecology* 68, 1479–1487.

Tauber, C.A., Johnson, J.B. & Tauber, M.J. (1992) Larval and developmental characteristics of the endemic Hawaiian lacewing *Anomalochrysa frater* (Neuroptera: Chrysopidae). *Annals of the Entomological Society of America* 85, 200–206.

Tauber, C.A., De León, T., Lopez Arroyo, J.I. & Tauber, M.J. (1998) *Ceraeochrysa placita* (Neuroptera: Chrysopidae): generic characteristics of larvae, larval description, and life cycle. *Annals of the Entomological Society of America* 91, 608–618.

Tauber, M.J. & Tauber, C.A. (1970a) Adult diapause in *Chrysopa carnea:* stages sensitive to photoperiodic induction. *Journal of Insect Physiology* 16, 2075–2080.

Tauber, M.J. & Tauber, C.A. (1970b) Photoperiodic induction and termination of diapause in an insect: response to changing day lengths. *Science* 167, 170.

Tauber, M.J. & Tauber, C.A. (1972a) Larval diapause in *Chrysopa nigricornis:* sensitive stages, critical photoperiod, and termination (Neuroptera: Chrysopidae). *Entomologia Experimentalis et Applicata* 15, 105–111.

Tauber, M.J. & Tauber, C.A. (1972b) Geographical variation in critical photoperiod and in diapause intensity of *Chrysopa carnea* (Neuroptera). *Journal of Insect Physiology* 18, 25–29.

Tauber, M.J. & Tauber, C.A. (1973a) Nutritional and photoperiodic control of the seasonal reproductive cycle in *Chrysopa mohave* (Neuroptera). *Journal of Insect Physiology* 19, 729–736.

Tauber, M.J. & Tauber, C.A. (1973b) Seasonal regulation of dormancy in *Chrysopa carnea* (Neuroptera). *Journal of Insect Physiology* 19, 1455–1463.

Tauber, M.J. & Tauber, C.A. (1974) Thermal accumulation, diapause, and oviposition in a conifer-inhabiting predator, *Chrysopa harrisii* (Neuroptera). *Canadian Entomologist* 106, 969–978.

Tauber, M.J. & Tauber, C.A. (1975) Natural daylengths regulate insect seasonality by two mechanisms. *Nature* 258, 711–712.

Tauber, M.J. & Tauber, C.A. (1976a) Developmental requirements of the univoltine species *Chrysopa downesi:* photoperiodic stimuli and sensitive stages. *Journal of Insect Physiology* 22, 331–335.

Tauber, M.J. & Tauber, C.A. (1976b) Environmental control of univoltinism and its evolution in an insect species. *Canadian Journal of Zoology* 54, 260–265.

Tauber, M.J. & Tauber, C.A. (1981) Seasonal responses and their geographical variation in *Chrysopa downesi:* ecophysiological and evolutionary considerations. *Canadian Journal of Zoology* 59, 370–376.

Tauber, M.J., Tauber, C.A. & Masaki, S. (1986) *Seasonal Adaptations of Insects*. Oxford University Press, New York.

Tauber, M.J., Tauber, C.A., Hoy, R.R. & Tauber, P.J. (1990) Life history, mating behavior, and courtship songs of the endemic Hawaiian *Anomalochrysa maclachlani* (Neuroptera: Chrysopidae). *Canadian Journal of Zoology* 68, 1020–1026.

Tauber, M.J., Tauber, C.A. & Gardescu, S. (1993) Prolonged storage of *Chrysoperla carnea* (Neuroptera: Chrysopidae). *Environmental Entomology* 22, 843–848.

Thierry, D., Cloupeau, R. & Jarry, M. (1994) Variation in the overwintering ecophysiological traits in the common green lacewing West-Palaearctic complex (Neuroptera: Chrysopidae). *Acta oecologica* 15, 593–606.

Toschi, C.A. (1965) The taxonomy, life history, and mating behavior of the green lacewings of Strawberry Canyon (Neuroptera: Chrysopidae). *Hilgardia* 36, 391–433.

Tyshchenko, V.P. & Kind, T.V. (1983) Neurohormonal mechanisms of the regulation of seasonal cycles. In *Hormonal Control of Insect Development*, ed. Gyljarov, M.C., pp. 82–117. Nauka, Leningrad. (in Russian)

Ushtchjekov, A.T. (1989) The rearing of insects with obligatory diapause. In *Proceedings of the 2nd Symposium on the Mass-Rearing of Insects*, pp. 69–70. Moscow. (in Russian)

Vannier, G. (1986) Accroissement de la capacité de surfusion chez les adultes de *Chrysoperla carnea* (Insectes, Névroptères) entrant en diapause hivernale. *Neuroptera International* 4, 71–82.

Vannier, G. (1988) Interruption expérimentale de la diapause dans deux populations de *Chrysoperla carnea* (Insectes, Névroptères): conséquences sur les températures de surfusion et de congélation. *Neuroptera International* 5, 25–37.

Vannier, G. & Canard, M. (1989) Cold hardiness and heat

tolerance in the early larval instars of *Nineta pallida* (Schneider) (Neuroptera: Chrysopidae). *Neuroptera International* 5, 231–238.

Volkovich, T.A. (1987) Role of light and temperature in the control of the active development and diapause of the lacewing *Chrysopa carnea* Stephens (Neuroptera: Chrysopidae). 1. Photoperiodic reaction under condition of constant and gradually changing daylength. *Entomologicheskoye Obozrenie* 66, 3–18. (in Russian with English summary)

Volkovich, T.A. (1996) Effects of temperature on diapause induction in *Chrysopa perla* (Linnaeus) (Insecta: Neuroptera: Chrysopidae). In *Pure and Applied Research in Neuropterology, Proceedings of the 5th International Symposium on Neuropterology*, ed. Canard, M., Aspöck, H. & Mansell, M.W., pp. 259–267. Sacco, Toulouse.

Volkovich, T.A. (1997) Effects of constant and variable temperature on diapause induction in the lacewing *Chrysopa phyllochroma* Wesmael (Neuroptera: Chrysopidae). *Entomologicheskoye Obozrenie* 76, 241–250. (in Russian with English summary)

Volkovich, T.A. (1998) Environmental control of seasonal cycles in green lacewings (Neuroptera: Chrysopidae) from forest–steppe zone of Russia. *Acta Zoologica Fennica* 209, 263–275.

Volkovich, T.A. & Arapov, V.V. (1993) Phenology and seasonal cycles of some species of lacewing Chrysopidae in the reservation "Forest on the Vorska river". *Vestnik St Petersburg University, Serie 3*, 4, 17–25. (in Russian with English summary)

Volkovich, T.A. & Arapov, V.V. (1994) Peculiarities of photoperiodic responses in lacewing *Chrysopa perla* L. (Neuroptera: Chrysopidae) under constant and changing temperature conditions. *Entomologicheskoye Obozrenie* 73, 506–520. (in Russian with English summary)

Volkovich, T.A. & Blumental, N.A. (1997) Photo-thermoperiodic responses in some species of lacewings (Chrysopidae): their role in diapause induction. *European Journal of Entomology* 94, 435–444.

Volkovich, T.A. & Orlova, N.A. (1998) Comparative study of seasonal strategies in two species of lacewings (Chrysopidae) in the forest–steppe zone of Russia. *Entomologicheskoye Obozrenie* 77, 3–15. (in Russian with English summary)

Wei, C.-s., Huang, B.-z. & Guo, C.-h. (1986) Studies on *Chrysopa boninensis* Okamoto in Guangzhou. *Acta Entomologica Sinica* 29, 174–180.

Whittington, A.E. & Brothers, D.J. (1991) Notes on the biology of *Mallada handschini* (Navás) and comparisons with other southern African species (Neuroptera: Chrysopidae). *Annals of the Natal Museum* 32, 215–220.

Withycombe, C.L. (1923) Notes on the biology of some British Neuroptera (Planipennia). *Proceedings of the Entomological Society of London* 1922, 501–594.

Zelený, J. (1963) Hemerobiidae (Neuroptera) from Czechoslovakia. *Acta Societatis Entomologicae Čechoslovenie* 60, 55–67.

PART 2

Lacewings in crops

Introduction to Part 2

The Editors

The species of lacewings found in the crop environment may, in some way, be 'pre-adapted' to that environment through their regular association with a given vegetation type, and by prey preferences. In seeking additional candidate species for use in IPM, surveys of the lacewings found 'naturally' on particular crops or vegetation may provide valuable leads on which may be useful, and thus targeted for increasing the collective impact on pests or – at least – for augmenting control. Much of the use of lacewings in IPM will rely increasingly on native species, perhaps those found naturally in relatively limited areas or microhabitats, and the study of such natural assemblages will also suggest combinations of species for complementary use. This may involve combinations of high impact generalist species (such as members of the *Chrysoperla carnea* group, so beneficial in a great variety of crop situations) and more specialised species whose more restricted predilections may favour their use in particular contexts.

This section helps us to understand the spectrum of lacewings which may be involved, and how the assemblages found on different crops vary. Despite the generalised information available, the essays collectively mention around 10% each of the global fauna of Chrysopidae and Hemerobiidae (113 species in 22 genera, and 64 species in 10 genera respectively), but also indicate the low involvement of Coniopterygidae. A general picture emerges of particular lacewing species being abundant on different crops, and a greater variety of low abundance (and, perhaps, less habitat-specific or habitat-dependent) taxa. In total, species from this diverse list have been found associated with 19 ground crops, 21 tree crops and 39 forestry and horticultural plants, with Europe and the Middle East having been more thoroughly surveyed than the remainder of the world.

Maintenance of such reservoirs of predators (and potential predators) is a key aspect of management, and involves steps to maintain surrounding areas of natural vegetation as habitats. This section therefore sets the background on lacewing abundance and diversity in crops for the practical developments discussed later in the book. Many of the cases described in Part 4 directly complement the information presented here.

CHAPTER 8
Lacewings in field crops

P. Duelli

8.1 INTRODUCTION

The best known aphidophagous predators in field crops are syrphid flies, coccinellid beetles, and chrysopid lacewings. Lacewings are likely to occur on almost all field crops all over the world. Where aphids occur on field crops, lacewings will also occur. There is one notable exception: there are no Chrysopidae on field crops in New Zealand, and all deliberate importations have been unsuccessful. However, there are some hemerobiid brown lacewings in field crops in New Zealand, mainly introduced from Australia (Wise, 1996).

Most lacewing species, green or brown, are arboreal or live on woody shrubs. Few species develop on field crops, but those species that do can be extremely abundant and widespread. The literature on predatory activity of lacewings in crops is dominated by one single species complex: *Chrysoperla carnea*. A literature survey (400 BIOSIS and CAB abstracts) of papers published since 1970 on lacewings in field crops indicated that 83% of all contributions refer to *C. carnea*. The problem with all published work on *C. carnea* is, that in the light of recent investigations it became clear that *C. carnea* is in fact a complex of at least 20 sibling species, some of which are at present only distinguishable behaviourally (see Chapter 3, this volume).

8.2 THE MAIN LACEWING GENERA PRESENT ON FIELD CROPS

The following genera are prioritised by the number of publications since 1970, with scientific names adapted to the new taxonomy proposed by Brooks & Barnard (1990) and Brooks (1994).

8.2.1 Chrysopidae

Genus Chrysoperla

The genus *Chrysoperla* is widespread in cultivated areas almost all over the world. The larvae are predatory, but adults feed on honeydew of homopteran insects and on pollen. The proportion of species inhabiting low vegetation, the herbaceous layer, is much higher than in any other lacewing genus. Therefore, many species of the genus *Chrysoperla* play a dominant role in the biological control of field-crop pests (New, 1975, 1988; Canard & Séméria, 1984).

Brooks (1994) divided the genus into four distinct species groups. Only two of them, the *carnea*-group and the *pudica*-group, contain well-known species considered as enemies of field-crop pests. The two other groups (the *comans*-group and the *nyerina*-group) contain only very few species, and are mainly arboreal.

The *carnea*-group: as pointed out in more detail by Henry *et al.* (see Chapter 3, this volume), the *carnea*-group contains at least 20 more or less cryptic sibling species. Among them, the most abundant species also are the most important in field crops: *C. carnea*, *C. plorabunda*, *C. lucasina*, *C. nipponensis*, and *C.c.5* (see below). All these at times had been synonymised with *C. carnea* (Aspöck *et al.*, 1980). Almost all publications using the name *C. carnea* in the last 50 years refer to one of these five species, which are common and widespread in parts of the northern hemisphere. The only representative of the *carnea*-group in the southern hemisphere is *C. zastrowi*, which is an important predator in field crops in southern Africa (Barnes, 1975).

The *pudica*-group: there are only a few, but important field crop species in this group: *C. congrua*, *C. rufilabris*, and *C. externa* (formerly *C. lanata*). In addition, this group contains some abundant arboreal species such as *C. pudica*. The *pudica*-group has been defined by Brooks (1994) on adult characters only. The larvae of *C. congrua* are completely different from the rest of this highly distinctive group, being pale and slender, and resemble larvae of the *carnea*-group. In contrast, the larvae of the other species of the *pudica*-group are humped, especially in resting position, with two dark lateral marks on the dorsal hump (e.g. *C. pudica, C. mutata, C. comanche, C. rufilabris*). Moreover, *C. congrua* is specific within the *pudica*-group in its habitat requirements. Adapted to living on tall grass, *C. congrua* is mainly found in monocotyledonous field crops such as wheat, corn, barley, and millet. *C. rufilabris* represents 5% of the BIOSIS and CAB abstracts examined (the highest number of citations since 1970 in the *pudica*-group), followed by *C. externa* with 1% and *C. congrua* with fewer than 1%.

Genus Chrysopa

Before 1977, when the genus *Chrysoperla* was raised from subgeneric status by Séméria (1977), most publications on lacewings in field crops used the genus name *Chrysopa* uncritically for any kind of green lacewing. Even today, many authors in applied entomology still use the genus name *Chrysopa* for genera such as *Chrysoperla* or *Brinckochrysa*. In this chapter, however, only the species of the true genus *Chrysopa s. str.* are included, which are characterised by predatory adults.

The genus is spread all over the world, but only in the northern hemisphere are there species of significance on field crops: *C. oculata* in North America, *C. pallens, C. formosa, C. phyllochroma*, and *C. commata* in Europe and Asia. Some other species such as *C. perla* are sometimes mentioned for field crops, but they rarely develop there, and invade from surrounding natural vegetation. None of the species in this genus reached 1% of the citations in CAB and BIOSIS.

Genus Mallada

Before the large genus *Dichochrysa* was split from the genus *Mallada* (formerly *Anisochrysa*) by Yang & Yang (1990, 1991), *Mallada* was by far the largest genus within the Chrysopidae, with more than 122 described species. Of all these, very few are known to be active in field crops. With 3% of all citations for chrysopids in field crops, *Mallada desjardinsi* (formerly *M. boninensis*) is the most important species. Furthermore, *M. signatus* is a common lacewing in field crops in Australia. Although *Dichochrysa* has the highest number of species in the family Chrysopidae, it is not of importance in field crops.

Genus Brinckochrysa

The genus *Brinckochrysa* is mainly arboreal, and in the tropics some species are important for biological control in plantations of citrus or mango. Only one species is repeatedly recorded from field crops, reaching 3% of all citations in BIOSIS and CAB since 1970: that is *Brinckochrysa scelestes*, mainly under the name *Chrysopa* or *Chrysoperla scelestes*, on the Indian subcontinent. Reports of Gautam (1994) of the two genera *Chrysoperla* (*C. carnea*) and *Brinckochrysa* (*B. scelestes*) interbreeding successfully cast some doubt on the taxonomic identity of the species used in India for biological control. So far, some closely related sibling species in the *carnea*-group have been reported to interbreed successfully in close confinement in the laboratory (Duelli, 1987a; Wells & Henry, 1994), but fertile hybrids between genera would be a surprise.

Genus Ceratochrysa

Only one species, *Ceratochrysa antica*, is known to occur regularly on field crops in tropical Africa.

Genus Ceraeochrysa

Only one species, *Ceraeochrysa cubana*, is known to play a major role in field crops in Central and South America.

8.2.2 Hemerobiidae

Many species of the Hemerobiidae are efficient predators of economic importance in agriculture, but there are few publications on their role in field crops. The genus *Hemerobius* contributes with the ubiquitous and common Palaearctic species *H. humulinus* and the Nearctic *H. pacificus* (Neuenschwander & Hagen, 1980). The genus *Micromus* is represented with *M. variegatus* in Eurasia, and *M. tasmaniae* in the Australian region (Samson & Blood, 1980). *Micromus timidus* is an extremely widespread tropical species ranging from Africa to the Pacific Islands.

8.3 DOMINANT LACEWING SPECIES IN FIELD CROPS IN DIFFERENT CONTINENTS

Seen on a world-wide scale, the species composition of lacewings in field crops does not depend on the type of field crop, but on the regional species composition of potential field-crop species on a continent. In other words, there are no particular species for cotton, wheat, alfalfa, or maize in any one country, but on each continent or subcontinent a different set of species occurs in field crops.

8.3.1 Europe and Mediterranean countries

Most work on lacewings in field crops has been done in Europe, but numerous contributions published in local journals and in various languages are often overlooked or simply ignored because there is no English translation. The main problem today, however, is the fact that in most older publications it is impossible to trace back what species in the *carnea*-complex was observed or investigated under the published name *Chrysoperla carnea*.

The original *C. carnea* described by Stephens (1846) was a British lacewing, and since most publications on lacewings in field crops in Europe and the Mediterranean area consider *C. carnea* to be the dominant and economically most important species, there is an urgent need to clarify what species in the *carnea*-group are actually present in field crops in these areas.

So far, at least seven species in the *carnea*-group (distinct either morphologically or as song morphs) are known to occur in Europe. Three of them, *C. mediterranea*, *C. renoni*, and *C. ankylopteriformis*, certainly have no affinities to field crops. *C. lucasina*, on the other hand, recently reinstated as a valid species by Lerault (1991) and verified by Henry *et al.* (1996), is the most important species in field crops in the southern half of Europe (Malet *et al.*, 1994) and in most Mediterranean countries. *C. lucasina* is morphologically distinct from all other European species of the *carnea*-group (recognisable at least when alive or in alcohol) by displaying a thin black stripe laterally on the pleura of the first abdominal segment. *Chrysoperla lucasina* is the only European *Chrysoperla* species in field crops that never changes colour in diapause.

The other three European sibling species in the *carnea*-group (*C.c.2*, *C.c.3*, and *C.c.4* – see Chapter 3, this volume) are clearly distinct song morphs, but sometimes difficult or impossible to identify morphologically.

C.c.2 can be distinguished from the two other sibling species by experts on morphological grounds, at least when alive or in alcohol. *C.c.2* so far has never been found in field crops or in other low vegetation, but it is a common species developing on deciduous trees in forests and in urban parks. It ranges from western and central Europe south to the Mediterranean coast in France and even Spain.

There is at present no easy way to separate *C.c.3* and *C.c.4* morphologically (see Chapter 3, this volume), but their precopulatory tremulation patterns are very distinct. *C.c.3* is a Mediterranean species, which has never been found north of the Alps, and *C.c.4* is the most common lacewing in central and northern Europe. *C.c.4* is the only species in Europe which often overwinters in large aggregations in barns and houses, displaying the characteristic carneous (flesh-red) diapause colouration which, it is said, has given the species the name *carnea*. Both *C.c.2* and *C.c.3* have a more yellowish-brown diapause colouration. *C.c.4* larvae are very frequent on tall grass and in field crops, but the adults can also be found on trees, mainly in spring, feeding on pollen, and in autumn before hibernation.

The conclusions for the interpretation of the species identity of former work on *C. carnea* are as follows. For publications on lacewings in field crops in northern Europe (especially Scandinavia and Russia) we can safely assume that the species called *C. carnea* were indeed *C.c.4*. If colour change or aggregations in winter are mentioned, it must have been *C.c.4* also in western and central Europe.

But for most work on *C. carnea* in western, central and eastern Europe it is hardly possible to decide retrospectively, whether *C. lucasina* or *C.c.4* had been investigated in the original publication, unless alcohol specimens were kept in reference collections.

In southern Europe and northern Africa the same applies to *C. lucasina* and *C.c.3*, in some places even *C.c.4* may occur sympatrically. In southern France (the district of Var) I was able in 1996 to collect all three species sympatrically in two neighbouring fields of maize and alfalfa. While the boreal and alpine populations of *C.c.4* in field crops only have one generation per year, the same species, as well as *C.c.3* and *C. lucasina*, have up to three or four generations in the south.

With the above-mentioned caveats on the identification of the true species name in former publications on *C. carnea*, we can interpret the literature on the lacewing fauna in the major field crops of Europe and the Mediterranean countries as follows. Lacewings in field crops identified as *C. carnea* which show diapause coloration or were found hibernating in 'green lacewing chambers' (Sengonca & Henze, 1992) are *C.c.4*. In all other cases, it could have been either *C.c.4* or *C. luca-*

sina, e.g. on wheat in Moldavia (Voicu & Nagler, 1987), on wheat in Britain (Vickerman & Sunderland, 1977), on cabbages for sauerkraut in Alsace, France (Herold & Stengel, 1994), on maize in Romania (Marin, 1988; Voicu, 1989), feeding on Colorado beetles in USSR, Ukraine and Moldavia (Shuvakhina, 1985), and on tobacco in Bulgaria (Kaitazov & Kharizanov, 1976; Kharizanov & Babrikova, 1978). In recent years, I have collected song-identified specimens of *C.c.4* on maize fields in Hungary, on alfalfa in Slovakia and Ukraina, and on various field crops in Switzerland and northern Italy.

In southern Europe and the Mediterranean countries the published species name *C. carnea* represents different species: in most cases in southern Europe (Miermont & Canard, 1975; Pantaleoni & Tisselli, 1986), *C. lucasina* would probably be the correct name today, as shown by Malet *et al.* (1994), but *C.c.3* or *C.c.4* can also be found in crop fields. The presently known distribution of *C. lucasina* in Europe and northern Africa is shown by Thierry *et al.* (1996) and Henry *et al.* (1996). *C.c.3* was also identified by the author from field crops in Israel and Algeria.

Further south and east of the Mediterranean basin, *C. carnea s. lat.* is considered to be by far the most important lacewing species on cotton in Egypt (Mostafa *et al.*, 1976; Maher-Ali *et al.*, 1983; Abdel-Fattah *et al.*, 1984; Salem & Matter, 1991), in Iraq (Awadallah & Khalill, 1979), in Israel (Gerling & Bar, 1985; Gerling *et al.*, 1997). On radish and potato plants, Al-Darkazly & Jabri (1989) regarded *C. carnea* as the most efficient predator. In all these cases, either *C. lucasina*, *C.c.3*, or most probably *C.c.5* was the species investigated, but certainly not the true *C. carnea* Stephens. *C.c.5* is the most common lacewing in irrigated alfalfa fields in the sandy deserts of Dubai, Oman and southern Israel (P. Duelli, unpublished data) and it might well be a common species in field crops all over the Sahel belt and on the Cape Verde Islands (H. Hölzel & P. Ohm, personal communication).

Apart from the genus *Chrysoperla*, very few other lacewing species have been reported from field crops in Europe and the Mediterranean regions. In central and eastern Europe, *C. carnea s. lat.* is often accompanied by *Chrysopa phyllochroma* and *C. commata*, e.g. on barley, rye, potatoes, and beans in Poland (Galecka & Zeleney, 1969), *C. formosa* on various pest insects in Romania (Voicu, 1988), *Micromus variegatus* and *Hemerobius*

humulinus on maize (Duelli, 1986*b*), and *M. angulatus* in alfalfa (Miermont & Canard, 1975). On tobacco in Bulgaria, *C. formosa*, *C. pallens*, and *C. perla* were found (Kharizanov & Babrikova, 1978).

8.3.2 Asia

In the Palaearctic part of Asia, the situation is similar to Europe, with a dominance of species of the *carnea*-group. Most publications from nations of the former USSR refer to *Chrysoperla carnea s. lat.* in cotton. Adashkevich *et al.* (1981) found in Uzbekistan that *C. carnea* appears too late in spring to have an impact on cotton aphids. But a number of other species were also collected on cotton in that country: *Chrysopa pallens*, *C. abbreviata*, and even the usually arboreal species *Cunctochrysa albolineata* and *Nineta vittata*. Inconsistent results were found for the impact of *Chrysoperla carnea* on spider-mites in cotton (Pavolva & Aripova, 1991). At present we have no idea what species of the *carnea*-group is meant by the name *C. carnea* in these Asian countries, but we know that the song morphs in central Asia (Kyrgyzstan) are different from the European ones (C.S. Henry & P. Duelli, unpublished data). *C. carnea s. lat.* is also reported from cotton in Tajikistan (Sukhoruchenko *et al.*, 1981), Azerbaijan (Aliev & Kurbanov, 1981; Rustamova, 1981; Gurbanov, 1982), Samarkand (Deryabin, 1986), and Afghanistan (Stolyarov *et al.*, 1974).

On the Indian subcontinent, unidentified sibling species of the *carnea*-group (maybe *C.c.5* or *C. sillemi*?) are reported under the name *C. carnea* as dominant on maize (Varm & Shenhmar, 1983) and cotton (Kapadia & Puri, 1990; Manjunath, 1993; Mannan *et al.*, 1995; Muralidharan & Chari, 1990; Natarajan, 1990; Singh *et al.*, 1993). Another very important green lacewing in Indian cotton fields is *Brinckochrysa scelestes* (often as *Chrysopa* or *Chrysoperla scelestes*). While Patel & Vyas (1985) report on seasonal abundance and oviposition sites, most papers are on pesticide resistance (Bhuvaneswari & Uthamasamy, 1994; Krishnamoorthy, 1985; Patel & Yadav, 1995). *B. scelestes* was mass-reared and released against potato pests in India (Ansari *et al.*, 1992) and can be hybridised successfully with *C. carnea s. lat.* according to Gautam (1994).

In China and Japan, *C. nipponensis* (in China under the name *C. sinica* and in Japan sometimes as *C. carnea*) is the most common species in crop fields. In Chinese cotton fields, *C. nipponensis* is often accompanied by

Chrysopa pallens (Huang *et al.*, 1990) and *C. formosa* (Li, 1987; Wang & Wang, 1996). *Chrysopa cognata*, possibly a synonym of *C. pallens*, is abundant in crop fields in Korea (Boo *et al.*, 1998). Cotton bollworms are also eaten by *C. phyllochroma* (Dai & Gua, 1992). Mass-reared *Chrysoperla nipponensis* were released in wheat fields in northern China (Zhou *et al.*, 1991).

8.3.3 North America
Species of the genus *Chrysoperla* are also dominant in crop fields of this continent. Most publications using the names *Chrysopa* or *Chrysoperla carnea* are most probably dealing with *C. plorabunda*. Since this species seems to be the only one in the *carnea*-group in America to display a reddish- or yellowish-brown diapause coloration (D. Johnson, personal communication), all publications mentioning such a colour change seem to refer to *C. plorabunda*. *Chrysoperla plorabunda* was manipulated with artificial honeydew in Californian alfalfa, safflower, and cotton fields (Hagen *et al.*, 1971; Duelli 1980*a*) and alfalfa in Utah (Evans & Swallow, 1993). They were investigated on cotton in Connecticut (Stark & Whitford, 1987), in Texas (Ables *et al.*, 1980; Rajakulendran & Plapp, 1982; Ridgway & Jones, 1969; Ridgway & Kinzer, 1974), on potatoes in Idaho (Ben Saad & Bishop, 1976) and California (Heimpel & Hough-Goldstein, 1992), wheat in Kansas (Messina *et al.*, 1995), cotton in Mississippi (Stark & Hopper, 1988), alfalfa in Michigan (Maredia *et al.*, 1992) and South Dakota (Micinski *et al.*, 1991), and maize in Iowa (Obrycki *et al.*, 1989). In field experiments in California, Cisneros & Rosenheim (1997) were able to show that *C. plorabunda* (as *C. carnea*) is a very efficient predator on cotton aphids, and that the reduviid bug *Zelus renardii* is not. But the bug very efficiently eats *C. plorabunda* larvae and thus releases aphid populations from regulation by lacewings. In Mexico also, *C. plorabunda* is the most frequent lacewing species in cotton (Jimenez Aragon, 1974).

From Texas to the east coast of the USA, *C. rufilabris* may take the position of *C. plorabunda* in field crops, e.g. on cotton in Texas (Treacy *et al.*, 1987). While *C. rufilabris* is an excellent predator on potato beetles (Nordlund *et al.*, 1991), the effect seems to be very limited on pests of water melon and leguminous forage crops (Legaspi *et al.*, 1996). In Florida, *C. rufilabris* is joined by *Ceraeochrysa cubana* (Dean & Schuster, 1995).

In the northern USA and Canada, *C. plorabunda* can be accompanied by *Chrysopa oculata* in maize (Coderre *et al.*, 1987) and in alfalfa (Lavallee & Shaw, 1969).

8.3.4 Central and South America
Here the dominant *Chrysoperla* species in field crops is *C. externa*. It feeds on cotton boll weevils in Nicaragua (Beije, 1993) and was successfully induced to oviposit in maize fields in Brazil by spraying molasses (Freitas & Giacheto Scaloppi, 1996). *Chrysoperla externa* is also reported to be common in maize in Peru (Nunez, 1988), while *Ceraeochrysa cincta* is more arboreal and often lives on citrus.

8.3.5 Africa south of the Sahara
The most widespread and locally most abundant species in field crops in sub-Saharan Africa is *Chrysoperla congrua*. It is known from the Cape region and Madagascar north to Ethiopia and the Arabian Peninsula and eastwards throughout Australasia to Easter Island (Brooks, 1994). In natural habitats, the species lives on tall grasses, and accordingly it is abundant on wheat and corn. Brettell (1982) also reports it as a less common species on cotton in Zimbabwe, together with *C. pudica*, which in turn is otherwise more common on shrubs and trees all over sub-Saharan Africa (Brooks, 1994). Kabissa *et al.* (1996) also reported *C. congrua* from cotton, but it seems to be active in the wrong season to eliminate cotton pests such as *Helicoverpa armigera*. Another lacewing species in cotton is *Mallada desjardinsi* (formerly *M. boninensis*) which ranges from eastern Africa, e.g. Tanzania (Kabissa *et al.*, 1995, 1996) and Zimbabwe (Brettell, 1979, 1984) to the Pacific Islands.

Another important species in field crops is *Chrysoperla zastrowi*, the only species of the *carnea*-group in the southern hemisphere. It abounds in cotton fields in South Africa (Barnes, 1975), and has its main distribution in southwestern Africa.

The genus *Ceratochrysa* with only three species is endemic to Africa, with *C. antica* known to feed on insect pests of cassava (Barnard & Brooks, 1984) and cotton (Kabissa *et al.*, 1989).

8.3.6 Australia
In Australia, the chrysopid *Mallada signatus* and the hemerobiid *Micromus tasmaniae* seem to be the most

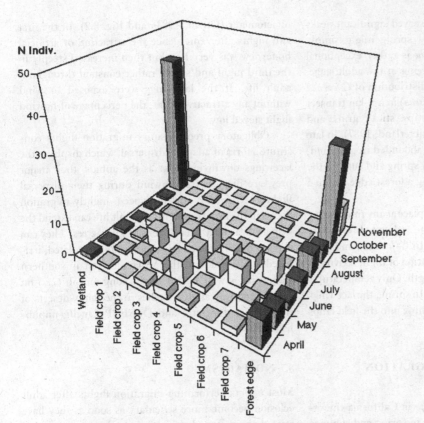

Fig. 8.1. Distribution of *Chrysoperla carnea s. lat.* (*C.c.4* and some *C. lucasina*) in space and time in a transect of trap stations extending over 5 km of intense agriculture (Field crop 1–7: wheat, maize, improved grassland), starting in an area of brushy wetland and ending on a semi-dry meadow along the edge of a mixed forest. Main dispersal flights to ecotonal overwintering sites occurred in October.

important lacewings in field crops, e.g. in cotton fields (Samson & Blood, 1980).

8.4 MIGRATION FLIGHTS AND NOMADISM IN FIELD CROPS

Field crops are extremely temporary habitats. For part of the year they are inhospitable, and often the kinds of crops change from year to year. Lacewing species as predators of soft-bodied, mostly homopterous, and often host-specific herbivores have to be mobile in order to follow their prey. Lacewings are predominantly arboreal, but a few species have adapted to living in temporary and patchy environments. These few species are the ones that are most successful in field crops, and they are abundant and widespread.

The strategy of field-crop specialists among the chrysopid and hemerobiid lacewings can be exemplified by the nomadic behaviour found in the genus *Chrysoperla*, particularly within the *carnea*-group. A detailed analysis of the dispersal and oviposition strategy was done with *C. plorabunda* in the California

Central Valley (Duelli, 1980*a*, *b*, 1984*b*, 1986*a*) and, later on, a similar study was performed on *C.c.4* or *C. lucasina* (not discriminated at that time) in Switzerland (P. Duelli, unpublished data). It is plausible to assume that other species in the *carnea*-group, such as *C. nipponensis* in eastern Asia, *C. externa* in Central America, *C.c.5* in northern Africa and the Arabian Peninsula, and *C. zastrowi* in southern Africa behave similarly.

The basic dispersal strategy consists of three types of flight behaviour: migration flights to overwintering sites after diapause induction in late summer and back into field crops in spring, preovipository migration flights to find new habitats with aphid colonies, and continued nomadism throughout the reproductive period, in order to spread the risk for the offspring in unpredictable, temporary, and patchy habitats.

8.5 MIGRATION FLIGHTS TO AND FROM OVERWINTERING SITES

At present we do not know whether migration flights to and from hibernation sites are a peculiarity of field-crop species. The two observed significant peaks of '*C. carnea*' in autumn and spring might simply reflect the fact that *Chrysoperla* is rather exceptional among lacewings by overwintering in the adult stage. Figure 8.1 shows the seasonal distribution of *C. carnea* (mainly *C.c.4* and a few *C. lucasina*) in a 5-km transect with window (interception) traps, sticky grids, and yellow pan traps (Limpach, Switzerland, 1987). In late spring and summer, the adults abounded in grassland, wheat, and maize fields, in early spring and autumn the peaks of trap catches were along a forest edge and in a brushy wetland.

Migration flights may take place at any time during and after diapause, as soon as the ambient temperature is above 10°C (Duelli, 1986a). In *C. lucasina* and *C.c.4* in Switzerland, there is no cessation of flight activity in winter induced by short daylength. Only temperatures below 10°C limit flight activity. In spring, the lacewings feed on tree pollen before returning into the field crops for oviposition.

8.6 PREOVIPOSITORY MIGRATION FLIGHTS

Adults of *Chrysoperla plorabunda* in California emerging from cocoons in field crops in spring and summer undergo an obligatory migration flight before they start to react to the scent of habitat kairomones (Duelli, 1980b). Only after these extended migration flights in the first two nights do the females copulate and, after another two to four days, start to oviposit (Duelli, 1980b). Preovipository migration flights are characterised by a downwind flight, which is not distracted by vegetative stimuli such as food or potential mating partners. Field experiments in alfalfa fields in the California Central Valley with strips sprayed with an artificial food mixture imitating aphid honeydew have shown that the numbers of prereproductive females caught in sticky grids were about the same upwind and downwind of the sprayed strips (Duelli, 1984a). The number of pregnant females, however, was much lower in the traps downwind of the food sprays than upwind. The interpretation is that pregnant females react to the food spray and land, while freshly emerged females continue their downwind migration flight.

In laboratory experiments, the age-dependence of the reaction to artificial food sprays was confirmed in an olfactometer (Duelli, 1984a and Fig. 8.2). In the first two nights after emergence the attraction of artificial honeydew was very low, but then increased steeply in the third night and stayed rather constant throughout adult life. If the lacewings were exposed to wind without any attractive scent, the percentage of upwind flight stayed low.

Obligatory preovipository migration flights constitute a form of adaptive dispersal, which displaces the lacewings downwind, just as the aphids, their major prey, are displaced downwind during their dispersal flights. Depending on wind speed, nightly migration flights in the first two nights of adult life can extend the dispersal capacity over dozens of kilometres. They can explain the presence of *C. plorabunda* on isolated, irrigated alfalfa fields in the Mojave Desert in southern California, as well as that of the song morph *C.c.5* on remote and tiny alfalfa patches in the sandy deserts of the Arabian Emirates and Dubai (P. Duelli, unpublished data).

8.7 NOMADISM

Most insects performing migration flights after adult eclosion become more sedentary as soon as they have started their reproductive period. But many species, particularly those of agricultural importance, go on dispersing throughout their reproductive period, depositing eggs in different fields. At least for *Chrysoperla plorabunda* in California, it could be shown that reproductively active females do not stay for more than one or two days in a crop field, even though there is plenty of food (Duelli, 1984a). In Switzerland, the weekly immigration and emigration rates of *C.c.4* or *C. lucasina* (not specifiable at that time) were investigated using 7-m-high sticky grid traps on each side of a 1-ha maize field (P. Duelli, unpublished data). Extrapolating the results from the four traps, each 1m wide, thus exposing a sticky surface of 7 m² on either side, yielded an average immigration of 1500 and an emigration of 1700 lacewings per hectare per night in July and August. Lacewings flying higher than the 7-m-high traps were considered to be migrants, which were not on an appetitive downwind flight (Duelli, 1980a) in search of food or mating partners. The average net emigration rate therefore is 200 individuals per hectare per night. At the same time, the

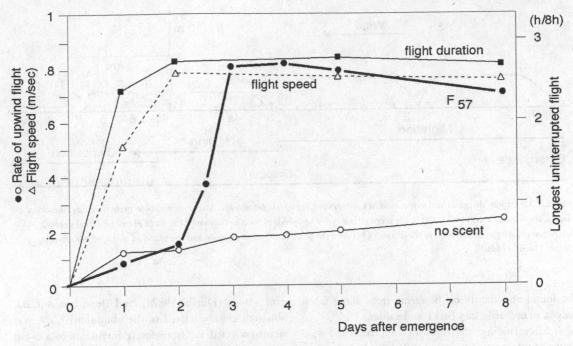

Fig. 8.2. Flight capacity of female *Chrysoperla plorabunda* after adult eclosion and the ontogenetic development of the attraction to the scent of artificial honeydew (F57: brewer's yeast with yeast hydrolysate) in an olfactometer. In a wind tunnel, the percentage of upwind flight was measured at different days after adult eclosion. Longest uninterrupted flight duration per night was recorded on a flight mill, and flight speed was measured on 2 m of phototactic free flight towards a light source. (From Duelli, 1984*a*.)

presence of adults in the maize field was counted weekly, as well as the number of larvae and pupae. Since the first two larval instars and the cocoons are difficult to find in the field, the number of third-instar larvae per 100 plants was used to estimate the production of adults per hectare and per day. In July and August, an average of 3500 adults was present in 1 ha of maize, and the production rate was 120 adults per hectare per day. A peak of maximum production was observed in the week of 12–19 August 1983, when 1800 adults emerged per hectare per day. Assuming that all immigrating lacewings flying lower than 7 m will land in a maize field which contains aphids, it can be concluded that an adult remains in that field for an average of only two days.

The spatial aspects of migration, dispersal, and oviposition behaviour in *C. plorabunda* are summarised in Fig. 8.3 (Duelli, 1984*a*). In the first two nights after emergence, the adults perform adaptive dispersal flights in the sense of preovipositional migration flights.

These are straight downwind flights mostly at elevations higher than 5 m above ground. The lacewings do not react to the scent of honeydew in the crop. While males start to be reproductively active already on day 2, the females do not copulate before day 3, when most of the migration period is over. In autumn, the migration flight from the crop fields to the hibernation sites is a similar kind of flight, while the lacewings are in a state of reproductive diapause.

After two nights of migration flights, which will bring the lacewings out of their native habitats into distant new breeding places, the scent of honeydew becomes a strong landing stimulus. The flight becomes an appetitive downwind flight, usually lower than 5 m above ground. The flight back from the overwintering sites into the crop fields is a similar kind of appetitive flight, where the lacewings are no longer in diapause and react to the scent of honeydew, but also of nectar, in search of pollen. Therefore, in late winter and early spring, field-crop specialists among the lacewings can

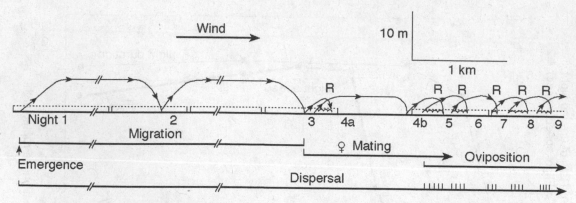

Fig. 8.3. Migration, dispersal and oviposition in Californian *Chrysoperla plorabunda*. After two nights of preovipository migration flights, the females start to react (R) to the scent of honeydew of aphids as soon as they enter the scent plume of a food source. Dispersal capacity is shown here as the distances between the place of origin of a female and the places of origin of her offspring. (From Duelli, 1984a.)

be found abundantly on flowering trees along forest edges, in orchards, city parks, and gardens.

After landing, the lacewings approach the food source in an appetitive upwind flight, which is a low, stepwise flight within the boundary layer. They feed on honeydew, nectar, and sometimes pollen. They mate after duetting with their specific vibrational calls (Henry, 1979). During the daytime, the lacewings rest on the underside of leaves and twigs. Towards the evening, they become active again, hopping with short flights from plant to plant. The lacewings do not remain at the food source for long. The females may deposit some eggs before dusk, but as soon as it gets darker than 10 lux (Duelli, 1986a), they take off and reach a new food source somewhere downwind. The eggs are deposited singly, mostly at dusk and dawn, and they are not as strictly placed close to aphid colonies as in predatory species (Duelli, 1987b).

Thus the dispersal of lacewing populations in crop fields is a continuous rolling downwind movement along the prevalent nightly wind directions – some kind of 'downwind nomadism'. At this point we can only speculate whether the dispersive behaviour observed in species of the *carnea*-group is a rather general feature of field crop species, or restricted to the few species observed, or whether it is a widespread trait in many lacewing genera. Obviously, large-scale population movements are not a characteristic of the *carnea*-group. *C. mediterranea* survives in isolated populations on pine in the Alps, with no apparent spread into neighbouring

pine stands (Duelli, 1987a), and the arboreal *C.c.2*, although closely related to the ubiquitous *C.c.4*, was never collected far from forest fragments in a 5-km transect through agricultural land in Switzerland (P. Duelli, unpublished data).

An evolutionary explanation for the dispersive behaviour of field-crop species could be their descent from forest-gap species. The phylogenetically very old lacewing families Chrysopidae and Hemerobiidae originally may only have consisted of forest species, which in an almost continuous virgin forest in temperate areas of the world had no reason to develop migratory behaviour. But some species adapted to living in forest gaps, and since those habitats were isolated and mostly temporary and unpredictable, the lacewings had to be mobile. Nowadays, these few original gap specialists profit from agriculture. Their habitats have extended drastically with human help, and since these new habitats also are temporary and unpredictable, the former gap specialists are now ideally adapted to living in field-crop habitats.

ACKNOWLEDGEMENTS

I would like to thank Alois Kempf for his competent help with the BIOSIS/CAB literature search, Herbert Hölzel, Peter Ohm and Ding Johnson for unpublished information, and Peter Wirz, Stephan Burkhart, and Mario Waldburger for their continuous support in rearing lacewings and their food.

REFERENCES

Abdel-Fattah, M.I., Hendi, A. & El-Said, A. (1984) Ecological studies on parasites of the cotton whitefly *Bemisia tabaci* Genn in Egypt. *Bulletin of the Entomological Society of Egypt Economic Series* 14, 95–106.

Ables, J.R., Jones, S.L. & House, V.S. (1980) Effect of diflubenzuron on entomophagous arthropods associated with cotton. *Southwestern Entomologist*, Supplement 1, 31–35.

Adashkevich, B.P., Adylov, Z.K. & Rasulev, F.K. (1981) The biomethod in action. *Zashchita Rastenii* 9, 9–10.

Al-Darkazly, T.A. & Jabri, N.M. (1989) Natural enemies of green peach aphid *Myzus persicae* Sulzer in middle of Iraq. *Journal of Agriculture and Water Resources Research Plant Production* 8, 105–114.

Aliev, A.A. & Kurbanov, G.G. (1981) Seasonal colonisation of chrysopids. *Zashchita Rastenii* 3, 35–45.

Ansari, M.A., Pawar, A.D. & Kumar, D.A. (1992) Possibility for biocontrol of tropical armyworm, *Spodoptera litura* (Lepidoptera: Noctuidae) on potato. *Plant Protection Bulletin (Faridabad)*, 44, 27–31.

Aspöck, H., Aspöck, U. & Hölzel, H. (1980) *Die Neuropteren Europas. Eine zusammenfassende Darstellung der Systematik, Ökologie und Chorologie der Neuropteroidea (Megaloptera, Raphidioptera, Planipennia)*, 2 vols. Göcke & Evers, Krefeld.

Awadallah, K.T. & Khalill, F.M. (1979) Insect predators in cotton and clover fields with special reference to the efficiency of *Coccinella septempunctata* L. *Mesopotamia Journal of Agriculture* 14, 173–182.

Barnard, P.C. & Brooks, S.J. (1984) The African lacewing genus *Ceratochrysa* new genus, new status (Neuroptera, Chrysopidae), a predator on the cassava mealybug *Phenacoccus manihoti* (Hemiptera, Homoptera, Pseudococcidae). *Systematic Entomology* 9, 359–371.

Barnes, B.N. (1975) The susceptibility of *Chrysopa zastrowi* Esb.-Pet. (Neuroptera: Chrysopidae) to two insecticides in the laboratory. *Phytophylactica* 7, 131–132.

Beije, C.M. (1993) Comparative susceptibility to two insecticides of the predator *Chrysoperla externa* Hagen and the boll weevil *Anthonomus grandis* Boh. in cotton in Nicaragua. In *Proceedings of the Section Experimental and Applied Entomology of the Netherlands Entomological Society*, vol. 4, ed. Sommeijer, M. J. & van der Blom, J., pp. 243–249. Netherlands Entomological Society, Amsterdam.

Ben Saad, A.A. & Bishop, G.W. (1976) Attraction of insects to potato plants through use of artificial honeydews and aphid juice. *Entomophaga* 21, 49–57.

Bhuvaneswari, K. & Uthamasamy, S. (1994) Toxicity of alphamethrin to beneficial insects in cotton ecosystem. *Journal of Insect Science* 7, 228–229.

Boo, K.S., Chung, I.B., Han, K.S., Pickett, J.A. & Wadhams, L.J. (1998) Response of the lacewing *Chrysopa cognata* to pheromones of its aphid prey. *Journal of Chemical Ecology* 24, 631–643.

Brettell, J.H. (1979) Green lacewings (Neuroptera, Chrysopidae) of cotton fields in Central Rhodesia 1. Biology of *Chrysopa boninensis* and toxicity of certain insecticides to the larva. *Rhodesian Journal Of Agricultural Research* 17, 141–150.

Brettell, J.H. (1982) Green lacewings (Neuroptera, Chrysopidae) of cotton fields in Central Zimbabwe 2. Biology of *Chrysopa congrua* and *Chrysopa pudica* and toxicity of certain insecticides to their larvae. *Zimbabwe Journal of Agricultural Research* 20, 77–84.

Brettell, J.H. (1984) Green lacewings (Neuroptera, Chrysopidae) of cotton fields in Central Zimbabwe 3. Toxicity of certain acaricides, aphicides and pyrethroids to larvae of *Chrysopa boninensis*, *Chrysopa congrua* and *Chrysopa pudica*. *Zimbabwe Journal of Agricultural Research*, 22, 133–139.

Brooks, S.J. (1994) A taxonomic review of the common green lacewing genus *Chrysoperla* (Neuroptera: Chrysopidae). *Bulletin of the British Museum (Natural History) Entomology* 63, 137–210.

Brooks, S.J. & Barnard, P.C. (1990) The green lacewings of the world: a generic review (Neuroptera: Chrysopidae). *Bulletin of the British Museum (Natural History) Entomology* 59, 117–286.

Canard, M., Séméria, Y. & New, T.R. (eds.) (1984). *Biology of Chrysopidae*. Dr W. Junk, Dordrecht.

Cisneros, J.J. & Rosenheim, J.A. (1997) Ontogenetic change of prey preference in the generalist predator *Zelus renardii* and its influence on predator–predator interactions. *Ecological Entomology* 22, 399–407.

Coderre, D., Provencher, L. & Tourneur, J.C. (1987) Oviposition and niche partitioning in aphidophagous insects on maize. *Canadian Entomologist* 119, 195–204.

Dai, X.F. & Gua, Y.Y. (1992) Predation of green lacewing (*Chrysopa phyllochroma*) on cotton bollworm (*Heliothis armigera* (*Helicoverpa armigera*)). *Acta Phytophylacica Sinica* 19, 23–28.

Dean, D.E. & Schuster, D.J. (1995) *Bemisia argentifolii* (Homoptera: Aleyrodidae) and *Macrosiphum euphorbiae* (Homoptera: Aphididae) as prey for two species of Chrysopidae. *Environmental Entomology* 24, 1562–1568.

Deryabin, V.I. (1986) Effects of defoliation on natural enemies. *Zashchita Rastenii* 8, 33.

Duelli, P. (1980a) Adaptive dispersal and appetitive flight in the green lacewing, *Chrysopa carnea*. *Ecological Entomology* 5, 213–220.

Duelli, P. (1980b) Preovipository migration flights in the green lacewing, *Chrysopa carnea* (Planipennia, Chrysopidae). *Behavioural Ecology and Sociobiology* 7, 239–246.

Duelli, P. (1984a) Dispersal and oviposition strategies in *Chrysoperla carnea*. In *Progress in World's Neuropterology*, *Proceedings of the 1st International Symposium on Neuropterology*, ed. Gepp, J., Aspöck, H. & Hölzel, H., pp. 133–146. Thalerhof, Graz.

Duelli, P. (1984b) Flight, dispersal, migration. In *Biology of Chrysopidae*, ed. Canard, M. Séméria, Y. & New, T.R., pp. 110–116. Dr. W. Junk, The Hague.

Duelli, P. (1986a) Flight activity patterns in lacewings (Planipennia: Chrysopidae). In *Recent Research in Neuropterology*, *Proceedings of the 2nd International Symposium on Neuropterology*, ed. Gepp, J., Aspöck, H. & Hölzel, H., pp. 165–170. Thalerhof, Graz.

Duelli, P. (1986b) A 'missing link' in the evolution of the egg pedicel in lacewings? *Experientia* 42, 624.

Duelli, P. (1987a) Eine isolierte Reliktpopulation von *Chrysoperla mediterranea* (Planipennia: Chrysopidae) in der Schweiz. *Mitteilungen der schweizerischen entomologischen Gesellschaft* 60, 301–306.

Duelli, P. (1987b) The influence of food on the oviposition-site selection in a predatory and a honeydew-feeding lacewing species (Planipennia: Chrysopidae). *Neuroptera International* 4, 205–210.

Evans, E.W. & Swallow, J.G. (1993) Numerical responses of natural enemies to artificial honeydew in Utah alfalfa. *Environmental Entomology* 22, 1392–1401.

Freitas, S.D. & Giacheto Scaloppi, E.A. (1996) Efeito da pulverização de melaço em plantio de milho sobre a população de *Chrysoperla externa* (Hagen) e distribuição de ovos na planta. *Revista de Agricultura (Piracicaba)* 71, 251–258.

Galecka, B. & Zeleney, J. (1969) The occurrence of predators of aphids of the genus Chrysopa spp. on crops growing on a 4 crop field and in the neighboring shelterbelts. *Ekologia Polska Seria* A 17, 351–360.

Gautam, R.D. (1994) Present status of rearing of chrysopids in India. *Bulletin of Entomology (New Delhi)* 35, 31–39.

Gerling, D. & Bar, D. (1985) Parasitization of *Chrysoperla carnea* (Neuroptera, Chrysopidae) in cotton fields of Israel. *Entomophaga* 30, 409–414.

Gerling, D., Kravchenko, V. & Lazare, M. (1997) Dynamics of common green lacewing (Neuroptera: Chrysopidae) in Israeli cotton fields in relation to whitefly (Homoptera: Aleyrodidae) populations. *Environmental Entomology* 26, 815–827.

Gurbanov, G.G. (1982). Effectiveness of the use of the common lacewing (*Chrysopa carnea* Steph.) in the control of sucking pests and the cotton moth on cotton. *Izvestiya Akademii Nauk Azerbaidzhanskoi SSR, Biologicheskikh Nauk.* 2 92–96.

Hagen, K.S., Sawall, E.F.J. & Tassan, R.L. (1971) The use of food sprays to increase effectiveness of entomophagous insects. *Komarek* E, 59–81.

Heimpel, G.E. & Hough-Goldstein, J.A. (1992) A survey of arthropod predators of *Leptinotarsa decemlineata* Say. in Delaware potato fields. *Journal of Agricultural Entomology* 9, 137–142.

Henry, C.S. (1979) Acoustical communication during courtship and mating in the green lacewing *Chrysopa carnea* (Neuroptera: Chrysopidae). *Annals of the Entomological Society of America* 72, 68–79.

Henry, C.S., Brooks, S.J., Johnson, J.B. (1996) *Chrysoperla lucasina* (Lacroix): a distinct species of green lacewing, confirmed by acoustical analysis (Neuroptera: Chrysopidae). *Systematic Entomology* 21, 205–218.

Herold, D. & Stengel, B. (1994) Les thrips. Nouveaux ravageurs du chou à choucroute en Alsace. *Infos (Paris)* 101, 31–35.

Huang, H., Yian, J.Z. & Li, D.Q. (1990) Predation model of *Chrysopa septempunctata* on cotton insect pests. *Natural Enemies of Insects* 12, 7–12.

Jimenez Aragon, J.G. (1974) Studies of population fluctuations of agricultural economically important insects in the Comarca Lagunera, Mexico. *Folia Entomologica Mexicana*, 56.

Kabissa, J.C.B., Selemani, E. & Fundi, F. (1989) New records of *Ceratochrysa antica*, *Mallada boninensis* and *Brinckochrysa* sp. (Neuroptera: Chrysopidae) from Tanzania. *Tropical Pest Management* 35, 206–207.

Kabissa, J.C., Kayumbo, H.Y. & Yarro, J.G. (1995) Comparative biology of *Mallada desjardinsi* (Navas) and *Chrysoperla congrua* (Walker) (Neuroptera. Chrysopidae), predators of *Helicoverpa armigera* (Hubner) (Lepidoptera: Noctuidae) and *Aphis gossypii* (Glover) (Homoptera: Aphididae) on cotton in eastern Tanzania. *International Journal of Pest Management* 41, 214–218.

Kabissa, J.C., Kayumbo, H.Y. & Yarro, J.G. (1996) Seasonal abundance of chrysopids (Neuroptera: Chrysopidae) preying on *Helicoverpa armigera* (Hubner) (Lepidoptera: Noctuidae) and *Aphis gossypii* (Glover) (Homoptera: Aphididae) on cotton in eastern Tanzania. *Crop Protection* 15, 5–8.

Kaitazov, A. & Kharizanov, A. (1976) The possibilities for using Chrysopidae. *Rastitelna Zashchita* 24, 22–25.

Kapadia, M.N. & Puri, S.N. (1990) Feeding behaviour of *Chrysoperla carnea* (Stephens) on the parasitized pupae of *Bemisia tabaci* (Gennadius). *Entomon* 15, 283–284.

Kharizanov, A. & Babrikova, T. (1978) Toxicity of

insecticides to certain species of chrysopids. *Rastitelna Zashchita* 26, 12–15.

Krishnamoorthy, A. (1985) Effect of several pesticides on eggs, larvae and adults of the green lace-wing *Chrysopa scelestes* Banks. *Entomon* 10, 21–28.

Lavallee, A.G. & Shaw, F.R. (1969) Preferences of golden-eye lacewing larvae for pea aphids, leafhopper and plant bug nymphs, and alfalfa weevil larvae. *Journal of Economic Entomology* 62, 1228–1229.

Legaspi, J.C., Correa, J.A. & Carruthers, R.I. (1996) Effect of short-term releases of *Chrysoperla rufilabris* (Neuroptera: Chrysopidae) against silverleaf whitefly (Homoptera: Aleyroididae) in field cages. *Journal of Entomological Science* 31, 102–111.

Leraut, P. (1991) Les *Chrysoperla* de la faune de France (Neuroptera Chrysopidae). *Entomologie Gallica* 2, 75–81.

Li, H.C. (1987) Augmentation of *Chrysopa* spp. to control cotton aphids by intercropping cotton and safflower. *Chinese Journal of Biological Control* 3, 109–111.

Maher-Ali, A., Moftah, S.A. & Rizk, G.A. (1983) Evaluation of the impact of certain predators on the population density of egg-masses of the cotton leafworm, *Spodoptera littoralis* (Boisd.) in cotton fields. *Bulletin de la Société Entomologique d'Egypte* 62, 111–116.

Malet, J.C., Noyer, C. & Maisonneuve, J.C. (1994) *Chrysoperla lucasina* (Lacroix) (Neur., Chrysopidae), prédateur potentiel du complexe Mediterranéen des *Chrysoperla* Steinmann: premier essai de lutte biologique contre *Aphis gossypii* Glover (Hom., Aphididae) sur melon en France méridionale. *Journal of Applied Entomology* 118, 429–436.

Manjunath, T.M. (1993) Biocontrol of *Helicoverpa armigera* and other cotton bollworms. *Indian Farming* 42, 18–23.

Mannan, V.D., Varma, G.C. & Brar, K.S. (1995) Seasonal fluctuations and host predator relationship of *Chrysoperla carnea* (Stephens) (Chrysopidae: Neuroptera). *Indian Journal of Ecology* 22, 21–26.

Maredia, K.M., Gage, S.H., Landis, D.A. & Wirth, T.M. (1992) Visual response of *Coccinella septempunctata* (L.), *Hippodamia parenthesis* (Say), (Coleoptera: Coccinellidae), and *Chrysoperla carnea* (Stephens), (Neuroptera: Chrysopidae) to colours. *Biological Control* 2, 253–6.

Marin, M. (1988) Prezenta speciei *Chrysopa carnea* Steph. (Neuroptera, Chrysopidae) in cultura porumbului. *Analele Institutului de Cercetari pentru Protectia Plantelor* 21, 77–83.

Messina, F.J., Jones, T.A. & Nielson, D.C. (1995) Host plant affects the interaction between the Russian wheat aphid and a generalist predator, *Chrysoperla carnea*. *Journal of the Kansas Entomological Society* 68, 313–319.

Micinski, S., Kirby, M.L. & Graves, J.B. (1991) Cotton *Gossypium hirsutum* L., tarnished plant bug *Lygus lineolaris* Palisot De Beauvois, cotton fleahopper *Pseudatomoscelis seriatus* Reuter, beneficial coccinellids (various spp.), beneficial hemipterans (various spp.), Neuroptera *Chrysopa* spp., spiders (various spp.): efficacy of selected insecticides for plant bug control 1990. *Insecticide and Acaricide Tests*, 197–198.

Miermont, Y. & Canard, M. (1975) Biology of the aphidophagous predator *Eumicromus angulatus* (Neuroptera; Hemerobiidae: laboratory studies and observations in the Southwest of France. *Entomophaga* 20, 179–191.

Mostafa, M.A., El-Bialy, S. & El-Berry, A.A. (1976) Fluctuation of populations of certain predators on flowers of cotton in Egypt. *Zeitschrift für angewandte Entomologie* 80, 225–228.

Muralidharan, C.M. & Chari, M.S. (1990) Effect of electrostatic and conventional spraying system on biocontrol agents in 'Hybrid 6' cotton. *Plant Protection Bulletin (Faridabad)* 42, 21–24.

Natarajan, K. (1990) Natural enemies of *Bemisia tabaci* Gennadius and effect of insecticides on their activity. *Journal of Biological Control* 4, 86–88.

Neuenschwander, P. & Hagen, K.S. (1980) Role of the predator *Hemerobius pacificus* in a non-insecticide treated artichoke field. *Environmental Entomology* 9, 492–495.

New, T.R. (1975) The biology of Chrysopidae and Hemerobiidae (Neuroptera) with reference to their usage as biocontrol agents: a review. *Transactions of the Royal Entomological Society of London* 127, 115–140.

New, T. R. (1988). Neuroptera. In *Aphids: Their Biology, Natural Enemies and Control*, ed. Minks, A.K.H., pp. 249–258. Elsevier, Amsterdam.

Nordlund, D.A., Vacek, D.C. & Ferro, D.N. (1991) Predation of Colorado potato beetle (Coleoptera, Chrysomelidae) eggs and larvae by *Chrysoperla rufilabris* (Neuroptera, Chrysopidae) larvae in the laboratory and field cages. *Journal of Entomological Science* 26, 443–449.

Nunez, E. (1988) Biological cycle and rearing of *Chrysoperla externa* and *Ceraeochrysa cincta* (Neuroptera, Chrysopidae). *Revista Peruana de Entomologia* 31, 76–82.

Obrycki, J.J., Hamid, M.N. & Sajap, A.S. (1989) Suitability of corn insect pests for development and survival of *Chrysoperla carnea* and *Chrysopa oculata* (Neuroptera, Chrysopidae). *Environmental Entomology* 18, 1126–1130.

Pantaleoni, R.A. & Tisselli, V. (1986) I Neurotteri delle colture agrarie: rilievi sui Crisopidi in alcune coltivazioni del forlivese. *Bollettino dell'Istituto di Entomologia dell'Università di Bologna* 40, 51–65.

Patel, I.S. & Yadav, D.N. (1995) Susceptibility of *Amrasca biguttula biguttula* and *Chrysopa scelestes* in cotton (*Gossypium* species) to three systemic insecticides. *Indian Journal of Agricultural Sciences* 65, 308–309.

Patel, K.G. & Vyas, H.N. (1985) Ovipositional site preference by green lacewing *Chrysopa scelestes* on cotton and green gram. *Gujarat Agricultural University Research Journal* 10, 79–80.

Pavlova, G.A. & Aripova, F.K. (1991) Predators of web mite on various cotton samples. *Zashchita Rastenii* 8, 41–42.

Rajakulendran, S.V. & Plapp, F.W. Jr (1982) Synergism of five synthetic pyrethroids by chlordimeform against the tobacco budworm (Lepidoptera: Noctuidae) and a predator, *Chrysopa carnea* (Neuroptera: Chrysopidae). *Journal of Economic Entomology* 75, 1089–1092.

Ridgway, R.L. & Jones, S.L. (1969) Inundative releases of *Chrysopa carnea* for control of *Heliothis zea* and *Heliothis virescens* on cotton-D. *Journal of Economic Entomology* 62, 177–180.

Ridgway, R.L. & Kinzer, R.E. (1974) Chrysopids as predators of crop pests. *Entomophaga, Mémoire hors Série* 7, 45–51.

Rustamova, M.R. (1981) Entomophagous insects on cotton. *Zashchita Rastenii*, 3, 36.

Salem, S.A. & Matter, M.M. (1991) Relative effects of neem seed oil and Deenate on the cotton leafworm, *Spodoptera littoralis* Boisd. and the most prevalent predators in cotton fields at Menoufyia Governorate. *Bulletin of Faculty of Agriculture, University of Cairo* 42, 941–952.

Samson, P.R. & Blood, P.R.B. (1980) Voracity and searching ability of *Chrysopa signata* (Neuroptera: Chrysopidae), *Micromus tasmaniae* (Neuroptera: Hemerobiidae), and *Tropiconabis capsiformis* (Hemiptera: Nabidae). *Australian Journal of Zoology* 28, 575–580.

Séméria, Y. (1977) Discussion de la validité taxonomique du sous-genre *Chrysoperla* Steinmann (Planipennia, Chrysopidae). *Nouvelle Revue d'Entomologie* 7, 235–238.

Sengonca, C. & Henze, M. (1992) Conservation and enhancement of *Chrysoperla carnea* Stephens (Neuroptera, Chrysopidae) in the field by providing hibernation shelters. *Journal of Applied Entomology* 114, 497–501.

Shuvakhina, E.Y. (1985) The lacewing against the Colorado beetle. *Zashchita Rastenii* 7, 40.

Singh, J., Sohi, A.S. & Dhaliwal, Z.S. (1993) Comparative incidence of *Helicoverpa armigera* (Hub.) and other insects pests on okra and sunflower intercrops in cotton under Punjab conditions. *Journal of Insect Science* 6, 137–138.

Stark, S.B. & Hopper, K.R. (1988) *Chrysoperla carnea* predation on *Heliothis virescens* larvae parasitized by *Microplitis croceipes*. *Entomologia Experimentalis et Applicata* 48, 69–72.

Stark, S.B. & Whitford, F. (1987) Functional response of *Chrysopa carnea* (Neuroptera, Chrysopidae) larvae feeding on *Heliothis virescens* (Lepidoptera, Noctuidae) eggs on cotton in field cages. *Entomophaga* 32, 521–528.

Stolyarov, M.V., Sugonyaev, E.S. & Umarov Sh, A. (1974) Dynamics of the arthropod community of cotton fields in northern Afghanistan (the basis of a system of integrated control of cotton pests). *Entomologicheskoe Obozrenie* 53, 245–257.

Sukhoruchenko, G.I., Smirnova, A.A. & Vikar, E.V. (1981) The effect of pyrethroids on arthropods of the cotton agrobiocoenosis. *Entomologicheskoe Obozrenie* 60, 5–15.

Thierry, D., Cloupeau, R. & Jarry, M. (1996) Distribution of the sibling species of the common green lacewing *Chrysoperla carnea* (Stephens) in Europe (Insecta: Neuroptera: Chrysopidae). In *Pure and Applied Research in Neuropterology, Proceedings of the 5th International Symposium on Neuropterology*, ed. Canard, M., Aspöck, H. & Mansell, M.W., pp. 233–240. Sacco, Toulouse.

Treacy, M.F., Benedict, J.H. & Lopez, J.D. (1987) Functional response of a predator (Neuroptera, Chrysopidae) to bollworm (Lepidoptera, Noctuidae) eggs on smoothleaf hirsute and pilose cottons. *Journal of Economic Entomology* 80, 376–379.

Varma, G.C. & Shenhmar, M. (1983) Some observations on the biology of *Chrysoperla carnea* (Stephens) (Chrysopidae: Neuroptera). *Journal of Research, Punjab Agricultural University* 20, 222–223.

Vickerman, G.P. & Sunderland, K.D. (1977) Some effects of dimethoate on arthropods in winter wheat. *Journal of Applied Ecology* 14, 767–778.

Voicu, M.C. (1988) Contributions to the knowledge of predatory Neuroptera and Coleoptera of agricultural pests in Romania. *Studii si Cercetari de Biologie Seria Biologie Animala* 40, 3–8.

Voicu, M.C. (1989) The role of the predatory insects in reducing the attack of the corn grain aphid (*Rhopalosiphum maidis* Fitch) (Hom., Aphididae) in Romania. *Revue Roumaine de Biologie* 34, 95–105.

Voicu, M.C. & Nagler, K. (1987) Chrysopidae, Coccinellidae and Syrphidae preying upon *Schizaphis graminum* Rond colonies in some agrobiocenoses in Moldavia, Romania. *Studii si Cercetari de Biologie Seria Biologie Animala* 39, 22–27.

Wang, A. & Wang, L. (1996) An exploratory study on cotton insect control by natural enemies. *China Cottons* 23, 31.

Wells, M.M. & Henry, C.S. (1994) Behavioural responses of hybrid lacewings (Neuroptera: Chrysopidae) to courtship songs. *Journal of Insect Behaviour* 7, 649–662.

Wise, K.A.J. (1996) Records concerning biological control of insect pests by Neuropteroidea (Insecta) in New Zealand. *Records of the Auckland Institute and Museum*, 32, 101–117.

Yang, X.-K. & Yang, C.-K. (1990). *Navasius*, a new genus of Chrysopinae (Neuroptera: Chrysopidae). *Acta Zootaxonomica Sinica* 15, 327–338.

Yang, X.-K. & Yang, C.-K. (1991). *Dichochrysa* nom. nov. for *Navasius* Yang & Yang, 1990 (Neuroptera: Chrysopidae) nec Esben-Petersen (Neuroptera: Myrmeleonidae). *Scientific Treatise on Systematic and Evolutionary Biology* 1, 150.

Zhou, W., Wang, R. & Qui, S. (1991) Field studies on the survival of *Chrysoperla sinica* (Neuroptera: Chrysopidae) mass reared and inoculatively released in wheat fields in Northern China. *Chinese Journal of Biological Control* 7, 97–100.

CHAPTER 9

Lacewings in fruit and nut crops

F. Szentkirályi

9.1 APPLE

Apple (*Malus pumila*) is one of the most widely culti-vated fruit-tree species. Different species from the *Malus* genus, such as *Malus sylvestris* (Europe), and *M. baccata* and *M. prunifolia* (southeast Asia, Japan, China) are endemic throughout Eurasia. *Malus pumila* has the greatest economic significance, and it is grown as a great number of cultivars all over the world from smaller house gardens to intensively managed large-sized commercial orchards mainly under temperate and Mediterranean climatic conditions.

9.1.1 Characteristics of lacewing assemblages in apple orchards

In apple orchards, species-rich arthropod communities can be built up consisting of hundreds of species (Szentkirályi & Kozár, 1991). Reviewing the published insect communities of apple orchards of the world, Szentkirályi & Kozár (1991) found that from total species-richness values, Neuropteroidea assemblages shared between only 0.6 and 5.1%. Homoptera, serving as the major potential prey source, ranged between 0.2% and 20.5%, meaning more than 100 species in certain cases. Due to its economic importance, apple is one of the cultivated plants, which has been very well studied regarding its pests. Thanks to numerous field investigations, a relatively high number of publications contain data on the natural enemies of apple pests. In the case of chrysopids and hemerobiids, most studies are from Europe and North America but some are from Asia, Australia, and New Zealand. Table 9.1 summar-ises lacewing records of European apple orchards, while all the others are shown on Table 9.2.

In European apple orchards (Table 9.1), a total of 15 green and 26 brown lacewing species have been detected. The frequency of occurrence and dominance values of the characteristic chrysopid species of apple orchards are as follow: *Chrysoperla carnea* (4%–91%), *Chrysopa pallens* (0.4%–75%), *C. formosa* (0.3%–47%), *C. phyllochroma* (1%–22%), *C. perla* (0.2%–4%), *Dichochrysa prasina* (0.2%–6%), *Nineta flava* (<2%). Although *N. flava* occurred with low dominance, it was found in about half of the orchards studied. The char-acteristic chrysopid assemblages for European apple orchards could be expected to be composed of four to eight species. The fact that in the majority of cases only one to two green lacewing species were recorded is due partly to lack of study and partly to heavy chemical treatments in commercial orchards. During the very detailed long-term (10 years) investigations of Hungarian apple ecosystem, even in apple orchards under intensive pesticide treatments, more than 10 chrysopid species were found (see columns 10*a–c* in Table 9.1).

Among the hemerobiids, 26 species have been recorded from European apple orchards. The com-monest brown lacewings were: *Wesmaelius subnebulosus* (4%–72%), *Hemerobius humulinus* (1%–39%), *Micro-mus angulatus* (3%–40%), *M. variegatus* (3%–40%), *Sympherobius pygmaeus* (5%–43%), *H. lutescens* (2%–14%), *W. nervosus* (1%–28%), *S. elegans* (<3%). From this the expected characteristic hemerobiid assemblage of apple orchards consists of four to six species. However, with long-term (many years) investi-gations, the number of brown lacewing species detected in orchards could be significantly raised (10–20 species) (Tolstova & Atanov, 1982, Szentkirályi, 1992). Detailed and long-term sampling is only one factor in the higher number of lacewing species recorded in Hungarian apple orchards. Another one is that in several cases, orchards are bordered by forested habitats (e.g. Table 9.1, column 10*c*). From forests, as lacewing reservoir habitats, many hemerobiids and chrysopids migrate to apple orchards. In particular more hemerobiid species are associated with conifers, consequently they occur

SOURCES AND METHODS OF EVALUATION

Literature sources were reviewed from 1910 until the present, although the overwhelming majority of usable data are derived from the publications of the last four decades. Most of the data are from European and North American publications. In order to have a better overview of the records of lacewing species, or rather of their assemblage in the given stands of cultivated crop plants, data have been arranged into tables. Within a table European lacewing data are shown first followed by the data of countries of other continents. If possible, the lacewing assemblages of European countries were arranged according to latitude, from northern areas to the southern and Mediterranean regions. In the case of those plants for which there were many literature sources from different geographical regions, two tables were constructed, usually one for the European region, and the other one for the remaining regions (America, Africa, Asia).

The list of species names of chrysopid and hemerobiid families can be found in the first column of the tables. For practical reasons, those names of species that were formerly published, but now are invalid, have been replaced according to the latest valid nomenclature. The correction of names was based upon the works of Aspöck & Hölzel (1996), Brooks & Barnard (1990), Monserrat (1990, 1996, 1998), and Penny *et al.* (1997). By this latter work concerning the newest nomenclature of North American lacewing species, the former name *Chrysoperla carnea* is synonymised with *Chrysoperla downesi* (Smith). Because the older invalid *carnea* is the most frequently used name in the literature, we did not replace it with the new one in the text and the related tables. However do not forget reading the text, that the American *Chrysoperla carnea* is not identical with the European one and its new valid name is *C. downesi*.

The columns of the tables contain either the lacewing assemblage of local stands of cultivated plants, or the data of faunistic collections of several localities altogether. Where the numbers of individuals collected were given and are sufficiently high, the percentage dominance values of each species have been calculated and are shown in the tables. If in a certain investigation, the samplings were made in several stands of crop plants, then the dominance distribution has been calculated for each site, and the minimal and maximal values (range) are presented for each species. For both Chrysopidae and Hemerobiidae, the percentage dominance distributions have been calculated separately. In the case of sites being close to each other, the numbers of lacewing individuals collected were summed, and from these data one dominance distribution was constructed. Where the calculation of dominance was impossible, because the numbers of individuals were too low or lacking, but the species was present, a '+' sign indicates the occurrence of the given species. The letter 'A' is put into the cell, if there was a reference that a certain species was dominant or abundant.

The number of literature references for each species can be found in the last column with an 'FR' sign (frequency of occurrence). The measure of these values refers to the closeness of the possible association between the given lacewing species and the given cultivated plant. The greater the FR and dominance percentage value of a certain species, the higher the probability that the lacewing is an important member of that Neuroptera assemblage found on a given cultivated plant.

Table 9.1. *Species composition and dominance distribution of chrysopid and hemerobiid assemblages found in apple (Malus pumila) orchards in Europe*

	1	2	3	4	5	6	7	8	9	10a	10b	10c	11	12	13	14	15	16	17	18	FR
Chrysopidae																					
Chrysoperla carnea s. lat. (Stephens)	+	+	+	+	A	+	+	+	+	10.4–89.1	38.2–62.9	51.8	90.8	+		+	4.2	+	+	+	18
Chrysopa formosa Brauer	+									0.3–44.8	13.1–47.0	22.8					20.8	+	+		6
Chrysopa pallens (Rambur)	+	25.0	+	+	+					0–3.2	0.4–8.0	1.7	9.2			+	75.0	+	+		11
Chrysopa perla L.	+									0.2–1.5	0.7–1.8	4.3						+	+		6
Chrysopa phyllochroma Wesmael	+									1.2–17.9	0.9–21.9	12.4									6
Chrysopa abbreviata Curtis										0–7.5	0.5–1.7	0.3									3
Chrysopa nigricostata Brauer										0–1.5											1
Chrysopa walkeri McLachlan												0.1									1
Dichochrysa prasina (Burmeister)										0–0.3	0.2–1.7	5.8						+			5
Dichochrysa ventralis (Curtis)										0–0.3		0.1									2
Dichochrysa flavifrons (Brauer)	+									0–0.3											2
Nineta flava (Scopoli)	+	1.3			+					0–1.5	0–0.5	0.3						+			8
Nineta vittata (Wesmael)		16.3										0.1									2
Chrysotropia ciliata (Wesmael)	+									0–0.3	0.3–0.4	0.5			+						3
Cunctochrysa albolineata (Killington)		57.5																			1
Hemerobiidae																					
Wesmaelius subnebulosus (Stephens)	+	66.7								4.4–10.0	12.5–71.8	29.3						+			7
Wesmaelius nervosus (Fabricius)		27.8									0–3.6	1.0						+	+		5
Wesmaelius malladai (Navás)											0–2.6	0.2									1
Wesmaelius concinnus (Stephens)										0–0.6		1.0									1
Wesmaelius mortoni (McLachlan)										0–0.5											1
Wesmaelius navasi (Andréu)																			+		1
Hemerobius humulinus L.	+	1.4		+						16.7–39.1	17.9–21.8	10.2					+	+	+		9
Hemerobius micans Olivier										0.6–1.9	0–3.6	2.9						+			4
Hemerobius nitidulus Fabricius										0–1.0	0–2.6	1.0									3
Hemerobius stigma Stephens												1.0						+			2
Hemerobius handschini Tjeder										0–0.1		0.8									2
Hemerobius lutescens Fabricius	+	4.2		+						1.9–3.6		13.7						+	+		6
Hemerobius perelegans Stephens																		+			1
Hemerobius marginatus Stephens										0–0.6								+			2

								FR
Hemerobius simulans Walker					+			1
Hemerobius pini Stephens	+	0–0.5	0.2		+			4
Hemerobius gilvus Stein			0.2					1
Hemerobius sp.				+				2
Micromus angulatus (Stephens)	+	10.7–20.4	2.6–40.0	13.9	+	+		6
Micromus variegatus (Fabricius)	+	7.4–39.5	2.6–9.1	10.6	+	+		5
Micromus lanosus (Zelený)		0–1.8	0.6					2
Psectra diptera (Burmeister)		1.0–3.0	0.2					2
Sympherobius pygmaeus (Rambur)	+	4.9–42.6	0–10.7	10.4	+	+		4
Sympherobius elegans (Stephens)	+	0–0.5	1.8–2.6	0.4	+	+		6
Sympherobius klapaleki Zelený			0.6					1
Megalomus tortricoides Rambur			0.6					1
Drepanepteryx phalaenoides (L.)	+	0–1.8	0–1.8	2.5				3
S_C	8 4 2 1 3	2 1 1	12 9	12	2 2 4	2	1 7	1
S_H	6 4	2 1	16 2	9 9	20	1 1 3	2 12	1

Note: The dominance values (%) are calculated from the number of individuals published in the reviewed papers.

Abbreviations: S_C, S_H, species richness of chrysopids and hemerobiids respectively; +, species present; FR, frequency of species occurrence in literature surveyed

Sources: 1, Europe (Principi & Canard, 1974); 2, England (New, 1967); 3, Norway (Baeschlin & Taksdal, 1979; Skånland, 1981); 4, Netherlands (Ravensberg, 1981); 5, Poland (Wiackowski & Wiackowska, 1968; Niemczyk et al., 1983); 6, Belarus (Koltun & Meleshko, 1994); 7, Germany (Krämer, 1961; Gottwald, 1991); 8, France (Malevez, 1976); 9, Switzerland (Mathys & Baggiolini, 1965; Baggiolini & Wildbolz, 1965; Wyss, 1995); 10a, b, c, Hungary (a: commercial, b: backyard, c: experimental–untreated orchards; Szabó & Szentkirályi, 1981, Szentkirályi, 1992; Szentkirályi, unpublished data); 11, Italy (Pantaleoni & Tisselli, 1985; Pantaleoni & Ticchiati, 1988); 12, Yugoslavia, Croatia (Ciglar & Schmidt, 1983; Ciglar, 1985; Markovic & Zivanovic, 1988; Injac & Dulic, 1992; Ciglar & Budinščak, 1993); 13, Romania (Polizu, 1932); 14, Bulgaria (Injac et al., 1978); 15, Moldova (Talitskaya, 1980; Zelený & Talitsky, 1966); 16, Russia former USSR (Semyanov, 1973); 17, Russia, former USSR (Tolstova & Atarov, 1982); 18, Lithuania (Zayantskauskas et al., 1982).

Table 9.2. *Species composition and dominance distribution of chrysopid and hemerobiid assemblages found in apple* (Malus pumila) *orchards in Asia, America, Australia, and New Zealand*

	1	2	3	4	5	6	7	8	9	10	11	12	13	FR
Chrysopidae														
Chrysoperla carnea s. lat. (Stephens)		49.1	+		A	+	28.0	+	+	+				8
Chrysoperla rufilabris (Burmeister)					+	+	52.0	+		+	+			6
Chrysoperla plorabunda (Fitch)						+								1
Chrysoperla harrisii (Fitch)						+								1
Chrysopa formosa Brauer	+	18.5												2
Chrysopa phyllochroma Wesmael		11.6												1
Chrysopa pallens (Rambur)		20.8												1
Chrysopa oculata Say				+	+	+	10.0	+	+		+			7
Chrysopa quadripunctata Burmeister						+	8.0	+		+	+			5
Chrysopa nigricornis Burmeister					+	+	+	+	A	+	+			7
Chrysopa sp.												+		1
Pseudomallada perfectus (Banks)							+							1
Chrysopa coloradensis Banks									+					1
Brinckochrysa scelestes (Banks)			+											1
Hemerobiidae														
Hemerobius humulinus L.		+			+	+	+	+		+	+			7
Hemerobius stigma Stephens				+		+		+						3
Hemerobius pacificus Banks									+					1
Hemerobius sp.						+					+			2
Micromus posticus (Walker)							A	+		+	+			4
Micromus subanticus (Walker)							+							1
Micromus tasmaniae (Walker)												+	+	2
Drepanacra binocula (Newman)												+	+	2
Sympherobius amiculus (Fitch)					+						+			2
S_C	1	4	2	1	4	7	6	5	4	4	4	1		
S_H		1		1	2	4	3	3	1	2	4	2	2	

Note: The dominance values (%) are calculated from the number of collected individuals published in the reviewed papers. *Abbreviations:* A, abundant; S_C, S_H, species richness of chrysopids and hemerobiids respectively; +, species present; FR, frequency of species occurrence in literature surveyed.

Sources: 1, China (Shi *et al.*, 1993); 2, China (Yan & Duan, 1986, 1988); 3, India (Thakur *et al.*, 1988; Pawar & Parry, 1989); 4, Canada, Quebec (LeRoux, 1960); 5, Canada, Ontario (Hagley, 1974, 1990 personal communication; Hagley & Allen, 1990); 6, USA, Wisconsin (Oatman *et al.*, 1964); 7, USA, Missouri (Childers & Enns, 1975); 8, USA, Pennsylvania (Asquith *et al.*, 1980; Brown *et al.*, 1988); 9, USA, Washington (Carroll & Hoyt, 1984; Grasswitz & Burts, 1995); 10, USA, Ohio (Holdsworth, 1968, 1970); 11, USA, Virginia (Haeussler & Clancy, 1944; Brown *et al.*, 1988; Parrella *et al.*, 1981); 12, Australia (MacLellan, 1973); 13, New Zealand (Collyer & van Geldermalsen, 1975).

only as transitory insects in apple orchards. Such species are *W. concinnus*, *W. mortoni*, *H. nitidulus*, *H. stigma*, *H. handschini*, or *H. pini*. These species were represented however with only one to two individuals in the lacewing fauna of apple orchards. It is also noticeable that at least half of the chrysopids and the majority of hemerobiids are broadleaf-forest inhabitants.

It can be stated from lacewing assemblages recorded in apple orchards in Table 9.2, that Asian (Chinese) chrysopid species are identical to European (e.g. *Chrysoperla carnea*, *Chrysopa formosa*, *C. pallens*). Hemerobiids are rather under-examined in Asian apple orchards, where only *H. humulinus* was found. *Drepanacra binocula* and *M. tasmaniae* are presumably characteristic of apple orchards in New Zealand and Australia (see Table 9.2, columns 12–13). Outside Europe, most studies of the Neuroptera fauna of apple trees have been carried out in North America. Overall, nine chrysopid and seven hemerobiid species have been discovered in US and Canadian apple orchards. On the basis of frequency values, characteristic green lacewing species in North American apple orchards are *Chrysoperla rufilabris*, *C. carnea* (by today's nomenclature = *Chrysoperla downesi* (Smith), see Penny *et al.*, 1997), *Chrysopa oculata*, *C. quadripunctata*, and *C. nigricornis*. Among hemerobiids, *H. humulinus*, *M. posticus*, and perhaps *H. stigma* are likely to be constant members of lacewing assemblages in American orchards. Data in Table 9.2 suggest that within these orchards, predicted assemblages contain five chrysopid and two to three hemerobiid species. In those European and American apple orchards where there is ground cover offering alternative prey sources, characteristic lacewing species can colonise. Such European species are *C. phyllochroma*, *C. abbreviata*, *C. walkeri*, *Micromus* spp. (*M. angulatus*, *M. variegatus*), or *Psectra diptera*. Typical low-vegetation-dwelling lacewings in North American apple orchards are *C. oculata* and the hemerobiid *Micromus* spp. as well as *M. posticus* and *M. subanticus*.

9.1.2 Apple aphids as prey for lacewings

Insect communities formed in apple orchards can be rich in species, with several hundred members. Within these Homoptera, as the principal food source for lacewings, ranged between 1 and 122 species in apple orchards world-wide (Szentkirályi & Kozár, 1991). Aphids are the most important lacewing prey group.

Despite the approximately 50 recorded aphid species associated with apple trees, the most abundant and harmful characteristic species are identical all over the world. Most studies on apple aphids have been in European and North American orchards. These typical apple aphids are as follows: the green apple aphid, *Aphis pomi*; rosy apple aphid, *Dysaphis devecta*; grey apple aphid, *D. plantaginea*; grass apple aphid, *Rhopalosiphum insertum*; woolly apple aphid, *Eriosoma lanigerum*.

Of the aphids listed, one of the most frequent species is *A. pomi*; consequently, most lacewing prey data are related to it. Many authors report lacewings from European apple orchards preying on this aphid. For example in a Belarus apple orchard, one of the dominant predators of this pest was *Chrysoperla carnea*, as indicated by Koltun & Meleshko (1994). In a Hungarian cottage garden, apple trees planted on sandy soil were infested by *A. pomi* exclusively. On their foliage the eggs laid by *Chrysopa formosa*, *C. pallens*, *C. perla*, and *Chrysoperla carnea* and second and third instar larvae of *Chrysopa formosa*, and *C. pallens* could be collected. The dominant species was *Chrysopa formosa*, while *Chrysoperla carnea* was represented with only a low number of individuals (F. Szentkirályi, unpublished data). A similar result was provided by Talitskaya (1980), who found *Chrysopa formosa* (20.8%) and *C. pallens* (75%) at a greater frequency than *Chrysoperla carnea* (4.2%). Having regard to these observations, it is likely that the earlier two chrysopid species prefer green apple aphids to a greater extent than *C. carnea*. In Romania, *C. carnea* and the hemerobiid *W. nervosus* were recorded on *A. pomi* colonies (Polizu, 1932).

In North America, *A. pomi* is especially troublesome among apple aphids, since it has become resistant to most insecticides. Developing continuously on apple trees, it can reach an extremely high density during summer and autumn. By infesting shoots on young trees (e.g. in apple tree nurseries), it may drastically reduce plant growth. Carroll & Hoyt (1984) studied the aphidophagous insect complexes associated with *A. pomi* on young apple trees in Washington State. They used exclusion cage experiments to determine the effectiveness of natural enemies. Thirty-nine predators were recorded, amongst them four chrysopids and a hemerobiid. Carroll & Hoyt investigated in field feeding trials the predation by adults and larvae of various lacewing species on *A. pomi* nymphs. The mean daily number of nymphs consumed per lacewing individual was as follows: *Chrysopa nigricornis*, male 33, female 105, larva 60; *C. oculata*, male 42, female 56, larva 140; *C. coloradensis*, female 44, larva 58; *Chrysoperla carnea*, larva 48. In field conditions, *Hemerobius pacificus* was only occasionally observed preying on *A. pomi*. The average predation of aphid individuals per day by this hemerobiid species was 23 for adults, and 18 for larvae. In Canadian apple orchards in Ontario, Hagley & Allen (1990) found chrysopid larvae, predominantly *C. carnea*, the most effective foliage-inhabiting predator of *A. pomi*. Using serological assays, they tested chrysopid larvae collected from foliage for predation on green apple aphids. Seropositive tests regarding chrysopid larvae ranged between 20% and 87%. According to laboratory studies, the mean number of *A. pomi* nymphs consumed by chrysopid larvae was 52.5 ± 24.6 per day. In Ohio (USA), *Chrysopa nigricornis* was observed to feed on *A. pomi* colonies, but this chrysopid species was uncommon in surveyed apple plantations (Holdsworth, 1970). Holdsworth found that the three aphid species were capable of damaging apple trees and aphidophagous predators including lacewings were unable to prevent the development of economic injury level.

Another aphid species, the pestiferous *D. devecta*, can be frequently recorded as curling apple leaves. In spite of this, there are few observations of lacewings preying on this pest. In a Hungarian experimental apple orchard, larvae of the hemerobiids *H. humulinus*, *Wesmaelius subnebulosus*, and *Drepanepteryx phalaenoides* were frequently found to be consuming rosy apple aphids (F. Szentkirályi, unpublished data). It was observed that *D. phalaenoides* larvae preferentially used the apple leaves rolled-up by this aphid to pupate inside them. In Moldova, *H. humulinus* proved also to be a predator of the aphid *Dysaphis devecta* (Talitskaya, 1980).

Woolly apple aphid, *E. lanigerum* originating from North America, was introduced into Europe in the 18th century. It has since spread across the world via transport of young apple trees from nurseries. It is one of the most serious apple pests. Even in early times, Standfuss (in Morton, 1910) reported from Switzerland that *E. lanigerum* colonies were the favourite food for larvae of hemerobiid *Drepanepteryx phalaenoides*, which were found to be active from the beginning of May till mid-July. Ravensberg (1981) in the Netherlands studied the natural control agents of this aphid. With low numbers of individuals, lacewings were present everywhere within apple orchards, and the predominant species were *Chrysoperla carnea* and the hemerobiids *H. humulinus*, and *H. lutescens* as predators of *E. lanigerum*. In three apple orchards in India, woolly apple aphid, infesting 48%–69% of twigs, peaked in July (Thakur *et al.*, 1988). During this investigation it was found that the decline of the aphid population following the peak in August, together with that of a syrphid species, was caused by the aphidophagous activity of *Brinckochrysa scelestes*. Thakur *et al.* concluded that the formerly active parasitoid *Aphelinus mali* was replaced with these two predators, and during this later period, these were the only efficient biological control agents. Holdsworth (1970) reported that in orchards in Ohio, *Chrysopa quadripunctata* also contributed to reducing the population level of *E. lanigerum*. In New Zealand, when this aphid species became more abundant on apple trees, *Micromus tasmaniae* and *Drepanacra binocula* were observed to feed on its colonies, but a hymenopteran parasitoid, *Anacharis zelandica*, restricted the numbers of the two hemerobiids (Collyer & van Geldermalsen, 1975). *E. lanigerum* was found in the following investigations, providing a further food source beside other aphid species for lacewings (see also Table 9.2) within apple stands: Switzerland (Baggiolini & Wildbolz, 1965), Hungarian experimental orchards (Mészáros, 1984), Croatia (Ciglar, 1985; Ciglar & Budinšcak, 1993), Canada, Quebec (LeRoux, 1960), and USA, Washington State (Carroll & Hoyt, 1984).

Generally, however, in orchards where lacewings were recorded, there were two to four species of aphid present. Other potential lacewing food sources were also found, including psyllids, mites, and mealybugs. For this reason, aphids in the following studies are only mentioned as potential prey, because the lacewings could also prey on other food sources. Baeschlin & Taksdal (1979) found *R. insertum* as a dominant species among aphids in Norwegian apple orchards, but *A. pomi* and *Dysaphis* spp. were also present. *Chrysoperla carnea* was the most dominant lacewing predator. *A. pomi*, the two *Dysaphis* spp., and *R. insertum* were also all present in Belgian apple orchards. The potential aphidophagous lacewings were *C. carnea*, occurring sporadically and a *Hemerobius* sp. (Malevez, 1976). In a one-year-old commercial apple-tree nursery in Poland three aphids served as prey for *C. carnea*: green apple aphid, rosy apple aphid, and a *Rhopalosiphum* sp. (Wiackowski & Wiackowska, 1968). Also in Polish apple orchards, Niemczyk *et al.* (1983), besides other prey, found the aphids *A. pomi*, *R. insertum*, and *D. plantaginea*. Within these orchards *C. carnea* was the dominant chrysopid species, while *Chrysopa pallens* and *C. phyllochroma* occurred only sporadically. Interestingly, the number of chrysopid larvae found was much lower than expected from the number of eggs present. Common lacewing larvae often walked down from apple-tree crowns to the grass or other plants of the understorey (Miszczak & Niemczyk, 1978). These authors reported that consequently, lacewing larvae were unable effectively to limit aphids and other pests in apple orchards. In Swiss apple orchards, where the individual number of detected *Chrysoperla carnea* larvae was very low, *E. lanigerum* was reported to be the most dominant aphid and *D. plantaginea* was an additional potential food source (Baggiolini & Wildbolz, 1965). In Hungarian experimental and commercial apple orchards, all the more important aphid species were discovered: *A. pomi*, *Dysaphis* spp., *R. insertum*, *E. lanigerum* (Mészáros, 1984). All the listed species served as prey for lacewings in these orchards (see Table 9.1, column 10*a*–*c*). In Croatian apple orchards, Ciglar (1985), Ciglar & Schmidt (1983), and Ciglar & Budinšcak (1993) found that the five aphid species mentioned above are associated regularly and frequently with apple. This was true both in managed commercial orchards and unmanaged apple stands, though in the latter case aphids were represented by lower individual numbers. Two aphidophagous lacewing species were recorded in their studies: *C. carnea* and *Chrysopa perla*.

In Quebec in North America, the aphids *A. pomi*, *E. lanigerum*, and *Rhopalosiphum* were recorded in apple orchards (LeRoux, 1960). Larvae and adults of *C. oculata* and *H. stigma* preyed upon these aphids. Three species of aphid (green, rosy, and woolly apple) occurred in apple orchards in Ohio (Holdsworth, 1970), and several lacewing species were observed preying on them. In New Zealand, three aphids infested apple trees, *Aphis spiraecola*, *Macrosiphum euphorbiae*, and *M. rosae*, which were preyed upon by two hemerobiids, *Micromus tasmaniae* and *D. binocula*.

9.1.3 Apple psyllids as prey for lacewings

Psyllids may build up another potential homopteran prey group in apple orchards. *Psylla mali* is the most harmful psyllid species. Semyanov (1973) in the St Petersburg region (Russia) detected the lacewing species *Chrysoperla carnea*, *Hemerobius perelegans*, and *H. marginatus* as the natural enemies of *P. mali*. This psyllid has also been detected in other investigations of apple orchards, but corresponding lacewing predation is not clear. Thus *P. mali* in Norway (Baeschlin & Taksdal, 1979) was described as having two population increases (second half of May to first half of June, and mid-August to late September). Lacewing larvae in this orchard were active mainly from August to November indicating the possibility that they preyed on psyllids in the autumn. *P. mali* has also been mentioned in Poland (Niemczyk *et al.*, 1983), in Hungary (Mészáros, 1984), in Switzerland (where *P. costalis* was also present) (Baggiolini & Wildbolz, 1965), and in Croatia (Ciglar & Schmidt, 1983; Ciglar & Budinšcak, 1993); corresponding potential predatory lacewing species are shown on Table 9.1.

9.1.4 Apple mealybugs as prey for lacewings

Mealybugs (Pseudococcidae) are also potential lacewing prey within the order Homoptera among apple pests, although few data are available. In Virginia (USA), Haeussler & Clancy (1944) found the Comstock mealybug (*Pseudococcus comstocki*) heavily infesting apple trees. Three lacewings were recorded within these orchards, *Chrysopa nigricornis*, *C. quadripunctata*, and the hemerobiid *Sympherobius amiculus*, but there was no noticeable reduction in the population level of this pest due to these predatory insects. However it is probable that the *S. amiculus* consumed mealybug larvae – similar predation has been frequently observed on other fruit trees. Collyer and van Geldermalsen (1975) reported *Phenacoccus graminicola* as a pest of apple orchards in New Zealand. Although this mealybug species is usually associated with grasses or white clover ground cover in apple orchards, these authors found that large numbers of individuals of this species infested the trunks and branches of apple trees. The hemerobiid *Micromus tasmaniae* was one of the reported natural enemies of *P. graminicola*.

9.1.5 Apple mites as prey for lacewings

Phytophagous mites are present in most apple orchards, supplying another food source, mainly for young larvae of lacewings. European red mite, *Panonychus ulmi*, is the most frequent apple mite pest, causing serious damage mainly in regularly and intensively pesticide-treated orchards. In Europe, Malevez (1976) has studied the seasonal dynamics of this mite in Belgian apple orchards. Two peaks of activity were found in the *P. ulmi* population (mid-May and late August). *Chrysoperla carnea* egg-laying began during the first peak (13–20 May) (Malevez, 1976). In Poland, Miszczak & Niemczyk (1978) analysed in detail the predation abilities of *Chrysoperla carnea* larvae at different densities of *P. ulmi* populations. They found that lacewing larvae were able to keep the European red mite population size below the economic threshold level when predator density was at least one chrysopid larva per 25 apple leaves. Furthermore, *Chrysoperla carnea* larvae were able to search out and capture mites, even when only one *P. ulmi* was present on 50 apple leaves. In Ohio (USA), Holdsworth (1968) examined *P. ulmi* control in an IPM apple orchard, where *C. rufilabris* and *Chrysoperla carnea* were present. In the case of high population densities of *P. ulmi* the number of *C. rufilabris* larvae was 2.7 times higher on apple foliage than at a low prey population level, while this ratio was 4.7 for *Chrysoperla carnea* larvae. Holdsworth also found that all chrysopid instars captured and readily accepted the mites. In Missouri (USA), Childers & Enns (1975) in five commercial and two abandoned apple orchards detected five lacewing species, *C. rufilabris*, *C. carnea*, *Chrysopa oculata*, *C. quadripunctata*, and *Micromus posticus* among 32 predaceous insects associated with tetranychid mites. They observed that *Chrysoperla rufilabris* larvae consumed *P. ulmi* in apple orchards, though this predator appeared at a very low population level. While other lacewing species are potential predators of this spider-mite, *Chrysopa oculata* and *M. posticus* larvae probably preferred lower vegetation, because larvae of these two lacewing species were regularly collected on water sprouts of apple trees, surface litter, and bark.

9.1.6 Apple moth as prey for lacewings

In Australian apple orchards, the light brown apple moth, *Epiphyas postvittana*, a polyphagous native pest, causes widespread damage on apples in the States of Victoria and Tasmania. MacLellan (1973), studying natural enemies of this moth, reported two hemerobiid predators of this pest, but they were present occasionally and only in small numbers in the examined orchards. At the same time, trash-carrier larvae of a chrysopid species gave a high percentage of positive serological reaction in precipitin tests. The seropositive reaction ratio was around 56%. In laboratory feeding experiments, chrysopid larvae readily consumed both first- and late-instar larvae of *E. postvittana*.

9.1.7 Seasonality of lacewings and their prey in apple orchards

In Europe and North America, only a small number of authors have paid attention to the synchrony between seasonal fluctuation of lacewings and patterns of their prey populations studied in apple orchards. There is some information regarding parallel sampling of lacewing predators and aphid seasonality. Such surveys started at the end of the 1960s and the early 1970s. Steiner *et al.* (1970) conducted the first such project in Europe (Germany). Sampling of phytophagous and zoophagous arthropods was done in six insecticide-treated and six untreated apple orchards. In these experimental orchards chemical treatments did not influence the seasonality of Neuroptera, especially chrysopids. Steiner *et al.* found two seasonal increases of abundance of chrysopids, which were well synchronised with the pattern of aphid numbers. The first lacewing peak was detected at the end of June, while the second peak was at the end of August or mid-September. Niemczyk *et al.* (1983) conducted another detailed phenological investigation on apple fauna in Poland. Within these orchards, larvae of *Chrysoperla carnea* were dominant and showed two small peaks, at the beginning and at the end of July. In this period *Aphis pomi* was most abundant in June and July, peaking in mid-July, while *Dysaphis plantaginea* showed two increases, the first peak in June, and a second smaller peak in September. In Italy, Pantaleoni & Ticchiati (1988) studying the Neuroptera fauna of fruit orchards monitored chrysopids and aphids with detailed samples of seasonal dynamics. Their collections were conducted in commercial apple orchards, where aphid populations could be collected on trees from mid-May to late July, with one peak at the beginning of July. Temporally delayed chrysopid egg-laying started during the peak abundance of aphids, peaked in mid-August, and declined by early October. Adults and larvae proved to be *C. carnea* species. Larval activity of *C. carnea* grew from mid-July, and peaked in early August. Adult *Chrysoperla carnea* expressed a similar seasonal activity pattern to that of their larvae.

In Hungary, during a 10-year (1976–85) agroeco-system research project, precise seasonality investigations were carried out regarding lacewings found in apple orchards. From the results of this research, Szabó & Szentkirályi (1981) published the seasonal flight-activity patterns of the most common chrysopid and hemerobiid species. They found that depending on orchard type and geographical region, the *C. carnea* flight pattern was varied. Generally two population increases characterised this species. The first period was from early June to mid-July, with a peak in late June or early July. The second period lasted from mid-September to mid-October with one peak in late September to early October. The other common species, *Chrysopa formosa*, also showed two flight-activity increases, one in June and another one in August. In Hungarian apple orchards, a third dominant species, *C. phyllochroma*, formed one flight peak in August. The most dominant hemerobiid species was *Hemerobius humulinus*, which had two flight-activity peaks, one in May or June, and the other one during September. The two flight periods of *Wesmaelius subnebulosus* occurred in June–July and in September–October. The next most frequent brown lacewing, *Micromus angulatus*, was active from the end of July to late September, and showed a peak during September. Szabó & Szentkirályi (1981) revealed, by counting eggs laid on apple leaves, that the main oviposition period of chrysopids occurred in June and July following mass propagation of apple aphids. The egg-laying pattern showed a small peak in late May and a larger one in early July. Egg-laying declined by the end of August. Based on long-term sampling conducted in apple orchards, the general seasonality patterns of the families Chrysopidae and Hemerobiidae showed the following characteristics (F. Szentkirályi, unpublished data). The seasonal flight activity of chrysopid species, calculated without *Chrysoperla carnea*, lasted from the beginning of May to the end of September, and it was divisible into two periods of increase. The first occurred in June and the second one built up during August and early September. Brown lacewing activity in Hungarian apple orchards lasted from April to October, with several peaks. Generally, the most intensive flight activity of hemerobiids was between mid-July and early October. Seasonally, four flight peaks of hemerobiids could be detected, a smaller one in mid-May, a greater one in late July, and the two highest peaks at the end of August and in late September. The activity pattern of these aphidophagous groups was well synchronised with the general seasonal dynamics of apple aphids found in Hungarian orchards.

Few authors provide information on seasonal changes in abundance of lacewings and their prey in North America. Asquith *et al.* (1980) give data on the seasonal occurrence of lacewings of Pennsylvanian (USA) apple orchards. Among chrysopids of these commercial orchards, *Chrysopa nigricornis* was active in July, *C. quadripunctata* in June and July, *C. oculata* from June to the end of August, *Chrysoperla rufilabris* from July to September, and *C. carnea* between June and late August. The following hemerobiid activities were detected in orchards: *H. humulinus* from June to August, *H. stigma* only in August, *M. posticus* during August and September. In Missouri (USA), Childers & Enns (1975) determined the following seasonality regarding lacewing larvae after studying seven apple orchards: *C. rufilabris* could be collected between mid-June and early September, *C. carnea* from late April to early September, *Chrysopa oculata* larvae from mid-June to the end of July, and *C. quadripunctata* from June through to the end of July. The hemerobiid *M. posticus* was found in these investigations from 13 March to 16 July. Other brown lacewing larvae were active on apple foliage from 16 May to 28 October (Childers & Enns, 1975). In Washington State (USA), Carroll & Hoyt (1984) made analyses to characterise chrysopid larvae and apple aphids phenologically. Lacewing larvae were active from late July to mid-October building up two or three peaks in mid-August and in mid-September, or in late August, mid-September, and early October. The aphid prey, *A. pomi*, was characterised by two increases in abundance during the season, the first period from mid-June to mid-July with one peak in early July, and the second one from late August to early October with one peak in the first half of September. These facts suggest that the second aphid increase corresponded with the second activity peak of chrysopid larvae. In Ohio (USA), Holdsworth (1970) reported that within apple orchards, the hemerobiids *M. posticus* and *H. humulinus* overwintered as adults and laid their eggs in April.

Seasonality patterns relating to lacewings inhabiting apple orchards have also been published in Canada. MacLellan (1977) conducted investigations on young and older semidwarf apple trees over five years

(1965–9). Counting egg numbers per tree in every 7–10-day interval throughout the season, green lacewing oviposition activity was recorded, and lasted from July to the end of August, peaking each year in late August. Hagley (1974) studied predacious arthropods over four years (1971–4) in an unsprayed apple orchard, where the seasonal distribution of two common lacewings, *C. oculata* and *H. humulinus*, was sampled. Lacewings on apple trees could be collected from May to the end of September. *C. oculata* populations showed two periods of increase, firstly in June or July, and secondly in late August to early September. *H. humulinus* expressed three or four peaks during season, basically one per month from June to September. The most characteristic peak of abundance of this hemerobiid species was between mid August and mid September.

9.1.8 Impacts of orchard management techniques on lacewings

Regular application of broad-spectrum pesticides and the maintenance of ground surface without weedy cover have caused numerous serious control problems in traditional pest management of commercial apple orchards. The development of pesticide resistance has limited the usefulness of regular insecticides or acaricides in the suppression of pests, whilst allowing several secondary phytophagous arthropod pests to become more important. Furthermore, natural enemies of pests, being more sensitive then their prey, are eradicated in orchards under intensive pesticide use. From the 1980s, in particular, these increasing problems have forced specialists to research alternative environmentally friendly pest management systems, such as IPM. One goal of IPM programmes is to amplify natural control factors via preservation and attraction of entomophagous arthropods. Methods include reduced spraying programmes, use of selective pesticides, selection of proper agricultural techniques, and increasing the diversity of surrounding vegetation.

Effects of ground cover

Ground cover with various crop plants or weeds may supply natural enemies with alternative food, shelter, oviposition sites, and overwintering microhabitats. Increase of plant diversity by cover cropping inside apple orchards can stabilise the populations of beneficial insects, because understorey vegetation maintains more aphids, and thus attracts and retains aphidoph-

aga. Extremely important is the sowing of plants – weeds – that are preferred by early-season predators during their colonisation.

As early as 1960, LeRoux drew attention to the importance of the ground cover in the preservation and survival of predatory insect fauna of apple orchards. He regarded plant cover as an integral part of apple ecosystems, which, together with bordering hedges, created a large reservoir for aphid predators.

Several studies have assessed the impact of ground-cover plant composition on predator conservation and augmentation. In Chinese apple orchards, Yan & Duan (1986, 1988) and Yan *et al.* (1997) made cover-cropping experiments. Various cover-cropping systems were developed that consisted of different herbaceous plant species. White sweet clover, *Melilotus albus*, and the weed *Lagopsis supina* (Labiatae) were the cover-crop plants, with alfalfa (*Medicago sativa*) planted between the rows. In orchards where white sweet clover was used as a between-rows ground-cover plant numbers of *Chrysoperla carnea*, *Chrysopa formosa*, *C. phyllochroma*, and *C. pallens* were higher in the apple tree canopy than in control orchards. The predominant lacewing species, *Chrysoperla carnea* (see Table 9.2, column 2) helped to suppress the mite pest *Tetranychus viennensis*.

Wyss (1995) tested effects of weed strips on aphids (*Dysaphis plantaginea*, *Aphis pomi*) and aphidophagous insects in a Swiss apple orchard. The flowering weed strip mixture sowed between rows consisted of 21 different herbaceous plant species. These weedy plants were selected by Wyss to offer food sources (pollen, nectar, and alternative aphid prey) for predators in those periods when aphids were absent from apple trees. Flowering weeds produced a significantly higher population level of chrysopids on tree foliage through the whole season. In addition the number of apple aphids was lower in the treated plot than in the control due to aphidophagous activity. In the early season, this was a very important factor in the control of the colonising *D. plantaginea*, because the herbaceous plant cover flowered at the right time to attract aphidophagous lacewings, like *Chrysoperla carnea*, into the apple orchard.

In Hungary the number of species in the weed flora within apple orchards ranged from 17 to 82 (Mészáros, 1984). With sweep-netting and pitfall trapping inside these apple orchards, numerous lacewings were collected from the ground-cover vegetation (F.

Szentkirályi, unpublished data). Combining larvae and adults, among the chrysopids, *C. carnea* (74.0%) and *Chrysopa phyllochroma* (10.2%) were the most dominant species, while *C. pallens*, *C. perla*, *C. formosa*, *C. abbreviata*, and *Dichochrysa prasina* were rare in the herbaceous layer. Of hemerobiids, the most frequent species were *Hemerobius humulinus* (42.3%) and *Wesmaelius subnebulosus* (40.8%). Two *Micromus* species (*M. angulatus* and *M. variegatus*) occurred with a common dominance value of 12.6%.

Effects of pest management programmes

Management type and agricultural techniques used in apple orchards can have a great impact on both phytophagous insects and their natural enemies. Cultivation methods, such as removal of weed cover, or pruning and the use of N_2-fertilisers (making leaves more vigorous, or resulting in the production of succulent shoots suitable for aphid development), may reduce the effectiveness of aphidophaga. However, in integrated pest management programmes, a reduced number of pesticide treatments may have a favourable influence on beneficial arthropods, increasing numbers within apple orchards. In British Columbia (Canada), Madsen & Madsen (1982) compared the beneficial arthropod fauna of an organic and a pesticide-treated apple orchard. They found that under organic management (nearly pesticide-free conditions), predatory insects and spiders were represented in larger numbers on tree foliage and ground-cover vegetation than in an orchard under a pesticide-spraying programme. In these field experiments, green and brown lacewings were recorded exclusively only in the organic apple orchard. However, the suppressive effects of pesticides on population size and species richness of lacewings can be potentially compensated by suitable surrounding vegetation acting as a reservoir for colonisation. Szentkirályi (1992) studied brown lacewings in seven Hungarian apple orchards under different pesticide regimes. The conclusion was that more diverse agriculture or surrounding woody vegetation encouraged species-rich hemerobiid assemblages even in intensively treated commercial apple orchards, according to the 'resource diversity hypothesis' (Szentkirályi & Kozár, 1991).

Recommended selective insecticides for IPM programmes may harm lacewings. For instance, Rumpf & Penman (1993) examined the effect of the insect growth regulator fenoxycarb on two lacewing species under laboratory and field conditions. Tests were carried out on *Chrysoperla carnea* and *Micromus tasmaniae* in Germany and New Zealand. In laboratory conditions, 0.01% fenoxycarb treatment reduced the pupation rate by 70%, and on apple trees 60% and 83% decreases were detected in the number of pupae of *Chrysoperla carnea* and *M. tasmaniae* respectively.

9.1.9 Biological control experiments with lacewings in apple orchards

At present there is intense interest by growers in biological control, due to the increasing insecticide resistance of apple aphids. This has resulted in a number of studies of augmentative release of natural enemies. Hagley (1989) applied *Chrysoperla carnea* as a biological control agent against *Aphis pomi* infesting dwarf apple trees in Ontario (Canada). The augmentative releases were undertaken during two seasons. Using eggs of *Chrysoperla carnea* at a density of 335000 individuals per hectare, the number of nymphs and apterous adults of *A. pomi* were reduced significantly on shoot terminals. Predator:prey ratios applied were 1:10 and 1:19, but lacewing larvae proved to be less effective in the last case because of the lower ratio. Grasswitz & Burts (1995) carried out augmentative-release-experiments with *C. rufilabris* against *A. pomi* on 1-year- and 2-year-old apple plots in Washington State (USA). During one trial they used *C. rufilabris* alone at a rate of 200 eggs or first-instar larvae per tree per week over a 6-week period. In another experiment, lacewings were released in combination with *Aphidoletes aphidimyza* at a rate of 400 eggs or first-instar stage larvae per tree per week for a 7-week-period. In these experiments, there was no significant suppressive effect on aphids due to *C. rufilabris* treatments and the number of *C. rufilabris* larvae found on trees was very low in comparison to numbers released. The reason for this may have been a low survival rate of lacewing larvae. Though larvae of this chrysopid species in laboratory feeding experiments readily accepted and consumed *Aphis pomi* nymphs, on apple trees lacewing larvae were observed to prey preferentially on alternative food sources, despite the presence of large number of green apple aphid colonies. So presumably, the failure of these release experiments with *C. rufilabris* was due to high mortality of young larvae caused by starvation. In India Thakur *et al.* (1992) conducted release experiments with the chrysopid *Brinckochrysa scelestes* against another apple pest, *Eriosoma lanigerum*. Population levels of the target aphid species were monitored before and after releases at four sites, to assess predatory efficiency. The rate of reduction in aphid populations due to *B. scelestes* larvae ranged between 23.8% and 34.8%, depending on the apple orchard.

9.2 PEAR

After apple, the common pear (*Pyrus communis*) is the second most intensively and widely cultivated fruit-tree species among pome fruits. Numerous arthropod pest groups are associated with pear, and their members consume different tree parts creating various phytophagous guilds. These guilds of herbivorous insects and mites together with their natural enemies can build up hierarchically organised subcommunities within pear orchards characterised by complicated structures (Gut *et al.*, 1991).

Among the phytophagous arthropods there are members of the sap-feeder guilds, serving as the main potential food source for lacewings, namely the homopteran insects (psyllids, aphids, coccids, cicadellids) and mites. Of these, pear psyllids are the most serious pests with the greatest economic importance. Aphids cause damage less frequently, although more than 40 species have been recorded on pear trees. This explains why there are fewer lacewing species documented in pear orchards than on apples. As the most important food source for most lacewings is aphids, in the case where aphids are rare, many lacewings prey on other homopteran insects, such as psyllids.

9.2.1 Characteristics of lacewing assemblages in pear orchards

The European and North American data of lacewing assemblages collected on cultivated and wild pear stands are presented in Tables 9.3 and 9.4, respectively. In Europe, the total species richness was 12 chrysopids and nine hemerobiids. By frequency of occurrence and percentage dominance values calculated from the few quantitative collections, the characteristic species of green lacewing assemblages associated with pear were *Chrysoperla carnea* (12%–100%), *Chrysopa formosa* (12%–67%), *C. pallens* (13%–18%). *Dichochrysa prasina* and *D. flavifrons* with smaller frequency and lower abundance can be regarded as further supple-

mentary species. The most characteristic brown lacewing species was *Hemerobius humulinus* (65%–100%). Additional species with lower occurrence were *W. subnebulosus* (10%), *H. micans*, and *Sympherobius pygmaeus*. In pear orchards with more developed herbaceous vegetation, *M. angulatus* (21%) could also be found. From Table 9.3 it can be concluded that the lacewing assemblage characteristic of pear orchards has three to four chrysopids and one to three hemerobiids. All of these species are ubiquitous and eurytopic without exception.

According to Table 9.4, lacewing assemblages in American pear orchards are relatively poor in number of species with a total of four chrysopids and six hemerobiids recorded. Among the green lacewings, three species can be considered as characteristic of pear trees by their frequency values. These are *Chrysoperla carnea*, *C. plorabunda* and *Chrysopa oculata*. For hemerobiids, it is likely that the members of the genus *Sympherobius* are characteristic of pear orchards, among which the most frequent was *S. angustus*. One reason for the low number of species may be that American orchards are less well studied. The scarce data suggest that both lacewing families form assemblages of one to three species in North American pear plantations.

9.2.2 Pear psyllids as prey for lacewings

The most serious pear pests belong to *Cacopsylla* (*Cacopsylla pyri*, *C. pyrisuga*, and *C. pyricola*) which are associated with both cultivated and wild *Pyrus* spp. *C. pyricola* was introduced to America from Europe in the 19th century, where it has become a serious pest. The above-mentioned pear psyllids are common all over Europe on different pear cultivars; the most abundant psyllid species on cultivated pear is *C. pyri*. Herard (1986) and Lyossoufi *et al.* (1994) have reviewed the rich entomophagous guilds associated with *Cacopsylla* species in European pear orchards. In the majority of investigations (Table 9.3) lacewings preyed on populations of the most abundant psyllid *C. pyri*. In the Czech Republic, the level of *C. pyri* populations has seriously increased in insecticide-treated orchards, because they have become resistant to classical insecticides (e.g. pyrethroids and organophosphates) (Kocourek & Beránková, 1996). This psyllid, which has two to three generations a year, is attacked in the Czech Republic by *Chrysoperla carnea* and *Hemerobius humulinus*

Table 9.3. *Species composition and dominance distribution of chrysopids and hemerobiids found in pear (Pyrus domestica) orchards of Europe*

	1	2	3	4	5a	5b	6	7	8	9	10	11	12	13	14	15	16	FR
Chrysopidae																		
Chrysoperla carnea s. lat. (Stephens)	+	+	+	+	70.6	11.6	+	86–100	+	+	+	A	+	98.8	+	+	+	16
Chrysopa formosa Brauer					11.8	66.8				+						+		3
Chrysopa pallens (Rambur)			+	+	17.6	16.6		0–13		+	+		+			+		8
Chrysopa perla (L.)													+		+			2
Chrysopa sp.															+			1
Dichochrysa prasina (Burmeister)								0–1				+				+	+	4
Dichochrysa flavifrons (Brauer)							+				+	+	+			+		5
Dichochrysa inornata (Navás)											+							1
Dichochrysa picteti (McLachlan)												+		1.2				2
Dichochrysa zelleri (Schneider)							+											1
Nineta flava (Scopoli)					+													1
Nineta vittata (Wesmael)	+																	1
Italochrysa italica (Rossi)										+								1
Hemerobiidae																		
Wesmaelius subnebulosus (Stephens)							+				+	+		10.3				4
Wesmaelius nervosus (Fabricius)											+							1
Hemerobius humulinus L.				+	+	100		100	+	+	+	+	+	64.7	+	+		12
Hemerobius micans Olivier													+	4.3	+			3
Hemerobius lutescens Fabricius													+	20.7				2
Micromus angulatus (Stephens)												+						1
Sympherobius pygmaeus (Rambur)										+	+		+					3
Sympherobius elegans (Stephens)												+						1
Drepanepteryx phalaenoides (L.)							+								+			2
S_C		1	1	2	4		2	3	1	2	5	3	4	2	2	6	2	
S_H	3			1	1		2	1		2	3	3	3	4	2	2	2	

Note: The dominance values (%) are calculated from the number of collected individuals published in the reviewed papers.

Abbreviations: A, abundant; S_C, S_H species richness of chrysopids and hemerobiids respectively; +, species present; FR, frequency of species occurrence in literature surveyed.

Sources: 1, UK (Hodgson & Mustafa, 1984, cited in Lyoussoufi et al., 1994); 2, Netherlands (Molleman et al., 1997); 3, Czech Republic (Kocourek & Beránková, 1996); 4, Hungary (Szentkirályi & G. Jenser, unpublished data); 5a, b, Moldova (a: *Dysaphis mali* infestation, b: *Dysaphis reaumuri* infestation, Talitskaya, 1980); 6, Moldova (Zelený & Talitsky, 1966); 7, Portugal (3 sites, Matias et al., 1988); 8, Spain (Franco, 1989, cited in Lyoussoufi et al., 1994); 9, France (Fauvel, 1982, cited in Lyoussoufi et al., 1994; Viollier & Fauvel, 1984); 10, France (Boujiou et al., 1984); 11, France (Herard, 1986); 12, Italy (Arzone, 1979; Giunchi, 1980); 13, Italy (Marzocchi & Pantaleoni, 1995); 14, former Yugoslavia (Rield, 1981; Grbić et al., 1990; Ciglar & Barić, 1992); 15, Greece (wild pear trees, Santas, 1987); 16, Russia, former USSR (Shalamberidze, 1980).

Table 9.4. *Species composition of chrysopids and hemerobiids found in pear (*Pyrus domestica*) orchards in America*

	1	2	3	4	5	6	7	8	9	10	11	FR
Chrysopidae												
Chrysoperla carnea s. lat. (Stephens)	+	+			+	+	+		+			6
Chrysopa oculata Say	+	+	+	+								4
Chrysoperla plorabunda (Fitch)			+		+			A		+		4
Chrysopa sp.											+	1
Hemerobiidae												
Hemerobius humulinus L.			+									1
Hemerobius pacificus Banks		+										1
Hemerobius ovalis Carpenter				+								1
Hemerobius sp.						+						1
Sympherobius angustus (Banks)			+						+	+		3
Sympherobius maculipennis Kimmins											+	1
Sympherobius californicus Banks								+				1
Sympherobius sp.			+									1
S_C	2	2	2		3	1	1	1	1	1	1	
S_H		1	3	1		1		1	1	1	1	

Abbreviations: A, abundant; S_C, S_H, species number of chrysopids and hemerobiids respectively; +, species present; FR, frequency of species occurrence in literature surveyed.

Sources: 1, Canada, British Columbia (Wilde, 1962); 2, Canada, British Columbia (McMullen & Jong, 1967); 3, Canada, Ontario (Hagley & Simpson, 1983); 4, USA, Washington (Fye, 1985); 5, USA, Washington (Burts, 1963); 6, USA, Oregon (Westigard *et al.*, 1968); 7, USA, Oregon (Gut *et al.*, 1991); 8, USA, California (Doutt, 1948; Doutt & Hagen, 1950); 9, USA, California (Nickel *et al.*, 1965); 10, USA, California (Madsen *et al.*, 1963); 11, Chile (Curkovic *et al.*, 1995).

(Kocourek & Beránková, 1996). In Hungary, *Cacopsylla pyri* is regularly associated with *Chrysoperla carnea* and *Chrysopa pallens* (F. Szentkirályi & G. Jenser, unpublished data) in commercial pear orchards. Most studies on lacewing predators of psyllids have been conducted in southern Europe. In Portugal, among the natural enemies of *Cacopsylla pyri*, three chrysopids were recorded, and *Chrysoperla carnea* was the most abundant of them (Matias *et al.*, 1988) (Table 9.3). Matias *et al.* (1990) investigating the predators of this psyllid also found that in pear orchards under IPM treatments, the abundance level of *Chrysoperla carnea* and *Chrysopa pallens* was ten times greater than in orchards intensively treated with insecticides. They recorded *H. humulinus* only from IPM orchards.

In southern France, *Cacopsylla pyri* produces five to seven generations each year (Herard, 1986). Among the entomophagous insects of *C. pyri*, rich lacewing assemblages were found in commercial pear orchards (Herard, 1986). A greater number of species can be a consequence of richer vegetation in the environment. Out of seven species, only *Chrysoperla carnea* was abundant; five of the other species (*Dichochrysa prasina*, *D. picteti*, *H. humulinus*, *Wesmaelius subnebulosus*, *Sympherobius pygmaeus*) were rare.

In Italian pear orchards, numerous lacewings participated in the natural control of *Cacopsylla pyri*. From eight pear orchards treated with pesticides at different concentrations, Arzone (1979) found only one in which *Chrysopa perla* was present in sufficient numbers to play an effective role in the suppression of *Cacopsylla pyri*. However, the number of lacewing species in orchards was not influenced by the number of insecticide treatments with two to three species in all cases, except where there were ten sprayings a year in which case none was present. *Chrysopa pallens* was the most common species in Arzone's orchards (recorded in six pear stands) while the other six lacewing species occurred only in one to three cases. During a survey of another Italian orchard, out of the six species of lacewings consuming *Cacopsylla pyri*, only two, *Chrysoperla carnea* and *H. humulinus*, were dominant species (Marzocchi & Pantaleoni, 1995).

In the pear orchards of the former Yugoslavia, *Cacopsylla pyri* is a major pest with at least five generations a year (Grbić *et al.*, 1990). In Croatian commercial orchards, Ciglar & Barić (1992) found *H. humulinus* among the natural enemies of this psyllid species. In Vojvodina, the univoltine *C. pyrisuga* could be found in small numbers with *C. pyri*, according to Grbić *et al.* (1990). They found that the larvae of *Chrysoperla carnea* and of *Hemerobius* spp. preyed on the nymphs and eggs of these psyllids, both under field and laboratory conditions. However, during the two years of this study, the neuropteran larvae were represented at such a low density that they could not have much influence on psyllid population levels in pear orchards.

On European wild pear (*Pyrus amygdaliformis*) the same psyllid species can be found as on cultivated pear, although different species can become dominant. In Greece large populations of *Cacopsylla notata* were detected on wild pear (Santas, 1987), while *C. pyrisuga* and *C. pyri* were rare. Santas's studies suggested that psyllids of wild pear can become alternative food sources for lacewings in pear orchards. In this way wild pear trees can serve as a reservoir for natural enemies. Of eight lacewings found, *Chrysoperla carnea* was the most abundant, followed by *D. flavifrons* and *D. prasina*. He also observed that *H. humulinus* larvae fed on psyllid nymphs. The larvae of the chrysopids *Chrysoperla carnea*, *D. flavifrons*, and *D. prasina* were able to develop successfully on psyllids until they became adults (Santas, 1987). In southern France Viollier & Fauvel (1984) collected the predators of psyllids by beating wild pear trees and abandoned pear orchards surrounded by garrigue vegetation type. The principal pest on wild pear was *Cacopsylla pyricola*, while *C. pyricola* and *C. pyri* were together the dominant phytophages in *Pyrus communis* stands. *C. pyrisuga* was represented at a low population level on both pear species. Viollier & Fauvel (1984) pointed out that wild pear trees were an important potential reservoir for the

predatory insects of psyllids, because they supplied alternative food prey. Among natural enemies, two lacewing species were found on the trees of both pear types, namely *Chrysoperla carnea* and *S. elegans*, the first one being a probable psyllid predator.

All data about the predation of lacewings on psyllids in North American pear orchards relate to the species *Cacopsylla pyricola*. This psyllid, after its introduction in the last century, has spread throughout the pear orchards of Canada and the United States. Studying the natural enemies of this psyllid, which is characterised by great abundance frequently exceeding the economic injury level, American authors have reported many field observations and laboratory experiments involving several lacewing species attacking larvae and eggs of this harmful psyllid. In Canada, Wilde (1962) and McMullen & Jong (1967) studied the natural enemies of *C. pyricola* infesting pear orchards in British Columbia. Their investigations were carried out in commercial orchards, where five psyllid generations developed per year. Sampling within pear orchards revealed that, of the green lacewings, *Chrysoperla carnea* and *Chrysopa oculata* adults were common. Larvae of both species were present in British Columbian orchards from early May to October and they consumed the eggs and nymphs of *Cacopsylla pyricola* as indicated by direct field observations. Lacewing larvae had the greatest abundance through July and August. Feeding experiments with larvae of the two lacewing species also supported their consuming of psyllids. However, the importance of *Chrysopa oculata* in the control of *Cacopsylla pyricola* populations is doubtful. It is commonly known that this lacewing is associated predominantly with the lower vegetation layer. Wilde (1962) and McMullen & Jong (1967) reported that in their experimental orchards there was ground cover, and the small cover-cropped plots in Wilde's orchards harboured more psyllid predators. So its preference for understorey vegetation guaranteed the regular presence of *Chrysopa oculata* in the pear groves.

Of brown lacewings, *Hemerobius pacificus* was the only species verified by McMullen & Jong (1967), as consuming the eggs and nymphs of *Cacopsylla pyricola* in British Columbian orchards. However this hemerobiid species was ineffective in the control of psyllids, because its abundance level was a tenth that of chrysopids. In Yakima Valley (Washington, USA), studying the natural enemies of *C. pyricola*, Fye (1985) detected the brown lacewing *H. ovalis* as one of the principal psyllid predators. He applied corrugated cardboard bands, as traps offering overwintering shelters, in order to record predatory insects associated with pear trees. It was determined from the collections of these bands that *H. ovalis* appeared in traps generally only late in the season, and that according to its vertical distribution this hemerobiid preferred the mid and upper crown of pear trees. In the absence of spraying, in Californian pear orchards, an anthocorid species and the hemerobiid, *Sympherobius angustus* were detected as major predators of pear psylla by Nickel *et al.* (1965). They observed that within the orchards examined, this brown lacewing species had greater importance than *Chrysoperla carnea*.

9.2.3 Pear aphids as prey for lacewings

Numerous investigations already mentioned here report on different aphid species in pear orchards in addition to psyllids. However, in most cases, aphid populations were low so that even if they were serving as lacewing prey their importance would be negligible. In pear orchards in Moldova lacewings were more abundant if aphids were present with psyllids (Talitskaya, 1980). According to the surveys of Cruz de Boelpaepe & Filipe (1990) carried out in Portugal, the aphid *Dysaphis pyri* was a common and injurious pest damaging the foliage and fruit of pear trees. In their studies, chrysopids were most active in the second half of May to early June. This period coincided with the second population build-up of *D. pyri*. The only lacewing mentioned as the natural enemy of this aphid within the examined orchards was *C. carnea*. Viollier & Fauvel (1984) found the same aphid species among the principal pests in an abandoned pear orchard southern France, where its colonies served as prey for the lacewings *C. carnea* and *S. elegans*. *D. pyri* was also discovered on wild pear trees in Greece by Santas (1987), though the aphid was present in low numbers and late in the autumn indicating the probability that this aphid would be eaten only by rather late-season active hemerobiids. Marzocchi & Pantaleoni (1995) observed *Melanaphis pyraria*, *Dysaphis* spp., and *Anuraphis* spp. aphids in Italian pear orchards. Despite the fact that out of three orchards sampled, two had a much lower aphid infestation level than that of psyllids (*Cacopsylla pyri*), it is nevertheless probable that they contributed to the survival of the five lacewing species found. In North

America, McMullen & Jong (1967) found *Aphis pomi* on pear trees of commercial orchards in British Columbia (Canada). With other additional prey types, this aphid species could be one of the food sources for the larvae of *Chrysoperla carnea*.

9.2.4 Pear scales as prey for lacewings

In pear orchards, a further homopteran prey group for lacewings is formed by scale insects. In particular mealybug (Pseudococcidae) populations may become a food source for certain lacewing species. In Californian orchards in Santa Clara Valley, the mealybug *Pseudococcus maritimus* infested pear trees and was attacked by *Sympherobius californicus* and *Chrysoperla plorabunda*. This latter lacewing species was a predominant enemy of mealybugs and it could suppress pest populations in the absence of insecticide treatments (Doutt & Hagen, 1950). In Chile, another mealybug species, *Pseudococcus affinis*, was discovered and claimed to be a harmful pest of pear orchards (Curkovic *et al.*, 1995). These authors recorded the hemerobiid *S. maculipennis* as a natural enemy of mealybugs, collected with corrugated bands fixed on trees. From wild pear in southern France Viollier & Fauvel (1984) recorded the coccid species *Epidiaspis leperii*. Out of the two lacewing species collected in their surveys, *S. elegans* may be a predator of this scale insect.

9.2.5 Pear mites as prey for lacewings

Increased phytophagous mite populations may become prey for young lacewing larvae and certain hemerobiid adults. The European red mite, *Panonychus ulmi*, can often be found on pear trees. Marzocchi & Pantaleoni (1995), working in Italian orchards, declared that populations of *P. ulmi* were present on trees in August and late summer. In the abandoned orchard and wild pear stands in southern France surveyed by Viollier & Fauvel (1984), five mite species belonging to the families Tetranychidae, Eriophyidae, and Tenuipalpidae were recorded, providing presumably a food source for the hemerobiid *Sympherobius elegans* found in these habitats. In house garden pear trees of Hungary infested with *P. ulmi*, larvae of *Hemerobius humulinus* were observed as they were feeding on mite individuals (F. Szentkirályi, unpublished data). In Canada, in pear orchards in British Columbia, the same mite species together with *Tetranychus telarius* were recorded

(McMullen & Jong, 1967). Considering their population level, mites could have been alternative prey for *Chrysoperla carnea* and *H. pacificus*. In the Californian pear orchards of Nickel *et al.* (1965), leaves were severely damaged by *P. ulmi* and pear rust mites, *Epitrimerus pyri*. Supposedly, these mites were prey for the brown lacewing *S. angustus*, which also appeared in great number in the same plantations.

Reviewing the data on lacewings and their potential prey species, the principal food sources of brown lacewings belonging to the genus *Sympherobius* characteristic for pear orchards are probably scale insects (mainly mealybugs) and mites. This is confirmed by the fact that *Sympherobius* species were always encountered in those pear orchards where the former prey arthropods were present as well (e.g. Doutt & Hagen, 1950; Nickel *et al.*, 1965; Viollier & Fauvel, 1984; Curkovic *et al.*, 1995).

9.2.6 Seasonality of lacewings and their prey in pear orchards

In addition to the ratio of entomophagous insects to pests, the synchrony between prey and predator is also important. Several authors have analysed the seasonal dynamics and the degree of synchrony between pear psyllids and lacewings in Europe. From this point of view, the most detailed sampling has been conducted in Portuguese orchards (Matias *et al.*, 1988, 1990; Matias, 1990; for aphids Cruz de Boelpaepe & Filipe, 1990). These studies indicated that chrysopid eggs and larvae were found in Portuguese pear orchards from February through to November. There was a tendency for chrysopid numbers to peak from the second half of May to the end of June (the peak was most frequently at the very end of May) and then again from late August to early October (the second peak was formed most frequently in mid-September). The number of eggs laid and of larvae present in the case of *Chrysoperla carnea* (see the high dominance value of *C. carnea* in Table 9.3, column 7) were relatively well synchronised with the pattern of the prey psyllid. Between the peaks of prey and predator, an approximately two-week lag for chrysopids could be detected. Sometimes there were not only two, but three lower or higher peaks in seasonal activity patterns of chrysopids, depending on the growth periods of *Cacopsylla pyri*. As a conclusion, Matias *et al.* (1988, 1990) reported that *Chrysoperla carnea* larvae had an outstanding role in the control of

psyllid populations. Marzocchi & Pantaleoni (1995), on the basis of their examinations in Italy, claimed that lacewing abundance is less synchronised with the changes in the number of individuals of pear psyllid populations. They found that *Cacopsylla pyri* was characterised by three periods of increase, occurring in late May to early June, in July, and in August–September. In their pear orchards, brown lacewings responded with an increase corresponding only to the second psyllid peak. Hemerobiids showed two peaks during the season, the first one in the second half of July and the other one in October. Chrysopids had two peaks as well, but the first one occurred a bit later, just before mid-August, while the autumn peak occurred in late-October. In Yugoslavia, Grbić *et al.* (1990) found that the seasonal dynamics of *C. pyri* eggs and nymphs were well synchronised with the pattern of chrysopid and hemerobiid larvae. *C. pyri* showed two periods of increase, with peaks in mid-May and in mid-July. Neuroptera larvae also peaked in mid-May, with a second larger peak in mid-July. This second increase lasted until mid-August. In Greece, Santas (1987) surveyed seasonal changes of psyllids and chrysopids associated with wild pear trees. Psyllids, predominantly eggs and younger larvae of *C. notata*, had two distinct population increases during the season, with peaks in June and September. Monthly sampling with the application of McPhail traps confirmed that the seasonal activity pattern of *Chrysoperla carnea* adults more or less followed that of psyllids, so its population growth continued from May through to October, with a low peak in July and a higher peak in September. The seasonal activity of *Dichochrysa flavifrons* and *D. prasina* was unimodal, the former peaked in August, the latter was evenly high during July and August. As a conclusion, chrysopids were not well synchronised with the dynamics of psyllid populations in the wild pear stands examined by Santas.

In North American pear orchards, lacewing seasonality has been less closely monitored than in Europe (Table 9.4). McMullen & Jong (1967) found that in British Columbia *Chrysoperla carnea* and *Chrysopa oculata* larvae preyed on eggs and nymphs of the psyllid *Cacopsylla pyricola* from early May through to October. They noticed that chrysopid larvae were most abundant in July and August. In California (Nickel *et al.*, 1965), in a semi-abandoned pear orchard with no insecticide spraying, found that nymphs and eggs of *C.*

pyricola showed three periods of increase, but that lacewings were represented only in low numbers, so a definite tendency regarding synchrony could not be determined. From August until the end of October, only a moderate increase in the abundance of lacewings was detected in these studies. Also in Californian pear orchards, during biological control experiments against the mealybug *Pseudococcus maritimus*, Doutt & Hagen (1949) sampled the natural activity of the green lacewing *Chrysoperla plorabunda* with trap banding on trunks. Mealybug populations had a bimodal distribution with a peak both in late July and in the first half of November. However, the number of lacewing pupae rose only from mid-August, and the pattern was characterised with a higher peak in mid-September and a smaller one in early November. According to these observations it is likely that the second lacewing peak was formed as a consequence of the autumn population increase in mealybugs.

9.2.7 Biological control experiments with lacewings in pear orchards
Despite the economic importance of pear cultivation, very few biological control experiments have been carried out with lacewings against pear pests. Doutt & Hagen (1949, 1950) conducted periodic release experiments with the green lacewing, *C. plorabunda*. This chrysopid proved to be a dominant natural enemy of the mealybug damaging Californian pear orchards, *P. maritimus*. Artificially mass-produced eggs of *C. plorabunda* were used in periodic releases on infested trees under DDT spraying control, because it was found earlier that this insecticide was toxic only for adults, while chrysopid larvae were able to survive the treatments (Doutt, 1948). These experiments made it clear that the timing of egg colonisation was very important in order to claim effectiveness. Results showed that smaller repeated colonisation using 250 eggs per tree, with proper time points and in two successive periods, could achieve a much better suppression of the mealybug than with a single release of a large number of chrysopid eggs.

9.3 PLUM

Although plum (*Prunus domestica*) is one of the most popular stone fruits, with many cultivars grown across a wide area of Europe, the data are still sparse on lacewing species from plum orchards (see Table 9.5). To date, Ward (1969) and Talitskaya (1980) have conducted the most accurate investigations in this respect. In plum orchards, the number of species collected ranges between one and five chrysopids, and one to two hemerobiids (Table 9.5). According to the records, the dominance values of greater than 10%, and the frequency-of-occurrence values, it is likely that three green lacewing species, *Chrysoperla carnea*, *Chrysopa formosa*, and *C. pallens*, and two brown lacewings, *Hemerobius humulinus* and *Wesmaelius subnebulosus*, might be the regular elements of aphidophagous guilds of plum trees. In European countries, a total of six chrysopid and six hemerobiid species were found in plum orchards, and the majority of these species were ubiquitous and eurytopic lacewings. In European plum orchards, the most common aphid prey of lacewings were *Phorodon humuli* and *Hyalopterus pruni*, while in Chile, a mealybug, *Pseudococcus affinis*, was the prey of the brown lacewing, *Sympherobius maculipennis* (Curkovic *et al.*, 1995). Although some investigations were conducted in untreated orchards, lacewings have a less important role in the suppression of aphid infestation (Ward, 1969; Hartfield, 1996). In Moldova, Talitskaya (1980) found that lacewings react differently according to the aphid prey present. She reported that five chrysopid species were present in the case of *H. pruni* infestation, while no green lacewings were recorded in orchards with the presence of the aphid *Brachycaudus amygdalinus*. On the other hand, hemerobiids appeared on both aphid prey, and *Hemerobius humulinus* was represented with a greater dominance in the case of the latter aphid species in Moldova. In Austria one individual of *H. humulinus* was also recorded from a plum tree (Ressl, 1974). Killington (1937) reported the occurrence on plum trees of another brown lacewing species, *Drepanepteryx phalaenoides*. The larvae of this brown lacewing species were often observed in great numbers amongst colonies of *Hyalopterus pruni* on infested plum trees (Standfuss, cited *in* Morton, 1910).

9.4 PEACH

Peach (*Prunus persica*) is a stone fruit tree indigenous in Asia, probably originating from China, where it has been kept growing for some 4000 years. It has been produced in numerous countries of the world for a long

Table 9.5. *Species composition and dominance distribution of chrysopids and hemerobiids found in plum* (Prunus domestica) *orchards*

	1	2	3	4	5	6a	6b	6c	7	8	FR
Chrysopidae											
Chrysoperla carnea s. lat. (Stephens)	+	+			+	12.4					4
Chrysopa formosa Brauer					+	15.2			+		3
Chrysopa pallens Rambur					+	67.6			+		3
Chrysopa perla (L.)						3.8					1
Chrysopa dubitans Mc Lachlan									+		1
Dichochrysa prasina (Burmeister)						1.0					1
Hemerobiidae											
Wesmaelius subnebulosus (Stephens)						50.0	80.0		+		3
Hemerobius humulinus L.	+					50.0	20.0				3
Hemerobius lutescens Fabricius	+										1
Hemerobius micans Olivier				+							1
Sympherobius elegans (Stephens)								+			1
Sympherobius maculipennis Kimmins										+	1
Drepanepteryx phalaenoides (L.)			+								1
S_C	1	1			3	5			3		
S_H	2		1	1		2	2	1	1	1	

Note: The dominance values (%) are calculated from the number of collected individuals published in the reviewed papers.
Abbreviations: S_C, S_H, species richness of chrysopids and hemerobiids respectively; +, species present; FR, frequency of species occurrence in literature surveyed.
Sources: 1, UK (Ward, 1969); 2, UK (Hartfield, 1996); 3, Switzerland (Standfuss, cited *in* Morton, 1910); 4, Poland (Wiackowski & Wiackowska, 1968); 5, Hungary (F. Szentkirályi, unpublished data); 6a, b, Moldova (Talitskaya, 1980); 6c, Moldova (Zelený & Talitsky, 1966); 7, Turkey (Sengonca, 1979); 8, Chile (Curkovic *et al.*, 1995).

time, but there are relatively few data in the literature about the lacewing species living in peach orchards.

9.4.1 Characteristic lacewings associated with peach orchards

Studies regarding lacewing species associated with peach trees have been carried out mainly in Europe and North America. Table 9.6 shows the related lists of species. Nine green lacewing species were found in European peach orchards. Of these three species appear characteristically in the studied orchards, namely *Chrysoperla carnea* (17%–93%), *Chrysopa pallens* (2%–50%), and *C. formosa* (3%–33%). In North America, ten chrysopid species appeared in peach orchards, and three of them were more dominant and

common: *Chrysoperla rufilabris, C. plorabunda,* and *Chrysopa oculata.* Of the brown lacewings four species in Europe and two in Canada were found in peach stands, but these were represented with only low population levels. The chrysopid and hemerobiid species that appear generally in peach orchards are common and ubiquitous. Based on this statement, the generally expected potential lacewing assemblages in peach orchards consist of three chrysopid and one to two hemerobiid species.

9.4.2 Peach aphids as prey for lacewings

Lacewings can prey upon numerous phytophagous arthropod groups associated with peach (aphids, aleyrodids, cicadellids, tetranychid mites, etc.). To date

Table 9.6. *Species composition and dominance distribution of chrysopid and hemerobiid assemblages found in peach* (Prunus persica) *orchards in Europe and Canada*

	1	2	3	4a	4b	5a	5b	6	7	8	FR
Chrysopidae											
Chrysoperla carnea s. lat. (Stephens)	+			92.6	88.5	+	74.4	67.0	16.7	+	8
Chrysoperla plorabunda (Fitch)										A	1
Chrysoperla rufilabris (Burmeister)										A	1
Chrysopa formosa Brauer					3.5		20.5	8.0	33.3		4
Chrysopa pallens (Rambur)		+		7.4	2.0		5.1	9.0	50.0		6
Chrysopa perla (L.)								5.0			1
Chrysopa phyllochroma Wesmael								5.0			1
Chrysopa abbreviata Curtis								2.0			1
Chrysopa walkeri McLachlan								1.0			1
Chrysopa oculata Say										A	1
Chrysopa quadripunctata Burmeister										+	1
Chrysopa nigricornis Burmeister										+	1
Chrysopa sp.			+								1
Dichochrysa prasina (Burmeister)					6.0						1
Dichochrysa ventralis (Curtis)						+		3.0			2
Meleoma signoretii Fitch										+	1
Meleoma emuncta (Fitch)										+	1
Ceraeochrysa lineaticornis (Fitch)										+	1
Hemerobiidae											
Wesmaelius subnebulosus (Stephens)						+					1
Hemerobius humulinus L.				+						+	2
Hemerobius stigma Stephens.										+	1
Micromus angulatus (Stephens)				+							1
Drepanepteryx phalaenoides (L.)			100								1
S_C	1	1	1	2	4	2	3	8	3	9	
S_H			1	2		1				2	

Note: The dominance values (%) are calculated from the number of collected individuals published in the reviewed papers. *Abbreviations:* S_C, S_H, species richness of chrysopids and hemerobiids respectively; +, species present; FR, frequency of species occurrence in literature surveyed.

Sources: 1, Portugal (Cruz de Boelpaepe *et al.*, 1988); 2, Spain (Monserrat & Marín, 1994); 3, France, South (Remaudière & Leclant, 1971); 4a, Italy, Padan Valley (Pantaleoni & Ticchiati, 1988); 4b, Italy, Po Valley (Pantaleoni & Tisselli, 1985); 5a, b, Hungary (a: house garden, b: commercial orchard, F. Szentkirályi, unpublished data); 6, Romania (Paulian, 1998); 7, Moldova (Talitskaya, 1980); 8, Canada, Ontario (Briand, 1931; Putman, 1932; Putman & Herne, 1966).

some 20 aphid species have been recorded on peach, the most important serious pest species (frequently virus vectors) being *Hyalopterus pruni*, *H. amygdali*, *Myzus persicae*, *M. varians*, *Brachycaudus schwartzi*, and *B. persicae*. It has been observed during some studies, which aphids from the above list served as prey for lacewings. In the south of France (valley of the River Rhone)

several *Drepanepteryx phalaenoides* larvae were found, feeding on aphid colonies of *H. pruni*, which infested peach trees (Remaudière & Leclant, 1971). Eggs, larvae, and adults of *Chrysoperla carnea*, *Chrysopa formosa*, and *C. pallens* were associated with *H. pruni* colonies in Moldova (Talitskaya, 1980) and also in Hungarian commercial peach orchards (F. Szentkirályi, unpublished data). In late October third-instar larvae of *Wesmaelius subnebulosus* were observed feeding on *B. schwartzi* colonies, which infested shoots of peach trees in a Hungarian house garden (F. Szentkirályi, unpublished data). In *H. amygdali* colonies in Egyptian peach groves, Darwish (1992) detected several *Chrysoperla carnea* larvae. Studying the predators of the same aphid species, Basky (1982) showed that within the aphidophagous complex, the participating proportion of Chrysopidae was only 11%–16%, in third place following coccinellids and syrphids. Other studies have also suggested that chrysopids and hemerobiids are not very important control factors in peach aphid populations (Remaudière & Leclant, 1971; Pantaleoni & Ticchiati, 1988). The pest management regime in operation is often responsible for the absence of lacewings. Pantaleoni & Ticchiati (1988) found that in commercial peach orchards in the aphid-increasing period (late April to mid-July), there was no lacewing activity on trees, because of five consecutive insecticide sprayings. After the termination of spraying, the egg-laying and larval activity of chrysopids (mainly *Chrysoperla carnea*) resumed immediately, peaked in mid-August, and strongly declined by the beginning of October. With a longer sampling period (four years), Paulian (1998) recorded a species-rich chrysopid assemblage in Romanian peach orchards; however only members of the *carnea*-group were abundant (*Chrysoperla lucasina* (Lacroix), 51%; *C. kolthoffi* (Navás), 47%; *Chrysoperla carnea s. str.*, 2%) and established in the central parts of the orchard.

In North America detailed studies started in the 1930s, in the commercial peach orchards of Ontario, Canada. These pioneer investigations are connected to the work done by Briand (1931), Putman (1932, 1963), Putman & Herne (1966), and Herne & Putman (1966) in orchard pest management. As a result of the investigations, the lacewing assemblages in the peach orchards of Niagara Peninsula (Ontario) were described, and their influence in the control of prey population was indicated. Of these assemblages, two species, *C. rufilabris* and *C. plorabunda* were by far the most abundant, with both their eggs and larvae present in all the orchards studied. *Chrysopa oculata* was usually common in most orchards, especially where weed or crop cover was in the herbaceous vegetation layer. *Chrysopa nigricornis* and *Chrysoperla carnea* were generally distributed everywhere, but they were never common. The hemerobiid *Hemerobius humulinus*

occurred in most peach orchards, but its density never exceeded one adult per 1000 leaves.

Studies between 1931 and 1935 shed light upon the fact that the abundance of peach aphids had little effect on lacewing fluctuations, and that one of the main food sources of larvae in these years was the eggs and larvae of the oriental fruit moth *Grapholita molesta* in the orchards. According to the observations of Briand (1931), lacewing larvae on the trees preyed on 50%–60% of the eggs of second-generation moths and 20%–45% of the eggs of third-generation moths. Putman (1932), under laboratory conditions, successfully reared many chrysopid larvae on the eggs of the oriental fruit moth. He determined that the average number of moth eggs consumed during the total developmental period of lacewing larvae was 535 in the case of *C. rufilabris* and 511 in the case of *C. plorabunda*. Field observation supplied evidence that further suitable potential prey sources for green lacewing larvae were present on the foliage of peach trees, because cicadellid nymphs and aleyrodid nymphs were also eaten, as well as a few thrips. For these predatory activities (mainly the consumption of *G. molesta* eggs and larvae), *C. rufilabris* was predominantly responsible. This typical arboreal species laid a great number of eggs on peach trees, and its larvae could frequently be found on the foliage. The oviposition of *C. rufilabris* was from early June to the end of August, with one peak in the first half of July. *C. plorabunda* had a smaller role in the control of *G. molesta*, because this species prefers lower tree-crowns in its egg laying. Furthermore, colonisation of peach orchards started later, in the last week of July, which suggests that the first generation of this species could have developed outside the orchards. The third most common chrysopid species, *Chrysopa oculata*, exclusively preferred for its egg-laying and larval activity the low, understorey vegetation infested by aphids. *C. oculata* eggs and larvae were found on a wide range of herbaceous plants, but none were collected from peach trees, so this lacewing species had no importance in the control of peach pests (Putman, 1932).

9.4.3 Peach mites as prey for lacewings
In Ontario, a detailed investigation of the natural enemies of another important pest group, the tetranychid and eriophyid mites, was conducted in seven peach orchards from 1946 to 1965 (Putman & Herne, 1966).

Amongst several mite pests, the European red mite, *Panonychus ulmi*, and the two-spotted spider-mite, *Tetranychus urticae*, were the most important on the peach trees. Only *P. ulmi* caused serious damage on peach. From the chrysopids, the larvae of *Chrysoperla carnea* and *C. rufilabris*, as generalist predators, fed freely on all stages of *P. ulmi*, *T. urticae*, and the brown mite *Bryobia arborea*. In those experimental plots, where the density of *P. ulmi* reached 10 or more per 100 peach leaves, chrysopid larvae consumed large quantities of mites. In conclusion, it was stated that although chrysopids could not completely suppress *P. ulmi*, they could keep it on an endemic level. In a mite-infested peach orchard, 90% of the hemerobiid larvae collected turned out to be *H. humulinus*. Both the larvae and adults of this species consumed all stages of *P. ulmi*. However, owing to their small population size, they had no importance in the control of mites.

9.4.4 Impacts of habitat characteristics and pest management on lacewings on peach

IPM systems established for peach must also include procedures that enhance the colonisation of lacewings in orchards. Putman (1963) reported that the secretion of extrafloral nectaries on peach leaves attracts predatory insects, like lacewings, that feed regularly on honeydew or nectar from blossom. It has been observed that adults of *Chrysoperla* species feed on peach leaf nectar at night in the laboratory.

Providing shelters for overwintering can also support the preservation of lacewing populations in agroecosystems. Tamaki & Halfhill (1968) put bands on peach-tree trunks in order to study how artificial overwintering sites affect beneficial arthropods, especially the predators of the green peach aphid, *Myzus persicae*. The evaluation of overwintering shelters over two years showed that the average number of individuals per tree was 0.7–0.9 for hemerobiid adults, while it was 0.04–0.3 for chrysopid larvae. These values correspond to 1.7%–2.6% and 0.1%–0.5% of predatory arthropods.

The application of pesticides that have no toxic effect on entomophagous arthropods is required within peach IPM programmes. Herne & Putnam (1966) checked the toxicity level of some insecticides on lacewings in Ontario peach orchards. They found that the larvae of *Chrysoperla carnea* were remarkably DDT-tolerant, but at the same time they reacted very sensitively to other chemicals, like parathion. Broadbent &

Pree (1984) evaluated the effects of insect growth regulators on predators associated with peach. The toxicity of growth regulators (e.g. diflubenzuron) was evident on the *Chrysopa oculata*. Both topical treatment and contact with treated leaves caused significant mortality rates, and restricted the moulting of the first-instar lacewing larvae.

On peach grown in a protected environment production system (under cover), the two-spotted spider-mite, *T. urticae*, became a principal pest, and was not readily suppressed by acaricides. Hagley & Miles (1987) made a mass-release experiment in a glasshouse with *Chrysoperla carnea* to assess the effects of biological control of *T. urticae*. The results showed that even eight weeks later, on all release trees, the number of mite-infested leaf clusters was significantly less than on control trees without lacewing treatment.

9.5 CHERRY

Studies of the aphidophages of sweet cherry (*Prunus avium*) and sour cherry (*Prunus cerasus*) orchards has also been much neglected. The few records regarding lacewings are from Europe. According to Table 9.7, four ubiquitous, eurytopic chrysopid species are the most common in cherry orchards, namely *Chrysoperla carnea*, *Chrysopa formosa*, *C. perla*, and *C. pallens*, the latter being the most dominant (14%–81%). By dominance values *Chrysoperla carnea* has the second rank (3%–76%), followed by *Chrysopa formosa* and *C. perla* (5%–20%). The brown lacewings recorded in cherry orchards are not numerous; they are represented by only one or two species. Data gathered to date suggest that the lacewing assemblages of cherry orchards consist of mainly three to four species. The prey of lacewings on cherry trees were the aphids *Myzus cerasi* (Hungary, Poland, Moldova) and *M. prunavium* (Poland). In Hungary, *Drepanepteryx phalaenoides* larvae were found more than once on cherry leaves curled by aphids, as they were consuming the prey colonies.

9.6 GRAPE

Vitis vinifera is the most commonly grown grape species, although others, such as *Vitis labrusca* and *Vitis rotundifolia*, are also cultivated to a smaller degree. The grape is one of the oldest cultivated plants.

Table 9.7. *Species composition and dominance distribution of chrysopid and hemerobiid assemblages found in orchards of cherry (*Prunus avium*) and sour cherry (*Prunus cerasus*) in Europe*

| | Prunus avium | | | | | Prunus cerasus | | | |
	1	2	3	4	5	6	7	8	FR
Chrysopidae									
Chrysoperla carnea s. lat. Stephens		+		3.2	25.0	76.2		9.5	4
Chrysopa formosa Brauer				9.7	12.0			4.8	2
Chrysopa pallens (Rambur)			+	67.7	63.0	14.3	+	81.0	5
Chrysopa perla (L.)		+		19.4			+	4.7	4
Chrysopa abbreviata Curtis			+						1
Dichochrysa prasina (Burmeister)						9.5			1
Hemerobiidae									
Wesmaelius subnebulosus (Stephens)	+								1
Hemerobius humulinus L.								+	1
Drepanepteryx phalaenoides (L.)			+						1
S_C		2	2	4	3	3	2	4	
S_H	1		1					1	

Note: The dominance values (%) are calculated from the number of collected individuals published in the reviewed papers.
Abbreviations: S_C, S_H, species richness of chrysopids and hemerobiids respectively; +, species present; FR, frequency of species occurrence in literature surveyed.
Source: 1, Spain (Monserrat & Marín, 1996); 2, Poland (tree nursery, Wiackowski & Wiackowska, 1968); 3, Hungary (F. Szentkirályi, unpublished data); 4, Moldova (Talitskaya, 1980); 5, Romania (Paulian & Andriescu, 1996); 6, Spain (Monserrat & Marín, 1994); 7, Hungary (F. Szentkirályi, unpublished data); 8, Moldova (Talitskaya, 1980).

Cultivation goes back at least 8000 years, and it has since spread all over the world. A large number of insect species inhabit grape plantations (e.g. Madsen & Morgan [1975] reported 122 species), forming various functional groups within grape agroecosystems (Remund *et al.*, 1989). There are few records about predatory insects associated with grapes although there is increasing interest in the use of biological control agents in integrated plant protection procedures of grapes.

9.6.1 Characteristics of lacewing assemblages in vineyards

Table 9.8 shows the lacewing assemblages recorded in vineyards. A total of 15 chrysopid and seven hemerobiid species have been found in European grape plantations to date. Based on few data it appears that the most common and most abundant species in vineyards was *Chrysoperla carnea*, which was also the most dominant species (39%–70%). *Dichochrysa prasina* (4%–22%), *C. formosa* (1%–15%), *C. pallens*, and *D. flavifrons* appear to be additional characteristic species. These chrysopid species occurred not only in commercial vineyards, but in smaller grape plantations, too. For example, on the grape plants in a Hungarian private garden, *Chrysoperla carnea* was often abundant, while *Chrysopa formosa* and *C. pallens* appeared sporadically. Numerous chrysopid species occurring with low abundance are not part of the characteristic fauna on vines, their presence depending only on the surrounding vegetation. Forest-dwelling species, such as those preferring conifers (*Nineta pallida*, *Peyerimhoffina gracilis*), or associated with oaks (*C. viridana*) have been found on vines (Schruft *et al.*, 1983; Pantaleoni & Tisselli, 1985). Other chrysopid species associated with the herbaceous vegetation layer, e.g. *C. phyllochroma*, could be characteristic species of

Table 9.8. *Species composition and dominance distribution within the local chrysopid and hemerobiid assemblages found on grape (*Vitis *spp.) in European, American, and Indian vineyards*

	1	2	3	4	5	6	7	8a	8b	9	10	11	FR
Chrysopidae													
Chrysoperla carnea s. lat. Stephens	69.8	+	+	+	A	39.1		23.3	11.1	+			9
Chrysoperla rufilabris (Burmeister)								10.5			+		2
Chrysoperla harrisii (Fitch)								3.8					1
Chrysoperla comanche (Banks)										+			1
Chrysopa formosa Brauer	0.8				+	15.3							3
Chrysopa pallens (Rambur)	0.4				+	7.9							3
Chrysopa perla (L.)		+											1
Chrysopa phyllochroma Wesmael		+			+								2
Chrysopa abbreviata Curtis					+								1
Chrysopa viridana Schneider						11.4							1
Chrysopa oculata Say							+	58.9	88.9	A			4
Chrysopa quadripunctata Burmeister								3.5					1
Dichochrysa prasina (Burmeister)	3.8					21.8							2
Dichochrysa clathrata (Schneider)						2.0							1
Dichochrysa ventralis (Curtis)	7.6												1
Dichochrysa flavifrons (Brauer)	11.8					2.5							2
Mallada desjardinsi (Navás)											+		1
Nineta pallida (Schneider)	0.4												1
Peyerimhoffina gracilis (Schneider)	0.4												1
Cunctochrysa albolineata (Killington)	2.7												1
Hypochrysa elegans (Burmeister)	2.3												1
Apertochrysa sp.											+		1
Hemerobiidae													
Wesmaelius subnebulosus (Stephens)	12.0												1
Hemerobius humulinus L.	20.0							13.9					2
Hemerobius micans Olivier	16.0												1
Hemerobius stigma Stephens	4.0							2.9					2
Hemerobius spp.								6.2					1
Micromus angulatus (Stephens)	8.0												1
Micromus variegatus (Fabricius)	36.0												1
Micromus paganus (L.)	4.0												1
Micromus posticus (Walker)								4.6					1
Micromus subanticus (Walker)								4.2					1
Micromus spp.								67.8	100				2
Sympherobius maculipennis Kimmins											+		1
Sympherobius amiculus (Fitch)								0.2					1
Sympherobius sp.								0.2					1
S_C	10	3	1	1	5	7	1	5	2	3	1	2	
S_H	7							8	1		1		

Table 9.8 (*cont.*)

Note: The dominance values (%) are calculated from the number of collected individuals published in the reviewed papers.
Abbreviations: S_C, S_H, species richness of chrysopids and hemerobiids respectively; +, species present; FR, frequency of species occurrence in literature surveyed.
Sources: 1, Germany (Schruft *et al.*, 1983); 2, Germany (Buchholz & Schruft, 1994); 3, Germany (Schirra, 1990); 4, Switzerland (Remund *et al.*, 1989); 5, Hungary (F. Szentkirályi, unpublished data); 6, Italy (Pantaleoni & Tisselli, 1985); 7, Canada, British Columbia (Madsen & Morgan, 1975); 8*a*, *b*, USA-Pennsylvania (*a*: commercial vineyards, *b*: abandoned vineyard; Jubb & Masteller, 1977); 9, USA, California (Daane *et al.*, 1994); 10, Chile (Ripa & Rojas, 1990); 11, India (Krishnamoorthy & Mani, 1989).

vineyard ecosystems, depending on the ground cover. For example, these species were quite frequent, and this latter chrysopid typically accompanied by *C. abbreviata* was represented by a large number of individuals in a Hungarian commercial grapeyard planted on sandy soil. Between grape rows the adults swarmed on aphid-infested weed plants (e.g. pigweed, *Amaranthus*) which covered the ground in many patches (F. Szentkirályi, unpublished data).

In European grape plantations, brown lacewings were detected by Schruft *et al.* (1983). Their sampling showed that out of the seven hemerobiid species found, the most characteristic were *Micromus variegatus* (36%), *Hemerobius humulinus* (20%), and *Wesmaelius subnebulosus* (12%). *H. micans* and *H. stigma* probably migrated to vineyards from surrounding forest. These few data suggest that lacewing assemblages associated with European vineyards are characterised by a low number of species, comprising one to three chrysopids and two to three hemerobiids, with a predominance of *Chrysoperla carnea*.

A total of six green and five brown lacewing species were found in North American vineyards. *Chrysopa oculata* (59%–89%) was dominant among chrysopids, followed by *Chrysoperla carnea* (11%–23%) and *C. rufilabris* (10.5%). These lacewing species are the characteristic members of the Neuroptera assemblages inhabiting the ecosystems of North American vineyards. Three hemerobiids, *H. humulinus* and two *Micromus* species (*M. posticus* and *M. subanticus*), may be regarded as characteristic of grapes, according to the studies of Jubb & Masteller (1977). The larvae of the two *Micromus* species were collected in greater numbers with pitfall trapping, indicating that they were associated with the herb layer. It seems that lacewing

assemblages inhabiting North American vineyards comprise one to three chrysopid and two to three hemerobiid species, as in Europe. *Micromus* species, preferring mainly herbaceous, low vegetation, occurred with higher abundance in those grape plantations where there was plant cover as well as aphid prey (e.g. in commercial vineyards in Jubb & Masteller's study). The appearance of additional species with low dominance in grape plantations usually depends on the adjacent vegetation type (e.g. forest patch, windbreaks) as a potential lacewing source.

9.6.2 Grape pests as prey for lacewings

Although a number of phytophagous arthropods can live on grape plants, the main food of predaceous lacewings, aphids, is rarely found. This means that weedy or crop-plant cover in grape plantations supporting alternative aphid prey is important in aphidophagous lacewing colonisation. Potential food sources associated with grape plants can be pest arthropods, such as spider-mites (Tetranychidae), aleyrodids, cicadellids, scale insects (coccids, pseudococcids), grape phylloxera, eggs and younger larvae of grape moths, and rarely the grapevine aphid, *Aphid illinoisensis*. Many of these pests have become resistant to pesticides, thus encouraging IPM programmes. The biology, seasonality, and preferred food sources of predators of grape pests needs to be studied before the development of such programmes. A few such studies on the arthropod fauna of grapes have been performed in both Europe and America. In California, Kinn & Doutt (1972) surveyed the phytophagous mites and their predators in treated and untreated commercial vineyards. They found chrysopids and hemerobiids in these grape plantations, but did not identify them. The potential prey of lacewings in these vineyards were *Tetranychus pacificus* and *Eutetranychus willamettei*, reaching highest densities on different grape varieties. In British Columbia, during a survey conducted in 14 vineyards by Madsen & Morgan (1975), only one lacewing species, *Chrysopa oculata*, was detected. The Virginia creeper leafhopper, *Erythroneura ziczac*, was the main potential homopteran prey for lacewings and was very abundant. In some cases the coccid *Lecanium* sp. and the grape phylloxera, *Phylloxera vitifoliae*, were also numerous.

Mealybugs may be an important food source for lacewings on grape. Ripa & Rojas (1990), in two vine-growing areas of Chile, often found the eggs of the hemerobiid *Sympherobius maculipennis* laid in bark crevices, near the egg clutches of the grape-damaging white vine mealybug, *Pseudococcus affinis*. This predatory lacewing was released against vine mealy bug as a biological control agent in grape-growing areas. In India, the mealybug *Maconellicoccus hirsutus* is a grape pest, and one of its main natural enemies is the green lacewing *Mallada desjardinsi* and an *Apertochrysa* sp. (Krishnamoorthy & Mani, 1989). The chrysopid *M. desjardinsi* is one of the biological control agent candidates of the region, and so its predatory potential was determined in feeding experiments, keeping its larvae on *M. hirsutus*. It was found that on average a total of 238 mealybug nymphs were consumed by an individual lacewing during its complete larval development (Mani & Krishnamoorthy, 1989).

There is evidence that grape moths (Tortricidae and Cochylidae) are also among the prey species of lacewings. In order to detect predation on the grape moth, *Eupoecilia ambiguella*, a harmful pest in European vineyards, an immunological test, ELISA, was applied by Buchholz *et al.* (1994). In these tests *Chrysoperla carnea* larvae showed positive reactions in 10.3% of cases indicating that they had been preying on the moth. The larvae of both *Chrysopa perla* and *Chrysoperla carnea* consumed eggs and larvae (first to third instars) of *E. ambiguella*. In addition adult *Chrysopa perla* and *C. phyllochroma* also consumed moth larvae (first to third instars) (Buchholz & Schruft, 1994).

In a study conducted by Jubb & Masteller (1977) in Pennsylvanian vineyards, several pests were found among potential arthropod prey of lacewings from July to the end of September. These included the red-banded leaf roller, grape berry moth, and mites. Direct predation by lacewing larvae on European red mite,

Panonychus ulmi, and on aphids was also observed in their studies. Every year, *C. oculata* proved to be the most abundant chrysopid species in their study vineyards. This species strongly prefers herbaceous and low shrubby vegetation. In grape plantations, its larvae could be collected by pitfall trapping, which also supports the close association of *Chrysopa oculata* with weedy cover. Similarly, Daane *et al.* (1994) found this species in the herbaceous vegetation of Californian grape plantations, whereas eggs and larvae collected from grape foliage consisted predominantly of *Chrysoperla carnea* and *C. comanche*. However, Jubb & Masteller (1977) reported that the predatory adults of *Chrysopa oculata* could also have a role in feeding on grape pests, because they occurred commonly on grape foliage. On wild grape (*Vitis riparia*), Jubb (personal communication) observed *C. oculata* larvae preying on grape phylloxera. Both the larvae and adults of the second most frequent and abundant chrysopid, *Chrysoperla carnea*, were associated with grape foliage. Consequently, this species is a potential predator of these pests, according to Jubb & Masteller (1977). Their surveys found that *C. rufilabris* was in close association with grape plants, so its larvae could probably participate in the control of phytophagous arthropods. During these samplings *Micromus* spp. were the most abundant brown lacewings and their larvae could be pitfall-trapped exclusively in the herbaceous weedy vegetation of vineyards. However the predaceous *Micromus* adults were sampled mainly from grape foliage, together with two *Hemerobius* spp. (see Table 9.8). Larvae of hemerobiid species were associated with both grape plants and vineyard ground vegetation. Based on these observations, these brown lacewing species could be expected to be among the natural enemies of grape pests (Jubb & Masteller, 1977). Other species occurring with low abundance like *C. harrisii* and *H. stigma* immigrated to the grape stand from the mixed coniferous woody hedgerow area, which closely bordered the studied vineyard.

9.6.3 Seasonality of lacewings and their prey in vineyards

The seasonal pattern of population fluctuation in the majority of chrysopid and hemerobiid species inhabiting temperate grape plantations indicates an increase in abundance in the period mid-August to the end of September or early October. The peaks of larval and adult activity occur in September (Jubb & Masteller, 1977). The larval population of *Chrysoperla carnea* in German vineyards rises in July, and peaks in mid-August (Schruft *et al.*, 1983). These characteristic increases in lacewing populations in late summer/early autumn may be caused by increasing prey abundance at the same time. Jubb & Masteller (1977) mentioned that in the commercial vineyard studied a heavy infestation of the mite *P. ulmi* developed. It is well known that populations of spider-mites increase in the second half of summer. Kinn & Doutt (1972) provided an example for this from vineyards in their studies, where *Tetranychus pacificus* mite populations increased rapidly from the end of July, peaked in the second half of August, then decreased until the end of September. Similarly, leafhopper pests on grapes showed three peaks during the season, and two of these developed in early August and in mid-September, respectively (Daane *et al.*, 1996). The late reproduction of these potential prey populations corresponded well with the described late-season activity patterns of lacewings.

9.6.4 Impacts of habitat structure and IPM programmes on lacewings in vineyards

Appropriate habitat manipulations to enhance natural enemies may play an important role in IPM programmes applied in commercial vineyards. The most significant way of influencing grape habitat structures is the establishment of ground cover. Remund *et al.* (1989, 1992) reported the effects of plant-cover management on the arthropod fauna of grapes. They found that the best method to increase the diversity and abundance of beneficial insects was in terraced vineyards with an alternating mowing of the green cover. Herbaceous plants and flowers over the vegetation period increased the abundance of beneficial arthropods by 220%. However, Buchholz & Schruft (1994) found no difference between predators living on the flowers and fruits of vine with two different types of undergrowth vegetation (*Geranio-Allietum vinealis* and *Lolio-Potentillion*) associations. According to their sampling series taken over four years from the flowers and fruits of grape plants, we could calculate the abundance proportions of lacewings within predatory arthropod assemblages which ranged from 1.0%–6.1% and 0.7%–5.6% for chrysopids to 0%–2.5% and 0%–2.0% for hemerobiids in each of the two vineyards.

In IPM programmes 'mating disruption' is applied to an increasing degree. The aim of this method is to inhibit the mating of the target pest with the application of species-specific sex pheromones. The advantage of this technique is that substituting insecticide treatments with pheromones may lower the pesticide load of the environment. In practice, sex pheromones are used mainly against pestiferous moths. An important question is the impact on predators, like lacewings, caused by the decrease of spraying frequency with this method. In Germany, Schirra (1990) examined the possible effects of mating disruption on the natural enemies of grape pests. The target pest was the grape moth, *Eupoecilia ambiguella*. In his experiment, there was a pheromone-controlled plot and a conventional insecticide-treated (Deltamethrin) plot. The results showed that within the predatory fauna, the abundance of the chrysopid *Chrysoperla carnea* became higher in the mating-disruption grape plot than in the insecticide-treated one. Delbac *et al.* (1996) and Stockel *et al.* (1997) studied the long-term effects of the same technique in the Bordeaux grapevine region of France. A grape moth pest was controlled using mating disruption in their experimental vineyards during the period 1989–95. Because of the use of pheromones, the frequency of insecticide and acaricide sprayings was greatly reduced. After a seven-year treatment, it was discovered that the cicadellid pest *Empoasca vitis*, which had previously been a problem, had ceased to be a pest. The conclusion was drawn that selective mating disruption enhanced the natural control of *E. vitis*, by a progressively increasing larval population of *Chrysoperla carnea* in insecticide-free vineyards during the experimental years.

9.6.5 Biological control experiments with lacewings in vineyards

In recent years, the application of lacewings as biological control agents against grape pests has been emphasised. Augmentation programmes carried out by Daane *et al.* (1996) and Daane & Yokota (1997) offer the chance to examine in detail the effect of different mass-release strategies on the success and effectiveness of lacewings. In their experiments inundative release tests were conducted in vineyards with *Chrysoperla carnea, C. comanche*, or *C. rufilabris*. These commercially available lacewings were tested and evaluated in release experiments against two cicadellid grape pests, *Erythroneura variabilis* and *E. elegantula*. The results showed that *C.*

carnea releases at a higher rate (~20.000 larvae/ha) were able to reduce the target leafhopper populations significantly. When all three lacewing species were tested, *C. rufilabris* achieved the best result with the above release rate. In commercial vineyards, the average reduction of cicadellids in *C. carnea* releases was 9.6%, which was not sufficient to suppress the leafhopper density when it was above the economic damage level. The treatments were most efficient if release timing was set to a 50%–70% hatching rate of leafhopper eggs. Larval releases had the best results when about 50% of individuals were able to survive until the third developmental stage, in contrast with egg-release plots, when there was found to be lacewing egg mortality of 70%.

9.7 OLIVE

The olive tree (*Olea europaea*) is one of the most important cultivated fruit species. The wild olive is endemic all around the Mediterranean Sea. This species prefers a Mediterranean climate characterised by dry, warm summers and mild, rainy winters. It has been cultivated since the Bronze Age in the eastern part of the basin of the Mediterranean Sea, from where it has spread all over the Mediterranean. The centre of world olive growing can be found in Spain, Italy, Greece, and Portugal. There is also olive production in several regions outside Europe, e.g. California (USA), Mexico, Peru, Australia, and South Africa.

9.7.1 Characteristics of lacewing assemblages in olive groves

Table 9.9 presents the lacewing assemblages found in olive groves. Investigations on lacewings associated with phytophagous insects on olive trees, as can be seen, have been conducted almost exclusively in Europe. It is also noticeable from Table 9.9 that the olive is one of the best-investigated fruit-tree species regarding Neuroptera. Nearly all the surveys have been quantitative, so the values of dominance distribution of lacewing assemblages can be calculated in most cases.

Over the surveyed Mediterranean regions 32 chrysopid and 6 hemerobiid species have been recorded in olive groves. Of these 13 chrysopids and 4 hemerobiids occur on the Iberian Peninsula, 16 chrysopids in southern France, 12 chrysopids in Italy including Sardinia, and 18 chrysopids and 4 hemerobiids in Greece including Crete and other islands. Characteristic green lacewing species in olive plantations should be as follows, according to frequency of occurrence (FR values) and range of dominance percentage: *Chrysoperla carnea* (1%–80%), *Dichochrysa flavifrons* (2%–80%), *D. prasina* (2%–58%), *D. zelleri* (1%–33%), *D. clathrata* (2%–14%), *Suarius nanus* (2%–22%), *Italochrysa italica* (1%–18%), *D. genei* (1%–3%), and *Chrysopa pallens* (0%–7%). These chrysopid species were detected in at least half of the investigated olive groves. Because of the sparse catches, only *Hemerobius humulinus* could be regarded as a characteristic hemerobiid species in olive orchards of Greece, while on the Iberian Peninsula it was *Wesmaelius subnebulosus* and *Sympherobius elegans* according to their higher percentage occurrence (Monserrat & Marín, 1996). These data show that in the basin of Mediterranean Sea, the expected lacewing assemblages associated with olive groves generally consisted of 7–10 chrysopid and 1–2 hemerobiid species. Although numbers of chrysopid species was similar in the surveyed countries and the more dominant and characteristic species were identical, there was little similarity between green lacewing assemblages within each south European growing region. Using the Jaccard index the similarity level ranged between 24%–50% (lowest: Greece vs. Iberian Peninsula = 24%; highest: Italy vs. southern France = 50%). Frequent occurrence of certain genera within lacewing assemblages may be detected. For example, *Dichochrysa* genus is represented with a relatively high (13 spp.) species richness. Other characteristic genera were *Italochrysa* and *Rexa*, and, in Greece, *Brinckochrysa*. It is also peculiar that the adults of the majority of recorded chrysopid species are non-predatory glyciphages or polliniphages. On the other hand predatory imagos are represented by only few species. The overwhelming majority (70%) of lacewing species associated with olive groves is xerothermophilous in their habitat preference; while about 20% are ubiquitous and eurytopic.

9.7.2 Monitoring methods for lacewings and major olive pests

The absence of hemerobiids and chrysopid species having predatory adults is likely to be due to the special collecting methods used in these cases. During most surveys the McPhail trap was applied, which was originally installed for monitoring adult olive fly, *Bactrocera (Dacus) oleae*. Usually 4%–5% diammonium phosphate was used in this olfactory trap type, often baited with protein hydrolysates as an attractant for non-predatory chrysopid adults. McPhail traps are not selective for this fly pest, but capture many other insects as well, including numerous predatory groups, such as green lacewings. The majority of lacewings in Table 9.9 were collected by this trapping method in the following investigations: 2, 3, 4, 7a, b, 8, 9a, 10, and 11. Direct collection (sweep-netting, beating, collections from fruits, insecticide fogging) from olive trees was used in the case of studies 1, 5, 6, and 9b. On this basis, it is possible that the trapping method led to bias in the species composition of lacewing assemblages, and that species characterised by predatory adults were underrepresented. The result that more than one hemerobiid species was found only where individuals were collected from the canopy (studies 1 and 9b) without the use of McPhail traps also supports this presumption.

Several serious insect pests attack olive trees. However, regular and widespread insecticide treatments against these pests have generated further plant protection problems. For instance, the eradication of beneficial insects and spiders by pesticides can result in outbreaks of secondary pests (e.g. the scale Aspidiotus nerii becomes a pest when its parasitoids are killed by frequent spraying). Furthermore, the number of insecticide-resistant strains of major pests has been rising. All these problems draw attention to the need for new control methods in olive production, such as reduced and localised treatments, use of selective pesticides, or bait sprayings to attract natural enemies into orchards. That is why demand has increased for the study of the use of lacewings in the control of major olive pests.

There are three traditional and very serious pests of olive orchards, belonging to the insect orders Diptera, Homoptera, and Lepidoptera. One of the most harmful pests is the olive moth, *Prays oleae* (Hyponomeutidae). Depending on the part of the olive tree (leaf, flower, and fruit) associated with moth larvae, there are phyllophagous, anthophagous, and carpophagous generations of *P. oleae*. The other main pest species is the olive black scale, *Saissetia oleae*. The black scale is a native, widespread species that strongly infests olive orchards all over the Mediterranean region. This pest causes considerable damage reaching such a high infestation level that trees are completely covered with sooty mould, a fungus mycelium growing on the

honeydew produced by scales on leaves and fruits. Another scale insect, which is normally a minor pest, the diaspidid *Aspidiotus nerii*, shows more frequent outbreaks as a consequence of pesticide treatments. The third main insect pest is the olive fruit fly, *B. oleae* (Tephritidae). Although lacewings have practically no role in control of this pest, the investigation of *B. oleae* and chrysopids is related because of the trapping methods used. It has become clear that trap types used for mass collection of the olive fly are not selective, they also capture other insect groups, such as lacewings.

Another type of trap is the yellow chromotropic trap, which attracts a large number of olive flies and other insects including Chrysopidae. The use of these traps allows an approximation to be made of green lacewing populations within olive insect communities. Raspi & Malfatti (1985), using yellow traps in Italian olive orchards, indicated that Neuroptera made up 0.3% of insect orders trapped. In Lebanon, Heim (1985) sampled arthropods from trees with a knock-down insecticide, using collecting sheets on the ground; he calculated the proportion of Neuroptera to be 0.1% of all insects found. In terms of lacewings as a percentage of natural enemies in olive orchards Raspi & Malfatti (1985) found that Neuroptera made up 1.5% of all predators and parasitoids. Raspi (1982) used delta yellow traps baited with ammonia carbonate to find that Chrysopidae made up 3.2% of all predatory and parasitoid insects. Neuenschwander (1982) also examined beneficial insect catches using yellow chromotropic traps during mass trapping of *B. oleae* in Crete, and found that Neuroptera (Chrysopidae and Coniopterygidae) made up 3.9% of all individual natural enemies. Despite these low proportions, neuropterological studies reveal that in moth and scale insect control in olive orchards, certain green lacewing species may be important oophagous or coccidophagous predators.

9.7.3 Olive pests as prey for lacewings

The olive moth, *P. oleae*, has been associated with chrysopids in Spain (Ramos *et al.*, 1978, 1983; Campos & Ramos, 1983, 1985), southern France (Alrouechdi *et al.*, 1980a, 1981), Italy (Liber & Niccoli, 1988; Pantaleoni *et al.*, 1993), and in Greece including Crete (Canard *et al.*, 1979; Canard & Laudého, 1980; Neuenschwander & Michelakis, 1980; Neuenschwander *et al.*, 1981). Lacewings have also been associated with the olive black

Table 9.9. *Species composition and dominance distribution within the local chrysopid and hemerobiid assemblages found in olive (Olea europaea) groves in the mediterranean parts of Europe and Near East*

	1	2	3	4	5	6	7a	7b	8	9a	9b	10	11	12	13	FR
Chrysopidae																
Chrysoperla carnea s. lat. (Stephens)	47.9	61.2	6.7–25.3	39.3	23.8–79.7	48.0	42.7	42.0	1.5	13.1	31.0	17.2	+			13
Chrysoperla mediterranea (Hölzel)			0–1.6													1
Chrysoperla mutata (McLachlan)																1
Chrysopa formosa Brauer	0.8	0.1	0–0.1		0–6.4								+			5
Chrysopa pallens (Rambur)		+	0–0.4		0–7.1		+					0.8	+			8
Chrysopa nigricostata Brauer													+			1
Chrysopa viridana Schneider	1.1	0.1	0–0.9							+		+	+			6
Chrysopa dubitans McLachlan											1.8		+			2
Dichochrysa prasina (Burmeister)	18.3	3.4	42.3–58.4	1.7	14.9–53.3	7.6	5.3	2.4					+			8
Dichochrysa flavifrons (Brauer)	15.6	24.2	11.6–28.4	22.2	0–9.5	37.2			65.8	50.2	37.1	13.7–80.0	+			13
Dichochrysa ariadne (Hölzel)												+				2
Dichochrysa picteti (McLachlan)	4.6	2.2	0.3–9.5	12.8												4
Dichochrysa zelleri (Schneider)			2.2–7.4	0.9			32.3	17.6	20.2	32.9	15.1	28.9		+	+	10
Dichochrysa clathrata (Schneider)				9.4	0–6.4		9.7	13.6	1.6		+		+			8
Dichochrysa granadensis (Pictet)	0.8															1
Dichochrysa subcubitalis (Navás)	1.5	0.1														2
Dichochrysa genei (Rambur)	3.4	3.4					1.7	1.2	+		1.1		+		+	10
Dichochrysa nachoi Monserrat	2.3															1
Dichochrysa baetica Hölzel			0–0.9													1
Dichochrysa iberica (Navás)			0–0.7													1
Dichochrysa venosa (Rambur)		1.6														2
Dichochrysa sp. (damaged individual)				0.9												1
Nineta flava (Scopoli)			0–0.4													1
Cunctochrysa albolineata (Killington)													+			1
Cunctochrysa baetica (Hölzel)	2.7	1.5				7.0							+			3
Italochrysa italica (Rossi)			+		0–17.8								+	+		6
Italochrysa vartianorum Hölzel														+		1
Notochrysa capitata (Fabricius)			0–1.1	9.4												2
Brinckochrysa michaelseni (Esben-Petersen)											4.1		+			6

Brinckochrysa nachoi (Monserrat)									0–1.3								1
Rexa lordina Navás					2.2				0–0.1								3
Rexa raddai (Hölzel)												+					2
Suarius nanus (McLachlan)		6.3	22.0	10.4	2.2	9.6						+		+			7
Unidentifiable (damaged individuals)						2.6											
Hemerobiidae																	
Wesmaelius navasi (Andréu)					9.1								+				1
Wesmaelius subnebulosus (Stephens)					63.6							+		+			3
Hemerobius humulinus L.												+	+	+			4
Micromus angulatus (Stephens)					9.1								+				1
Sympherobius pygmaeus (Rambur)					18.2												2
Sympherobius elegans (Stephens)																	1
S_C	12	12	16		10	8	7	4	10	7	10	8	7	8	11	17	1	4
S_H	4		1				1		1	1	1		1	3	1	

Note: The dominance values (%) are calculated from the number of collected individuals published in the reviewed papers.

Abbreviations: S_C, S_H, species richness of chrysopids and hemerobiids respectively; +, species present; FR, frequency of species occurrence in literature surveyed.

Sources: 1, Spain (Monserrat & Marín, 1994, 1996); 2, Spain (Campos & Ramos, 1983); 3, Southern France (Alrouechdi *et al.*, 1980*a*, *b*; Alrouechdi, 1984); 4, Italy (Liber & Niccoli, 1988); 5, Southern Italy (Pantaleoni & Curto, 1990); 6, Italy, Sardinia (Pantaleoni *et al.*, 1993); 7*a*, *b*, Greece, Aguistri (*a*, *b*: two orchards; Canard & Laudého, 1977); 8, Greece, Akrefnion (Canard & Laudého, 1980); 9*a*, *b*, Crete (*a*: McPhail trapping, *b*: knock-down by insecticide fogging; Neuenschwander & Michelakis, 1980; Neuenschwander *et al.*, 1981); 10, Western Crete (Canard *et al.*, 1979); 11, Greece (Santas, 1984); 12, Lebanon (Heim, 1985); 13, Turkey (Sengonca, 1981).

scale, *S. oleae*, in Spain (Granada) (McEwen *et al.*, 1994), southern France (Alrouechdi *et al.*, 1980*a*, *b*; 1981), Italy including Sardinia (Delrio, 1983; Liber & Niccoli, 1988; Pantaleoni *et al.*, 1993), and in Greece including Crete (Canard *et al.*, 1979; Canard & Laudého, 1980; Neuenschwander & Michelakis, 1980; Neuenschwander *et al.*, 1981). Another scale insect, the diaspidid *Aspidiotus nerii*, was also present on olive trees in Cretan groves (Canard *et al.*, 1979; Neuenschwander & Michelakis, 1980; Neuenschwander *et al.*, 1981) and may serve as a food source for lacewings. The same is true for the psyllid *Euphyllura olivina*, a minor pest on olive, which has been found in olive orchards in southern France (Alrouechdi *et al.*, 1980*a*, *b*, 1981) and Spain (McEwen *et al.*, 1994).

9.7.4 The olive moth (*Prays oleae*) as prey for lacewings

Ramos *et al.* (1978), Campos (1989), and Campos & Ramos (1983, 1985) investigated the natural enemies of *Prays oleae* eggs in olive orchards under an IPM programme in southern Spain between 1970 and 1986. They found that chrysopid larvae were the most active among oophagous predators. The ratio of attacked *P. oleae* eggs was studied in all the three generations. They showed that egg predation rate was lowest in the phyllophagous generation with an average of 0.8% mortality (range: 0%–1.5%), higher in the anthophagous generation at 6.8% (range: 0%–34.3%), while in case of the most harmful carpophagous generation it had the highest value with an average of 64% (range: 8.4%–96.3%). According to Ramos *et al.* (1983), the real predatory efficiency of chrysopid larvae on moth eggs laid on olive fruits ranged between 26% and 97% during the period 1970–81. In these olive orchards of Granada, Campos & Ramos (1985) found *Chrysoperla carnea* the most important oophagous predator of *P. oleae*. They carried out a long-term investigation to analyse the functional (predatory) response of *C. carnea* larvae to the increasing numbers of eggs laid by *P. oleae* moth on the surface of olive fruits. The number of eggs laid by olive moths per fruit was most frequently between 1 and 6, but fruits having more than 6 eggs occurred too. Campos and Ramos observed that *C. carnea* larvae indiscriminately attacked both isolated (1 egg/fruit) and grouped (more than 1 egg/fruit) moth eggs, however it was easier for them to discover fruits with many eggs because of the higher encounter rate.

Mean percentages of consumed *P. oleae* eggs were not different and ranged between 72.1% and 76.6% in case of fruits with 1 to 5 laid eggs. A small decrease appeared only at egg numbers of 6 or more, in which case predation was 69%–70%. They concluded from these results that chrysopid activity remained constant regardless of the number of prey eggs per fruit, however real larval efficiency decreased when egg density increased on fruits. Alrouechdi (1981) also examined under laboratory conditions the predatory behaviour of *C. carnea* larvae as a function of number of eggs laid by the olive moth on olive fruits. He demonstrated that all three larval stages of *C. carnea* consumed *P. oleae* eggs of any age, no matter whether they were single or laid in groups. The higher the number of moth eggs, the greater was the attack rate of chrysopid larvae in each instar. In all three larval stages functional response was positive, and consumption rate doubled when the number of eggs was doubled from 3 to 6. He also observed that the third instar *C. carnea* frequently attacked and consumed even mature larvae of *P. oleae*. In olive orchards in southern France, the predation effect of *C. carnea* larvae was investigated in relation to the *P. oleae* carpophagous generation. Alrouechdi *et al.* (1981) found that under natural conditions the rate of predation increased with increasing number (between 1 and 3) of moth eggs per fruit in contrast with the results of Campos & Ramos (1985). The mean predation rate, depending on orchard site and year, varied at 8.4%–26.8% at 1 egg per fruit, 9.2%–25.8% at 2 eggs per fruit, and 13.3%–45.0% at 3 eggs per fruit. In experiments under semi-natural conditions Alrouechdi *et al.* (1981) showed that the predation rate of chrysopid larvae increased from 12.5% to 50.8%, when the number of *P. oleae* eggs per olive fruit changed from 1 to 3 or higher. In the third olive growing region (Italy), Liber & Niccoli (1988) carried out similar investigations regarding the predation effect of chrysopid larvae. They demonstrated that predation by *C. carnea* larvae gradually rose during the season together with increasing infestation level (egg number) of *P. oleae* on fruits. The percentage of moth eggs consumed by chrysopid larvae ranged between 7% and 34%.

In these studies on the predation efficiency of chrysopid larvae, their protective effects on fruit infested by olive moth were also measured. The rate of fruit protection varied depending on season, year, and site. In Italian olive orchards, Liber & Niccoli (1988) showed that the percentage of fruits protected by chrysopid predation varied from 6% to 22% depending on *P. oleae* infestation level. The mean rate of fruit protection in southern France ranged between 7.7% and 25.2%, and in southern Spain between 28.2% and 67.7%, depending on sampling site and year (Alrouechdi *et al.*, 1981; Campos & Ramos, 1985).

9.7.5 Olive black scale (*Saissetia oleae*) as prey for lacewings

In southern France the distribution of *S. oleae* within orchards influenced the number of chrysopid adults and eggs. Alrouechdi *et al.* (1981) detected that the greater the infestation level of *S. oleae* and the amount of honeydew produced, the higher the rate of chrysopid adults and eggs. In orchards infested by olive scale, 92% of chrysopid eggs were laid by *Chrysoperla carnea*, although two *Dichochrysa* species (*D. prasina* and *D. flavifrons*) were dominant according to adult sampling. Alrouechdi *et al.* (1981) reported that the number of adult chrysopids captured in all three species was highest at high infestation levels of olive black scale and lowest at low infestation levels. Most studies support the opinion that honeydew produced by *S. oleae* and *E. olivina* is attractive to chrysopid adults, and acts as a food supply for them, contributing to fecundity and increases in oviposition rate. At the same time, the pre-imaginal larval stages of these homopteran species are the most important components of the diet of chrysopid larvae within olive orchards (Canard *et al.*, 1979; Neuenschwander & Michelakis, 1980; Alrouechdi *et al.*, 1981; Delrio, 1983; Liber & Niccoli, 1988; Pantaleoni *et al.*, 1993). However even the complete coccidophagous insect complex is unable to suppress and keep *S. oleae* under control, because warmer and drier climatic conditions favour the rapid population growth of this scale pest.

9.7.6 Seasonality of lacewings and their prey in olive groves

To understand the role of chrysopids in olive pest control, it is necessary to review the seasonal activity patterns of adults and their oviposition, and also the synchrony of these patterns with potential prey.

In the Iberian Peninsula Campos (1986, 1989) and Campos & Ramos (1983) investigated the seasonality of chrysopids associated with olive orchards. *Chrysoperla carnea* (61%) was the dominant species followed by *Dichochrysa flavifrons* (24%). Campos (1986, 1989) reported that the reproductive period of *Chrysoperla carnea* extended from March to November, during which three generations per year developed. By sampling adults, three activity periods could be detected, the first in March–April, the second in June–July, and the third in August to October. There were two peaks in the number of eggs on olive trees deposited by *Chrysoperla carnea*, first in June, and then in August or September. There were two periods of increase in the number of the adult *D. flavifrons*, first in June–July, and then during August–September (Campos & Ramos, 1983; Campos, 1986). *D. genei*, *D. venosa*, *Rexa lordina*, and *Cunctochrysa baetica* showed one peak through the season, while *D. prasina* and *D. picteti* formed two smaller peaks in the season of 1982 (Campos & Ramos, 1983).

In olive orchards in Provence (southern France) Alrouechdi *et al.* (1980*a*, *b*, 1981) and Alrouechdi (1984) reported that the dominant lacewings as concluded from trapping of adults were *D. prasina* (42–58%), *D. flavifrons* (12–28%), and *Chrysoperla carnea* (6–25%). On the other hand, the following dominance order was gained with rearing of eggs collected in groves: *Chrysoperla carnea* (91%–92%), *D. prasina* (1%–5%), *Chrysopa pallens* (2%–3%), and *D. flavifrons* (0.5%–4%) (Alrouechdi, 1984). McPhail trap catches showed that chrysopid adults had six to seven population peaks from May to mid-October. *Chrysoperla carnea* was characterised by three peaks, the first late May to early June, the second in late June to mid-July, and the smaller third one in mid- to late August. Capture peaks for *D. prasina* adults were mid-May, early June, mid-July, end of August to early September, and the end of September. Flight peaks of *D. flavifrons* were in early June, July, early August, and in October (Alrouechdi, 1980 *a*, *b*; Alrouechdi, 1984). The total number of lacewing eggs peaked in September, but a high number of eggs was also found in July and August. Alrouechdi *et al.* (1981) studied the coincidence of the temporal pattern of *Chrysoperla carnea* eggs with its prey. The number of chrysopid eggs was highest from June to September. Peaks in egg numbers varied depending on year and site. In one of years the maximum was in September, when *Chrysopa pallens* laid an extremely high number of eggs on olive trees. The peak of egg-laying occurred during July and August in olive orchards in southern France (Alrouechdi, 1984).

Alrouechdi *et al.* (1981) investigated *Chrysoperla carnea* egg-laying activity over two years in olive orchards. Oviposition started in mid-May and finished in early October. Egg-laying was synchronised with the pre-imaginal stages of *P. oleae*, *S. oleae*, and *E. olivina*. Egg numbers showed three to four peaks per season: the first was in mid-June, the second in mid-July, and the third largest in the second half of August. A smaller egg-laying peak was detected at the beginning of August (in 1978) and in the first half of September (in 1978 and 1979). Based on these patterns, it seems that in both years the appearance of three generations of *E. olivina* coincided with peaks of chrysopid egg-laying. The effect of *E. olivina* on lacewing egg-laying was probably influenced by the abundance of the other two major olive pests over the season. Thus in the development of the mid-July peak, the psyllid and the anthophagous generation of the olive moth were important, while in the mid-June egg-laying peak the second generation of the psyllid plus the carpophagous generation of *P. oleae* could be important. The smaller lacewing peak in late September followed closely the third-generation period of *E. olivina*. Populations of young *S. oleae* present on olive trees from late June to the end of the season could also contribute to chrysopid egg-laying. Larvae of *Chrysoperla carnea* were present from mid-May to the end of September in olive orchards examined by Alrouechdi *et al.* (1981) in Provence.

In Italy, several authors have carried out sampling in olive orchards to monitor seasonal changes in the abundance levels of chrysopids. In the region of Pisa, northern Italy, Raspi & Malfatti (1985) trapped adult chrysopids using yellow traps. They recorded two activity periods, the first from the second half of June to early July, the other one in the second half of October. Pantaleoni & Curto (1990) in Apulia, southern Italy, investigated seasonal activity patterns of imagos of three dominant green lacewing species, *Chrysoperla carnea*, *Dichochrysa prasina*, and *Italochrysa italica*, using sweep-netting from olive tree foliage during 1983–84. In the case of *Chrysoperla carnea*, a bimodal pattern was characteristic with two peaks, the first and largest in mid-June, and the second in late July to early August. According to the authors, this chrysopid species had three generations a year in olive orchards: the first in May–June, the second in July–August, and the third, smaller, generation in August-September. Seasonality of *D. prasina* imagos was varied depending

on site, and it was characterised by more peaks. This species could be collected with greatest abundance from olive trees during August, sometimes in early September, as well. A further activity peak of *D. prasina* was detected in mid-June or the end of June to early July. Seasonality of the third dominant species, the univoltine *I. italica*, had one well-developed characteristic peak. However, this chrysopid probably did not take part in the control of olive pests, because its larvae prey upon preimaginal stages of an ant, *Crematogaster scutellaris*. Delrio (1983), monitoring natural enemies of *S. oleae*, found that *Chrysoperla carnea* adults were present in a Sardinian olive orchard with two periods of flight, the first in February and March, and the second, larger flight period, between mid-July and early October, peaking in the first half of September. Maximum larval density of *C. carnea* developed in July, coinciding with greatest abundance of first-instar larvae of *S. oleae*. Pantaleoni *et al.* (1993) also made detailed investigations regarding lacewings in Sardinian olive orchards. They showed that approximately half of the eggs laid on olive shoots consisted of *C. carnea* (48%), with another 45% laid by *D. flavifrons* and *D. prasina* together. The remaining eggs were produced by *Cunctochrysa baetica* (7.4%). Larval species composition was as follows: *Chrysoperla carnea* (91%), *Cunctochrysa baetica* (7%), and *Dichochrysa* spp. (2%). Pantaleoni *et al.* (1993) examined the seasonal distribution of eggs and larvae of these lacewing species on olive trees. The oviposition of *Chrysoperla carnea* extended from March to November, with most eggs deposited from June to the end of August. The egg-laying pattern showed two smaller peaks, in early April and in early May, respectively. Over the main oviposition period the highest peak was found in mid-July. *C. carnea* larvae corresponded well with chrysopid egg-laying pattern. The *Dichochrysa* spp. egg-laying pattern was bimodal. The first increase in egg numbers was from mid-May to late June with a peak at the beginning of June. The second egg-laying period was from mid-July to early September, and peaked in late August. A low level of larval activity was observed in the case of *Dichochrysa* spp. in March–April, in early July, and in October. The oviposition pattern of *Cunctochrysa baetica* was bimodal increasing in June, and then again from mid-July to late August. The larval activity pattern was also bimodal, characterised by a higher peak in mid-June and a smaller one in mid-August.

In Greece Canard & Laudého (1977, 1980) reported different green lacewing seasonal patterns from different regions and orchards. On the island of Aguistri, McPhail trapping indicated that adult *Chrysoperla carnea* showed a bimodal or trimodal pattern depending on site, with peaks at the beginning of August and early September, or in June, mid-August and the first half of September. *D. zelleri* adults were commonest in June and September. *D. clathrata* and *Suarius nanus* adult patterns were bimodal. Adults of *Dichochrysa* spp. peaked in late July to early August and in mid-September. *Suarius nanus* peaked at the beginning of August and in late August. In the Akrefnion region, *D. flavifrons* (66%) had the highest dominance followed by *D. zelleri* and *S. nanus*, while *Chrysoperla carnea* was represented with only 1.5%. The greatest number of *D. flavifrons* adults was present in the first half of September. *D. zelleri* and *S. nanus* had highest ratio in samples at the end of July to early August. Furthermore, a smaller rise in abundance of this species was observed at the end of September. In olive orchards in Crete, Canard *et al.* (1979), Neuenschwander & Michelakis (1980), and Neuenschwander *et al.* (1981) carried out investigations on lacewing seasonality. In Cretan olive groves, *D. flavifrons* and *D. zelleri* were also the most dominant chrysopid species, followed by *Chrysoperla carnea* with 17%. McPhail traps indicated that adults were mainly active from the end of May to October, with low catches between December and April (Canard *et al.*, 1979). Neuenschwander & Michelakis (1980) used pesticide-sondage (pyrethroid fogging) to study seasonal lacewing activity pattern. The presence of adults and larvae of the more dominant chrysopid species on olive trees coincided well during the season. *Dichochrysa* spp. (*flavifrons* and *zelleri*) had two adult activity periods: the first, smaller, one in May and the greater one in July–September with a peak in July. Larvae of *Dichochrysa* spp. were also present in these periods on olive trees, although their pattern indicated only one peak in July. Adults of *Chrysoperla carnea* showed two periods of increase; firstly during May–July and secondly in September. The seasonal activity pattern of larvae of this species was unimodal with a high peak in June. *Chrysoperla carnea* larvae were recorded with greatest abundance on trees between May and August. During the period of greatest oviposition (June to July), the potential food sources for chrysopid larvae were larvae of the anthophagous

generation and eggs of the carpophagous generation of *P. oleae*. Using the sondage technique, data were gathered on the presence of adults of the chrysopid species *S. nanus* on olive trees, which was detected between mid-June and mid-September with a peak in August.

Reviewing what is known regarding the seasonality of the most abundant chrysopid species recorded in olive orchards around the Mediterranean Basin, we can draw some conclusions: (1) *Chrysoperla carnea* is the main oophagous and larval chrysopid predator of olive pests (dominance values above 90%); (2) in most olive-growing areas (Spain, southern France, Italy) *Chrysoperla carnea* can develop at least three or four generations per year; (3) the major predatory activity of polyphagous *Chrysoperla carnea* larvae coincides with abundance patterns of preimaginal stages of olive scale, psyllid, and moth pests; (4) the larvae of *Dichochrysa* spp. are characterised by low dominance but can also take part in control of olive pests; however, they prey mainly on populations of *Saissetia oleae*.

9.7.7 Impacts of pest management programmes and non-olive vegetation on lacewings

Olive orchards must be sprayed regularly because of the presence of harmful pests. Irrigation is also required in certain olive orchards in dry weather. In order to protect the soil from desiccation, herbicide applications are used to keep the ground surface weed-free. In several olive orchards, other fruit trees are planted as intercrops. All of these management techniques affect lacewing assemblages in olive orchards including their predatory activity level. A number of authors report on these influences within olive orchards. Insecticide treatments reduced the predatory efficacy of chrysopid larvae on the black olive scale in Sardinia (Delrio, 1983). Ramos *et al.* (1978) also found that the number of predated eggs of *P. oleae* decreased on insecticide-sprayed olive trees compared to untreated blocks, because of lowered oophagous activity of chrysopid larvae. Such effects are often combined in orchards. For example Pantaleoni & Curto (1990) reported that in an insecticide-free olive orchard with irregular inter-row spacing and intercropping with other fruit trees, a species-rich chrysopid assemblage could be detected, in contrast with an orchard characterised by chemical treatments and large inter-row spacing without intercropping, where there was a species-poor lacewing assemblage.

Another important factor having a considerable

influence on the characteristics of olive lacewing assemblages is the non-olive vegetation type within and around orchards. Canard *et al.* (1979) and Canard & Laudého (1980) reported that the presence of intercrops and non-crop vegetation influenced the number of species of chrysopids within olive orchards. Natural garrigue vegetation or legume fields around groves produced chrysopid assemblages of highest number of species and abundance. Within the experimental olive orchard, many more chrysopid adults were captured near non-crop bushes than in other inner parts of the orchard (Canard & Laudého, 1980). In a non-irrigated olive orchard bordered with non-cultivated characteristic native vegetation with bushes and trees (e.g. *Spartium, Cistus, Pistatia, Pinus, Quercus*), a species-rich chrysopid assemblage was detected by Alrouechdi *et al.* (1980*b*). McEwen & Ruiz (1993) demonstrated that non-olive vegetation could play an important role in enhancing chrysopid egg-laying in olive orchards. Surveying many natural and weedy plants, they encountered chrysopid eggs laid on 15 mainly dicotyledonous species. The overwhelming majority of recorded eggs belonged to the genus *Chrysoperla* and a smaller part to *Dichochrysa* spp. Interestingly, these authors raised the question as to whether the influence of these plants on lacewings is beneficial (i.e. as alternative food sources for chrysopids they enhance their colonisation in groves), or rather disadvantageous for biological control of olive pests (e.g. as trap plants, they can reduce the lacewing oviposition rate on olive tree foliage). Answering these questions requires further investigations in the future.

9.7.8 Lacewing manipulation experiments with attractants in olive groves

An additional way to increase the control efficacy of chrysopids is to manipulate their population distribution with attractants in olive orchards. Such manipulation experiments were carried out by Liber & Niccoli (1988) in an Italian olive orchard, in order to test the possibility of improving the effectiveness of chrysopids by applying attractant artificial honeydew as food. This food comprised autolysed brewer's yeast, sucrose, and water in a ratio of 1:7:20, to which acid(HCl)-hydrolysed L-tryptophan was added. They found that significantly more adult *Chrysoperla carnea* individuals were trapped in the treated plot compared with an untreated one. This was true for not only *C. carnea*, but

for other chrysopid species as well. Liber & Niccoli (1988) found that percentages of predated eggs of olive moth and the protection rate of fruits were also higher on treated trees. However, there was an uncertainty in the judgement of the attractive effect, because within treated plots, infestation of olive black scale was on a much higher level than on untreated olive trees. So the possibility could not be excluded that the attractiveness of natural honeydew produced by *Saissetia oleae* also contributed to the concentration of chrysopid adults.

In southern Spain McEwen *et al.* (1994) carried out detailed experiments with manipulative spraying of L-tryptophan solutions to concentrate *C. carnea* adults onto olive-tree foliage, and thus improve lacewing oviposition rate. They sprayed the acid-hydrolysed L-tryptophan solution once a week over a five-week period, in an amount of 1 g/tree and 2 g/tree. They demonstrated that numbers of adult *C. carnea* trapped after applications of tryptophan solution were significantly higher in treated plots than in control ones. The number of captured chrysopid adults was doubled in 2 g/tree treatments compared with 1 g/tree sprayings. At the same time, they did not find any increase in the number of *C. carnea* eggs laid on foliage in treated blocks. The reason for this was claimed to be the lack of proteinaceous food components in the sprayed solution, which are necessary for increasing and maintaining of fecundity of lacewing females. Presumably that is why the expected growth of egg production did not occur in blocks treated with only tryptophan (McEwen *et al.*, 1994).

9.8 CITRUS

More than 30 species belong to the genus *Citrus* (family Rutaceae). Over thousands of years they have become among the best-known and economically most important cultivated fruit trees. Citrus species seem to have originated in southeast Asia and the Indonesian archipelago and have spread from that geographical area to other areas of the world with a subtropical or Mediterranean climate. For example by 3000 BC, lemon was cultivated in Persia (now Iran). The recent Latin nomenclature, *Citrus*, derives from the ancient Egyptian name 'chitri'.

The economic importance of citrus species is considerable and we know that they have also been used in medicine since ancient times. Today they are not only fruits of world-wide importance but also serve as raw materials or ingredients for the food industry (beverages, essences, juices, etc.), cosmetics (e.g. aromas, perfumes, citric acid, bergamot oil), medicines (volatiles, vitamin specifics). In parks and gardens numerous citrus species are planted as evergreen ornamental plants. The best-known citrus species are lemons, oranges, mandarins, and grapefruits, all having several cultivated varieties.

The wild ancestor of today's cultivated lemon (*Citrus limon*) lives on the southern slopes of the Himalayas. Cultivated lemon species originate from China. The orange tree (*C. aurantium*) is indigenous to India. The sweet orange cultivar (convar. *sinensis*) spread from southeast Asia, mainly from China, and India. In the 15th century, the orange was introduced into Europe (Italy) from the Middle and Near East (Persia). During the 16th century, it was also introduced to Florida. The mandarin (*C. reticulata*) is endemic in China, Vietnam, and the Philippines. Its cultivars have been grown mainly in East Asian countries, but since the 19th century have also been planted in the Mediterranean area. Today citrus fruits are intensively cultivated in subtropical and Mediterranean regions all over the world. Outside Asia, the principal growing countries are around the Mediterranean Sea, e.g. Italy, Spain, and Greece. In Africa, northern Mediterranean areas plus South Africa are the main citrus-producing regions. On the American continent following introduction during the 16th–18th centuries, most citrus fruit orchards are found in the USA (California, Florida), Mexico, Argentina, and Brazil.

Because of the relatively high number of *Citrus* species and their long history of cultivation, numerous phytophagous arthropods have adapted to these fruit trees and have attained pest status. Citrus has an extremely species-rich assemblage of arthropods suitable for lacewing prey. For example some 20 species of aphids have been recorded on citrus fruit trees, 60% of which are cosmopolitan and four of which are harmful pests. Armoured scale pests (Diaspididae: 42 spp.) and non-armoured scale pests (Coccidae: 34 spp.) are also important pests on citrus in the Palaearctic region. Only two aleyrodid insect pests occur on citrus. Mites are found in very high numbers on citrus, with about 160 species recorded to date, belonging to 26 families. Of all these homopteran and mite groups, at least 20 species are considered to be serious citrus pests all over the world.

9.8.1 Characteristics of lacewing assemblages in citrus orchards

This high number of potential prey species suggests that species-rich lacewing assemblages could be expected to be associated with citrus orchards. Lacewing species recorded from citrus are presented in Tables 9.10 and 9.11, arranged by geographical region. Table 9.10 shows chrysopid and hemerobiid species of European, Near Eastern, and Southeast Asian regions. In Europe and the Near East, the numbers of species of green and brown lacewings recorded were 12 and 2, respectively. In countries of Southeast Asia, chrysopid and hemerobiid species number were 9 and 6, respectively. The most frequent and generally dominant or abundant species in the Palaearctic region was *Chrysoperla carnea*. This was followed by *Chrysopa pallens* in frequency order, also showing a widespread geographical presence in citrus orchards. Further species occurred in only one or two investigations (countries), so it is difficult to confirm whether these species are characteristic for green lacewing assemblages living in citrus groves. It is remarkable that in both Europe and Southeast Asia, *Dichochrysa* species occurred in a relatively higher number; out of a total 17 species recorded, 10 belonged to this genus. *Mallada desjardinsi* is probably a characteristic chrysopid of Asian lacewing assemblages, because this species was regularly found in stands of other cultivated plants within this region. The number of hemerobiid species was very low; in Europe and the Near East a total of 2, in Southeast Asian regions, thanks to detailed, long-term investigations in Japan, 6 species were recorded on citrus. The potential hemerobiid assemblages are characterised by a predominance of *Micromus* and *Sympherobius* species. It is hard to gain a real picture of expected lacewing assemblages living in citrus orchards because in this respect, these fruit species are understudied. For example, quantitative samples have not been carried out (it was impossible to determine dominance structure of lacewing assemblages), and in the case of more detailed investigations (Table 9.10, columns 3 and 14) much higher lacewing species numbers were recorded. These data inform us that in European regions, the expected characteristic chrysopid assemblages of citrus consist of 2–4 more dominant and 5–6 low-abundance species. In the same region, there are characteristically 1–2 species of hemerobiids probably belonging to the *Sympherobius*

Table 9.10. *Species composition of chrysopid and hemerobiid assemblages found in European and Asian citrus (*Citrus spp.*) groves*

	1	2	3	4	5	6	7	8	9	10	11	12	13	14	FR
Chrysopidae															
Chrysoperla carnea s. lat. (Stephens)		+	A	A	+	+		A	+	+	A		+		10
Chrysoperla shansiensis (Kuwayama)											A				1
Chrysopa formosa Brauer			+								+				2
Chrysopa pallens (Rambur)			+		+				+		+	+		+	6
Chrysopa viridana Schneider			+												1
Chrysopa perla (L.)								+							1
Chrysopa sp.													+		1
Dichochrysa prasina (Burmeister)			+												1
Dichochrysa flavifrons (Brauer)			A												1
Dichochrysa picteti (McLachlan)			+												1
Dichochrysa zelleri (Schneider)			+												1
Dichochrysa inornata (Navás)			+												1
Dichochrysa venosa (Rambur)		+													1
Dichochrysa venusta (Hölzel)			+												1
Mallada desjardinsi (Navás)											+		+		2
Mallada basalis (Walker)													+		1
Dichochrysa cognatella (Okamoto)														+	1
Plesiochrysa lacciperda (Kimmins)													+		1
Hemerobiidae															
Hemerobius sp.												+			1
Micromus numerosus Navás														+	1
Micromus multipunctatus Matsumura														+	1
Micromus variegatus (Fabricius)														+	1
Micromus dissimilis (Nakahara)														+	1
Sympherobius gayi Navás		+													1
Sympherobius fallax Navás								+	+						2
Notiobiella subolivacea Nakahara														+	1
S_C		2	10	1	2	1	1	1	2	1	5	1	4	3	
S_H	1							1			1		1	5	

Abbreviations: A, abundant; S_C, S_H, species richness of chrysopids and hemerobiids respectively; +, species present; FR, frequency of species occurrence in literature surveyed.

Sources: 1, Portugal (Monserrat, 1991); 2, Spain (Ros *et al.*, 1988; Monserrat & Marín, 1994; Sechser *et al.*, 1994); 3, Italy-Calabria (Pantaleoni & Lepera, 1985); 4, Italy-Sardinia (Delrio, 1986); 5, Greece (Santas, 1984); 6, Cyprus (Orphanides, 1991); 7, Malta (Borg, 1932); 8, Turkey (Sengonca, 1979; Uygun *et al.*, 1990); 9, Georgia–former USSR (Kokhreidze, 1982); 10, Israel (Swirski *et al.*, 1985); 11, China (Li *et al.*, 1996; Wei *et al.*, 1986; Xu & Zhong, 1988; Zhou *et al.*, 1991); 12, Korea (Catling *et al.*, 1977); 13, India (Satpute *et al.*, 1986; Krishnamoorthy & Mani, 1989); 14, Japan (Nakao, 1962, 1964, 1968; Nakao *et al.*, 1972, 1977, 1985).

Table 9.11. *Species composition of chrysopid and hemerobiid assemblages found in American and African citrus* (Citrus *spp.*) *groves*

	1	2	3a	3b	4	5	6	7a	7b	8	FR
Chrysopidae											
Chrysoperla externa (Hagen)		+	+	+							3
Chrysoperla rufilabris (Burmeister)		+									1
Chrysoperla pudica (Navás)									+		1
Chrysoperla plorabunda (Fitch)	+	+									2
Chrysoperla congrua (Walker)									+		1
Chrysopa sp.		+			+						2
Dichochrysa nicolaina (Navás)									+		1
Dichochrysa handschini (Navás)									+		1
Dichochrysa sjoestedti (Weele)									+		1
Mallada desjardinsi (Navás)								+		+	2
Dichochrysa sp.									+	+	2
Chrysopodes lineafrons Adams & Penny			+								1
Ceraeochrysa caligata (Banks)			+								1
Ceraeochrysa everes (Banks)			+								1
Ceraeochrysa smithi (Navás)			+								1
Ceraeochrysa cubana (Hagen)		+	+								2
Ceraeochrysa cincta (Schneider)			+		+						2
Ceraeochrysa valida (Banks)		+									1
Ceratochrysa antica (Walker)								+		+	2
Brinckochrysa lauta (Esben-Petersen)							+				1
Borniochrysa squamosa (Tjeder)									+	+	2
Plesiochrysa brasiliensis (Schneider)			+								1
Plesiochrysa elongata (Navás)			+								1
Leucochrysa floridana Banks		+									1
Leucochrysa sp.			+								1
Hemerobiidae											
Hemerobius pacificus Banks	+										1
Hemerobius bolivari Banks				+							1
Hemerobius sp.				+							1
Micromus sjöstedti Weele									+		1
Sympherobius subcostalis Monserrat						+					1
Sympherobius californicus Banks	+										1
Sympherobius barberi (Banks)	+										1
Nomerobius argentinensis Olazo					+						1
Megalomus sp.					+						1
S_C	1	7	10	1	2		1	2	7	4	
S_H	3			2	2	1			1		

Abbreviations: S_C, S_H, species richness of chrysopids and hemerobiids respectively; +, species present; FR, frequency of species occurrence in literature surveyed.

Sources: 1, USA, California (Woglum *et al.*, 1947; DeBach *et al.*, 1950; DeBach & Bartlett, 1951; Bartlett, 1957); 2, USA, Florida (Muma, 1955, 1959; Childers *et al.*, 1992); 3a, Brazil, Jaboticabal region (Freitas & Fernandes, 1992); 3b, Brazil (Chagas *et al.*, 1982); 4, Argentina (Olazo-Gonzales, 1987); 5, Central & South America (Monserrat, 1998); 6, Africa, Angola (Barnard & Brooks, 1984); 7a, South Africa, Mpumalanga (Van den Berg & De Beer, 1996); 7b, South Africa, Transvaal (Van den Berg *et al.*, 1987); 8, Mascarene Islands, La Réunion (Ohm & Hölzel, 1997).

genus. In Southeast Asia, the characteristic chrysopid assemblages of citrus groves have 3–5 species, while hemerobiid assemblages consist of 5 species, characteristically from the genus *Micromus*.

Table 9.11 shows lacewing species collected in citrus orchards in the American continent (USA, Brazil, Argentina; columns 1–5), or in Africa. There was no quantitative sampling regarding lacewings in these regions, so it was impossible to calculate dominance values. Chrysopid species numbers associated with American citrus orchards totalled 14, of which 7 occurred in the USA, while 10 were found in South America. In American citrus orchards the expected species richness of chrysopid assemblages would be 7–10. Characteristic species in North America are likely to belong to the genus *Chrysoperla* (mainly *C. externa*, *C. plorabunda*) in most cases, while South American species are characteristically from the *Ceraeochrysa*. The number of brown lacewing species reported from the Americas is 8. Here, the most characteristic hemerobiids are probably species from the genera *Hemerobius* and *Sympherobius*. Expected species number in hemerobiid assemblages is 2–3.

African lacewing collections have been carried out in Angola, South Africa, and Madagascar (Table 9.11, columns 6–8). In Africa, a total of 9 chrysopid species have been recorded on citrus. Expected species number in chrysopid assemblages in these regions is 2–7. Characteristic species are of the genus *Dichochrysa*. Ohm and Hölzel (1997) found many green lacewing species in citrus orchards or gardens on islands of the Madagascar region, possibly indicating a close association with prey on citrus trees. For example, on the island of La Réunion, the following characteristic species were detected: *Mallada desjardinsi* (= *M. boninensis*), *Borniochrysa squamosa*, *Brinckochrysa lauta*, and *Ceratochrysa antica*. Only one hemerobiid species, *Micromus sjöstedti*, occurred in African citrus plantations.

9.8.2 Citrus aphids as prey for lacewings

Among homopteran citrus pests, aphids represent the insect group with the greatest economic importance. They not only cause direct damage to citrus trees, but also indirectly, because they are able to transfer the tristeza virus (CTV), a highly dangerous disease for citrus. The recorded number of aphid species living on citrus is about 20, of which four species are major pests,

although others are known secondary or minor pests. The four most widespread species in the Eurasian region are: *Aphis citricola*, *A. gossypii*, *Toxoptera aurantii*, and *T. citricidus*. The first three are cosmopolitan polyphagous species, whilst the latter infests citrus orchards mainly in the Far East and the southern hemisphere. All four species are involved in CTV transmission. Three species of aphid cause heavy losses in the Mediterranean region. The black citrus aphid, *T. aurantii*, occurs on all citrus species and on their hybrids. The green citrus aphid, *A. citricola* (= *A. spiraecola*) appeared in the 1960s in this region, supposedly introduced from Nearctic citrus plantations. This species prefers orange, mandarin, and clementine trees. The third species, the cotton aphid, *A. gossypii*, has a similar host plant preference to *A. citricola*.

Santas (1984) found *T. aurantii* and *Myzus persicae* to be aphid pests in Greek citrus orchards. Examining natural enemies of these, he found two lacewing species. The chrysopid *Chrysopa pallens* preyed on both species of aphid, while *Chrysoperla carnea* did so only on *M. persicae*. In the region of Georgia-Adjaria (former USSR), two of the most serious citrus pests are *T. aurantii* and *A. citricola*. Kokhreidze (1982) studied the aphidophagous complex of these aphids, and reported 20 species, including *Chrysopa pallens* and *Chrysoperla carnea*. In Southeast Asia, Catling *et al.* (1977) examined aphid-infested Korean citrus orchards. They observed moderate aphid outbreaks in the summer shoot-growing period. *A. gossypii* was a minor pest in these orchards in the period from late July to late September. They found *Chrysopa pallens* and a *Hemerobius* species as natural enemies of this aphid. Surveying Chinese citrus orchards, Li *et al.* (1996) recorded seven aphid species: *A. citricola*, *A. gossypii*, *T. citricidus*, *T. aurantii*, *T. odinae*, *M. persicae*, and *Sinomegaura citricola*. More than 100 aphidophagous natural enemies were found and three green lacewing species were mentioned among the predators: *C. carnea*, *C. pallens*, and *C. formosa*. In Japanese citrus plantations, Nakao carried out studies over 20 years on assemblages of insect pests and their associated natural enemies (Nakao, 1962, 1964, 1968; Nakao *et al.*, 1972, 1977, 1985). In pesticide-free citrus orchards, an insect community of 133 species was identified, from which 22 injurious pests and 18 predators were recorded. Certain brown lacewing species of genus *Micromus*, like *M. numerosus*, were among the more significant predators. This hemerobiid, together with *M. multipunctatus*, preyed on populations of the dominant aphid, *T. citricidus*.

9.8.3 Citrus whiteflies as prey for lacewings

The next major homopteran pest group, which can act as a food source for lacewings in citrus groves, is the whiteflies (Aleyrodidae). Whiteflies are among the most serious citrus pests in several countries within the Mediterranean regions of Europe and the Near East. Two aleyrodid species, *Parabemisia myricae* and *Dialeurodes citri*, have been found among the economically most important pests on citrus trees. *D. citri* is definitely a key pest in countries of the eastern part of the Mediterranean Basin, e.g. Cyprus, Turkey, and Israel (Uygun *et al.*, 1997). In these regions, two to three *D. citri* generations per year infest well-developed citrus leaves, while seven to nine generations of *P. myricae* damage young shoots. Aleyrodids produce honeydew, attracting numerous indigenous predators (Uygun *et al.*, 1990). In Cyprus Orphanides (1991) and in Turkey Uygun *et al.* (1990, 1997) examined natural enemies of this whitefly in citrus groves. Both of them detected *Chrysoperla carnea* among the most common polyphagous predatory insects of *P. myricae*. However larvae of this lacewing had little impact on high aleyrodid populations, in contrast with whitefly parasitoids, *Eretmocerus* spp., which had a significant control effect (Orphanides, 1991; Uygun *et al.*, 1990). From Israel, Swirski *et al.* (1985) published data on whitefly predators of *P. myricae*, and mentioned the chrysopid *C. carnea* and the hemerobiid *Sympherobius fallax*. Aleyrodid pests may occur on citrus trees in Asian regions as well. For example in the Nagpur region of India, Satpute *et al.* (1986) carried out a two-year investigation on the entomophagous insects of the whitefly *Aleurocanthus woglumi*. Among the three most important predators, the green lacewing *Mallada desjardinsi* was recorded.

9.8.4 Citrus scales as prey for lacewings

The most harmful pests of citrus plantations are scale insects. Although numerous species cause serious economic damage (e.g. *Aonidiella aurantii, A. citri, Aspidiotus nerii, Chrysomphalus* spp., *Parlatoria* spp., *Lepidosaphes beckii, Unaspis* spp., *Saissetia oleae, Coccus* spp., *Ceroplastes* spp., *Planococcus citri, Pseudococcus* spp., and *Icerya purchasi*), there are surprisingly few

studies on lacewing natural enemies of scale insects in Asia. In Indian citrus plantations, Krishnamoorthy & Mani (1989) studied the natural enemies of mealybugs in a four-year series of investigations. According to their observations, populations of *Planococcus citri* served as prey for several chrysopid species, namely: *Mallada basalis, M. desjardinsi, Plesiochrysa lacciperda*, and *Chrysoperla carnea*. In Korean mandarin orchards the key pests were red wax scale, *Ceroplastes rubens*, and white wax scale, *C. ceriferus*, while *Coccus hesperidum* and *Pseudococcus comstocki* were minor pests (Catling *et al.*, 1977). Catling *et al.* reported *Chrysopa pallens* and a *Hemerobius* species as among the ten predatory insects of these scales. In Japan, Nakao (1964, 1968) observed two lacewings that are associated with citrus scales; *C. pallens* preyed on the dominant scales *Aulaecorthum magnoliae* and *I. purchasi*, whilst *Micromus numerosus* consumed only the latter species.

9.8.5 Citrus psyllids and moths as prey for lacewings

Psyllids rarely cause serious damage in citrus groves and there has been only one study on the citrus psyllid, *Diaphorina citri*, and its natural enemies. Liu (1989) declared the chrysopid *Mallada desjardinsi* an important predator of *Diaphorina citri* under open field conditions and studied, under laboratory conditions, the effect of psyllids on the larval development, survival, and reproduction of this chrysopid.

A number of species of moth also attack citrus. For example, the citrus leaf miner, *Phyllocnistis citrella*, is an important pest in Chinese citrus groves. Chen *et al.* (1989) investigated the predatory efficiency of the most abundant chrysopid species, *M. desjardinsi*, in citrus plantations on citrus leaf miner in laboratory experiments. He found that 149 second- and third-instar leaf miners were consumed on average by one chrysopid larva during its development. The functional response of lacewing larvae to increasing density of leaf miners followed the Holling type II (disc equation).

9.8.6 Citrus mites as prey for lacewings

Although several species of mite are abundant and cause economic damage on citrus trees, the available data on lacewings as acariphagous control agents are limited in the Asian region. The citrus red mite, *Panonychus citri*, is a major citrus pest in Korea due in part to the fact that orchards are sprayed regularly

(7–15 times) per season using broad-spectrum insecticides against other pests (Catling *et al.*, 1977). Thus,
natural enemies of *P. citri* are killed and mite populations increase. Lacewing species were mentioned by
Catling *et al.* (1977) as among mite predators. Zhou *et
al.* (1991) found 62 species of natural enemies of *P. citri*,
and among the major five predators, larvae of
Chrysoperla carnea were recorded. Density of the lacewing population was 0.25–0.8 individuals per 1000
leaves, and larvae on average consumed 98 mites daily.
According to long-term surveys, the citrus red mite can
be considered as one of the most harmful pests in
orange groves in Japan (Nakao, 1964, 1968; Nakao *et al.*,
1972, 1977). Nakao and his team, on the basis of long
observation, described the food-chain connections
between certain citrus pests and their natural enemies.
The hemerobiid *Micromus numerosus* was detected regularly among the predators of *P. citri. Chrysopa pallens*
and *Dichochrysa cognatella* were also found as *P. citri*
predators in some orchards.

9.8.7 Citrus pests as prey for lacewings in America

The connection between aphids and predators on
citrus orchards in the Americas has only rarely been
examined, with data on aphidophagous lacewings only
from Brazil. Chagas *et al.* (1982) detected a whitefly
species, *Aleurothrixus flocculus* and an aphid, *Toxoptera
citricidus*, living on citrus with three species of lacewing,
belonging to the genera *Chrysopa*, *Megalomus*, and
Hemerobius recorded among their predators, without
further identification. Souza *et al.* (1989, 1990) investigated the hemerobiid *Nusalala uruguaya* (Navás), a
potential predator of aphids living on citrus in Brazil,
in the laboratory. They collected three species of
aphids, *T. citricidus*, *Brevicoryne brassicae* and
Dactynotus species, and fed them to larvae and adults of
N. uruguaya. Larvae mainly consumed the *Dactynotus*
sp., and this aphid resulted in the most rapid lacewing
development and viability. *Toxoptera citricidus* was not
considered a suitable prey for the hemerobiid – all the
larvae died in the first instar. In the case of *B. brassicae*
the hemerobiids died before the pupal ecdysis.

Due to the regular use of insecticides, development
of resistance of certain pests on citrus such as scale
insect started in the 1920s in North America. At the
same time insecticide use killed the more sensitive entomophagous insects. For example, up to the 1950s the
citrus mealybug, *Planococcus citri*, caused severe problems in citrus orchards resulting from the application of
DDT. Previously, predators of *P. citri* (*Sympherobius
californicus*, *S. barberi*, and *Chrysoperla plorabunda*)
were detected in California (Woglum *et al.*, 1947).
Bartlett (1957) took 13 consecutive monthly samplings
on the populations of *P. citri* and its entomophagous
complex in treated and untreated orange trees. He
found that larvae of *C. plorabunda* were resistant to
DDT but that the chrysopid population level was very
low, together with that of the hemerobiid *S. californicus*,
so their effectiveness in the control of mealybug was
questionable. Moreover, DeBach and Bartlett (1951)
observed earlier that mealybug populations in citrus
orchards were increasing as a consequence of chlordane
spraying and that resistant *C. plorabunda* reacted by
increasing their numbers in treated trees.

Another important harmful species in the USA,
the Florida red scale, *Chrysomphalus aonidum*, was
examined by Muma (1959) who conducted a detailed
survey on its coccidophagous natural enemies in orange
and grapefruit orchards in Florida, recording six chrysopid species (Table 9.11, column 2, except *Chrysoperla
externa*). The most common lacewing species were
Ceraeochrysa valida, *C. cubana*, and *Chrysoperla rufilabris*; in particular the two latter chrysopids proved to be
abundant, and there is no doubt that they played a role
in scale control. The other chrysopid species were represented only in small numbers. On another citrus scale
pest, *Lepidosaphes beckii*, a feeding generalist *Chrysopa
(Ceraeochrysa) lateralis* (Guérin) was found (Muma,
1955). This green lacewing was able to complete its life
cycle preying on this purple scale. On citrus in Central
and South America, Monserrat (1998) published data
on the hemerobiid species *Sympherobius subcostalis*
preying on mealybugs *Planococcus citri* and *Phenacoccus
gossypii* scale pests.

The citrus red mite, *Panonychus citri*, is one of the
most important mite pests living on citrus groves in
North America, but citrus rust mite (*Phyllocoptruta
oleivora*) and six-spotted mite (*Tetranychus sexmacula-
tus*) are also injurious in Florida. The chrysopid *C. lateralis* was associated with the heavy infestations of the
citrus red mite (Muma, 1955). In South California *P.
citri* is also a major pest in citrus plantations. Among the
principal predators of this mite larvae of *Chrysoperla
plorabunda* are recorded as the third most effective
natural enemy (DeBach & Bartlett, 1951). These

authors also detected the brown lacewings *Sympherobius* spp. and *Hemerobius pacificus* consuming citrus red mite as a secondary food source. These hemerobiids had limited relevance in comparison with other predators. DeBach & Bartlett (1951) reported a high degree of parasitisation of green lacewing larvae that reduced their predatory efficiency on citrus trees.

9.8.8 Citrus psyllids and aleyrodids as prey for lacewings in South Africa

South Africa is another area where surveys of the natural enemies of citrus pests have been conducted. The major homopteran pests found were psyllids and aleyrodids. In South Africa the citrus psyllid *Trioza erytreae* is an indigenous insect pest that has eight overlapping generations yearly in groves. *T. erytreae* has been considered the major pest since it is a vector of greening disease on citrus trees caused by the South African heat-sensitive strain of *Liberobacter africanum*. The citrus psyllid as a vector can transmit this disease producing a very poor yield with green, unripe fruits. Several predators of the citrus psyllid were recorded by Van den Berg *et al.* (1987) during a three-year study conducted in orchards in areas of the former Transvaal. Among the predators seven chrysopids and a hemerobiid were observed preying on *T. erytreae* (Table 9.11, column 6*b*). Green lacewings were frequently abundant, and in order of importance they followed spiders in second place. *Dichochrysa* and *Chrysoperla* species consumed the nymphal stage, while *Suarius* attacked psyllid eggs. Although the larvae of *Micromus sjöstedti* fed on psyllid nymphs they frequently preferred aphids. The high rate of parasitisation of chrysopids (>50%) in the larval stage significantly reduced their control ability against the pest. Although predatory arthropods reduced the citrus psyllid populations, they were unable to keep the pest below the economic damage level (Van den Berg *et al.*, 1987).

The spiny blackfly, *Aleurocanthus spiniferus*, is known as the most destructive aleyrodid attacking citrus species in southern Africa. In cases of heavy infestation of *A. spiniferus* tree vigour and fruit production are reduced because of sooty mould cover on leaves, fruits, and branches. Since 1988 the spiny blackfly has spread to many citrus trees of gardens and commercial plantations in South Africa (Van den Berg & De Beer, 1996). These authors surveyed the natural enemies of *A. spiniferus* infesting citrus trees in gardens.

Two chrysopid species were recorded, and *Mallada desjardinsi* larvae were very active and voracious in field conditions preying on blackfly nymphs. Although numerous predators contributed to suppression of the aleyrodid populations, they were unable to reduce it below an economically acceptable level (Van den Berg & De Beer, 1996).

9.8.9 Seasonality of lacewings and their prey in citrus orchards

Although there are relatively numerous published data on lacewings living on citrus, their seasonal activity and synchrony with the temporal distribution pattern of pests has been reported only in a few cases. From the European region only Pantaleoni & Lepera (1985) have published monitoring data on the seasonality of green lacewings living on citrus. They recorded ten chrysopid species through sweep-netting collection in orange groves in southern Italy. *Chrysoperla carnea* was the most dominant species followed by *Dichochrysa flavifrons*, while *D. prasina* and *Chrysopa formosa* shared the third dominance rank. No data were reported on potential or real prey of these lacewings. From their sample the seasonal abundance pattern of chrysopid species was described. Trimodal and synchronous seasonal patterns represented the abundance of adults of both *Chrysoperla carnea* and *Chrysopa formosa*. The highest peak occurred in mid-June, the second and third smaller ones in mid-July and late August. The activity pattern of *D. flavifrons* and *D. prasina* was bimodal, characterised by two peaks. There was a smaller peak of *D. flavifrons* in mid-June and a higher one in late August to early September. The first moderate peak of adults of *D. prasina* was in mid-July with the second one at the end of August to early September. The synchronous seasonal activity patterns of *D. inornata* and *D. venusta* were very similar and characterised with two peaks: the earlier one was detected in the first half of June and the second one was found in mid-to late August. The seasonal activity dynamics of the chrysopid *D. zelleri* was unimodal and peaked at the end of August or early September. *Chrysopa viridana* and *Chrysopa pallens* were represented by low numbers, and so the seasonality pattern of these species was not given.

In Southeast Asia Catling *et al.* (1977) found that outbreaks of aphids (*Aphis gossypii* and *A. citrina*) living in citrus groves overlapped with the second flushing characterised by many young shoots in August. They also observed that the abundance of the predator complex involving *C. pallens* and a *Hemerobius* species peaked in August or September following a period of aphid increase. In citrus plantations in Japan the aphid *T. citricidus* preyed on by the hemerobiid *Micromus numerosus* expressed two periods of rising population, in early June and during the end of July to early September. The aphidophagous *M. numerosus* was active from June to September having three peaks, in the first half of June, in mid-August, and in mid-September (Nakao, 1962, 1964). Chagas *et al.* (1982) investigated the seasonality of *T. citricidus* and *A. flocculus* citrus aphids and their natural enemies in Brazil. The former aphid species showed three, and the latter four population increases during the season. These increasing periods of citrus aphids were more or less overlapping with the activity period of a *Chrysopa*, *Megalomus*, and *Hemerobius* species. Due to small numbers of individuals these lacewings produced no characteristic peak over the seasons.

DeBach *et al.* (1950) studied seasonal population changes of the citrus red mite, *P. citri*, and its predators living on citrus plantations in California. This mite pest was characterised by two periods of population growth. Regularly the first one occurred in June, rarely in May, with a peak in mid-June. The second growth in mite abundance lasted from the end of September to late December mostly represented by a peak in October. Taking into account the order of importance, *Chrysoperla plorabunda* was in the third place among the predators. According to published diagrams it seems that the seasonality of this chrysopid more or less followed the population changes of citrus red mite. Over five seasons the major activity period of *C. plorabunda* was in July, and a peak was reached between late June and mid-July. In late September a second, lower peak of this chrysopid was recorded in two of five seasons which overlapped with the beginning of the second period of increase of *P. citri*. Briefly, the increased abundance of *C. plorabunda* in July could play a role in the control of the citrus red mite because this activity peak coincided with a period of decreasing population of the pest.

9.8.10 Impacts of IPM programmes on lacewings in citrus orchards

The worldwide occurrence of growing citrus fruits, their high economic importance, and increasing pesticide resistance problems explain the great interest in IPM technologies in citrus orchards (Cavalloro &

DiMartino, 1986; Vacante, 1997). Numerous experiments have been made in citrus plantations in order to introduce environmentally sound and tolerable insecticides in citrus pest control. However there may be undesirable impacts of these chemicals to non-target beneficials, such as lacewings. For example a new insect growth regulator, diofenolan, was used against moth and scale pests in citrus orchards in Spain (Sechser *et al.*, 1994). The impact assessment showed that in field trials the populations of *Chrysoperla carnea* could survive diofenolan sprays both in treated and in control plots. However in laboratory tests larvae of *C. carnea* did not moult to the pupal stage after treatment with this insect-growth regulator. Poisoned protein baits have been applied for control of the citrus pest *Ceratitis capitata* in a Spanish experiment by Ros *et al.* (1988). The results showed that a great number of adult *Chrysoperla carnea* were also captured and killed by attractant hydrolysed protein bait traps treated with insecticide.

9.8.11 Biological control experiments with lacewings in citrus orchards

In the framework of IPM programmes biological control methods have been implemented against citrus pests in many cases but studies using green or brown lacewings are few. This may be explained by the successful suppression of serious pests through introduced exotic parasitoids rather than predators. One of the earliest implementations of lacewing species for biological control was carried out in Palestine during 1926/7. The classical mass-rearing and releasing method was applied by using a local brown lacewing, *Sympherobius fallax*, against the citrus mealybug (Bodenheimer & Guttfeld, 1929). Since the 1980s some new reports have been published on releasing lacewings as biological control agents against citrus pests. For example mass-rearing and seasonal colonisation of *Chrysoperla carnea* resulted in the best protection against lemon pests infesting citrus trees growing in greenhouses in the Republic of Georgia (Nikolaishvili & Mekvabishvili, 1990). In China *C. shansiensis* is used against citrus and other pests (Peng, 1988), while *Mallada desjardinsi* (Wei *et al.*, 1986) and *M. basalis* (Wu, 1992, 1995) have been released against citrus red mite, in field conditions.

9.9 TROPICAL AND SUBTROPICAL FRUITS

In the subtropical and tropical regions of the world, many kinds of fruit are grown, which can be host plants for a number of homopteran and mite pests, serving as prey for lacewings. However, due to the lack of research, there are only a few records of chrysopids and hemerobiids as predators of these pests. Data available are shown in Table 9.12, suggesting that in the case of the plantations investigated, only one or two neuropteroid species were found for any one specific site. Two species were recorded most frequently, namely *Chrysoperla carnea* and *Plesiochrysa lacciperda*, although the latter occurred only in India.

Among the numerous pests of the date palm (*Phoenix dactylifera*), one of the most important noxious insects is the armoured scale, *Parlatoria blanchardii*. The parlatoria date scale originated from the oases of Mesopotamia, and it took 30 years to spread into most date-growing countries of the world, heavily infesting the palms. In spite of this, only a few surveys have been conducted on the lacewing natural enemies of this scale pest. In African date-palm plantations, from coccidophagous lacewings, Smirnoff (1957, in Morocco) and Bitaw & Bin-Saad (1988, in Libya) both recorded *Chrysoperla carnea*, the only chrysopid species found to be an active natural enemy of parlatoria scale. According to Smirnoff's (1957) surveys, *C. carnea* appeared regularly in the Moroccan oases, producing five generations a year, whereas in Israel green lacewings were found only infrequently on date palms (Kehat, 1967). Recently from Spain, Monserrat & Marín (1994) reported lacewings on date palm, namely one specimen of *Chrysopa formosa* and eight specimens of *Dichochrysa venosa*.

Avocado (*Persea americana*) grown as a subtropical fruit in various regions of the world under suitable climatic conditions, has many serious insect pests, such as whitefly pseudococcids, diaspidids, and leaf rollers. In Israel, among the native natural enemies of the whitefly *Parabemisia myricae* and the long-tailed mealybug, *Pseudococcus longispinus*, the green lacewing *Chrysoperla carnea* was the dominant species, while the brown lacewing *Symperobius fallax* was rare (Swirski *et al.*, 1980, 1985). According to these investigations, lacewings proved to be less important as biological control agents, in contrast with parasitoids, because of their polyphagism, low level of population, and a high rate of parasitisation. The green lacewing species *C. plorabunda* was one of the more important natural enemies of the principal pest of Californian avocado plantations, the diaspidid *Hemiberlesia lataniae* (Chua & Wood, 1990). The leaf roller, *Homona spargotis* is a very harmful pest on

avocado in Queensland, Australia. Pinese & Brown (1986) detected that a *Chrysopa* species preyed on the eggs of this leaf roller, but it had little controlling influence on the pest population.

In mango (*Mangifera indica*) orchards a total of six chrysopids and two hemerobiid species have been reported in the literature. In Israel, *Chrysoperla carnea* preyed on the mixed population of two diaspidid species living on mango trees, the oriental red scale, *Aonidiella orientalis*, and the Californian red scale, *A. aurantii* (Ofek *et al.*, 1997). Fasih & Srivastara (1990) reported from India that two chrysopid species might be potential biological control agents of mangohoppers. They found all the developmental stages of *Mallada desjardinsi* and *Plesiochrysa lacciperda* on infested mango trees from February to September. The two green lacewings fed on all the three cicadellid species *Idioscopus clypealis*, *I. nitidulus*, and *Amritodus atkinsoni*. In Africa, *Brinckochrysa* and *Dichochrysa* species have been found on mango trees (Hölzel & Ohm, 1991, Ohm & Hölzel, 1997, 1998; Hölzel *et al.*, 1994). Among the chrysopid species, *D. nicolaina* is probably most closely associated with mango plantations, because it was frequently collected on trees. Monserrat (1998) reported the only brown lacewing species, *Symperobius barberi* that preyed on the pseudococcid *Planococcus citri*, which is a pest on mango planted in America.

Lacewings recorded on guava (*Psidium guajava*) have been reported only from Indian orchards by Krishnamoorthy & Mani (1989) and Mani & Krishnamoorthy (1990). Presently, in India, the mealybugs *Planococcus citri* and *Ferrisia virgata* are the major pests on guava. The chrysopids *Chrysoperla carnea* and *Plesiochrysa lacciperda* are commonly associated in large numbers of individuals with these mealybug pests. The authors found that the two chrysopid species had a significant role in the suppression of the severe infestation level of mealybugs.

Litchi (*Litchi chinensis*) has been produced for 2000 years in Southeast Asia, and orchards can also be found in many other tropical regions. In spite of this, lacewing records on litchi have only been published by Hölzel & Ohm (1991) and Ohm & Hölzel (1997) from the islands of Mauritius and La Réunion. Three green lacewing species were collected: *Chrysoperla brevicollis*, *Mallada desjardinsi*, and *M. mauriciana*, and a hemerobiid species, *Micromus timidus*.

There are barely any data regarding the presence

Table 9.12. *Chrysopid and hemerobiid species found in plantations of various tropical and subtropical fruits*

	Date palm		Avocado				Mango				Guava		Litchi	Banana	FR
	1	2	3	4	5	6	7	8	9	10	11a	11b	12	13	
Chrysopidae															
Chrysoperla carnea s. lat. (Stephens)	+	+	+		+						+				??
Chrysoperla plorabunda (Fitch)				+											??
Chrysoperla brevicollis (Rambur)													+		??
Mallada desjardinsi (Navás)						+			+				+		??
Dichochrysa mauriciana (Hölzel & Ohm)													+		??
Dichochrysa nicolaina (Navás)							+		+						??
Ceratochrysa antica (Walker)									+						??
Brinckochrysa beninensis Hölzel & Duelli								+							??
Brinckochrysa decaryella (Navás)						+			+		+	+			??
Plesiochrysa lacciperda (Kimmins)										+					??
Hemerobiidae															
Micromus timidus (Hagen)									+	+			+	+	??
Sympherobius barberi (Banks)	+														??
Sympherobius fallax Navás			+												??
S_C	1	1	1	1	1	2	1	1	4	1	2	1	3		
S_H	1		1						1	1			1	1	

Abbreviations: S_C, S_H, species number of chrysopids and hemerobiids respectively; +, species present; FR, frequency of species occurrence in literature surveyed.

Sources: 1, Africa, Libya (Bitaw & Bin-Saad, 1988); 2, Africa, Morocco (Smirnoff, 1957); 3, Israel (Swirski *et al.*, 1980, 1985); 4, USA, California (Chua & Wood, 1990); 5, Israel (Ofek *et al.*, 1997); 6, India (Fasih & Srivastava, 1990); 7, Africa, Senegal, Namibia, and The Gambia (Hölzel *et al.*, 1994, 1997); 8, Africa, Benin (Hölzel & Duelli, 1994); 9, Mascarene Islands, Mauritius and La Réunion; Comoro Islands (Hölzel & Ohm, 1991; Ohm & Hölzel, 1997, 1998); 10, USA (Monserrat, 1998); 11*a*, India (Krishnamoorthy & Mani, 1989); 11*b*, India (Mani & Krishnamoorthy, 1990); 12, Mascarene Islands, Mauritius (Hölzel & Ohm, 1991, Ohm & Hölzel, 1997); 13, Tonga Islands (Stechmann & Völkl, 1990)

Table 9.13. *Chrysopid and hemerobiid species found on various herbaceous and bushy small fruits*

	Strawberry				Blackcurrant	Blueberry	Melon
	1	2	3	4	5	6	7
Chrysopidae							
Chrysoperla carnea s.l. (Stephens)		+			90.0	+	✓
Chrysoperla lucasina (Lacroix)	+						+
Chrysopa phyllochroma Wesmael					10.0		
Mallada basalis (Walker)				+			
Hemerobiidae							
Wesmaelius subnebulosus (Stephens)					43.8		
Wesmaelius nervosus (Fabricius)					43.8		
Micromus tasmaniae (Walker)				+			
Sympherobius pygmaeus (Rambur)					12.5		
S_C	1	1	1		2	1	1
S_H				1	3		

Note: The dominance values (%) are calculated from the number of collected individuals published in the reviewed paper.
Abbreviations: S_C, S_H, species number of chrysopids and hemerobiids respectively; +, species present.
Sources: 1, France (Villeneuve *et al.*, 1997); 2, Italy (Bonomo *et al.*, 1991; Pari *et al.*, 1993); 3, Taiwan (Chang & Huang, 1995; Tzeng & Kao, 1996); 4, New Zealand (Butcher *et al.*, 1988); 5, Poland (Wiackowski & Wiackowska, 1968); 6, USA, Michigan (Whalon & Elsner, 1982); 7, France (Malet *et al.*, 1994).

of lacewings in other kinds of tropical fruit stands. On Tonga, the aphid *Pentalonia nigronervosa* was found to be a regular pest in banana plantations by Stechmann & Völkl (1990). They stated that the most frequent aphidophagous species was the hemerobiid *Micromus timidus*, which was able to prey on the ant-guarded banana-aphid colonies, as the only specialised predator. There is also one report on lacewings on papaya (*Carica papaya*). Cheng & Chen (1996) reported that against the papaya-damaging mite species *Tetranychus cinnabarinus*, the green lacewing *Mallada basalis* (Walker) has recently been applied as a biological control agent in Taiwan.

9.10 SMALL FRUITS AND MELONS

Lacewings preying on pests living on small fruits, such as berries or currants and fruits like melon, have been neglected in the literature. There are only sporadic data available as can be seen from Table 9.13.

The best-studied crop is strawberry (*Fragaria* spp.) grown either in the field or under glass. Aphids and mites as prey of lacewings regularly infest strawberry plantations. The impact of predators on the two-spotted spider-mite (*Tetranychus urticae*) was studied through field samplings and by serological methods on strawberry plantations in New Zealand (Butcher *et al.*, 1988). Only one lacewing species, a hemerobiid (*Micromus tasmaniae*) found in small numbers, was represented among the predatory arthropods found. According to immunological determinations of the dietary composition of *M. tasmaniae* aphids made up the largest single prey component (13.3%) while mites only accounted for 2% of prey consumed. Consequently, the impact of this brown lacewing on this strawberry pest was minimal.

Tetranychid mites cause damage everywhere that strawberries are grown. In the 1990s a number of initiatives has been launched using various lacewing species in IPM and biological control programmes, in Europe as well as in Southeast Asia. For instance, *Mallada desjardinsi* and *M. basalis* chrysopids have been mass-reared and released in control experiments against *Tetranychus urticae* and *T. kanzawai* mite pests of strawberry plantations in Taiwan (Lo *et al.*, 1990; Chang & Huang, 1995; Tzeng & Kao, 1996). In these biological control experiments involving releasing eggs or larvae of lacewings periodically, colonies of *T. kanzawai* and of *T. urticae* were decreased by 60%–90% and 50%–90% respectively. The lacewing *Chrysoperla lucasina* has been released against aphids living on greenhouse and field strawberries in IPM programmes in France (Villeneuve & Trottin-Caudal, 1997). Release of *Chrysoperla carnea* produced an effective control of strawberry aphids in Italy (Bonomo *et al.*, 1991). Release of larvae of *Chrysoperla carnea* against *Macrosiphum euphorbiae* and *Chaetosiphon fragaefolii* aphids (Pari *et al.*, 1993) resulted in the conclusion that biological control was an effective technique to control these two pests. Easterbrook (1998) reported that the relative abundance of aphidophagous insects such as chrysopids in field strawberries of southeast England is normally low indicating that naturally occurring populations are unlikely to exert economic levels of pest control.

Wiackowski & Wiackowska (1968) recorded five lacewing species on blackcurrant (*Ribes* sp.) bushes. Of these *Chrysoperla carnea*, *Wesmaelius subnebulosus*, and *W. nervosus* were dominant (Table 9.13). They found three aphid species serving as prey for lacewings on the bushes: *Aphidula schneideri*, *Hyperomyzus lactucae*, and *Cryptomyzus ribis*. Whalon & Elsner (1982) studied an aphid vector (*Illinoia pepperi*) of blueberry shoestring virus and its natural enemies on blueberry (*Vaccinium corymbosum*) plantings in Michigan (USA). *Chrysoperla carnea* was found among predators at lower density (2 ± 1.2 individuals/10 shoot terminals in average). When they applied acephate, an aphicide, only a few chrysopids were able to survive the treatment.

Recently *C. lucasina* (part of the common green lacewing [*carnea*] species complex) was used in a biological control experiment to suppress the aphid *Aphis gossypii* living on melon in the south of France (Malet *et al.*, 1994). Second-instar larvae of *C. lucasina* were released in a predator:prey ratio of 1:20. Aphid populations increased at a slower rate in treated as opposed to control plots. The results indicated that the impact of the predator was sufficient to reduce significantly the *A. gossypii* colonies.

9.11 NUTS

9.11.1 Almond

Almond (*Prunus dulcis*) is indigenous in central and western Asia, but is grown in other parts of the world with a warm climate. Commercial production is carried out mainly in the Mediterranean regions of Europe and in the USA (California).

Several phytophagous groups of arthropods (e.g. aphids, mites, scales) live on almond and are potential food sources for lacewings. The most important almond aphids are *Brachycaudus amygdalinus*, *B. helichrysi*, *B. persicae*, *Hyalopterus amygdali*, *Myzus amygdalinus*, and *M. persicae*. These species can also be found on other fruit trees belonging to the *Prunus* genus such as peach and plum. Despite the extensive growing area there are scarcely any data available in the literature regarding lacewing species associated with almond orchards. To monitor natural enemies, Triggiani (1973) took weekly samples for four years in an insecticide-free almond orchard from aphid-infested trees. Five of the above-mentioned aphid species were found, which infested the orchards during April and May. Only one lacewing species, *Chrysoperla carnea*, was recorded as an aphidophagous insect. Lacewing adults and larvae were represented but only in very low numbers. The majority of *Chrysoperla carnea* specimens appeared in May, immediately after the aphid peak. The conclusion was drawn that the aphidophagous complex was not efficient in the control of aphid infestations on almond.

Monserrat & Marín (1994, 1996) have provided the most records regarding green and brown lacewing species associated with almond, from their faunistic collections conducted in Spain between 1976 and 1993. The following seven green lacewing species appeared on almond (the percentage values in parentheses relate to the species dominance): *Dichochrysa prasina* (39.7%), *Chrysoperla carnea s. lat.* (28.8%), *Chrysopa pallens* (12.3%), *C. formosa* (11.0%), *D. flavifrons* (5.5%), *D. granadensis* (1.4%), and *C. viridana* (1.4%). According to these dominance data, it is likely that mainly the most common, ubiquitous chrysopid species can be found on almond, among which the most frequent are the *D. prasina* and *Chrysoperla carnea* complex. Only one brown lacewing, *Wesmaelius subnebulosus*, occurred on almond trees, represented by only three specimens.

9.11.2 Hazelnut

Hazelnut or filbert (*Corylus avellana*), as a tree-nut crop, is widely distributed throughout Europe, Asia, and North Africa, and has also been introduced into North America. Hazelnuts are among the major nut crops cultivated commercially. Among the numerous minor and major insect pests living on hazelnut, aphids can occasionally reach the economic damage level (AliNiazee, 1998). In non-commercial forest-type habitats, communities of large numbers of insects, such as aphids and associated natural enemies, can build up on hazelnut trees.

An important pest of cultivated hazelnuts is the filbert aphid, *Myzocallis coryli*, which is a widespread species in Europe, North America (AliNiazee, 1998), and in South America (Aguilera & Pacheco, 1995). Another but not so harmful aphid species, *Corylobium avellanae*, can also commonly occur on hazelnut. While in Europe insecticides are rarely necessary to control these aphids, in Oregon (USA), where hazelnut is produced in commercial orchards, *M. coryli* was a serious pest in the 1980s. In recent years the filbert aphid has become a secondary pest due to the introduction of parasitoids which have proved to be a successful method of biological control (Viggiani, 1982; AliNiazee, 1998).

In Oregon, Messing & AliNiazee (1985) studied the predators of *M. coryli* in hazelnut orchards. Fifty-five predaceous insect species were associated with this aphid in hazelnut agroecosystems. Four chrysopid and four hemerobiid species were found (Table 9.14) on hazelnut trees; however these lacewings were only found in small numbers.

Table 9.14 shows those lacewing species that have been recorded on hazelnut (from commercial orchards to forested habitats). The majority of the data (columns 1–9) are from European regions, however, with a few exceptions (Hungary, Italy), these data did not originate from hazelnut orchards, but from collections performed in natural vegetation. In European countries, a total of 13 chrysopid and 19 hemerobiid species were found on hazelnut. The largest numbers of species ($S_C = 11$, $S_H = 13$) were gathered in Spain, as a result of long-term and numerous local collections (Monserrat & Marín, 1994, 1996). Aphidophagous species, like lacewings, colonise or abandon hazelnut groves, strongly depending on the presence of food sources and on the type of the surrounding vegetation with

alternative prey (AliNiazee, 1998). For example, woody environments can be an important reservoir for lacewings, too. This is supported by the greater numbers of species of chrysopids and hemerobiids originating from collections on hazelnut trees and bushes in forest habitats (e.g. Switzerland, Austria). In these sites, 7–8 green and 9–11 brown lacewing species were collected from hazelnut foliage. The greater number of chrysopid species in Hungarian hazelnut groves was also due to adjacent oak forests (see Table 9.14, column 5).

The characteristic species of green lacewing assemblages associated with hazelnut, according to frequency of occurrence and range of dominance values are: *Chrysoperla carnea* (20%–38%), *Dichochrysa prasina* (10%–33%), *Chrysotropia ciliata* (4%–55%), *D. ventralis* (4%–13%), *Cunctochrysa albolineata* (2%–17%), *Nineta flava* (1%–17%), and with lower dominance values *Chrysopa perla* and *D. flavifrons*. The characteristic brown lacewings are as follow: *Hemerobius humulinus* (24%–64%), *H. micans* (10%–54%), *H. lutescens* (7%–11%), *Micromus lanosus* (15%–17%), and less frequently *H. marginatus*. Based upon these data, the most probable hazelnut lacewing assemblages in Europe are formed from 6–7 chrysopids and 3–4 hemerobiids. By comparisons of the species compositions in different countries, it can be stated that most species associated with hazelnut (especially the common and dominant ones) are identical with the characteristic lacewings of broadleaf forests (*Quercus*, *Fagus*). It supports the statement that surrounding forests may be important colonisation reservoirs for hazelnut stands. For example, in Hungarian deciduous forests, the hemerobiid *H. micans* could be regularly collected in great numbers from the foliage of native hazelnut bushes (F. Szentkirályi, unpublished data).

9.11.3 Walnut

European walnut (*Juglans regia*) is an Asian endemic tree, planted throughout the world. In spite of this, according to Table 9.15, there have been surprisingly few investigations regarding lacewing assemblages associated with walnuts. The majority of data in the table represents records derived from several sites, rather than quantitatively sampled assemblages of local orchards. A total of 11 chrysopid and 5 hemerobiid species were collected on walnut. From Table 9.15 the expected lacewing assemblage associated with walnut

Table 9.14. *Species composition and dominance distribution of chrysopids and hemerobiids found on trees or orchards of hazelnut (*Corylus avellana*)*

	1	2	3	4	5	6	7	8	9	10	11	FR
Chrysopidae												
Chrysoperla carnea s.l. (Stephens)		+	+		20.8	37.7	+	20.8	+	+		8
Chrysoperla rufilabris (Burmeister)										+		1
Chrysopa formosa Brauer					4.2							1
Chrysopa pallens (Rambur)			10.5		16.7							2
Chrysopa perla (L.)		+	7.9			2.6		8.3	+			5
Chrysopa nigricornis Burmeister										+		1
Dichochrysa prasina (Burmeister)		A	10.5		33.3	31.7	+	25.0	+			7
Dichochrysa ventralis (Curtis)		+	7.9		4.2	8.3		12.5				5
Dichochrysa flavifrons (Brauer)	+	+	2.6			8.8						4
Dichochrysa clathrata (Schneider)						0.5						1
Dichochrysa inornata (Navás)						2.3						1
Cunctochrysa albolineata (Killington)		+	5.3			1.6		16.7				4
Nineta pallida (Schneider)						0.3						1
Nineta flava (Scopoli)		+			16.7	1.0		16.7				4
Ceraeochrysa placita (Banks)										+		1
Chrysotropia ciliata (Wesmael)			55.3		4.2	5.2						3
Hemerobiidae												
Wesmaelius subnebulosus (Stephens)		+				5.5						2
Wesmaelius malladai (Navás)		+										1
Megalomus tortricoides Rambur		+	1.2						+			3
Hemerobius humulinus L.		+	30.4	+	64.0	26.7		24.4	+	+		8
Hemerobius micans Olivier	+	+	54.0	+	25.0	25.6		9.8	+			8
Hemerobius lutescens Fabricius		+	9.3	+	11.0	6.7			+			6
Hemerobius gilvus Stein						3.3			+			2
Hemerobius simulans Walker	+		0.6									2
Hemerobius pini Stephens			0.6			1.1						2
Hemerobius marginatus Stephens			2.5					48.8				2
Hemerobius atrifrons McLachlan		+										1
Hemerobius stigma Stephens										+		1
Hemerobius ovalis Carpenter										+		1
Hemerobius pacificus Banks										+		1
Hemerobius sp.											+	1
Micromus angulatus (Stephens)		+	0.6			3.3						3
Micromus variegatus (Fabricius)			0.6			1.1						2
Micromus paganus (L.)		+				1.1						2
Micromus lanosus (Zelený)		+				15.5		17.1	+			4
Sympherobius pellucidus (Walker)		+				1.1						2
Sympherobius klapaleki Zelený						4.4						1
Sympherobius elegans (Stephens)				+		4.4						2
S_C	1	7	8		7	11	2	6	3	4		
S_H	2	11	9	4	3	13		4	6	4	1	

Table 9.14 (*cont.*)

Note: The dominance values (%) are calculated from the number of collected individuals published in the reviewed papers.
Abbreviations: A, abundant; S_C, S_H, species richness of chrysopids and hemerobiids respectively; +, species present; FR, frequency of species occurrence in literature surveyed.
Sources: 1, Norway (Greve, 1967); 2, Switzerland (Eglin-Dederding, 1980); 3, Austria (Ressl, 1971, 1974); 4, Czechoslovakia (Zelený, 1963); 5, Hungary (F. Szentkirályi, unpublished data); 6, Spain (Monserrat & Marín, 1994, 1996); 7, Italy (Pantaleoni, 1996); 8, Yugoslavia, Montenegro (Devetak, 1991); 9, Bulgaria (Popov, 1991);10, USA (Messing & AliNiazee, 1985); 11, Chile (Aguilera & Pacheco, 1995)

consists of 4–5 chrysopid and 2–4 hemerobiid species as a maximum. The majority of the species on walnut trees are common and ubiquitous, among which the most frequent species with the greatest dominance values are *Chrysopa pallens*, *C. viridana*, and *Dichochrysa prasina* from the chrysopids, and *Wesmaelius subnebulosus* and *Hemerobius humulinus* from the hemerobiids. With the exception of *C. formosa*, the lacewing species are also deciduous-forest-dwellers and they are probably the members of the real walnut neuropteroid assemblages.

The most important and most common potential prey of these lacewings is the walnut aphid, *Chromaphis juglandicola*, which may occur in any region where walnut is grown, from Eurasia to North America. In the USA, this aphid is one of the most harmful pests of the genus *Juglans*. The walnut aphid has many natural enemies, and several lacewing species can be found among them. In northern California Sluss & Hagen (1966) recorded the common green lacewing *Chrysoperla carnea* in all the walnut orchards surveyed. However, only in one case did they find larvae of this species to respond to densities of *Chromaphis juglandicola* and to have a significant suppressive effect on aphid population levels. Also in California Frazer & van den Bosch (1973) studied the possibilities of biological control of *C. juglandicola* in detail. They found the chrysopids *Chrysoperla carnea* and *Chrysopa nigricornis* to be the most dominant components within the aphidophagous assemblages of the walnut aphid. During a season, *Hemerobius* sp. with other predatory species occurred only sporadically. The level of synchrony between the aphid population and the seasonal fluctuations of aphidophages was monitored with weekly sampling. They found that the seasonal fluctuation pattern in walnut aphid numbers was trimodal. The three peaks

were in mid-May, late July to early August, and late September to the first half of October. The aphidophagous groups, like chrysopids, also expressed characteristic trimodal activity patterns, following the aphid seasonal dynamics. Aphid and chrysopid numbers were well synchronised during both the second and third population peaks, but the first lacewing peak followed the aphid one at least three weeks later in early June.

Of the natural enemies of *Chromaphis juglandicola*, *Chrysopa pallens* has been recorded from China (Li, 1992) while in Hungary nine lacewing species were collected. In this latter case, another aphid species, *Callaphis juglandis*, also infested the walnut trees. Besides imagos of *Chrysopa pallens*, *W. subnebulosus*, and *H. humulinus*, eggs and larvae could also be found among aphid colonies on leaves (F. Szentkirályi, unpublished data).

9.11.4 Pecan

Pecan (*Carya illinoinensis*, Juglandaceae) is an important commercial tree-nut crop in Mexico and in the southern USA; however it is also produced in other parts of the world, for instance in South Africa (Van den Berg & Maritz, 1995), and Israel (Mansour, 1993). Pecan is a long-season crop (requiring approximately seven months to mature) which can be attacked by a large complex of pests. Among several important pecan pests can be found the foliar-feeding aphids. Evidence is growing that the regular use of insecticides in traditional pest management has resulted in the development of resistance in pecan aphids. In order to solve this problem, there is an increasing interest in enhancing the controlling effects of natural enemies. To achieve this goal, intensive investigations were conducted in the 1980s to study the aphidophagous guilds associated with pecan trees.

Table 9.15. *Species composition and dominance distribution of chrysopids and hemerobiids found on walnut (*Juglans regia*) trees or groves*

	1	2a	2b	3a	3b	4	5	6	7	FR
Chrysopidae										
Chrysoperla carnea s. lat. (Stephens)									+	1
Chrysopa formosa Brauer		7.7								1
Chrysopa pallens (Rambur)		30.8		+			+	+		4
Chrysopa nigricornis Burmeister									+	1
Chrysopa viridana Schneider				+	+		+			3
Dichochrysa prasina (Burmeister)		57.7								1
Dichochrysa flavifrons (Brauer)					+					1
Dichochrysa subflavifrons (Tjeder)							+			1
Cunctochrysa albolineata (Killington)				+			+			2
Nineta flava (Scopoli)		3.8		+						2
Chrysotropia ciliata (Wesmael)		+					+			2
Hemerobiidae										
Wesmaelius subnebulosus (Stephens)		80.0	20.0	+		92.6				4
Hemerobius humulinus L.	+	20.0	20.0							3
Hemerobius micans Olivier			30.0	+						2
Hemerobius sp.									+	1
Sympherobius pygmaeus (Rambur)			30.0			7.4				2
S_C		1	4	4	2		5	1	2	
S_H	1	2	4	1	1	2		1		

Note: The dominance values (%) are calculated from the number of collected individuals published in the reviewed papers.
Abbreviations: S_C, S_H, species richness of chrysopids and hemerobiids respectively; +, species present; FR, frequency of species occurrence in literature surveyed.
Sources: 1, Spain (Monserrat & Marín, 1996); 2a, b, Hungary (a, b, house gardens, F. Szentkirályi, unpublished data); 3a, Moldova (Talitskaya, 1980); 3b, Moldova (Zelený & Talitsky, 1966); 4, Turkey (Sengonca, 1979); 5, Turkey (Sengonca, 1981); 6, China (Li, 1992); 7, USA (Frazer & van den Bosch, 1973).

Characteristics of lacewing assemblages in pecan orchards

Lacewing assemblages recorded in pecan orchards are shown in Table 9.16. In the USA, a total of 10 chrysopid and 9 hemerobiid species were found, while in South Africa, only two green lacewings occurred on pecan groves, *Chrysoperla pudica* and *Borniochrysa squamosa*. Within green lacewing assemblages, the most frequent and most dominant species was *C. rufilabris* (66%–91%). Three other subdominant and frequent chrysopids were found: *Chrysopa quadripunctata* (2%–23%), *C. nigricornis* (0.4%–12%), and *Chrysoperla carnea* (1%–16%). It is likely that, with lower occurrence and abundance, *Ceraeochrysa lineaticornis* and species from *Leucochrysa* (*Nodita*) could also be considered to be characteristic components of pecan lacewing assemblages.

Micromus posticus (60%–84%) proved to be the most dominant and most frequent hemerobiid species of pecan orchards, followed by the subdominant *H. humulinus* (6%–37%) and *M. subanticus* (4%–6%). Based on the further species found in Table 9.16 it

Table 9.16. *Species composition and dominance distribution within the local chrysopid and hemerobiid assemblages found in American and South African pecan* (Carya illinoinensis) *orchards*

	1	2a	2b	3	4	5	6	7	FR
Chrysopidae									
Chrysoperla carnea s. lat. (Stephens)	3.7	14.1	15.6	1.1					4
Chrysoperla rufilabris (Burmeister)	70.2	80.2	65.7	90.6	A	+	+		7
Chrysoperla cf. pudica (Navás)								+	1
Chrysoperla plorabunda (Fitch)				1.4					1
Chrysopa quadripunctata Burmeister	22.9	4.4	6.3	1.8	A	+			6
Chrysopa nigricornis Burmeister	0.5	0.9	12.4	0.4	+	A			6
Chrysopa oculata Say		0.3		1.4					2
Ceraeochrysa lineaticornis (Fitch)		0.1		0.7	+				3
Leucochrysa insularis (Walker)					+				1
Leucochrysa pavida (Hagen)	0.9			2.5					2
Leucochrysa sp.	1.8								1
Borniochrysa squamosa (Tjeder)								+	1
Hemerobiidae									
Hemerobius humulinus L.	18.2	6.5	36.7	6.3	+				5
Hemerobius stigma Stephens		1.2					+		2
Hemerobius sp.			12.5						
Micromus posticus (Walker)	75.8	83.7	59.5	75.0	A	+	+		7
Micromus subanticus (Walker)		4.1	3.8	6.3					3
Micromus variolosus Hagen		2.0							1
Sympherobius barberi (Banks)	6.0								1
Sympherobius amiculus (Fitch)		2.0							1
Sympherobius occidentalis (Fitch)		0.4							1
S_C	6	6	4	8	5	3	1	2	
S_H	3	7	3	4	2	1		2	

Note: The dominance values (%) are calculated from the number of collected individuals published in the reviewed papers.
Abbreviations: A, abundant; S_C, S_H, species richness of chrysopids and hemerobiids respectively; +: species present; FR, frequency of species occurrence in literature surveyed.
Sources: 1, USA, Texas (Liao *et al.*, 1984); 2a, USA, Kansas (Dinkins *et al.*, 1994); 2b, USA, Georgia (Dinkins *et al.*, 1994); 3, USA, Oklahoma (Smith *et al.*, 1996); 4, USA, Alabama (Edelson & Estes, 1987); 5, USA, Georgia (Mizell & Schiffhauer, 1987a); 6, USA, Florida (Mizell & Schiffhauer, 1987b); 7, South Africa (Van den Berg & Maritz, 1996).

seems that *Micromus* and *Sympherobius* spp. can be characteristic components of brown lacewing assemblages in pecan orchards. Surveys tell us that the most probable typical lacewing assemblages consist of 4–6 chrysopid and 3–4 hemerobiid species in North American pecan orchards.

Pecan aphids as prey for lacewings

The most important aphids attacking and damaging pecan trees are the black-margined aphid, *Monellia caryella*, black pecan aphid, *Melanocallis caryaefoliae*, and yellow pecan aphid, *Monelliopsis pecanis*. These aphid species serve as prey for lacewings in pecan orchards and can cause infestations on trees either alone, as with *Monellia caryella* (Liao *et al.*, 1984, 1985; Mansour, 1993) and *Monelliopsis pecanis* (Van den Berg & Maritz, 1996), or combined together, as with *Monellia caryella* and *Monelliopsis pecanis* (Dinkins *et al.*, 1994; Edelson & Estes, 1987; Mizell & Schiffhauer, 1987*b*), or *Monellia caryella* with the aphid *Longistigma caryae* (Smith *et al.*, 1996). The fluctuations of pecan aphid populations can be generally characterised by a bimodal seasonal pattern with an early and a late peak (Edelson & Estes, 1983; Mizell & Schiffhauer, 1987*b*; Mansour, 1993), or with mid- and late-season peaks (Liao *et al.*, 1984, 1985). Sometimes, the pecan aphid had only one greater seasonal peak (Liao *et al*, 1984; Mansour, 1993; Van den Berg & Maritz, 1995). Many investigations have been conducted to measure how well the seasonal population dynamics of pecan aphids are synchronised with the abundance patterns of lacewings, and to indicate the degree to which lacewings control the aphids. Liao *et al.* (1984, 1985) made detailed studies in order to characterise the predator–prey relations. Monitoring all the developmental stages they found that numbers of chrysopid larvae and eggs and of hemerobiid adults showed a significant positive correlation to *M. caryella* densities in most cases. In particular, numbers of chrysopid eggs laid, as a measure of the reproductive numerical response, apparently increased with increasing aphid density. With the aid of experiments involving exclusion methods, evidence was supported that chrysopids play a major role in the collapse of pecan aphid populations. It was found that the most voracious species was *Chrysoperla rufilabris*, having an average daily larval consumption of 39.3 nymphs from *M. caryella* during its developmental time. The larvae of *Micromus posticus* and then *Chrysopa quadripunctata* followed, with 37.7 and 26.6 as a mean number of individuals consumed per day, respectively. Dinkins *et al.* (1994) reported that in Kansas and Georgia (USA), the population fluctuations of lacewing species responded to the increase of aphids with a one-week lag. At the same time, in Kansas, individual numbers of lacewings correlated well with the population increases of two primary leafhopper groups *Erythroneura* and *Empoasca* spp., that could develop on trees when aphids were mainly absent. In Alabama pecan orchards, Edelson & Estes (1987) found *Chrysoperla rufilabris*, *Chrysopa quadripunctata*, and *M. posticus* as the most abundant aphidophagous species. These predators were in synchrony with their prey, and they showed a maximum abundance when the aphid population reached the greatest level. Contrary to the observation that in these studies lacewing species had a major influence on the control of pecan aphids, it seems that in other parts of the world their effect is negligible. In Israel, chrysopids followed aphid fluctuations, but because of their low abundance, their impact was minor (Mansour, 1993). Green lacewing eggs and pupae were found among colonies of yellow pecan aphids in South Africa (Van den Berg & Maritz, 1996). Lacewings were the second most abundant group after coccinellids, but still, because of their low density, they probably make a limited contribution to the suppression of pecan aphids.

Impacts of habitat manipulations and IPM programmes on lacewings in pecan orchards

Due to the developing resistance of aphids, and the suppressive effects of pesticides on natural enemies, alternative tactics are required in the augmentation and the preservation of aphidophages in pecan orchards. Habitat manipulation and biological control experiments have been conducted in order to enhance the predatory effects of lacewings. One of the habitat manipulation methods is in the use of a cover-crop management system to enhance beneficial insects in pecan orchards. Certain legumes, like *Trifolium* and *Vicia*, are potential plant species for use as cool-season cover crops. Smith *et al.* (1994) studied the effect of such cool-season annual and perennial ground cover in pecan orchards. Chrysopids and hemerobiids were present with low abundance, and their density was not closely correlated with the size of aphid populations of the legume cover. Smith *et al.* (1996) used legumes and a mixture of these legumes and grasses as ground cover in experimental pecan orchards. They found that on pecan trees, out of the 13 indicated lacewing species, only the density of *Chrysoperla rufilabris* was influenced by a cover crop. The individual numbers of *C. rufilabris* were three to five times higher on pecan trees with legume cover (crimson clover/vetch), than on trees with grass cover. Eleven warm-season cover crops, or their mixtures, were evaluated in experimental plots situated in the understorey of pecan orchards (Bugg & Dutcher, 1989). Sesbania plant species appeared to be the best source of cowpea aphid as prey for aphidophagous insects.

Tree intercropping inside pecan orchards in order

to ensure an alternative food source for lacewings is another way of habitat manipulation. Mizell & Schiffhauer (1987b) indicated that aphids living on crapemyrtle tree were a food supplement for the lacewings *C. rufilabris* and *Micromus posticus* in the period of low population levels of pecan aphids. A further lacewing conservation method is to increase the number of refuges or overwintering places in pecan orchards. Mizell & Schiffhauer (1987a) could promote the successful overwintering of lacewings with the use of 'trunk traps' put on pecan trees, which gave shelter to them.

Several IPM programmes have been instigated in the last 15 years, as a pest control procedure in pecan orchards. To achieve reductions in pesticide applications, biological controls have been attempted with lacewings. The most important biological control agent against pecan aphids was the chrysopid *C. rufilabris*. Pyrethroids have no toxic effect on the adults and larvae of *C. rufilabris*, as indicated with laboratory tests (Mizell & Schiffhauer, 1990). This characteristic feature of *C. rufilabris* is advantageous, because the use of insecticides cannot be totally neglected against pests on pecan. For lacewing releases applied in biological control, mass-rearing is necessary. Studying such mass-rearing techniques, Elkarmi *et al.* (1987) found that *C. rufilabris* produced twice as many eggs per day as *Chrysoperla carnea*. As a biological control against pecan aphids, within an IPM programme, practical experiments using inoculative release of *C. rufilabris* have been made (LaRock & Ellington, 1996). Natural enemies of lacewings can interfere with such programmes. Tedders *et al.* (1990) found that the honeydew produced by pecan aphids attracted a high number of red imported fire ants (*Solenopsis invicta*) to the trees, which proved to be a major predator of lacewing eggs, larvae, and pupae within the foliage of pecan trees.

ACKNOWLEDGEMENTS

I am very grateful to Dr Peter McEwen (Insect Investigations Ltd, School of Biosciences, Cardiff University, UK) for his careful editorial work, useful comments, and critical language correction on the manuscript. I also thank Dr Ferenc Kozár (Department of Zoology, Plant Protection Institute of Hungarian Academy of Sciences) for reviewing and corrections of nomenclature of scale insects and whiteflies used in this chapter, and Ferenc Kádár (Department of Zoology, Plant Protection Institute of Hungarian Academy of Sciences) for help in the typing of bibliography. This work was supported, in part, by the Hungarian National Science Foundation (grant number: OTKA T 023284).

REFERENCES

Aguilera, P.A. & Pacheco, V.C. (1995) Determinación de depredadores del pulgón del avellano europeo, *Myzocallis coryli* (Goeze) (Homoptera: Aphididae) en la IX region de Chile. *Revista Chilena de Entomologia* 22, 17–19.

AliNiazee, M.T. (1998) Ecology and management of hazelnut pests. *Annual Review of Entomology* 43, 395–419.

Alrouechdi, K. (1981) Relations comportementales et trophiques entre *Chrysoperla carnea* (Stephens) (Neuroptera; Chrysopidae) et trois principaux ravageurs de l'olivier. I. La teigne de l'olivier *Prays oleae* Bern. (Lep. Hyponomeutidae). *Neuroptera International* 1, 122–134.

Alrouechdi, K. (1984) Les chrysopides (Neuroptera) en oliveraie. In *Progress in World's Neuropterology, Proceedings of the 1st International Symposium on Neuropterology*, ed. Gepp, J., Aspöck, H. & Hölzel, H., pp. 147–152. Thalerhof, Graz.

Alrouechdi, K., Canard, M. & Pralavorio, R. (1980a) Répartition des adultes et des pontes de chrysopides (Neuroptera) récoltés dans une oliveraie de Provence. *Neuroptera International* 1, 65–74.

Alrouechdi, K., Lyon, J.-P. & Canard, M. (1980b) Les chrysopides (Neuroptera) récoltés dans une oliveraie du sud-est de la France. *Acta Oecologica, Oecologia Applicata* 1, 173–180.

Alrouechdi, K., Pralavorio, R. & Canard, M. (1981) Coincidence et relations prédatrices entre *Chrysopa carnea* (Stephens) (Neur., Chrysopidae) et quelques ravageurs de l'olivier dans le sud-est de la France. *Bulletin de la Société Entomologique Suisse* 54, 281–290.

Arzone, A. (1979) Indagini sui limitatori naturali di *Psylla pyri* (L.) in Piemonte. *Bollettino del Laboratorio di Entomologia Agraria Portici* 36, 131–149.

Aspöck, H. & Hölzel, H. (1996) Neuropteroidea of North Africa, Mediterranean Asia and of Europe: a comparative review. In *Pure and Applied Research in Neuropterology, Proceedings of the 5th International Symposium on Neuropterology*, ed. Canard, M., Aspöck, H. & Mansell, M.W., pp. 31–86. Sacco, Toulouse.

228   F. SZENTKIRÁLYI

Asquith, D., Croft, B.A. & Hoyt, S.C. (1980) The systems
approach and general accomplishments toward better
insect control in pome and stone fruits. In *New
Technology of Pest Control*, ed. Huffaker, C.B., pp.
249–317. Wiley New York.

Baeschlin, R. & Taksdal, G. (1979) Die Fauna einer
Obstanlage in Ostnorwegen. 1. Die
Populationsentwicklung von Arthropodengruppen in
den Kronen von Apfelbäumen. *Meldinger fra Norges
landbrukshøgskole* 58, 1–44.

Baggiolini, M. & Wildbolz, T. (1965) Comparaison de
différentes méthodes de recensement des populations
d'arthropodes vivant aux dépens du pommier.
Entomophaga 10, 247–264.

Barnard, P.C. & Brooks, S.J. (1984) The African lacewing
genus *Ceratochrysa* (Neuroptera: Chrysopidae): a
predator on the cassava mealybug, *Phenacoccus manihoti*
(Hemiptera: Pseudococcidae). *Systematic Entomology* 9,
359–371.

Bartlett, B.R. (1957) Biotic factors in natural control of
citrus mealybugs in California. *Journal of Economic
Entomology* 50, 753–756.

Basky, Z. (1982) Predators and parasites of *Hyalopterus pruni*
and *Hyalopterus amygdali* populations living on peach,
plum and reed. *Acta Phytopathologica Academiae
Scientiarum Hungaricae* 17, 311–316.

Bitaw, A.A. & Bin-Saad, A.A. (1988) Natural enemies of
date palm pests in Jamahiriya. *Arab and Near East
Plant Protection Newsletter* 7, 26.

Bodenheimer, F.S. & Guttfeld, M. (1929) Über die
Möglichkeiten einer biologischen Bekämpfung von
Pseudococcus citri Risso (Rhy. Cocc.) in Palestina. Eine
monographische Studie. *Zeitschrift angewandte
Entomologie* 15, 67–136.

Bonomo, G., Catalano, G., Maltese, V. & Sparta, S. (1991)
Esperienze di lotta biologica e integrata nella
fragolicoltura marsalese. *Informatore Agrario* 47,
97–100.

Borg, J. (1932) *Scale insects of the Maltese Islands.*
Government Printing Press, Valletta.

Bouyjou, B., Canard, M. & Nguyen, T.X. (1984) Analyse
par battage des principaux prédateurs et proies
potentielles en vergers de poirier non traités.
IOBC/WPRS Bulletin 7, 148–166.

Briand, L.J. (1931) Notes on *Chrysopa oculata* Say and its
relation to the oriental peach moth (*Laspeyresia molesta*
Busck.) infestation in 1930. *Canadian Entomologist* 63,
123–126.

Broadbent, A.B. & Pree, D.J. (1984) Effects of diflubenzuron
and BAY SIR 8514 on beneficial insects associated with
peach. *Environmental Entomology* 13, 133–136.

Brooks, S.J. & Barnard, P.C. (1990) The green lacewings of

the world: a generic review (Neuroptera: Chrysopidae).
*Bulletin of the British Museum (Natural History)
(Entomology)* 59, 117–286.

Brown, M.W., Adler, C.R.L. & Weires, R.W. (1988) Insects
associated with apple in the mid-Atlantic states. *New
York's Food and Life Sciences Bulletin* 124, 1–31.

Buchholz, U. & Schruft, G. (1994) Räuberische
Arthropoden auf Blüten und Früchten der Weinrebe
(*Vitis vinifera* L.) als Antagonisten des Einbindigen
Traubenwicklers (*Eupoecilia ambiguella* Hbn.) (Lep.,
Cochylidae). *Journal of Applied Entomology* 118, 31–37.

Buchholz, U., Schmidt, S. & Schruft, G. (1994) The use of
an immunological technique in order to evaluate the
predation on *Eupoecilia ambiguella* (Hbn.)
(Lepidoptera: Cochylidae) in vineyards. *Biochemical
Systematics and Ecology* 22, 671–677.

Bugg, R.L. & Dutcher, J.D. (1989) Warm-season cover crops
for pecan orchards: horticultural and entomological
implications. *Biological Agriculture and Horticulture* 6,
123–148.

Burts, E.C. (1963) The pear psylla in Central Washington.
Washington Agricultural Experimental Station Circular
416, 1–11.

Butcher, M.R., Penman, D.R. & Scott, R.R. (1988) Field
predation of two spotted spider mites in a New Zealand
strawberry crop. *Entomophaga* 33, 173–185.

Campos, M. (1986) Influencia del compleijo parasitario
sobre las poblaciones de *Chrysoperla carnea*
(Neuroptera, Chrysopidae) en olivares del sur de
España. *Neuroptera International* 4, 97–105.

Campos, M. (1989) Observaciones sobre la bioecología de
Chrysoperla carnea (Stephens) (Neuroptera:
Chrysopidae) en el sur de España. *Neuroptera
International* 5, 159–164.

Campos, M. & Ramos, P. (1983) Chrisópidos (Neuroptera)
capturados en un olivar del sur de España. *Neuroptera
International* 2, 219–227.

Campos, M. & Ramos, P. (1985) Some relationships between
the number of *Prays oleae* eggs laid on olive fruits and
their predation by *Chrysoperla carnea*. In *Integrated Pest
Control in Olive Groves, Proceedings of the
CEC/FAO/IOBC International Joint Meeting, Pisa*, ed.
Cavalloro, R. & Crovetti, A., pp. 237–241. Balkema,
Rotterdam.

Canard, M. & Laudého, Y. (1977) Les névroptères capturés
au piège de McPhail dans les oliviers en Grèce. 1: L'Île
d'Aguistri. *Biologia Gallo-Hellenica* 7, 65–75.

Canard, M. & Laudého, Y. (1980) Les névroptères capturés
au piège de McPhail dans les oliviers en Grèce. 2: La
région d'Akrefnion. *Biologia Gallo-Hellenica* 9,
139–146.

Canard, M., Neuenschwander, P. & Michelakis, S. (1979)

Les névroptères capturés au piège de McPhail dans les oliviers en Grèce. 3: La Crète occidentale. *Annales de la Société Entomologique de France* (N.S.) 15, 607–615.

Carroll, D.P. & Hoyt, S.C. (1984) Natural enemies and their effects on apple aphid, *Aphis pomi* DeGeer (Homoptera: Aphididae), colonies on young apple trees in Central Washington. *Environmental Entomology* 13, 469–481.

Catling, H.D., Lee, S.C., Moon, D.K. & Kim, S. H. (1977) Towards the integrated control of Korean citrus pests. *Entomophaga* 22, 335–343.

Cavalloro, R. & Di Martino, E. (1986) *Integrated Pest Control in Citrus Groves. Proceedings of the Experts' Meeting, Commission of the European Communities,* A.A. Balkema, Rotterdam.

Chagas, E.F. das, Silveira Neto, S., Braz, A.J.B.P., Mateus, C.P.B. & Coelho, I.P. (1982) Flutuação populacional de pragas e predadores em citros. *Pesquisa Agropecuária Brasileira* 17, 817–824.

Chang, C.P. & Huang, S.C. (1995) Evaluation of the effectiveness of releasing green lacewing, *Mallada basalis* (Walker) for the control of tetranychid mites on strawberry. *Plant Protection Bulletin Taipei* 37, 41–58. (in Chinese)

Chen, R.T., Chen, Y.H. & Huang, M.D. (1989) Biology of the green lacewing, *Chrysopa boninensis* and its predation efficiency on the citrus leaf miner, *Phyllocnistis citrella*. In *Studies on the Integrated Management of Citrus Insect Pests,* pp. 96–105. Academic Book and Periodical Press, Guangzhou, Guangdong, China.

Cheng, W.Y. & Chen, S.M. (1996) Utilization of green lacewing in Taiwan. *Taiwan Sugar* 43, 20–22.

Childers, C.C. & Enns, W.R. (1975) Predaceous arthropods associated with spider mites in Missouri apple orchards. *Journal of the Kansas Entomological Society* 48, 453–471.

Childers, C.C., Futch, S.H. & Stange, L.A. (1992) Insect (Neuroptera; Lepidoptera) clogging of a microsprinkler irrigation system in Florida citrus. *Florida Entomologist* 75, 601–604.

Chua, T.H. & Wood, B.J. (1990) Other tropical fruit trees and shrubs. In *Armored Scale Insects, their Biology, Natural Enemies and Control,* vol. B, ed. Rosen, D. pp. 453–552. Elsevier, Amsterdam.

Ciglar, I. (1985) Natural enemies in apple orchards in SR Croatia, Yugoslavia. *Agriculturae Conspectus Scientificus* 68, 131–139.

Ciglar, I. & Barić, B. (1992) Control of pear psylla *Psylla pyri* L. (Homoptera: Psyllidae) in commercial orchards in North-East of Croatia, Yugoslavia. *Acta Phytopathologica et Entomologica Hungarica* 27, 155–163.

Ciglar, I. & Budinšcak (1993) Dynamic of fauna population in managed and unmanaged apple orhards. *Agronomski Glasnik* 1–2, 63–72. (in Croatian)

Ciglar, I. & Schmidt, L. (1983) Insect fauna of apple orchard 'Borinci' – Vinkovci, Croatia, Yugoslavia. *Acta Entomologica Jugoslavica* 19, 83–90. (in Croatian)

Collyer, E. & van Geldermalsen, M. (1975) Integrated control of apple pests in New Zealand. 1. Outline of experiment and general results. *New Zealand Journal of Zoology* 2, 101–134.

Cruz de Boelpaepe, M.O. & Filipe, M.N. (1990) Les pucerons du poirier. Dynamique des populations de *Dysaphis piri* (Boyer Fons) en vergers dans la province de Beira-Baixa (Portugal). *Bulletin SROP* 13, 31–35.

Cruz de Boelpaepe, M.O., Peixoto, E.S. & Saraiva, T. (1988) Dynamique des populations aphidiennes en vergers de pêchers, dans la region de Ribatejo e Oeste. Rôle des facteurs de l'environnement. *Bulletin SROP* 11, 17–19.

Curkovic, S.T., Barria, P.G. & Gonzalez, R.R. (1995) Observaciones preliminares sobre insectos y acaros presentes en vides, perales, ciruelos y kakis detectados con trampas de agregación. *Acta Entomologica Chilena* 19, 143–154.

Daane, K.M. & Yokota, G.Y. (1997) Release strategies affect survival and distribution of green lacewings (Neuroptera: Chrysopidae) in augmentation programs. *Environmental Entomology* 26, 455–464.

Daane, K.M., Yokota, G.Y., Hagen, K.S. & Zheng, Y. (1994) Field evaluation of *Chrysoperla* spp. in augmentative release programs for the variegated grape leafhopper. *Erythroneura variabilis*. In *Quality Control of Mass-Reared Arthropods, Proceedings of the 7th Workshop of the IOBC Working Group,* Rimini, Italy, July 1993 ed. Nicoli, G., Benuzzi, M. & Leppla, N.C., pp. 105–120.

Daane, K.M., Yokota, G.Y., Zheng, Y. & Hagen, K. S. (1996) Inundative release of common green lacewings (Neuroptera: Chrysopidae) to suppress *Erythroneura variabilis* and *E. elegantula* (Homoptera: Cicadellidae) in vineyards. *Environmental Entomology* 25, 1224–1234.

Darwish, E.T.E. (1992) Bionomics of the mealy peach aphid *Hyalopterus amygdali* (BlanC.) on peach trees in Menoufia, Egypt. *Acta Phytopathologica et Entomologica Hungarica* 27, 205–210.

DeBach, P. & Bartlett, B. (1951) Effects of insecticides on biological control of insect pests of citrus. *Journal of Economic Entomology* 44, 372–383.

DeBach, P., Fleschner, C.A. & Dietrick, E.J. (1950) Studies of the efficacy of natural enemies of citrus red mite in southern California. *Journal of Economic Entomology* 43, 807–819.

Delbac, L., Fos, A., Lecharpentier, P. & Stockel, J. (1996) Mating disruption against *Eudemis*. Impact on the 'green cicadellid in the Bordeaux vineyard. *Phytoma* 488, 36–39.

Delrio, G. (1983) The entomophagous insects of *Saissetia oleae* (Oliv.) in Sardinia. In *Entomophagous Insects and Biotechnologies against Olive Pests, Proceedings of the European Commission Experts' Meeting on Entomophages and Biological Methods in Integrated Control in Olive Orchards*, ed. Cavalloro, R. & Piavaux, A., pp. 15–23.Chania, Greece.

Delrio, G. (1986) Studies on citrus red mite in Sardinia. In *Proceedings of the Experts' Meeting: Integrated Pest Control in Citrus-Groves*, ed. Cavalloro, R. & Di Martino, E., pp. 189–197.

Devetak, D. (1991) Neuropteroidea. Megaloptera, Raphidioptera, Planipennia (Insecta). Fauna Durmitora 4. *Crnogorska akademija nauka i umjetnosti Posebna izdanja, knjiga 24, Odjeljenje prirodnih nauka, knjiga 15.* A.A. Balkema, Rotterdam. 135–159.

Dinkins, R.L., Tedders, W.L. & Reid, W. (1994) Predaceous neuropterans in Georgia and Kansas pecan trees. *Journal Entomological Science* 29, 165–175.

Doutt, R.L. (1948) Effect of codling moth sprays on natural control of the Baker mealybug. *Journal of Economic Entomology* 41, 116–117.

Doutt, R.L. & Hagen, K.S. (1949) Periodic colonization of *Chrysopa californica* as a possible control of mealybugs. *Journal of Economic Entomology* 42, 560–561.

Doutt, R.L. & Hagen, K.S. (1950) Biological control measures applied against *Pseudococcus maritimus* on pears. *Journal of Economic Entomology* 43, 94–96.

Easterbrook, M.A. (1998) The beneficial fauna of strawberry fields in south-east England. *Journal of Horticultural Science and Biotechnology* 73, 137–144.

Edelson, J.V. & Estes, P.M. (1983) Intracanopy distribution and seasonal abundance of the yellow pecan aphids *Monellia caryella* and *Monelliopsis nigropunctata* (Homoptera: Aphididae). *Environmental Entomology* 12, 862–867.

Edelson, J.V. & Estes, P.M. (1987) Seasonal abundance and distribution of predators and parasites associated with *Monelliopsis pecanis* Bissell and *Monellia caryella* (Fitch) (Homoptera: Aphidae). *Journal of Entomological Science* 22, 336–347.

Eglin-Dederding, W. (1980) Die Netzflügler des Schweizerischen Nationalparks und seiner Umgebung (Insecta: Neuropteroidea). In *Ergebnisse der wissenschaftlichen Untersuchungen im Schweizerischen Nationalpark*, vol. 15, pp. 281–351. Lüdin AG, Liestal.

Elkarmi, L.A., Harris, M.K. & Morrison, R.K. (1987) Laboratory rearing of *Chrysoperla rufilabris* (Burmeister), a predator of insect pests of pecans. *Southwestern Entomologist* 12, 73–78.

Fasih, M. & Srivastava, R.P. (1990) Parasites and predators of insect pests of mango. *International Pest Control* 32, 39–41.

Frazer, B.D. & van den Bosch, R. (1973) Biological control of the walnut aphid in California: the interrelationship of the aphid and its parasite. *Environmental Entomology* 2, 561–568.

Freitas, S. De. & Fernandes, O.A. (1992) A preliminary statement on green lacewings in citrus in the Jaboticabal region of Brazil (Insecta: Neuroptera: Chrysopidae). In *Current Research in Neuropterology, Proceedings of the 4th International Symposium on Neuropterology*, ed. Canard, M., Aspöck, H. & Mansell, M.W., pp. 147–150. Sacco, Toulouse.

Fye, R.E. (1985) Corrugated fiberboard traps for predators overwintering in pear orchards. *Journal of Economic Entomology* 78, 1511–1514.

Giunchi, P. (1980) Possibilities of biological control of pear insects. *IOBC/WPRS Bulletin* 3, 48–49.

Gottwald, R. (1991) Wichtige Antagonisten in Apfelanbau der DDR. In *Verhandlungen des XII. Internationalen Symposiums über Entomofaunistik Mitteleuropas*, ed. Dolin, W.G., pp. 74–77. Naukova Dumka, Kiev.

Grasswitz, T.R. & Burts, E.C. (1995) Effect of native natural enemies and augmentative releases of *Chrysoperla rufilabris* Burmeister and *Aphidoletes aphidimyza* (Rondani) on the population dynamics of the green apple aphid, *Aphis pomi* DeGeer. *International Journal of Pest Management* 41, 176–183.

Grbić, M., Lakic, B. & Mihajlović, L. (1990) Predators and parasitoids of *Psylla pyri* L. (Hom.: Psyllidae) in Vojvodina (Yu). *Bulletin SROP* 13, 44–54.

Greve, L. (1967) Faunistical notes on Neuroptera from Southern Norway. *Norsk Entomologisk Tidsskrift* 14, 37–43.

Gut, L.J., Liss, W.J. & Westigard, P.H. (1991) Arthropod community organization and development in pear. *Environmental Management* 15, 83–104.

Haeussler, G.J. & Clancy, D.W. (1944) Natural enemies of Comstock mealybug in the eastern states. *Journal of Economic Entomology* 37, 503–509.

Hagley, E.A.C. (1974) The arthropod fauna in unsprayed apple orchards in Ontario II. Some predacious species. *Proceedings of the Entomological Society of Ontario* 105, 28–40.

Hagley, E.A.C. (1989) Release of *Chrysoperla carnea* Stephens (Neuroptera: Chrysopidae) for control of the green apple aphid, *Aphis pomi* DeGeer (Homoptera: Aphididae). *Canadian Entomologist* 121, 309–314.

Hagley, E.A.C. & Allen, W.R. (1990) The green apple aphid, *Aphis pomi* DeGeer (Homoptera: Aphididae), as prey of polyphagous arthropod predators in Ontario. *Canadian Entomologist* 122, 1221–1228.

Hagley, E.A.C. & Miles, N. (1987) Release of *Chrysoperla*

carnea Stephens (Neuroptera: Chrysopidae) for control of *Tetranychus urticae* Koch (Acarina: Tetranychidae) on peach grown in a protected environment structure. *Canadian Entomologist* 119, 205–206.

Hagley, E.A.C. & Simpson, C.M. (1983) Effects of insecticides on predators of the pear psylla, *Psylla pyricola* (Hemiptera: Psyllidae), in Ontario. *Canadian Entomologist* 115, 1409–1414.

Hartfield, C.M. (1996) Aphid natural enemies in United Kingdom plum orchards. In *IOBC/WPRS Working Group Meeting on Integrated Plant Protection in Stone Fruit*, ed. Cravedi, P., Hartfield, C.M. & Mazzoni, E., pp. 87–92. IOBC/WPRS, Avignon.

Heim, G. (1985) Effect of insecticidal sprays on predators and indifferent arthropods found on olive trees in the north of Lebanon. In *Integrated Pest Control in Olive Groves, Proceedings of the CEC/FAO/IOBC International Joint Meeting, Pisa*, ed. Cavalloro, R. & Crovetti, A., pp. 456–465. Balkema, Rotterdam.

Herard, F. (1986) Annotated list of the entomophagous complex associated with pear psylla, *Psylla pyri* (L.) (Hom.: Psyllidae) in France. *Agronomie* 6, 1–34.

Herne, D.H.C. & Putman, W.L. (1966) Toxicity of some pesticides to predacious arthropods in Ontario peach orchards. *Canadian Entomologist* 98, 936–942.

Holdsworth, R.P. Jr (1968) Integrated control: effect on European red mite and its more important predators. *Journal of Economic Entomology* 61, 1602–1607.

Holdsworth, R.P. Jr (1970) Aphids and aphid enemies: effect of integrated control in an Ohio apple orchard. *Journal of Economic Entomology* 63, 530–535.

Hölzel, H. & Duelli, P. (1994) *Brinckochrysa beninensis* n. sp. – eine neue Chrysopiden-Spezies aus Westafrika (Neuroptera: Chrysopidae). *Entomologische Zeitschrift* 104, 54–58.

Hölzel, H. & Ohm, P. (1991) Chrysopidae der Mascarene-Inseln (Neuropteroidea: Planipennia). *Entomologische Zeitschrift* 101, 333–352.

Hölzel, H., Ohm, P. & Stelzl, M. (1994) Chrysopidae aus Senegal und Gambia II. Belonopterygini und Chrysopini (Neuroptera). *Entomofauna* 15, 377–396.

Hölzel, H., Ohm, P. & Stelzl, M. (1997) Chrysopidae von Namibia (Neuroptera, Chrysopinae). *Mitteilungen münchener entomologische Gesellschaft* 87, 47–71.

Injac, M. & Dulic, K. (1992) Population of phytophagous mites and occurrence of predators in apple orchards. *Acta Phytopathologica et Entomologica Hungarica* 27, 299–304.

Injac, M., Sivčev, I. & Vukovic, M. (1978) Development of *Chrysopa carnea* Steph. in the program of the integral apple protection. *Zastita Bilja* 29, 371–379. (in Croatian)

Jubb, G.L. Jr & Masteller, E.C. (1977) Survey of arthropods in grape vineyards of Erie county, Pennsylvania: Neuroptera. *Environmental Entomology* 6, 419–428.

Kehat, M. (1967) Survey and distribution of common lady beetles (Col., Coccinellidae) on date palm trees in Israel. *Entomophaga* 12, 119–125.

Killington, F.J. (1937) *A monograph of the British Neuroptera*, vol. 2. Ray Society, London.

Kinn, D.N. & Doutt, R.L. (1972) Initial survey of arthropods found in North Coast vineyards of California. *Environmental Entomology* 1, 508–513.

Kocourek, F. & Beránková, J. (1996) Regulations of *Psylla pyri* (L.) by selective insecticides and natural antagonists in IPM of pear orchards. *IOBC/WPRS Bulletin* 19, 339–340.

Kokhreidze, G.G. (1982) Natural enemies of tea and green citrus aphids on citrus plants in Adjaria. *Bulletin of the Academy of Sciences of the Georgian SSR* 105, 609–611. (in Russian)

Koltun, N.E. & Meleshko, N.I. (1994) Predators of the green apple aphid (*Aphis pomi* De Geer) and their effectiveness in apple nurseries in Belarus. *Vestsi Akademii Agrarnykh Navuk Belarusi* 4, 57–59. (in Russian)

Krämer, P. (1961) Untersuchungen über den Einfluss einiger Arthropoden auf Raubmilben (Acari). *Zeitschrift angewandte Zoologie* 48, 257–311.

Krishnamoorthy, A. & Mani, M. (1989) Records of green lacewings preying on mealybugs in India. *Current Science* 58, 155–156.

LaRock, D.R & Ellington, J.J. (1996) An integrated pest management approach, emphasizing biological control, for pecan aphids. *Southwestern Entomologist* 21, 153–166.

LeRoux, E.J. (1960) Effects of 'modified' and 'commercial' spray programs on the fauna of apple orchards in Quebec. *Annales de la Société Entomologique du Québec* 6, 87–121.

Li, B.C., Ye, Q.M., Li, Z., Ye, X.X., Jin, L.F. & Ye, Z. (1996) The food-web pattern of citrus insect communities in Zhejiang Province II. The sub-community of aphids. *Acta Agriculturae Zhejiangensis* 8, 271–273. (in Chinese)

Li, J.P. (1992) Morphology and bionomics of *Chromaphis juglandicola* Kaltenback and its control. *Entomological Knowledge* 29, 345–347.

Liao, H.T., Harris, M.K., Gilstrap, F.E., Dean D.A., Agnew, C.W., Michels, G.J. & Mansour, F. (1984) Natural enemies and other factors affecting seasonal abundance of the blackmargined aphid on pecan. *Southwestern Entomologist* 9, 404–420.

Liao, H.T., Harris, M.K., Gilstrap, F.E. & Mansour, F. (1985) Impact of natural enemies on the blackmargined pecan aphid, *Monellia caryella* (Homoptera: Aphididae). *Environmental Entomology* 14, 122–126.

Liber, H. & Niccoli, A. (1988) Observations on the effectiveness of an attractant food spray in increasing chrysopid predation on *Prays oleae* (Bern.) eggs. *Redia* 71, 467–482.

Liu, Z.M. (1989) Studies on the interaction system of citrus psylla, *Diaphorina citri* KuW. and its natural enemies, *Tetrastichus* sp. and *Chrysopa boninensis* Omamoto. In *Studies on the Integrated Management of Citrus Insect Pests*, pp. 144–164. Academic Book and Periodical Press, Guangzhou, Guangdong, China.

Lo, K.C., Lee, W.T. & Wu, T.K. (1990) Use of predators for controlling spider mites (Acarina, Tetranychidae) in Taiwan, China. In *The Use of Parasitoids and Predators to Control Agricultural Pests, FFTC NARC International Seminar*, National Agricultural Research Centre, Tukuba gun, Japan.

Lyoussoufi, A., Rieux, R., Armand, E., Faivre d'Arcier, F. & Sauphanor, B. (1994) La faune entomophage des psylles du poirier (Homoptera: Psyllidae) en Europe. Revue bibliographique: II – Insectes Oligonéoptéres prédateurs. *IOBC/WPRS Bulletin* 17, 93–98.

MacLellan, C.R. (1973) Natural enemies of the light brown apple moth, *Epiphyas postvittana*, in the Australian Capital Territory. *Canadian Entomologist* 105, 681–700.

MacLellan, C.R. (1977) Populations of some major pests and their natural enemies on young and semidwarf apple trees in Nova Scotia. *Canadian Entomologist* 109, 797–806.

Madsen, H.F. & Madsen, B.J. (1982) Populations of beneficial and pest arthropods in an organic and a pesticide treated apple orchard in British Columbia. *Canadian Entomologist* 114, 1083–1088.

Madsen, B.J. & Morgan, C.V.G. (1975) Mites and insects collected from vineyards in the Okanagan and Similkameen Valleys, British Columbia. *Journal of the Entomological Society of British Columbia* 72, 9–14.

Madsen, H.F., Westigard, P.H. & Sisson, R.L. (1963) Observations on the natural control of the pear psylla, *Psylla pyricola* Forster, in California. *Canadian Entomologist* 95, 837–844.

Malet, J.C., Noyer, C., Maisonneuve, J.C. & Canard, M. (1994) *Chrysoperla lucasina* (Lacroix) (Neur., Chrysopidae), a potential predator of the Mediterranean *Chrysoperla* Steinmann complex: first experiment to control *Aphis gossypii* Glover (Hom., Aphididae) on melon in France. *Journal of Applied Entomology* 118, 429–436.

Malevez, N. (1976) Observations sur l'entomofaune d'un verger de pommiers dans une perspective de lutte intégrée. *Parasitica* 32, 109–140.

Mani, M. & Krishnamoorthy, A. (1989) Feeding potential and development of green lacewing *Mallada boninensis* (Okamoto) on the grape mealybug, *Maconellicoccus hirsutus* (Green). *Entomon* 14, 19–20.

Mani, M. & Krishnamoorthy, A. (1990) Natural suppression of mealybugs in guava orchards. *Entomon* 15, 245–247.

Mansour, F.A. (1993) Natural enemies and seasonal abundance of the blackmargined aphid (*Monellia caryella*) in pecan orchards in Israel. *Phytoparasitica* 21, 329–332.

Markovic, M. & Zivanovic, M. (1988) Ecological selectivity of some acaricides to the predators of *Panonychus ulmi* KoC. *Zastita Bilja* 39, 183–195.

Marzocchi, L. & Pantaleoni, R.A. (1995) Indagine sui principali entomofagi predatori (Insecta Heteroptera, Neuroptera et Coleoptera) in pereti della Pianura Padana. *Bolletino dell'Istituto di Entomologia dell'Università di Bologna* 49, 21–40.

Mathys, G. & Baggiolini, M. (1965) Méthodes de recensement d'insectes ravageurs dans des vergers soumis à des essais de lutte intégrée. *Mitteilungen der schweizerischen entomologischen Gesellschaft* 38, 120–141.

Matias, C. (1990) Bilan de la protection intégrée en vergers de poiriers au Portugal depuis 1983. *Bulletin SROP* 13, 8–12.

Matias, C., Nguyen, T.X. & M. Canard, M. (1988) Role prédateur possible des chrysopes (Neuroptera: Chrysopidae) à l'encontre des psylles (Homoptera: Psyllidae) du poirier au Portugal. *Neuroptera International* 5, 93–101.

Matias, C., Bouyjou, B., Avelar, J. & Domingues, V. (1990) Faune prédatrice et proies potentielles en vergers de poiriers dans deux situations (lutte chimique et lutte intégrée) au Portugal. *Bulletin SROP* 13, 11–16.

McEwen, P.K. & Ruiz, J. (1993) Relationship between non-olive vegetation and green lacewing eggs in a Spanish olive orchard. *Antenna* 18, 148–150.

McEwen, P.K., Jervis, M.A. & Kidd, N.A.C. (1994) Use of a sprayed L-tryptophan solution to concentrate numbers of the green lacewing *Chrysoperla carnea* in olive tree canopy. *Entomologia Experimentalis et Applicata* 70, 97–99.

McMullen, R.D. & Jong, C. (1967) New records and discussion of predators of the pear psylla, *Psylla pyricola* Forster, in British Columbia. *Journal of the Entomological Society of British Columbia* 64, 35–40.

Messing, R.H. & AliNiazee, M.T. (1985) Natural enemies of *Myzocallis coryli* (Hom.: Aphididae) in Oregon hazelnut orchards. *Journal of the Entomological Society of British Columbia* 82, 14–18.

Mészáros, Z. (ed.) (1984) Results of faunistical and floristical studies in Hungarian apple orchards (Apple Ecosystem Research No. 26). *Acta Phytopathologica Academiae Scientiarum Hungaricae* 19, 91–176.

Miszczak, M. & Niemczyk, E. (1978) Green lacewing (*Chrysopa carnea* Steph.) (Neuroptera – Chrysopidae), as a predator of European mite (*Panonychus ulmi* Koch) on apple trees. Part II. The effectiveness of *Chrysopa carnea* larvae in control of *Panonychus ulmi* KoC. *Fruit Science Reports* 4, 21–31.

Mizell, R.F. & Schiffhauer, D.E. (1987a) Trunk traps and overwintering predators in pecan orchards: survey of species and emergence times. *Florida Entomologist* 70, 238–244.

Mizell, R.F. & Schiffhauer, D.E. (1987b) Seasonal abundance of the crapemyrtle aphid, *Sarucallis kahawaluokalani*, in relation to the pecan aphids, *Monellia caryella* and *Monelliopsis pecanis* and their common predators. *Entomophaga* 32, 511–520.

Mizell, R.F. & Schiffhauer, D.E. (1990) Effects of pesticides on pecan aphid predators *Chrysoperla rufilabris* (Neuroptera: Chrysopidae), *Hippodamia convergens*, *Cycloneda sanguinea* (L.), *Olla v-nigrum* (Coleoptera: Coccinellidae), and *Aphelinus perpallidus* (Hymenoptera: Encyrtidae). *Journal of Economic Entomology* 83, 1806–1812.

Molleman, F., Drukker, B. & Blommers, L. (1997) A trap for monitoring pear psylla predators using dispensers with the synomone methylsalicylate. *Proceedings of the Section Experimental and Applied Entomology of the Netherlands Entomological Society* 8, 177–182.

Monserrat, V.J. (1990) A systematic checklist of the Hemerobiidae of the world (Insecta: Neuroptera). *Advances in Neuropterology, Proceedings of the 3rd International Symposium on Neuropterology*, ed. Mansell, M.W. & Aspöck, H., pp. 215–262. South African Department of Agricultural Development, Pretoria.

Monserrat, V.J. (1991) Nuevos datos sobre algunas especies del genero *Hemerobius* L., 1758 (Insecta, Neuroptera: Hemerobiidae). *Graellsia* 47, 61–70.

Monserrat, V.J. (1996) Revisión del genero *Hemerobius* de Latinoamérica (Neuroptera, Hemerobiidae). *Fragmenta Entomologica* 27, 399–523.

Monserrat, V.J. (1998) Nuevos datos sobre los hemeróbiidos de América (Neuroptera: Hemerobiidae). *Journal of Neuropterology* 1, 109–153.

Monserrat, V.J. & F. Marín (1994) Plant substrate specificity of Iberian Chrysopidae (Insecta: Neuroptera). *Acta Oecologica* 15, 119–131.

Monserrat, V.J. & F. Marín (1996) Plant substrate specificity of Iberian Hemerobiidae (Insecta: Neuroptera). *Journal of Natural History* 30, 775–787.

Morton, K.J. (1910) Life-history of *Drepanepteryx phalaenoides*, Linn. *Entomologist's Monthly Magazine* 46, 54–62.

Muma, M.H. (1955) Factors contributing to the natural control of citrus insects and mites in Florida. *Journal of Economic Entomology* 48, 432–438.

Muma, M.H. (1959) Natural control of Florida red scale on citrus in Florida by predators and parasites. *Journal of Economic Entomology* 52, 577–586.

Nakao, S. (1962) A list of insects collected in a citrus grove near Fukuoka City (Ecological studies on the insect community of citrus groves, IV). *Kontyu* 30, 50–70.

Nakao, S. (1964) The interspecific relations among insects in a citrus grove (Ecological studies on the insect community of citrus groves, V). *Kontyu* 32, 490–503.

Nakao, S. (1968) Ecological studies on the insect community of citrus groves I–V. *Review of Plant Protection Research* 1, 97–106.

Nakao, S., Nohara, K. & Ono, T. (1972) Fundamental study on the integrated control of citrus red mite in the summer orange grove. *Mushi* 46, 1–27.

Nakao, S., Nohara, K. & Nagatomi, A. (1977) Studies on pests and their predators at two groves of 'Kuroshima-mikan,, a native citrus on Nagashima, Kagoshima Pref. *Applied Entomology and Zoology* 12, 334–346.

Nakao, S., Nohara, K. & Nagatomi, A. (1985) Effect of insecticide treatments on the fauna of a natural growth of citrus of Japan. *Mushi* 50, 91–114.

Neuenschwander, P. (1982) Beneficial insects caught by yellow traps used in mass-trapping of the olive fly, *Dacus oleãe. Entomologia Experimentalis et Applicata* 32, 286–296.

Neuenschwander, P. & Michelakis, S. (1980) The seasonal and spatial distribution of adult and larval Chrysopids on olive-trees in Crete. *Acta Oecologica* 1, 93–102.

Neuenschwander, P., Canard, M. & Michelakis, S. (1981) The attractivity of protein hydrolysate baited McPhail traps to different chrysopid and hemerobiid species (Neuroptera) in a Cretan olive orchard. *Annales de la Société Entomologique de France* (N.S.) 17, 213–220.

New, T.R. (1967) Trap-banding as a collecting method for Neuroptera and their parasites, and some results obtained. *Entomologist's Gazette* 18, 37–44.

Nickel, J.L., Shimizu, J.T. & Wong, T.Y. (1965) Studies on natural control of pear psylla in California. *Journal of Economic Entomology* 58, 970–976.

Niemczyk, E., Olszak, R. & Pruska, M. (1983) *The Role and Exploitation of Predatory and Parasitic Insects in Limiting the Population of the more Important Pests in Apple Orchards. Final Report, Project PL–AR-78, Research Institute of Pomology and Floriculture Skierniewice, Poland*, 1–104.

Nikolaishvili, A.A. & Mekvabishvili, S.S. (1990) Entomophages of pests occurring on lemon trees and their effectiveness in the greenhouse. *Subtropicheskie Kul'tury* 6, 102–110. (in Russian)

Oatman, E.R., Legner, E.F. & Brooks, R.F. (1964) An ecological study of arthropod populations on apple in northeastern Wisconsin: insect species present. *Journal of Economic Entomology* 57, 978–983.

Ofek, G., Huberman, G., Yzhar, Y., Wysoki, M. Kulitsky, W., Reneh, S. & Inbal, Z. (1997) The control of the oriental red scale, *Aonidiella orientalis* Newstead and the California red scale, *A. aurantii* (Maskell) (Homoptera: Diaspididae) in mango orchards in Hevel Habsor (Israel). *Alon Hanotea* 51, 212–218.

Ohm, P. & Hölzel, H. (1997) Beitrag zur Kenntnis der Neuropterenfauna der Maskarenen (Neuroptera: Coniopterigidae, Hemerobiidae, Chrysopidae, Myrmeleontidae). *Zeitschrift für Entomologie* 18, 221–236.

Ohm, P. & Hölzel, H. (1998) A contribution to the knowledge of the neuropterous fauna of the Comoros: the Neuroptera of Mayotte. *Acta Zoologica Fennica* 209, 183–194.

Olazo-Gonzales, E.V. (1987) Neuroptera associated with citrus crops in the Tucuman Province and the description of a new species of *Nomerobius* (Hemerobiidae). *Revista de Investigación, Centro de Investigaciones para la Regulación de Poblaciones de Organismos Nocivos* 5, 37–54. (in Spanish)

Orphanides, G.M. (1991) Biology and biological control of *Parabemisia myricae* (Kuwana) (Homoptera: Aleyrodidae) in Cyprus. *Technical Bulletin of the Cyprus Agricultural Research Institute* 135, 6.

Pantaleoni, R.A. (1996) Distribuzione spaziale di alcuni Neurotteri Planipenni su piante arboree. *Bollettino dell'Istituto di Entomologia dell'Università di Bologna* 50, 133–141.

Pantaleoni, R.A. & Curto, G.M. (1990) I Neurotteri delle colture agrarie: Crisopidi in oliveti del Salento (Italia meridionale). *Bollettino dell'Istituto di Entomologia dell'Università di Bologna* 45, 167–179.

Pantaleoni, R.A. & Lepera, D. (1985) I Neurotteri delle colture agrarie: indagine sui Crisopidi in agrumeti della Calabria. In *Estratto dagli Atti XIV Congresso Nazionale Italiano di Entomologia*, Palermo, Erice, Bagheria, pp. 451–457. Accademia Nazionale Italiana di Entomologia, Palermo.

Pantaleoni, R.A. & Ticchiati, V. (1988) I Neurotteri delle colture agrarie: osservazioni sulle fluttuazioni stagionali di popolazione in frutteti. *Bollettino dell'Istituto di Entomologia dell'Università di Bologna* 43, 43–57.

Pantaleoni, R.A. & Tisselli, V. (1985) I Neurotteri delle colture agrarie: rilievi sui Crisopidi in alcune coltivazioni del forlivese. *Bollettino dell'Istituto di Entomologia dell'Università di Bologna* 40, 51–65.

Pantaleoni, R.A., Lentini, A. & Delrio, G (1993) Crisopidi in oliveti della Sardegna. Risultati preliminari. In *Atti del Convegno: Tecniche, Norme e Qualità in Olivicoltura*, Potenza, 15–17 December 1993, pp. 879–890. Universià degli Studi della Basilicata, Potenza, Italy.

Pari, P., Lucchi, C. & Brigliadori, M. (1993) Application of biological control techniques to strawberries in protected cultivation. *Informatore Agrario* 49, 49–54.

Parrella, M.P., McCaffrey, J.P. & Horsburgh, R.L. (1981) Population trends of selected phytophagous arthropods and predators under different pesticide programs in Virginia apple orchards. *Journal of Economic Entomology* 74, 492–498.

Paulian, M. (1998) Occurrence of chrysopids (Neuroptera, Chrysopidae) and moving activity of their populations within a peach orchard agro-ecosystem in Romania. *Acta Zoologica Fennica* 209, 207–210.

Paulian, M. & Andriescu, I. (1996) Chrysopidae and Hemerobiidae recorded from crops and adjacent natural habitats in the Danube Delta, Romania (Insecta: Neuroptera). In *Current research in Neuropterology, Proceedings of the 4th International Symposium on Neuropterology*, ed. Canard, M., Aspöck, H. & Mansell, M.W., pp. 203–206. Sacco, Toulouse.

Pawar, A.D. & Parry, M. (1989) Record of natural enemies of important fruit pests in Ladakh (J & K). *Indian Journal of Plant Protection* 17, 291–292.

Peng, Y.K. (1988) Mass rearing and field release of *Chrysopa shansiensis* (Neuroptera: Chrysopidae) against citrus red spider mites (Tetranychidae). *Chinese Journal of Biological Control* 4, 137. (in Chinese)

Penny, N.D., Adams, P.A. & Stange, L.A. (1997) Species catalog of the Neuroptera, Megaloptera, and Raphidioptera of America North of Mexico. *Proceedings of the California Academy of Sciences* 50, 39–114.

Pinese, B. & Brown, J.D. (1986) The avocado leafroller: a pest of increasing significance. *Queensland Agricultural Journal* 112, 289–292.

Polizu, S. (1932) Contribution to the biology of the green apple aphid, *A. pomi*. *Buletinul Muzeului National de Istorie Naturala din Chisinău* 4, 39–44. (in Romanian)

Popov, A.K. (1991) Baum- und strauchbewohnende Neuropteren in Bulgarien. *Acta Zoologica Bulgarica* 41, 26–36.

Principi, M.M. & Canard, M. (1974) Les Névropteres. Les organismes auxiliaires en verger de pommiers. *OILB/SROP* 3, 151–162.

Putman, W.L. (1932) Chrysopids as a factor in the natural control of the oriental fruit moth. *Canadian Entomologist* 64, 121–126.

Putman, W.L. (1963) Nectar of peach leaf glands as insect food. *Canadian Entomologist* 95, 108–109.

Putman, W.L. & Herne, D.H. (1966) The role of predators and other biotic agents in regulating the population density of phytophagous mites in Ontario peach orchards. *Canadian Entomologist* 98, 808–820.

Ramos, P., Campos, M. & Ramos, J.M. (1978) Osservazioni biologiche sui trattamenti contro la tignola dell'olivo (*Prays oleae* Bern., Lep. Plutellidae). *Bollettino del Laboratorio di Entomologia Agraria Portici* 35, 16–24.

Ramos, P., Campos, M. & Ramos, J.M. (1983) Present status of research on biological control of the olive moth in Spain. In *Entomophagous Insects and Biotechnologies against Olive Pests, Proceedings of the European Commission Experts' Meeting*, ed. Cavalloro, R. & Piavaux, A., pp. 127–135. A.A. Balkema, Rotterdam.

Raspi, A. (1982) Considerazioni preliminari sulla cattura di entomofauna utile mediante l'impiego di trappole chemiocromotropiche nell'oliveto. *Frustula Entomologica* 5, 103–109.

Raspi, A. & Malfatti, P. (1985) The use of yellow chromotropic traps for monitoring *Dacus oleae* (Gmel.) adults. In *Integrated Pest Control in Olive Groves, Proceedings of the CEC/FAO/IOBC International Joint Meeting*, Pisa, ed. Cavalloro, R. & Crovetti, A., pp. 428–440. Balkema, Rotterdam.

Ravensberg, W.J. (1981) The natural enemies of the woody apple aphid, *Eriosoma lanigerum* (Hausm.) (Homoptera: Aphididae), and their susceptibility to difenbenzuron. *Medelingen van de Faculteit Landbouwwetenschappen Rijksuniversiteit Gent* 46, 437–441.

Remaudière, G. & Leclant, F. (1971) Le complexe des ennemis naturels des aphides du pêcher dans la Moyenne Vallée du Rhone. *Entomophaga* 16, 255–267.

Remund, U., Niggli, U. & Boller, E.F. (1989) Faunistische und botanische Erhebungen in einem Rebberg der Ostschweiz. Einfluss der Unterwuchsbewirtschaftung auf das Ökosystem Rebberg. *Landwirtschaft Schweiz* 2, 393–408.

Remund, U., Gut, D. & Boller, E. (1992) Influence of flora and management of permanent green cover on the arthropod fauna – preliminary conclusions of a 4-year investigation in 21 vineyards in Eastern Switzerland. *IOBC/WPRS Bulletin* 15, 107.

Ressl, F. (1971) Untersuchungen über die Chrysopiden des Bezirkes Scheibbs (Niederösterreich). Beitrag zur Kenntnis der Ökologie, Phänologie und Verbreitung der Chrysopiden Mitteleuropas. *Beiträge zur Entomologie* 21, 597–607.

Ressl, F. (1974) Untersuchungen über die Hemerobiiden (Neuroptera, Planipennia) des Bezirkes Scheibbs (NOE). Ein Beitrag zur Kenntnis der Verbreitung, Ökologie und Phänologie der Hemerobiiden Mitteleuropas. *Entomologische Gesellschaft Basel* 24, 10–28.

Rield, H. (1981) *Importation of natural enemies for control of pear psylla*, Psylla pyricola *Foerster in the Pacific Northwest and California*. Progress Report for Cooperative Agreement No. 58-9AH2-0-510 between USDA/SEA Agricultural Research and the University of California.

Ripa, S.R. & Rojas, P.S. (1990) Manejo y control biológico del chanchito blanco de la vid. *Revista Frutícola* 11, 82–87.

Ros, J.P., Moner, P., Roig, V., Castillo, E. & Lorite, P. (1988) Efficiency of hydrolyzed protein in baited sprays against *Ceratitis capitata* Wied. *Boletín de Sanidad Vegetal Plagas* 14, 5–9.

Rumpf, S. & Penman, D. (1993) Effects of the insect growth regulator fenoxycarb on two lacewing species in the laboratory and field. In *Proceedings of the 46th New Zealand Plant Protection Conference*, pp. 97–101. New Zealand Plant Protection Society, Rotorua, New Zealand.

Santas, L.A. (1984) On some Chrysopidae of Greece. In *Progress in World's Neuropterology, Proceedings of the 1st International Symposium on Neuropterology*, ed. Gepp, J., Aspöck, H. & Hölzel, H., pp. 167–172. Thalerhof, Graz.

Santas, L.A. (1987) The predators' complex of pear-feeding psyllids in unsprayed wild pear trees in Greece. *Entomophaga* 32, 291–297.

Satpute, P.P., Taley, Y.M., Nimbalkar, S.A. & Kadu, N.R. (1986) Natural enemies of citrus black fly, *Aleurocanthus woglumi* Ashby, in Nagpur region. *Bulletin of Entomology New Delhi* 27, 128–131.

Schirra, K.J. (1990) Effects of applications through several years without using insecticides on beneficial arthropods in viticulture. *IOBC/WPRS Bulletin* 13, 282–285.

Schruft, G., Wegner, G., Müller, R.-D. & Sampels, J. (1983) Das Auftreten von Florfliegen (Chrysopidae) und anderen Netzflüglern (Neuroptera) in Rebanlagen. *Weinwissenschaft* 38, 186–194.

Sechser, B., Reber, B. & Wesiak, H. (1994) Selectivity of diofenolan (CGA 59 205) and its potential for integrated scale control. In *Pest and Diseases, Proceedings of the Brighton Crop Protection Conference*, pp. 1193–1198. BCPC Publications.

Semyanov, V.P. (1973) Entomophages of *Psylla mali* and increase of their role. *Zashchita Rastenii* 5, 19–20. (in Russian)

Sengonca, Ç. (1979) Beitrag zur Neuropterenfauna der Türkei. *Nachrichtenblatt der bayerischen Entomologen* 28, 10–15.

Sengonca, Ç. (1981) Die Neuropteren Anatoliens I. Chrysopidae. *Mitteilungen der münchener entomologischen Gesellschaft* 71, 121–137.

Shalamberidze, N. (1980) The pear psyllid. *Zashchita Rastenii* 7, 27. (in Russian)

Shi, W.C., Li, J.R. & Liu, X. (1993) A study of the diversity of insect communities using gray relational grade analysis in some apple orchards in Maoxian County. *Journal of Southwest Agricultural University* 15, 137–143. (in Chinese)

Skånland, H.T. (1981) Studies on the arthropod fauna of Norwegian apple orchard. *Fauna Norvegica* Series B 28, 25–34.

Sluss, R.R. & Hagen, K.S. (1966) Factors influencing the dynamics of walnut aphid populations in northern California. In *Ecology of Aphidophagous Insects, Symposia ČSAV*, ed. Hodek. I., pp. 243–248. Academia, Prague.

Smirnoff, W. (1957) La cochenille du palmier dattier (*Parlatoria blanchardi* Targ.) en Afrique du Nord. Comportement, importance économique, prédateurs et lutte biologique. *Entomophaga* 2, 1–98.

Smith, M.W., Eikenbary, R.D., Arnold, D.C., Landgraf, B.S., Taylor, G.G., Barlow G.E., Carroll, B.L., Cheary, B.S., Rice, N.R. & Knight, R. (1994) Screening cool-season legume cover crops for pecan orchards. *American Journal of Alternative Agriculture* 9, 127–135.

Smith, M.W., Arnold, D.C., Eikenbary, R.D., Rice, N.R., Shiferaw, A., Cheary, B.S. & Carroll, B.L. (1996) Influence of ground cover on beneficial arthropods in pecan. *Biological Control* 6, 164–176.

Souza, B., Ciociola, A.I. & Matioli, J.C. (1989) Biologia comparativa de *Nusalala uruguaya* (Navás, 1923) (Neuroptera; Hemerobiidae) alimentada com diferentes especies de afideos. II. Fases de pre-pupa, pupa e adulta. *Anais da Sociedade Entomologica do Brasil* 18 (Supplement), 43–51.

Souza, B., Matioli, J.C. & Ciociola, A.I. (1990) Biologia comparativa de *Nusalala uruguaya* (Navás, 1923) (Neuroptera; Hemerobiidae) alimentada com diferentes especies de afideos. I. Fase de larva. *Anais da Escola Superior de Agricultura 'Luiz de Queiroz'* 47, 283–300.

Stechmann, D.-H. & Völkl, W. (1990) A preliminary survey of aphidophagous insects of Tonga, with regards to the biological control of the banana aphid. *Journal of Applied Entomology* 110, 408–415.

Steiner, H., Immendoerfer, G. & Bosch, J. (1970) The arthropods occurring on apple-trees throughout the year and possibilities for their assessment. *EPPO Publications Series A* 57, 131–146.

Stockel J.P., Lecharpentier P., Fos, A., Delbac, L., Witzgall, P. & Arn, H. (1997) Effects de la confusion sexuelle contre l'eudemis *Lobesia botrana* sur les populations d'autres ravageurs et d'auxiliaires dans le vignoble Bordelais. *Bulletin OILB/SROP* 20, 89–94

Swirski, E., Izhar, Y., Wysoki, M., Gurevitz, E. &

Greenberg, S. (1980) Integrated control of the long-tailed mealybug, *Pseudococcus longispinus* (Hom.: Pseudococcidae), in avocado plantations in Israel. *Entomophaga* 25, 415–426.

Swirski, E., Blumberg, D., Wysoki, M. & Izhar, Y. (1985) Data on phenology and biological control of the Japanese bayberry whitefly, *Parabemisia myricae*, in Israel. *Phytoparasitica* 13, 73.

Szabó, S. & Szentkirályi, F. (1981) Communities of Chrysopidae and Hemerobiidae (Neuroptera) in some apple orchards. *Acta Phytopathologica Academiae Scientiarum Hungaricae* 16, 157–169.

Szentkirályi, F. (1992) Brown lacewing (Neuropteroidea: Hemerobiidae) assemblages in various types of apple orchards. *Acta Phytopathologica et Entomologica Hungarica* 27, 601–604.

Szentkirályi, F. & Kozár, F. (1991) How many species are there in apple insect communities?: testing the resource diversity and intermediate disturbance hypotheses. *Ecological Entomology* 16, 491–503.

Talitskaya, N.V. (1980) Chrysopids and hemerobiids (Chrysopidae, Hemerobiidae, Neuroptera) – aphidophaga of fruit cultures. *Entomofagi Vreditelei Rasteniy, Kishinev, Stiinca* 56–63. (in Russian)

Tamaki, G. & Halfhill, J.E. (1968) Bands on peach trees as shelters for predators of the green peach aphid. *Journal of Economic Entomology* 61, 707–711.

Tedders, W.L., Reilly, C.C., Wood, B.W., Morrison, R.K. & Lofgren, C.S. (1990) Behavior of *Solenopsis invicta* (Hymenoptera: Formicidae) in pecan orchards. *Environmental Entomology* 19, 44–53.

Thakur, J.N., Pawar, A.D. & Rawat, U.S. (1988) Observations on the correlation between population density of apple woolly aphid and its natural enemies and their effectiveness in Kullu Valley (H.P.). *Plant Protection Bulletin, India* 40, 13–15.

Thakur, J.N., Pawar, A.D. & Rawat, U.S. (1992) Apple woolly aphid, *Eriosoma lanigerum* Hausmann (Hemiptera: Aphididae) and post-release impact of its natural enemies in Kullu Valley (H.P.). *Plant Protection Bulletin Faridabad* 44, 18–20.

Tolstova, Y.S. & Atanov, N.M. (1982) Effect of the chemical means of plant protection on the fauna of arthropods in the orchards. I. Long-term application of the pesticides to the agrobiocenosis. *Entomologicheskoe Obozrenie* 61, 441–453.

Triggani, O. (1973) Contributo alla conoscenza dell'azione svolta dai nemici naturali degli afidi del mandorlo (*Amygdalus communis*) in agro di Bari. *Istituto di Entomologia Agraria dell'Università di Bari, Entomologica* 9, 119–135.

Tzeng, C.C. & Kao, S.S. (1996) Evaluation on the safety of pesticides to green lacewing, *Mallada basalis* larvae.

Plant Protection Bulletin Taipei 38, 203–213. (in Chinese)

Uygun, N., Ohnesorge, B. & Ulusoy, R. (1990) Two species of whiteflies on citrus in Eastern Mediterranean: *Parabemisia myricae* (Kuwana) and *Dialeurodes citri* (Ashmead). *Journal of Applied Entomology* 110, 471–482.

Uygun, N., Ulusoy, M.R., Karaca, Y. & Kersting, U. (1997): Approaches to biological control of *Dialeurodes citri* (Ashmead) in Turkey. *IOBC/WPRS Bulletin*, 20, 52–62.

Vacante, V. (1997) *Integrated Control in Citrus Fruit Crops.* IOBC/WPRS Bulletin, 20(7).

Van den Berg, M.A. & De Beer, M.S. (1996) Natural enemies of the spiny blackfly, *Aleurocanthus spiniferus* (Hem.: Aleyrodidae), in Mpumalanga South Africa. *Proceedings of the International Society of Citriculture* 1, 667–669.

Van den Berg, M.A. & Maritz, M. (1995) Seasonal occurrence and age specific mortality of the yellow pecan aphid, *Monelliopsis pecanis* (Hemiptera: Aphididae), in an orchard in the eastern Transvaal Lowveld. *Journal of the South African Society for Horticultural Science* 5, 107–109.

Van den Berg, M.A. & Maritz, M. (1996) Predators of the yellow pecan aphid, *Monelliopsis pecanis* (Homoptera: Aphididae), in the eastern parts of South Africa. *African Plant Protection* 2, 123–125.

Van den Berg, M.A., Deacon, V.E., Fourie, C.J. & Anderson, S.H. (1987) Predators of the citrus psylla, *Trioza erytreae* (Hemiptera: Triozidae), in the Lowveld and Rustenburg areas of Transvaal. *Phytophylactica* 19, 285–289.

Viggiani, G. (1982) Natural enemies of the filbert aphids in Italy. In *Aphid Antagonists, Proceedings of a Meeting of the European Commission Experts' Group*, A.A. Balkema, ed. Cavalloro, R., pp. 109–113. Rotterdam.

Villeneuve, F. & Trottin-Caudal, Y. (1997) Integrated control of vegetable crops in France: achievements and perspectives. *Infos Paris* 135, 40–44.

Viollier, B. & Fauvel, G. (1984) Comparaison de la faune vivant sur 2 espèces de poiriers, *Pyrus amygdaliformis* Vill. et *P. communis* L., en garrigue et dans un verger abandonné de la région de Montpellier. *Agronomie* 4, 11–18.

Ward, L.K. (1969) A survey of the arthropod fauna of plum trees at East Malling in 1966. *Bulletin of Entomological Research* 58, 581–599.

Wei, C.S., Huang, B.Z. & Guo, C.H. (1986) Studies on *Chrysopa boninensis* Okamoto in Guangzhou. *Acta Entomologica Sinica* 29, 174–180. (in Chinese)

Westigard, P.H., Gentner, L.G. & Berry, D.W. (1968) Present status of biological control of the pear psylla in southern Oregon. *Journal of Economic Entomology* 61, 740–743.

Whalon, M.E. & Elsner, E.A. (1982) Impact of insecticides on *Illinoia pepperi* and its predators. *Journal of Economic Entomology* 75, 356–358.

Wiackowski, S.K. & Wiackowska, I. (1968) Investigations on the entomofauna accompanying aphids occurring on fruit trees and bushes. *Polskie Pismo Entomologiczne* 38, 255–283.

Wilde, W.H.A. (1962) Bionomics of the pear psylla, *Psylla pyricola* Foerster in pear orchards of the Kootenay Valley of British Columbia, 1960. *Canadian Entomologist* 94, 845–849.

Woglum, R.S., LaFollette, J.R., Landon, W.E. & Lewis, H.C. (1947) The effect of field-applied insecticides on beneficial insects of citrus in California. *Journal of Economic Entomology* 40, 818–820.

Wu, T.K. (1992) Feasibility of controlling citrus red spider mite, *Panonychus citri* (Acarina: Tetranychidae) by the green lacewing, *Mallada basalis* (Neuroptera: Chrysopidae). *Chinese Journal of Entomology* 12, 81–89. (in Chinese)

Wu, T.K. (1995) Integrated control of *Phyllocnistis citrella*, *Panonychus citri*, and *Phyllocoptruta oleivora* with periodic releases of *Mallada basalis* and pesticide applications. *Chinese Journal of Entomology* 15, 113–123. (in Chinese)

Wyss, E. (1995) The effects of weed strips on aphids and aphidophagous predators in an apple orchard. *Entomologia Experimentalis et Applicata* 75, 43–49.

Xu, G.J. & Zhong, S.T. (1988) A preliminary study on the biology of *Ricania sublimbata* Jacobi and its control. *Insect Knowledge* 25, 93–95.

Yan, Y. & Duan, J. (1986) Some observations on biological control of apple tree spider mites 1. Preliminary report on the conservation and augmentation of natural enemies by cover cropping in apple orchards. *Acta Agriculturae Boreali-Sinica* 1, 1–7. (in Chinese)

Yan, Y. & Duan, J. (1988) The effect of cover cropping in apple orchards on the predator community on the apple tree. *Acta Phytophylacica Sinica* 15, 23–27. (in Chinese)

Yan, Y., Yu, Y., Du, X. & Zhao, B. (1997) Conservation and augmentation of natural enemies in pest management of Chinese apple orchards. *Agriculture Ecosystems and Environment* 62, 253–260.

Zayantskauskas, P.A., Yonaytis, V.P. & Yakimavitsus, A.B. (1982) *Complex of Beneficial Organisms in the Biocenoses of Orchards in Lithuanian SSR*. Institute of Zoology and Parasitology of Academy of Sciences, Lithuanian SSR, Vilnius. (in Russian)

Zelený, J. (1963) Hemerobiidae (Neuroptera) from Czechoslovakia. *Acta Societatis Entomologicae Czechoslovaka* 60, 55–67.

Zelený, J. & Talitsky, V.I. (1966) To the knowledge of snakeflies (Raphidioptera), lacewings (Neuroptera) and scorpionflies (Mecoptera) of Moldavian SSR. *Trudy Moldavian Nauch-Issledovatelskogo Instituta Sadovodstva Vinogradstva i Vinodelia Entomologia* 13, 85–91. (in Russian)

Zhou, C.A., Zou, J.J., Peng, J.C., Ouyang, Z.Y., Hu, L.C., Yang, Z.L. & Wang, X.B. (1991) Predation of major natural enemies on *Panonychus citri* and its comprehensive evaluation in citrus orchards in Hunan, China. *Acta Phytophylactica Sinica* 18, 225–229. (in Chinese)

CHAPTER 10
Lacewings in vegetables, forests, and other crops

F. Szentkirályi

10.1 VEGETABLES

The vegetable crops discussed here are grown mainly in smaller plots, in house gardens, and greenhouses although some of them are also field crops. Many pests live on vegetables, but knowledge of associated lacewings as natural enemies of these is sparse.

Various cultivars of bean (*Phaseolus vulgaris*) are produced world-wide. Both larvae and adults of the following lacewing species were collected from aphid (*Aphis fabae* complex) infested bean plants in a garden in Hungary: *Chrysoperla carnea*, *Chrysopa formosa*, *Micromus angulatus*, and *M. variegatus* (F. Szentkirályi, unpublished data). *Chrysoperla carnea* was a predator of the mite *Tetranychus urticae* on beans in Turkey (Aydemir & Toros, 1990). The brown lacewing *M. timidus* occurs in colonies of *A. craccivora* infesting bean plantations in Western Samoa (Stechmann & Semisi, 1984). Pantaleoni & Tisselli (1985) sampled potential biological control agents by sweep-netting on stands of *Vicia faba* in Italy. Five chrysopid species were recorded with percentage of species incidence as follows: *Dichochrysa prasina* (44%), *Chrysopa formosa* (27.7%), *Chrysoperla carnea* (23.4%), *D. clathrata* (2.8%), and *Chrysopa pallens* (2.1%). The presence of arboreal *Dichochrysa* spp. on beans reflects that surrounding vegetation acts as a lacewing reservoir, while the presence of *Chrysoperla carnea* and *Chrysopa formosa* may be related to lacewing species associated with phytophagous insects and mites of bean plantations.

Cabbage (*Brassica oleracea*) also has numerous aphid pests. *Chrysopa perla* was one of the most important natural enemies of *Brevicoryne brassicae* in cabbage fields in the Moscow area (Russia) (Ter-Simonjan *et al.*, 1982). Somen-Singh *et al.* (1994) demonstrated that the aphidophagous *M. timidus* was among the natural enemies of the key pest *Myzus persicae* on cabbage in India. Weires & Chiang (1973), studying the possibilities of integrated control against major cabbage pests in Minnesota, found *Micromus posticus* to be the most

abundant brown lacewing, while a second species, *M. subanticus*, appeared only occasionally.

Growing of cultivated amaranths (*Amaranthus* spp.) originating from South America has spread not only to tropical–subtropical areas but also to temperate areas of some European countries. In Hungary, the following lacewings were recorded from the plant canopy in a period of aphid infestation from June to July: *Chrysoperla carnea*, *Chrysopa formosa*, *C. perla*, and *M. angulatus*. Two further chrysopid species, *C. phyllochroma* and *C. abbreviata*, were frequently captured from an experimental amaranth field planted on sandy soil (F. Szentkirályi, unpublished data).

Artichoke (*Cynara scolymus*) (Compositae) occurs generally in warmer areas of the world. Neuenschwander & Hagen (1980) studied the predatory role of *Hemerobius pacificus* in an insecticide-free artichoke field in the USA, where it appeared to be valuable for control of early seasonal aphid infestations when the prey population was still low.

Root crops are the most important food plants in the tropics. Hölzel *et al.* (1994) collected the chrysopid species *Ceratochrysa antica* from plantations of manioc or cassava (*Manihot esculenta*) in Nigeria. Barnard & Brooks (1984) reported that larvae of *C. antica* fed on the cassava mealybug, *Phenacoccus manihoti*, an introduced serious pest in southeast Nigeria. Boussienguet (1986) recorded another chrysopid, *Mallada desjardinsi*, which consumed this mealybug on manioc in Gabon. Taro (*Colocasia esculenta*) is another typical root crop produced for its tuber. Surveying its aphid pests Stechmann & Völkl (1990) recorded *Pentalonia nigronervosa* from the Tonga Islands, and Stechmann & Semisi (1984) reported *Aphis gossypii* from Western Samoa. Larvae of the brown lacewing *M. timidus* were frequently recorded in colonies of both aphid species. *Dialeurodes cardamomi*, a whitefly, is a sporadic minor pest on its host plant, the aromatic cardamom (*Elettaria cardamomum*) in South India. Selvakumaran *et al.*

(1996) monitored 17 cardamom plantations, finding *M. desjardinsi* for the first time among natural enemies of this whitefly.

Data on lacewings on vegetable crop plants cover releases for biological control in field or glasshouse experiments. In recent years *M. angulatus* has often been applied as a biological control agent to suppress aphids living on vegetables. It can control aphids under greenhouse conditions (Yarkulov, 1986; Stelzl *et al.*, 1992; Tverdyukov *et al.*, 1993). Examples of protection of greenhouse vegetables by applying lacewings against aphids are described below.

Lacewings were used to control two aphid species living on cucumber (*Cucumis sativus*). Against *A. gossypii*, Lezhneva & Anisimov (1995) used *M. angulatus*, while Ter-Simonjan *et al.* (1982) and Ushchekov (1989) released *Chrysoperla carnea* and *Chrysopa perla* respectively. Larvae of *M. angulatus* combined with *Verticillum lecanii* fungus treatment were also effectively applied by Potemkina & Kovalenko (1990) against *A. frangulae* on cucumber. Under glass the green peach aphid, *Myzus persicae*, is a major pest of green pepper. Lezhneva & Anisimov (1995) tried to decrease populations by releasing *Micromus angulatus*, and Tulisalo *et al.* (1977) applied *Chrysoperla carnea* against the same pest. By releasing *C. carnea*, infestation by *Aulacortum solani* could be kept below the economic threshold level in field tomatoes (Ter-Simonjan *et al.*, 1982). Lezhneva & Anisimov (1995) also used *M. angulatus* against *Aphis nasturtii* causing damage to greenhouse tomato plots. Due to the susceptibility of *M. angulatus* to high temperatures in the greenhouse, its predatory effectiveness was limited.

Aubergine (*Solanum melongena*) hosts a very common, harmful virus vector, the green peach aphid, *Myzus persicae*. Hassan (1978) successfully used *Chrysoperla carnea* against this aphid in a greenhouse. The optimal result could be reached by using a ratio of predator to prey of 1:5 releasing them four to seven times.

10.2 BEVERAGE AND SUGAR CROP PLANTS

10.2.1 Cocoa and coffee

Cocoa (*Theobroma cacao*), an endemic plant in South America, is widely produced in countries with a warm, humid, tropical climate in equatorial Africa. Some records of lacewings on cocoa are derived from this region. Hölzel & Monserrat (1992) detected seven

chrysopid species on cocoa in Equatorial Guinea including members of the genera *Ankylopteryx*, *Parankylopteryx*, *Ceratochrysa* and *Chrysoperla* (Table 10.1; for notes on the methodology of compiling the tables in this chapter, see p. 291). A brown lacewing, *Hemerobius tolimensis*, was collected from cocoa in Colombia, South America (Monserrat, 1996). It seems that among hemerobiids more species of *Notiobiella* preferred the cocoa plantations, as in Cameroon (Monserrat, 1984).

Coffee (*Coffea* spp.), an endemic bush in Africa and the Arabian Peninsula, is grown widely under tropical conditions. Although it has numerous insect pests, such as pseudococcid or diaspidid scales, their lacewing natural enemies have rarely been reported. Only three records of lacewing species are shown in Table 10.1, two from Africa and one from India. *Chrysoperla congrua* was captured on *Coffea liberica* growing in Equatorial Guinea (Hölzel & Monserrat, 1992). *Micromus timidus* was documented from *Coffea arabica* in Zaïre (Monserrat, 1992). A *Dichochrysa* species from the *Coffea* canopy infested by the pseudococcid *Ferrisia virgata* is mentioned from India (Balakrishnan *et al.*, 1991).

10.2.2 Tea

Commercial varieties of tea (*Camellia sinensis*) are grown as a perennial monoculture in diverse agroecological conditions between latitudes 41° N and 16° S. Over the long life period (40–50 years) of some plantations, the tea bushes are consumed by aphids, scales, and mites, which can be prey food for lacewings. According to recent estimations 1034 arthropod species infest tea plants world-wide (Muraleedharan & Chen, 1997). These potential prey characteristically colonise different age-groups of tea plants. Banerjee (1983) found that sap-feeding mites and aphids accumulated on younger tea plants (aged below 15 years), while mature plants (aged between 22 and 36 years) were colonised by scale insects (Coccidae, Diaspididae) and mites, possibly related to change in nitrogen content during the life span of tea. However, other factors, in particular changes in the architectural and morphological complexity of tea plants, should also be taken into consideration.

Published records of lacewings on tea growing in China, India and Florida are summarised in Table 10.2. Six species of *Chrysoperla*, *Ceraeochrysa*, and *Chrysopa* were found. Only one hemerobiid (*Micromus timidus*)

Table 10.1. *Chrysopid and hemerobiid species found on cocoa and coffee trees*

	Theobroma cacao		Coffea spp.			
	1	2	3	4	5	FR
Chrysopidae						
Chrysoperla congrua (Walker)		+	+			2
Ceratochrysa antica (Walker)		+				1
Dichochrysa sp.		+			+	2
Ankylopteryx tristicta Navás		+				1
Ankylopteryx splendidissima Gerstaecker		+				1
Parankylopteryx polysticta (Navás)		+				1
Parankylopteryx tenuis Hölzel, Stelzl & Ohm		+				1
Hemerobiidae						
Hemerobius tolimensis Banks	+					1
Micromus timidus Hagen				+		1
S_C		7	1		1	
S_H	1			1		

Abbreviations: S_C, S_H, species number of chrysopids and hemerobiids respectively; +, species present; FR, frequency of species occurrence in literature surveyed.
Sources: 1, South America, Columbia (Monserrat, 1996); 2, Africa, Equatorial Guinea (Hölzel & Monserrat, 1992); 3, Africa, Equatorial Guinea (Hölzel & Monserrat, 1992); 4, Africa, Zaïre (Monserrat, 1992); 5, India (Balakrishnan *et al.*, 1991).

was mentioned from India. Further research may increase the number of lacewing species in tea plantations.

As well as records of lacewings in tea plantations, there are some data on their potential role in biological control. The diaspídid *Fiorinia theae*, a serious tea pest, is widely distributed in the warmer parts of the world, and occurs also on related plants (camellias and hollies). Munir & Sailer (1985) sampled its natural enemies on *Camellia japonica* in Florida. *Chrysopa* spp. preying on nymphs had no significant role in the control of tea scale because of their extremely low population level and irregular appearance. However some lacewings can be successful in biological control of certain tea pests. In Southeast Asia tetranychid mites are major tea pests. Chrysopids such as *Mallada basalis* and *M. desjardinsi* have been released for biological control in Taiwan (Lo *et al.*, 1990; Cheng & Chen, 1996). Releases were conducted with *Chrysoperla nipponensis* against the pink mite *Acaphylla theae* because this lacewing was dominant among its natural enemies in China (Zhao & Hou, 1993).

Micromus timidus was an important predator of tea insect pests in India. The leaf-folding caterpillar of the tortricid moth *Cydia leucostoma* is one of the most harmful tea pests; *M. timidus* is a major predator of this pest (Selvasundaram & Muraleedharan, 1987). The tea aphid, *Toxoptera aurantii*, is common on tea shoots. Colonies were attacked by *M. timidus* in all plantations surveyed in South India (Radhakrishnan & Muraleedharan, 1989). Seasonal fluctuations of the hemerobiid species closely followed the population changes of *T. aurantii*. In the course of larval development *M. timidus* consumed 203 aphids on average. Chrysopids have also been recorded as natural enemies of *T. aurantii*; Dai-Xuan (1995) recorded *Chrysopa pallens* attacking *T. aurantii* in China.

10.2.3 **Sugarcane**
Sugarcane (*Saccharum officinarum*) originated from tropical Asia and has long been grown world-wide. It is hosts numerous insect pests (for instance aphids, scales, leafhoppers, and mites) that are potential prey for lacewings. The limited data available on green and brown

Table 10.2. *Chrysopid and hemerobiid species found in tea (*Camellia sinensis*) plantations*

	1	2	3	4	5	FR
Chrysopidae						
Chrysoperla carnea s.l. (Fitch)			+			1
Chrysoperla rufilabris (Burmeister)	+					1
Chrysoperla harrisii (Fitch)	+					1
Chrysopa pallens (Rambur)		+				1
Ceraeochrysa claveri (Navás)	+					1
Ceraeochrysa cincta (Schneider)	+					1
Hemerobiidae						
Micromus timidus Hagen				+	+	2
S_C	4	1	1			
S_H				1	1	

Abbreviations: S_C, S_H, species number of chrysopids and hemerobiids respectively; +, species present; FR, frequency of species occurrence in literature surveyed.

Sources: 1, USA, Florida (Collins & Whitcomb, 1974); 2, China (Dai-Xuan, 1995); 3, China (Zhao & Hou, 1993); 4, India (Selvasundaram & Muraleedharan, 1987); 5, India (Radhakrishnan & Muraleedharan, 1989).

lacewings associated with sugarcane can be seen in Table 10.3. The major pest is the sugarcane woolly aphid, *Ceratovacuna lanigera*. Its lacewing predators include *Italochrysa aequalis*, *Dichochrysa alcestes*, *Chrysoperla furcifera*, and *Micromus timidus* (Azuma & Oshiro, 1971; Lim & Pan, 1979; Arakaki, 1992*a*, *b*). Arakaki found eggs and larvae of *M. timidus* in colonies of *Ceratovacuna lanigera* among 17 species of aphidophaga associated with sugarcane on Okinawa Island. It seems that this lacewing is a specialised predator of Cerataphidini, including this pest. Analysing seasonality of *C. lanigera*, Arakaki (1992*b*) reported two population peaks: a larger peak in spring and a smaller one in autumn. Seasonal fluctuation was mirrored by changing number of larvae of *M. timidus*. *Mallada innotata* was recorded among the principal natural enemies of another sugarcane pest, the pseudococcid *Saccharicoccus sacchari* (De Barro, 1990).

Research on the possible role played by lacewings in biological control of sugarcane pests has been initiated. The feeding potential of *Brinckochrysa scelestes* on eggs of the sugarcane leafhopper, *Pyrilla perpusilla*, was studied by Mishra *et al.* (1996). Single larvae consumed a mean of 170 eggs during development. *Mallada*

basalis was released in field trials against mite pests on sugarcane plantations in Taiwan (Cheng & Chen, 1996).

10.3 ORNAMENTAL PLANTS

Ornamental plants, which can be woody (trees and shrubs) or herbaceous, have several functions. They are planted in urban areas for amenity purposes (e.g. in parks), or for landscaping. In the latter case, they are used as windbreaks, hedgerows, or lines of trees along roads. Landscape plants offer food (nectar, pollen, alternative prey), shelter, egg-laying surfaces, and overwintering microhabitats for predatory insects, thus enhancing the natural biological control of agricultural pests. Ornamental plants are host to numerous phytophagous arthropods, which can cause considerable damage. Lacewings, as predatory insects, are associated with ornamental plants in both ways.

Table 10.4 contains available data on lacewings collected on woody ornamental plants. In urban areas, within parks, forest tree species are often planted as ornamentals, on which woodland-inhabiting lacewings can frequently be found, for example *Hemerobius micans*

Table 10.3. *Chrysopid and hemerobiid species found in sugarcane (*Saccharum officinarum*) stands*

	1	2	3	4	5	6	FR
Chrysopidae							
Chrysoperla furcifera (Okamoto)					+		1
Italochrysa sp. near aequalis (Walker)						+	1
Brinckochrysa scelestes (Banks)			+				1
Mallada innotata (Walker)		+					1
Dichochrysa alcestes (Banks)				+			1
Chrysopa sp.						+	1
Hemerobiidae							
Micromus timidus Hagen	+			+			2
Micromus sp.						+	1
S_C		1	1	1	1	2	
S_H	1			1		1	

Abbreviations: S_C, S_H, species number of chrysopids and hemerobiids respectively; +, species present; FR, frequency of species occurrence in literature surveyed.

Sources: 1, Taiwan (Cheng *et al.*, 1992); 2, Australia, Queensland (De Barro, 1990); 3, India, Orissa (Mishra *et al.*, 1996); 4, Japan, Okinawa (Arakaki, 1992*a, b*); 5, Japan, Okinawa (Azuma & Oshiro, 1971); 6, Malaysia (Lim & Pan, 1979).

on *Quercus robur* (Lammes, 1993), and *Dichochrysa prasina* on the conifer *Tsuga canadensis*. This latter lacewing preys on hemlock woolly adelgid, *Adelges tsugae*, in Japan (McLure, 1995). On ornamental plants, lacewings can exploit psyllids, coccids, or aphids. For instance in California, the larvae of *H. pacificus* fed on a psyllid, *Acizzia uncatoides*, living on an *Acacia* landscape plant. The abundance of this hemerobiid increased significantly in response to increasing psyllid densities (Dreistadt & Hagen, 1994). Miyanoshita & Kawai (1992) studied predators of *Ceroplastes japonicus*, a coccid damaging the ornamental bush *Euonymus japonicus*, using exclusion experiments. The main mortality factor in summer was the larva of the *M. desjardinsi*.

Bamboo species (*Phyllostachys* spp., *Bambusa* sp.) are used as ornamental plants all over the world. According to Table 10.4, a diverse lacewing assemblage is associated with bamboo. On Mauritius, Hölzel & Ohm (1991) collected four chrysopid species, while in Georgia (former USSR), Agekjan (1973) recorded three chrysopid and three hemerobiid species among the predators of bamboo aphids *Takecallis taiwanus* and *Melanaphis bambusae* in the region of the Black Sea coast. In Japan, *Micromus numerosus* was a natural enemy of *Pseudoregma bambucicola* (Morimoto & Shibao, 1993). On the ornamental shrub *Physocarpus opulifolius* in North America, Wheeler & Hoebeke (1985) recorded a diverse insect assemblage, including larval and adult *H. humulinus* associated with *Aphis neilliae* and *Utamphorophora humboldti*.

Ornamental plants supply alternative food sources for lacewings. Pecan aphids have lacewing predators that overwinter on bark, but the available prey is insufficient to sustain them in the spring in the USA. Crape myrtle, *Lagerstroemia indica*, an exotic shrub from Asia planted in the USA, hosts the crapemyrtle aphid, *Sarucallis kahawaluokalani* (Mizell & Schiffhauer, 1987). Predatory lacewings on the pecan aphid complex used crapemyrtle aphids as alternative food when other prey were at low levels. Adults and larvae of *Chrysoperla rufilabris* responded to peaks in the aphid population. Bugg (1987) monitored nectar-feeding insects visiting the soapbark tree, *Quillaja saponaria*. Among the flower-visiting entomophagous insects, a great number of lacewings, namely adults of

Table 10.4. *Chrysopid and hemerobiid species found on various ornamental plants*

	1	2	3	4	5	6	7	8	9	10	11	FR
Chrysopidae												
Chrysoperla carnea s.l. (Stephens)					+			+			+	3
Chrysoperla rufilabris (Burmeister)			+									1
Chrysoperla brevicollis (Rambur)							+					1
Chrysopa pallens (Rambur)								+				1
Mallada desjardinsi (Navás)							+			+		2
Dichochrysa prasina (Burmeister)		+										1
Cunctochrysa albolineata (Killington)								+				1
Apertochrysa eurydera (Navás)							+					1
Brinckochrysa decaryella (Navás)							+					1
Hemerobiidae												
Wesmaelius subnebulosus (Stephens)								+				1
Hemerobius humulinus L.				+				+				2
Hemerobius micans Olivier	+											1
Hemerobius stigma Stephens			+									1
Hemerobius pacificus Banks						+						1
Hemerobius ovalis Carpenter					+							1
Micromus numerosus Navás										+		1
Micromus variegatus (Fabricius)								+				1
Micromus posticus (Walker)			+									1
S_C		1	1		1		4	3		1	1	
S_H		1		2	1	1		3	1			

Abbreviations: S_C, S_H, species number of chrysopids and hemerobiids respectively; +, species present; FR, frequency of species occurrence in literature surveyed.

Sources: 1, Finland, *Quercus robur* and *Sorbus aucuparia* (Lammes, 1993); 2, Japan, *Tsuga canadensis* (McClure, 1995); 3, USA, *Lagerstroemia indica* (Mizell & Schiffhauer, 1987); 4, USA, *Physocarpus opulifolius* (Wheeler & Hoebeke, 1985); 5, USA, *Quillaja saponaria* (Bugg, 1987); 6, USA, *Acacia longifolia* (Dreistadt & Hagen, 1994); 7, Mauritius, *Phyllostachys bambusoides*, bamboo (Hölzel & Ohm, 1991); 8, Georgia, bamboo (Agekjan, 1973); 9, Japan, bamboo (Morimoto & Shibao, 1993); 10, Japan, *Euonymus japonicus* (Miyanoshita & Kawai, 1992); 11, Czech Republic, *Lupinus polyphyllus* (Stary & Havelka, 1991).

C. carnea and H. ovalis, fed on floral nectar. Weekly suction samples from the tree yielded a significantly higher number of individuals for both species during the flowering period.

There are many examples of biological control using lacewings against aphids infesting ornamentals. In the USA, Raupp et al. (1994) carried out an augmentative release of C. carnea and C. rufilabris against bean aphid, A. fabae, on hawthorn, Crataegus phaenopyrum, but there was no evidence of a reduction in the aphid population. Inundative release of Chrysoperla carnea was also unsuccessful in suppressing Illinoia liriodendri populations damaging the tulip tree, Liriodendron tulipifera. The main reason for failure was that the tree-visiting ant, Linepithima humile, removed 98% of the lacewing eggs placed on the trees (Dreistadt et al., 1986).

The principal pests of greenhouse flowering and ornamental crops are Aphididae, Aleyrodidae, Coccidae, Thripidae, and Agromyzidae. A frequent greenhouse ornamental, chrysanthemum, can be infested by many aphid species. Release experiments have been performed using *C. carnea* against *Brachycaudus helichrysi*, *Macrosiphoniella sanborni*, and *A. gossypii* (Del Bene *et al.*, 1993), and with *C. rufilabris* against *Myzus persicae* and *A. gossypii* (Hesselein *et al.*, 1993). *Chrysoperla rufilabris* is frequently used for biological control in closed production systems. It was evaluated in inundative releases for the control of *Heliothis virescens*, a noctuid pest on *Petunia* sp. (Davidson *et al.*, 1992), and for the aleyrodid *Bemisia tabaci* on *Hibiscus rosa-sinensis* (Breene *et al.*, 1992). *Chrysoperla carnea* was released by Heinz & Parrella (1990) for the suppression of two honeydew-excreting aphids, *A. gossypii* and *M. persicae* on greenhouse marigolds, *Tagetes erecta*. Successful control by *C. carnea* was achieved when releases were started immediately after the aphid infestation.

10.4 BROAD-LEAVED FORESTS

It is appropriate to discuss forests in a book that focuses on lacewings in the crop environment. Surveying the literature, it clear that the overwhelming majority of green and brown lacewings occurring in agricultural and horticultural habitats are also forest inhabitants. Forests, with their various habitat types (inner forest, clearings, forest edges) serve as reservoirs for lacewings that may have a role in control of pests of cultivated plants. In numerous cases, field crop colonisation of lacewings is enhanced by the presence of forests nearby. Most forests of both temperate and tropical zones are under sylvicultural management and these procedures have a significant effect on forest-inhabitant lacewing assemblages.

From the literature the overwhelming majority of investigations on forest lacewing fauna are from Europe. Few data on forest inhabiting lacewings exist for other areas, and those published are mainly simply collections from certain tree species. For this reason lacewing species and assemblages are presented only from studies in European countries. Tables 10.5–10.11 show chrysopid and hemerobiid species from the most frequently studied broad-leaved and coniferous tree species. Data from faunistic collections carried out in several places in a country, from a certain tree species, are separated from those related to assemblages of given forest stands. In cases where sufficient collections or

samplings were available, the dominance values of lacewing species are given. In other cases, only the presence of species is indicated.

10.4.1 Beech

European beech (*Fagus silvatica*) (Fagaceae) prefers a wet, Atlantic climate. Extensive forests occur, particularly in the plains of northern regions of central Europe, while on mountainous regions, on average 600–1200 m a.s.l., and in southern Europe 1000–1700 m a.s.l., the beech forms isolated vegetation zones. In south Europe *F. moesiaca* grows in the Balkans and *F. orientalis* can be found in the east (Asia Minor, Crimea, Caucasus). Species of the genus *Nothofagus* are indigenous in temperate rainforests of the southern hemisphere.

Table 10.5 summarises dominance distributions, frequency of occurrence, and species richness values of chrysopid and hemerobiid species and assemblages relating to the European beech. Columns 1–6 represent faunistic collections, and columns 7–13 show assemblages found in local *Fagetum* associations. Among green lacewings, a total of 17 species were recorded from *Fagus silvatica* and *F. moesiaca* (column 5) in all European countries. The most frequent and dominant chrysopid was the generalist *Chrysoperla carnea* (30%–98%). This was followed by *Dichochrysa flavifrons* (0.3%–39%), *D. prasina* (1%–14%), *Chrysa perla* (0.2%–10%), *D. ventralis* (0.4%–6%), and *C. pallens* (0.3%–3%). In one case the proportion of *Hypochrysa elegans* also exceeded 10% (0.2%–13%). In nearly half the cases, *Chrysotropia ciliata* and *Cunctochrysa albolineata* were also represented in low numbers. *Chrysoperla carnea* and *Dichochrysa* species are characteristic chrysopids of beech forests. *Hypochrysa elegans* prefers and associates specifically with *F. silvatica* (Séméria, 1984; Pantaleoni, 1990*b*; Monserrat & Marín, 1994). In Hungarian mountainous beech stands, *H. elegans* adults were collected with sweepnetting in great numbers from foliage of sprouts growing out from trunks (F. Szentkirályi, unpublished data), supporting the association mentioned above. The expected chrysopid assemblages of beech stands consist of 7–11 species in faunistic collections and 6–13 species in local samplings.

Among hemerobiids there are 21 species on the basis of faunistic collections and 22 species according to assemblage records in beech forests. From all the

Table 10.5. *Species composition and dominance distribution of chrysopid and hemerobiid assemblages in European beech (Fagus silvatica) forests*

	From records on individual faunistic collections							Local assemblage in *Fagetum* association						FR
	1	2	3	4	5	6	7	8	9	10	11	12	13	
Chrysopidae														
Chrysoperla carnea s. lat. (Stephens)	+	+	+	94.9	30.6	+	85.7	+	+	+	69.9	97.8	78.3–79.6	11
Chrysopa pallens (Rambur)		+	+		2.8	+	0.8		+		0.7		0–0.3	6
Chrysopa perla (L.)			+	0.2	2.8	+	0.1				7.8		0–10.2	7
Chrysopa phyllochroma Wesmael							4.7							1
Chrysopa viridana Schneider					2.8	+							0–1.9	2
Dichochrysa prasina (Burmeister)			+	2.9	13.9	+						1.1	1.0–3.5	6
Dichochrysa ventralis (Curtis)			+	0.4	5.6	+	2.6						1.6–3.1	6
Dichochrysa flavifrons (Brauer)			+	0.4	38.9	+		+			0.7	0.3	0–1.6	7
Nineta flava (Scopoli)				0.3	2.8								0–1.0	4
Nineta vittata (Wesmael)						+	0.1				4.6			3
Nineta pallida (Schneider)				0.2									0–1.3	2
Nineta guadarramensis (Pictet)				0.3								0.5		2
Peyerimhoffina gracilis (Schneider)		+					4.7				16.3		0–1.0	2
Chrysotropia ciliata (Wesmael)		+	+			+	1.2					0.2	+	5
Cunctochrysa albolineata (Killington)				0.2			0.2					0.2	0.8–3.1	5
Nothochrysa fulviceps (Stephens)				0.1			0.1					0.2		3
Hypochrysa elegans (Burmeister)				0.2		+							1.0–12.8	3
Hemerobiidae														
Wesmaelius subnebulosus (Stephens)	+			6.5		+	1.4				3.9	8.6	1.2–1.8	6
Wesmaelius nervosus (Fabricius)			0.8				16.5						0–0.5	4
Wesmaelius quadrifasciatus (Reuter)							0.7				1.3			2
Hemerobius humulinus L.	+	+	19.1	1.9	7.1	+	1.4			+	39.5	1.1	1.8–9.6	11
Hemerobius micans Olivier	A	A	65.6	86.6	74.8	+	25.9			+		88.2	62.8–81.4	10
Hemerobius nitidulus Fabricius							2.2							1
Hemerobius stigma Stephens				1.9			6.5						0–0.6	3
Hemerobius handschini Tjeder					0.8									1
Hemerobius lutescens Fabricius			3.8		1.6	+	7.9			+	2.6		9.6–17.9	8
Hemerobius marginatus Stephens	+		4.6		9.4	+	0.7				1.3			6

Species	1	2	3	4	5	6	7	8	9	10	11	12	13	FR
Hemerobius simulans Walker												1.1		1
Hemerobius fenestratus Tjeder						0.8								1
Hemerobius pini Stephens	1.2–4.6	5.3	5.0											3
Hemerobius gilvus Stein	1.4–3.6													2
Hemerobius contumax Tjeder													+	1
Micromus angulatus (Stephens)		6.6										1.5		1
Micromus variegatus (Fabricius)	0–0.5	11.8	1.4										+	5
Micromus paganus (L.)		22.4	12.2										+	4
Micromus lanosus (Zelený)						3.1						2.3		2
Sympherobius pygmaeus (Rambur)	0–0.9					1.6								1
Sympherobius elegans (Stephens)	0–0.6		2.9										+	6
Sympherobius klapaleki Zelený											1.1			2
Sympherobius fuscescens (Wallengren)		1.3									1.1			2
Sympherobius pellucidus (Walker)													+	3
Megalomus hirtus (L.)						0.8						0.8		1
Megalomus tortricoides Rambur			0.4										+	1
Drepanepteryx phalaenoides (L.)		3.9	14.4			0.8						1.5	+	6
S_C	2	11	7	9	9	8	9	9	1	2	1	6		13
S_H	5	3	9	9	9	15	9	9	0	6	11	5		12

Note: The dominance values (%) are calculated from the number of collected individuals published in the reviewed papers.

Abbreviations: S_C, S_H, species richness of chrysopids and hemerobiids respectively; +, species present; A, abundant; FR, frequency of species occurrence in literature surveyed.

Sources: 1, Czech Republic and Slovakia (Zelený, 1963, 1978); 2, Austria (Ressl, 1971, 1974); 3, Austria (Gepp, 1974); 4, Spain (Monserrat & Marín, 1994, 1996); 5, Yugoslavia-Montenegro, *Fagus moesiaca* (Devetak, 1991); 6, Bulgaria (Popov, 1991); 7, Denmark (Nielsen, 1977); 8, France (Baylac, 1980); 9, Switzerland (Eglin, 1967); 10, Germany (Rabeler, 1962); 11, Slovakia (Vidlička, 1994); 12, Spain (Marín & Monserrat, 1991); 13, Italy (Pantaleoni, 1990*a*, *b*; signed with +).

European regions, 27 brown lacewing species were collected on beech trees. Dominance values and frequency of occurrence reveal that the following brown lacewings are characteristic in European beech woods: *Hemerobius micans* (26%–88%), *H. humulinus* (1%–40%), *H. lutescens* (2%–18%), *Drepanepteryx phalaenoides* (0.4%–14%), *H. marginatus* (1%–9%), *Wesmaelius subnebulosus* (1%–9%), *Micromus variegatus* (0.5%–12%), and *Sympherobius elegans* (0.6%–3%). In several cases the proportions of two further hemerobiid species exceeded 10%: *M. paganus* (12%–22%) and *W. nervosus* (0.5%–17%). However, these two species did not occur frequently on beech and even when they were present, the surroundings included other types of forests (e.g. conifers) from where adults could have migrated. Among the listed brown lacewings, the most characteristic abundant species was *H. micans*, which is closely associated with beech (Zelený, 1963; Ressl, 1974; Nielsen, 1977; Pantaleoni, 1990*a*, *b*; Marín & Monserrat, 1991; Monserrat & Marín, 1996). In these collections, *H. micans* and *Chrysoperla carnea* together represented more than 90% of lacewing adults. Tree preference of *H. micans* was also supported by cluster analysis, showing that it was a main component of the hemerobiid group specifically characteristic for beech forests (Pantaleoni, 1990*a*; Monserrat & Marín, 1996). The second most common hemerobiid species on beech was the ubiquitous *H. humulinus*. According to characteristic species, species richness of most predictable assemblages is nine, although S_H values of Table 10.5 suggest that up to 15 hemerobiids might be present, for example because of immigrants from nearby coniferous forests.

Only Monserrat (1991, 1996, 1998) has reported data on lacewings associated with other species of beech outside Europe. He collected brown lacewings living in *Nothofagus* forests of Chile and Argentina, namely *H. bolivari* Banks (Monserrat, 1996) and *H. chilensis* Nakahara. The latter brown lacewing was recorded from *N. dombeyi* (Monserrat, 1998).

It seems that the number of characteristic species living in local lacewing assemblages of *Fagetum* associations tends to decrease with increasing altitude. Based on Table 10.5 the number of typical beech forest lacewing species ranked by altitude level in certain locations are as follows. In a 90-year-old Danish lowland beech forest 6 characteristic chrysopids and 8 characteristic hemerobiids were detected (Nielsen, 1977). Five chry-

sopid and 6 hemerobiid species were collected as typical in a calciphilous beech wood (*Cephalentero-Fagion*) located at 600 m a.s.l. in Slovakia (Vidlička, 1994). Pantaleoni (1990*a*, *b*) recorded 6 typical chrysopid and 4–5 hemerobiid species in three stands of Apennine beech forests (*Trochischanto-Fagetum* with *Tilio-*, *Abieti-Aceri* and *Luzulo-Fagetum* sub-associations) at 1000–1400 m a.s.l. in Italy. There were only 2–4 green and 2–3 brown lacewing species in three locations at around 1400–1700 m a.s.l. in a mountainous relic beech wood (*Galio-Fageto* series: Supramediterranean Iberian Ayllon series) in Spain (Marín & Monserrat, 1991).

Few homopterans associated with beech are potential prey for lacewings. Two aphid species, *Lachnus pallipes* and *Phyllaphis fagi*, live on *F. silvatica* in Europe. The latter is a pest because its population totally covers the leaves of beech trees during outbreaks. The scale insect *Cryptococcus fagisuga* settles in crevices of the trunk and branches and is a pest in beech forests of Europe. Larvae of *H. micans* and *Chrysotropia ciliata* consumed *P. fagi* in Denmark (Nielsen, 1977). No other data on lacewing predators of *P. fagi* are available. There are limited data on predation of beech scales by lacewings. Baylac (1980) studied *Cryptococcus fagisuga* and its natural enemies in beech stands in northern France. *Dichochrysa flavifrons* and *S. elegans* were recorded among predators, but were uncommon. The average individual number of chryopid larvae sampled in 0–2 m on the trunk varied between 0.04 and 0.33 per trunk (range: 0.01–0.62) in accordance with the increasing infestation level of *C. fagisuga*. In Danish beech forests, Nielsen (1977) reported that *C. fagisuga* was consumed by larvae of *D. ventralis*.

Nielsen (1977) elucidated the pattern of beech lacewing flight activity by using light traps in Denmark. *Chrysoperla carnea* showed three activity periods; a small peak in May–June, another higher activity in August to early September, peaking in late August, and the last sharp and high peak in early October indicating activity before overwintering. The other dominant species, *Chrysotropia ciliata*, showed a high activity level in the period of May to October. Its mass flight occurred and peaked in July. The third dominant chrysopid, the herb layer inhabitant *Chrysopa phyllochroma*, swarmed through July and August and peaked in late July. Because of low abundance of other lacewings, Nielsen only gave the length of activity period. *Dichochrysa ventralis* was recorded from late July to the end of August. Adults of *Drepanepteryx phalaenoides* were active from the end of August to early October. Flights of *H. micans* were detected between early June and early October, and mid-August to mid-October in the case of *H. lutescens*. Marín & Monserrat (1991) also reported the seasonal pattern of the two most dominant lacewing species collected in three beech stands in Spain. *Chrysoperla carnea* showed three seasonal activity periods. The abundance was low in June, a second main activity period was detected between late July and mid-September with a peak in August. The third activity peak was in late September to mid-October, characterised by few individuals. *Hemerobius micans* showed a trivoltine pattern. Its overlapping populations peaked in May to mid-June, late July to early August, and mid-August to mid-September depending on beech stands. In Italian beech forests Pantaleoni (1990*a*) monitored three typical hemerobiids associated with beech during two years in two localities. *Hemerobius micans* was active from May to early October with a smaller peak in late June and a higher peak in mid-July. A third characteristic lacewing species, *Hemerobius humulinus*, was represented by one high activity period in July. Two characteristic beech chrysopids were documented by Pantaleoni (1990*a*). *Hypochrysa elegans* was active during May and June with a sharp larger peak in mid-June. In one of two forests the mass flight of *Chrysoperla carnea* was recorded in August and September with a unimodal pattern. In the other forest stand this lacewing species showed a bimodal seasonality with peaks in mid-June and in August, respectively.

10.4.2 Oaks

Importance of *Quercus* species
About 450 species of *Quercus* (Fagaceae) occur. Most oak species are indigenous to the continental and subtropical regions of the northern hemisphere.

Lacewing assemblages in oak forests
Despite the large number of species, information on lacewing assemblages has to date been limited to European forests. Lacewings have been sampled from *Quercus robur*, *Q. petraea*, *Q. cerris*, *Q. pubescens*, *Q. ilex*, *Q. suber*, *Q. coccifera*, *Q. faginea*, *Q. rotundifolia*, and *Q. pyrenaica*. Of these *Q. robur* and *Q. petraea* prefer a rather humid climate, providing mesic habitats for lacewings in northern, western, and central Europe. Both

species also have forms adapted to Atlantic, Mediterranean and continental climates. The other oaks mentioned are thermophilous and xerophilous species creating xerothermous forest stands in the areas of south Europe and warmer regions of central–eastern Europe (*Q. cerris*, *Q. pubescens*). Depending on climatic factors, soil type (acidophilous, basophilous), and other conditions of habitats, there are many mountainous and lowland oak-forest associations. Due to the long evolutionary development of *Quercus*, the number of herbivorous insects associated with oaks is very high. For instance, 630 phytophagous insects have been recorded living on oak forests in Hungary (central Europe) (Csóka, 1998) and of these 9% (56 spp.) belong to the Homoptera – the major food resource for lacewings.

Table 10.6 summarises the recorded chrysopids and hemerobiids from each *Quercus* species studied in Europe.

Chrysopid assemblages
Altogether 33 chrysopid species have been recorded in European oak forests. Seven species are directly associated with coniferous trees while the other 26 species are potential members of characteristic assemblages on each species of *Quercus*. About half of the 26 species (46%) are xerothermophilous, 27% are ubiquitous (or euryoecious), and only a few prefer meso- or hygrophilous habitats. The highest numbers of chrysopid species were found on the following oaks: *Quercus rotundifolia* (23 spp.), *Q. pyrenaica* (20 spp.), *Q. robur* (19 spp.), *Q. faginea* (17 spp.), and *Q. cerris* (12 spp.). These large numbers of species reflect long-term, intensive collections made in the Iberian Peninsula in numerous localities by Monserrat & Marín (1994). By the values of dominance and frequency of occurrence the following chrysopid species would be generally associated with oaks (Table 10.6): *Chrysoperla carnea* (3%–67%), *Dichochrysa prasina* (11%–40%), *D. flavifrons* (4%–41%), *Chrysopa viridana* (0.3%–19%), *D. ventralis* (1%–20%), *Chrysotropia ciliata* (1%–34%), *Nineta flava* (0.03%–6%), *C. pallens* (0.3%–6%), *D. picteti* (0.1%–10%), *D. granadensis* (0.3%–5%), *D. clathrata* (0.1%–3%), and *Cunctochrysa albolineata* (0.03%–2%). All these are mentioned as oak-forest inhabitants. *Dichochrysa flavifrons*, *Chrysopa viridana*, and *N. flava* have a stronger preference for oaks (Aspöck *et al.*, 1980; Monserrat & Marín, 1994). According to Aspöck *et al.* (1980) other chrysopids may

Table 10.6. *Species composition and dominance distribution of chrysopids and hemerobiids found on different oaks (Quercus spp.) in European countries*

	1	2	3	4	5	6	7	8	9	10	11	12	13	14	FR
Chrysopidae															
Chrysoperla carnea s. lat. (Stephens)	+	A	+	66.9	2.5	+		+	28.9	42.2	46.8	58.3	37.8	+	12
Chrysoperla mediterranea (Hölzel)											0.1				1
Chrysoperla mutata (McLachlan)											0.03				1
Chrysopa formosa Brauer	+			0.3						0.1	0.3	0.3		+	3
Chrysopa pallens (Rambur)			5.7	0.3						1.8	0.6	0.3	0.7	A	9
Chrysopa perla (L.)			11.4									0.1			3
Chrysopa viridana Schneider				0.3	3.8					18.8	5.6	2.0	6.8		7
Chrysopa nigricostata Brauer				0.1						0.02	0.02	0.4			4
Chrysopa dorsalis Burmeister														+	1
Chrysopa regalis Navás										0.1	0.03				2
Chrysopa nierembergi Navás										0.1					1
Dichochrysa prasina (Burmeister)	+	+	11.4	13.5	13.9	+	+		39.5	27.6	33.5	25.7	27.0	+	12
Dichochrysa ventralis (Curtis)		+	20.0	1.0								1.2			4
Dichochrysa flavifrons (Brauer)			5.7	11.9	40.5	+	+		18.4	4.5	4.1	3.7	17.6	A	11
Dichochrysa clathrata (Schneider)				0.1			+		2.6		0.1	0.1			5
Dichochrysa zelleri (Schneider)							+								1
Dichochrysa granadensis (Pictet)				0.3					5.3	0.4	2.6	0.3			5
Dichochrysa inornata (Navás)				1.2						0.1	0.4	0.1			4
Dichochrysa picteti (McLachlan)				0.2			+			1.4	2.1	0.05	10.1		5
Dichochrysa venosa (Rambur)											0.1				1
Dichochrysa subcubitalis (Navás)										0.3	0.2	0.3			3
Dichochrysa iberica (Navás)				0.1							1.1	0.7			3
Dichochrysa genei (Rambur)											0.2				1
Dichochrysa spp.					39.2										
Nineta flava (Scopoli)			5.7	0.5				+	5.3	0.03	0.03	0.5		+	7
Nineta vittata (Wesmael)								+							1
Nineta pallida (Schneider)			5.7												1
Nineta guadarramensis (Pictet)				1.2						1.2		4.6			3
Chrysotropia ciliata (Wesmael)		+	34.3	1.4			+	+							4
Cunctochrysa albolineata (Killington)		+		1.7						0.1	0.03	0.4			6
Cunctochrysa baetica (Hölzel)										0.4	1.3	0.4			3

Species	1	2	3	4	5	6	7	8	9	10	11	12	13	14	FR
Nothochrysa fulviceps (Stephens)				0.2						0.3			0.4	0.6	3
Italochrysa italica (Rossi)				0.4						0.4			0.4	0.1	4
Brinckochrysa nachoi Monserrat				0.5						0.3			0.3		1
Hemerobiidae															
Wesmaelius subnebulosus (Stephens)	14.9			5.1						42.1				10.2	6
Wesmaelius nervosus (Fabricius)	+			+											1
Wesmaelius malladai (Navás)	0.5									0.3					1
Hemerobius humulinus L.	33.0	A	A	0.8									1.1		10
Hemerobius micans Olivier	3.2	A	A							2.9			0.6		8
Hemerobius stigma Stephens	0.5			0.4						2.9				1.7	4
Hemerobius handschini Tjeder	0.6														1
Hemerobius lutescens Fabricius	13.4			25.0											4
Hemerobius perelegans Stephens	2.7														1
Hemerobius marginatus Stephens	+			+											2
Hemerobius gilvus Stein	4.8			1.8									24.4		4
Micromus angulatus (Stephens)	1.1			0.4											2
Micromus variegatus (Fabricius)	7.6			0.5											2
Micromus paganus (L.)	1.3														1
Micromus lanosus (Zelený)	1.3			0.5											3
Sympherobius pygmaeus (Rambur)	1.9			6.9						100	92.4	41.3	45.5		8
Sympherobius elegans (Stephens)	+			8.7									+		4
Sympherobius klapaleki Zelený	6.4			0.4						2.9			10.2		5
Sympherobius fuscescens (Wallengren)	0.6			+											1
Sympherobius riudori Navás				0.8											1
Megalomus hirtus (L.)										5.7					1
S_C	17	9	4	2	5	2	4	12	6	17	23	20	6		6
S_H	13	8	2	2	8	4	1	8	5	9	2				2

Note: The dominance values (%) are calculated from the number of collected individuals published in the reviewed papers.

Abbreviations: S_C, S_H, species richness of chrysopids and hemerobiids respectively; +, species present; A, abundant; FR, frequency of species occurrence in literature surveyed.

Oak species: 1–5, *Quercus robur*; 6–7, *Q. cerris*; 8–9, *Q. faginea*; 10, *Q. petraea*; 11, *Q. rotundifolia*; 12, *Q. pyrenaica*; 13, *Q. suber*; 14, *Q. ilex*.

Sources: 1, 8, Czech Republic & Slovakia (Zelený, 1963, 1971, 1978); 2, Austria (Gepp, 1974); 3, Austria (Ressl, 1971, 1974); 4, Spain (Monserrat & Marín, 1994, 1996); 5, Italy (Pantaleoni, 1996); 6, Yugoslavia (Devetak, 1991); 7, Bulgaria (Popov, 1991); 9–14, Spain (Monserrat & Marín, 1994, 1996).

also occur in oak forests. Such species are as follows: *N. guadarramensis, N. vittata, D. inornata, D. iberica, Cunctochrysa baetica, Italochrysa italica,* and *Nothochrysa fulviceps* (Table 10.6). Numerous *Dichochrysa* species in Table 10.6 inhabit bushy garrigue vegetation in Mediterranean areas in which xerophilous oaks can also be found. Monserrat & Marín (1994) produced a dendrogram analysing plant substrate specificity in chrysopids, and certain well-separated lacewing clusters coincided with preferences to some *Quercus* species. Such clusters were shown for *Q. robur* and *Q. pyrenaica*: *Nothochrysa fulviceps, Nineta guadarramensis, N. flava, Cunctochrysa albolineata, Dichochrysa ventralis, Chrysotropia ciliata,* and *Chrysopa perla*. Most of these species are not xerothermous. They prefer more humid, mesic vegetation types, such as the forests of *Q. robur* throughout Europe. Monserrat & Marín (1994) could detect further xerothermous lacewing clusters, which were associated with *Quercus rotundifolia* and with *Q. robur.*

Seven overwhelmingly ubiquitous, eurytopic chrysopid species are common and frequent in European oaks (see Table 10.6). These are generally accepted as stable members of local lacewing assemblages of *Quercetum* associations.

Table 10.7 shows chrysopid assemblages found by local samples in different European oak-forest associations. Columns 1–9 of the table show western and central European oak forests, while columns 10–16 show south European ones. *Quercus robur* or *Q. petraea* (6, 7) dominated the western and central European oak forests, while south European *Quercetum* (10–16) associations were characterised by *Q. ilex, Q. pubescens, Q. cerris,* or *Quercus rotundifolia.*

From chrysopid assemblages in European oak forests, 30 species have been recorded, 3 associated with conifers. Of the remaining 27 species, in western and central Europe 20, and in south Europe 16 green lacewings were detected in oak forest types. Table 10.7 reveals that in European oak stands, the common and most frequent or most dominant chrysopids within local assemblages are the following: *Chrysoperla carnea* (7%–88%), *Dichochrysa prasina* (1%–74%), *Chrysopa viridana* (0.2%–29%), *C. pallens* (0.3%–18%), and *D. flavifrons* (0.1%–11%). At the same time, *Nineta flava* (0.5%–38%), *Chrysotropia ciliata* (0.2%–14%), *D. ventralis* (1.5%–8%), and *Cunctochrysa albolineata*

(0.2%–25%) were characteristic only in western and central European forests, although Table 10.6 suggests that they also occur in south European oaks. Characteristic (dominant and frequent) chrysopid species associated with oak trees or oak forests based on Tables 10.6 and 10.7 are identical; consequently, these lacewings may be considered to be the expected constant and common elements of chrysopid assemblages within European oak forests.

According to dendrograms of our cluster analyses, applied in order to analyse similarity relations between chrysopid assemblages in Table 10.7, three well separated groups were detected. Regarding species composition, the total south European chrysopid assemblages (10–16) formed one group. Within this cluster similar, sclerophyllous oak forests with a dry, warm climate took part. For instance, *Quercetum ilicis* (12–15) was such a typical Mediterranean forest association. The other group (1–5, 8–9) was created by chrysopid assemblages of western and central European oak forests, dominated by *Q. robur*. A third, two-member group (6–7), well separated from the former ones, was represented by lacewings in forests of Hungarian *Quercetum petraeae*. There are ecological reasons for the separation between chrysopid assemblages living in the warmer and drier oak forests of southern Europe and those living in the temperate, more humid ones of western and central Europe. The level of euryoecious species is 25%, of xerothermophilous is 56%, and of hygrothermophilous is only 6%. The rate of xerothermophilous species in oak stands of western and central Europe is significantly lower, at 30%. The rate of mesohygrophilous species is higher, at 25%, and eurytopic lacewings were represented at 35%. The difference between the two regions was in the frequency of occurrence and dominance values of some chrysopid species, derived from exclusivity of numerous lacewing species to the Mediterranean area in oak forests of southern Europe, such as *Dichochrysa* spp. As shown in Table 10.7 certain chrysopids were present only in samples from the western and central European region. For instance, the somewhat mesohygrophilous *Chrysopa perla, D. ventralis, N. flava, N. vittata, Chrysotropia ciliata,* and *Cunctochrysa albolineata* prefer less dry, mesic habitats. This tendency is shown in higher dominance values of these chrysopids calculated from faunistic collections presented in Table 10.6. The average number of characteristic species in oak arboreal chrysopid assemblages is eight to nine in west and central Europe while it is six to seven in south Europe.

Hemerobiid assemblages

Twenty-one hemerobiid species were recorded in faunistic collections on eight species of European oaks (Table 10.6). Five are characteristically associated with coniferous trees. The other 16 would be real components of assemblages dwelling in oak forests. The majority of species – as with chrysopids – are ubiquitous (43.8%) or thermophilous (37.5%) while fewer were hygrothermophilous (18.7%). Due to extended sampling efforts (Monserrat & Marín, 1996) the highest number of hemerobiids was recorded on oaks of the Iberian Peninsula (columns 4, 11, and 12). The total number of hemerobiid species on oaks in Europe is: *Quercus robur*: 14, *Q. cerris*: 9, *Q. pyrenaica*: 7, *Quercus rotundifolia*: 6, *Q. petraea*: 5, *Q. faginea*: 4, *Q. suber*: 3, and *Q. ilex*: 2. According to their frequency of occurrence and range of dominance values (%), the following seven species are associated with European oaks as constant components of assemblages: *Hemerobius humulinus* (1%–49%), *H. micans* (1%–25%), *H. lutescens* (13%–25%), *H. gilvus* (2%–24%), *Sympherobius pygmaeus* (2%–100%), *S. klapaleki* (0.4%–10%), and *Wesmaelius subnebulosus* (5%–42%). These species are euryoecious or thermophilous, and the stronger preference of some of them for oaks is emphasised (Aspöck *et al.*, 1980; Monserrat & Marín, 1996). These are: *H. micans, H. lutescens, H. gilvus, S. pygmaeus*, and *S. klapaleki*. In their similarity analysis Monserrat & Marín (1996) found separate clusters for hemerobiids preferring certain oak species. According to this, *H. humulinus, H. lutescens*, and *H. perelegans* were associated with *Q. robur, S. pygmaeus* with *Q. faginea*, and *H. gilvus* and *S. klapaleki* with *Q. pyrenaica*. Moreover, other species from Table 10.6 are also oak-forest inhabitants, for example *S. elegans, Drepanepteryx phalaenoides*, and *S. pellucidus* (Aspöck *et al.*, 1980). Adults of the last species were collected in abundance using insecticide-fogging from foliage of *Q. robur* (Barnard *et al.*, 1986). Aspöck *et al.* (1980) also mentioned in connection with this species that its larvae have so far been found only on *Quercus*. These data supported the idea that *S. pellucidus* is associated with oaks much more intensively than previously supposed. Table 10.6 shows that *Micromus* species are associated with *Q. robur*. However, they are probably not associated with oaks but rather with the richer herbaceous understorey or bushy vegetation layer as their preferred habitat.

On the basis of a preliminary similarity analysis of composition of hemerobiid assemblages on oaks (Table 10.6), some characteristic groups can be detected. By using binary similarity indices the following clusters occur: <*Q. faginea, Q. suber*, and *Q. petraea*> (in Spain); <*Q. cerris, Quercus rotundifolia*, and *Q. pyrenaica*> (in Spain and Bulgaria); <*Q. robur*> (hemerobiid patterns were separated country by country). By using quantitative similarity indices the following well-separated oak clusters were seen to be formed by hemerobiids: <*Q. pyrenaica* and *Quercus rotundifolia*> (in Spain); <*Q. faginea* and *Q. petraea*> (in Spain); <*Q. robur* and *Q. robur*> (Spain and Austria).

Table 10.8 covers recorded assemblages sampled in local European oak associations or urban forests. Columns 1–10 refer to results of collections in western and central Europe whilst columns 11–14 cover southern Europe. *Q. robur* was the dominant tree species in oak associations by studies in west and central Europe, but *Q. petraea* and *Q. cerris* also participated in the canopy level in certain cases (columns 4, 6, 7). The typical tree species in oak forests of southern Europe (columns 11–14) were *Q. pubescens, Q. petraea, Q. cerris, Q. ilex, Quercus rotundifolia*, and *Q. suber*.

In total 29 hemerobiid species were recorded in forests as indicated in Table 10.8. Only 18 species are truly associated with oak trees, because the other 11 species are known as exclusive inhabitants of coniferous trees. All of these 18 species were found in western and central Europe in oak hemerobiid assemblages while only seven species were found in southern Europe. From the former region 50% of the species were ubiquitous, 39% were thermophilous, and 11% were hygrothermophilous. Hemerobiids with a hygrothermous habitat preference were absent from oak forests in southern Europe with a drier and warmer climate, and the proportions of xerothermous and ubiquitous species were slightly higher, 43% to 57% respectively.

By values of dominance range and frequency of occurrence (Table 10.8), the most characteristic species in hemerobiid assemblages were as follows: *H. humulinus* (1%–64%), *S. pygmaeus* (0.4%–80%), *H. micans* (4%–30%), *W. subnebulosus* (0.6%–24%), *S. klapaleki*

Table 10.7. Species composition and dominance distribution of the local chrysopid assemblages found in various types and associations of oak (Quercus) forests in European countries

	1	2	3	4	5	6	7	8	9	10	11	12	13	14	15	16	FR
Chrysopidae																	
Chrysoperla carnea s. lat. (Stephens)	22.1–62.5	25.8	A	+	88.1	7.4	31.0	31.4	33.5	21.0	18.9	+	18.2	12.0	+	64.5	16
Chrysoperla mediterranea (Hölzel)													17.7				1
Chrysopa formosa Brauer						+							0.2	15.0			4
Chrysopa pallens (Rambur)	0.9–1.0	0.8	+		1.6	1.4	4.8	+	18.0	1.1	0.5			10.7	+	0.3	13
Chrysopa perla (L.)						0.1	7.1	0.5	12.4								4
Chrysopa phyllochroma Wesmael						+	2.4		0.5								3
Chrysopa abbreviata Curtis																	1
Chrysopa viridana Schneider						0.2	4.8			4.7	12.6		1.2	29.2		10.1	7
Chrysopa nigricostata Brauer						+			12.9								2
Chrysopa dorsalis Burmeister	0–0.2										0.5						2
Chrysopa walkeri McLachlan						+				0.2							2
Dichochrysa prasina (Burmeister)	34.3–73.8	14.1			3.3	42.1	1.2		10.8	57.3	42.1	+	14.7	23.2	+	20.8	13
Dichochrysa ventralis (Curtis)	0–1.5	1.6				+	8.3	2.1									5
Dichochrysa flavifrons (Brauer)		0.1			0.4	11.1	10.7			2.1	5.8	+	3.2	3.9	+	2.3	11
Dichochrysa clathrata (Schneider)										1.3							1
Dichochrysa zelleri (Schneider)										7.5	15.8	+	4.3		+		5
Dichochrysa picteti (McLachlan)										2.6			9.0				2
Dichochrysa iberica (Navás)												+	25.5				2
Nineta flava (Scopoli)	0.5–2.3	37.5	+		2.5	9.3	1.2	9.3	3.6								8
Nineta vittata (Wesmael)	0–0.5	5.5						2.6									3
Nineta guadarramensis (Pictet)						17.7										1.4	2
Nineta carinthiaca (Hölzel)						2.4											1
Nineta inpunctata (Reuter)							1.2										1
Chrysotropia ciliata (Wesmael)	0–0.2	14.1		+	4.1		14.3	12.4	2.1								8
Cunctochrysa albolineata (Killington)	0–0.2	0.8				6.2	11.9	25.3									5
Cunctochrysa baetica (Hölzel)											1.1		2.8	3.9		0.3	5
Nothochrysa capitata (Fabricius)			+					16.5									2
Nothochrysa fulviceps (Stephens)						2.0	1.2										2

Italochrysa italica (Rossi)									2.1		3.2		+		3
Hypochrysa elegans (Burmeister)					0.2					2.6					2
Species richness of chrysopids	9	9	5	2	6	17	13	8	9	10	9	6	11	8	7

Note: The dominance values (%) are calculated from the number of collected individuals published in the reviewed papers.

Abbreviations: A, abundant; +, species present; FR, frequency of species occurrence in literature surveyed.

Sources: 1, Poland, *Pino–Quercetum* (Czechowska, 1985); 2, Poland, *Potentillo albae–Quercetum* (Czechowska, 1990); 3, Switzerland, mixed oak forest (Eglin, 1967); 4, Germany, *Querco–Carpinetum* (Rabeler, 1962); 5, Germany, urban oak forest, *Quercus robur*, *Q. petraea* (Saure & Kielhorn, 1993); 6, Hungary, *Quercetum petraeae-cerris* (F. Szentkirályi, unpublished data); 7, Hungary, *Quercetum petraeae*, mixed with smaller plantations of *Picea abies* and *Pinus silvestris* (F. Szentkirályi, unpublished data); 8, UK, park forest, *Quercus robur* (Barnard *et al.*, 1986; New, 1967, signed by +); 9, Hungary, gallery forest, *Quercetum roboris* (Szentkirályi, 1998); 10, Italy, pre-Appeninic oak forest, *Q. pubescens*, *Q. petraea* (Pantaleoni, 1990a); 11, Italy, hill oak forest, *Q. pubescens*, *Q. cerris* (Pantaleoni, 1990a); 12, Southern France, *Quercetum ilicis* (Du Merle, 1983a); 13, Southern France, *Quercetum ilicis* (Canard, 1987); 14, Italy, *Quercetum ilicis* (Pantaleoni, 1984); 15, Croatia, *Orno–Quercetum ilicis* (Devetak, 1998); 16, Spain, *Bupleuro rigidi–Querceto rotundifoliae* (Marín & Monserrat, 1987).

Collecting methods: Sweep-netting from tree foliage (columns 3, 4, 10, 11, 14–16); beating of branches (12, 13); pyrethroid-fogging (8); yellow and red pan trapping (1, 2, 5); light trapping (7, 9); Malaise, light, and pan trapping (6).

Table 10.8. *Species composition and dominance distribution of the local hemerobiid assemblages found in various types and associations of oak (Quercus) forests in European countries*

	1	2	3	4	5	6	7	8	9	10	11	12	13	14	FR
Hemerobiidae															
Wesmaelius subnebulosus (Stephens)	0–1.9					0.6	7.4	1.0	14.3	3.5				24.3	7
Wesmaelius nervosus (Fabricius)	0–3.8	0.8				0.6		7.7		0.3					5
Wesmaelius quadrifasciatus (Reuter)										0.3					1
Wesmaelius concinnus (Stephens)	20.8–22.9	16.0			4.3										3
Hemerobius humulinus L.	34.4–45.3	64.0	+	+	14.9	27.4	11.1	25.8	23.8	19.1	1.3	13.3	13.5		13
Hemerobius micans Olivier	3.8–7.5	8.0	+	+		29.3	16.7	21.5	4.8	13.7	7.5	17.8	5.4		12
Hemerobius nitidulus Fabricius	3.8–18.3	0.8			2.1			0.3		0.6					5
Hemerobius stigma Stephens	13.2–15.3	0.8			6.4					1.3					4
Hemerobius atrifrons McLachlan										1.7					1
Hemerobius handschini Tjeder							3.7			1.5		3.3			3
Hemerobius lutescens Fabricius						1.3	13.0	10.1		4.1					4
Hemerobius marginatus Stephens						+				0.1					2
Hemerobius simulans Walker										0.1					1
Hemerobius fenestratus Tjeder										0.8					1
Hemerobius pini Stephens	0–7.6									0.8					2
Hemerobius gilvus Stein						6.4				0.4	70.0	48.9	1.4		5
Hemerobius contumax Tjeder							1.9								1
Micromus angulatus (Stephens)						2.5	11.1		38.1	7.7				1.4	5
Micromus variegatus (Fabricius)						5.7	22.2		14.3	18.4					4
Micromus paganus (L.)			+							0.1					2
Micromus lanosus (Zelený)						3.2									1
Psectra diptera (Burmeister)										0.1					1
Sympherobius pygmaeus (Rambur)	0–0.4	0.8			2.1	14.6		2.0	4.8	12.0	21.3	16.7	79.7	72.8	11
Sympherobius elegans (Stephens)						+				3.4				1.4	3
Sympherobius klapaleki Zelený		8.0			63.8	7.0				1.4					4
Sympherobius fuscescens (Wallengren)	0–3.8									0.3					2
Sympherobius pellucidus (Walker)							3.7	31.5		0.3					3

	1	2	3	4	5	6	7	8	9	10	11	12	13	14
Megalomus tortricoides Rambur	0–0.4				7.4									1
Drepanepteryx phalaenoides (L.)	0.8		6.4	1.3	1.9					8.1				6
Species richness of hemerobiids	11	9	3	2	7	14	11	8	6	25	4	5	4	4

Note: The dominance values (%) are calculated from the number of collected individuals published in the reviewed papers.

Abbreviations: +, species present; FR, frequency of species occurrence in literature surveyed.

Sources: 1, Poland, *Pino–Quercetum* (Czechowska, 1985); 2, Poland, *Potentillo albae–Quercetum* (Czechowska, 1990); 3, Switzerland, mixed oak forest (Eglin, 1967); 4, Germany, *Querco–Carpinetum* (Rabeler, 1962); 5, Germany, urban oak forest, *Quercus robur*, *Q. petraea* (Saure & Kielhorn, 1993); 6, Hungary, *Quercetum petraeae–cerris* (F. Szentkirályi, unpublished data); 7, Hungary, *Quercetum petraeae*, mixed with smaller *Picea abies* and *Pinus silvestris* plantations (F. Szentkirályi, unpublished data); 8, UK, park forest, *Q. robur* (Barnard et al., 1986); 9, Hungary, gallery forest, *Querco-ulmetum* with *Q. robur* (Szentkirályi, 1998); 10, Hungary, *Querceto petraeae–Carpinetum* and *Quercetum pubescents*, mixed with *Picea abies*, *Pinus silvestris*, *P. nigra* and *Larix decidua* plantations (F. Szentkirályi, unpublished data); 11, Italy, pre-Appeninic oak forest, *Quercus pubescens*, *Q. petraea* (Pantaleoni, 1990a); 12, Italy, hill oak forest, *Quercus pubescens*, *Q. cerris* (Pantaleoni, 1990a); 13, Italy, *Quercetum ilicis* (Pantaleoni, 1984); 14, Spain, *Bupleuro rigidi–Querceto rotundifoliae* (Marín & Monserrat, 1987).

Collecting methods: Sweep-netting from foliage (columns 3, 4, 11–14); insecticide-fogging (8); yellow and red pan trapping (1, 2, 5); light trapping (7, 9, 10); Malaise, light, and pan trapping (6).

(1%–64%), *H. gilvus* (0.4%–70%), *D. phalaenoides* (0.4%–8%), *H. lutescens* (1%–13%). With one exception *(D. phalaenoides)*, hemerobiids are the same as those noted in Table 10.6. This coincidence confirms that these eight species can be considered as standard components of lacewing assemblages in European oak forests.

According to dendrograms of our similarity analysis made from Table 10.8 the assemblages in southern Europe (11–14) are separated from those in western and central Europe (1–9). Due to the high number of species the assemblage in column 10 is isolated from those previously mentioned. This latter case gives a typical example of how characteristic hemerobiids of coniferous stands scattered among oak forests can migrate to the *Quercus* canopy. It can be observed from the higher dominance values of Table 10.8 that certain thermophilous species (*H. gilvus* and *S. pygmaeus*) prefer warmer, dry forests in the Mediterranean region. The data in Table 10.6 also reflect this tendency. On the basis of their higher dominance percentage or exclusive occurrence the following brown lacewings prefer oak forests of western and central Europe, characterised by a more humid climate: *H. humulinus, H. micans, H. lutescens, S. klapaleki,* and *D. phalaenoides. Micromus* species (*M. angulatus* and *M. variegatus*) also belong to the group associated with richer herbaceous vegetation.

Lacewings on oaks outside Europe

Investigations on lacewing assemblages associated with *Quercus* species outside Europe are rare. Monserrat (1996, 1998) published data on hemerobiids – without indication of oak species – collected from *Quercus* of the Neotropical area, in particular in Central and South America. The following lacewings were recorded from *Quercus* spp.: *Hemerobius martinezae* Monserrat (Mexico and Central America), *H. bolivari* Banks (in Central and South America), *Hemerobius jucundus* Navás (Mexico and Central America), and *Micromus subanticus* (Walker) from Turkey oak (North and Central America).

Aphids as prey for lacewings in oak forests

About 225 aphid species from more than 40 genera have been recorded using *Quercus* spp. as a host plant (Blackman & Eastop, 1994). The recorded numbers of aphid species on the nine *Quercus* spp. surveyed for

lacewings in Europe are as follows: *Q. robur*: 22, *Q. ilex*: 21, *Q. petraea*: 20, *Q. pubescens*: 17, *Q. cerris*: 14, *Q. pyrenaica*: 9, *Q. coccifera*: 8, *Q. suber*: 5, and *Q. faginea*: 1 (Blackman & Eastop, 1994). The major groups of aphids and some important pestiferous species infesting the above mentioned oaks that are potential prey for lacewings as follows:

1. Drepanosiphinae (colonies on leaves): *Myzocallis castanicola, Hoplocallis microsiphon* (mainly in the Mediterranean region), *Diphyllaphis mordvilkoi* (southern Europe), *Hoplochaetaphis zachvatkini* (Mediterranean region), *Tuberculatus annulatus, T. eggleri, T. neglectus, T. querceus.*
2. Thelaxinae (colonies on young shoots and stems, leaves, developing acorns): *Thelaxes suberi, T. dryophila.*
3. Lachnidae (aphids on twigs, older stems or branches, in crevices in trunks of oaks): *Lachnus pallipes* (in northwest and central Europe), *L. roboris, Stomaphis quercus.*
4. Phylloxeridae (infestation mainly on leaves, sometimes younger twigs): *Phylloxera glabra, P. quercus* (in particular in southern Europe), *P. quercina* (in the Mediterranean region), *Acanthochermes quercus.*

Despite the great number of Homoptera associated with oaks, few data on their natural enemies are available. In the case of aphids the only data on lacewings are for Phylloxeridae. Steffan (1972) mentioned lacewing larvae eating two *Phylloxera* species. *Chrysotropia ciliata, Nineta flava, Dichochrysa prasina, Chrysopa perla,* and *Sympherobius pygmaeus* consumed individuals of *Phylloxera glabra* infesting *Q. robur*. The other aphid species, *P. coccinea*, living on *Q. robur, Q. petraea,* or *Q. pubescens*, was eaten by *Chrysotropia ciliata, N. flava,* and *Chrysopa perla.* Pantaleoni (1982) found freshly emerged adults of *C. dorsalis* in high numbers on seedlings of *Q. ilex* infested by *P. quercus.* It is notable that certain lacewing species associated with coniferous trees consume prey on oaks. We may suppose that this explains why numerous lacewing species dwelling in coniferous forest occur frequently in oak stands (see Tables 10.6–10.8). In particular, there are some hemerobiids present with a high dominance range in lacewing assemblages of oak forests (e.g. *Wesmaelius concinnus, Hemerobius nitidulus,* or *Hemerobius stigma* in Table 10.8).

Outside Europe, only Shantibala *et al.* (1994) in

India have reported predation by a lacewing on oak aphid. They detected *Micromus timidus* consuming *Cervaphis quercus*, which infests new growth and leaves of *Quercus* spp. in Southeast Asia and belongs to Greenideinae.

Du Merle (1983*a*) found *Thelaxes confertae* aphids on *Q. ilex* trees in southern France. In laboratory conditions larvae of the chrysopids *Cunctochrysa baetica, Dichochrysa prasina* and *D. zelleri* consumed individuals collected from infested oak trees. These lacewings were also recorded on trees of *Q. ilex* and *Q. pubescens* in the same places where the aphid was collected.

Scale insects as prey for lacewings on oaks

More than 140 coccoid species have been recorded on various *Quercus* spp. from the Palaearctic region (Kozár, 1998). Scale insect families are rich in species associated with oaks. Their species richness and the major oak pests are as follow: Asterolecanidae (20 spp.): *Asterodiaspis variolosa;* Beesonidae (2 spp.); Coccidae (15 spp.): oak soft scale, *Parthenolecanium rufulum*; Diaspididae (50 spp.): *Quadraspidiotus zonatus, Targionia vitis, Diaspidiotus wuenni*; Eriococcidae (5 spp.); Kermesidae (35 spp.): including *Kermes quercus*; Lecanodiaspididae (5 spp.); Margarodidae (9 spp.): *Palaeococcus fuscipennis* (in Mediterranean region and Switzerland); Ortheziidae (2 spp.).

Numerous scale insects (such as Kermesidae) feed exclusively on *Quercus* spp. Some species, particularly in southern Europe, cause severe damage. *Targionia vitis* and *Parthenolecanium rufulum* cause high mortality of oak seedlings in nurseries. Data on lacewings consuming oak scales are rare, but it is supposed that many lacewing species feed on them, at least when aphids are absent. Schmutterer (1972) published data on *Dichochrysa prasina* that preyed upon the oak soft scale *P. rufulum*. In connection with crop plants and fruit trees (see Chapter 9, this volume) there is clear evidence that certain species of chrysopids and hemerobiids specialise on scales. Such species have trash-carrier larvae, for instance *Dichochrysa, Cunctochrysa,* and *Chrysotropia* spp. *Symperobius* spp. are also coccidophagous. Many representatives of these genera have a high dominance ratio on oaks, for example, *Dichochrysa prasina, D. flavifrons, C. ciliata, S. klapaleki,* and *S. pygmaeus* (see Tables 10.6–10.8). Adults of more than 80% of recorded chrysopid species are not predatory (Tables 10.6–10.7) so it would be useful to know what species

of lacewing larvae are found on oaks. There are however few data, because lacewing adults have predominantly been sampled in the studies carried out. Applying the insecticide-fogging method, Barnard *et al.* (1986) collected many more chrysopid and hemerobiid larvae than adults from the canopy level of *Q. robur*. The dominance rate of chrysopid larvae differs from that of adults in almost every species. Surprisingly the dominance rates of trash-carrier larvae of *Cunctochrysa albolineata* (35.6%) and *Nothochrysa capitata* (29.2%) were higher than for *Chrysoperla carnea*. It is probable that not only adults but also larvae of the dominant *S. pellucidus* are present in great numbers on *Q. robur* (see Aspöck *et al.*, 1980). These species are potential predators of scale insects. Barnard *et al.* (1986) found adults of *N. capitata* and *S. pellucidus* in within-canopy distributions that showed strong preference towards the area around the trunk. Examining the distribution of chrysopid species within *Q. robur* trees Pantaleoni (1996) supports our presumption that lacewings may consume scale insects on oak twigs and bark in great numbers. He found that eggs and larvae of *Dichochrysa* species (*Dichochrysa prasina* and *D. flavifrons*) were predominant (above 80%). He also surveyed the proportional distributions of eggs and larvae within total lacewing assemblages both on the foliage and trunk. The proportion of *Dichochrysa prasina* eggs laid on the foliage was 6% and on the trunk 36%, while for larvae this proportion was foliage 3% and trunk 97%. Eggs, and especially larvae, of *Dichochrysa prasina* were mainly on the trunk. It is supposed that larvae of *Dichochrysa prasina* preyed upon scales or *Lachnus* spp. inhabiting bark crevices and older branches

Whiteflies (Aleyrodidae) as prey for lacewings on oaks

In their systematic world catalogue, Mound & Halsey (1978) mentioned altogether 36 aleyrodid species living on *Quercus* spp. The number of species per whitefly genus on oaks are as follow: *Aleuroplatus*: (5), *Asterobemisia*: (3), *Bemisia*: (2), *Pealius*: (5), *Trialeurodes*: (4), *Tetraleurodes*: (4), *Setaleyrodes*: (8). The major, sometimes pestiferous aleyrodid species serving as potential food sources for lacewings on oaks are as follows: *Asterobemisia carpini* (in particular on *Q. robur*), *Bemisia tabaci, Pealius quercus* (on *Q. ilex* and *Q. robur*), *Trialeurodes vaporariorum,* and *Setaleyrodes quercicola*. Mound & Halsey (1978) reported some data on predatory activity of lacewings upon aleyrodid species of oaks. *B. tabaci* was preyed on by *Nineta flava* and *Brinckochrysa scelestes* while *T. vaporariorum* was attacked by *Chrysoperla rufilabris*.

Moths as prey for lacewings on oaks

Oaks are food plants for numerous moth species, many of which are serious defoliating pests. In the absence of Homoptera prey, eggs and young larvae of certain moth groups can become food sources for lacewings. Such alternative prey among oak moth pests include the tortricids, geometrids, and noctuids (mainly in tree nurseries). Du Merle (1983*a*, *b*) examined in detail the predatory activity of lacewings and their possible role in the control of a moth species in oaks in southern France. He investigated egg mortality of the green oak tortricid *Tortrix viridana* caused by its predators and parasitoids in various oak forests in ten places in the Mediterranean and central regions of France. Chrysopids played a significant role in natural control of this pest. Eggs of *T. viridana* were deposited on the bark of the oak twigs, generally in groups with two layers of eggs. Surveying numerous egg clusters Du Merle found that the upper eggs were predated by chrysopid larvae in higher proportion (range: 17%–38%) than the lower ones (range: 3%–18%). Mortality rates (%) of moth eggs caused by green lacewing larvae on oaks from ten sites were as follow: *Quercus robur* and *Q. petraea*: 22.4%–36.9%, *Q. suber* (only in one place): 26%, *Q. ilex* (at 700 and 800 m a.s.l.): 16.8%–27.7%, *Q. pubescens* (at 800 and 900 m a.s.l.): 15%–36.8%, *Q. pubescens* (at 1150 and 1350 m a.s.l.): 4.2%–8.7%. From the latter data, the range of larval predation of lacewings depended on altitude, being between 15% and 37% below 900 m, while it was lower than 9% above 1100 m. Du Merle (1983*a*) recorded six chrysopid species from *Q. pubescens* and the evergreen oak *Q. ilex* (see Table 10.7, column 12). Larvae of *Cunctochrysa baetica* and *Dichochrysa prasina* were observed in the field eating eggs of *T. viridana* deposited on the bark of the oak. Du Merle supposed that larvae of other chrysopid species also consumed tortricid eggs. In my opinion this is very probable because trash-carrier larvae of *Dichochrysa* and *Cunctochrysa* (but not *Chrysoperla carnea*) hunt aphids and other prey on bark. Du Merle (1983*a*) also examined the voracity of the chrysopid larvae in the laboratory. The average number of moth (*T. viridana*) eggs

consumed by an individual larva of chrysopid species was: *Dichochrysa prasina*: 47 eggs/day, *D. zelleri*: 32 eggs/day, *Cunctochrysa baetica* = 50 eggs/day.

Seasonal activity patterns of chrysopids in western and central European regions

There are no accounts of synchrony between seasonal patterns of green lacewings and their potential prey on oaks. We can demonstrate only the seasonality of the most predominant oak lacewings in Europe.

Barnard *et al.* (1986) examined the presence of lacewings between April and October in oak canopy in Great Britain. Adults of the dominant species *Chrysoperla carnea* showed no peak during this period. Larvae of *C. carnea* were represented in smaller numbers from mid-June up to mid-July reflecting the minor importance of their predatory role on oak trees. The flight activity of *Cunctochrysa albolineata* was from mid-June to the end of July while larvae were recorded from mid-July to mid-October. Larvae were present with greatest dominance (35.6%) in the canopy of *Quercus robur*. Adults of *Nothochrysa capitata*, third in rank of dominance, were active from late June to late July and larvae were active between mid-July and mid-October. The dominance value of larvae of *N. capitata* was 29.2% and its importance for predation followed that of *C. albolineata*. Flight activity of *Chrysotropia ciliata* was from mid-June to the end of July. Larvae were present in samples between late July and late September, except from late August to early September. Numbers of *Dichochrysa ventralis* were limited, and adults were observed from late June to mid-July.

Saure & Kielhorn (1993) examined the temporal occurrence of chrysopid adults on *Q. robur* in Germany from April to the end of October. The dominant *Chrysoperla carnea* was present during the whole of this period. *Nineta flava* occurred from the second half of July to early September. *Chrysotropia ciliata* was present from early June to early August and *Dichochrysa prasina* from early July to the beginning of September. *Chrysopa pallens* could be trapped during a short period in mid-July. Saure & Kielhorn (1993) could not define activity peaks for each lacewing species due to the low number of individuals in their samples.

Czechowska (1990) monitored the seasonality of lacewings in oak forests in Poland by yellow pan trapping (Table 10.7, column 2). *Nineta flava* was the predominant species, its seasonal pattern varying with place and year. Individuals appeared from early June to mid- or late September showing two peaks, the first in the first half of July, and the second between late July and mid-August. The size of the peaks varied according to the forest stand and year. *Chrysoperla carnea*, the second most abundant species, was present from early April to early June with a peak in the second part of May, and between August and September with another peak in early September. Low abundance characterised this species in June–July. *Chrysotropia ciliata* occurred from June to early September, but showed mass swarming in July–August with two peaks, in the first part of July and in early August respectively. *Dichochrysa prasina* was found from June to September, with a peak in late July to early August. *Chrysopa pallens* showed a very short occurrence period in late July to the end of August. Gepp (1974) also gave information on lacewings in the Kaiserwald forest (Austria), but defined only the length of seasonality of each species. *Chrysoperla carnea* appeared from April to October with the exception of July. *Chrysotropia ciliata* could be collected between early May and mid-July while the swarming flight of *D. prasina* lasted longer, until early August. *Dichochrysa ventralis* was present for only two months from the end of May to the end of July. Predatory adults of *Chrysopa pallens* were active from mid-July to mid-August. Szentkirályi (1984, 1998) published data on the seasonality of chrysopids from Hungary. Flight pattern data were based on intensive light trapping, and on long-term monitoring conducted in oak forests. Seasonal flight-activity patterns of *Chrysoperla carnea* and other chrysopids were analysed separately. Depending on locality and year the flight patterns of *Chrysoperla carnea* varied but could occur from March to the end of October. Mass flights occurred from June to late August. The first activity increase was in April (probably representing activity of overwintered adults), the second in May, a third in late June and early July, a fourth at the start of August, and the last in early October. However, the greatest predatory activity of *Chrysoperla carnea* larvae is limited to a short period, from early June to mid-August. Flight activity of other chrysopids lasted from mid-May to the end of October in general in Hungarian oak forests (Szentkirályi, 1984). Two peaks of flight activity were observed during the season, the first one in early June and the latter in the first part of August.

Seasonal activity patterns of chrysopids in southern Europe

Seasonality of lacewings in oak forests has also been examined in Mediterranean areas of Europe. The composition of lacewing assemblages living in dry and warm oak forests (for instance, *Quercetum ilicis*) was similar in these areas. For this reason, features of lacewing seasonality are given for each species. Data on seasonal activity patterns of the most dominant chrysopid species in oak forests are derived from Spain (Marín & Monserrat, 1987), southern France (Canard, 1987), and Italy (Pantaleoni, 1982, 1984, 1990*a*). In the Mediterranean region, *Chrysoperla carnea* showed great plasticity regarding possible number of generations per annum. Because it overwinters as adults, adult activity can be detected from March to December. The seasonal flight pattern of *Chrysoperla carnea* has five peaks in Spain as follows: at the end of June, end of July, first half of September, and in early October; prior to these, during April, individuals of the overwintering generation also form a distinct period of increasing activity. In France, the activity of overwintered adults was in March. From May to December three further peaks were recorded, the first in the second half of June, the largest one in mid-October, with a small peak in early August. A bimodal seasonality pattern was characteristic for *C. carnea* in Italy with one peak in the second half of June and another smaller peak in mid-September.

Chrysopa viridana is one of the most typical and abundant lacewing species in Mediterranean oak forests. In Spain four activity periods occur from the end of April to late October. The first peak occurs in early May, the second one in the first half of June, the next one after mid-July, and the last, greatest peak in middle of September. In Italy, Pantaleoni (1990*a*) also found four flight-activity periods in the case of *C. viridana*, though these occurred over a shorter season, in late June, in mid-July, in early August, and in early September. In other investigations, seasonality of *C. viridana* was characterised by a bimodal pattern (Pantaleoni, 1984), the first smaller peak in May, the second in the first half of August.

Dichochrysa prasina in Mediterranean regions showed the longest seasonal activity period in Spain, from late April to mid-October, while in southern France and Italy this period lasted from May to October or the end of September. In Spain, two large and two small activity peaks were found, while in the other two countries bimodal seasonality patterns were observed. On this basis, in Mediterranean regions, four activity periods could be separated relating to *Dichochrysa prasina* adults: an early peak in late April, a second one between May and the end of June with a larger peak in early June (Italy, Spain) or late June (France), the third a mass flight lasting from July to the end of September with a large peak in early August (Italy) or in late August (Spain), and finally a fourth smaller peak in late September (Spain and France).

Dichochrysa flavifrons was frequent in southern Europe, with longer seasonality and a trimodal pattern in Spain, so differing from other regions where bimodal activity patterns were observed. Adult *D. flavifrons* occurred in the various areas as follows: peak in mid-June (France, Italy) or end of June (Spain), in late July to mid-August (only in Spain), in early September (Italy) or in the second half of September (Spain) and in mid-October (France).

Data on seasonality of *D. zelleri* in oak stands are available only from France and Italy. Flight-activity patterns in both countries were similar and bimodal. The two periods of activity were from May to the end of June and from mid-July to the end of September. The first peaks in the two countries were concurrent in mid-June. The second peak was in the first part of September (France, Italy) or, depending on locality, it could be detected in the first ten days of August (Pantaleoni, 1990*a*).

Dichochrysa picteti was the other typical chrysopid species living in oak forests in the Mediterranean region, and its seasonality has been examined only in southern France and Italy. Adults of *D. picteti* were active between May and October with a more abundant peak in October. In Italian oak forests the first peak was reached in late May to early June, while the other activity period was from the end of July till mid-October with a peak in the second part of August.

The third *Dichochrysa* species was *D. iberica*, preferring oak trees, which has been monitored only in southern France (Canard, 1987). It was the most dominant species in the forest showing trimodal seasonal activity as a major character. The first flight period was between May and late July with a peak in late June. The second activity period was between August and October with a peak in early September. The last sea-sonal flight-activity period took place from October till mid-December with a peak in mid-October.

Cunctochrysa baetica occurred with low but regular frequency in oak forests and seems to be a univoltine species, although Canard (1987) reported two activity periods in June and in August–September. It could be collected in Spain in late September for a short period and in Italy from May to the end of July (Pantaleoni, 1984). In the latter case the peak of activity was recorded between the end of May and early or mid-June (Pantaleoni, 1990*a*).

Individuals of *Chrysopa pallens* were only collected in low numbers because of dry habitats in the region. Its short flight activity period is in the first half of August in Spain or in mid-August in Italy.

Seasonal activity patterns of hemerobiids in oak forests

Due to limited numbers in samples seasonal activity patterns of hemerobiids can only be analysed in a few cases. *Hemerobius humulinus* is one of the most constant members of hemerobiid assemblages in oak forests and occurs Europe-wide (Table 10.8). According to Barnard *et al.* (1986) *H. humulinus* appeared on oaks from May to mid-September, and showed a clear peak at the end of July. In Germany flight activity was detected from late April till mid-October with the major activity between mid-June and early August (Saure & Kielhorn, 1993). Seasonality of *H. humulinus* in thermophilous oak forests in Poland was from April to September. This species was most abundant in May and early June but had a second smaller peak of activity in July (Czechowska, 1990). According to Gepp (1974), in Kaiserwald (Austria) the seasonality patterns of *H. humulinus* consisted of two periods: the first between early April and mid-June, and the second from the first ten days of June to early September. On the basis of long-term light-trapping data in Hungary the seasonal flight period of *H. humulinus* was from early April to the end of October (Szentkirályi, 1992, 1997). During April and May activity of the overwintered generation was detected in a first smaller peak in June, followed by three further activity peaks with mass flights in July, August, and the second half of September. Swarming activity gradually reduced till late October. There are at least four generations of this species in Hungary, both in forests and in cultivated fields. In southern Europe only Pantaleoni (1990a) has investigated the seasonality of this species. He detected three yearly generations of H. humulinus in Italy. The swarming activity of the first generation covered the period from mid–May to mid–June, and the mass flight occurred in July with a sharp peak at the end of the month. The third period of activity was recorded during September with a smaller peak in the middle of the month.

The other significant, common hemerobiid species in European oak forests is Sympherobius pygmaeus, which has a shorter seasonality in northern than in southern regions. Flight activity in Britain is mid–June to mid–August (Barnard et al., 1986). A short peak of flight activity was detected in June in oak forests in Germany and Poland. Due to the limited number of individuals, the flight patterns in these oak forests could not be assessed (Czechowska, 1990; Saure & Kielhorn, 1993). Using long-term light-trapping data (Szentkirályi, 1992, 1997, 1998) found the flight pattern of S. pygmaeus in central Europe to be similar to that detected in Mediterranean regions. Swarming took place from early May to mid–October in Hungary, Spain, and Italy, with three activity periods. The first was from mid–May till mid–June with a lower peak in early June. The second mass flight covered the period from early July to late August with two peaks; the first in the second half of July, the second in mid–August. The third flight period was in September–October with a smaller maximum in mid–September. Sympherobius pygmaeus is supposed to have three to four generations yearly in Hungarian oak forests. In Spain, Marín & Monserrat (1987) found four flight activity periods with one peak in each. During the season gradually increasing peaks occurred; the first low peak in early May, the second in the first part of June, the third in the second half of July to early August, while the last and highest peak was recorded in mid–September. Three flight periods were recognised in Italy, their peaks also increasing as the season progressed (Pantaleoni, 1990a). The first low peak was found at the end of May. During the next swarming period the second peak, mid–June to the end of July, produced maximum catches in the first half of July. The period of mass flight of S. pygmaeus took place from mid–August to the last ten days of September, and its peak was registered in the first half of September. According to these detailed seasonal activity patterns S. pygmaeus has three to four

generations yearly in oak forests located in the warmer regions of Europe with a major predation period from mid–July to mid–August and in mid–September.

The duration of Hemerobius micans activity in oak forests varied, lasting from May to mid–August in Great Britain and in Poland (Barnard et al., 1986, Czechowska, 1990), from mid–May to late July in Austria (Gepp, 1974), from mid–April to the end of October in Hungary (Szentkirályi, 1997, and unpublished data), and from early June to early August in Italy (Pantaleoni, 1990a). In the latter case one individual of H. micans occurred in mid–September. During the swarming period no distinctive peak was observed but most individuals of H. micans were collected in May and August in Poland and these peaks were similar to maximum captures in Hungarian light traps. A small peak occurred in the second half of May, one sharp peak in early July, a further one in late July to early August, and the last but also smallest peak in the first half of September. Mass flights took from early July to the end of August. Pantaleoni (1990a) showed two peaks of H. micans flight patterns in oak forests in Italy, the first at the turn of June to July and another in late July. These two maxima of H. micans coincided with the two peaks of mass flight found in Hungary.

The fourth hemerobiid species living in oak forests is Wesmaelius subnebulosus. Examination of its seasonality have only been made in Hungary (Szentkirályi, 1992, 1997, 1998), Italy (Pantaleoni, 1982), and Spain (Marín & Monserrat, 1987). Seasonal activity of W. subnebulosus was detected from March to mid–November in Hungary and Italy. Adults are adapted to lower temperatures and remain active during the early and late season and, moreover, can overwinter in mild winters. The number of individuals captured was low in oak forests in Spain and seasonality took from early April to the end of July. Wesmaelius subnebulosus had three activity periods in Hungarian oak forests, the first in May–June with a limited peak in mid–May. The second period covered late June to mid–August coinciding with mass flight, with a greater peak in late July to early August. The third period was from mid–September to mid–October with a smaller peak in early October. The highest number of individuals appeared in the first part of April in Spain with a smaller peak of activity also in early June, and later on limited activity from the end of June to mid–July. Pantaleoni detected a trimodal seasonal pattern of *W. subnebulosus* in Italy. During the season three well-separated peaks of flight activity could be detected. The first covered mid-April, the second, greater one, at the turn of June to July (the period of mass flight), the third smallest one in mid-October. According to these seasonality patterns *W. subnebulosus* has three to four generations per year in oak forests of central and southern Europe.

Sympherobius klapaleki is another lacewing species preferring oak, being active in Germany from the first ten days of May to the last ten days of June as well as from mid-August to mid-September. Saure & Keilhorn (1993) considered these two flight-activity periods as two generations. In Poland *S. klapaleki* had only one long activity period in May, then swarming intensity gradually decreased until the end of August.

Hemerobius gilvus reaches a high population level in warmer Mediterranean oak forests, and its seasonality can be observed better there. Pantaleoni (1990*a*) collected this hemerobiid species from *Quercus pubescens*. A characteristic bimodal seasonal pattern consisted of two separated periods, the first from the beginning of May to the end of June, the second from mid-August to the end of September. Peaks of population were recorded in mid-May and in the first part of September.

Surveying these seasonality patterns suggests that predatory activity of hemerobiids in European oak forests may be expected from mid-June to the end of August. Characteristic maximum mean activity patterns of hemerobiids can be defined in general: thus for example at least three generations of most species is likely in central and south European oak forests. These periods were detected in Hungary in May–June, mid-June to the end of August, and mid-September to mid-October (Szentkirályi, 1984). Marín & Monserrat (1987) described similar peaks in the seasonal hemerobiid patterns, as follows: in June, in early July to mid-August, and in September.

10.4.3 Other broad-leaved trees
Numerous data on lacewing fauna of other deciduous tree species are available in the neuropterological literature but the overwhelming majority of collections have been made in Europe. These broad-leaf tree species rarely form separate stands; they are elements rather of mixed forests. Given their economic importance a short overview on associated lacewing assemblages to these trees is relevant. There are both hardwood- and soft-

wood producing species among these trees, which are used for industrial tools, agricultural timber (for instance, *Ulmus, Alnus* spp.), furniture (*Tilia, Ulmus, Carpinus*), building timber (such as *Alnus*), shipbuilding (for instance *Ulmus*), paper- and match-making (*Populus* spp.), as well as for other industrial raw materials (*Fraxinus, Acer, Tilia* spp.). *Populus* and *Salix* species that grow rapidly produce a huge amount of biomass during a short period and are grown in monoculture as agricultural crop plants. Due to these advantageous features softwood species are used as new sources of bio-energy in order to extract alcohol, glucides, or chips for burning. Many of the broad-leaf tree species discussed here are planted along highways and in the streets of cities or as ornamental plants in urban parks, for example hornbeams, lime, maple, or birch.

The European lacewing fauna recorded on other broad-leaf tree species is reviewed on the base of the following papers: UK: Killington (1937); Poland: Czechowska (1990); Germany: Ohm (1965); Switzerland: Eglin (1980*a, b*); Austria: Ressl (1971, 1974), Gepp (1974, 1988/9); Czech Republic and Slovakia: Zelený (1963, 1971, 1978), Vidlička (1994); Hungary: F. Szentkirályi (unpublished data); Spain: Monserrat & Marín (1994, 1996); Italy: Pantaleoni (1982, 1984); Croatia: Devetak (1991); Bulgaria: Popov (1991). Tables are not given separately, and the method of evaluation is the same as applied elsewhere in this chapter.

Hornbeams
Two hornbeam species, *Carpinus betulus* and *C. orientalis* (Corylaceae) are distributed in natural plant associations in Europe. The first is usually found in mixed forests with oak and beech, while the second can be found in the Mediterranean area of southern Europe. Altogether 15 green lacewing species have been recorded from European hornbeams. The number of recorded species from surveyed countries ranges from four to ten. On the basis of frequency of occurrence and dominance values (%) the following nine chrysopid species may be considered as characteristic lacewings related to *Carpinus* stands in decreasing order of dominance: *Chrysotropia ciliata, Nineta flava, Chrysoperla carnea, Chrysopa pallens, Dichochrysa prasina, D. ventralis, D. flavifrons, C. perla,* and *Hypochrysa elegans*. The first three species are the most frequent and abundant. Further chrysopid species represented by low

numbers of individuals and only one location were as follows: *Dichochrysa inornata, N. vittata, Cunctochrysa albolineata, Nothochrysa fulviceps, N. capitata* and *Nineta inpunctata*.

Only nine brown lacewing species have been recorded on hornbeam but five can be considered to be characteristic: *Hemerobius humulinus* (14%–67%), *H. micans* (17%–79%), *H. lutescens, Drepanepteryx phalaenoides*, and *Sympherobius klapaleki*. Individuals of *S. pygmaeus, Micromus angulatus, H. marginatus*, and *Wesmaelius nervosus* were captured from hornbeam trees only in one country. Numbers of hemerobiid species per country ranged from three to six.

The most probable prey of lacewings living on hornbeams are Homoptera. Thirteen species of aphid are associated with *Carpinus* species. In Europe and the Middle East the undersides of leaves of *C. betulus* are infested by *Myzocallis carpini* aphids. According to observations in oak–hornbeam forests in Hungary larvae and adults of *H. micans* as well as *H. humulinus* prey on this aphid (F. Szentkirályi, unpublished data). Scale insects are not specialised on *Carpinus* species although the oak pest *Parthenolecanium rufulum* as a polyphagous species can be detected from time to time. Schmutterer (1972) mentioned from the forestry literature that larvae of *Dichochrysa prasina* prey on this coccid species. Seven whitefly species were found on hornbeam trees. Populations of *Asterobemisia carpini* and *Pealius quercus* associated with *C. betulus* may also be prey of lacewings.

Research on seasonality of potential prey of lacewings living on hornbeam has been neglected. Populations of the aphid *Myzocallis carpini* appear to infest the foliage of hornbeam at a high density in springtime (April–May) and in autumn (September to early November) (F. Szentkirályi, unpublished data).

Limes

In the case of European lime trees (Tiliaceae) lacewings have been recorded only on *Tilia cordata* (small-leafed lime) and *T. platyphyllos* (large-leafed lime). Altogether 12 chrysopid species have been collected from *Tilia* species. On the basis of frequency of occurrence and dominance values (%) the following six green lacewings seem to be characteristic species on lime trees: *Nineta flava* (28%–59%), *Chrysotropia ciliata* (35%), *Chrysoperla carnea* (27%), *Dichochrysa prasina* (7%–21%), *N. vittata*, and *Cunctochrysa albolineata*

(21%). In addition a few individuals of *Chrysopa pallens, C. perla, D. flavifrons, Nothochrysa fulviceps, N. inpunctata*, and *Hypochrysa elegans* were collected on lime in certain countries. The number of green lacewing species found on *Tilia* trees was about two to eight from country to country.

Seven species of hemerobiids have been recorded on European *Tilia* species. Five species may be regarded as characteristic, as follows: *Hemerobius humulinus* (38%–68%), *H. micans* (11%–36%), *H. lutescens* (25%), *Sympherobius klapaleki* (17%), and *H. marginatus* (1%). From *T. cordata* trees, *Wesmaelius nervosus* and *Micromus angulatus* have also been collected in Poland (Czechowska, 1990). The number of hemerobiid species recorded on limes in various countries ranges from two to five.

From the potential homopteran prey of lacewings 13 species of aphids are represented on 16 American and Palaearctic *Tilia* species. In Europe *Eucallipterus tiliae* and *Patchiella reaumuri* aphids infest *T. cordata* and *T. platyphyllos. Tiliaphis* species occur on limes in east Asia. Populations of *E. tiliae* can be found in high numbers on lime foliage in Hungary during May and July and from late August to the end of October. During this latter autumnal period (in late August to late September), third-instar larvae of *H. humulinus* were observed on strongly infested leaves of *T. cordata* as they were attacking *E. tiliae*.

From coccids, the polyphagous *Eulecanium tiliae, Lepidosaphes ulmi* and *Quadraspidiotus ostreaeformis* infest *Tilia* species. Four species of whiteflies are recorded from *Tilia*. There are no data relating to lacewings attacking the latter two homopteran groups on lime trees.

Maples

In Europe the maples (Aceraceae) most frequently examined for lacewings are *Acer campestre* and the European sycamore *A. pseudoplatanus*. Less frequently collections of lacewings have been made from *A. platanoides, A. tataricum* (Hungary), and *A. monspessulanum* (Spain). Altogether 15 chrysopid species have been detected in maples and ten might be considered potential members of characteristic lacewing assemblages. These chrysopids are as follows: *Dichochrysa prasina* (59%), *Chrysotropia ciliata* (27%), *Chrysopa pallens* (18%), *Nothochrysa fulviceps* (14%), *Nineta flava, Chrysoperla carnea* (s. lat.), *Chrysopa perla, C. viridana*,

D. ventralis, and *D. flavifrons*. It seems that these more frequently occurring lacewing species have no preference for any *Acer* species, possibly due to the similar prey on each. *Dichochrysa picteti*, *N. vittata*, *Cunctochrysa albolineata*, *Italochrysa italica*, and *Hypochrysa elegans* are rare on *Acer*. Aspöck *et al.* (1980) also mentioned *D. ariadne* among green lacewings from maples. The recorded number per country of green lacewings living on *Acer* ranged from three to nine.

Thirteen hemerobiid species have been found on European maple trees. On the basis of frequency of occurrence and dominance percentage (%) those associated with *Acer* are: *Hemerobius humulinus* (18%–58%), *H. micans* (18%–23%), *H. lutescens* (9%–10%), *H. marginatus* (3%–27%), *Sympherobius elegans*, *S. pygmaeus*, *Wesmaelius subnebulosus*, *Micromus lanosus*, and *W. nervosus*. A limited number of individuals of *M. variegatus*, *M. paganus*, and *S. pellucidus* as well as *H. stigma* are also present on maple trees.

Several aphid genera are specific to *Acer* trees and are represented by many monophagous aphids (Blackman & Eastop, 1994). Aphids using maples as host plants mostly belong to *Periphyllus* (35 spp.) and *Drepanaphis* (20 spp.).

Periphyllus aceris mainly infests the hedge maple, *A. campestre* and *A. platanoides*, while *P. acericola* and *P. testudinaceus* attack *A. pseudoplatanus*. *Drephanaphis acerinum* and *D. platanoidis* prefer *A. pseudoplatanus*, and *A. platanoides* is preferred by *D. acerifoliae*. According to investigations in Hungary, *Periphyllus* aphids living on *Acer* trees are most abundant on foliage during May–July and from late September to early November (Balázs, 1963; F. Szentkirályi, unpublished data). At high levels of infestation the aphid may be the preferred food for lacewings during these periods. In early May adults of *H. humulinus*, *H. micans*, and *H. lutescens* visited *A. platanoides* infested by *Periphyllus* spp. in Hungary. During the period of late September and October egg clusters of *Nineta flava* were laid on aphid-infested leaves of *A. platanoides*, while young larvae of *H. humulinus* preying on aphids were observed on *A. tataricum* trees in Hungary (F. Szentkirályi, unpublished data).

Two species of scale insects, namely *Acanthococcus aceris* and *Phenacoccus aceris*, frequently infest *Acer* species in large numbers. Their presence may explain the preference of coccidophagous lacewings (such as *Dichochrysa* and *Sympherobius*) for maples.

Fourteen species of aleyrodids are associated with the *Acer* genus; for instance the monophagous *Aleyrodes aceris* infests *Acer platanoides*, and *Aleyrodes acerinus* infests mainly *Acer campestre*. *Rhinocola aceris* is most frequent psyllid on maple trees. Data on lacewings preying on these two Homoptera groups on maples are lacking.

Ashes

Lacewings living on ashes (Oleaceae) have been sporadically examined. There are data available on lacewings collected from *Fraxinus excelsior*, *F. ornus*, and *F. angustifolia* trees in Europe only from the Czech Republic, Slovakia, Hungary, Spain, and Croatia. Altogether, seven chrysopid species have been found on *F. ornus* and *F. angustifolia* trees in the Mediterranean region. Of these only five may be considered as characteristic lacewing species associated with ashes: *Chrysoperla carnea* (36%), *Dichochrysa prasina* (32%), *D. flavifrons* (15%), *D. subcubitalis* (16%), and *Chrysopa viridana*. Sometimes *Nineta flava* and *C. pallens* are also collected on ashes. Five species were recorded on ash trees in Spain and Croatia.

Nine hemerobiids were found on ashes, with *Hemerobius humulinus* and *H. micans* characteristic. Further hemerobiid species collected from *Fraxinus* were: *Wesmaelius subnebulosus*, *H. lutescens*, *H. marginatus*, *H. gilvus*, *Micromus lanosus*, *Sympherobius elegans*, and *M. tortricoides*. Aspöck *et al.* (1980) also mentioned *S. pygmaeus* on *Fraxinus*.

Of the potential homopteran prey of lacewings numerous species of Pemphigini aphids are specialised to ashes. In particular certain representatives of *Prociphilus* aphids use ashes as primary host plants. For instance, *Prociphilus fraxini*, *P. americanus* or *P. oriens* are associated with *F. excelsior*. The fact that species belonging to other aphid genera are not associated with European ashes may explain why so few species of lacewings have been found on *Fraxinus* trees. The spring generations of *Prociphilus* species producing leaf nest colonies may serve as prey for *Sympherobius* hemerobiids as observed for other gall-forming aphids.

Of the scale insects, *Pseudochermes fraxini* is associated with *Fraxinus* species. This coccid is widely dispersed on ashes in Europe. In sites with a warmer climate it frequently becomes a forest pest due to its high abundance. Populations of this pest may offer a food source on *Fraxinus* trees for *Dichochrysa* lacewing species preferring scale insects. Individuals of *P. fraxini* infesting the trunks of *F. ornus* are regularly preyed on by larvae of *Dichochrysa prasina*. This may explain why *Dichochrysa* species (*prasina, flavifrons,* and *subcubitalis*) were recorded with higher total dominance (63%) on trees of *Fraxinus angustifolia* in Spain.

In Psyllidae, the monophagous *Phyllopsis fraxini* attacks *F. excelsior* across Europe. It may be eaten by lacewings on ashes, but there are no relevant data about lacewing predation.

Elms

Lacewing collections have been conducted on *Ulmus minor, U. glabra, U. campestris,* and *U. scabra* elms (Ulmaceae) growing in Europe. Altogether 13 chrysopid species were recorded on elms. Of these 12 were found on trees of *U. minor* according to long-term collecting in Spain by Monserrat & Marín (1994). Only three chrysopid species were present on elms in each country studied. On the basis of frequency of occurrence and dominance value (%) four green lacewings seem to be characteristic for elms: *Chrysoperla carnea* (51%), *Dichochrysa prasina* (28%), *D. flavifrons* (8%), and *Chrysopa pallens*. Further chrysopid species were recorded in Spain as follow: *C. viridana* (3%), *D. ventralis, D. granadensis, D. picteti* (4%), *D. venosa, D. subcubitalis* (2%), *D. iberica,* and *Italochrysa italica*. Moreover, *Nineta flava* was found on elms on a north German island (Amrum) (Ohm, 1965).

Only five species of hemerobiid have been observed on elms. Two species may be considered as typical of elm, namely *Sympherobius pygmaeus* (92%) and *Wesmaelius subnebulosus*. Other hemerobiids collected from elm trees in central Europe are: *Hemerobius humulinus, H. micans,* and *H. lutescens*. Aspöck *et al.* (1980) indicated that *Drepanepteryx phalaenoides* is also associated with elms.

Of potential homopteran prey 68 aphid species have been recorded on elms. *Ulmus* is a primary host plant of 44 species of Eriosomatini, gall-forming aphids. Of the other important Homoptera groups only the elm woody coccid, *Gossyparia spuria,* infests *Ulmus* species. This pestiferous scale is widely spread and mainly infests young elm trees in Europe. Although concrete observations of lacewing predation on this aphid are absent, the high dominance of *S. pygmaeus* and numerous *Dichochrysa* species, 47% combined, is indicative of predation in the case of *U. minor* in Spain.

Seven whitefly species from the Aleyrodidae live on *Ulmus* species, such as *Asterobemisia carpini*. Data on predation of lacewings on whitefly are not available.

Alders

Data relating lacewings to on alders (Betulaceae) have been published only on three species, namely the European alder *Alnus glutinosa, A. incana,* and *A. viridis*. Forest associations (*Alnetum*) created by alders provide special habitats with wet and cool microclimate for lacewings. These associations are located along the edges of brooks, in flood plains, in other wetlands (e.g. *A. glutinosa*), and in high mountainous regions (*A. incana, A. viridis* in the Alps; Eglin, 1980a, b, c).

Of the green lacewings 12 species have been detected on European alders. The following eight chrysopid species can be regarded as typical elements of lacewing assemblages in *Alnetum* associations: *Chrysoperla carnea* (42%–88%), *Chrysopa perla* (4%–29%), *Chrysotropia ciliata* (53%), *Dichochrysa flavifrons* (3%–28%), *Dichochrysa prasina* (4%–22%), *Chrysopa pallens* (6%), *Nineta flava,* and *Cunctochrysa albolineata*. Limited numbers of individuals of *Chrysopa formosa, C. viridana, N. vittata,* and *Hypochrysa elegans* have also been captured from *Alnus* trees in single countries. The number of chrysopid species per country recorded on alder ranges between three and seven.

Altogether 12 hemerobiid species have been recorded from alder trees. Six species of brown lacewings are characteristic of the *Alnus* genus as follows: *Hemerobius humulinus* (78%), *H. lutescens* (2%), *H. micans, Micromus angulatus* (12%), *Drepanepteryx phalaenoides,* (3%) and *Wesmaelius subnebulosus* (2%). Other hemerobiids such as *W. nervosus, H. perelegans, H. marginatus, M. paganus, M. lanosus,* and *Psectra diptera* have also been collected in some countries. The number of species by country ranges from two to seven.

The presence and high abundance of certain chrysopids and hemerobiids indicate the wet and cooler microclimatic conditions of *Alnetum*. Examples are *Chrysopa perla* (in Switzerland and Austria), *Chrysotropia ciliata* (Austria), and *H. lutescens* or *H. micans* (Switzerland, Austria, and Slovakia). Association between *Sympherobius elegans* and alders was mentioned by Aspöck *et al.* (1980).

Fifty aphid species in 24 genera have been recorded on alders. Many of these feed exclusively on *Alnus*. Eight species infest European alder, *A. glutinosa,* including *Pterocallis alni, P. maculata, Glyphina betulae, Betacallis alnicolens,* and *Clethrobius comes.* According to Eglin (1980*b*), in the Swiss Alps *Chrysoperla carnea, Chrysopa perla, Dichochrysa prasina, H. humulinus,* and *H. micans* prey on *P. albidus* infesting *A. incana. Hemerobius humulinus* consumed the *P. alni* on *A. viridis.*

Psylla alni specialises on *Alnus* in Europe. All the above-mentioned lacewing species fed on this psyllid and the aphids on *A. incana* and *A. viridis* in Switzerland. Only three aleyrodids and nine coccids live on alders in Europe, but data are not available on whether they are consumed by lacewings or not.

Birch

Lacewing assemblages associated with *Betula pendula* have been examined both in Europe and in North America. The species composition is close to that on alders, probably reflecting a similar prey spectrum. Twelve chrysopid species have been recorded on birch, and six of these are frequent and dominant in Europe. These are: *Chrysoperla carnea* (41%), *Dichochrysa prasina* (25%), *D. ventralis* (16%), *Chrysopa perla, Nineta flava,* and *Cunctochrysa albolineata.* In Europe *Chrysopa pallens, D. flavifrons, D. inornata, Chrysotropia ciliata, Nothochrysa fulviceps,* and *N. inpunctata* were sporadically found on silver birch. Hajek & Dahlsten (1988) found three species on *B. pendula* in northern California (USA) as follows: *Chrysoperla plorabunda* (Fitch), *C. rufilabris* (Burmeister), and *Chrysopa nigricornis* Burmeister. These are all common and abundant in North America. The repeated occurrence on birch and strong association with it of such species as *Chrysopa perla* or *Cunctochrysa albolineata* may be the result of the wet and cooler climate of habitats created by *B. pendula* forest stands in Europe. The number of green lacewing species recorded on birches varied from one to eight per country.

Altogether 12 species of brown lacewings have been recorded on European birches. Seven species seem to be typical on *B. pendula.* They are: *Wesmaelius nervosus, Hemerobius micans, H. perelegans, H. marginatus, H. humulinus, H. lutescens,* and *Micromus paganus.* The most frequent were *W. nervosus* and *H. micans.* The number of species on birch ranges from two to five from country to country. *Hemerobius pacificus* was the only hemerobiid recorded on *B. pendula* by Hajek & Dahlsten (1988) in California.

Sixty-one aphid species in 18 genera feed on *Betula.* Only a small number of aphid species restrict their feeding to single species of birches (Blackman & Eastop, 1994). Sixteen species are associated with *B. pendula.* The most important genera recorded on birch are: *Betulaphis, Calaphis, Callipterinella, Euceraphis,* the gall-forming *Hamamelistes, Monaphis, Stomaphis,* and *Betacallis.* In Europe in the Alps Eglin (1980*a, b*) found that *Chrysoperla carnea, H. micans, H. humulinus,* and *M. paganus* preyed on *Euceraphis betulae* on birch trees. Hajek & Dahlsten (1988) studied natural enemies of aphids living on *B. pendula* introduced as an ornamental plant into northern California. They recorded three species: *Callipterinella calliptera, Betulaphis brevipilosa,* and *Euceraphis betulae* that commonly occur also in Europe. All the above-mentioned lacewings were found as predators of these aphids on birch trees (Hajek & Dahlsten, 1988).

Poplars

There are many species of poplars (Salicaceae) and, in particular, varieties and hybrids of *Populus canadensis* are widely cultivated. However, lacewing records are only available for *P. tremula, P. alba,* and *P. nigra.* Altogether 11 chrysopid species have been collected from poplars in the central and southern regions of Europe. Six species are typical and common on *Populus* trees as follows: *Chrysoperla carnea* (16%–55%), *Dichochrysa prasina* (16%–50%), *D. flavifrons* (6%–16%), *Chrysopa pallens* (6%), *C. formosa* (5%), and *C. perla.* Further species of green lacewings on poplars are found occasionally: *D. picteti* (4%), *C. viridana* (3%), *D. ventralis, C. abbreviata* (4%), *Nineta flava,* and *C. dorsalis.* The last is strongly associated with pines. The number of chrysopid species collected per country ranges between one and eight.

Nine hemerobiid species have been detected on poplars, but only five can be considered typical, namely *Sympherobius pygmaeus* (67%), *Hemerobius humulinus* (22%), *Wesmaelius subnebulosus* (6%), *H. micans,* and *H. lutescens.* In various countries *S. elegans, S. pellucidus, H. handschini,* and *S. gratiosus* are also collected occasionally from *Populus.* The last is strongly associated with poplars in Spain (Monserrat & Marín 1996). The number of species on *Populus* varies from one to four per site. Aspöck *et al.* (1980) also indicated *W. malladai* living on poplars.

About 120 poplar-feeding aphid species in 24

genera are known from all over the world. Some species seem to be host-specific. Poplar aphids may be free-living species or pseudogall- or gall-formers. The latter all belong to the Pemphiginae. Of the aphid species 22 live on the white poplar, *P. alba*, while 30 species infest the black poplar, *P. nigra*. European aspen, *P. tremula*, hosts 12 aphid species. Aphids infesting *Populus* belong to the genera *Chaitophorus*, *Doraphis*, *Pachypappa*, *Pemphigus*, *Phylloxerina*, *Pterocomma*, and *Thecabius*. Of the aphidophagous lacewings, egg batches (17–38 eggs per cluster) deposited by *Chrysopa pallens* were observed on the foliage of *Populus alba* with strong aphid infestations in Hungary. Adults of *H. micans* and *W. subnebulosus* were also collected on trees of *P. canadensis*. From this latter poplar in Hungary heavily infested by *Pemphigus* sp. larvae of *S. pygmaeus* occur inside galls formed on leaf petioles (F. Szentkirályi, unpublished data). Eglin (1980*b*) collected numerous lacewings, *Chrysoperla carnea*, *Dichochrysa flavifrons*, *D. ventralis*, *Dichochrysa prasina*, *H. humulinus*, *H. micans*, *H. lutescens*, and *S. pellucidus* from aphid-infested trees of *Populus tremula* in high-mountainous habitats in the Swiss Alps. Pantaleoni (1982, 1984) recorded many adults of *S. pygmaeus* from *P. nigra* ssp. *pyramidalis* trees in Italy infested by populations of *Chaitophorus leucomelas* and *Pemphigus spyrothecae*. He also found larvae of this brown lacewing species inside galls created by *Pemphigus*. Both Italian and Hungarian observations suggest that *Sympherobius* lacewings may be strongly associated with *Populus* in order to exploit populations of *Pemphigus*.

From the polyphagous scale insects 15 species infest *Populus* in central Europe (Kozár & Kosztarab, 1980). Some coccids (*Diaspidiotus gigas*, *D. perniciosus*, and *Lepidosaphes ulmi*) reach high population densities and may be potential alternative food for lacewings. The continuous presence of larvae and adults of these coccids during the whole season on the trunk and branches may greatly contribute to maintenance of *Dichochrysa* and *Sympherobius* lacewings in *Populus* stands.

Other groups of phytophagous insects also appear on poplars and are eaten by lacewings. *Chrysomela populi* and *C. tremulae* (Chrysomelidae) are major pests on poplars. *Chrysoperla carnea* and *Chrysopa formosa* were among the principal predators of both *Chrysomela* beetles in Turkey (Zeki & Toros, 1990). Augustin & Lévieux (1993) studied the life history of *Chrysomela tremulae* on stands of six poplar species in the central region of France. The leaf beetle produced two to three generations yearly, preyed on by various insects. Among them larvae of *Chrysoperla carnea* were observed to attack the first two larval stages of *Chrysomela tremulae*.

Willows

About 200 species of willows (Salicaceae) grow in the continental temperate northern hemisphere. Data have only been published on lacewings from *Salix alba*, *S. caprea*, *S. fragilis*, *S. repens*, and *S. purpurea* willow trees in Europe. Altogether 14 chrysopid species have been found on willows. Seven are characteristic and common on *Salix*: *Chrysoperla carnea* (53%), *Chrysopa abbreviata*, *Dichochrysa prasina* (14%), *D. flavifrons* (8%), *C. perla*, *Chrysopa pallens* (8%), and *Nineta flava*. Further chrysopids also occur, namely: *C. formosa*, *C. phyllochroma*, *D. ventralis*, *D. picteti*, *N. vittata*, *Nothochrysa fulviceps*, and *C. walkeri*.

Only seven hemerobiid species have been recorded in European willow stands, in certain countries and in low abundance. The number of species in any country ranged from one to three. These are: *Wesmaelius subnebulosus*, *Hemerobius humulinus*, *H. lutescens*, *Sympherobius pygmaeus*, *Drepanepteryx phalaenoides*, *Micromus variegatus*, and *Psectra diptera*.

Of these lacewings recorded on *Salix*, many species prefer the herbaceous plant layer. These are *C. abbreviata*, *C. phyllochroma*, *C. walkeri*, and to some extent *C. formosa*, and the hemerobiids *P. diptera* and *M. variegatus*. Numerous authors have found *C. abbreviata* characteristic and abundant on willows (Ohm, 1965; Eglin 1980*b*; Gepp, 1988/9; Popov, 1991). The association of lacewing species preferring low vegetation with the *Salix* genus is reasonable because willows are regularly bushes, or the canopy is kept low in order to harvest willow twigs easily. Lacewings preferring wet and cooler habitats can be found on the list of species, such as *C. perla*, *H. lutescens*, and *D. ventralis*.

Aphids are the most numerous prey on *Salix*. More than 120 aphid species use willows as food. The most important species belong to *Chaitophorus* (48 spp.), Pterocommatini (29 spp. in four genera), and *Cavariella* (20 spp.). For instance, white willow, *S. alba*, is associated with 28 aphid species, goat willow, *S. caprea*, with 25 species, and 15 species live on grey willow, *S. cinerea* (Blackman & Eastop, 1994). Despite this rich aphid fauna on willows, few data are available on how lacewings prey on them. Eglin (1980*b*) observed individuals of *Chrysopa abbreviata* consuming *Aphis farinosa* that infested *Salix* species in the Swiss Alps. *Aphis saliceta* on *S. purpurea* was eaten by *Chrysoperla carnea*.

More or less the same species of scale insect infest *Salix* as poplars. From the 13 scale species, *Chionaspis salicis* and *Lepidosaphes ulmi* may be regarded the most harmful pests for willows in central Europe (Kozár & Kosztarab, 1980). Reaching high infestation levels these two scale insects may be important food for *Dichochrysa* in willow stands in Europe. *Salix* serve as host plants for 15 species of whiteflies. Data are not available on lacewing predation on scales or whiteflies on *Salix*.

Eucalypts

Eucalyptus (Myrtaceae) forms natural dry forest in Australia and Indonesia. Because of their extremely rapid growth and considerable water consumption/transpiration they have been introduced into tropical and subtropical areas in order to dry out marshlands as well as to gain higher timber production. They serve as an important source of wood as well as ornamental trees planted in public parks or along alleys. Moreover they provide curative volatile oils and other medicines. *Eucalyptus globulus* has been introduced and cultivated in the Mediterranean region of Europe.

Only Monserrat & Marín (1994, 1996) working in southern Spain have collected lacewings associated with *E. globulus*. They recorded 11 chrysopid species on eucalypt foliage. The characteristic lacewings with greater dominance values on *E. globulus* are: *Brinckochrysa nachoi* (43%), *Chrysoperla carnea* (32%), *Dichochrysa flavifrons* (7%), *D. picteti* (5%), *D. subcubitalis* (5%), and *D. genei* (5%). A few individuals of further lacewing species were also detected: *Italochrysa italica*, *Chrysopa formosa*, *Cunctochrysa baetica*, *Chrysoperla mediterranea*, and *Chrysopa mutata*.

Monserrat & Marín collected only two hemerobiid species, *Wesmaelius subnebulosus* (three individuals) and *Megalomus tineoides* (one individual) from *E. globulus*.

It is probable that polyphagous Homoptera species are the main food sources for lacewings on eucalypt trees in the Mediterranean region. The presence of more species and higher total dominance (22%) of *Dichochrysa* and the outstanding abundance of *B. nachoi* suggest that scale insects may be the major prey of green lacewings on *E. globulus*. According to the

Palaearctic scale insect catalogue (Kozár, 1998), 13 coccids, 12 diaspidids, and 1 phoenicococcid species are associated with *Eucalyptus*. There are many scale insect species among them that are also considered as harmful pests for other cultivated plants (for instance, *Olea* or *Oleander*) in the Mediterranean region, such as *Coccus hesperidum*, *Saissetia oleae*, or *Aspidiotus nerii*. Populations of the monophagous *Pseudaulacaspis eucalypticola* or the polyphagous *P. eugeniae* more frequently infest eucalypts in Asia and Australia. The suspicion that *B. nachoi* (in the Iberian Peninsula) is a coccidophagous or psylloidophagous species, is supported by regular occurrence of *B. michaelseni* (in Greece and Crete) on olive trees, where coccids *S. oleae* and *A. nerii* and the psyllid *Euphyllura olivina* are the major food source.

The possibility of lacewings preying upon psyllids on eucalypts is also supported by the work of Majer & Recher (1988), conducted in natural forest stands in Australia. They collected and compared arthropod assemblages of four eucalypt species (*Eucalyptus wandoo*, *E. marginata*, *E. accedens*, and *E. calophylla*) using the chemical knockdown and branch-clipping method in Victoria. In each of the forests sampled the dominance of Neuroptera was at a very low level (below 1%). However, our calculations from their results show that lacewings increased their abundance level up to three to four times in response to a strong outbreak of a psyllid *Eucalyptolyma* sp. (dominance range 40%–80%) on sampled trees of *Eucalyptus calophylla*, compared to other eucalypt species with much lower dominance (range 8%–24%) of this potential prey.

10.5 CONIFERS

10.5.1 Importance, composition, and distribution of coniferous forests

Conifers are represented by about 550 species, and are distributed worldwide. The highest number of species is in East Asia and the temperate areas of North America. The trees have economic importance in the following fields: industrial, building, mining, and piling timber (for instance, *Pinus silvestris*, *Picea abies*, *Abies alba*, *Larix* spp., cedars); paint industry (*Araucaria*); joinery (*Pinus silvestris*, *Picea abies*, *Pinus strobus*, *Larix decidua*); resin production (*Pinus nigra*); shipbuilding (*Picea abies*); cellulose industry (*P. abies*); and the chemical industry (*P. abies*, *Abies alba*, *A. balsamea*). Due to their aesthetic qualities, fast-growing conifers are frequently planted as ornamental trees in urban parks (e.g., *Pinus mugo*, *P. strobus*, *Picea pungens*, or *Cedrus* spp.). Growing in poor soil and tolerating drought, certain Pinus species (*P. nigra*, *P. silvestris*) are planted to forest bare slopes of lime or dolomite hills and to stabilise sandy soils or dunes.

Studies on lacewings associated with conifers have been made exclusively in the areas of continental and Mediterranean climate of the northern hemisphere. For this reason in this review I focus on species from these areas.

The most important forest-forming coniferous trees in Eurasia are the pines (*Pinus*), spruces (*Picea*), true firs (*Abies*), and larches (*Larix*). Besides *Pinus* and *Abies*, hemlock (*Tsuga*) and Douglas fir (*Pseudotsuga*) species form forests in North America. Of about 90 species of *Pinus*, the most important is the Scots pine, *P. silvestris*, in Europe and North Asia. It grows mainly in highland areas. The mountain pine, *P. mugo*, inhabits the sub-alpine zone of high mountains in central Europe and the Balkans. The Austrian pine, *P. nigra*, prefers the warmer sub-Mediterranean and Mediterranean areas of Europe. Italian stone pine, *P. pinea*, is the characteristic conifer of the Mediterranean coastal zone. From 50° N, and also in high mountains, the Norway spruce, *Picea abies* (= *excelsa*) forms closed forests. Amongst *Abies* spp., silver fir, *A. alba*, is a highland species of central and southern Europe. The deciduous European larch, *Larix decidua*, forms typical mixed forests with *Pinus cembra* in the Alps.

The major species of coniferous forests in North America is Douglas fir, *Pseudotsuga menziesii*. Further conifer species also participate in forest stands, such as the loblolly pine *Pinus taeda*, white fir *Abies concolor*, hemlock *Tsuga canadensis*, and balsam fir *A. balsamea*.

Highland coniferous forests serve as habitats for lacewings in European high mountains. The mountainous coniferous zone consists of *Picea abies* stands. At its lower border it forms mixed forests with *A. alba*, oak, or hornbeam, and at its upper border with *P. cembra* and *L. decidua*. Other species create hilly coniferous forests in southern Europe. *Abies alba* and *P. uncinata* are present in the Pyrenees, while *P. peuce* or *Picea amorica* are found in the Balkan mountains as characteristic conifer species. At the upper border of the coniferous zone, sub-alpine bushy vegetation mainly comprises *Pinus mugo*.

The continuous zone ranging between 50° and 70° N is the taiga. Mixed forests are found at the southern edge of taiga and include Scots pine, Norway spruce, oak, beech, and hornbeam in Europe. *Pinus silvestris* forests grow on poor, sandy, and pebbly soil (Poland, Baltic areas). *Pinus strobus*, *Picea glauca*, and *P. rubra* represent the coniferous taiga belt of North America, mixed with *Acer* species around the Great Lakes. The major characteristic coniferous forest species are *Tsuga canadensis* or *Pseudotsuga menziesii* elsewhere in North America. In the coolest areas of eastern Siberia, *Larix* species are associated with *Pinus silvestris* at the northern edge of the boreal coniferous forest belt.

10.5.2 Lacewing assemblages in European coniferous forests

Characteristic life conditions in coniferous habitats

Specific environmental conditions exist in coniferous forests that strongly influence the composition of lacewing assemblages. Due to an almost closed canopy it is dark inside conifer forests. The trunks become bald and leafless and green foliage is found only in the upper crown level, thus aphids and other homopteran prey groups are within this tree layer (frequently over 30 m above the ground). Due to the strong shading and slow decomposition of fallen needles, the understorey as well as the bush level is practically absent (except at gaps in the trees). For these reasons a monoculture-like vegetation develops in both planted and natural coniferous forest. The repetitive architecture of conifer trees, the baldness of trunks, the development of young shoots and cones at the same time, and the continuous falling of leaves produces many similar microhabitats. These provide lacewings with shelters and with egg-laying, cocooning, and overwintering microsites. Homopteran food sources associated with evergreen conifer foliage are present continuously over the season – a fact that may promote increased abundance of lacewing species.

Secondary plant substances produced in conifers affect the food quality of potential phytophagous prey of lacewings. Among the secondary organic compounds in conifers, resins are characteristic. Conifer resins consist of mixtures of monoterpenoids, sesquiterpenoids, diterpenoid acids, and phenolic compounds produced in cavities or ducts in all parts of the tree. Terpenoids are thought to have ecological functions serving as a defensive agent against herbivores. Many diterpenes have been reported to have toxic effects in conifers against attacking phytophagous insects and their predators. Consequently the metabolism of both sap-feeding Homoptera (aphids, scales) and their lacewing predators have had to adapt to terpenoids in the diet.

Several generalist species of lacewings from conifers immigrate to lower deciduous zones to avoid disadvantageous environmental circumstances. The geographical location of coniferous forests may to a large extent define the seasonality and population dynamics of lacewings. The cool climate in coniferous forests of high mountains or the northern taiga belt limits the number of lacewing generations, while the number of generations may increase in southern parts of Europe, in particular in stands of *Pinus* spp. having drier and warmer microclimatic conditions.

Lacewings have been recorded in European stands of *Abies alba*, *Picea abies*, *Pinus silvestris*, *P. nigra*, *P. cembra*, *P. pinaster*, *P. pinea*, *P. halepensis*, *P. peuce*, *P. mugo*, and *Larix decidua*, and North American stands of *Pseudotsuga menziesii*, *Abies concolor*, *A. balsamea*, and *Pinus taeda*.

Chrysopid assemblages

Table 10.9 shows green lacewings collected from eight species of conifer in Europe. Columns refer to faunistic records made in different sites and times in each country. In cases where dominance values (%) could be calculated their range and frequency of occurrence value also reflect tree species preference.

Twenty-nine chrysopid species have been collected from European conifers. Species associated exclusively with conifers (Killington, 1937; Ohm, 1973; Aspöck *et al.*, 1980) are *Nineta pallida*, *Peyerimhoffina gracilis*, *Chrysopa dorsalis*, *C. regalis*, *Dichochrysa genei*, *Chrysoperla mediterranea*, and *Notochrysa capitata*. Besides these specialist chrysopids on conifers the above-mentioned works list generalist chrysopids that appear on broad-leaf trees as well as on conifers, namely: *Nineta flava*, *N. vittata*, *Chrysopa perla*, *Dichochrysa prasina*, *D. flavifrons*, *D. iberica*, *D. zelleri*, *D. ventralis*, and *Chrysoperla carnea* (*s. lat.*). Most chrysopids (62%) recorded from coniferous species prefer thermophilous or xerothermous habitats. On the basis of Table 10.9 I reviewed the characteristic chrysopids on each coniferous species.

The number of green lacewing species varied between one and five on *Albies alba* (columns 1–4). *Nineta pallida* and *P. gracilis* from the conifer specialist species and *C. carnea* from the generalists were the most frequent. Moreover, *Chrysopa perla*, *Dichochrysa prasina*, *D. flavifrons*, and *D. ventralis* occurred on this conifer in some countries.

Fourteen chrysopid species from *Picea abies* were recorded (columns 5–9), and the number of species per country ranged from two to thirteen. The richest lacewing fauna (13 spp.) was registered on Norway spruce in Austria (Ressl, 1971). According to values of dominance percentage and frequency of occurrence the most common lacewing species associated with *P. abies* were as follows: specialists – *N. pallida* (51%), *Peyerimhoffina gracilis* (16%); generalists – *Dichochrysa prasina* (15%), *Chrysoperla carnea*, *Chrysopa pallens*, and *Chrysopa perla* (8%). Further species were recorded in some countries, such as *C. nigricostata*, *D. ventralis*, *N. flava*, *Chrysotropia ciliata*, *Cunctochrysa albolineata*, *Nothochrysa capitata*, and *Hypochrysa elegans*.

Most lacewing collections have been made in relation to pine species (columns 10–21), in particular to *Pinus silvestris* (columns 10–15). Altogether 17 chrysopid species have been recorded on Scots pine, while the number of species per country varied. Most species (15) were recorded in Spain (Monserrat & Marín, 1994). From the conifer-specialist chrysopids four species were present. *Chrysopa dorsalis* (14%) and *C. regalis* (<1%) associate exclusively with pines, whilst *Nineta pallida* (<1%) and *Peyerimhoffina gracilis* (<1%) prefer spruce. Of the generalist lacewings on Scots pine, *Chrysoperla carnea* (74%) and *Chrysopa perla* (68%) had high dominance, with lower dominance for *Chrysopa pallens* (0.2 – 14%), *Dichochrysa prasina* (7%), and *D. flavifrons* (4–5%). Association of *D. iberica* (4%) and *C. nigricostata* (4%) with *Pinus* is uncertain. Substrate specificity of the first species is unknown but it has been collected in monocultures or mixed stands of pine forests in Spain (Aspöck *et al.*, 1980). Higher dominance of *D. iberica* was found on *Pinus pinaster* and *P. halepensis* trees. Association of *C. nigricostata* with conifers is controversial in the literature. Numerous authors state that this species lives on low vegetation (e.g. Aspöck *et al.*, 1980), whilst others have collected it from deciduous trees and on *Pinus* species. Individuals of *C. nigricostata* have been trapped near *P. nigra* stands on sandy soil as well as in

Table 10.9. *Species composition and dominance distribution of chrysopids found on different conifers (Abies, Picea, Pinus spp.) in European countries*

	1	2	3	4	5	6	7	8	9	10	11	12	13	14	15	16	17	18	19	20	21	FR
Chrysoperla carnea s. lat. (Stephens)	+	+	+	+	+	+	+		+		+	+	73.8	+	+	+	+	50.5	+	29.0	20.2	18
Chrysoperla mediterranea (Hölzel)		+													+	+	+	8.8	+	5.8	41.2	4
Chrysopa formosa Brauer							+						0.3			+			+	4.3	1.3	5
Chrysopa pallens (Rambur)				+			+	0.5	+			14.3	0.2								0.1	7
Chrysopa perla (L.)		+			+			8.1				67.9										6
Chrysopa viridana Schneider													1.3		+	+		1.1		8.7	1.5	5
Chrysopa nigricostata Brauer									+				4.3		+	+		2.2				5
Chrysopa dorsalis Burmeister										+		14.3			+	+			+			4
Chrysopa regalis Navás								14.7					0.5		+	+		12.1			0.4	4
Dichochrysa prasina (Burmeister)								1.9	+				7.0		+	+		8.8	+	5.8	2.1	11
Dichochrysa ventralis (Curtis)								1.9					1.9									3
Dichochrysa flavifrons (Brauer)												3.6	5.2			+		2.2	+	24.6	2.8	10
Dichochrysa clathrata (Schneider)													1.9			+						2
Dichochrysa alarconi (Navás)																					0.9	1
Dichochrysa granadensis (Pictet)													0.5			+					0.6	2
Dichochrysa picteti (McLachlan)																				11.6	5.8	2
Dichochrysa venosa (Rambur)									+												0.7	1
Dichochrysa subcubitalis (Navás)													0.3			+			+		0.4	2
Dichochrysa iberica (Navás)													3.7					11.0		7.2	13.9	4
Dichochrysa genei (Rambur)																				2.9	7.4	2
Nineta flava (Scopoli)						+		0.9					0.2									3
Nineta pallida (Schneider)							+	51.2	+		+		0.6									8
Peyerimhoffina gracilis (Schneider)							+	16.1			+		0.3									7
Chrysotropia ciliata (Wesmael)								2.8														1
Cunctochrysa albolineata (Killington)								0.9	+								+					3
Cunctochrysa baetica (Hölzel)																					0.3	1
Nothochrysa capitata (Fabricius)								0.9	+													2
Italochrysa italica (Rossi)																		3.3			0.4	2
Hypochrysa elegans (Burmeister)								0.5									+					2
Number of chrysopid species	4	5	1	2	4	2	3	13	8	2	2	5	15	1	6	8	6	9	5	9	17	

Note: The dominance values (%) are calculated from the number of collected individuals published in the reviewed papers.

Abbreviations: +, species present; FR, frequency of species occurrence in literature surveyed

Coniferous tree species: 1–4, *Abies alba*; 5–9, *Picea abies*; 10–15, *Pinus silvestris*; 16–17, *Pinus nigra*; 18–19, *Pinus pinaster*; 20, *Pinus pinea*; 21, *Pinus halepensis*.

Sources: 7, 10, Czech Republic & Slovakia (Zelený, 1963, 1978); 1, 8, 12, Austria (Ressl, 1971); 5, 11, Austria (Gepp, 1974); 2, 13, 16, 18, 20, 21, Spain (Monserrat & Marín, 1994); 19, Italy (Pantaleoni, 1982); 3, 6, 14, Yugoslavia, Montenegro (Devetak, 1991); 4, 9, 15, 17, Bulgaria (Popov, 1991).

Collecting methods: 1–7, yellow and red pan trapping; 9, light trapping; 8, 10, 12–18, sweep-netting or branch clipping; 11, Malaise trapping.

deciduous forest without any surrounding conifers in Hungary. It is probable that *C. nigricostata* lives in both types of forest. The approximate number of species of lacewings is 11 in *P. silvestris* stands. From Scots pine trees in Spanish stands further additional chrysopids were also recorded (Table 10.9, column 13) although they were represented by few individuals and cannot be considered as typical.

Collections from other pine species (*P. nigra, P. pinaster,* and *P. pinea*) have been made only in Mediterranean countries (Table 10.9, columns 16–21). On the basis of dominance and frequency of occurence values the above mentioned *Pinus* species are preferred by the following lacewings: *Chrysoperla carnea* (20%–50%), *C. mediterranea* (6%–41%), *D. flavifrons* (2%–25%), *D. iberica* (7%–14%), *D. picteti* (6%–12%), *Dichochrysa prasina* (2%–9%), *Chrysopa regalis* (0.4%–12%). *C. viridana* (1%–9%) a typical species preferring oaks occurs regularly on pine trees as does *C. formosa* (1%–4%). Besides *C. regalis,* the presence of high numbers of *Dichochrysa* species in pine forests in the Mediterranean region is characteristic, and the majority of them may be feeding on Homoptera. Such lacewings are: *D. genei, D. venosa, D. granadensis, D. alarconi,* or *D. clathrata.* This opinion is supported by the cluster analysis of Monserrat & Marín (1994). According to this analysis the above *Dichochrysa* species together with *Chrysoperla mediterranea, Chrysopa regalis,* and *C. nigricostata* formed well-defined separated clusters in conifer assemblages.

Table 10.10 shows local lacewing assemblages encountered in natural or planted conifer forests in Europe. A total of 15 species were collected in local assemblages, while the local species richness ranged from 1 to 10. A total of 13 chrysopids were detected in local pine forests. Expected species richness of these local green lacewing assemblages associated with pine stands is 6–7. The following species can be regarded as typical of inhabitants of Scots pine forests: *Chrysoperla carnea* (4%–89%), *Chrysopa dorsalis* (0.03%–8%), *Dichochrysa prasina* (2%–82%), *Chrysopa pallens* (2%–50%), and *Chrysopa perla* (1%–6%). *Cunctochrysa albolineata* (0.03%–21%), *D. ventralis* (<5%), *N. vittata* (1%–5%), *Nineta flava* (1%–2%), *Chrysotropia ciliata, Chrysopa formosa, C. phyllochroma,* and *D. flavifrons* were not constant elements in assemblages, arriving rather from surrounding oak, hornbeam, or birch stands. According to Czechowska's

studies (1985, 1994, 1995), these broad-leaf tree species mix with the dominant *Pinus silvestris* in Scots pine forest associations (columns 1–6). In the *P. mugo* stand, only one chrysopid species, *Chrysoperla carnea,* appeared in Switzerland (Eglin, 1980*a, b, c*).

In the *Piceetum* associations (Table 10.10 columns 11–13), seven chrysopid species occurred. Species richness of lacewing assemblages ranged from 1 to 4. Of lacewing species preferring *Picea abies, Nineta pallida* and *Peyerimhoffina gracilis* were recorded in Hungarian investigations. A further characteristic species on spruce was *C. carnea.* Other species detected only in one country on spruce are *Chrysopa pallens, Chrysopa perla, N. vittata,* and *Chrysotropia ciliata.*

Six chrysopid species have been recorded from larch trees in *Laricetum* associations (Table 10.10, columns 14–16). Species number in individual locations varied from 3 to 6. Green lacewings expressing strong preference for *Larix decidua* have not been discovered. Some eurytopic lacewing species were present in all the investigations, namely: *Chrysoperla carnea, Chrysopa pallens,* and *Nineta flava.* In a few stands, *Chrysopa perla, Dichochrysa prasina,* and *D. flavifrons* were also captured.

Species richness of chrysopid assemblages within mixed coniferous forests (Table 10.10, columns 17–18) may vary widely depending on composition of vegetation and geographical location. In high mountains, such as the Alps (Eglin, 1980*a, b, c*), only *Chrysoperla carnea* was found in mixed forests of *Picea, Pinus,* and *Larix.* In Mediterranean mixed conifer forests of Italy (*Pinus, Picea,* and *Abies* spp.) Pantaleoni (1990*a*) detected six chrysopid species. Among these, besides both spruce specialists *N. pallida* (12%) and *Peyerimhoffina gracilis* (30%), the pine-preferring *Chrysopa dorsalis* (5%) has also been found. *Dichochrysa flavifrons* was represented with a significant dominance (33%); this species was also most abundant on *Pinus pinea* according to Table 10.9.

Summarising all the above facts, six species are known conifer specialists, according to Tables 10.9 and 10.11. *Nineta pallida* and *Peyerimhoffina gracilis* are mainly associated with Norway spruce, but can also appear with lower abundance in *Abies alba* and *Pinus silvestris* stands. A definite preference is expressed by *Chrysopa dorsalis, C. regalis, Chrysoperla mediterranea,* and *D. genei* toward *Pinus.* Values of their frequency of occurrence also support the idea that *Chrysopa nigricos-*

tata and *D. iberica* should be included in pine lacewing assemblages. On the basis of higher dominance values and frequent presence the following species may be considered as functional components of coniferous assemblages from the generalist (euryoecious) chrysopids: *Chrysoperla carnea, Dichochrysa prasina, D. flavifrons, Chrysopa perla*, and *Chrysopa pallens*. Further additional species can often be found in coniferous forests mixed with broad-leaf tree species.

Hemerobiid assemblages

Table 10.11 presents hemerobiid faunistic collections on conifers in Europe. Half the recorded species (49%) were xerothermic lacewings preferring dry and warm habitats. A few hemerobiids were euryoecious (17%), hygrothermic (~ 9%) or hygrophilous (11%). The following typically prefer various conifers:

- *Hemerobius fenestratus* (mainly on *Picea* although it has also been found on *Larix* and *Pinus*);
- *H. atrifrons* (showing a defined preference for *Larix* but also on other conifer species);
- *H. nitidulus, H. handschini, H. stigma, Sympherobius fuscescens, Wesmaelius concinnus, W. mortoni*, and *W. ravus* (all of them have strong preference to *Pinus* spp.);
- *H. simulans* (preference for *Larix* but it also appears on spruce and pine trees);
- *H. pini* and *H. contumax* (collected from *Picea* in particular but on other conifer species also);
- *Drepanepteryx algida* (exclusively associated with *Larix*);
- *W. quadrifasciatus* and *W. fassnidgei* (with defined preference to *Larix* but also occurring on other conifer species);
- *S. pellucidus* (living on *Picea, Abies*, and *Pinus* species);
- *S. riudori* and *H. schedli* (known from *Pinus*).

Authors such as Aspöck *et al.* (1980) report certain species from conifers that are also known from broadleafed trees, for instance, *S. elegans, D. phalaenoides, W. nervosus, Megalomus tortricoides, W. malladai, W. subnebulosus, H. humulinus*, or *H. perelegans*.

Referring to Table 10.11, 14 species have been recorded on *Abies alba* (columns 1–5). Four species occur frequently and are possible constant elements of the lacewing assemblages. Of these, three are conifer specialists (*H. pini, H. contumax*, and *W. quadrifascia-*

tus), while *H. micans* is a generalist usually appearing on broad-leaf trees. *Hemerobius simulans, S. fuscescens*, and *M. tortricoides* prefer conifers, but were rarely found on *A. alba* trees. Certain, mainly euryoecious, hemerobiids were also collected that could live both on broad-leaf and coniferous trees, such as *W. subnebulosus, H. humulinus*, and *W. nervosus*. A further three lacewing species found on *A. alba* but which can be found on either broad-leaf trees or conifers are *W. malladai, M. paganus*, and *S. pellucidus*. The number of recorded species on *A. alba* ranges between three and nine per country.

Twenty-two species of hemerobiid have been collected on *Picea abies* (columns 6–10). The range of recorded species per site is between 10 and 17 indicating a relatively species-rich assemblage. Brown lacewings preferring *P. abies* ranked by frequency of occurrence and dominance values are *H. pini* (31%), *H. contumax* (7%), and *H. fenestratus* (8%). The other conifer specialists with lower dominance were *W. quadrifasciatus* (4%), *W. ravus* (<1%), *H. nitidulus* (4%), and *H. stigma* (2%). Further brown lacewings living on other conifers have also been recorded on Norway spruce in limited numbers and in some places only. These are *H. atrifrons, H. handschini, S. fuscescens, W. mortoni, W. concinnus, M. paganus*, and *S. pellucidus*. It is very probable that *H. humulinus* (17%), *H. micans* (18%), *W. subnebulosus*, and *W. nervosus* from the generalist, eurytopic hemerobiids (also found in broad-leaf trees) are the real elements of spruce lacewing assemblages.

The number of collections of hemerobiids on *Larix decidua* in Europe is limited (columns 11–12). All the known larch specialist brown lacewings, *H. atrifrons, H. simulans, W. quadrifasciatus*, and *D. algida*, were recorded. Lacewings preferring other conifers were also collected on *Larix* trees, such as *H. nitidulus, H. stigma*, and *H. pini*. Genuine connections of these latter species with *L. decidua* have not yet been proved, as immigration from other bordering conifer stands to the larch canopy may explain their presence.

Most data were gathered from hemerobiids living on *Pinus* species. According to columns 13–26, 29 brown lacewing species have been recorded in pine stands in Europe. A total of 22 hemerobiid species (columns 13–18) have been collected from *P. silvestris*. The number of lacewing species per country varied from 8 to 11. From the pine specialists the more

Table 10.10. *Species composition and dominance distribution within the local green and brown lacewing assemblages found in various types and associations of coniferous forests in European countries*

	1	2	3	4	5	6	7	8	9	10	11	12	13	14	15	16	17	18	FR
Chrysopidae																			
Chrysoperla carnea s. lat. (Stephens)	4.5	8.2	12.9	25.9	6.3	7.0	88.9	+	68.6	+	+	+	+	+	+	+	+	13.7	18
Chrysopa pallens (Rambur)	18.1	16.4	3.9		50.0	21.0	2.2	+	7.5			+	+	+	+	+			12
Chrysopa perla (L.)	0.5	0.8				0.7			6.0							+			6
Chrysopa formosa Brauer									0.7										1
Chrysopa phyllochroma Wesmael									3.0										1
Chrysopa dorsalis Burmeister				0.03					8.2									5.5	3
Dichochrysa prasina (Burmeister)	58.8	65.6	81.9	72.2	37.5	49.0	6.7		2.2							+		5.5	10
Dichochrysa ventralis (Curtis)	4.5		0.03	0.03		0.7		+								+			6
Dichochrysa flavifrons (Brauer)									0.7									32.9	2
Nineta flava (Scopoli)		0.6	0.6	1.8		0.7	2.2												6
Nineta vittata (Wesmael)	4.5	0.8												+	+	+			4
Nineta pallida (Schneider)											+		+					12.3	3
Peyerimhoffina gracilis (Schneider)											+		+					30.1	3
Nothochrysa capitata (Fabricius)													+						1
Chrysotropia ciliata (Wesmael)			0.03						0.7				+						3
Cunctochrysa albolineata (Killington)	9.0	8.2	0.6	0.03	6.3	21.0			2.2										7
Hemerobiidae																			
Wesmaelius subnebulosus (Stephens)	0.5								2.1			+		+		+			5
Wesmaelius nervosus (Fabricius)		0.4	1.0	6.3		0.2	1.4						+			+	+		8
Wesmaelius quadrifasciatus (Reuter)					0.5		1.4			+					+	A	A		8
Wesmaelius fassnidger (Killington)										+					+	+	+		4
Wesmaelius malladai (Navás)								+		+			+			+	+		4
Wesmaelius concinnus (Stephens)	37.4	31.8	54.8	19.0	67.4	53.9	60.9		6.3			+		+		+			9
Wesmaelius mortoni (McLachlan)	0.5	0.1	0.1	0.5	0.5	2.2	1.4												5
Wesmaelius razus (Withycombe)										+		+		+	+	+	+		4
Hemerobius humulinus L.	4.7	3.5	36.2	47.6	2.1	2.2	2.9	+	14.6			+	+	+	+	+	+	2.9	15
Hemerobius micans Olivier	0.5	2.9	2.9	1.6													+	8.6	5
Hemerobius nitidulus Fabricius	4.7	21.2	3.9	14.3	4.3	15.7	14.5	+	18.8	+									10
Hemerobius stigma Stephens	42.1	38.9	1.0	3.2	20.3	24.7	8.7	+	8.3	A		+				A		28.6	15

Species	1	2	3	4	5	6	7	8	9	10	11	12	13	14	15	FR
Hemerobius atrifrons McLachlan						0.2	+	+	+				A	+		6
Hemerobius handschini Tjeder						18.8	+	A	+			+	A	+	60.0	6
Hemerobius lutescens Fabricius						6.3							+			2
Hemerobius fenestratus Tjeder	0.5						+			+			+			2
Hemerobius pini Stephens	4.7	3.5	3.2	0.4	0.2	2.1	+	+	A	+			+	+		11
Hemerobius contumax Tjeder							+		+							1
Hemerobius simulans Walker							+			+						1
Micromus variegatus (Fabricius)						8.3										1
Micromus angulatus (Stephens)						12.5										1
Micromus paganus (L.)							+	+		+			+	+		3
Sympherobius pygmaeus (Rambur)				1.4												1
Sympherobius elegans (Stephens)			1.6	4.3			+									3
Sympherobius klapaleki Zelený				2.9												1
Sympherobius fuscescens (Wallengren)	4.7	3.2	4.8		0.2		+			+			+			7
Sympherobius pellucidus (Walker)						2.1		+								1
Megalomus tortricoides Rambur													+			1
Drepanepteryx algida (Erichson)													A			1
Drepanepteryx phalaenoides (L.)	0.4	0.4			0.2	0.2	+						A			3
S_C	7	6	7	6	4	7	4	3	1	1	7	3	6	1	6	
S_H	10	8	7	9	7	10	10	9	12	7	1	1	15	12	4	

Note: The dominance values (%) are calculated from the number of collected individuals published in the reviewed papers.

Abbreviations: S_C, S_H, species richness of chrysopid and hemerobiid assemblages respectively; A, abundant; +, species present; FR, frequency of species occurrence in literature surveyed.

Forest type/association: 1–5, *Peucedano–Pinetum* (dominated by *Pinus silvestris*, mixed with *Picea abies*, *Betula pendula*, sporadically *Quercus robur* and *Carpinus betulus*); 6, *Leucobryo–Pinetum* (*Pinus silvestris*); 7, urban pine forest (*P. silvestris*); 8, *Pinetum* (*P. silvestris*); 9, pine forest plantation (*P. silvestris*); 10, *Mugetum* (*P. mugo*); 11, *Picea abies* plantation; 12, *Athyrio–Piceetum* (*P. abies*); 13, *Piceetum montanum* and *subalpinum* (*P. abies*); 14, larch trees in urban habitat (*Larix decidua*); 15–16, *Laricetum* (*L. decidua*); 17, *Larici–Pinetum cembrae* (*L. decidua*, *Pinus mugo*, *P. cembra*, *Picea abies* mixed); 18, non-spontaneous coniferous forest (*Pinus*, *Picea*, *Abies* spp. mixed).

Sources: 1, 2, 6, Poland (Czechowska, 1994); 3, 4, Poland (Czechowska, 1985); 5, Poland (Czechowska, 1995); 7, Germany (Saure & Kielhorn, 1993); 9, 11, Hungary (F. Szentkirályi, unpublished data); 12, Slovakia (Vidlička, 1994); 8, 10, 13, 16, 17, Switzerland (Eglin, 1980*a*, *b*, *c*; 1981); 14, 15, Austria (Schremmer, 1962*b*); 18, Italy (Pantaleoni, 1990*a*).

Table 10.11. *Species composition and dominance distribution of hemerobiids found on different conifers (Abies, Picea, Larix, Pinus spp.) in European countries*

	1	2	3	4	5	6	7	8	9	10	11	12	13	14	15	16	17	18	19	20	21	22	23	24	25	26	FR
Wesmaelius subnebulosus (Stephens)				+		+	+		+	+			+			11.0	+	+	+		7.8	+	+	+			14
Wesmaelius nervosus (Fabricius)		+	+			+	+	0.6							0.8												4
Wesmaelius quadrifasciatus (Reuter)	+	+	+			+	+	3.9	+		+		+					+								+	10
Wesmaelius fassnidgei (Killington)																		+								+	2
Wesmaelius malladai (Navás)				+																							3
Wesmaelius concinnus (Stephens)						+	+		+				+		5.9					+							6
Wesmaelius mortoni (McLachlan)						+	+						+					+		+							4
Wesmaelius navasi (Andréu)																								+			1
Wesmaelius ravus (Withycombe)						+		0.6		+			+			0.2	+	+				+					7
Wesmaelius reisseri Aspöck & Aspöck																0.1											1
Hemerobius humulinus L.		+		+		+		17.4	+	+	+			+	3.4	0.1	+	+	+			+					14
Hemerobius micans Olivier			+		+	+		18.3	+	+					5.9		+	+								+	11
Hemerobius nitidulus Fabricius			+		+	+		4.2	+	+	+	+	+	+	43.7	31.2		+	+	+	3.1		+			+	13
Hemerobius stigma Stephens					+	+		2.1	+	+	+	+	+	+	16.8	37.3		+	+	+	82.4					+	15
Hemerobius atrifrons McLachlan						+		1.8	+				+														5
Hemerobius handschini Tjeder						+		0.3	+	+	+				2.5	1.4		+	+	+	2.1		+		+	+	10
Hemerobius lutescens Fabricius																											2
Hemerobius simulans Walker				+					+		+																4
Hemerobius fenestratus Tjeder							+	7.8					+		0.8												3
Hemerobius pini Stephens	+	+	+	+	+	+	+	30.9	A	+	+	+	+		1.7		+	+	+	+		+					18
Hemerobius contumax Tjeder						+		6.9	+				+					+		+							9
Hemerobius schedli Hölzel																									+		1
Micromus variegatus (Fabricius)							+	1.5																			2
Micromus paganus (L.)				+						+																	2
Micromus lanosus (Zelený)								0.6	+																		2
Sympherobius pygmaeus (Rambur)																0.5					3.1						2
Sympherobius elegans (Stephens)								1.2							10.9	3.9			+		1.0			+			6
Sympherobius fuscescens (Wallengren)				+		+		0.3					+		7.6	14.3	+		+		0.5					+	9
Sympherobius pellucidus (Walker)			+					1.5																			2
Sympherobius riudori Navás																			+								1
Megalomus hirtus (L.)																								+			1
Megalomus pyraloides Rambur				+																							1
Megalomus tortricoides Rambur																										+	2
Drepanepteryx algida (Erichson)												+															1
Drepanepteryx phalaenoides (L.)	+												+														2
Number of hemerobiid species	4	5	8	9	3	13	4	17	10	11	7	3	11	3	11	10	9	8	7	6	7	3	2	6	4	9	

Note: The dominance values (%) are calculated from the number of collected individuals published in the reviewed papers.

Abbreviations: A, abundant; +, species present; FR, frequency of species occurrence in literature surveyed.

Coniferous tree species: 1–5, *Abies alba*; 6–10, *Picea abies* (Zelený, 1963, 1978); 2, 8, 12, 15, *Larix decidua*; 11–12, *Larix decidua*; 13–18, *Pinus silvestris*; 19–20, *P. nigra*; 21–22, *P. pinaster*; 23, *P. pinea*; 24, *P. halepensis*; 25, *P. peuce*, 26, *P. cembra*.

Sources: 1, 6, 11, 13, Czech Republic and Slovakia (Zelený, 1963, 1978); 2, 8, 12, 15, Austria (Ressl, 1974); 7, 14, Austria (Gepp, 1974); 3, 16, 19, 21, 23, 24, Spain (Monserrat & Marín, 1996); 22, Italy (Pantaleoni, 1982); 4, 9, 17, Yugoslavia, Montenegro (Devetak, 1991); 5, 10, 18, 20, 25, Bulgaria (Popov, 1991); 26, Switzerland (Eglin, 1980b).

common species were *H. nitidulus* (31%–44%), *H. stigma* (17%–37%), *S. fuscescens* (8%–14%), *H. handschini* (1%–3%), and *W. concinnus* (6%), while *W. ravus* (<1%) and *W. mortoni* were occasionally sampled. Of hemerobiids living on other conifers *H. pini* (<2%) was represented in limited numbers. *Sympherobius elegans* (4%–11%), *W. subnebulosus* (11%), *H. humulinus* (<4%), and *H. micans* (<6%) from generalist species also living on broad-leaf trees were frequently found on *P. silvestris*. The aforementioned four hemerobiids together with the more common and dominant five pine specialists (plus *H. pini*) may be considered as possible components of lacewing assemblages on Scots pine forests.

A limited amount of data on lacewings living on Austrian pine (columns 19–20) supports the idea that the same hemerobiid species are associated with *P. nigra* as with *P. silvestris*. The total recorded 10 species possibly belonging to assemblages of *P. nigra* are as follows: from conifer specialists *W. concinnus*, *W. mortoni*, *H. nitidulus*, *H. stigma*, *H. handschini*, *H. pini*, *S. fuscescens*, and *S. riudori* (in Spain); from generalists *W. subnebulosus* and *S. elegans*.

Altogether 9 hemerobiid species have been collected from *Pinus pinaster* living in the Mediterranean region (columns 21–22). *Hemerobius stigma* (82%), *H. nitidulus* (3%), *H. handschini* (2%), and *S. fuscescens* from 7 recorded species in Spain are specialists on pine. Pantaleoni (1982) could not find conifer specialist hemerobiids on this pine species from Italy, only euryoecious *W. subnebulosus*, *H. humulinus*, and *H. micans*. Besides these broad-leaf tree inhabitants, Monserrat & Marín (1996) found a further 2 generalist brown lacewing species, *S. pygmaeus* (3%) and *S. elegans* (1%), in the canopy of *P. pinaster*.

Monserrat & Marín (1996) recorded only 2 hemerobiid species, *H. stigma* and *W. subnebulosus* on *P. pinea* (column 23) among conifer species in the Mediterranean area.

Six brown lacewing species were detected on *P. halepensis* in Spain (column 24). *Hemerobius stigma* was the only conifer specialist, whilst the others were either generalists (*W. subnebulosus*, *S. elegans*) or rare hemerobiids (*W. navasi*, *M. hirtus*, and *M. pyraloides*).

Popov (1991) found 4 hemerobiids on *P. peuce* trees (column 25) in mountainous areas of the Balkans: *H. handschini*, *H. pini*, *W. malladai*, and the very rare *H. schedli*.

In the Swiss Alps Eglin (1980b) collected 9 brown lacewing species from *P. cembra* (column 26). The majority of them were conifer specialists, such as *W. quadrifasciatus*, *W. fassnidgei*, *H. stigma*, or *H. pini*.

Information on forest associations and collecting methods involving Table 10.10 have been explained in the case of discussion of chrysopids above. Each important coniferous hemerobiid was found in higher dominance within local samplings. A total of 29 brown lacewing species are listed in Table 10.10 belonging to local assemblages.

Typical members of brown lacewing assemblages (in total 25 species) of *Pinetum* forests (columns 1–10) ranked by frequency of occurrence and dominance values are as follows: *W. concinnus* (6%–67%), *H. stigma* (1%–42%), *H. humulinus* (2%–48%), *H. nitidulus* (4%–21%), *H. handschini* (19%), *W. nervosus*, (0.2%–6%), *H. pini* (<5%), *S. fuscescens* (<5%), and *W. mortoni* (0.1%–2%). On the basis of occurrence, *H. micans*, *H. atrifrons*, *W. quadrifasciatus*, *S. pygmaeus*, and *S. elegans* may be regarded as further functional and occasional components of lacewing assemblages of *Pinetum* stands, in accordance with Table 10.11. The expected number of species in brown lacewing assemblages in European *Pinus* forests is 9–11 according to Table 10.10.

Despite extensive areas of *Piceetum* forest the number of studies on lacewings is limited (Table 10.10, columns: 11–13). Taking into consideration the data of Table 10.11, the following species may be components of hemerobiid assemblages (total number of species 13) associated with Norway spruce forests: *H. pini*, *H. stigma*, *H. humulinus*, *W. quadrifasciatus*, *W. subnebulosus*, *W. ravus*, *H. contumax*, and *H. fenestratus*.

Investigations on hemerobiid assemblages living in local *Laricetum* associations were carried out in Austria and Switzerland (Table 10.10, columns 14–16). Eglin (1980b) detected species-rich assemblages collecting 15 species from *Larix decidua* stands. Of the species preferring European larch he found *W. quadrifasciatus*, *H. atrifrons*, and *D. algida* as abundant brown lacewings while *W. fassnidgei* was recorded in low numbers. The other hemerobiids belong to species that appear on different conifers or both equally on broad-leaf trees and conifers.

Columns 17–18 (Table 10.10) give examples on hemerobiid assemblages from mixed forests created by various conifer species. The Swiss example (Table

10.10, column 17) demonstrates that each of the brown lacewings living on differing conifer species can be represented in the assemblage, while in the Italian case (column 18) only species characteristic on *Pinus* trees in higher abundance were present (*H. handschini* 60%, *H. stigma* 29%).

The cluster analysis of Monserrat & Marín (1996) supports the existence of brown lacewing assemblages associated with various coniferous species. They detected two separate characteristic groups through analysis: one of them consisted of hemerobiids living on *Pinus* trees (*P. silvestris*, in particular), the other one was formed by species collected mainly from *Abies alba*.

Lacewings on conifers outside Europe

There are only sporadic data on lacewings living on conifers outside Europe. Monserrat (1991) mentioned that an endemic brown lacewing, *Hemerobius eatoni* Morton, was collected from *Pinus canariensis* trees in the Canary Islands. In the same article he reported two hemerobiids also found on *Pinus* tree species, *H. harmandinus* Navás in India and *H. nairobicus* Navás in Africa.

Studies on lacewings in conifers in America are also limited. Certain references on brown lacewings living on conifers are available in Garland (1978), Deyrup & Deyrup (1978), and Monserrat (1996, 1998), while on green lacewings only in the study of Ehler & Kinsey (1995). Deyrup & Deyrup reared adults of *H. stigma* Stephens, *H. pacificus* Banks, *H. kokaneeanus* Currie, *H. bistrigatus* Currie, and *Wesmaelius coloradensis* (Banks) from cocooned larvae living on Douglas fir trees (*P. menziesii*) in the USA. *Hemerobius humulinus*, *H. stigma*, and *H. simulans* had been found earlier on balsam fir (*Abies balsamea*) (Garland, 1978). The following hemerobiid species are mentioned by Monserrat on *Pinus* tree species: *H. stigma* in USA, *Micromus variolosus* Hagen in USA, and *Wesmaelius magnus* (Kimmins) in Guatemala. Moreover, *H. elongatus* Monserrat was collected from *A. religiosa* in Mexico. Ehler & Kinsey (1995) studying white fir aphids found the green lacewings *Chrysoperla carnea*, *Chrysopa coloradensis* Banks, and *C. nigricornis* Burmeister as potential natural enemies living on *A. concolor*.

Aphids as prey for lacewings on conifers

Two large aphid groups, namely the families Lachnidae and Adelgidae have a great number of representatives that are known as conifer-feeding species and which are potential food sources for conifer forest inhabiting lacewings. The more frequent aphids using conifers as host plants belong to the following genera: *Cinara*, *Eulachnus*, *Schizolachnus*, *Adelges*, *Pineus*, *Essigella*, *Prociphilus*, *Elatobium*, *Mindarus*, and *Pachypappa*. Among conifers the *Pinus* genus is associated with the most aphids, with some 170 pine-feeding species. From these, more than 100 species belong to the genus *Cinara*. The following European pines are preferred by the given number of aphid species: *P. mugo* (16 spp.), *P. nigra* (19 spp.), *P. silvestris* (29 spp.), *P. pinea* (10 spp.), *P. pinaster* (9 spp.), and *P. peuce* (2 spp.).

From the Lachnidae, 38 *Cinara* species are associated with the genus *Picea*. *Picea* species are primary hosts of Adelgidae, which have certain host specificity. Within the Pemphiginae, for *Pachypappa*, *Prociphilus*, and *Pachypapella* species, spruce species are only secondary hosts, and they live in white wax on the roots of these trees. According to Blackman & Eastop (1994), 29 aphid species infesting trees of Norway spruce have been recorded.

Some 25 *Cinara* species have been detected on the genus *Abies*. The true firs are secondary host trees for all *Dreyfusia* spp. subgenus *Adelges*, which have economic importance, migrating from galls to their primary host *Picea*. From silver firs (*Abies alba*) 13 aphid species, and from balsam-fir (*A. balsamea*) 8 aphid species have been described.

It is likely that the host tree choice of representatives of the aphid fauna living on the genus *Larix* is much more specific than in case of the other conifers. However, more common widely distributed larch-feeding aphids (e.g. *Cinara laricis*, *Adelges laricis*) did not discriminate between *Larix* species. From trees of European larch (*L. decidua*), 10 aphid species have been found so far belonging to the genera *Cinara* and *Adelges*.

Aphid prey of lacewings living on conifers was first reviewed in Killington's (1937) book. Among predatory lacewings, he collected data from the literature on the aphid food of three hemerobiids associated with conifers: *Hemerobius stigma* consumed *Cinara pini* and a *Eulachnus* species, while *H. nitidulus* predated *C. pini*, *Eulachnus agilis* and *Cinara* (*Lachnus*) *tomentosus*. *Sympherobius fuscescens* fed on *Pineus strobi*.

In Austria, Schremmer (1962a, b) made detailed investigations on aphid prey of lacewings on *Larix decidua*. He mainly observed predation of lacewing larvae in the larch canopy. The chrysopids *Chrysoperla carnea*, *Chrysopa pallens*, and *Nineta flava*, and hemerobiids *Wesmaelius subnebulosus* and *W. quadrifasciatus* were recorded. From the Adelgidae and Lachnidae prey of these lacewings included *Adelges laricis*, *A. viridis*, *Cinara laricis*, *C. kochiana*, and *C. laricicola*. Among aphids infesting *L. decidua*, Eglin (1980*a*, *b*) found *C. cuneomaculata* and *C. laricis* species in the Swiss Alps. According to his observations, the following lacewings appeared as natural enemies of these two larch-associated aphids: *Chrysoperla carnea*, *Chrysopa perla*, *W. fassnidgei*, *W. quadrifasciatus*, *H. atrifrons*, and *Drepanepteryx algida*.

In Poland, Kolodziejak (1994) investigated potential aphid prey on *Pinus silvestris* within three pine forest areas. On Scots pine trees, seven species were recorded from the Lachnidae: *Eulachnus agilis*, *Schizolachnus pineti*, *Cinara pinea*, *C. pilosa*, *C. pini*, *C. pinihabitans*, and *C. nuda*. The first two were the most abundant in all the pine stands. Within the same pine stands in Poland Czechowska (1985, 1994, 1995) sampled lacewing assemblages (see Table 10.10, columns 1–6). These lachnid aphids could therefore be potential prey of the following lacewings detected on canopy of *P. silvestris* in Czechowska's studies: *Dichochrysa prasina*, *D. ventralis*, *W. concinnus*, *H. nitidulus*, *H. stigma*, *H. humulinus*, *S. fuscescens*, and *W. mortoni*.

From Italy, Pantaleoni's (1982) investigations provide data on the feeding connection between aphidophagous lacewings and aphids associated with *Pinus nigra* trees. Larvae of *W. subnebulosus* and adults of *Chrysopa dorsalis* fed on *Cinara schimitscheki* and *Eulachnus rileyi*.

Eglin (1980*b*) studied aphid–lacewing prey–predator connections under high mountainous conditions. His investigations were conducted on stands of *Pinus cembra* and *P. mugo* in the sub-alpine zone of the Alps. On *P. cembra*, *Chrysoperla carnea*, *W. quadrifasciatus*, *W. ravus*, *W. fassnidgei*, *H. stigma*, and *H. handschini* preyed on the aphid *Cinara cembrae*. On *P. mugo* trees Eglin (1980*b*) recorded infestation of two aphids, *Cinara pini* and *C. neubergi*, which were consumed by *Chrysoperla carnea*, *W. malladai*, *H. stigma*, *H. handschini*, and *H. pini*.

Eglin (1980*b*) also reported on lacewings that feed on aphids infesting *Picea abies*. He found that *W. quad-rifasciatus*, *H. stigma*, *H. pini*, and *H. humulinus* fed on the aphids *Cinara pinicola* and *C. pilicornis* associated with Norway spruce stands. An additional prey source could be the widely distributed green spruce aphid, *Elatobium abietinum*, which is known as a serious pest frequently exceeding the economic damage level on various *Picea* species (e.g. *P. alba*, *P. glauca*, or *P. pungens*).

Pschorn-Walcher & Zwölfer (1956) studied the aphidophaga of the balsam woolly aphid, *Adelges piceae*, infesting silver fir trees in the Swiss Alps. From subdominant predators the chrysopid *D. ventralis* was found regularly with lower abundance, and from incidental aphidophaga *Chrysopa pallens*, *D. flavifrons*, *Nothochrysa capitata*, *W. quadrifasciatus*, and *Drepanepteryx phalaenoides* were occasionally recorded as preying on the woolly aphid. Fully-grown larvae of *Dichochrysa ventralis* captured both diapausing larvae and adults of *A. piceae*. Garland (1978) and Ehler & Kinsey (1995) published data on aphidophagous lacewings associated with American true firs. Garland reviewed former Canadian publications, which reported *H. stigma*, *H. humulinus*, and *H. simulans* as natural enemies of the balsam woolly aphid, *Adelges piceae*. Eight species of aphids infesting balsam fir stands are recorded (Blackman & Eastop, 1994), from which *Adelges nordmannianae*, *A. pectinatae*, *Cinara confinis*, *C. curvipes*, or *Mindarus abietinus* are additional potential prey for lacewings.

In the Holarctic region the widespread balsam twig aphid, *Mindarus abietinus*, is associated primarily with *Abies* species, such as the white fir, *A. concolor*. Ehler & Kinsey (1995) made detailed studies in California between 1989 and 1992 of a new aphid, *Mindarus kinsey*, infesting one- and two-year-old seedlings of *Abies concolor* in forest nurseries. They monitored potential lacewing predators with suction traps in the nursery. *Chrysoperla carnea* was dominant, while other lacewings collected were *Chrysopa nigricornis* and *C. coloradensis*. Despite the relatively high abundance of lacewing adults detected by trappings, lacewing larvae were seldom found on aphid-infested white fir seedlings.

Regarding lacewings that prey on aphids living on Douglas fir (*Pseudotsuga menziesii*), early information was published by Laidlaw (1936) who observed *W. subnebulosus* feeding on *Adelges cooleyi* populations. Deyrup & Deyrup (1978) examining lacewings associated with Douglas fir found that from cocooned larvae collected on trees heavily infested with *A. cooleyi*, several hundred *H. stigma* adults could be reared out. On this basis, they supposed this hemerobiid to be an important predator of *A. cooleyi*. Another fir, *Tsuga canadensis*, is a food-plant of the hemlock woolly adelgid, *A. tsugae*. In Japan, the chrysopid *Dichochrysa prasina* was found on this aphid on hemlock trees (McClure, 1995).

Scale insects as prey for lacewings on conifers

Among potential homopteran prey, scale insects come next in importance after aphids. It is characteristic of coniferous tree species that there are many mono- and oligophagous scale insects living on them all over the world (Kozár & Kosztarab, 1980; Kozár, 1998). For lacewings, the nearly constant presence of scale larvae on needles gives an alternative source of food even in periods without aphids.

Some scale insects damaging *Picea abies* stands are potential prey for lacewings. On Norway spruce, individuals of the pseudococcid *Paroudablis piceae* can be found in the majority of stands. Presumably, Eglin (1980*b*) referred to this scale insect in his study, in which he mentioned a *Pseudococcus* species infesting *P. abies* and *Pinus silvestris*. This scale insect associated with spruce was eaten by larvae of *Chrysoperla carnea* and *Chrysotropia ciliata* and by *Hemerobius pini*, *H. stigma*, and *Wesmaelius quadrifasciatus*. This pseudococcid species also infested Scots pine and was consumed by *Chrysoperla carnea*, *H. nitidulus*, and *Sympherobius elegans* as reported by Eglin.

Members of the family Diaspididae, namely *Leucaspis loewi*, *L. pini*, and *L. pusilla*, are characteristic on *Pinus*. The last spreads through southern and central Europe. In a case in Hungary of infestation with *L. pusilla* shoots and needles of *P. nigra* had a length of half or one-third of healthy ones. *Matsucoccus feytaudi* (South France, Italy), *M. josephi* (Israel), *Marchalina hellenica* (Italy, Greece, and Cyprus) live on *Pinus* tree species, exclusively in the Mediterranean regions. Covassi *et al.* (1991) studied the spread of the margarodid *Matsucoccus feytaudi* on *P. pinaster* in Italy. They found that eggs and young larvae of this scale was also consumed by hemerobiids (*H. stigma* and *H. simulans*) of the predatory guild on pine. In Chinese pine forests, Wang & Hu (1987) discovered among natural enemies of scale insects, that the lacewing *Chrysopa kulingensis* (Navás) is an important predator on the population of a *Matsucoccus* species.

Other arthropod prey types on conifers

The Nantucket pine-tip moth, *Rhyacionia frustrana*, is a serious forest pest of loblolly pine, *Pinus taeda*, causing retarded growth and multiple or crooked stems. Eikenbary & Fox (1968) made field investigations in South Carolina (USA) in order to describe the predators of *R. frustrana* in plantations of young loblolly pine. Larvae of *Chrysoperla plorabunda* preyed actively on larvae of *R. frustrana*, which were consumed within 5–10 minutes. Chrysopid larvae occurred commonly from May to September, walking on needles or shoots and hunting for prey. Eikenbary & Fox (1968) reared adults of *Hemerobius stigma* from *R. frustrana* infested tips. Hemerobiid larvae were observed to enter the tunnels of pine-tip moth although predation by this lacewing was not detected in field conditions.

Psocids (Psocoptera) are also among potential prey of lacewings in conifer forests. Borkowski (1986) studied arthropods living in the tree crown in a 50-year-old Scots pine forest. Psocids were present in his samples both in spring and autumn but in lower numbers than two other greater potential food sources, namely Homoptera and eggs or larvae of Lepidoptera. Investigations of Turner (1984) proved that lacewings could consume individuals of Psocoptera living on conifers. He studied predatory activity on epiphytic herbivores of the larch tree canopy in southern England. Turner made serological determinations of the main larch-dwelling epiphyte grazing groups, namely Psocoptera and Collembola in diets of predators. Homoptera (principally psyllids) was the third group of prey examined. Collected from *Larix* trees by beating, 50% of chrysopid larvae, and 81% of hemerobiid larvae gave positive serological reactions to antisera of the three prey orders. According to positive reactions of lacewings the proportion of each prey type in the diet was as follows. From hemerobiid larvae 76.5% consumed Psocoptera, 59% preyed upon Collembola, and 53% consumed psyllids (mainly *Psylla melanoneura*). According to serological analysis *H. humulinus* fed on each prey type. From chrysopid larvae 50% preyed upon Collembola while 75% did so upon psocids and psyllids. Hemerobiid larvae produced positive reactions to antisera mainly in June–July and in September.

Among potential prey of lacewings inhabiting

conifers there are also mites. Schremmer (1962*b*) found among phytophagous mites the tetranychid *Paratetranychus ununguis* and the eriophyid *Eriophyes pini* var. *laricis* that infested *Larix decidua* trees in Austria. *Paratetranychus ununguis* is well known from spruce trees but its occurrence on other conifer species was also detected. *Eriophyes pini* causes major loss through gall formation. Schremmer (1962*b*) mentioned young larvae of chrysopids and hemerobiids as well as Coniopterygidae as predators of this mite species. Gut-content analysis of adult central European lacewings made by Stelzl (1991) indicated the probability of lace-wings inhabiting conifer forests preying upon mites and consuming them. Several species of chrysopids and hemerobiids specialising on conifers were collected from their typical forest habitats. Stelzl (1991) found mites among the intestinal contents of the following lacewing species: *H. stigma* and *H. pini* (conifer special-ists), *Chrysopa perla*, *Chrysopa pallens*, *H. humulinus*, and *H. micans* (generalists from conifers as well as broad-leaf trees).

With the exception of *Sympherobius fuscescens* all predatory adults of lacewings examined in conifer forests consumed aphids. Adults of certain conifer lace-wings prey upon Diptera, or even Lepidoptera. There were Diptera remains in the gut contents of *Drepanepteryx algida*, *Wesmaelius quadrifasciatus*, *H. stigma*, *H. pini*, *H. nitidulus*, *H. micans*, *H. humulinus*, *C. dorsalis*, and *Chrysopa perla*. Lepidoptera remains were discovered among food items of *H. stigma*, *H. nitidulus*, *H. micans*, *Chrysopa perla*, and *Chrysopa pallens*. Lacewings associated closely with conifers cannot therefore be considered exclusively as aphidophagous species, as they are able to attack and consume a broad range of prey types.

Lacewings as biological control agents against pests on conifers

Insecticide treatments may reduce populations of lace-wings in large-sized forest stands. Evidence of this harmful effect has already been obtained through pest management during the 1950s. For instance, Zoebelein (1957) studied the impact of DDT on an insect commu-nity in a 60-year-old spruce stand using the insecticide-fogging technique. He ascertained that after treatment numbers of chrysopid larvae were reduced by 66% in comparison with their number before treatment. According to this experiment mortality of green lace-

wing larvae closely depended on timing of insecticide treatment. Treatment in the evening (20.00–22.00 hours) resulted in seven times more lacewing larvae dying than treatment in the morning (04.00–06.00 hours). This result was explained by greater activity of lacewing larvae during the evening. In order to prevent harmful impact of pesticides on natural enemies various biological control programmes should be developed in the framework of forest IPM. Lacewings as biological control agents against conifer pests were applied in the 1930s in Canada (Garland, 1978). Two exotic hemerob-iid species, *Hemerobius nitidulus* and *H. stigma*, were imported from Europe and introduced against the balsam woolly aphid, *Adelges piceae*, in New Brunswick. Altogether 920 eggs of *H. nitidulus* were released in 1935, and 1561 eggs of *H. stigma* in 1935, 2600 eggs in 1937, and 4260 eggs in 1938. After release, these heme-robiid species were not recovered, for several reasons. The low number of eggs in the case of *H. nitidulus* may have been inadequate for a successful colonisation. The number of *H. stigma* eggs was much higher, but even so there was a lack of detection later, because the endemic species *H. stigmaterus* inhabiting the same area proved to be identical to the imported *H. stigma* (Garland, 1978).

In China, Wang & Hu (1987) carried out lacewing release experiments against *Matsucoccus* sp. infesting pine trees. Compared to controls, a significant reduc-tion in numbers of scales was gained within five weeks of releasing *Chrysopa kulingensis* in predator-exclusion experiments. Ehler & Kinsey (1995) published another example of an aphid-management programme in a conifer tree nursery. *Mindarus kinseyi*, a new aphid species, became a serious pest of white fir seedlings in tree nurseries. Ehler & Kinsey (1995) made augmenta-tive release experiments with two commercially avail-able lacewings, *Chrysoperla rufilabris* and *C. carnea* (= *C. downesi*). In every case, one-day-old lacewing larvae were applied to aphid-infested experimental seedlings. Results proved that *M. kinseyi* is a suitable prey for both chrysopids, from which a larval individual could consume on average 100 aphids within 10–12 days. For successful biological control, at least five individuals of *C. rufilabris* larvae per infested seedling were required. In this case chrysopid larvae eliminated aphids within two weeks. Larvae developed well and pupated after release, but imagos and further generations of any experimental chrysopid species in tree nurseries have not been detected. Ehler & Kinsey (1995) concluded from this that releases of these lacewing species should not to be expected to provide season-long control of *M. kinseyi*.

Seasonality of prey of lacewings in coniferous forests

On the basis of sampling to detect seasonal changes of aphid species being prey for lacewings, it seems that two periods of growing abundance level occur generally in conifer stands. The first happens in the period of May–July and the second in late August to the end of November under temperate climatic conditions. Seasonal patterns of *Adelges piceae* (Pschorn-Walcher & Zwölfer, 1956), *Cinara laricis* (Schremmer, 1962*b*), *Eulachnus agilis, Schizolachnus pineti, C. pinea*, and *C. pilosa* (Kolodziejak, 1994) were found in Europe. In the USA *Mindarus kinseyi* was represented more or less by higher abundance and its seasonal patterns were characterised by many peaks during the season, except in the period of February–May, when aphid population levels dropped (Ehler & Kinsey, 1995). Several aphid and scale species overwintering in larval or egg stages in the evergreen canopy of conifer trees supply potential food sources for those lacewings that overwinter as predatory adults or larvae and remain active in early spring and in late autumn to early winter (e.g. larvae of *Dichochrysa* spp., adults of *Hemerobius stigma* or *Wesmaelius subnebulosus*).

Seasonal activity pattern of chrysopids

On the basis of Malaise trapping in Slovakia, *Peyerimhoffina gracilis* was active between early August and mid-September, and its peak was in early August (Vidlička, 1994). Ressl (1971) and Gepp (1974, 1988/9) published data on seasonal characteristics of *P. gracilis* in Austria. It seems that a bimodal activity pattern characterises this chrysopid species in the Alps. Adults appeared in early February to March. The first period of seasonal activity took from mid-June to mid-July while the second was from mid-August to the end of October. Gepp (1988/9) considered this species as univoltine. In his experience overwintered adults lay eggs early during February – March. Feeding activity of larvae occurs between March and late July. In Italy, by light trapping, Pantaleoni (1990*a*) detected a bimodal flight pattern. The first swarming took from late May to mid-July while the second was from mid-July to early October. The flight peaks were recorded in early June and in mid-August respectively.

From the viewpoint of flight activity *Nineta pallida* was investigated in Austria (Ressl, 1971; Gepp, 1974, 1988/9). Seasonal activity of *N. pallida* lasted from early July to the end of October, with the swarming peak in mid-August. The same unimodal seasonal pattern of this chrysopid species was found also in Italy by Pantaleoni (1990*a*) although flight started in late July. Gepp (1988/9) proved that egg-laying of the monovoltine *N. pallida* occurred in late summer, and larvae overwintered in the second instar stage. The third-instar larvae were active on spruce in the period of mid-April to end of May.

There are limited data on seasonality of *Chrysopa dorsalis*. Gepp (1988/9) reported that adults were active between June and mid-September, and their mass flight took place between mid-July and mid-August in Austria. In Italy *C. dorsalis* flew from early May to early September, and its peak activity occurred during the period of late June to early July.

Seasonal activity pattern of hemerobiids

Short descriptions of seasonality are given below only for those conifer-specialist brown lacewings that have been studied at least in three European countries.

Hemerobius stigma is known to overwinter in the adult stage. In England its seasonal activity took place from January to December (Killington, 1937), in Poland from May to August (Czechowska, 1995), in Switzerland from May to late October (Eglin, 1980*b*), in Austria from January to September (Ressl, 1974), in Hungary from June to October (Szentkirályi, 1997), and in Italy from May to September (Pantaleoni, 1990*a*). In northern parts and cooler mountainous regions of Europe *H. stigma* showed an unimodal distribution of its activity and peaked in July (Chechowska, 1995; Eglin, 1980*b*) or in June (Ressl, 1974). In Hungary Szentkirályi (1997) found the mass flight of this hemerobiid in the period of July–August characterised with a peak in early July. However Pantaleoni (1982) detected a bimodal pattern of seasonal flight of *H. stigma* in Italy. The first period of flight activity lasted from mid-May to early July with a peak in mid-June. The second period was found between early July and late September, and the maximal activity occurred at the end of July to early August.

According to the above-mentioned authors, the seasonal activity of *H. nitidulus* lasted between April and September in England and Austria, and from early May to mid-October in Hungary. The activity pattern was unimodal with a peak in July in Austria, but tri-modal flight distribution was detected by light trapping in Hungary. In the latter case the mass flight occurred in late-June to mid-August, characterised by a peak in early July.

The length of the seasonal activity period showed by *H. pini* was as follows in different countries: May to July in England, May to October in Switzerland, and April to September in Austria. In the last two cases *H. pini* expressed its greatest activity in July.

The seasonality pattern of *Sympherobius fuscescens* was detected in the period of June to September in England, June to July in Poland, and June to August in Austria. The unimodal peak of adult activity occurred in June in each of the above countries.

Adults of *Wesmaelius quadrifasciatus* were active from May to August in England, from June to September in Switzerland, and from May to September in Austria. In the latter country the activity peak was in June.

Seasonal activity of *W. concinnus* began in May in England, in June in Poland, and finished in August in both countries. The maximal activity was shown by this hemerobiid in June both in Poland and Austria.

ACKNOWLEDGEMENTS

I am very grateful to Dr Peter McEwen (Insect Investigations Ltd, School of Biosciences, Cardiff University, UK), to Andy Whittington (National Museums of Scotland, Edinburgh, UK), and to Dr Tim New (School of Zoology, La Trobe University, Bundoora, Australia) for their careful editorial work and critical language corrections on the manuscript. Many thanks to Dr Ferenc Kozár (Department of Zoology, Plant Protection Institute of Hungarian Academy of Sciences) for corrections of nomenclature of scale insects used in this chapter, to Ferenc Kádár (Department of Zoology, Plant Protection Institute of Hungarian Academy of Sciences) for help in typing the bibliography. This work was supported, in part, by the Hungarian National Science Foundation (grant number: OTKA T 023284).

REFERENCES

Agekjan, N.G. (1973) Neuroptera feeding on bamboo aphids in Adzharia and their parasites. *Entomologicheskoe Obozrenie* 52, 549–564. (in Russian)

Arakaki, N. (1992*a*) Predators of the sugar cane woolly aphid, *Ceratovacuna lanigera* (Homoptera: Aphididae) in Okinawa and predator avoidance of defensive attack by the aphid. *Applied Entomology and Zoology* 27, 159–161.

Arakaki, N. (1992*b*) Seasonal occurrence of the sugar cane woolly aphid, *Ceratovacuna lanigera* (Homoptera: Aphididae), and its predators in sugar cane fields of Okinawa Islands. *Applied Entomology and Zoology* 27, 99–105.

Aspöck, H. & Hölzel, H. (1996) Neuropteroidea of North Africa, Mediterranean Asia and of Europe: a comparative review (Insecta). In *Pure and Applied Research in Neuropterology, Proceedings of the 5th International Symposium on Neuropterology*, ed. Canard, M., Aspöck, H. & Mansell, M.W., pp. 31–86. Sacco, Toulouse.

Aspöck, H., Aspöck, U. & Hölzel, H. (1980) *Die Neuropteren Europas. Eine zusammenfassende Darstellung der Systematik, Ökologie und Chorologie der Neuropteroidea (Megaloptera, Raphidioptera, Planipennia) Europas*, 2 vols. Göcke & Evers, Krefeld.

Augustin, S. & Lévieux, J. (1993) Life history of the poplar beetle *Chrysomela tremulae* F. in the central region of France. *Canadian Entomologist* 125, 399–401.

Aydemir, M. & Toros, S. (1990) Natural enemies of *Tetranychus urticae* Koch. (Acarina, Tetranychidae) on bean plants in Erzincan. In *Proceedings of the 2nd Turkish National Congress of Biological Control*, pp. 261–271. Ege Universitesi. Izmir, Turkey.

Azuma, S. & Y. Oshiro (1971) Studies on the insect pests of sugar cane woolly aphid, *Ceratovacuna lanigera* Zehntner. *Bulletin of the Okinawa Agricultural Experimental Station* 5, 9–27.

Balakrishnan, M.M., Sreedharan, K., Venkatesha, Krishnamoorty, P. & Bhat, P. K. (1991) Observations on *Ferrisia virgata* (Ckll.) (Homoptera: Pseudococcidae) and its natural enemies on coffee, with new records of predators and host plants. *Journal of Coffee Research* 21, 11–19.

Balázs, K. (1963) Einige Beobachtungen über die Blattläuse der Acer- und Tilia Arten. *Folia Entomologica Hungarica* 16, 195–210. (in Hungarian)

Banerjee, B. (1983) Arthropod accumulation on tea in young and old habitats. *Ecological Entomology* 8, 117–123.

Barnard, P.C. & Brooks, S.J. (1984) The African lacewing genus *Ceratochrysa* (Neuroptera: Chrysopidae): a predator on the cassava mealybug, *Phenacoccus manihoti* (Hemiptera: Pseudococcidae). *Systematic Entomology* 9, 359–371.

Barnard, P.C., Brooks, S.J. & Stork, N.E. (1986) The seasonality and distribution of Neuroptera, Raphidioptera and Mecoptera on oaks in Richmond

Park, Surrey, as revealed by insecticide knock-down sampling. *Journal of Natural History* 20, 1321–1331.

Baylac, M. (1980) Faune associée à *Cryptococcus fagi* (Baer.) (Homoptera: Coccoidea) dans quelques hetraies du nord de la France. *Acta Oecologica, Oecologia Applicata* 1, 199–208.

Blackman, R.L. & Eastop, V.F. (1994) *Aphids on the World's Trees: An Identification and Information Guide*. CAB International, Wallingford UK.

Borkowski, K. (1986) Contribution to the knowledge of the insect fauna of Scotch pine tree crowns. *Polskie Pismo Entomologiczne* 56, 667–676. (in Polish)

Boussienguet, J. (1986) The natural enemy complex of cassava mealybug, *Phenacoccus manihoti* (Hom., Coccoidea, Pseudococcidae) in Gabon. I. Faunistic inventory and relationships. *Annales de la Société Entomologique de France* 22, 35–44. (in French)

Breene, R.G., Meagher, R.L. Jr, Nordlund, D.A. & Wang, Y. T. (1992) Biological control of *Bemisia tabaci* (Homoptera: Aleyrodidae) in a greenhouse using *Chrysoperla rufilabris* (Neuroptera: Chrysopidae). *Biological Control* 2, 9–14.

Brooks, S.J. & Barnard, P.C. (1990) The green lacewings of the world: a generic review (Neuroptera: Chrysopidae). *Bulletin of the British Museum (Natural History) Entomology* 59, 117–286.

Bugg, R.L. (1987) Observations on insects associated with a nectar-bearing Chilean tree, *Quillaja saponaria* Molina (Rosaceae). *Pan-Pacific Entomologist* 63, 60–64.

Canard, M. (1987) Cycle annuel et place de *Chrysoperla mediterranea* (Hölzel) (Neuroptera, Chrysopidae) en forêt mediterranéenne. *Neuroptera International* 4, 279–285.

Cheng, W.Y. & Chen, S.M. (1996) Utilization of green lacewing in Taiwan. *Taiwan Sugar* 43, 20–22.

Cheng, W.Y., Wang, Z.T., Hung, T.H. & Hung, J.K. (1992) Seasonal occurrence of the sugarcane woolly aphid and its predators on autumn planted canes. *Report of the Taiwan Sugar Research Institute* 139, 19–31.

Collins, F.A. & Whitcomb, W.H. (1974) Preliminary survey for the natural enemies of tea scale, *Fiorinia theae* Green in Florida, Unpublished report. Entomology Department, University of Florida, Gainesville.

Covassi, M., Binazzi, A. & Toccafondi, P. (1991) Studies on the entomophagous predators of a scale of the genus *Matsucoccus* Cock. in Italy. I. Faunistical–ecological notes on species observed in pine forests in Liguria and Tuscany. *Redia* 74, 575–597. (in Italian)

Csóka, G. (1998) Insect herbivore guild of the oaks native to Hungary. *Forestry Researches* 88, 311–318. (in Hungarian)

Czechowska, W. (1985) Neuropteran (*Planipennia* and *Raphidioptera*; *Neuropteroidea*) communities of coniferous forests in the Kampinoska Forest and in Białoęka Dworska near Warsaw. *Fragmenta Faunistica* 29, 391–403.

Czechowska, W. (1990) Neuropterans (Neuropteroidea) of linden–oak–hornbeam and thermophilous oak forests of the Mazovian Lowland. *Fragmenta Faunistica* 34, 95–119.

Czechowska, W. (1994) Neuropterans (Neuropteroidea: Raphidioptera, Planipennia) of the canopy layer in pine forests. *Fragmenta Faunistica* 36, 459–467.

Czechowska, W. (1995) Neuropteroidea and Coccinellidae (Coleoptera) of pine canopies of the pine forests in the Berezinsky Biosphere Reserve in Byelorussia. *Fragmenta Faunistica* 38, 159–163.

Dai-Xuan (1995) Preliminary study on the ecological niches of the black citrus aphid and its natural enemies. *Journal of Tea Science* 15, 79–80.

Davidson, N.A., Kinsey, M.G., Ehler, L.E. & Frankie, G.W. (1992) Tobacco budworm, pest of petunias, can be managed with Bt. *California Agriculture* 46, 7–9.

De Barro, P.J. (1990) Natural enemies and other species associated with *Saccharicoccus sacchari* (Cockerell) (Hemiptera: Pseudococcidae) in the Bundaberg area, southeast Queensland. *Journal of the Australian Entomological Society* 29, 87–88.

Del Bene, G., Gargani, E. & Landi, S. (1993) Lotta biologica e integrata contro insetti dannosi alle piante da fiore e ornamentali: risultati preliminari. *Colture Protette* 22, 13–18.

Devetak, D. (1991) Neuropteroidea. Megaloptera, Raphidioptera, Planipennia (Insecta). Fauna Durmitora 4. *Crnogorska Akademija Nauka i Umjetnosti Posebna Izdanja, Knjiga 24, Odjeljenje Prirodnih Nauka, Knjiga 15, Titograd*, 135–159.

Devetak, D. (1998) Neuroptera in different habitats in Istria and Quarnero (NW Balkan). *Acta Zoologica Fennica* 209, 95–98.

Deyrup, M. & Deyrup, N. (1978) Pupation of *Hemerobius* in Douglas-fir cones. (Neuroptera: Hemerobiidae). *Pan-Pacific Entomologist* 54, 143–146.

Dreistadt, S.H. & Hagen, K.S. (1994) Classical biological control of the acacia psyllid, *Acizzia uncatoides* (Homoptera: Psyllidae), and predator–prey–plant interactions in the San Francisco Bay area. *Biological Control* 4, 319–327.

Dreistadt, S.H., Hagen, K.S. & Dahlsten, D.L. (1986) Predation by *Iridomyrmex humilis* (Hym.: Formicidae) on eggs of *Chrysoperla carnea* (Neu.: Chrysopidae) released for inundative control of *Illinoia liriodendri* (Hom.: Aphididae) infesting *Liriodendron tulipifera*. *Entomophaga* 31, 397–400.

Du Merle, P. (1983a) Les facteurs de mortalité des oeufs de *Tortrix viridana* L. (Lep., Tortricidae). I. Le complexe des prédateurs (Hym., Formicidae; Derm., Forficulidae; Orth., Phaneropteridae; Neur., Chrysopidae). *Agronomie* 3, 239–246.

Du Merle, P. (1983b) Les facteurs de mortalité des oeufs de *Tortrix viridana* L. (Lep., Tortricidae). III. Action régulatrice de chacun des facteurs et examen de la mortalité totale. *Agronomie* 3, 429–434.

Eglin, W. (1967) Die Mecopteren und Neuropteren des Kantons Tessin/Südschweiz. *Mitteilungen der schweizerischen entomologischen Gesellschaft Basel* 17, 41–58.

Eglin, W. (1980a) Auf Netzflüglerfang in den Gebirgswäldern des Schweizerischen Nationalparks (Insecta, Neuropteroidea), 1938–78. *Acta Musei Reginaehradecensis, Série A: Scientiae Naturales Supplementum*, 31–34.

Eglin, W. (1980b) Die Netzflügler des Schweizerischen Nationalparks und seiner Umgebung (Insecta: Neuropteroidea). *Ergebnisse der wissenschaftlichen Untersuchungen im schweizerischen Nationalpark* 15, 281–351.

Eglin, W. (1980c) Die Insektenfauna des Hochmoores Balmoos bei Hasle, Kanton Luzern. III. Neuropteroidea und Mecoptera (Netzflügler and Schnabelfliegen). *Entomologische Berichte Luzern* 3, 76–85.

Eglin, W. (1981) Zur Insektenfauna des Siedereiteiches bei Hochdorf, Kanton Luzern. III. Neuropteroidea (Netzflügler). *Entomologische Berichte Luzern* 5, 68–70.

Ehler, L.E. & Kinsey, M.G. (1995) Ecology and management of *Mindarus kinseyi* Voegtlin (Aphidoidea: Mindaridae) on white-fir seedlings at a California forest nursery. *Hilgardia* 62, 1–62.

Eikenbary, R.D. & Fox, R.C. (1968) Arthropod predators of the Nantucket pine tip moth, *Rhyacionia frustrana*. *Annals of the Entomological Society of America* 61, 1218–1221.

Garland, J.A. (1978) Reinterpretation of information on exotic brown lacewings (Neuroptera: Hemerobiidae) used in a biocontrol programme in Canada. *Manitoba Entomologist* 12, 25–28.

Gepp, J. (1974) Die Netzflügler (Megaloptera, Raphidiodea, Planipennia) des Kaiserwaldes südwestlich von Graz (mit einer zoogeographischen Analyse). *Mitteilung der Abteilung für Zoologie Landesmuseum Joanneum* 3, 11–28.

Gepp, J. (1988/9) Zur ökologischen Differenzierung der präimaginalen Stadien baumbewohnender Chrysopiden im Alpenraum (Planipennia, Chrysopidae). *Sitzungsberichten der österreichischen Akademie der Wissenschaften mathematische-naturwissenschaftliche Klasse I* 197, 1–73.

Hajek, A.E. & Dahlsten, D.L. (1988) Distribution and dynamics of aphid (Homoptera: Drepanosiphidae) populations on *Betula pendula* in northern California. *Hilgardia* 56, 1–33.

Hassan, S.A. (1978) Releases of *Chrysopa carnea* Steph. to control *Myzus persicae* (Sulzer) on eggplant in small greenhouse plots. *Zeitschrift für Pflanzenkrankheiten und Pflanzenschutz* 85, 118–123.

Heinz, K.M. & Parrella, M.P. (1990) Biological control of insect pests on greenhouse marigolds. *Environmental Entomology* 19, 825–835.

Hesselein, C., Robb, K., Newman, J., Evans, R. & Parrella, M.P. (1993) Demonstration/integrated pest management program for potted chrysanthemums in California. *Bulletin OILB/SROP* 16, 71–76.

Hölzel, H. & Monserrat, V.J. (1992) Chrysopidae from Equatorial Guinea. *Entomofauna* 13, 465–476.

Hölzel, H. & Ohm, P. (1991) Chrysopidae der Mascarene-Inseln (Neuropteroidea: Planipennia). *Entomologische Zeitschrift* 101, 333–352.

Hölzel, H., Ohm, P. & Stelzl, M. (1994) Chrysopidae aus Senegal und Gambia II. Belonopterygini und Chrysopini (Neuroptera). *Entomofauna* 15, 377–396.

Killington, F. J. (1937) *A Monograph of the British Neuroptera*, vol. 2. Ray Society, London.

Kolodziejak, E. (1994) Communities of *Lachnidae* (*Aphidoidea*) inhabiting pine canopies in Polish pine forests situated in three forest health zones. *Fragmenta Faunistica* 36, 377–385.

Kozár, F. (1998) *Catalogue of Palearctic Coccoidea*. Plant Protection Institute, Hungarian Academy of Sciences, Budapest.

Kozár, F. & Kosztarab, M. (1980) Coccoidea of Central European forests and their host relationships. *Acta Musei Reginaehradecensis, Série A: Scientiae Naturales Supplementum*, 203–211.

Laidlaw, W.B.R. (1936) The brown lacewing flies (Hemerobiidae): their importance as controls of *Adelges cooleyi* Gillette. *Entomologist's Monthly Magazine* 72, 164–174.

Lammes, T. (1993) Faunistic rarities, *Hemerobius micans*. *Entomologica Fennica* 4, 15.

Lezhneva, I.P. & Anisimov, A.I. (1995) Predatory aphidophages for protection of greenhouse crops. *Zashchita Rastenii* 11, 39–40. (in Russian)

Lim, G.T. & Pan, Y.C. (1979) A survey of the natural enemies of sugar-cane aphids in Perak, Malaysia. *International Organisation for Biological and Integrated Control of Noxious Animals and Plants (IOBC) Newsletter* 11-12, 5.

Lo, K.C., Lee, W.T., Wu, T.K. & Ho, C.C. (1990) Use of predators for controlling spider mites (Acarina, Tetranychidae) in Taiwan, China. *FFTC National*

Agicultural Research Centre International Seminar on 'The Use of Parasitoids and Predators to Control Agricultural Pests', National Agricultural Research Centre, Tukuba gun, Japan.

Majer, J.D. & Recher, H.F. (1988) Invertebrate communities on Western Australian eucalypts: a comparison of branch clipping and chemical knockdown procedures. *Australian Journal of Ecology* 13, 269–278.

Marín, F. & Monserrat, V.J. (1987) Los neurópteros del encinar ibérico. *Boletín de Sanidad Vegetal Plagas* 13, 347–359.

Marín, F. and Monserrat, V.J. (1991) The community of Neuropteroidea from Iberian southern beechwoods. In *Behaviour and Impact of Aphidophaga*, ed. Polgár, L., Chambers, R.J., Dixon, A.F.G. & Hodek, I., pp. 187–198. SPB Academic Publishing, The Hague.

McClure, M.S. (1995) Using natural enemies from Japan to control hemlock woolly adelgid. *Frontiers of Plant Science* 47, 5–7.

Mishra, B.K., Jena, B.C. & Mishra, P.R. (1996) Biology and feeding potential of *Chrysopa scelestes* Banks feeding on the eggs of *Pyrilla perpusilla* Walker. *Indian Sugar* 45, 757–759.

Miyanoshita, A. & Kawai, S. (1992) Influence of predation by *Mallada boninensis* (Okamoto) (Neuroptera: Chrysopidae) and autumn movement of female adults on survival of *Ceroplastes japonicus* Green (Homoptera: Coccidae). A model experiment with cages. *Japanese Journal of Applied Entomology and Zoology* 36, 196–199.

Mizell, R.F. & Schiffhauer, D.E. (1987) Seasonal abundance of the crapemyrtle aphid, *Sarucallis kahawaluokalani*, in relation to the pecan aphids, *Monellia caryella* and *Monelliopsis pecanis* and their common predators. *Entomophaga* 32, 511–520.

Monserrat, V.J. (1984) Sobre las especies africanas del género *Notiobiella* Banks, 1909, II. (Neuroptera, Planipennia, Hemerobiidae). In *Progress in World's Neuropterology, Proceedings of the 1st International Symposium on Neuropterology*, ed. Gepp, J., Aspöck, H. & Hölzel, H. pp. 99–124. Thalerhof, Graz.

Monserrat, V. (1990) A systematic checklist of the Hemerobiidae of the world (Insecta: Neuroptera). In *Advances in Neuropterology, Proceedings of the 3rd International Symposium on Neuropterology*, ed. Mansell, M.W. & Aspöck, H., pp. 215–262. South African Department of Agricultural Development, Pretoria.

Monserrat, V.J. (1991) Nuevos datos sobre algunas especies del género *Hemerobius* L., 1758 (Insecta, Neuroptera: Hemerobiidae). *Graellsia* 47, 61–70.

Monserrat, V.J. (1992) New data on the Afrotropical brown lacewings (Neuroptera: Hemerobiidae). *Journal of the Entomological Society of South Africa* 55, 123–136.

Monserrat, V.J. (1996) Revision del género *Hemerobius* de Latinoamerica (Neuroptera, Hemerobiidae). *Fragmenta Entomologica* 27, 399–523.

Monserrat, V.J. (1998) Nuevos datos sobre los hemeróbidos de América (Neuroptera: Hemerobiidae). *Journal of Neuropterology* 1, 109–153.

Monserrat, V.J. & Marín, F. (1994) Plant substrate specificity of Iberian Chrysopidae (Insecta: Neuroptera). *Acta Oecologica* 15, 119–131.

Monserrat, V.J. & Marín, F. (1996) Plant substrate specificity of Iberian Hemerobiidae (Insecta: Neuroptera). *Journal of Natural History* 30, 775–787.

Morimoto, M. & Shibao, H. (1993) Predators of the bamboo aphid *Pseudoregma bambucicola* (Homoptera: Pemphigidae) in Kagoshima, Southern Japan. *Applied Entomology and Zoology* 28, 246–248.

Mound, L.A. & Halsey, S.H. (1978) *A Systematic Catalogue of the Aleyrodidae (Homoptera) with Host Plant and Natural Enemy Data*. British Museum (Natural History), London.

Munir, B. & Sailer, R.I. (1985) Population dynamics of the tea scale, *Fiorinia theae* (Homoptera: Diaspididae), with biology and life tables. *Environmental Entomology* 14, 742–748.

Muraleedharan, N. & Chen, Z.M. (1997) Pests and diseases of tea and their management. *Journal of Plantation Crops* 25, 15–43.

Neuenschwander, P. & Hagen, K.S. (1980) Role of the predator *Hemerobius pacificus* in a non-insecticide treated artichoke field. *Environmental Entomology* 9, 492–495.

New, T.R. (1967) Trap-banding as a collecting method for Neuroptera and their parasites, and some results obtained. *Entomologist's Gazette* 18, 37–44.

Nielsen, E.S. (1977) Studies on lacewings (Neuroptera s.str.) in a Danish beech stand. *Entomologiske Meddelelser* 45, 45–64. (in Danish)

Ohm, P. (1965) Zusammensetzung und Entstehungsgeschichte der Neuropterenfauna der Nordfriesischen Insel Amrum. *Verhandlung Vereinigung naturwissenschaftliche Heimatforschung, Hamburg* 36, 81–101.

Ohm, P. (1973) Durch die Forstwirtschaft ermöglichte Vergrößerung der Verbreitungsareale nadelholzbewohnender Netzflügler. *Faunistische-ökologische Mitteilungen* 4, 299–304.

Pantaleoni, R.A. (1982) Neuroptera Planipennia del comprensorio delle Valli di Comacchio: indagine ecologica. *Bollettino dell'Istituto di Entomologia dell'Università di Bologna* 37, 1–73.

Pantaleoni, R.A. (1984) Neuroptera Planipennia del comprensorio delle Valli di Comacchio: le neurotterocenosi del *Quercetum ilicis* e del *Populus nigra pyramidalis. Bollettino dell'Istituto di Entomologia dell'Università di Bologna* 39, 61–74.

Pantaleoni, R.A. (1990a) I Neurotteri (Neuropteroidea) della valle del Bidente-Ronco (Appennino Romagnolo). *Bollettino dell'Istituto di Entomologia dell'Università di Bologna* 44, 89–142.

Pantaleoni, R.A. (1990b) Neurotteri e fasce di vegetazione in Romagna. *Bollettino dell'Istituto di Entomologia dell'Università di Bologna* 44, 143–154.

Pantaleoni, R.A. (1996) Distribuzione spaziale di alcuni Neurotteri Planipenni su piante arboree. *Bollettino dell'Istituto di Entomologia dell'Università di Bologna* 50, 133–141.

Pantaleoni, R.A. & Tisselli, V. (1985) I Neurotteri delle colture agrarie: rilievi sui Crisopidi in alcune coltivazioni del forlivese. *Bollettino dell'Istituto di Entomologia dell'Università di Bologna* 40, 51–65.

Penny, N.D., Adams, P.A. & Stange, L.A. (1997) Species catalog of the Neuroptera, Megaloptera, and Raphidioptera of America North of Mexico. *Proceedings of the California Academy of Sciences* 50, 39–114.

Popov, A.K. (1991) Baum- und strauchbewohnende Neuropteren in Bulgarien. *Acta Zoologica Bulgarica* 41, 26–36.

Potemkina, V.I. & Kovalenko, T.K. (1990) The use of aphid predator *Micromus angulatus* in the integrated vegetable protection system. *Nauchno Tekhnicheskii Byulleten, VASKhNIL, Sibirskoe Otdelenie* 2, 24–32. (in Russian)

Pschorn-Walcher, H. & Zwölfer, H. (1956) The predator complex of the white-fir woolly aphids (genus *Dreyfusia, Adelgidae*). *Zeitschrift für angewandte Entomologie* 39, 63–75.

Rabeler, W. (1962) Die Tiergesellschaften von Laubwäldern (Querco-Fagetea) im oberen und mittleren Wesergebiet. *Mitteilungen floristische-soziolische Arbeitsgemeinschaft* N.F. 9, 200–229.

Radhakrishnan, B. & Muraleedharan, N. (1989) Life history and population dynamics of *Micromus timidus* Hagen a predator of the tea aphid, *Toxoptera aurantii* (Boyer de Fonscolombe). *Journal of Plantation Crops* 16 (Supplement), 189–194.

Raupp, M.J., Hardin, M.R., Braxton, S.M. & Bull, B.B. (1994) Augmentative releases for aphid control on landscape plants. *Journal of Arboriculture* 20, 241–249.

Ressl, F. (1971) Untersuchungen über die Chrysopiden des Bezirkes Scheibbs (Niederösterreich). Beitrag zur Kenntnis der Ökologie, Phänologie und Verbreitung der Chrysopiden Mitteleuropas. *Beiträge zur Entomologie* 21, 597–607.

Ressl, F. (1974) Untersuchungen über die Hemerobiiden (Neuroptera, Planipennia) des Bezirkes Scheibbs (NOE). Ein Beitrag zur Kenntnis der Verbreitung, Ökologie und Phänologie der Hemerobiiden

Mitteleuropas. *Entomologische Gesellschaft Basel* 24, 10–28.

Saure, C. & Kielhorn, K.H. (1993) Netzflügler als Bewohner der Kronenregion von Eiche und Kiefer (Neuroptera: Coniopterygidae, Hemerobiidae, Chrysopidae). *Faunistische-ökologische Mitteilungen* 9, 391–402.

Schmutterer, H. (1972) Unterordnung Coccoidea, Schildläuse. Coccidae (Lecaniidae), Napfschildläuse. In *Die Forstschädlinge Europas, Würmer, Schnecken, Spinnentiere, Tausendfüssler und hemimetabole Insekten*, vol.1, ed. Schwenke, W., pp. 405–418. Paul Parey, Hamburg.

Schremmer, F. (1962a) Beobachtungen und Untersuchungen über die Insektenfauna der Lärche (*Larix decidua*) im östlichen Randgebiet ihrer natürlichen Verbreitung, mit besonderer Berücksichtigung einer Grosstadtlärche, part I. *Zeitschrift für angewandte Entomologie* 45, 1–48.

Schremmer, F. (1962b) Beobachtungen und Untersuchungen über die Insektenfauna der Lärche (*Larix decidua*) im östlichen Randgebiet ihrer natürlichen Verbreitung, mit besonderer Berücksichtigung einer Grosstadtlärche, part II. *Zeitschrift für angewandte Entomologie* 45, 113–153.

Selvakumaran, S., David, B.V. & Kumaresan, D. (1996) Observations on the natural enemies of the whitefly *Kanakarajiella cardamomi* (David and Subramaniam) a pest on cardamon. *Indian Journal of Environment and Toxicology* 6, 26–27.

Selvasundaram, R. & Muraleedharan, N. (1987) Natural enemies of certain leaf folding caterpillar pests of tea in southern India. *Journal of Coffee Research* 17, 118–119.

Séméria, Y. (1984) Savannah: 5.3. Mediterranean Climates. In *Biology of Chrysopidae*, ed. Canard, M. Séméria, Y. & New, T.R., pp. 176–180. Dr W. Junk, The Hague.

Shantibala, K., Somen, L., Debaraj, Y. & Singh, T.K. (1994) Development and predatory efficiency of the larvae of *Micromus timidus* Hagen (Neuroptera: Hemerobiidae) on an oak aphid, *Cervaphis quercus* Takahashi. *Indian Journal of Hill Farming* 7, 212–214.

Somen-Singh, L.S., Devjani, P., Debaraj, Y. & Singh, T.K. (1994) Studies on the seasonal incidence of *Myzus persicae* Sulzer (Homoptera: Aphididae) on cabbage in relation to abiotic and biotic factors. *Proceedings of the National Academy of Sciences, India, Section B, Biological Sciences* 64, 95–98.

Stary, P. & Havelka, J. (1991) *Macrosiphum albifrons* Essig, an invasive lupin aphid and its natural-enemy complex in Czechoslovakia (Homoptera, Aphididae). *Acta Entomologica Bohemoslovaka* 88, 111–120.

Stechmann, D.-H. & Semisi, S.T. (1984) Insektenbekämpfung in West-Samoa unter besonderer Berücksichtigung des Standes biologischer und integrierter Verfahren. *Anzeiger für Schädlingskunde Pflanzenschutz Umweltschutz* 57, 65–70.

Stechmann, D.-H. & Völkl, W. (1990) A preliminary survey of aphidophagous insects of Tonga, with regards to the biological control of the banana aphid. *Journal of Applied Entomology* 110, 408–415.

Steffan, A.W. (1972) Unterordnung Aphidina, Blattläuse. In *Die Forstschädlinge Europas, Würmer, Schnecken, Spinnentiere, Tausendfüssler und hemimetabole Insekten*, vol. 1, ed. Schwenke, W., pp. 162–386. Paul Parey, Hamburg.

Stelzl, M. (1991) Untersuchungen zu Nahrungsspektren mitteleuropäischer Neuropteren-Imagines (Neuropteroidea, Insecta). Mit einer Diskussion über deren Nützlichkeit als opponenten von Pflanzenschädlingen. *Journal of Applied Entomology* 111, 469–477.

Stelzl, M., Hassan, S.A. & Gepp, J. (1992) Zuchtversuche an Hemerobiiden (Neuroptera, Planipennia) als Antagonisten von Gewachshausschädlingen. *Mitteilungen der deutschen Gesellschaft für allgemeine und angewandte Entomologie* 8, 187–192.

Szentkirályi, F. (1984) Analysis of light trap catches of green and brown lacewings (Neuropteroidea: Planipennia, Chrysopidae, Hemerobiidae) in Hungary. In *Verhandlungen des X. International Symposiums über Entomofaunistik Mitteleuropas*, ed. Kaszab, Z. pp. 177–180. Budapest.

Szentkirályi, F. (1992) Spatio-temporal patterns of brown lacewings based on the Hungarian light-trap network (Insecta: Neuroptera: Hemerobiidae). In *Current Research in Neuropterology, Proceedings of the 4th International Symposium on Neuropterology*, ed. Canard, M., Aspöck, H. & Mansell, M.W. pp. 349–357. Sacco, Toulouse.

Szentkirályi, F. (1997) Seasonal flight patterns of some common brown lacewing species (Neuroptera, Hemerobiidae) in Hungarian agricultural regions. *Biologia (Bratislava)* 52, 291–302.

Szentkirályi, F. (1998) Monitoring of lacewing assemblages (Neuroptera: Chrysopidae, Hemerobiidae) by light trapping in region of Körös-Maros National Park. *Crisicum* 1, 151–167. (in Hungarian)

Ter-Simonjan, L.G., Bushtshik, T.N. & Voskresenskaja, V.N. (1982) Biological control of pests in vegetable plantings. *Acta Entomologica Fennica*, 40, 33–35.

Tulisalo, U., Tuovinen, T. & Kurppa, S. (1977) Biological control of aphids with *Chrysopa carnea* on parsley and green pepper in the greenhouse. *Annales Entomologici Fennici* 43, 97–100.

Turner, B.D. (1984) Predation pressure on the arboreal epiphytic herbivores of larch trees in southern England. *Ecological Entomology* 9, 91–100.

Tverdyukov, A.P., Nikonov, P.V. & Yushchenko, N.P. (1993) Micromus. *Zashchita Rastenii* 7, 15. (in Russian)

Ushchekov, A.T. (1989) *Chrysopa perla* for aphid control. *Zashchita Rastenii* 11, 20–22. (in Russian)

Vidlička, L. (1994) Flight activity of some Planipennia species. *Biologia (Bratislava)* 49, 729–737.

Wang, L.Y. & Hu, H.L. (1987) Studies on *Chrysopa kulingensis* Navás. *Natural Enemies of Insects* 9, 25–28. (in Chinese)

Weires, R.W. & Chiang, H.C. (1973) Integrated control prospects of major cabbage insect pests in Minnesota – based on the faunistic, host varietal, and trophic relationships. *Minnesota Agricultural Experimental Station Technical Bulletin* 291, 42.

Wheeler, A.G. & Hoebeke, E.R. (1985) The insect fauna of ninebark, *Physocarpus opulifolius* (Rosaceae). *Proceedings of the Entomological Society of Washington* 87, 356–370.

Yarkulov, F.Y. (1986) The biological method in greenhouses in the Maritime Territory. *Zashchita Rastenii* 12, 20–21. (in Russian)

Zeki, H. & Toros, S. (1990) Determination of natural enemies of *Chrysomela populi* L. and *Chrysomela tremulae* F. (Coleoptera, Chrysomelidae) harmful to poplars and the efficiency of their parasitoids in Central Anatolia region. In *Proceedings of the 2nd Turkish National Congress of Biological Control*, pp. 251–260. Ege Universitesi, Izmir, Turkey. (in Turkish)

Zelený, J. (1963) Hemerobiidae (Neuroptera) from Czechoslovakia. *Acta Societatis Entomologicae Čechosloveniae*, 60, 55–67.

Zelený, J. (1971) Megaloptera-, Raphidioptera-, Neuroptera- und Mecoptera-Arten des Gebirges Novohradské hory. *Sborník Jihočeského muzea v Českých Budějovicích Přírodní Vědy (Supplementum)* 9, 39–43. (in Czech)

Zelený, J. (1978) Les fluctuations spatio-temporelles des populations de Névroptères aphidiphages (*Planipennia*) comme élément indicateur de leur spécificité. *Annales de Zoologie et Écologie Animale*, 10, 359–366.

Zhao, Y.F. & Hou, J.W. (1993) Investigation and feeding of natural enemies of *Acaphylla theae* Watt. *Journal of Tea* 19, 35–37. (in Chinese)

Zoebelein, G. (1957) Zur Beeinflussung der Insektenfauna des Waldes durch chemische Grossschädlingsbekämpfungen. *Zeitschrift für angewandte Entomologie* 41, 320–331.

SOURCES AND METHODS OF EVALUATION

Literature sources were reviewed from 1910 until the present, although the overwhelming majority of usable data are derived from the publications of the last four decades. In order to have a better overview of the records of lacewing species, or rather of their assemblage in the given stands of cultivated plants, ornamentals, or forest trees, data have been arranged into tables. Within the tables of the forest lacewings the European data are arranged by geographical regions (from northern areas to the southern and Mediterranean regions) and by the tree species of forest associations.

The list of species names of chrysopid and hemerobiid families can be found in the first column of the tables. For practical reasons, those names of species that were formerly published, but now are invalid, have been replaced according to the latest valid nomenclature. The correction of names were based upon the works of Aspöck & Hölzel (1996), Brooks & Barnard (1990), Monserrat (1990, 1996, 1998), and Penny *et al.* (1997). By this latter work concerning the newest nomenclature of North American lacewing species, the former name *Chrysoperla carnea* is synonymised with *Chrysoperla downesi* (Smith). Because the older invalid *carnea* is the most frequently used name in the literature, we did not replace it with the new one in the text and the related tables. However do not forget reading the text, that the American *C. carnea* is not identical with the European one and its new valid name is *C. downesi*.

The serial number of columns in the tables refers to the publication sources, also to the corresponding countries, and in some cases to tree species or associations, as indicated in the notes at the bottom of the tables. The columns contain either the lacewing assemblages of local stands of cultivated plant, forest, or the data of faunistic collections of several localities altogether. Where the numbers of individuals collected were given and are sufficiently high, the percentage dominance values of each species have been calculated and are shown in the tables. If in a certain investigation, the samplings were made in several stands of a forest type, then the dominance distribution has been calculated for each site, and the minimal and maximal values (range) are presented for each species. For both Chrysopidae and Hemerobiidae, the percentage dominance distributions have been calculated separately. In the case of sites being close to each other, the numbers of lacewing individuals collected were summed, and from these data one dominance distribution was constructed. Where the calculation of dominance was impossible, because the numbers of individuals were too low or lacking, but the species were present, a '+' sign indicates the occurrence of the given species. An 'A' letter is put into the cell, if there was a reference that a certain species was dominant or abundant.

The number of literature references for each species can be found in the last column with an 'FR' sign (frequency of occurrence). The measure of these values refers to the closeness of the possible association between the given lacewing species and the given cultivated plant. The greater the FR and dominance percentage value of a certain species, the higher the probability that the lacewing is an important member of that Neuroptera assemblage found on a given cultivated plant. In the tables, the number of species and the species richness of assemblages are shown in the last two rows, separately for chrysopids (S_C) and hemerobiids (S_H).

PART 3

Principles

Introduction to Part 3

The Editors

This section of the book gives an outline of the principles behind the successful use of lacewings in crop protection. As such it provides practical information on how best to use lacewings, what happens to lacewings once they are in the field, and how success might be measured.

To these ends the section outlines the use of lacewings in biological control in an historical context, covering failures as well as successes, details ways in which to produce large numbers of lacewings in the laboratory for augmentation purposes, and how to improve the effectiveness of lacewings already in the field. Features of the nutrition of lacewing larvae and adults are covered in more detail as are ecological relations between lacewings and the environment in which they exist.

A large part of this section of book is devoted to the relevant area of the relationship between lacewings and pesticides and the highly topical area of the effects of *Bacillus thuringiensis* via the ingestion of transgenic corn-fed prey on lacewings is discussed.

CHAPTER 11

The use of lacewings in biological control

L.J. Senior and P.K. McEwen

11.1 HISTORICAL DEVELOPMENT

Lacewings have long been recognised as effective predators of aphids and other arthropod pests. More than 250 years ago, Réaumur (1742) discussed the use of lacewings for biological control of aphids in greenhouses (Stiling, 1985). Many early authors noted the large range and huge numbers of insect and mite pests consumed by lacewing larvae (Killington, 1936). Balduf (1939) stated that the Chrysopidae and Hemerobiidae were very effective biological control agents, feeding on some of the world's most important agricultural and horticultural pests. However, it was not until the 20th century that studies began on their potential for biological control. According to Ridgway & Murphy (1984) the first studies of the release of lacewings for control of a pest were those of Doutt & Hagen (1949, 1950), who examined the use of lacewings for the control of mealybugs. Later, Dunn (1954) investigated the potential of the hemerobiid *Micromus variegatus* Fabricius as a predator of aphids.

The principal factor limiting the manipulative use of lacewings for biological control has been that of producing large numbers economically. Finney (1948, 1950) began the first studies on the mass culture and distribution of *Chrysoperla carnea*. During the 1960s and early 1970s, work was carried out to devise an artificial diet for lacewings, to facilitate mass-rearing (Hagen & Tassan, 1965, 1966, 1970; Vanderzant, 1969; Butler & Ritchie, 1971). Improvements in rearing methods precipitated an increase in the number of trials of field releases of lacewings, particularly *C. carnea*, during the 1960s and 1970s. For example, Ridgway & Jones (1968, 1969) used *C. carnea* to control lepidopterous pests (*Helicoverpa zea* (Boddie) and *H. virescens* Fabricius), Miszczak & Niemczyk (1978) used the same species for control of European red mite (*Panonychus ulmi* Koch), & Adashkevich & Kuzina (1971) and Shuvakhina (1977, 1978) used *C. carnea* to control Colorado potato beetle [*Leptinotarsa decemlineata*

(Say)]. However, the majority of trials have targeted aphids either in the greenhouse (Scopes, 1969; Tulisalo & Tuovinen, 1975; Tulisalo *et al.*, 1977; Hassan, 1978), or on field crops (Shands *et al.*, 1972; Hagley, 1989). Field releases have been reviewed in greater detail by Ridgway & Murphy (1984) and Wang & Nordlund (1994).

In the latter part of the last century (1970s onwards), with the introduction of integrated pest management (IPM), the role of biological control has gained more credibility. As the disadvantages of chemical controls are recognised, they are increasingly becoming subordinate to other methods, such as biological control, and numerous studies have investigated interactions between pesticides and beneficial species. A few species of lacewings, mainly chrysopids, are now well studied and accepted as important biological control agents. This is well illustrated by the increase over the last 30 years in the number of commercial companies selling biological controls, including lacewings (van Lenteren *et al.*, 1997). Methods for their use continue to be developed and refined, and more species investigated.

11.2 LACEWING SPECIES USED IN BIOLOGICAL CONTROL

Most lacewing species used for biological control are members of either the Chrysopidae or the Hemerobiidae. To date, most research has been directed towards the former family, particularly species of the genus *Chrysoperla*, several of which are available commercially. This is largely because the adults of several species within this genus (e.g. *C. carnea*) are not predaceous, feeding instead on nectar, pollen and honeydew, so they are relatively easy to mass-produce. The *C. carnea* species complex is common in Europe, *C. rufilabris* (Burmeister) in North America, and *C. sinica* (Tjeder) in China (Wang & Nordlund, 1994). Hence these are the species that have been studied most intensively.

The hemerobiids have received less attention, despite the fact that they are at least as diverse as the chrysopids in some local assemblages (New, 1975). Two genera in particular have been studied: *Micromus* [e.g. *M. angulatus* (Stephens) and *M. tasmaniae* Walker] and *Hemerobius* (e.g. *H. pacificus* Banks). Species in these two genera are the most common, agriculturally important hemerobiids in North America (Krakauer & Tauber, 1996).

11.3 THE EFFECTIVENESS OF LACEWINGS AS BIOLOGICAL CONTROL AGENTS

Larval lacewings fulfil many of the requirements of an effective biological control agent. They are voracious, active predators, with an excellent searching capacity (Sundby, 1966; Bond, 1980). Afzal & Khan (1978) found that one *Chrysoperla carnea* larva could consume an average of 487 aphids (*Aphis gossypii* Glover) or 511 whitefly pupae [*Bemisia tabaci* (Gennadius)] before pupation. Lingren *et al.* (1968) stated that larval *Chrysopa* spp. were among the most efficient predators of immature *Heliothis zea* and *H. virescens*. Lacewings are also polyphagous, feeding on a large range of small, soft-bodied arthropods (Principi & Canard, 1984). Their prey include pests of economic importance such as aphids, whitefly, mites, mealybugs, Colorado potato beetle, and various lepidopteran pests such as bollworms (literature reviewed by Wang & Nordlund, 1994). This is an important commercial consideration; as they may be used to control several arthropod pests, lacewings have a broad potential market.

Lacewings are abundant, with a world-wide distribution (Zelený, 1984) and a broad habitat range, covering a variety of agroecosystems (New, 1984; Wang & Nordlund, 1994). Therefore, they may be used in virtually any cropping system. They are often already present in the general area of a crop, and may be manipulated to increase their numbers, either by augmentation or conservation. When released as eggs or larvae, as is common practice, these stages are unable to disperse, remaining in the target area.

Because of their resistance to several common insecticides, lacewings are suitable for use in integrated pest management programmes, giving them an advantage over other, more sensitive predators and parasitoids. For instance, Grafton-Cardwell & Hoy (1985) compiled data on the susceptibility of *C. carnea* to a range of pesticides. They found that eggs were extremely tolerant to the majority of pesticides tested, and that all stages were tolerant to many synthetic pyrethroids, botanicals, microbial insecticides, insect growth regulators, fungicides, herbicides, acaricides and the formamidine insecticide chlordimeform. The interactions of lacewings with pesticides are discussed by Vogt *et al.* (Chapter 16, this volume).

Lacewings have a high reproductive rate (Elkarmi *et al.*, 1987), a relatively long adult life, and a long oviposition period (New, 1975). Advances in rearing techniques have made it possible for several species to be cultured in large numbers on simple diets (Nordlund & Morrison, 1992). Furthermore, recent developments in the production of cheap, artificial food for larval rearing could significantly reduce lacewing production costs (Senft, 1997).

11.4 GENERAL SCHEMES OF USE

There are four methods which may be employed in the use of lacewings for biological control. Definitions are taken from van Driesche & Bellows (1996).

Introduction of a new natural enemy species (classical biological control)
This method is used to control introduced pest species. New enemies are typically natural enemies of the pest in its native homeland.

Lacewings are not generally introduced using this method. As at least one lacewing species is often already in the crop, it is more desirable to enhance the numbers of those present rather than to introduce exotic species (New, 1975). The sibling species issue complicates matters; although '*Chrysoperla carnea*' is widely distributed, different sibling species predominate in different areas. As *C. carnea* has, until recently, been treated as a single species, many releases may in fact have been unintentional introductions of new sibling species.

Augmentation using inoculative releases
'Small numbers of a natural enemy are introduced early in the crop cycle with the expectation that they will reproduce in the crop and their offspring will continue to provide pest control for an extended period of time.'

The principal disadvantage of this method is that many lacewings (including *C. carnea*) disperse after adult emergence, and consequently do not remain in the crop. However, Larock & Ellington (1996) used inoculative releases of *C. rufilabris* successfully, as part of an integrated pest management programme to control pecan aphids. This method may also be suitable for protected crops, where the lacewing adults are unable to disperse. For an inoculative release to provide long-term control, the lacewings must be able to reproduce. Therefore it is necessary that the pest or pests present in the crop, or alternative prey, provide a sufficiently nutritious diet for the predators, which will enable them to complete their development.

Augmentation using inundative releases

'Used when insufficient reproduction of the released natural enemies is likely to occur, and pest control will be achieved exclusively by the released individuals themselves.'

This method is used more commonly. It may be employed to control pests which do not support development of the lacewings, and in situations where immediate control is desired, for example on short-term crops, or when numbers of the pest have already reached damaging levels. Several companies now produce lacewings commercially, and techniques have been developed for the mass release of eggs and larvae (Nordlund *et al.*, Chapter 12, this volume). However, mass releases are expensive; until lacewings can be produced cheaply, this method will not be economically viable for low-value crops. For instance, King *et al.* (1985) concluded that, with each release costing $385/ha, control of *Heliothis* spp. with *C. carnea* was not economically viable.

Conservation

'This approach seeks to identify and rectify negative influences that suppress natural enemies and to enhance agricultural fields (or other sites) as habitats for natural enemies.'

This technique assumes that natural enemies already exist in the vicinity of the crop and that, given the opportunity, they have the potential to adequately suppress the pest. Although lacewings are widespread, efficient predators, the natural population present in a crop is often insufficient to effectively control pests. Conservation techniques can be used to increase numbers. Examples include:

1. Attractants, arrestants, and food supplements. Attempts have been made to identify chemicals which may be sprayed on to crops to manipulate field populations of lacewings (Hagen *et al.*, 1976; McEwen *et al.*, 1994; Dodds & McEwen, 1998). Artificial honeydews play a dual role; as well as acting as attractants, they provide a source of protein for adults, thus increasing their fecundity.
2. Hibernation shelters. These may be used to protect overwintering adults, thereby potentially increasing the population of lacewings present in a crop the following spring (McEwen & Sengonca, Chapter 27, this volume).
3. Cropping patterns. Strip harvesting, intercropping, and other techniques may be used to provide food, overwintering sites, and shelter from adverse weather conditions and from pesticides.
4. Minimising the impact of pesticides. More selective pesticides may be used, such as insect growth regulators. Also, pesticides can be used more selectively, by reducing the dosage, limiting the areas treated, limiting the numbers of applications, and using selective formulations.
5. Managing natural enemies of lacewings. For example, Itioka & Inoue (1996) described how lacewings failed to control mealybug in a crop in which ants were present. When the ants were removed, the pest population decreased by 94%.

Conservation techniques have the advantage that they do not rely on mass production of the predators, and hence may be simpler and cheaper. However, these methods alone may not provide sufficient control of a pest. It is more likely that they will be used as part of an integrated pest management programme, or combined with augmentative releases.

11.5 FAILURE OF BIOLOGICAL CONTROL USING LACEWINGS

Many studies have been undertaken to study predation by lacewings on a huge variety of arthropod pests in a number of crops. Many laboratory tests produce favourable results, and several examples of successful field releases are given in this volume (e.g. Tommasini & Mosti, Chapter 26, this volume; Maisonneuve, Chapter 30, this volume). However, despite the excellent qualifications of lacewings as biological control agents, failures often occur. It is important that the reasons for these failures are identified, in order to optimise the use of lacewings for biological control.

Although lacewings prey upon a large variety of pests, their nutritional requirements for growth, development, and reproduction are much more specific. For example, Legaspi *et al.* (1994) found that all immature stages of the whitefly *Bemisia tabaci* were preyed upon by *Chrysoperla rufilabris*, but none of the lacewing larvae survived to adulthood. Similarly, *C. carnea* larvae were unable to survive on a diet of *Trialeurodes vaporariorum* (Westwood) (the greenhouse whitefly) alone, despite consuming large numbers of this pest (Senior & McEwen, 1998). Cohen & Brummett (1997) compared the biomass and methionine (the least abundant essential amino acid) content of *Ephestia kuehniella* Zeller eggs and *B. tabaci* pupae. They found that although an exclusive diet of whitefly pupae satisfied the biomass requirements of lacewing larvae, it was nutritionally inadequate. Even related prey species may give different responses. For example, Osman & Selman (1996) found that diets of three different aphid species produced significantly different juvenile mortality in *C. carnea* larvae. Furthermore, a pest may be ignored in favour of a preferred prey species; Adashkevich *et al.* (1972) found that when lacewing larvae were offered several prey simultaneously, aphids were attacked first, followed by thrips, then tetranychids.

The suitability of the prey may also be affected by the host plant. Legaspi *et al.* (1996) found that whitefly (*B. argentifolii* Bellows & Perring) reared on cucumber or cantaloupes were better quality prey for *C. carnea* and *C. rufilabris* than whitefly reared on poinsettia or lima bean. Aphids may acquire toxicity through feeding on certain alkaloid-rich plants (Emrich, 1991). Moreover, host plants may have physical characteristics which compromise the efficacy of predators. Qualities such as the trichome density or waxiness of leaves may affect the mobility, and hence the searching capacity of lacewing larvae (Treacy *et al.*, 1987; Eigenbrode *et al.*, 1995).

Failure frequently occurs if the population of the prey species is too large. Daane *et al.* (1996) found that when leafhopper densities were high, lacewings failed to reduce their numbers to below the economic injury threshold. Similarly, timing of release is important; releases should take place near the beginning of the season, before the pest population peaks.

The stage at which the insect is released is another important factor. Daane & Yokota (1997) determined that lacewings released as larvae are more likely to survive and reduce pest numbers than those released as eggs. The same authors also compared two commercial delivery systems, and found significant differences in their efficacy.

The obligate dispersal present in many species, including *C. carnea*, may make these species unsuitable for augmentative releases; Trouvé *et al.* (1996) failed to control hop aphids despite three releases of *C. carnea*, because the lacewings did not settle in the hop field.

Finally, the recent discovery that *C. carnea*, a recognised biological control agent, is actually a sibling species complex, may have important implications for its use in biological control. For instance, Thierry *et al.* (1994) found that sibling species overwintered in different sites, and that there was variation in the recovery of reproductive activity. The authors concluded that differences between the sibling species could well have practical implications for the use of *C. carnea* for biological control.

More field trials are needed to assess the efficacy of lacewings as biological control agents of particular pests, and to analyse the reasons for failure.

11.6 CONCLUSION

Lacewings have the potential to be effective biological control agents. They are suitable for use in a huge variety of crops, against a large number of pests, and for integration into IPM programmes. Since their first use in biological control programmes over 50 years ago, much progress has been made, and continuing advances in mass-rearing techniques will make their use more economically attractive. However, it is clear that more research is necessary in order for lacewings to be regarded as efficient, reliable biological control agents.

REFERENCES

Adashkevich, B.P. & Kuzina, N.P. (1971) *Chrysopa* against the Colorado potato beetle. *Zashchita Rastenii* 12, 23. (in Russian)

Adashkevich, B.P., Kuzina, N.P. & Shijko, E.S. (1972) Rearing, storage and usage of *Chrysopa carnea* Steph. *Vop. biol. Zashch. Rast., Kishinev* 1, 8–13. (in Russian)

Afzal, M. & Khan, M.R. (1978) Life history and feeding behaviour of green lacewing *Chrysopa carnea* Steph. (Neuroptera, Chrysopidae). *Pakistan Journal of Zoology* 10, 83–90.

Balduf, W.V. (1939) *The Bionomics of Entomophagous Insects, Part 3.* John S. Swift, New York. (reprinted 1974 by E.W. Classey, London).

Bond, A.B. (1980) Optimal foraging in a uniform habitat: the search mechanism of the green lacewing. *Animal Behaviour* 28, 10–19.

Butler, G.D. Jr & Ritchie, P.L. Jr (1971) Feed Wheast and the abundance and fecundity of *Chrysopa carnea*. *Journal of Economic Entomology* 64, 933–934.

Cohen, A.C. & Brummett, D.L. (1997) The non-abundant nutrient (NAN) concept as a determinant of predator-prey fitness. *Entomophaga* 42, 85–91.

Daane, K.M. & Yokota, G.Y. (1997) Release strategies affect survival and distribution of green lacewings (Neuroptera: Chrysopidae) in augmentation programs. *Environmental Entomology* 26, 455–464.

Daane, K.M., Yokota, G.Y. & Zheng, Y., Hagen, K.S. (1996) Inundative release of common green lacewings (Neuroptera, Chrysopidae) to suppress *Erythroneura variabilis* and *Erythroneura elegantula* (Homoptera, Cicadellidae) in vineyards. *Environmental Entomology* 25, 1224–1234.

Dodds, C. & McEwen, P.K. (1998) Electroantennogram responses of green lacewings (*Chrysoperla carnea*) to plant volatiles: preliminary results. *Acta Zooogica Fennica* 209, 99–102.

Doutt, R.L. & Hagen, K.S. (1949) Periodic colonization of *Chrysopa californica* as a possible control of mealybugs. *Journal of Economic Entomology* 42, 560.

Doutt, R.L. & Hagen, K.S. (1950) Biological control measures applied against *Pseudococcus maritimus* on pears. *Journal of Economic Entomology* 43, 94–96.

van Driesche, R.G. & Bellows, T.S. Jr (1996) *Biological Control.* Chapman & Hall, London.

Dunn, J.A. (1954) *Micromus variegatus* Fabricius (Neuroptera) as a predator of the pea aphid. *Proceedings of the Royal Entomological Society, London A* 29, 76–80.

Eigenbrode, S.D., Moodie, S. & Castagnola, T. (1995) Predators mediate host plant resistance to a phytophagous pest in cabbage with glossy leaf wax. *Entomologia Experimentalis et Applicata* 77, 335–342.

Elkarmi, L.A., Harris, M.K. & Morrison, R.K. (1987) Laboratory rearing of *Chrysoperla rufilabris* (Burmeister), a predator of insect pests of pecans. *Southwestern Entomologist* 12, 73–78.

Emrich, B.H. (1991) Acquired toxicity of the lupin aphid, *Macrosiphum albifrons*, and its influence on the aphidophagous predators *Coccinella septempunctata*, *Episyrphus balteatus* and *Chrysoperla carnea*. *Zeitschrift für Pflanzenkrankheiten und Pflanzenschutz* 98, 398–404.

Finney, G.L. (1948) Culturing *Chrysopa californica* and obtaining eggs for field distribution. *Journal of Economic Entomology* 41, 719–721.

Finney, G.L. (1950) Mass-culturing *Chrysopa californica* to obtain eggs for field distribution. *Journal of Economic Entomology* 43, 97–100.

Grafton-Cardwell, E.E. & Hoy, M.A. (1985) Intraspecific variability in response to pesticides in the common green lacewing, *Chrysoperla carnea* (Stephens) (Neuroptera: Chrysopidae). *Hilgardia* 53, 1–32.

Hagen, K.S. & Tassan, R.L. (1965) A method of providing artificial diets to *Chrysopa* larvae. *Journal of Economic Entomology* 58, 999–1000.

Hagen, K.S. & Tassan, R.L. (1966) Artificial diet for *Chrysopa carnea* Steph. In *Ecology of Aphidophagous Insects*, ed. Hodek, I., pp. 83–87. Academia, Prague.

Hagen, K.S. & Tassan, R.L. (1970) The influence of the food Wheast, and related *Saccharomyces fragilis* yeast products on the fecundity of *Chrysopa carnea* (Neuroptera, Chrysopidae). *Canadian Entomologist* 102, 806–811.

Hagen, K.S., Greany, P., Sawall, E.F. Jr & Tassan, R.L. (1976) Tryptophan in artificial honeydews as a source of an attractant for adult *Chrysopa carnea*. *Environmental Entomology* 5, 458–468.

Hagley, E.A.C. (1989) Release of *Chrysoperla carnea* Stephens (Neuroptera: Chrysopidae) for control of the green apple aphid, *Aphis pomi* DeGeer (Homoptera: Aphididae). *Canadian Entomologist* 121, 309–314.

Hassan, S.A. (1978) Releases of *Chrysopa carnea* Steph. to control *Myzus persicae* (Sulzer) on eggplant in small greenhouse plots. *Zeitschrift für Pflanzenkrankheiten und Pflanzenschutz* 85, 118–123.

Itioka, T. & Inoue, T. (1996) The role of predators and attendant ants in the regulation and persistence of a population of the citrus mealybug *Pseudococcus citriculus* in a satsuma orange orchard. *Applied Entomology and Zoology* 31, 195–202.

Killington, F.J. (1936) *A Monograph of the British Neuroptera*, Vol. I, Ray Society, London.

King, E.G., Hopper, K.R. & Powell, J.E. (1985). Analysis of systems for biological control of crop arthropod pests in the U.S. by augmentation of predators and parasites. In *Biological Control in Agricultural IPM Systems*, ed. Hoy, M.A. & Herzog, D.C., pp. 201–228. Academic Press, London.

Krakauer, A.H. & Tauber, C.A. (1996) Larvae of *Micromus* – generic characteristics and a description of *Micromus subanticus* (Neuroptera, Hemerobiidae). *Annals of the Entomological Society of America* 89, 203–211.

Larock, D.R. & Ellington, J.J. (1996) An integrated pest management approach, emphasizing biological control, for pecan aphids. *Southwestern Entomologist* 21, 153–166.

Legaspi, J.C., Carruthers, R.I. & Nordlund, D.A. (1994) Life history of *Chrysoperla rufilabris* (Neuroptera: Chrysopidae) provided sweetpotato whitefly *Bemisia tabaci* (Homoptera: Aleyrodidae) and other food. *Biological Control* 4, 178–184.

Legaspi, J.C., Correa, J.A., Carruthers, R.I., Legaspi, B.C. & Nordlund, D.A. (1996) Effect of short-term releases of *Chrysoperla rufilabris* (Neuroptera, Chrysopidae) against silverleaf whitefly (Homoptera, Aleyrodidae) in field cages. *Journal of Entomological Science* 31, 102–111.

van Lenteren, J.C., Roskam, M.M. & Timmer, R. (1997) Commercial mass production and pricing of organisms for biological control of pests in Europe. *Biological Control* 10, 143–149.

Lingren, P.D., Ridgway, R.L. & Jones, S.L. (1968) Consumption by several arthropod predators of eggs and larvae of two *Heliothis* species that attack cotton. *Annals of the Entomological Society of America* 61, 613–618.

McEwen, P.K., Jervis, M.A. & Kidd, N.A.C. (1994) Use of sprayed L-tryptophan solution to concentrate numbers of the green lacewing *Chrysoperla carnea* in olive tree canopy. *Entomologia Experimentalis et Applicata* 70, 97–99.

Miszczak, M. & Niemczyk, E. (1978) Green lacewing *Chrysopa carnea* Steph. (Neuroptera, Chrysopidae) as a predator of the European mite *Panonychus ulmi* Koch on apple trees. 2. The effectiveness of *Chrysopa carnea* larvae in biological control of *Panonychus ulmi* Koch. *Fruit Science Reports* 5, 21–31.

New, T.R. (1975) The biology of Chrysopidae and Hemerobiidae (Neuroptera), with reference to their usage as biocontrol agents: a review. *Transactions of the Royal Entomological Society, London* 127, 113–140.

New, T.R. (1984) The need for taxonomic revision in Chrysopidae. In *Biology of Chrysopidae*, ed. Canard, M., Séméria, Y. & New, T.R., pp. 37–42. Dr W. Junk, The Hague.

Nordlund, D.A. & Morrison, R.K. (1992) Mass rearing *Chrysoperla*. In *Advances in Insect Rearing for Research and Pest Management*, ed. Anderson, T.E. & Leppla, N.C., pp. 427–439. Westview Press, Boulder CO.

Osman, M.Z. & Selman, B.J. (1996) Effect of larval diet on the performance of the predator *Chrysoperla carnea* Stephens (Neuropt., Chrysopidae). *Zeitschrift für angewandte Entomologie* 120, 115–117.

Principi, M.M. & Canard, M. (1984) Feeding habits. In *Biology of Chrysopidae*, ed. Canard, M., Séméria, Y. & New, T.R., pp. 76–92. Dr W. Junk, The Hague.

Ridgway, R.L. & Jones, S.L. (1968) Field-cage releases of *Chrysopa carnea* for suppression of populations of the bollworm and the tobacco budworm on cotton. *Journal of Economic Entomology* 61, 892–898.

Ridgway, R.L. & Jones, S.L. (1969) Inundative releases of *Chrysopa carnea* for control of *Heliothis* on cotton. *Journal of Economic Entomology* 62, 177–180.

Ridgway, R.L. & Murphy, W.L. (1984) Biological control in the field. In *Biology of Chrysopidae*, ed. Canard, M., Séméria, Y. & New, T.R., pp. 220–228. Dr W. Junk, The Hague.

Scopes, N.E.A. (1969) The potential of *Chrysopa carnea* as a biological control agent of *Myzus persicae* on glasshouse chrysanthemums. *Annals of Applied Biology* 64, 433–439.

Senft, D. (1997) Mass-reared insects get fast-food. *Agricultural Research* 45, 4–7.

Senior, L.J. & McEwen, P.K. (1998) Laboratory study of *Chrysoperla carnea* (Stephens) (Neuropt., Chrysopidae) predation on *Trialeurodes vaporariorum* (Westwood) (Hom., Aleyrodidae). *Journal of Applied Entomology* 122, 99–101.

Shands, W.A., Simpson, G.W. & Brunson, M.H. (1972) Insect predators for controlling aphids on potatoes. 1. In small plots. *Journal of Economic Entomology* 65, 511–514.

Shuvakhina, E.Y. (1977) Criteria of biological effectiveness of *Chrysopa carnea* in the control of Colorado potato beetle *Leptinotarsa decemlineata* on potato crops. *Byull. vses. nauchno-issled. Inst. Zashch. Rast. Vredit.* 41, 3–6. (in Russian)

Shuvakhina, E.Y. (1978) Experience of using *Chrysopa carnea* for control of Colorado potato beetle *Leptinotarsa decemlineata* in the Voronezh region Byull. vses. nauchno-issled. Inst. *Zashch. Rast. Vredit.* 42, 3–9. (in Russian)

Stiling, P.D. (1985) *An Introduction to Insect Pests and their Control.* Macmillan, London.

Sundby, R.A. (1966) A comparative study of the efficiency of three predatory insects *Coccinella septempunctata* L. (Coleoptera, Coccinellidae), *Chrysopa carnea* Steph. (Neuroptera, Chrysopidae) and *Syrphus ribesii* L. (Diptera, Syrphidae) at two temperatures. *Entomophaga* 11, 395–404.

Thierry, D., Cloupeau, R. & Jarry, M. (1994) Variation in the overwintering ecophysiological traits in the common green lacewing West-Palearctic complex (Neuroptera, Chrysopidae). *Acta Oecologica* 15, 593–606.

Treacy, M.F., Benedict, J.H., Lopez, J.D. & Morrison, R.K. (1987) Functional-response of a predator (Neuroptera,

Chrysopidae) to bollworm (Lepidoptera, Noctuidae) eggs on smoothleaf, hirsute, and pilose cottons. *Journal of Economic Entomology* 80 , 376–379.

Trouvé, C., Lédée, S., Brun, J. & Ferran, A. (1996) Biological control of the hop aphid. A review of three years of tests in northern France. *Phytōma* 486, 41–44.

Tulisalo, U. & Tuovinen, T. (1975) The green lacewing *Chrysopa carnea* Steph. (Neuroptera, Chrysopidae) used to control the green peach aphid *Myzus persicae* Sulz., and the potato aphid *Macrosiphum euphorbiae* Thomas (Homoptera, Aphididae), on greenhouse green pepper. *Annales Entomologici Fennici* 41, 94–102.

Tulisalo, U., Tuovinen, T. & Kurppa, S. (1977) Biological

control of aphids with *Chrysopa carnea* on parsley and green pepper in the greenhouse. *Annales Entomologici Fennici* 43, 97–100.

Vanderzant, E.S. (1969) An artificial diet for larvae and adults of *Chrysopa carnea*, an insect predator of crop pests. *Journal of Economic Entomology* 62, 256–257.

Wang, R. & Nordlund, D.A. (1994) Use of *Chrysoperla* spp. (Neuroptera: Chrysopidae) in augmentative release programmes for control of arthropod pests. *Biocontrol News and Information* 15), 51N–57N.

Zelený, J. (1984) Chrysopid occurrence in west palearctic temperate forests and derived biotopes. In *Biology of Chrysopidae*, ed. Canard, M., Séméria, Y. & New, T.R., pp. 151–160. Dr W. Junk, The Hague.

CHAPTER 12

Mass-rearing, release techniques, and augmentation

D.A. Nordlund, A.C. Cohen, and R.A. Smith

12.1 INTRODUCTION

The family Chrysopidae contains some 90 genera or subgenera and almost 2000 species (New, 1984) of predaceous insects, including some commercially available species that are widely used in augmentative biological control. These predators feed on a variety of slow-moving, soft-bodied arthropods (Hydorn, 1971; Hydorn & Whitcomb, 1979). Hunter (1997) lists 65, 20, and 54 North American suppliers of *Chrysoperla carnea* (Stephens), *C. comanche* (Banks), and *C. rufilabris* (Burmeister), respectively. Also, *C. carnea* is used in Europe and *C. sinica* (Tjeder) is used in China (Wang & Nordlund, 1994). Large numbers of chrysopids are required for augmentative biological control programmes. Thus cost-effective mass-rearing systems for lacewings are necessary for their successful use. An effective artificial diet and/or innovations that will facilitate rearing process automation are the first steps to development of an effective mass-rearing system. Current methods for commercial rearing of lacewings use, primarily, *Sitotroga* or *Ephestia* eggs for larval diet. An artificial diet (Cohen & Smith, 1998) for larvae, which we expect will replace the expensive insect eggs, has been developed. The rearing process, though, still consists of labour-intensive operations. Automation of one or more of these processes could considerably decrease the cost of rearing large numbers (i.e. millions to trillions) of lacewings. Effective release systems must also be available to fulfil the objective of augmentative biological control programs. Success in periodic release programs requires that agents of sufficient quality be available in sufficient numbers at a reasonable cost.

12.2 MASS-REARING

Mass-rearing techniques for *Chrysoperla* spp. were reviewed recently by Nordlund & Morrison (1992). Several needs for improving existing techniques were identified. More recently, the status of chrysopid rearing in India was reviewed by Gautam (1994). The processes needing improvements include larval food presentation, adult feeding and oviposition units, mechanised egg collection and de-stalking, a mechanised system for larval rearing unit preparation, and field application techniques.

12.3 LARVAL REARING

Over the years, several methods have been used for rearing lacewings with a cannibalistic tendency. Finney (1948, 1950) reared *Chrysoperla* larvae by using paper sheets covered with *Phthorimaea operculella* (Zeller) eggs as prey and *Chrysoperla* eggs sprinkled on top in a wood tray covered with muslin. Ridgway *et al.* (1970) introduced Hexcel® (Hexcel Products Inc., Dublin CA), which has a honeycomb shape, to separate the larvae into individual cells. *Sitotroga cerealella* (Olivier) eggs were provided as food for the larvae. Morrison *et al.* (1975) began using ornamental Masonite® separated by cotton organdie for larval rearing. When production of Masonite was discontinued, Morrison (1977*a*, *b*) proposed using Verticel® (Hexacomb, University Park IL), which consists of triangular cells *c.* $1 \times 0.5 \times 0.5$ cm (Fig. 12.1). The larvae were fed *Sitotroga* eggs. Several other materials, including plastic light diffusing grids and shredded paper, have been used to reduce larval cannibalism. With Verticel, the process of larval rearing unit preparation involves several steps. The Verticel is first cut to size, which varies from insectary to insectary. A polyvinyl acetate glue and water mixture is applied to one side (usually with a paint roller or brush), a piece of organdie is pressed into place, and the assembly is set aside to dry (Morrison, 1977*a*). Lacewing eggs, generally mixed with *Sitotroga* or *Ephestia* eggs, are then sprinkled into the cells and the process of applying glue and organdie is repeated for the other side of the unit. After the eggs

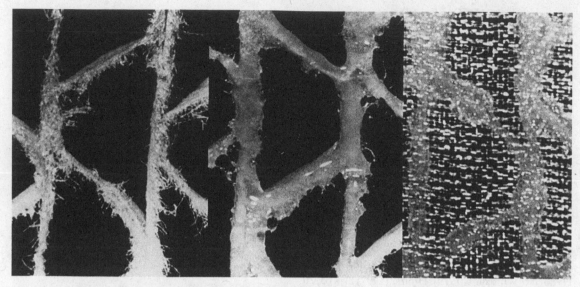

Fig. 12.1. Verticel (left), Verticel coated with Quick Pack hot melt glue using a Hysol Hot Wheelcoater (centre), and coated Verticel to which organdie has been attached (right). (After Nordlund, 1993.)

hatch, food is supplied by sprinkling prey eggs on top of the unit, with the larvae feeding through the organdie.

Verticel is still used widely in Western countries for larval rearing units, and it has a variety of attributes (lightweight, inexpensive, easy disposal, and space efficient) that make it useful for this purpose. Nordlund (1993) described an improved technique for attaching organdie to Verticel larval rearing units. The technique involved application of a hot melt glue (Quik Pak) to the Verticel, using a Hysol Hot Wheelcoater 600 WC (both from Dexter Hysol Aerospace & Industrial Products Div., Seabrook NH) (Fig. 12.2). The organdie was attached to the unit with a flat iron. Since that time, R.A. Smith (unpublished data) has developed an improved technique for cutting the organdie using a rotary cutter (Fig. 12.3), and a pneumatic heat press for attaching the organdie to the Verticel. Assuming a supply of glue-coated Verticel, this process could be automated, including the placement of eggs in the individual cells and packages of artificial diet on the unit.

12.4 ADULT HOLDING AND EGG HARVESTING

For the adults, 1-gallon (3.78 l), cylindrical, waxed-cardboard cartons were introduced by Finney (1948)

for use in emergence and oviposition. Since that time, adult feeding and oviposition units (AFOUs) have generally been cylinders of various sizes and materials (Fig.12.4). Paper liners were commonly used to facilitate egg harvesting (Fig. 12.5). Finney (1948) anaesthetised adults with carbon dioxide to change the AFOUs. Morrison & Ridgway (1976) used a screen bottom on their units and a vacuum device to facilitate change of units and egg collection. A vacuum device is used to pull adults to the bottom of the unit, to permit collection of the liners or transfer of adults from one AFOU to another.

Lacewings deposit their stalked eggs on the organdie top and paper liners, or sides, of the AFOUs, making removal, collection, and handling of the eggs difficult. Finney (1950) used a sodium hypochlorite solution to separate eggs from the substrate and remove stalks. Extreme care must be taken when using this method because underexposure to the solution results in too few eggs being removed, and overexposure damages the eggs (Morrison, 1977*b*). Ridgway *et al.* (1970) developed a method for removing eggs using nylon netting loosely rolled into a ball. The netting was rubbed across the substrate on which the eggs were attached. This method did not completely remove the stalks and often damaged the eggs. Also, to reduce damage to the eggs, the liners and tops of the AFOUs

Fig. 12.2. Hysol Hot Wheelcoater 600 WC from Dester Hysol Aerospace & Industrial Products Division, which can be used to apply hot melt glue to Verticel larval rearing units.

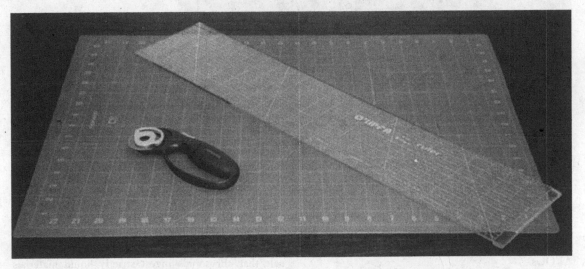

Fig. 12.3. A 45-mm rotary cutter from Fiskars Inc., a 18×24 in (45.72×60.96 cm) gridded cutting mat from Fiskars Inc, and a 24-in (60.96 cm) lip-edge ruler from O'LIPFA used for more efficient cutting of organdie for larval rearing units.

must be held for *c.* 24 hours to allow the eggs to harden before harvest. Morrison (1977*b*) developed a new method which utilised a hot-wire-vacuum removal device. This device was successful, in terms of egg removal; however, it limited the efficiency of the adult cages and was easily misaligned.

Nordlund & Correa (1995*a*) described a hot-wire egg-harvesting system and an AFOU, which did result in improved harvesting efficiency. However, it was a complicated system and the AFOU did not improve the efficiency of adult holding. Nordlund & Correa (1995*b*) described a cubic AFOU with 9.3 times the internal surface area, allowing the unit to hold 9 times the number of adults, of the traditional 3-gallon (11.36 l)

Fig. 12.4. A traditional 3-gallon (11.361 l) cardboard adult feeding and oviposition unit.

cardboard AFOU used in many rearing facilities in the USA (Figs. 12.6, 12.7). They also described a sodium–hypochlorite–based-egg harvesting system (Figs. 12.8, 12.9), which could easily be automated using a robotic arm and commercially available conveyor/accumulator systems. This system (without the automation) provided approximately 89% savings in labour for handling the AFOU (adult transfer, feeding and watering, harvest of eggs, and cleaning) and an 82% reduction in shelf-space requirements, compared to the cylindrical AFOUs.

12.5 ARTIFICIAL DIET FOR LACEWING LARVAE

Simmonds (1944) argued that development of artificial diets for entomophages would greatly reduce cost and enhance the potential for success of augmentative biological control programs. Yet, Grenier *et al.* (1994) pointed out that there was still no artificial-diet-based commercial rearing of entomophages, including chrysopids. The paucity of artificial-diet-based rearing,

after over 50 years of effort, must be construed as a failure.

The pioneering work on artificial diets for *Chrysoperla* spp. was conducted by K.S. Hagen and E.S. Vanderzant. Remarkably, Hagen's (1950) work on the requirement for protein in adult chrysopid diets is still the basis for adult diets. All of the larval artificial diets described by Hagen & Tassan (1966, 1970), Vanderzant (1969, 1973), Ridgway *et al.* (1970), and Yazlovetsky (1992) were liquid diets based on the rationale that chrysopids ingest strictly liquid materials (Vanderzant, 1969, 1973; Yazlovetsky, 1992). There has been considerable effort to develop chemically defined diets, though often the most expensive and least nutritionally useful ingredients are chemically pure.

Cohen (1995, 1998*a*, *b*) and Cohen & Smith (1998) discussed the need for recognising the importance of macromolecular components in diets for entomophagous insects. The fact that most entomophagous arthropods, including chrysopid larvae, use extra-oral digestion means that they are geared to processing complex foods whose major nutrients occur in suitably

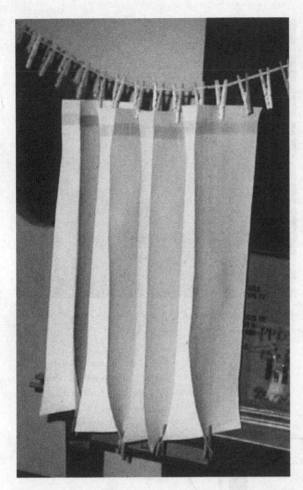

Fig. 12.5. Liners and tops from a traditional cardboard adult feeding and oviposition unit being held for eggs to mature before harvest.

high concentrations. Attempts to present nutrients such as defined free amino acids or defined sterols forces the nutritionist to put these ingredients into forms and concentrations that are alien to entomophages.

For example, it is impossible to dissolve or emulsify cholesterol in sufficient amounts to rival those found in prey (host) eggs or larval tissues. The solubility of cholesterol is 0.2 μg/g of water, and with an emulsifier such as Tween®, about 50 μg could be suspended in 1 g of water. However, lepidopteran eggs and larvae commonly carry 1,000 μg of cholesterol per g of biomass (Cohen, 1998a). Similarly, the protein

content of prey eggs or bodies is commonly 5%–15% of the biomass. The low solubility of many amino acids prevents anywhere near this amount from being dissolved. The essential amino acids leucine and isoleucine are practically insoluble in water. Furthermore, large amounts of free amino acids that are soluble and simple sugars, along with salts, create a condition of excess osmolality. Finally, the palatability of the diet composed of defined ingredients is a great departure from that of whole foods.

Cohen & Smith (1998) reasoned that a solid diet for larvae would be more suitable than liquids, which are inherently dilute, and would improve performance. The presentation of solids also allows the predators to recapture their digestive enzymes for further use in their gut, a most important aspect of their normal feeding process (Cohen, 1998a). The ingredients for the Cohen & Smith (1998) diet are listed in Table 12.1. The rationale for each of these ingredients was discussed by Cohen & Smith (1998). At 400× magnifica'" tion, it was evident that the diet was a mixture of stringy solids in a slurry of 5–40-mm particles (Fig. 12.10), giving the diet a consistency similar to the insides of insect larvae or partially developed insect eggs.

The diet resembles the insides of insect prey both in texture and in composition. As a semi-solid, the diet presented fewer leakage problems compared to liquid diets. This diet was packaged in stretched Parafilm® (American National Can, Chicago IL) (Cohen & Smith, 1998) (Fig. 12.11), and the predators fed readily and apparently normally. An improved packaging system is being developed.

The protein, lipid, carbohydrate, and water content of the diet were about 17%, 15%, 5%, and 62% respectively. The cholesterol content was about 2600 mg/kg of diet. These values are similar to those for the nutrient profile of lepidopteran eggs (Fig. 12.12) (Cohen, 1992) and gross composition of lepidopteran larvae (Cohen, 1998a) (Table 12.2). 'Winnowing' effects often observed when insects are first subjected to feeding on artificial diet (Bartlett, 1984; Cohen, 1992) were not seen in the insects reared on this artificial diet. Hens' egg provides the desirable nutrients and puts them in a concentrated form. Raw hens' eggs have been used in diets for predatory insects for three decades (e.g. Atallah & Newsom, 1966; Hassan & Hagen, 1978), but raw eggs dilute nutrients and have the disadvantages of liquid diets. Cooking the eggs reduces the

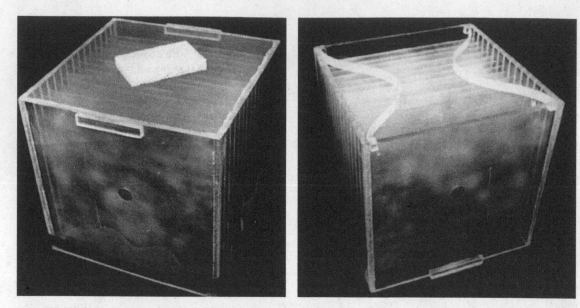

Fig. 12.6. Cubic adult feeding and ovipostion unit. Normal view (left) shows the placement of the watering sponge on top. Upside down view (right), with bottom screen removed, shows placement of feeding strips. (From Nordlund & Correa, 1995*b*.)

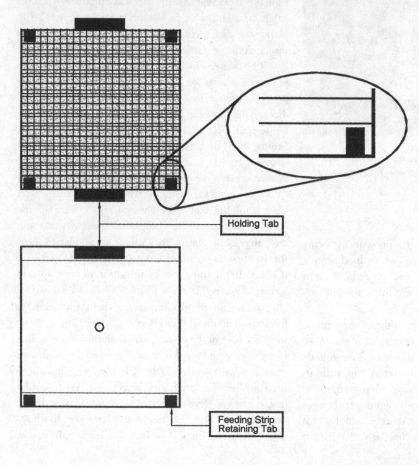

Fig. 12.7. Schematic diagram showing the top and side views of the cubic adult feeding and ovipositon unit. The fibreglass insect screen bottom and elastic retainer are not shown. (From Nordlund & Correa, 1995*b*.)

Holding Tab

Feeding Strip Retaining Tab

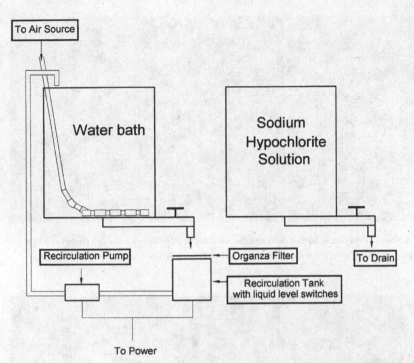

Fig. 12.8. Schematic diagram of the equipment used in the sodium–hypochlorite-based green lacewing egg-harvesting system. (From Nordlund & Correa, 1995*b*.)

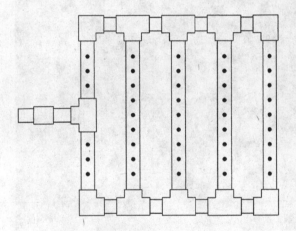

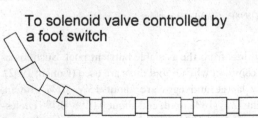

Fig. 12.9. Schematic diagram of the pneumatic agitation device for the sodium-hypochlorite-based green lacewing egg harvesting system. (From Nordlund & Correa, 1995*b*.)

Table 12.1. *Diet ingredients (in grams) in original formulation of the Cohen & Smith (1998) artificial diet and a scaled-up formulation*

Ingredient	0.5 kg batch formulation	30 kg batch formulation
Ground beef	100	6000
Beef liver	100	6000
Sucrose	15	900
Brewer's yeast	10	600
Honey solution (20%)	25	1500
Hens' eggs	100	6000
Sorbic acid	0.6	36
Propionic acid	0.6	36
Streptomycin[a]	0.1	6
Chlortetracycline[a]	0.1	6
Water	65	3900
Acetic acid (10% in water)	5	300

Note:
[a] Antibiotics are not currently used in the *Chrysoperla* colony at the BCMRRU. Research efforts are currently being undertaken to reduce microbial load by non-chemical means.

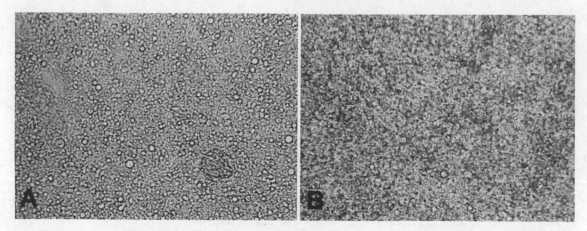

Fig. 12.10. Micrograph of the Cohen & Smith (1998) artificial diet (left) and the contents of a *Helicoverpa zea* egg (right) at 400 ×, showing the similarity of appearance and texture. The artificial diet does contain some larger particles, which are not visible in this micrograph.

Fig. 12.11. A *Chrysoperla rufilabris* larva feeding on the Cohen & Smith (1998) artificial diet packaged in Parafilm®.

microbial load and simulates the natural concentrations and chemical form of nutrients (i.e. lipoproteins and glycolipoproteins).

The use of solid or semi-solid diet permits presentation of a very nutrient-rich mixture of dietary components, especially lipids (which include essential nutrients such as fatty acids, sterols, and lipid-soluble vitamins), without separation of these components and

their loss from the available nutrient pool. Such losses are common when liquid diets are used (Cohen, 1992). Prey larvae and eggs are about 15%–20% protein, about 12%–18% lipid, and about 0.01%–0.1% cholesterol (Cohen, 1992, 1998*a*). By contrast, prey haemolymph is about 4%–5% protein and about 2%–3% lipid (Florkin & Jeuniaux, 1974). Earlier diets for *Chrysoperla* spp. were about 3%–5% protein hydroly-

Table 12.2. *Composition of two components of insect haemolymph, total body composition, and artificial medium*

Component	Total larva	Larval haemolymph[a]	Artificial medium
Protein	16%–20%	2%–5%	19%
Fat	12%–16%	1%–5%	17%
Cholesterol	~1000 µg/g	~50 µg/g	~1260 µg/g

Note:

[a] Amounts determined for *Trichoplusia ni* larvae (Cohen, 1998).

sates and less than 2% lipid (Vanderzant, 1973; Hassan & Hagen, 1978; Nepomnyashchaya *et al.*, 1979). These diets were developed under the assumption that so-called 'liquid-feeding' predators were obligated to feed on body fluids (i.e. haemolymph) that were already liquids before feeding had begun.

During the development of the diet, it was prepared in 0.5 kg batches using a small food processor, which is inadequate for a mass-rearing system. Recently, the batch size was increased to 30 kg, using a Hobart Cutter Mixer HCM (Troy OH) (Fig. 12.13), with the result that cost of labour for diet preparation has decreased from about $32.00/kg to about $0.50/kg, from 84% to about 8 % of total diet production costs. The larger scale of production is adequate to produce 30 000 adult *C. rufilabris* per batch. We expect future cost reductions as the preparation scale increases, permitting wholesale/bulk purchases of the diet components.

12.6 ADULT DIET

Finney (1948) fed the adults a combination of honey and honeydew from the common mealybug, *Planococcus citri* (Risso). The honeydew was found to enhance oviposition. However, since production of the honeydew required rearing *P. citri*, using a synthetic substitute would be more economical. Hagen (1950) used a protein hydrolysate of brewer's yeast to obtain high fecundity. Later Hagen & Tassan (1966, 1970) used brewer's yeast or Wheast™ and sucrose (1:1 by weight) and water for feeding adult *Chrysoperla*.

Far less research has been undertaken on development and improvement of diets for *Chrysoperla* adults

than on larval diets. We assume that it was the early and impressive success with adult diets such as those reported by Hagen & Tassan (1965, 1970) that seemed to obviate the need for extensive further research. A combination of yeast products, sometimes mixed with the dairy product whey, sugars, and water have served as the basic adult diet both in the laboratory and for field-based enhancement of chrysopid populations. Various substitutions have been made, for example, using yeast hydrolysates or autolysates, yeast extracts, or whole yeast products such as brewer's yeast or torula yeast. However, McEwen & Kidd (1995) showed that the different components of artificial diets play an enormous role in egg productivity by adult green lacewings. Their point that far more should be learned about the role of these materials for adult chrysopids is well taken.

This point is reinforced by the studies of Chang *et al.* (1995) where the goal of storage of different life stages of chrysopids overlaps with the nutritional preparation for insects to be stored. These authors showed that post-diapause adult *C. carnea* produced far more eggs if they were fed sugars and no proteins before being put into diapause. This finding was unexpected in light of Hagen's (1950) and Hagen & Tassan's (1966, 1970) findings of the need for protein in the adult diet to enhance egg production. The work of Chang *et al.* (1995) showed that although protein-containing diets were still necessary before high levels of egg production could be achieved, there are subtleties of timing of protein intake that have strong bearing on egg production. Their work suggested that there are probably other subtle interactions of components in adult nutrition that could have a great bearing on adult fecundity. For example, in nature, adult chrysopids consume pollen (rather than yeast). Pollens contain (along with proteins) large amounts of lipids, including polyunsaturated fatty acids and sterols (McCaughey *et al.*, 1980; Standifer, 1966*a*, *b*; Standifer *et al.*, 1968).

Although the nutritional quality of pollen has been well studied in the domain of honey-bee nutrition, almost no attention has been paid to the nutritional value of different kinds of pollens for chrysopids. This is especially unfortunate in light of the dramatic differences found in the qualities and quantities of pollens from different plants (McCaughey *et al.*, 1980; Standifer 1966*a*, *b*; Standifer *et al.*, 1968). This point is illustrated by a consideration of the sterol composition of pollens from 15 different plant species (Standifer *et*

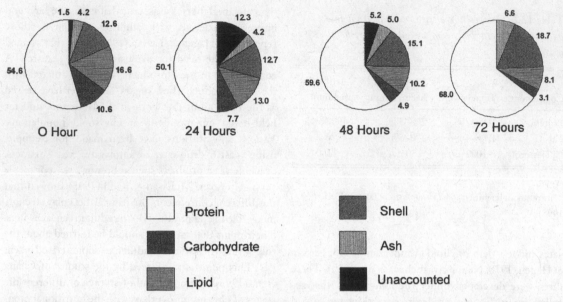

Fig. 12.12. Chemical composition of dry material in *Heliothis virescens* (F.) eggs during four periods of development. (After Cohen & Patana, 1985.)

Fig. 12.13. A Hobart Cutter Mixer (HCM) 450, which can be used for mixing 30 kg batches of the Cohen & Smith (1998) artificial diet.

al., 1968) where one of four sterols (cholesterol, 24–methylene-cholesterol, β–sitosterol, or stigmasterol) predominated; and the sterols represented three distinct classes as far as insect metabolic pathways are concerned: C-27, C-28, or a C-29. Standifer (1967) found that protein qualities from different pollens also had a major impact on honey-bee nutrition. Pollen from dandelion was about six times as effective at promoting hypopharyngeal gland development as sorghum pollen on a percentage protein basis. Such differences in the nutritional quality of pollens and their very powerful effects on honey-bee nutrition should be taken as an impetus for further study of chrysopid nutrition, especially with respect to lipids, which have been sorely neglected in studies of entomophages, including chrysopids.

12.7 RELEASE TECHNIQUES

Use of lacewings in biological control programs requires cost-effective systems for release of large numbers into the target area. Lacewings are typically released as eggs or larvae, but can also be released as adults. Field release of adult lacewings is somewhat problematic, because many species have a migratory preoviposition period (Duelli, 1984). When young adults are released there is a strong likelihood that they will leave the target field before they begin to oviposit. The use of young adults in area-wide programmes, involving early season inoculative releases, may hold some promise. However, predators and parasitoids can destroy a large number of eggs, particularly late in the growing season. We are aware that at least one company markets mature adults, which have been fed in the insectary, based on the theory that the adults are ready to oviposit and will locate and oviposit in sites with potential prey within the target field. This approach, though seemingly expensive, deserves some additional study.

Release of lacewing eggs also has problems. Mortality of the eggs can be affected by high temperatures, placement (on or off the plant), and predators (Daane *et al.*, 1993). Also, after hatching, young lacewing larvae must find their first meal quickly or perish. Generally, we want to release lacewings when the pest populations are low, so that they can keep them low. This leads to suggestions that it would be useful to devise a system for attaching a prey egg to each lacewing egg before it is placed on a plant. At the least, it might be useful to apply prey material to plants at the same time the lacewing eggs are applied. However, either process could add significantly to the cost of using lacewings.

Traditionally, *Chrysoperla* eggs have been released manually by sprinkling eggs or a mixture of eggs and some type of filler (vermiculite, rice hulls, saw dust, etc.) on the plant. However, in recent years, there has been increasing research into development of mechanical methods of dispersal. Daane *et al.* (1993) reported a mechanical method of releasing eggs that were mixed with corn-cob grit. The mixture was placed in a 5-gallon (18.93 l) container with a funnel that had an adjustable opening. The release rate could be controlled by the funnel opening or by the speed of the tractor on which the container was mounted.

Tests have been conducted to determine the viability of eggs that have been released using mechanical methods. Gardner & Giles (1996) tested how *Chrysoperla* eggs were affected by extraction from vermiculite, tractor vibration, discharge through a field applicator, and environmental conditions. There was no significant difference in egg hatch from these procedures or a control. There has also been interest in a release technique for spraying eggs that are in a liquid

solution. Tests on the viability of eggs in certain solutions have been conducted. Jones & Ridgway (1976) found that eggs could be suspended in a 0.125% agar solution for more than 60 minutes with no reduction in hatch rate. McEwen (1996) wished to determine if longer storage times in spray media could be accomplished. It was determined that compared to eggs held in air, eggs could be kept in a 0.125% agar solution or water for up to one day (at 4 °C) with no reduction in hatch rate. Sengonca & Lochte (1997) used water to suspend lacewing eggs and found that the eggs are not damaged when deflector jet nozzles are used for spraying applications. They also tested several substances to increase adhesion of the eggs to the plants.

The 'Bio-Sprayer' (Beneficial Insectary, Oak Run CA and Smucker Mfg Inc., Harrisburg OR), developed by W. L. Tedders Jr of the US Department of Agriculture, Agricultural Research Service, in conjunction with Beneficial Insectary and Smucker Mfg Inc., is a mechanised device used for release of lacewing eggs (Fig. 12.14). It utilises low pressure air and BioCarrier™ (Smucker Mfg Inc.), which is a solution that is used to evenly distribute the eggs and help adhere them to the plant. Sengonca & Lochte (1997) reported the development of spray and atomiser techniques for applying *C. carnea* eggs using water as the carrier, and noted that eggs could remain in water for up to 12 hours with no negative impact on hatch rate. They suggest use of deflector jet nozzles with a bore diameter of at least 0.9 mm and spray pressure up to 3 bar.

Aerial application is another method that can be used to release *Chrysoperla*. With this method, fixed-wing aircraft, helicopters, or even radio-controlled model aeroplanes, combined with a release hopper, can be used to release the *Chrysoperla* eggs . One helicopter pilot was able to cover 200 acres with green lacewings in 30 minutes (Grossman, 1990). One radio-controlled model aeroplane (Fig. 12.15), called a Beneficial Insect Release Device or B.I.R.D. (Arizona Biological Control Inc. (ARBICO), Tucson AZ) can cover 50 acres in 10 minutes (McClintic, 1992).

Daane *et al.* (1993) reported that the release of larvae instead of eggs allows for better control of the numbers released. Also, the larvae are not as susceptible to predation as the eggs are. Currently, most releases of larvae are performed manually with a camel-hair paintbrush or by placing rearing units in the plant. Both

Fig. 12.14. The Bio-Sprayer from Smucker Mfg Inc. can be used to release *Chrysoperla* eggs that are suspended in a solution called BioCarrier™ .

methods are extremely time-consuming and labour-intensive. Some commercial producers provide larvae mixed with rice hulls in a bottle with an applicator tip. The bottle is squeezed to release the larvae and rice hulls (Planet Natural, Bozeman MT). Automation of larvae release is also being studied. Research has been conducted (Morisawa & Giles, 1995) to determine the effect of rotation and vibration on *Chrysoperla* larvae when mixed with vermiculite or rice hulls. It was concluded that larvae can sustain mechanical agitation with no significant loss of viable larvae. One problem with dispensing larvae is their tendency to migrate to the sides of the container.

12.8 AUGMENTATION

Classical biological control and augmentative biological control (via inoculative or inundative releases) have a critical theoretical difference (Nordlund, 1996). In classical biological control the goal is the establishment of a regulatory mechanism to maintain the pest population below economic levels for a long period. Regulation requires, by definition, a density-dependent biological

control agent, meaning one that is essentially host/prey specific. With many inoculative and particularly inundative releases, regulation is not the goal – therapeutic control is. Therapeutic control whether one uses a conventional pesticide or a biological control agent, does not require regulation or a host/prey-specific agent (Nordlund & Legaspi, 1996).

The lacewings that are available from commercial sources are generalist predators. There are several characteristics of *Chrysoperla* spp. that make them attractive as agents for inundative releases. Larvae of *Chrysoperla* spp. feed on a wide range of insect and mite pests in a variety of agroecosystems (Bickley & MacLeod, 1956; Hydorn, 1971; Agnew *et al.*, 1981). This broadens the potential market for these insects and permits increased investment in rearing capacity. Adults are not predatory and can be easily cultured on relatively simple diets (Nordlund & Morrison, 1992). They have a high reproductive potential (Elkarmi *et al.*, 1987) and high searching rates (Fleschner, 1950). When released as eggs or larvae, *Chrysoperla* spp. are unable to leave the target area and will feed on any prey that they can capture. *Chrysoperla* larvae exhibit tolerance to some insecti-

Fig. 12.15. The B.I.R.D. – Beneficial Insect Release Device from Arizona Biological Control Inc. (ARBICO) is a radio-controlled model aeroplane that can be used to release chrysopid eggs.

cides (Plapp & Bull, 1978; Shour & Crowder, 1980; Hassan *et al.*, 1985; Pree *et al.*, 1989).

Chrysoperla spp. have been used against a variety of aphid species in cabbage, pepper, tomato, eggplant, peas, potato, and cotton (Adashkevich & Kuzina, 1974; Beglyarov & Smetnik, 1977; Radzivilovskaya & Daminova, 1980; Shands *et al.*, 1972; Anon., 1982). They have been used against Colorado potato beetle [*Leptinotarsa decemlineata* (Say)] in eggplant and potato (Adashkevich & Kuzina, 1971; Shuvakhina, 1974; Nordlund *et al.*, 1991); European red mite [*Panonychus ulmi* (Koch)] in apple (Miszczak & Niemczyk, 1978); the grape mealybug [*Pseudococcus maritimus* (Ehrhorn)] in pear (Doutt & Hagen, 1949, 1950); the Comstock mealybug [*Pseudococcus comstocki* (Kuwana)] in mulberry and catalpa (Beglyarov & Smetnik, 1977); and the Western grape leafhopper (*Erythroneura elegantula* Osborne) and variegated leafhopper (*E. variabilis* Beamer) on grapes (Daane *et al.*, 1993; Daane & Yokota,

1997). Daane *et al.* (1993) suggested releases in corn-cob grit mixtures, when 25%–50% of the eggs have hatched and demonstrated that, for variegated grape leafhopper control, releases should be synchronised with leafhopper egg hatch. Daane & Yokota (1997) reported that distribution of eggs in a relatively even pattern, as opposed to clumps (in cups) reduces cannibalism; however, scratching of egg surfaces during the mixing process could result in increased mortality due to desiccation. The chrysopid *Mallada basalis* (Walker) is used in Taiwan to control several tetranychid mites on strawberry and screen-house grown papaya (Chang, 1994; Chang & Huang, 1995). *Chrysoperla* spp. have also been used against bollworm [*Helicoverpa zea* (Boddie)], tobacco budworm [*Heliothis virescens* (Fabricius)], and *Helicoverpa armigera* (Hübner) in cotton (Lingren *et al.*, 1968; Ridgway & Jones, 1968, 1969; Kinzer, 1977; Ridgway *et al.*, 1977; Beijing Plant Protection Institute, 1979; Anon., 1982).

A variety of chrysopids can be used in augmentative biological control programs, in many different cropping systems against many different hosts. We do not believe that we have really begun to use these valuable predaceous insects to their full potential.

12.9 ENVIRONMENTAL MANIPULATION TO INCREASE THE EFFECTIVENESS OR NUMBER OF LACEWINGS

Several efforts have been made to enhance or manipulate the numbers of *Chrysoperla* spp. in the field. Pioneering work in this subject came out of Hagen's laboratory where the use of food sprays (such as whey and yeast combinations) was demonstrated as useful in building up populations of chrysopids (Hagen & Tassan, 1970). The rationale behind this work is that nutrients and chemicals that attract or arrest chrysopids could either contribute to population build-up or redistribution of predators in sites where they are needed. In early studies, Hagen *et al.* (1976) and van Emden & Hagen (1976) showed that products of the breakdown of the amino acid tryptophan were highly attractive to *C. carnea*. More recently, McEwen *et al.* (1994) showed the efficacy of tryptophan solutions to concentrate numbers of green lacewings in olive groves. These studies incorporated acid hydrolysed L-tryptophan and omitted sugar and yeast products that were included in earlier studies (reviewed by McEwen *et al.*, 1994). These studies raise interesting questions about promising strategies for concentrating lacewings in strategic places and inducing them to lay eggs in microhabitats where they are needed to control pests.

12.10 CONCLUSION

Chrysopidae include several species of predators that are used for augmentative biological control. However, the use of these beneficial insects is far from reaching its full potential. Some factors contributing to this limited use are (real or perceived) high production costs, limited production capacity, inefficient release technologies/strategies, an institutionalised bias against generalist predators, and inadequate data from large-scale studies on the most effective approaches for their use. We believe that chrysopids have an important place in our arsenal of pest-management tools, but admit that the above issues must be addressed before they will take their rightful place in that arsenal. We have discussed a number of advances in rearing technologies that can contribute to reduced costs and increased capacity. Increased availability of inexpensive chrysopids will result in increased practical and experimental use and further development of improved release technology and an increased understanding of how, when, and where to use augmentative releases of these potentially very important biological control agents.

ACKNOWLEDGEMENTS

The text of this chapter has been approved for publication as a Journal Article No. J-9422 of the Mississippi Agricultural and Forestry Experiment Station, Mississippi State University. Mention of a commercial or proprietary product does not constitute endorsement by the United States Department of Agriculture.

REFERENCES

Adashkevich, B.P. & Kuzina, N.P. (1971) *Chrysopa* against the Colorado potato beetle. *Zashchita Rastenii* 12, 23. (in Russian)

Adashkevich, B.P. & Kuzina, N. P.(1974) Chrysopids on vegetable crops. *Zashchita Rastenii* 9, 28–29.

Agnew, C.W., Sterling, W.L. & Dean, D.A. (1981) Notes on the Chrysopidae and Hemerobiidae of eastern Texas with keys for their identification. *Supplement to the Southwestern Entomologist* No. 4.

Anon. (1982) *Biological Control of Pests in China*. China Program, Science and Technology Exchange Division, Office of International Cooperation and Development, United States Department of Agriculture, Washington DC.

Atallah, Y.H. & Newsom, L.D. (1966) Ecological and nutritional studies on *Coleomegilla maculata* DeGeer (Coleoptera: Coccinellidae). I. The development of an artificial diet and a laboratory rearing technique. *Journal of Economic Entomology* 57, 1173–1179.

Bartlett, A.C. (1984) Genetic changes during insect domestication. In *Advances and Challenges in Insect Rearing*, ed. King, E.G. & Leppla, N.C., pp. 2–8. United States Department of Agriculture, New Orleans, LA.

Beglyarov, G.A. & Smetnik, A.I. (1977) Seasonal colonization of entomophages in the USSR. In *Biological Control by Augmentation of Natural Enemies*, ed. Ridgway, R.L. & Vinson, S.B., pp. 283–328. Plenum Press, New York.

Beijing Plant Protection Institute (1979) A preliminary report on the use of *Chrysoperla sinica* Tjeder to control agricultural and forest pests. *Natural Enemies of Insects* 1, 11–18. (in Chinese)

Bickley, W.E. & MacLeod, E.G. (1956) A synopsis of the nearctic Chrysopidae with a key to the genera. *Proceedings of the Entomological Society of Washington* 58, 177–202.

Chang, C.P. (1994) *Biological Control in Screen-house Papaya – Control of Spider Mites with Green Lacewing*. Taiwan Apicultural and Sericultural Experimental Station. (in Chinese).

Chang, C.P. & Huang, S.C. (1995) Evaluation of the effectiveness of releasing green lacewings, *Mallada basalis* (Walker) for the control of tetranychid mites on strawberry. *Plant Protection Bulletin* 37, 41–58. (In Chinese with English summary)

Chang, Y.F., Tauber, M.J. & Tauber, C.A. (1995) Storage of the mass-produced predator *Chrysoperla carnea* (Neuroptera: Chrysopidae): influence of photoperiod, temperature, and diet. *Environmental Entomology* 24, 1365–1374.

Cohen, A.C. (1992) Using a systematic approach to develop artificial diets for predators. In *Advances in Insect Rearing for Research and Pest Management*, ed. Anderson, T. E. & Leppla, N.C., pp. 77–92. Westview Press, Oxford.

Cohen, A.C. (1995) Extra-oral digestion in predaceous terrestrial arthropoda. *Annual Review of Entomology* 40, 85–103.

Cohen, A.C. (1998*a*) Solid-to-liquid feeding: the inside(s) story of extra-oral digestion in predaceous arthropoda. *American Entomologist* 44, 103–117.

Cohen, A.C. (1998*b*) Artificial media for rearing entomophages comprising cooked, whole egg. U. S. Patent No. 5834177. U. S. Patent and Trademark Office, 10 November 1998.

Cohen, A.C. & Patana, R. (1985) Chemical composition of tobacco budworm eggs during development. *Comparative Biochemistry and Physiology* 18, 165–169.

Cohen, A.C. & Smith, L.K. (1998) A new concept in artificial diets for *Chrysoperla carnea*: the efficacy of solid diets. *Biological Control* 13, 49–54.

Daane, K.M. & Yokota, G.Y. (1997) Release strategies affect survival and distribution of green lacewings (Neuroptera: Chrysopidae) in augmentation programs. *Environmental Entomololgy* 26, 455–464.

Daane, K.M., Yokota, G.Y., Rasmussen, Y.D., Zheng, Y. & Hagen, K.S. (1993) Effectiveness of leafhopper control varies with lacewing release methods. *Californian Agriculture* 47, 19–23.

Doutt, R.L. & Hagen, K.S. (1949) Periodic colonization of *Chrysopa californica* as a possible control of mealybugs. *Journal of Economic Entomology* 42, 560.

Doutt, R.L. & Hagen, K.S. (1950) Biological control measures applied against *Pseudococcus maritimus* on pears. *Journal of Economic Entomology* 43, 94–96.

Duelli, P. (1984) Flight, dispersal, migration. In *Biology of Chrysopidae*, ed. Canard, M., Séméria, Y. & New, T.R., pp. 110–116. Dr W. Junk, The Hague.

Elkarmi, L.A., Harris, M.K. & Morrison, R.K. (1987) Laboratory rearing of *Chrysoperla rufilabris* (Burmeister), a predator of insect pests of pecans. *Southwestern Entomologist* 12, 73–78.

van Emden, H.F. & Hagen, K.S. (1976) Olfactory reactions of the green lacewing *Chrysopa carnea* to tryptophan and certain breakdown products. *Environmental Entomology* 53, 469–473.

Finney, G.L. (1948) Culturing *Chrysopa californica* and obtaining eggs for field distribution. *Journal of Economic Entomology* 41, 719–721.

Finney, G.L. (1950) Mass culturing *Chrysopa californica* to obtain eggs for field distribution. *Journal of Economic Entomology* 43, 97–100.

Fleschner, C.A. (1950) Studies on searching capacity of the larvae of three predators of the citrus red mite. *Hilgardia* 20, 223–265.

Florkin, M. & Jeuniaux, C. (1974) Hemolymph composition. In *The Physiology of Insects*, vol. 5, ed. Rockstein, M., pp. 256–307. Academic Press, New York.

Gardner, J. & Giles, K. (1996) Handling and environmental effects on viability of mechanically dispensed green lacewing eggs. *Biological Control* 7, 245–250.

Gautam, R.D. (1994) Present status of rearing of chrysopids in India. *Bulletin of Entomology, New Delhi* 35, 31–39.

Grenier, S., Greany, P.D. & Cohen, A.C. (1994) Potential for mass release of insect parasitoids and predators through development of artificial culture techniques. In *Pest Management in the Subtropics: Biological Control – A Florida Perspective*, ed. Rosen, D., Bennett, F.D. & Capinera, J.L., pp. 181–206. Intercept, Andover, UK.

Grossman, J. (1990) Biological control takes off. *Agrichemical Age* Oct. 1990, 12–15, 18.

Hagen, K.S. (1950) Fecundity of *Chrysopa californica* as affected by synthetic foods. *Journal of Economic Entomology* 43, 101–104.

Hagen, K.S. & Tassan, R.L. (1965) A method of providing artificial diets to *Chrysopa* larvae. *Journal of Economic Entomology* 58, 999–1000.

Hagen, K.S. & Tassan, R.L. (1966) The influence of protein hydrolysates of yeasts and chemically defined diets on the fecundity of *Chrysopa carnea* Stephens (Neuroptera). *Vestník Ceskoslovenské Spolecnosti Zoologiské* 30, 219–227.

Hagen, K.S. & Tassan, R.L. (1970) The influence of food Wheast and related *Saccharomyces fragilis* yeast products on the fecundity of *Chrysopa carnea* (Neuroptera: Chrysopidae). *Canadian Entomologist* 102, 806–811.

Hagen, K.S., Greany, P., Sawall, E.F. Jr. & Tassan, R.L. (1976) Tryptophan in artificial honeydews as a source of an attractant for adult *Chrysoperla carnea*. *Environmental Entomology* 5, 458–468.

Hassan, S.A. & Hagen, K.S. (1978) A new artificial diet for rearing *Chrysopa carnea* larvae (Neuroptera: Chrysopidae). *Zeitschrift für angewandte Entomologie* 86, 315–320.

Hassan, S.A., Klingauf, F. & Shahin, F. (1985) Role of *Chrysopa carnea* as an aphid predator on sugar beet and the effect of pesticides. *Zeitschrift für angewandte Entomologie* 100, 163–174.

Hunter, C.D. (1997) *Supplies of Beneficial Organisms in North America*. California Environmental Protection Agency, Sacramento.

Hydorn, S.B. (1971) Food preferences of *Chrysopa rufilabris* Burmeister in central Florida. MSc thesis, University of Florida, Gainesville.

Hydorn, S.B. & Whitcomb, W.H. (1979) Effects of larval diet on *Chrysopa rufilabris*. *Florida Entomologist* 62, 293–298.

Jones, S.L. & Ridgway, R.L. (1976) *Development of Methods for Field Distribution of Eggs of the Insect Predator* Chrysopa carnea *(Stephens)*. Agricultural Research Service, United States Department of Agriculture, ARS-S-124.

Kinzer, R.E. (1977) Development and evaluation of techniques for using *Chrysopa carnea* Stephens to control *Heliothis* Species that Attack Cotton. PhD thesis, Texas A&M University College Station.

Lingren, P.D., Ridgway, R.L. & Jones. S.L. (1968) Consumption by several common predators of eggs and larvae of two *Heliothis* species that attack cotton. *Annals of the Entomological Society of America* 61, 613–618.

McCaughey, W.F., Gilliam, M. & Standifer, L. N. (1980) Amino acids and protein adequacy for honey bees of pollens from desert plants and other floral sources. *Apidologie* 11, 75–86.

McClintic, D. (1992) Bug blitzers. *The Furrow* 97, 8.

McEwen, P.K. (1996) Viability of green lacewing *Chrysoperla carnea* Steph. (Neuropt., Chrysopidae) eggs stored in potential spray media, and subsequent effects on survival of first instar larvae. *Journal of Applied Entomology* 120, 171–173.

McEwen, P.K. & Kidd, N.A.C. (1995) The effects of different components of an artificial food on adult green lacewing (*Chrysoperla carnea*) fecundity and longevity. *Entomologia Experimentalis et Applicata* 77, 343–346.

McEwen, P.K., Jervis, M.A. & Kidd, N.A.C. (1994) Use of a sprayed L-tryptophan solution to concentrate numbers of the green lacewing *Chrysoperla carnea* in olive tree canopy. *Entomologia Experimentalis et Applicata* 70, 97–99.

Miszczak, M. & Niemczyk, E. (1978) Green lacewing *Chrysopa carnea* Steph. (Neuroptera, Chrysopidae) as a predator of the European mite *Panonychus ulmi* Koch on apple trees. 2. The effectiveness of *Chrysopa carnea* larvae in biological control of *Panonychus ulmi* Koch. *Fruit Science Reports* 5, 21–31.

Morisawa, T. & Giles, D.K. (1995) Effects of mechanical handling on green lacewing larvae (*Chrysoperla rufilabris*). *Transactions of the ASAE* 11, 605–607.

Morrison, R.K. (1977a) A simplified larval rearing unit for the common green lacewing. *Southwestern Entomologist* 2, 188–190.

Morrison, R.K. (1977b) New developments in mass production of *Trichogramma* and *Chrysopa* spp. In *Proceedings of 1977 Beltwide Cotton Production Research Conference*, pp. 149–151. National Cotton Council, Memphis TN.

Morrison, R.K. & Ridgway, R.L. (1976) *Improvements in Techniques and Equipment for Production of a Common Green Lacewing,* Chrysopa carnea. Agricultural Research Service, United States Department of Agriculture, ARS-S-143.

Morrison, R.K., House, V.S. & Ridgway, R.L. (1975) Improved rearing unit for larvae of a common green lacewing. *Journal of Economic Entomology* 68, 821–822.

Nepomnyashchaya, A.M., Abashkin, A.S. & Yazlovetsky, I.G. (1979) *Methodical indication of rearing lacewing in artificial diet*. All-Union Institute of Biological Control in Plant Protection, Kishinev. (in Russian)

New, T.R. (1984) The need for taxonomic revision in Chrysopidae. In *Biology of Chrysopidae*, ed. Canard, M, Séméria, Y. & New, T.R., pp. 37–42. Dr W. Junk, The Hague.

Nordlund, D.A. (1993) Improvements in the production system for green lacewings: a hot melt glue system for preparation of larval rearing units. *Journal of Entomological Science* 28, 338–342.

Nordlund, D.A. (1996) Biological control, integrated pest management and conceptual models. *Biocontrol News and Information* 17, 35–44.

Nordlund, D.A. & Correa, J.A. (1995a) Improvements in the production system for green lacewings: an adult feeding and oviposition unit and hot wire egg harvesting system. *Biological Control* 5, 179–188.

Nordlund, D.A. & Correa, J.A. (1995*b*) Description of green lacewing adult feeding and oviposition units and a sodium hypochlorite-based egg harvesting system. *Southwestern Entomologist* 20, 293–301.

Nordlund, D.A. & Legaspi, J.C. (1996) Whitefly predators and their potential for use in Biological Control. In *Bemisia: 1995 Taxonomy, Biology, Damage, Control and Management*, ed. Gerling, D. & Mayer, R.T., pp. 499–513. Intercept, Andover, UK.

Nordlund, D.A. & Morrison, R.K. (1992) Mass rearing of *Chrysoperla* species. In *Advances in Insect Rearing for Research and Pest Management*, ed. Anderson, T.E. & Leppla, N.C., pp. 427–439. Westview Press, Boulder CO.

Nordlund, D.A., Vacek, D.C. & Ferro, D.N. (1991) Predation of Colorado potato beetle (Coleoptera: Chrysomelidae) eggs and larvae by *Chrysoperla rufilabris* (Neuroptera: Chrysopidae) larvae in the laboratory and field cages. *Journal of Entomological Science* 26, 443–449.

Plapp, F.W. Jr & Bull, D.L. (1978) Toxicity and selectivity of insecticides to *Chrysopa carnea*, a predator of the tobacco budworm. *Environmental Entomology* 7, 431.

Pree, D.J., Archibald, D.E. & Morrison, R.K. (1989) Resistance to insecticides in the common green lacewing *Chrysoperla carnea* (Neuroptera: Chrysopidae) in southern Ontario. *Journal of Economic Entomology* 82, 29–34.

Radzivilovskaya, M.A. & Daminova, D.B. (1980) Prospects of using the aphis lion (*Chrysopa vulgaris*) against aphids in Dzhizak region, Uzbek SSR. In *Protecting Cotton in the Dzhizakskiy Oblast: Proceedings of a Scientific Research Conference*, ed. Davletskina, A.D., Eremenko, T.S. & Zhurauskaya, S.A, p. 87. Fan, Tashkent. (in Russian)

Ridgway, R.L. & Jones, S.L. (1968) Field-cage releases of *Chrysopa carnea* for suppression of populations of the bollworm and tobacco budworm on cotton. *Journal of Economic Entomology* 61, 895–898.

Ridgway, R.L. & Jones, S.L. (1969) Inundative releases of *Chrysopa carnea* for control of *Heliothis* on cotton. *Journal of Economic Entomology* 62, 177–180.

Ridgway, R.L., Morrison, R.K. & Badgley, M. (1970) Mass rearing green lacewing. *Journal of Economic Entomology* 63, 834–836.

Ridgway, R.L., King, E.G. & Carrillo, J.L. (1977) Augmentation of natural enemies for control of plant pests in the Western Hemisphere. In *Biological Control by Augmentation of Natural Enemies*, ed. Ridgway,

R.L.& Vinson, S.B., pp. 379–416. Plenum Press, New York.

Sengonca, Ç. Lochte, C. (1997) Development of a spray and atomizer technique for applying eggs of *Chrysoperla carnea* (Stephens). *Journal of Plant Diseases and Protection* 104, 214–221.

Shands, W.A., Simpson, G.W. & Brunson, M.H. (1972) Insect predators for controlling aphids on potatoes. 1. In small plots. *Journal of Economic Entomology* 65, 511–514.

Shour, M.H., & Crowder, L.A. (1980) Effects of pyrethroid insecticides on the common green lacewing. *Environmental Entomology* 9, 306–309.

Shuvakhina, E.Y. (1974) Lacewings and their utilization in controlling pests on agricultural crops. In *Biological Agents for Plant Protection*, Kolos, Moscow. pp 185–199. (in Russian)

Simmonds, F.J. (1944) The propagation of insect parasites on unnatural hosts. *Bulletin of Entomological Research* 35, 219–226.

Standifer, L.N. (1966*a*) Some lipid constituents of pollens collected by honeybees. *Journal of Apicultural Research* 5, 93–98.

Standifer, L.N. (1966*b*) Fatty acids in dandelion pollen gathered by honey bees, *Apis mellifera* (Hymenoptera: Apidae). *Annals of the Entomological Socitey of America* 59, 1005–1008.

Standifer, L.N. (1967) A comparison of the protein quality of pollens for growth-stimulation of the hypopharyngeal glands and longevity of honey bees, *Apis mellifera* L. (Hymenoptera: Apidae). *Insectes Sociaux* 14, 415–426.

Standifer, L.N., Deveys, M. & Barbier, M. (1968) Pollen sterols – a mass spectrographic survey. *Phytochemistry* 7, 1361–1365.

Vanderzant, E. (1969) An artificial diet for larvae and adults of *Chrysopa carnea*, an insect predator of crop pests. *Journal of Economic Entomology* 62, 256–257.

Vanderzant, E. (1973) Improvements in the rearing diet for *Chrysopa carnea* and the amino acid requirements for growth. *Journal of Economic Entomology* 66, 336–338.

Wang, R. & Nordlund, D.A. (1994) Use of *Chrysoperla* spp. (Neuroptera: Chrysopidae) in augmentative release programs for control of arthropod pests. *Biocontrol News and Information* 15, 51N–57N.

Yazlovetsky, I.G. (1992) Development of artificial diets for entomophagous insects by understanding their nutrition and digestion. In *Advances in Insect Rearing for Research and Pest Management*, ed. Anderson, T.E. & Leppla, N.C., pp 41–62. Westview Press, Boulder CO.

CHAPTER 13

Features of the nutrition of Chrysopidae[1] larvae and larval artificial diets

I.G. Yazlovetsky

13.1 INTRODUCTION

The potential for the use of predaceous Chrysopidae larvae as biological agents to control the density of noxious arthropods has been noted by many authors (New, 1975, 1988; Principi, 1983; Nordlund & Morrison, 1992). Appropriate technologies and equipment have been developed and continue improving which utilise eggs of the Angoumois grain moth *Sitotroga cerealella* and some other Lepidoptera as food for larvae under mass rearing of the most widely distributed representative of this family, *Chrysoperla carnea*. However, the high cost of *C. carnea* in mass production under these technologies (about $12.50 per 1000 larvae) does not currently allow wide application of this predator in the field (King & Coleman, 1989). The rationale of economic expediency and effectiveness of mass-rearing of Chrysopidae with their subsequent release against noxious insects has for over 50 years been promoting active research for different types of artificial food for these predators, in particular artificial nutritional diets.

For insects, Singh (1977) defined artificial nutritional diets as any food which is prepared by a human being using various methods and on which insects can develop in captivity. Artificial food substrates for predaceous and parasitic insects include artificial nutritional diets (including alternative or substitute preys and hosts), specially treated with the aim of improving conditions of entomophagous nutrition. These include thermal treatment, making coverage, impregnation of inert materials with liquid artificial nutritional diets, their jellying, encapsulation, etc.

Hodek (1978a) introduced the concept of adequate prey (food) for predaceous insects, under which he considered only food that provided normal development and reproduction for the predator. Among the main reasons for inadequacy he mentioned lack in food of necessary ingredients in required quantities. In other works, this author noted that diverse aspects of entomophagous insect nutrition were much more complex and much less studied than that of phytophages which influenced negatively development of techniques for the rearing of insect predators and parasites (Hodek, 1978b, 1993).

Over 20 years ago we came to conclusion that the multiple, contradictory, and often mutually exclusive demands of artificial nutritional diets as the base of artificial food substrates of complete value for mass-rearing of entomophages could not be met by standard methods of empirical selection, which were usually used when compiling artificial nutritional diets of acceptable composition. To solve this problem, it is necessary to use improved methodology, which is based on the knowledge of the trophic biology of insect predators and parasites as well as on maximising their nutritional requirements. The work at this level evidently assumes more complete use of information on prey and host acceptability and utilisation, morphology of the alimentary canal, and the physiology of absorption, digestion, and food assimilation (Nepomnyashaya *et al.*, 1979; Yazlovetsky *et al.*, 1979a; Yazlovetsky, 1992). Similar methodology has been developed by Grenier *et al.* (1986) and Thompson (1986) in their prolonged investigations of artificial nutritional diets for parasitoids. Cohen (1992) formulated a 'systematic approach to develop artificial diets for predators' on similar principles.

High demands for artificial food substrates and

---

[1] The author of this chapter considers it necessary to use predominantly the species names given in the original publications on artificial nutritional diets for Chrysopidae.

Table 13.1. *Classification of artificial nutritional diets for entomophagous insects*

Number	Name of artificial nutritional diet category	Purpose	Technique for development and preparation
1	Subnatural	a) Revealing of principal possibility for *in vitro* growth and development b) Laboratory rearing for scientific purposes[a]	Minimum treatment of natural food substrates for entomophages (quick killing of host insects and prey, homogenisation and lyophilisation of their tissues, haemolymph separation, etc.)
2	Holidic = chemically defined	a) Determination of nutritional requirements, study of metabolism b) Study of nutrition and digestion	Nutritional mixtures are prepared from individual inorganic and organic highly purified ingredients of non-complex and well-studied composition on the basis of the general conception of insect nutritional requirements
3	Meridic	a) Study of nutrition and digestion b) Laboratory rearing for other scientific purposes	They are prepared on the basis of holidic artificial nutritional diet by means (if possible) of equivalent replacement of many purified substances of known structure for considerably fewer number of components of more complex poorly known composition and low degree of purity (e.g. mixture of highly purified amino acids and vitamins is replaced by protein hydrolysates, brewer's yeast and autolysates)
4	Oligidic	a) Rearing for scientific purposes b) Mass-rearing	Mixtures are made mainly from unpurified natural substances of vegetable and animal origin and insufficiently studied composition through empiric selection of the most suitable ratio
5	Suboptimal	a) Rearing for scientific purposes b) Provision of programmes for mass-rearing of entomophagous insects with artificial diet of acceptable quality and cost	The results of study of trophic behaviour, nutritional requirements, features of nutrition and digestion provide optimal ratio of selected components of either known or specially studied composition as well as necessary physical and mechanical properties of artificial food substrates under development

Note:

[a] Scientific purposes mean study of behaviour, physiology, mechanism of resistance, toxicological evaluation of pesticides, isolation and identification of pheromones, evaluation of efficiency of entomophage releases, etc.

artificial nutritional diets, suitable for the majority of insect predators and, in particular, parasites (as compared, for example, with phytophages), complicate their development. It widens the spectrum of the methods used and the set of ingredients applied when compiling artificial nutritional diets, complicating systematisation of the latter. Currently there is no convenient common system of classification of artificial nutritional diets for entomophagous insects. The Dougherty classification (Dougherty, 1959) for arthropods, which splits all artificial nutritional diets into three groups on the basis of compositional data, degree of study, and purity of ingredients applied, is most commonly used. The special classification of artificial nutritional diets for entomophagous insects (Bratti, 1990), adding the category of 'subnatural artificial nutritional diets' to the Dougherty terminology, suggested by Grenier (1986), and supplement and with the category of 'suboptimal artificial nutritional diets' introduced by us (Table 13.1), is used for the first time in this review. The latter category is defined by specific peculiarities of nutrition and digestion of insect predators and parasites as well as by the actual development of alternative technologies for their mass-rearing. The main principles for creation of suboptimal artificial nutritional diets were earlier described by Yazlovetsky (1986, 1992).

Amazing modesty has often been noted in the results associated with the development of artificial nutritional diets for insect predators, as compared to similar achievements for parasitoids (Mellini, 1975; Waage et al., 1985; Yazlovetsky, 1986; Cohen, 1992, 1996). To quote Cohen, 'the progress in dietetics for predators has been overstated, and the difficulties have been underestimated' (Cohen, 1992: 78). As a result, even artificial nutritional diets that were described for predaceous insects over 20 years ago, have not yet been applied for mass-rearing, including for Chrysopidae.

It is known that the quality of food in larval stages strongly influences both preimaginal development and reproduction capacity of adult insect predators, including Chrysopidae (New, 1975; Principi, 1983). Hence, creation of artificial nutritional diets and artificial food substrates of full value, adequate to the nutritional requirements, peculiarities of nutrition, and digestion of chrysopid larvae, is one of the key problems in development of alternative (i.e. not using natural food) technologies for mass-rearing of predators of this family. Studies of nutrition and nutritional requirements of Chrysopidae larvae, in particular, C. carnea, which have been carried our during the last 15–20 years, place this family as one of the most investigated predators in this respect (Hagen, 1987). However, absence until recently of the acceptable alternative technologies for chrysopid mass-rearing testifies to the insufficiency of knowledge currently accumulated for successful development of adequate artificial nutritional diets and artificial food substrates.

The aim of our review is to evaluate the current state of development of artificial nutritional diets and artificial food substrates for Chrysopidae larvae, as well as to project optimal ways for development of investigations in this field.

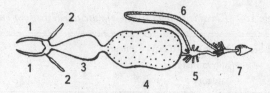

Fig. 13.1. Schematic diagram of the alimentary tract of Chrysopidae larvae. 1, mandibles; 2, salivary glands; 3, foregut; 4, midgut; 5, hindgut; 6, Malpighian tubules; 7, silk-separating reservoir. (After Ermicheva et al., 1987.)

13.2 NUTRITION MECHANISMS OF CHRYSOPERLA CARNEA LARVAE

13.2.1 Morphology of the alimentary canal of Chrysopidae larvae

A schematic structure of the alimentary tract of Chrysopidae larvae is presented in Fig. 13.1.

Larvae have a pair of mandibles (1), linked by canals with salivary glands (2). The intestine consists of the foregut (stomodeum) (3) and midgut (mesenteron) (4). The midgut is not continuous with the cavity of the poorly developed hindgut (5) which is a characteristic peculiarity of Chrysopidae larvae. Mesenteron content is incorporated into a multilayer peritrophic membrane. The hindgut does not directly participate in the nutrition processes, but only transports proteinaceous substances produced by the Malpighian tubules into the silk-separating reservoir and used by larvae for construction of the cocoon. The lack of communication between the mid- and hindguts does not allow excretion of undigested remnants of food absorbed by the larvae. These remnants are accumulated in the distal part of the midgut in the form of black and brown viscous mass (meconium), during the whole larval stage of ontogenesis. Defecation takes place only once after imaginal moulting. Filling of the midgut with undigested remnants may result in its hypertrophy and untimely completion of larval nutrition (Spiegler, 1962; Yazlovetsky, 1986, 1992; Ermicheva et al., 1987). Evidently, this morphological peculiarity of the alimentary canal is one of the reasons for the distinctions between the physiological mechanisms of nutrition regulation in Chrysopidae larvae and other well-studied insect species (Bond, 1978). Respectively, it is this peculiarity which causes special demands on digestion of food ingredients, including artificial nutritional diets.

According to our observations, the meconium's consistency and properties depend considerably on artificial nutritional dietary composition. Rapid mortality up to 30% of emergent adults due to the sticking together of wing edges and reproductive organs by meconium, preventing mating, results in a deviation from optimum results.

13.2.2 Extra-oral mechanisms of food digestion in Chrysopidae larvae

As follows from the above, the study of feeding and digestion features of Chrysopidae larvae is very useful for the development of artificial nutritional diets for mass-rearing. It is worth mentioning the study of the role of extra-intestinal or extra-oral digestion in larval nutrition. Extra-oral digestion is a chemical preliminary treatment of the food, which increases the degree of assimilation and nutrient value. This mechanism considerably widens predator capacity for assimilation of food substrates, including artificial ones. However, to proceed from this, the role of extra-oral digestion in predator nutrition must be ascertained when developing artificial nutritional diets.

Cohen (1995) reviewed the importance of extra-oral digestion in predaceous terrestrial Arthropoda in detail. Considerable attention in the study of extra-oral digestion is paid to biochemical, mainly enzymological, methods of investigation. Applied to Neuroptera, the author particularly notes the insufficient study and knowledge of digestive chemistry (biochemistry) of this impressive group of insect predators (Cohen, 1995).

The first attempts to study the composition and to localise enzymes of the alimentary canal of green lacewing larvae were made in France (Ferran et al., 1976). This was the purpose for determination of hydrolytic activity in homogenates of the salivary glands as well as intestines of last-instar larvae of Chrysoperla carnea and Chrysopa perla. Standard set Apizym for visual determination of enzyme microquantities was used as a tool for investigations. Approximate estimation of the activity of revealed enzymes was carried out using five-point scale. Potentials of this method allowed the authors to establish presence of 19 different hydrolytic enzymes in the alimentary canal. Their identification may be doubtful only in case of lipases. The idea is that revealing of homogenate activity to β-naphtyl-myristate does not allow us to conclude presence of lipases in the predator's alimentary canal. Enzymes demonstrating such activity (tissue C_{14}-esterases) do not participate in digestion and, hence can not be considered as lipases (acylhydrolases of the highest triglycerides) which obtained the systematic number 3.1.1.3 in the classification of enzymes recommended by the International Biochemical Union (Brockerhof & Jensen, 1974).

In the study of the nature of active distribution of the investigated hydrolases in the alimentary canal of larval predators, the semiquantitative method utilised did not provide unambiguous results. However, Ferran et al. (1976) managed to make rather fixed conclusions about slight proteolytic (except leucinaminopeptidase) activity of homogenates in Chrysoperla carnea and Chrysopa perla larvae salivary glands.

During the past 15 years, we have attempted experimentally to prove or disprove the existence of extra-oral digestion mechanism in Chrysopidae larvae. Thus, we investigated the composition, properties, and distribution of the main hydrolytic enzyme activity in the alimentary tract of intensively feeding third-instar larvae of Chrysoperla carnea (Ermicheva et al., 1987; Sumenkova et al., 1989; Moontyan & Yazlovetsky, 1988, 1992, 1993; Yazlovetsky, 1992; Yazlovetsky et al., 1993). Some results of these investigations are given in Fig. 13.2. The share (%) of activity of every hydrolase group investigated, found in the appropriate organ, in its total activity in the alimentary canal of the C. carnea larvae, is given in the ordinate axis of the diagram.

As can be seen in Fig. 13.2, the prevailing share of hydrolytic enzyme activity is concentrated in the midgut, with minimal manifestation in salivary glands and the foregut. For example, endopeptidase activity of the homogenate of the pair of the salivary glands equals 0.02%, and aminopeptidase 0.03% of the values of the corresponding activities of the total homogenate of the alimentary tract. Only 4.8% of endopeptidase and 1.5% of leucinaminopeptidase activity was revealed in the foregut. However, one cannot completely exclude the possibility of mechanical introduction of these enzymes into the foregut from the midgut during the preparation process. Practically all the endopeptidases, about 80% of carbohydrases, and 85% of lipases are localised in the midgut content, while aminopeptidases sare distributed between its content and the wall. Such localisation of hydrolases in the intestine of C. carnea

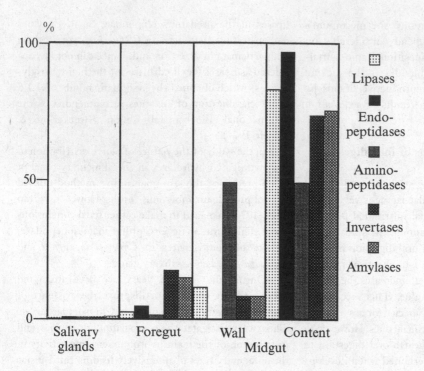

Fig. 13.2. Distribution of the activity of hydrolytic enzymes in the alimentary tract of *Chrysopa carnea* larvae.

larvae is well known for spatial organisation and dividing of highly molecular food components described for many groups of insects having a peritrophic membrane (Santos & Terra, 1986). The presence of the peritrophic membrane, as well as sufficient size of the crop, was not characteristic of predaceous insects with extraoral digestion (Terra & Ferreira, 1981).

It is interesting to note that Hemiptera are the only group of insects which lack the peritrophic membrane (Terra, 1990). They also utilise extra-oral digestion most intensively (Cohen, 1995). This was confirmed by our investigations on localisation, composition, and some characteristics of the main digestive enzymes of predaceous bugs *Podisus maculiventris* and *Perillus bioculatus* (Hemiptera: Pentatomidae). However, in the salivary glands of larvae of the older instars of the above two species we observed high activity of proteolytic and amylolytic enzymes, but almost complete absence of lipase activity (Moontyan & Yazlovetsky, 1994; Sumenkova & Yazlovetsky, 1994; Yazlovetsky & Kaplan, 1995; Yazlovetsky *et al.*, 1996). This suggests that predaceous Pentatomidae utilise extra-oral digestion for extra-intestinal digestion of proteins and polysaccharides, except fats.

We also observed nutrition features of *C. carnea*

larvae when feeding them with coloured liquid artificial nutritional diets, packed in transparent microcapsules. It is distinctly visible that artificial nutritional diets enter the mandible canal soon after the piercing of the capsule; sucking is continuous, quick, and without regurgitation. Dense excrements were clearly visible forming in the distal part of the midgut. These excrements often consisted of layers with different colours depending of the type of food absorbed.

Hence, we have sufficient data to describe the nutrition and digestion process of *C. carnea* larvae the following way. Digestion of nutrients may commence even in the prey body as affected by small quantities of enzymes (mainly carbohydrases), which are produced by the predator salivary glands. However, only partial destruction of large molecules of nutrient substances takes place here, which eases their subsequent movement and digestion. At this stage polymerous and oligomerous carbohydrates may be hydrolysed, while highly molecular proteins are practically unbroken. The macerated food moves into the stomodeum, where its hydrolysis continues under the influence of carbohydrases, proteases, and lipases, which came from salivary glands together with the food, and, probably, penetrated the foregut from the midgut. However, the contribution of salivary glands and stomodeum in digestion of proteins, fats, and carbohydrates is relatively small. This process takes place mainly in the midgut, covered by a peritrophic membrane. Necessary digestive enzymes are produced in the wall of the mesenteron and excreted into its aperture; this secretion is stimulated by the food, which enters the midgut (Ermicheva *et al.*, 1987; Yazlovetsky, 1992; Yazlovetsky *et al.*, 1993).

Recently, Cohen (1998) attempted to apply the concept of the leading role of extra-oral digestion in nutrition of the majority of predaceous arthropods to *C. carnea* larvae. This application looks rather disputable when compared with the results of our investigation of the nutrition mechanism of this predator.

We can not completely deny the presence of an extra-oral digestion mechanism in *C. carnea* larvae. However, we have experimental data sufficient to exclude the leading role of *external* food digestion in their nutrition. That is why we suppose that *C. carnea* is not included among the number of predaceous terrestrial arthropods which use extra-oral digestion as a main method of utilising their prey. This conclusion may considerably influence further development of artificial nutritional diets and artificial food substrates for mass-rearing of Chrysopidae.

13.3 CURRENT STATE OF ARTIFICIAL NUTRITIONAL DIETS FOR CHRYSOPIDAE LARVAE

In 1948 and 1950 Finney was the first to report rearing of *Chrysopa californica* on coddled larvae of the potato tuber moth, *Phthorimaea operculella* (Finney, 1948, 1950). Since then, 14 species of Chrysopidae have been grown on different artificial food substrates and the number of the publications describing artificial food substrates and artificial nutritional diets for Chrysopidae larvae has risen to over 70. These publications mainly pay attention to artificial food substrates and artificial nutritional dietary development for the most widely distributed Chrysopidae. They are *Chrysoperla carnea* (over 50 publications), *Chrysopa septempunctata* (7), *C. sinica* (4), and *C. perla* (4).

Table 13.2 provides a bibliographic summary of the main publications on the preparation, composition and techniques of artificial nutritional diets for Chrysopidae larvae. When compiling it we used some data from well-known bibliographic works on artificial nutritional diets for insects (Singh, 1977; Gomez *et al.*, 1982, 1985). The application of the classification chosen by us, to the reported artificial nutritional diets for green lacewing larvae, are given in the right-hand column of this table.

13.3.1 Subnatural artificial nutritional diets

Ingredients and preparation techniques typical for this artificial nutritional diet category, firstly applied by Finney (1948, 1950), are being used and developed (Table 13.2). An event, which is worth noting, was the first use of artificial nutritional diets for Chrysopidae larvae, consisting of powder from lyophilised homogenate of larval drone honey-bee brood, which was simultaneously and rather successfully tested on 24 species of coccinellids (Matsuka & Niijima, 1985). Six generations of *Chrysopa septempunctata* and *Chrysoperla carnea*, four generations of *Chrysopa formosa* and one generation of *Chrysopa intima*, *Chrysoperla furcifera*, and *Mallada formosanus* were reared with good developmental parameters on drone powder, which was called the 'substitutive' diet by its authors, and water (Okada & Niijima, 1971; Okada *et al.*, 1974; Niijima, 1989; Niijima & Matsuka, 1989). Ferran *et al.*, (1981) used a mixture of honey and powder, made from lyophilised homogenate of larval worker honey-bee brood, as the nutrition diet for Chrysopidae larvae. Five generations of *Chrysoperla carnea*, were reared, while breeding of *Chrysopa septempunctata* and *C. perla* ceased on the second or third generation. From 50 larvae of *C. formosa* it was possible to obtain only two disfigured adults, which soon died (Ferran *et al.*, 1981). Though drone powder is undoubtedly the best diet among subnatural artificial nutritional diets for some Chrysopidae larvae, it has one serious disadvantage, which hinders its use in technologies of mass-rearing. Under conditions that are optimal for Chrysopidae maintenance (70%–80% of relative air humidity and 25%–7°C) the hygroscopic drone powder becomes a syrup, which is quickly oxidised by polyphenoloxidases and covered by mould (Niijima & Matsuka, 1989). At the same time the semi-liquid state of artificial nutritional diets completely suits the feeding method of Chrysopidae larvae, which in the opinion of Niijima (1989) explains the better results obtained by rearing on drone powder, compared to less hygroscopic powder from lyophilised aphids.

Table 13.2. *Published reports on advances in artificial nutritional diet development for* Chrysopidae *larvae*

Chrysopidae species, references	Numbers of categories[a] of artificial nutritional diet	Chrysopidae species, references	Numbers of categories of artificial nutritional diet
Chrysoperla carnea = *Chrysopa carnea* = *Chrysopa californica* = *Chrysoperla plorabunda* = *Chrysoperla mohave* = *Chrysoperla nipponensis* = *Chrysopa vulgaris*		*Chrysoperla furcifera*	
		Niijima & Matsuka, 1989	1
		Chrysopa septempunctata	
Ageeva *et al.*, 1988	3,5[b]	Ferran *et al.*, 1981	1,4,5
Ageeva *et al.*, 1990	3,5	Kaplan *et al.*, 1989	3,5
Babrikova & Radeeva, 1990	1,3–3[c]	Niijima, 1989	2
Babyi, 1979	1,3	Niijima & Matsuka, 1989	1–2,5
Bigler *et al.*, 1976	1,3–3	Okada *et al.*, 1974	1
Cohen, 1983	1,3	Okada & Niijima, 1971	1
Cohen, 1998	4,5	*Chrysopa sinica*	
Ferran *et al.*, 1981	1,4,5	Cai *et al.*, 1983	3,4
Finney, 1948	1	Yazlovetsky *et al.*, 1992	3,5
Finney, 1950	1	Ye *et al.*, 1979	3,4
Hagen & Tassan, 1965	3	Zhang & Zhou, 1998	3,4
Hagen & Tassan, 1966*a*	3	*Chrysopa perla*	
Hagen & Tassan, 1966*b*	3	Canard, 1973	1,4
Hasegawa *et al.*, 1983	1	Bigler *et al.*, 1976	1,3–3
Hasegawa *et al.*, 1989	1–2,5	Ferran *et al.*, 1981	1,4,5
Hassan & Hagen, 1978	3,4	Yazlovetsky *et al.*, 1992	3,5
Kariluoto, 1980	3,4	*Chrysopa rufilabris*	
Keiser *et al.*, 1991	3,5	Hydorn & Whitcomb, 1979	3,4
Letardi & Caffarelly, 1989	3,4	*Chrysopa lanata lanata*	
Martin *et al.*, 1978	3	Botto & Crouzel, 1979	4
Nepomnyashaya *et al.*, 1979	3,5	*Chrysopa albolineata*	
Niijima & Matsuka, 1989	1–2,5	Yazlovetsky *et al.*, 1992	3,5
Ponomareva, 1971	1,3–3,4	*Chrysopa scelestes*	
Ponomareva & Beglyarov, 1973	1,3–3,4	Krishnamoorthy & Nagarkatti, 1983	3,4
Sogoyan & Lyashova, 1971	1,3–4		
Tartarini, 1983	3,4	*Chrysopa formosa*	
Vanderzant, 1969	3	Ferran *et al.*, 1981	1,4,5
Vanderzant, 1973	3	Niijima & Matsuka, 1989	1
Wuhrer & Hassan, 1990	3,4	*Chrysopa intima*	
Yazlovetskij & Nepomnyashaya, 1981	3	Niijima & Matsuka, 1989	1
Yazlovetsky *et al.*, 1979*a*	3,5	*Mallada prasina* = *Anisochrysa prasina* = *Chrysopa prasina*	
Yazlovetsky *et al.*, 1979*b*	3	Yazlovetsky *et al.*, 1992	3,5
Yazlovetsky *et al.*, 1992	3,5		
Yazlovetsky *et al.*, 1990	3,5	*Mallada flavifrons* = *Anisochrysa flavifrons*	
Yazlovetsky, 1992	3,5	Cava & Sgobba, 1982	3,4
Yazlovetsky & Abashkin, 1977	3		
Yazlovetsky & Nepomnyashaya, 1977	3	*Mallada formosanus*	
Zaki & Gestrana, 1990	4	Niijima & Matsuka, 1989	1

Notes:

[a] Categories are explained in Table 13.1.

[b] The artificial nutritional diet described in the article, can be referred to two or several categories.

[c] The article describes two artificial nutritional diets, which refer to different categories.

The solution of this problem can be achieved through the encapsulation of liquid artificial nutritional diets containing drone powder (Cohen, 1983).

Addition of a stew of crickets to meridic artificial nutritional diets has helped to complete the developmental cycle of *Chrysopa perla* larvae. Only three adults were reared (2 males and 1 female which failed to lay eggs without feeding on aphids), although it indicated that in principle it was possible to rear Chrysopidae (being predators in all nutrition stages of their development) without natural prey (Canard, 1973).

Other examples of the use of techniques for preparation of subnatural artificial nutritional diets are inconsequential, because they serve as attempts to improve artificial nutritional diets of other categories, through the addition of minimally treated insects. Thus, nutritional diet containing honey, milk, alkaline casein hydrolysate, pepton (a product of pepsin hydrolysis with a mixture of meat wastes obtained from slaughter-houses), brewer's yeast hydrolysate, mixture of water and fat, soluble vitamins, ascorbic acid and water, differs from meridic and oligidic artificial nutritional diets only by addition of considerable quantities (over 30% by weight) of powder made from *Sitotroga cerealella* moths. The moths were killed by heating them up to 90°C, then finely crushed and dried. Ponomareva (1971) and Ponomareva & Beglyarov (1973) reared one generation of *Chrysoperla carnea* with good development parameters on this diet. However, three years later, in France, Bigler *et al.* (1976) reported the unsuccessful testing of artificial nutritional diets of this composition, on larvae of *Chrysoperla carnea* and *Chrysopa perla*. Only 5% of *Chrysoperla carnea* larvae reached the third-instar, while all larvae of *Chrysopa perla* died at the first-instar. Additives to artificial nutritional diets of meridic composition of up to 20% of the homogenate of *S. cerealella* eggs (Sogoyan & Lyashova, 1971; Babrikova & Radeeva, 1990) or 10% mixture of pea aphid *Acyrthosiphon pisum* homogenate with the summed extract of fats from *S. cerealella* moths (Babyi, 1979) also failed to give any advantages over artificial nutritional diets for chrysopids similarly composed.

13.3.2 Holidic (chemically defined) artificial nutritional diets

Data on the development of holidic artificial nutritional diets for green lacewing larvae were found only in three articles; all of them were made by one team of Japanese authors and published more than 10 years ago

(Hasegawa *et al.*, 1989; Niijima, 1989; Niijima & Matsuka, 1989). It is worth noting that it is these three publications which primarily defined progress in the study of chrysopid nutritional requirements. The first article describes four recipes of holidic artificial nutritional diets, which differ in amino acid composition. On the best diet two generations of *Chrysoperla carnea* were obtained, slightly retarded in their development compared to those in the control sample. Survival of adults in F_2 constituted 44.8%, while females laid more than 1000 eggs for two months. The diet is complex in its composition, which is typical for artificial nutritional diets of this category. It consists of 65 individual components, including 23 amino acids, 2 carbohydrates, 5 organic acids, 6 fatty acids, cholesterol, 11 mineral salts, and 17 vitamins. A peculiarity of this holidic artificial nutritional diet was the use of literature data (to find its optimal composition) about the composition of amino acids and mineral substances of aphids, honey-bees, and other insects. This diet allowed the determination of optimum zones for the ratio of amino acids and carbohydrates as well as mineral substances. It was also shown that exclusion of five organic acids from artificial nutritional diets does not prevent larvae from reaching the imaginal stage, while absence of three out of six fatty acids and three fat soluble vitamins considerably decreased their survival (Hasegawa *et al.*, 1989).

Niijima (1989) applied the same artificial nutritional diets after slight modification of the preparation procedure to investigate nutrition requirements of *Chrysopa septempunctata* larvae. He established that holidic artificial nutritional diets, consisting of 65 components, provide normal development and oviposition of this chrysopid with slight reduction of adult body weight and increase of the developmental period, compared to the control sample (i.e. feeding of larvae and adults on drone powder). The author explained the appearance of some adults with poorly spread and deformed wings, by the influence of nutritional and physical factors, i.e. low humidity and lack of space at adult emergence. Nutrition value of certain components for *C. septempunctata* larvae was determined by their gradual exclusion from artificial nutritional diet composition. Exclusion of all mineral salts resulted in death of the majority of larvae already at the first instar. Withdrawal of many vitamins also confirmed their importance. Larvae also developed on artificial nutritional diets without cholesterol; however, fertility of eggs laid by emergent adults was considerably reduced.

At the same time replacement of trehalose by sucrose, elimination of β-alanin and α-aminobutyric acid, all five organic acids, five inorganic salts, three water-soluble vitamins, fatty acids, and fat-soluble vitamins did not considerably influence their developmental parameters. Lowered requirement for fatty acids and fat-soluble vitamins, on one hand, and a narrower interval of the optimal ratio of carbohydrates and amino acids, on the other hand, are almost the only considerable distinctions in nutritional requirements of *Chrysoperla carnea* and *Chrysopa septempunctata*. These two species also differ from each other by nutrition regimes in the imaginal phase.

When it was clarified that some components were not needed, it was possible to simplify considerably the initial holidic artificial nutritional diets. Two simplified diets, consisting of 54 and 37 ingredients respectively, resulted in normal development of *C. septempunctata* for three generations, while the fourth generation exhibited a conspicuous sharp reduction of the percentage of emergence, reduced adult body weight, and prolongation of the developmental period. Only two adults were obtained on the best of these two diets (54 ingredients) in the fifth generation and their development lasted over 50 days as compared to 30 in the first generation (Niijima, 1989).

The last of these three Japanese works is of interest because data are given for the general chemical composition of drone powder and three aphid species, which were used at optimisation of the holidic artificial nutritional diets discussed above. As a result, these artificial nutritional diets were proposed as basic ones for the study of nutritional requirements of aphidophagous insects (Niijima & Matsuka, 1989).

13.3.3 Meridic artificial nutritional diets

Research into artificial nutritional diets of this category for chrysopid larvae began in 1965/6. Hagen & Tassan (1965) were the first to report two recipes of liquid artificial nutritional diets, which did not contain insect material and were made of enzymatic hydrolysates of yeast and casein, fructose, ascorbic acid, choline chloride, and water. Though simple, these artificial nutritional diets provided complete development of *Chrysoperla carnea* from newly emerged larvae until adults, which laid fertile eggs. Hence, in principle the possibility for the development of Chrysopidae on meridic artificial nutritional diets had been proved.

Besides, the authors were the first who tried rather successfully to improve conditions for the nutrition of predaceous larvae on liquid artificial nutritional diets through their encapsulation in paraffin capsules by means of simple devices (Hagen & Tassan, 1965, 1966a, b). The promising results of this pioneer work, as well as the evident need to improve the composition of the first meridic artificial nutritional diets (there was a considerable time-lag in larval development and cocoon mass compared to the control sample), resulted in improvements to the Hagen & Tassan recipes. Vanderzant (1969) managed full success, reporting rearing on meridic artificial nutritional diets, which consisted of casein and soy enzymatic hydrolysates, soya-bean oil and lecithin, cholesterol, fructose, four mineral salts, eight vitamins from B group, ascorbic acid, choline, and water. Increased complexity of the composition, compared to Hagen & Tassan's artificial nutritional diets, evidently allowed the nutritional requirements of this green lacewing to be more completely met. This is particularly so for fats, which resulted in almost halving the duration of larval development stage, with adult emergence equal to 55%–65% of the initial number of first-instar larvae (Vanderzant, 1969). This artificial nutritional diet was the first artificial diet of acceptable cost; its physical properties and test results allowed its use for development of methods for *C. carnea* mass-rearing. Consequently, Vanderzant's recipe was successfully reproduced many times in other countries (Bigler *et al.*, 1976; Yazlovetsky & Abashkin, 1977; Babrikova & Radeeva, 1990). This artificial nutritional diet was the basis for development of devices for the production of microencapsulated artificial food substrates for chrysopids (Yazlovetsky & Abashkin, 1977; Martin *et al.*, 1978; Yazlovetsky *et al.*, 1979b; Yazlovetskij & Nepomnyashaya, 1981). Later, Vanderzant used modifications of her artificial nutritional diets to reveal amino acid, fat, and ascorbic acid requirements of *C. carnea*. She reported continuous *in vitro* rearing of 18 generations of this predator (Vanderzant, 1973). Modifications of this artificial nutritional diet was also carried out by other authors with the aim to make it cheaper and to adapt it for microencapsulation conditions (Yazlovetsky *et al.*, 1979a, b; Yazlovetskij & Nepomnyashaya, 1981). The green lacewing *Chrysopa perla* was also reared on Vanderzant's diet; however, the yield of adults was only 2.5%. At the same time a much more complex meridic artificial nutritional diet by French authors demonstrated poorer results in comparative studies with *Chrysoperla carnea* (adult yield was 7.5% as compared to 35% on Vanderzant's diet) (Bigler *et al.*, 1976).

13.3.4 Oligidic artificial nutritional diets

Hassan & Hagen (1978) attempted to improve artificial nutritional diets for Chrysopidae by adding to meridic compositions such nutritionally valuable ingredients as honey, chicken eggs, and dried cow's milk. From seven recipes tested on *Chrysoperla carnea* larvae, the best was the mixture of honey, sugar, food yeast flakes, enzymatic hydrolysates of yeast and casein, chicken egg yolk, and water. Three generations of the predator were reared on it. Developmental duration and cocoon mass were lower than in control insects – 33% and 6% respectively – while adult yield varied within 90% and 100% of the initial number of larvae. The authors concluded that a new nutritional diet can be used as 'a supplementary larval food in the mass production of this beneficial arthropod' (Hassan & Hagen, 1978). This work has been further developed in Germany by Wuhrer & Hassan (1990) with the results of testing 24 types of artificial nutritional diets, composed on the base of honey, fructose, yeast, and egg yolk. It revealed the positive effect of soya-bean oil and plant pollen on *C. carnea* larvae development (Wuhrer & Hassan, 1990). Slightly earlier work (Cava & Sgobba, 1982) reported *in vitro* rearing of the green lacewing *Anisochrysa flavifrons*. They fed larvae with Hassan & Hagen's artificial nutritional diet, jellied through the addition of agar-agar, and which ensured satisfactory developmental parameters (Cava & Sgobba, 1982). Using this method for feeding of *C. carnea* larvae with Hassan & Hagen's artificial nutritional diet Tartarini (1983) came to the conclusion that this artificial food substrate had some advantages compared with the larvae of *Galleria mellonella*. Letardi & Caffarelly (1989) introduced ascorbic acid into Hassan & Hagen's artificial nutritional diet, excluded food flakes and almost doubled the content of enzymatic hydrolysates of yeasts. Tests of this liquid artificial nutritional diet has shown that in spite of prolongation of the larval development period (about 29%), it can substitute eggs of *Ephestia kuehniella* for feeding *C. carnea* larvae. The authors noticed an interesting detail: meconium, usually viscous though not sticky when feeding larvae with natural or substituting prey, was often liquid and sticky when larvae were fed on artificial nutritional diets. This resulted in death and failure to copulate in some adult insects.

Concurrent with the Hassan & Hagen article, Ye *et al.* (1979) published results of breeding 10 generations of *Chrysopa sinica* on an artificial nutritional diet, which consisted of chicken eggs, brewer's yeast, honey, cane sugar, ascorbic acid, and water. The cocoon yield (from the initial number of larvae) and adult yield (from cocoons) varied within 58%–77% and 77%–94% respectively; average adult fertility varied from 464 to 1562 eggs. Similar approaches were used when making artificial nutritional diets for this species in another Chinese publication (Cai *et al.*, 1983). Good development parameters for *Chrysoperla carnea* and *Chrysopa septempunctata* were reported on oligidic artificial nutritional diets which were microencapsulated with the help of specially designed highly efficient devices (1000 g of microcapsules per hour). The yield of cocoons from newly emerged larvae of both species reached 95% (Zhang & Zhou, 1998).

Kariluoto added about 14% of beef liver and 1% of Bacto-agar Difco in an artificial nutritional diet of meridic composition. Artificial nutritional diets jellied by this method resulted in a yield of 80% for *Chrysoperla carnea* adults from the initial quantity of the first-instar larvae as compared to 90% in control (feeding of larvae on aphid *Myzus persicae*) (Kariluoto, 1980).

Composition of the oligidic artificial nutritional diet prepared by the addition of dried milk and pig grease to components which were usually applied in meridic nutritional diets was tested by Ponomareva (1971). On this artificial nutritional diet, 60.8% of *C. carnea* larvae completed their development. Butter was used (without evident success) as a source of fat in an oligidic artificial nutritional diet for larvae of this green lacewing (Sogoyan & Lyashova, 1971), and crushed walnut kernels instead of enzyme casein hydrolysate (Ponomareva & Beglyarov, 1973).

An attempt was made to use homogenised biomass of the alga *Chlorella vulgaris* as the only protein source in artificial nutritional diets for *Chrysoperla carnea* larvae. All larvae fed on such artificial nutritional diet died, being unable to complete their development. Only the addition of usual ingredients of the meridic artificial nutritional diet resulted in the rearing of adults with the yield of 66.7% of the initial number of first-instar larvae (Zaki & Gestrana, 1990).

13.3.5 Suboptimal artificial nutritional diets

The main purpose of artificial nutritional diets of this category is the provision of programs for mass-rearing of insect entomophages with the artificial food of acceptable quality and cost (Table 13.1). This trend is considered as one of the most promising, when developing economically expedient and efficient technologies for the production of beneficial insects (Mellini, 1975; Singh, 1982; Waage *et al.*, 1985; Bratti, 1990). However, until recently no alternative technologies have been developed for mass-rearing of any species of predators or parasites, which is evidently explained by the difficulty of finding a solution to this problem. Details on successful implementation of big projects for Chrysopidae rearing, using artificial food substrates for feeding larvae, prepared on the basis of artificial nutritional diets, were absent in publications reviewed by us. Available Chinese publications on this theme are laconic (Ma *et al.*, 1986; Zhang & Zhou, 1998).

To be the base for economically expedient and efficient technology of Chrysopidae mass production, artificial food substrates for larvae should meet the following requirements:

1. Provide development of lacewings with the main biological parameters, which do not produce the same results on natural prey.
2. Be considerably (10 or more times) cheaper than the food used in existing rearing systems.
3. Not impede production mechanisation.

In turn, the following main requirements are necessary for artificial nutritional diets as the base for artificial food substrates:

1. To have a set of substances in the available form and necessary concentration for more completely meeting Chrysopidae nutrition requirements.
2. To consist of cheap ingredients.
3. To include a minimum quantity of indigestible admixtures.
4. To contain antiseptics, which retard multiplication of microorganisms and do not harm the insect.
5. To correspond in its parameters (homogeneity, viscosity, capability of forming resistant emulsions) to the conditions of the packing process.

The microencapsulation process of the liquid also requires rather strict technical conditions, with particular reference to its viscosity, homogeneity, and ability to form stable emulsions.

Since 1970 in the USA and since 1972 in the USSR special devices have been developed for the production of microencapsulated artificial food substrates for Chrysopidae (Yazlovetsky & Abashkin, 1977; Martin *et al.*, 1978; Keiser *et al.*, 1986; Yazlovetsky *et al.*, 1990; Nordlund & Morrison, 1992; Yazlovetsky, 1992). These devices pack liquid artificial nutritional diets into thin-walled, ball microcapsules. On the one hand, these allow isolation of the artificial nutritional diet from undesirable contact with air, microorganisms, and the larval body, and on the other hand, provide the predator with the possibility of freely sucking the content of the capsule, by piercing its envelope with the mandibles. Such artificial food substrates currently provide the most adequate method for predaceous insects with a piercing and sucking mouth, such as Chrysopidae larvae (Bond, 1978; Principi, 1983).

Vanderzant's artificial nutritional diet, which was specially adapted to packing conditions (Vanderzant, 1973; Yazlovetsky & Abashkin, 1977; Martin *et al.*, 1978; Yazlovetsky *et al.*, 1979*b*), was applied to get the microencapsulated artificial food substrate. However, the results obtained did not exceed the level of laboratory production of kilogram quantities of microcapsules and those of field testing of reared larvae on small plots. This was explained by imperfect microcapsules and insufficient adequacy of artificial nutritional diets (Yazlovetskij & Nepomnyashaya, 1981; Yazlovetsky *et al.*, 1990; Nordlund & Morrison, 1992). However, accumulated experience resulted in progress in this field.

We suppose that for the development of suboptimal artificial nutritional diets for entomophages, targeted research is necessary for cheap ingredients of known or specially studied composition, which will allow the reduction of the cost of artificial nutritional diets. Such research should include a preliminary evaluation of the availability of the main nutritional substances of selected ingredients for their assimilation by the digestion system of the insect (Yazlovetsky, 1986, 1992).

An artificial nutritional diet for *Chrysoperla carnea* larvae, which we developed in 1979 with the use of some elements of this methodology, can be referred to the suboptimal diet category. It is characterised by the inclusion of water extracts of wheat germ and brewer's yeasts with amino acid composition studied by us. Optimal content of these amino acids, water-soluble vitamins, and microelement sources in artificial nutritional diets was determined on the basis of data from the literature and results of investigation of the composition of three aphid species (*Myzodes persicae, Megoura viciae,* and *Aphis fabae*), as well as *Sitotroga cerealella* eggs. Mathematical methods were used to optimise amino acid composition. The required persistence of water and lipid emulsion has been reached through introduction of Tween 80 emulsifier into artificial nutritional diets (Yazlovetsky *et al.,* 1979*a,b*; Nepomnyashaya *et al.,* 1979; Yazlovetsky, 1992). The artificial nutritional diet suited the microencapsulation conditions and the artificial food substrates obtained ensured larval development from hatching until the first moult, with biological parameters exceeding those of the control. Expenses were 25–30 times lower for production of the second-instar larvae. High effectiveness during their release against aphids in greenhouses was demonstrated, as compared to the control rearing on *S. cerealella* eggs (Yazlovetskij & Nepomnyashaya, 1981). However, the yield of adults on this microencapsulated artificial food substrate did not exceed 30%, which was mainly explained by difficulties of the first-instar larvae in feeding through the microcapsule envelope. It is worth noting that, unlike artificial nutritional diets of other categories, adequacy of suboptimal artificial nutritional diets and preparation on their basis of artificial food substrates should be evaluated mainly according to indexes of insect development from hatching until the targeted developmental stage. Therefore, in the case of *C. carnea,* the yield of adult insects is not considered as the principal criterion, since expenses of maintaining stock culture are relatively small even in large-scale rearing, and it can be more profitable to maintain cultures on natural food. Besides, one should note that multiple reproduction on artificial food substrates would inevitably affect the quality of predators produced (Cohen, 1992, 1996).

Enrichment of this artificial nutritional diet with myristic acid (as trimyristate) and sitosterol, resulted, for the first time, in a completed developmental cycle of *Chrysopa septempunctata* on liquid artificial nutritional diets that contained no insect material. Grounds for this enrichment were the result of analysis of data from the literature and our own data on lipid composition of several aphid species (Kaplan *et al.,* 1986; Keiser & Yazlovetsky, 1988). However, the yield of adults did not exceed 10%, indicating the need for further improvement of artificial nutritional diets (Kaplan *et al.,* 1989).

Our work on improvement of suboptimal artificial nutritional diets for Chrysopidae larvae was done in three main directions.

1. Further accumulation and use of information about composition of the main nutrients of natural food.

    New data were obtained on the composition and utilisation of sterols (Keiser & Yazlovetsky, 1989, 1991), nucleic acids (Sekirov *et al.,* 1990), and carbohydrates (I.G. Yazlovetsky, unpublished data) by *Chrysoperla carnea* larvae.

2. Research into the composition of cheap sources of protein, amino acids, lipids, and carbohydrates.

    Studies of the origin of commercial protein and vitamin concentrates of yeast (eight preparations), fungal (seven preparations), and bacterial (three preparations) resulted in the choice of protein and vitamins concentrate from the biomass of the bacteria *Acinetobacter calcoaceticus.* Besides the mixture of amino acids and proteins, which is similar in its nutritional value to animal proteins, this concentrate contains 9% of fats and 8% of carbohydrates (Ageeva *et al.,* 1979, 1989). Preference was given to sunflower oil and lecithin after their comparison with soya-bean products. Among seven studied carbohydrate sources, sucrose was the most suitable for artificial nutritional diets for *C. carnea* larvae. Satisfactory results were also obtained, however, on cheaper starch syrup, potato, and maize starch, explained by the higher activity of both invertase and α-amylase in the alimentary canal of this predator (Moontyan & Yazlovetsky, 1988, 1992, 1993).

    It resulted in the development of a rather perfect suboptimal artificial nutritional diet, which consisted of 11 ingredients, including extracts of biomass of *Acinetobacter calcoaceticus,* sucrose, mineral salts, vitamins, mixture of sunflower oil and lecithin, Tween 80 emulsifier, antiseptic (sorbic acid), and water. To increase solubility of the bacterial biomass and to formulate a persistent emulsion, the artificial nutritional diet was treated with ultrasound (Keiser *et al.*, 1991). Tests were made on five Chrysopidae species by moistening the artificial nutritional diet with sterile pine sawdust. Survival of the respective developmental stages and egg fertility of the predators, reared on this artificial food substrate, yielded better than the control individuals, which were fed on *Sitotroga cerealella* eggs. The proportion of adults with wings and genitalia stuck with meconium decreased considerably; the yield of fully valuable adults was 60%, 57%, 44%, 47%, and 48% for *Chrysoperla carnea, Chrysopa sinica, C. prasina, C. albolineata*, and *C. perla*, respectively (Ageeva *et al.*, 1990; Yazlovetsky *et al.*, 1992). Unfortunately this artificial nutritional diet has not been tested in microencapsulated form.

3. Accumulation and use of information on peculiarities of food absorption and digestion by Chrysopidae larvae.

We tried to estimate suitability of artificial nutritional diet ingredients both on the basis of results of their compositional study as well as by hydrolysing their components with corresponding digestive enzymes of *Chrysoperla carnea* larvae. As already mentioned, it was absolutely necessary for compiling suboptimal artificial nutritional diets for Chrysopidae, because of architectural peculiarities of the alimentary canal which result in specificity of mechanisms of nutrition regulation in chrysopid larvae (Bond, 1978; Ermicheva *et al.*, 1987). Thus, for example, the choice of the biomass of the bacterium *Acinetobacter calcoaceticus* as the only source of amino acids and proteins in the composition of the above-mentioned suboptimal artificial nutritional diet, which was successfully tested on larvae of five Chrysopidae species, was to a greater extent determined by the results of the comparative evaluation of hydrolysation of this and other protein concentrates of microbial origin with proteolytic enzymes of the predator's alimentary canal (Ageeva *et al.*, 1989). Such enzymological evaluation of suitability of protein components of artificial nutritional diets became possible thanks to the information and methodical experience acquired during studies made in our laboratory of the mechanism of Chrysopidae larvae nutrition and digestion (Yazlovetsky, 1992).

Some elements of developmental methodology of suboptimal artificial nutritional diets for Chrysopidae larvae were previously used in one French (Ferran *et al.*, 1981) and two Japanese (Hasegawa *et al.*, 1989, Niijima & Matsuka, 1989) publications (Table 13.2). Cohen (1998) recently reported a new concentrated artificial nutritional diet for *C. carnea*, which was composed of beef products, chicken eggs, sugar, honey, yeasts, antiseptics, and water. These selected ingredients are categorised as oligidic, while in its developmental and preparation technique it is a typical suboptimal artificial nutritional diet. The chemical composition of the prey's body was used as a model, which allowed the concentration of steroids, phospholipids, and proteins, adequate to their content in the natural food. More than 35 generations of *C. carnea* were obtained with the body size, fecundity, voracity, and developmental period equivalent or even exceeding parameters of control groups of predators reared on lepidopterous eggs (Cohen, 1998). The concept of 'concentrated nutrients' is of considerable interest for the development of suboptimal artificial nutritional diets. This short communication lacks data about the cost of this artificial nutritional diet and its suitability for microencapsulation.

13.4 CONCLUSION

At the end of this review we come to the conclusion, that in the past 30–35 years noticeable advances have been achieved in development of artificial nutritional diets for larvae of no fewer than 14 Chrysopidae species. The level of knowledge achieved is to some extent sufficient for the development of different artificial nutritional diets, suitable for laboratory rearing of green lacewings for investigation purposes. It is of special reference to the widely distributed and best-studied *Chrysoperla carnea*. However, the creation of artificial food substrates for economically efficient mass production of Chrysopidae still requires the development of better special artificial nutritive diets, belonging to the suboptimal category. The main criteria for estimation of such artificial nutritional diets should be the quantity of insects of acceptable quality reproduced on them, associated with labour, time, and financial expenses.

REFERENCES

Ageeva, L.I., Sumenkova, V.V. & Yazlovetsky, I.G. (1979) Amino acid composition of some commercial protein concentrates. In *Biochemistry and Physiology of Insects*, ed. Popushoi, I.S., pp. 63–70. Shtiintsa Publishing, Kishinev. (in Russian with English Summary).

Ageeva, L.I., Keiser, L.S. & Yazlovetsky, I.G. (1988) Artificial diet for *Chrysopa carnea* larvae. Authors certificate of the USSR N1405126. *Bulletin of Inventions of USSR* 23, 247.

Ageeva, L.I., Ermicheva, F.M., Nepomnyashaya, A.M., Sekirov, I.A., Sumenkova, V.V. & Yazlovetsky I.G. (1989) *Selection and Determination of Biological Value of Protein Components for Artificial Nutritional Diets for Entomophagous Insects (Methodical Guidelines)*. All–Union Research Institute of Biological Methods for Plant Protection, Kishinev. (in Russian)

Ageeva, L.I., Yazlovetsky, I.G. & Keiser L.S. (1990) Artificial diet for *Chrysopa* larvae. Authors certificate of the USSR N1552410. *Bulletin of Inventions of USSR* 11, 264.

Babrikova, T. & Radeeva, K. (1990) Development of artificial diets for mass-rearing of *Chrysopa carnea* Steph. (Neuroptera, Chrysopidae). *Bulletin of High Agricultural School* (Plovdiv, Bulgaria) 35, 91–98. (in Bulgarian with English summary)

Babyi, V.S. (1979) Liquid artificial diet for *Chrysopa carnea* larvae. In *Biochemistry and Physiology of Insects*, ed. Popushoi, I.S., pp. 73–76. Shtiintsa Publishing, Kishinev. (in Russian with English summary)

Bigler, F., Ferran, A. & Lyon, J.P. (1976) L'élevage larvaire de deux prédateurs aphidiphages (*Chrysopa carnea* Steph., *Chrysopa perla* L.) à l'aide de différents milieux artificiels. *Annales de Zoologie – Écologie Animale* 8, 551–558.

Bond, A.B. (1978) Food deprivation and the regulation of meal size in larvae of *Chrysopa carnea*. *Physiological Entomology* 3, 27–32.

Botto, E.N. & Crouzel, I.S. (1979) Dietas artificiales y capacidad de postura de *Chrysopa lanata lanata* (Banks) en conditiones de laboratorio. *Acta Zoologica Lilloana* 35, 745–758.

Bratti, A. (1990) Techniche di allevamento *in vitro* per gli stadi larvali di insetti entomophagi parassitoidi. *Bollettino dell'Istituto di Entomologia dell'Università di Bologna* 44, 169–220.

Brockerhoff, H. & Jensen, R.G. (1974) *Lipolytic Enzymes*. Academic Press, New York.

Cai, C.R., Zhang, X.D. & Zhao, J.Z. (1983) Studies on artificial diet for rearing larvae of *Chrysopa sinica* Tjeder. *Natural Enemies of Insects* 5, 82–85.

Canard, M. (1973) Influence de l'alimentation sur le développement, la fécondité et la fertilité d'un prédateur aphidiphage: *Chrysopa perla* (L.). PhD thesis, Université Paul-Sabatier, Toulouse.

Cava, F. & Sgobba, D. (1982) Prove di allevamento in ambiente condizionato di *Anisochrysa flavifrons* (Brauer) (Neuroptera, Chrysopidae). *Bollettino dell'Istituto di Entomologia dell'Università di Bologna* 36, 227–244.

Cohen, A.C. (1983) Improved method of encapsulating artificial diet for rearing predators of harmful insects. *Journal of Economic Entomology* 76, 957–959.

Cohen, A.C. (1992) Using a systematic approach to develop artificial diets for predators. In *Advances in Insect Rearing for Research and Pest Management*, ed. Anderson, T.E. & Leppla, N.C., pp. 77–91. Westview Press, Boulder CO.

Cohen, A.C. (1995) Extra-oral digestion in predaceous terrestrial Arthropoda. *Annual Review of Entomology* 40, 85–103.

Cohen, A.C. (1996) Technology transfer potential of artificial media for entomophages. *Information Bulletin of West Palaearctic Section International Organization for Biological Control of Noxious Animals and Plants* (WPRS IOBC) 19, 80.

Cohen, A.C. (1998) Application of the nutrient concentration concept to development of diet for mass-rearing. In *Proceedings of International Conference of Integrated Pest Management*, Guangzhou, 15–20 June 1998, p. 215.

Dougherty, E.C. (1959) Introduction to axenic culture of invertebrate metazoa: a goal. *Annals of the New York Academy of Sciences* 77, 27–54.

Ermicheva, F.M., Sumenkova, V.V. & Yazlovetsky, I.G. (1987) Localization of protease in guts of *Chrysopa carnea* larvae. *Izvestia Academy of Science Moldavian SSR, Seria of Biology and Chemistry Sciences* 4, 49–52. (in Russian)

Ferran, A., Bigler, F. & Lyon, J.-P. (1976) Étude des activités enzymatiques des glandes salivaires et des intestins de trois insectes prédateures de pucerons: *Chrysopa carnea* Steph., *Chrysopa perla* L.(*Neuroptères: Chrysopidae*) et *Semiadalia 11 notata* Sch. (*Coleoptères, Coccinellidae*). *Annales de Zoologie–Ecologie Animale* 8, 513–521.

Ferran, A., Lyon, J.-P., Larroque, M.-M. & Formento, A. (1981) Essai d'élevage de différents prédateurs aphidiphages (*Coccinellidae, Chrysopidae*) à l'aide de poudre lyophilissée de couvain de reines d'abeilles. *Agronomie* 1, 579–586.

Finney, G.L. (1948) Culturing *Chrysopa californica* and

obtaining eggs for field distribution. *Journal of Economic Entomology* 41, 719–21.

Finney, G.L. (1950) Mass-culturing *Chrysopa californica* to obtain eggs for field distribution. *Journal of Economic Entomology* 43, 97–100.

Gomez, A.N., Estrada, L.C. & Galan, R.B. (1982) *Dietas Artificiales en Insecta: Un Compendio de Referencias.* Instituto Nacional de Investigaciones Agrarias, Madrid.

Gomez, A.N., Estrada, L.C., Galan, R.B. & Gonzales, C.I. (1985) *Dietas Artificiales en Insecta. Un Compendio de Referencias (1980-1984).* Instituto Nacional de Investigaciones Agrarias, Madrid.

Grenier, S. (1986) Biologie et physiologie des relations hôtes–parasitoides chez 3 tachinaires (*Diptera: Tachinidae*) d'interêt agronomique. Développement en milieux artificiels. Lutte biologique. PhD thesis, l'Institut National des Sciences Appliquées de Lyon et l'Université Claude-Bernard, Lyon.

Grenier, S., Delobel, B. & Bonnot, G. (1986) Physiological considerations of importance to the success of *in vitro* culture: an overview. *Journal of Insect Physiology* 32, 403–408.

Hagen, K.S. (1987) Nutritional ecology of terrestrial insect: Predators. In *Nutritional Ecology of Insects, Mites, Spiders and Related Invertebrates* ed. Slansky, F. & Rodriguez, J.G., pp. 533–538. Wiley, New York.

Hagen, K.S. & Tassan, R.L. (1965) A method of providing artificial diets to *Chrysopa* larvae. *Journal of Economic Entomology* 58, 999–1000.

Hagen, K.S. & Tassan, R.L. (1966a) Artificial diet for *Chrysopa carnea* Stephens. In *Ecology of Aphidophagous Insects*, ed. Hodek, I., pp. 83–87. Academia, Prague.

Hagen, K.S. & Tassan R.L. (1966b) A method of coating droplets of artificial diets with paraffin for feeding *Chrysopa* larvae. In *Ecology of Aphidophagous Insects*, ed. Hodek, I., pp. 89–90. Academia, Prague.

Hasegawa, M., Saeki, Y. & Sato, Y. (1983) Artificial rearing of some beneficial insects on drone powder and the possibility of their application. *Honeybee Science* 4, 153–156.

Hasegawa, M., Niijima, K. & Matsuka, M. (1989) Rearing *Chrysoperla carnea* (*Neuroptera, Chrysopidae*) on chemically defined diet. *Japanese Journal of Applied Entomology and Zoology* 24, 96–102.

Hassan, S.A. & Hagen, K.S. (1978) A new artificial diet for rearing *Chrysopa carnea* larvae (Neuroptera, Chrysopidae). *Zeitschrift für angewandte Entomologie* 86, 315–320.

Hodek, I. (1978a) Causes de la non-adéquation des proies chez les Coccinelles aphidiphages. *Annales de Zoologie–Écologie Animale* 10, 423–432.

Hodek, I. (1978b) Specificité alimentaire des entomophages vis-à-vis de leur proie. *Annales de Zoologie–Écologie Animale* 10, 407–413.

Hodek, I. (1993) Habitat and food specificity in aphidophagous predators. *Biocontrol Science and Technology* 3, 91–100.

Hydorn, S.B. & Whitcomb, W.H. (1979) Effects of larval diet on *Chrysopa rufilabris*. *Florida Entomologist* 62, 293–300.

Kaplan, P.B., Nepomnyashaya, A.M. & Yazlovetsky, I.G. (1986) Fatty acid composition of the predatory larvae of aphidophagous insects and its dependence on the diet. *Entomologicheskoe obozrenie* 65, 262–268.

Kaplan, P.B., Keiser, L.S. & Yazlovetsky, I.G. (1989) The artificial diet for larvae of *Chrysopa septempunctata*. *Zoologichesky Zhurnal (Moscow)* 68, 118–122. (in Russian with English summary).

Kariluoto, K.T. (1980) Survival and fecundity of *Adalia bipunctata* L. (Coleoptera: Coccinellidae) and some other predatory insect species on an artificial diet and a natural prey. *Annales Entomologici Fennici* 46, 101–106.

Keiser, L.S. & Yazlovetsky, I.G. (1988) Comparative studies on sterols of *Chrysopa carnea* larvae and different types of their food. *Journal of Evolutionary Biochemistry and Physiology (Leningrad)* 24, 157–164.

Keiser, L.S. & Yazlovetsky, I.G. (1989) The influence of sterols of artificial nutritional media on the development of *Chrysopa carnea* larvae. *Agricultural Biology (Moscow)* 1, 119–123 (in Russian with English summary)

Keiser, L.S. & Yazlovetsky, I.G. (1991) The content of total $\Delta^{5,7}$-sterols and ecdisteroids at various stages of development of the aphis lion *Chrysopa carnea*. *Journal of Evolutionary Biochemistry and Physiology (Leningrad)* 27, 417–421.

Keiser, L.S., Yazlovetsky, I.G., Nepomnyashaya, A.M., Abashkin, A.S. & Zernow, V.S. (1986) Composition for microencapsulation of artificial diets. Authors certificate of the USSR N1249725. *Bulletin of Inventions of USSR* 29, 265.

Keiser, L.S., Yazlovetsky, I.G. & Ageeva, L.I. (1991) The method of diet preparing for *Chrysopa carnea* Steph. larvae. Authors certificate of the USSR N1664246. *Bulletin of Inventions of USSR* 27, 21.

King, E.G. & Coleman, R.J. (1989) Potential for biological control of *Heliothis* species. *Annual Review of Entomology* 34, 53–75.

Krishnamoorthy, A. & Nagarkatti, S. (1983) A mass-rearing technique for *Chrysopa scelestes* Banks (Neuroptera, Chrysopidae). *Journal of Entomology Research* 5, 93–98.

Letardi, A. & Caffarelly, V. (1989) Impredo di una dieta semi-artificiale allo stato liquido per l'allevamento di larve di *Chrysoperla carnea* (Stephens) (Planipennia, Chrysopidae). *Redia* 72, 195–203.

Ma, A., Zhang, X. & Zhao, J. (1986) A machine for making encapsulated diet for rearing *Chrysopa* spp. *Chinese Journal of Biological Control* 2, 145–147.

Martin, P.B., Ridgway, R.L. & Schuetze, C.E. (1978) Physical and biological evalutions of an encapsulations diet for rearing *Chrysopa carnea. Florida Entomologist* 61, 145–152.

Matsuka, M. & Niijima, K. (1985) Harmonia axyridis. In *Handbook of Insect Rearing,* vol. 1, ed. Singh, P. & Moore, R.F., pp. 265–268. Elsevier, Amsterdam.

Mellini, E. (1975) Possibilità de allevamento de insetti entomophagi parassiti su diete artificiali. *Bollettino dell'Istituto di Entomologia dell'Università di Bologna* 32, 257–290.

Moontyan, E.M. & Yazlovetsky, I.G. (1988) Digestive carbohydrases of *Chrysopa carnea* larvae. *Izvestia Academy of Science Moldavian SSR, Seria of Biology and Chemistry Sciences* 3, 56–59. (in Russian)

Moontyan, E.M. & Yazlovetsky, I.G. (1992) Comparative studies on digestive carbohydrases of lacewings. *Journal of Evolutionary Biochemistry and Physiology (Leningrad)* 28, 563–539. (in Russian with English summary)

Moontyan, E.M. & Yazlovetsky, I.G. (1993) Characterization of the midgut amylase in larvae of *Chrysopa carnea. Journal of Evolutionary Biochemistry and Physiology (Leningrad)* 29, 236–243. (in Russian with English summary)

Moontyan, E.M. & Yazlovetsky, I.G. (1994) Digestive carbohydrases in larval predaceous bugs *Podisus maculiventris* and *Perillus bioculatus. Journal of Evolutionary Biochemistry and Physiology (Leningrad)* 30, 161–167. (in Russian with English summary).

Nepomnyashaya, A.M., Mencher, E.M. & Yazlovetsky, I.G. (1979) A new approach to the development of artificial nutritive diets for mass-rearing of beneficial insects. Optimization of nutritive diets by simplex lattice method. In *Biochemistry and Physiology of Insects,* ed. Popushoi, I.S., pp. 29–36. Shtiintsa Publishing, Kishinev. (in Russian with English summary)

New, T.R. (1975) The biology of *Chrysopidae* and *Hemerobidae* (Neuroptera), with reference to their usage as biocontrol agents: a review. *Transactions of the Royal Entomological Society of London* 127, 115–140.

New, T.R. (1988) *Neuroptera.* In *Aphids, Their Biology, Natural Enemies and Control,* vol. 2, ed. Minks, A.K. & Harrewijin, P., pp. 249–258. Elsevier, Amsterdam.

Niijima, K. (1989) Nutritional studies on an aphidophagous chrysopid, *Chrysopa septempunctata* Wesmael. 1. Chemically-defined diets and general nutritional requirements. *Bulletin of the Faculty of Agriculture, Tamagawa University* 29, 22–30.

Niijima, K. & Matsuka, M. (1989) Artificial diets for mass-production of chrysopids (Neuroptera). In *Proceedings of FFTC-NARC International Seminar on Agricultural Research Centre, 'The Use of Parasitoids and Predators to Control Agricultural Pests',* pp. 1–15. Tukuba Science City, Ibaraki-ken.

Nordlund, D.A. & Morrison, R.K. (1992) Mass rearing of *Chrysoperla* species. In *Advances in Insect Rearing for Research and Pest Management,* ed. Anderson, T.E. & Leppla, N.C., pp. 427–439. Westview Press, Boulder CO.

Okada, I. & Niijima, K. (1971) Artificial rearing of lacewing, *Chrysopa septempunctata* Wesmael, with special reference to a new diet using drone honey bee brood. *Heredity* 25, 41–44.

Okada, I., Matsuka, M. & Tani, M. (1974) Rearing a green lacewing *Chrysopa septempunctata* Wesmael on pulverized drone honey bee brood. *Bulletin of the Faculty of Agriculture, Tamagawa University* 14, 26–32.

Ponomareva, I.A. (1971) Artificial diet for the larvae of *Chrysopa carnea* Steph. In *Proceedings of Conference of Biological Pest Control on Vegetable and Fruit Crops,* ed. Sikura, A.I. *et al.,* pp. 77–79. All-Union Research Institute of Biological Methods for Plant Protection, Kishinev. (in Russian)

Ponomareva, I.A. & Beglyarov, G.A. (1973) Development of artificial diets for rearing of green lacewing, *Chrysopa carnea* Steph. In *Problems of Plant Protection (1969–1971),* vol. 2, ed. Shapa, V.A., pp. 67–68. Shtiintsa Publishing, Kishinev. (in Russian with English summary)

Principi, M.M. (1983) I Neurotteri Crisopidi e la possibilità della loro utilizzazione in lotta biologica e in lotta integrata. *Bollettino dell'Istituto di Entomologia dell'Università di Bologna* 38, 231–262.

Santos, C.D. & Terra, W.R. (1986) Distribution and characterization of oligomeric digestive enzymes from *Erinnyis ello* larvae and inferences concerning secretory mechanisms and the permeability of the peritrophic membrane. *Insect Biochemistry* 16, 691–700.

Sekirov, I.A., Yazlovetsky, I.G. & Dobrikov, M.I. (1990) Utilization of dietary nucleic acids and some of their precursors by larval aphis lion *Chrysopa carnea. Journal of Evolutionary Biochemistry and Physiology (Leningrad)* 26, 35–40 (in Russian with English summary)

Singh, P. (1977) *Artificial Diets for Insects, Mites and Spiders.* Plenum Press, New York.

Singh, P. (1982) The rearing of beneficial insects. *New Zealand Entomologist* 7, 304–310.

Sogoyan, L.N. & Lyashova, L.V. (1971) Artificial diets for some predator insects. In *Proceedings of Conference of Biological Pest Control on Vegetable and Fruit Crops*, ed. Sikura, A.I. *et al.*, pp. 92–93. All–Union Research Institute of Biological Methods for Plant Protection, Kishinev. (in Russian)

Spiegler, P.E. (1962) The origin and nature of the adhesive substance of larvae of the genus *Chrysopa* (Neuroptera, Chrysopidae). *Annals of the Entomological Society of America* 55, 69–77.

Sumenkova, V.V. & Yazlovetsky, I.G. (1994) Digestive proteases from predaceous bugs *Podisus maculiventris* and *Perillus bioculatus*. *Journal of Evolutionary Biochemistry and Physiology (Leningrad)* 30, 632–642. (in Russian with English summary)

Sumenkova, V.V., Ermicheva, F.M. & Yazlovetsky, I.G. (1989) Characterization of gut proteases from aphis lions with different alimentary patterns. *Journal of Evolutionary Biochemistry and Physiology (Leningrad)* 25, 436–442. (in Russian with English summary)

Tartarini, E. (1983). Influenza di differenti metodi di allevamento larvale sullo sviluppo e sulla fecondità di *Chrysoperla carnea* (Stephens) (Neuroptera, Chrysopidae). *Bollettino dell'Istituto di Entomologia dell'Università di Bologna* 38, 1–24.

Terra, W.R. (1990) Evolution of digestive systems of insects. *Annual Review of Entomology* 35, 181–200.

Terra, W.R. & Ferreira, C. (1981) The physiological role of the peritrophic membrane and trehalase: digestive enzymes in the midgut and excreta of starved *Rhynchosciara*. *Journal of Insect Physiology* 27, 325–331.

Thompson, S.N. (1986) Nutrition and *in vitro* culture of insect parasitoids. *Annual Review of Entomology* 31, 197–219.

Vanderzant, E.S. (1969) An artificial diet for larvae and adults of *Chrysopa carnea*, an insect predator of crop pests (Neuroptera, Chrysopidae). *Journal of Economic Entomology* 62, 256–257.

Vanderzant, E.S. (1973) Improvements in the rearing diet for *Chrysopa carnea* and the amino acid requirements for growth. *Journal of Economic Entomology* 66, 336–8.

Waage, J.K., Carl, K.P., Mills, N.J. & Greathead, D.J. (1985) Rearing entomophagous insects. In *Handbook of Insect Rearing*, vol. 1, ed. Singh, P. & Moore, R.F., pp. 45–66. Elsevier, Amsterdam.

Wuhrer, B. & Hassan, S. (1990) Möglicher Einzatz von Kunstfutter in Massenzuchten von Prädatoren am Beispiel von *Chrysoperla carnea* and *Coccinella septempunctata*. In *Mitteilungen aus der Biologischen Bundesanstalt fur Land-und Forstwirtschaft, Berlin–Dahlem*, number 266, p. 312. Parey Buchverlag, Berlin.

Yazlovetsky, I.G. (1986) Current tendencies in the development of artificial nutritional media for mass-rearing of insects. *Information Bulletin of East Palaearctic Section International Organization for Biological Control of Noxious Animals and Plants* (EPS IOBC) 15, 21–42. (in Russian with English summary)

Yazlovetsky, I.G. (1992) Development of artificial diets for entomophagous insects by understanding their nutrition and digestion. In *Advances in Insect Rearing for Research and Pest Management*, ed. Anderson, T.E. & Leppla, N.C., pp. 41–62. Westview Press, Boulder, CO.

Yazlovetsky, I.G. & Abashkin, A.S. (1977) Microencapsulated artificial diet for *Chrysopa carnea* larvae. *Zaschita Rastenii*, 3, 32. (in Russian)

Yazlovetsky, I.G. & Kaplan, P.B. (1995) Comparative studies on lipid nutrition in larvae of predaceous bugs *Podisus maculiventris* and *Perillus bioculatus*. *Journal of Evolutionary Biochemistry and Physiology (Leningrad)* 31, 21–28. (in Russian, Summary in English)

Yazlovetsky, I.G. & Nepomnyashaya, A.M. (1977) The liquid diet for mass-rearing of *Chrysopa* larvae. In *Entomophages in Plant Protection*, ed. Popushoi, I.S., pp. 72–75. Shtiintsa Publishing, Kishinev. (in Russian with English summary)

Yazlovetskij, I.G. & Nepomnyashaya, A.M. (1981) Essai d'élevage massif de *Chrysopa carnea* sur milieux artificiels micro-encapsulés et de son application dans la lutte contre les pucerons en serre. In *Proceedings of Colloque Franco-soviétique 'Lutte Biologique et Intégrée contre les Pucerons'*, pp. 51–58. Institut National de la Recherche Agronomique, Paris.

Yazlovetsky I.G., Mencher, E.M., Nepomnyashaya, A.M. & Sumenkova, V.V. (1979*a*) New approach to elaboration of artificial nutritive diets for mass-rearing of entomophagous insects. *Izvestia Academy of Science Moldavian SSR, Seria of Biology and Chemistry Sciences* 1, 55–63. (in Russian)

Yazlovetsky, I.G., Nepomnyashaya, A.M., Sumenkova, V.V. & Mencher, E.M. (1979*b*) Artificial diet for mass-rearing of predatory insects. Authors certificate of the USSR N 688161. *Bulletin of Inventions of USSR* 36, 6.

Yazlovetsky, I.G., Sumenkova, V.V. & Nepomnyashaya, A.M. (1979*c*) Biochemical investigation of nutritional requirements of aphidophagous insects. General characterization and amino acid composition of some aphid species and the Angoumois grain moth eggs. In *Biochemistry and Physiology of Insects*, ed. Popushoi, I.S., pp. 19–28. Shtiintsa Publishing, Kishinev. (in Russian with English summary).

Yazlovetsky, I.G., Abashkin, A.S. & Keiser, L.S. (1990) Mass rearing of *Chrysopa carnea* Steph. *Zaschita Rastenii*, 1, 27–28. (in Russian)

Yazlovetsky, I.G., Ageeva, L.I. & Keiser, L.S. (1992) Artificial diet for five species of Chrysopids. *Zoologichesky Zhurnal (Moscow)* 71, 123–129. (in Russian with English summary)

Yazlovetsky, I.G., Lupu, E.I., Kaplan, P.B. & Aimert, K.M. (1993) Properties of intestinal lipases and new data on nutrition mechanism of larval aphis lion *Chrysopa carnea*. *Journal of Evolutionary Biochemistry and Physiology (Leningrad)*, 29, 139–145. (in Russian with English summary)

Yazlovetsky, I.G., Sumenkova, V.V. & Moontyan, E.M. (1996) Predaceous pentatomids on artificial nutritional diets: enzymological aspects. In: *Proceedings of the 20th*

International Congress of Entomology, Firenze, 25–31 August 1996, p. 614.

Ye, Z.-C., Han, Y.-M., Wang, D.-G., Liang, S.-Z. & Li, S.-Y. (1979) Studies on the artificial diets for larvae and adults of *Chrysopa sinica* Tjeder. *Acta Phytophylacica Sinica* 6, 11–16.

Zaki, F.N. & Gestrana, M. (1990) Rearing the aphid lion, *Chrysopa carnea* Steph. on different preparations of diet based on algae. In *Abstracts Volume of the 4th Conference on Aphidophaga,* Godollo, 3–7 September 1990, p. 68.

Zhang, X. & Zhou, H. (1998) Mass rearing green lacewing in China. In *Proceedings of International Conference of Integrated Pest Management,* Guangzhou, 15–20 June, 1998, p. 225.

CHAPTER 14

Ecological studies of released lacewings in crops

K.M. Daane

14.1 INTRODUCTION

The effectiveness of augmentative releases of lacewings in the field has varied greatly, with experimental studies reporting from 0% to 100% pest reduction (for reviews, see Ridgway & Murphy, 1984; King *et al.*, 1985; Daane *et al.*, 1998). Numerous factors can influence lacewing release effectiveness, and while some can be controlled (e.g. release rate), the success or failure of many release programmes is contingent upon the ecological interactions at the release site. For this reason, to realise fully the commercial potential of lacewings, the ecological interactions among lacewings, their prey, and the release environment must be better understood. Crucial areas of ecological study discussed here include: (1) the influence of environmental or host plant conditions on lacewing performance, (2) the role of other natural enemies on released lacewing density, (3) the effect of pesticide residues in the release environment, and (4) the importance of predator–prey interactions in the field. Given the above concerns, research has recently been conducted that helps to categorise the ecological relationship of released lacewings with respect to target pest(s) and the biotic or abiotic factors in the release environment.

14.2 HABITAT INFLUENCE ON RELEASE EFFECTIVENESS

The impact of predators can vary between habitats, as described by Hodek (1993) for aphidophagous predators, and many predators may be rendered ineffective without leaving the release site. Not surprisingly, different species of lacewings appear to be associated with certain habitats, such as conifers or broad-leafed trees (New, 1975), and lacewings released into less desirable habitats may have reduced effectiveness against an otherwise acceptable prey species. Nevertheless, only two lacewing species [*Chrysoperla carnea* (Stephens) *s. lat.* and *C. rufilabris* (Burmeister)] are primarily used in augmentation programmes. Briefly reviewed here are habitat factors that can influence lacewing performance: climate, plant structure, and ground cover.

Climatic conditions

Lacewings are released into environments with widely varying temperatures and humidities, most often depending on geographic location, seasonal release period, and agroecosystem. Lacewing release sites range from cotton fields in Texas, to apple orchards in Washington, to greenhouses in Europe. Often, such differences in environmental conditions are not fully considered in the development of release tactics. For example, under relatively humid conditions (75% RH), *C. carnea* and *C. rufilabris* are similar in most life-history traits (Tauber & Tauber, 1983). However, under low to moderate humidity (35%–55% RH), *C. rufilabris* has a prolonged preoviposition period, reduced fecundity, increased preimaginal mortality, and a slower developmental rate. Consequently, while *C. rufilabris* is widespread in the USA, it is virtually absent from the dry, southwestern states. Given the above, inoculative release of *C. rufilabris* into dry areas would be unlikely to persist in subsequent generations.

Temperature-induced mortality of *C. carnea* can occur in hot, dry areas. Daane & Yokota (1997) evaluated release of *C. carnea* eggs in vineyards in California's Central Valley, where ambient summer temperatures are commonly between 34°C and 40°C; their field studies found extensive egg mortality. Laboratory experiments showed that temperatures >37°C resulted in mortality of insectary-reared *C. carnea* spp. eggs, which are typically produced at temperatures closer to 27°C, and may have been a contributing factor in the field mortality of the *C. carnea* eggs. Differences in temperature and humidity tolerance become all the more important given the recent advancements in *Chrysoperla* taxonomy and biology. The *C. carnea* species-complex in the USA is recognized by distinct entities, e.g. *C. carnea*, *C. plorabunda* (Fitch), and *C. downesi* (Banks). Separate populations within the *C. carnea* species-complex have diverse and variable seasonal cycles, patterns of habitat choice and seasonal movement, polymorphisms (Tauber & Tauber, 1986) and courtship songs (e.g. Henry *et al.*, 1993). A call here is made for a better understanding of the biology of each of these populations, biotypes, or species in order to better match the mass-reared natural enemy with the climatic conditions common in the more basic geographic regions or crop environments (e.g. hot and dry climates, glasshouse systems).

Habitat-lacewing interactions

Once viable lacewing larvae have been successfully released onto or near the targeted pest, they must immediately begin to search for food. Naturally, the success of the lacewing larvae is influenced by their ability to forage upon the pest's host plant or in the pest habitat and successfully find and capture the prey. The ecological interactions between predators and prey habitat can be so disruptive that variations in plant morphology between closely related plant species or even cultivars can affect predator performance.

One of the best-known examples is the abundance or type of trichomes on the leaf surface, which has been shown to affect the searching ability of many species (e.g. Obrycki, 1986; Messina *et al.*, 1995). *Chrysoperla rufilabris*, for example, is an important predator of boll-worms in southeastern USA (Lingren *et al.*, 1968), and has been used in augmentative release programmes against this pest. However, on some cotton cultivars, leaf trichomes act as a mechanical barrier to *C. rufilabris* larvae, reducing their mobility as compared with smooth-leaf cultivars (Treacy *et al.*, 1987). Similarly, in field and glasshouse experiments, Eigenbrode *et al.* (1995, 1996) found that *C. carnea* significantly reduced diamondback moth, *Plutella xylostella* L., densities on glossy-leaf cabbage but did not produce similar results on normal-wax plants. A closer examination showed lacewings spent less time searching for prey and more time in other activities, including grooming and scrambling, on standard cabbage than on the glossy-leaf cultivar. Scanning electron micrographs showed plant debris, probably wax, from the standard cabbage impeded *C. carnea* larval mobility, thereby reducing its effectiveness against diamondback moth larvae on normal-wax versus glossy-leaf cabbage cultivars.

Habitat–herbivore interactions

Another way the targeted plant species can disrupt lacewing release is by influencing the distribution of herbivores, which is likely to influence the ability of lacewings to locate and capture pests. Messina *et al.* (1997) and Clark & Messina (1998) quantified the behaviour of *Chrysoperla plorabunda*, a species closely related to *C. carnea*, on two perennial grasses (crested wheatgrass and Indian rice-grass) with different leaf structures. In the absence of prey, lacewing larval residence times were similar on the two grasses. On plants infested with aphids, lacewing larvae captured significantly more aphids on Indian rice-grass than on crested wheatgrass. Comparisons between aphid-free and aphid-infested plants suggest that differences in plant architecture, rather than predator movement, modified prey accessibility. Aphids on seedlings and mature plants of crested wheatgrass frequently occurred in concealed locations; aphids on Indian rice-grass were more likely to feed in exposed locations.

On individual plants, both prey and predator species have preferences for different plant structures (e.g. leaves, flowers), which may influence release timing, depending on the distribution of predators. Atakan *et al.* (1996) studied predator distribution on two cotton cultivars (cv. Pima and Acala) and showed *C. carnea* were found on the leaves of both varieties, whereas predatory *Orius* spp. colonized mainly the flowering structures on Pima plants but the leaves on Acala. This variability in prey and predator distribution resulted in a significant and positive correlation between thrips and predator populations on Pima but not on Acala plants, and suggests that *C. carnea*, in some instances, is the preferred release species. These examples show the influence of tritrophic interactions among the lacewing, targeted herbivore, and host plant, and illustrate the necessity of matching the released predator's biological characteristics not only with the climatic conditions and prey species (see below), but also with the crop.

The between-plant spatial distribution of lacewing prey (or prey host plant) can also influence release success. Heinz (1998) used matrices of potted plants within greenhouses to generate spatially subdivided habitats and test the spread of arthropod pests and the abilities of natural enemies to locate them. *Chrysoperla rufilabris* larvae did not disperse as well as adult parasitoids (*Aphidius colmani* Viereck). This resulted in natural-enemy displacement distances and diffusion constants that were significantly lower for released lacewings, which located 49% of the aphid-infested pots, than released parasitoids, which located 97% of aphid-infested pots. Similarly, Breene *et al.* (1992) showed lacewing releases used to suppress the sweet-potato whitefly, *Bemisia tabaci* (Gennadius), on *Hibiscus* were more effective when host plants were in contact. In this instance, the plant-to-plant contact provided bridges for released lacewings to disperse from plants with low to high prey density. In essence, plant spacing improved the predators' search capacity and prey capture efficiency.

Ground covers and polycultures

A diverse plant community can influence arthropod natural-enemy populations by providing critical food or habitat resources that might not be found in a simple plant community (Andow, 1991; Bugg & Waddington, 1994). In agroecosystems, the plant community can be manipulated through the addition of cover crops or allowing resident (weedy) vegetation to grow. This increased diversity in the plant community sometimes results in an increase in the numbers of entomophagous insects and a decrease in the numbers of herbivores (reviewed by Altieri & Letourneau, 1984; Andow, 1991). Some IPM practitioners have suggested that lacewing releases in perennial crops are improved if ground covers are present because the alternative prey in ground covers may help to establish a larger, more permanent lacewing population (e.g. Altieri & Schmidt, 1985). Here, a physical aspect of the release ecosystem may affect not only the microclimatic conditions of the release site (e.g. temperature, humidity, increased plant biomass) but also has the potential to influence a biotic component (e.g. predaceous and/or phytophagous arthropods) of the ecosystem as well.

Nevertheless, there are few studies that clearly indicate that ground covers have a positive influence on lacewing release effectiveness or lacewing density in the perennial crop above. Costello & Daane (1998*a*, *b*) reported little or no difference in the number of predaceous arthropods collected on vines with or without ground covers, although lacewing density in the ground-cover was often high (Daane *et al.*, 1994). Smith *et al.* (1996) surveyed arthropods in pecan orchards with different types of ground-cover vegetation. They found that ground cover had little effect on the density or type of arthropods present in the pecan canopy, except that densities of *Chrysoperla rufilabris* were greater during July in pecans with a legume ground cover. Ground cover may, in some instances, reduce release effectiveness. For example, *C. carnea* prefers low-lying vegetation and may move from the perennial crop to feed on non-targeted hosts in the ground cover. Similar results were reported in polycultures, or agroecosystems with two or more cash crops. Schultz (1988) made counts of green lacewing eggs on cotton intercropped with corn, beans, or weeds, and in cotton monocultures and found significantly fewer eggs on cotton plants in polycultures, compared with monocultures. Schultz suggested mechanisms to explain the reduction in egg numbers in the intercrops, including a decrease in the numbers of prey (aphids) and the presence of other predators (earwigs).

14.3 NATURAL-ENEMY INTERACTIONS

Predators

Some combinations of natural-enemy species released together have been shown to work effectively. For example, laboratory experiments were conducted to quantify the behaviour and effectiveness of two predators (*Chrysoperla plorabunda* and the seven-spotted lady beetle, *Coccinella septempunctata* L.), used either singly or in combination, to control bean aphids, *Aphis fabae* Scopoli (Chang, 1996). In that study, releases of both lacewing and beetle larvae slowed the growth of aphid populations and there was no evidence of a negative intraguild interaction between these predator species. The lack of intraguild competition may be explained by the finding that *Chrysoperla plorabunda* and *Coccinella septempunctata* larvae differed significantly in their search distribution on the plant. However, not all interactions among natural enemies enhance release effectiveness and, for that reason, the natural-enemy complex at the release site must be considered as an ecological component of the release environment. In fact, earlier laboratory work with *Chrysoperla carnea* and *Coccinella septempunctata* showed interference and competitive behaviour between these generalist predators (Sengonca & Frings, 1985). Indeed, studies in cotton fields of interspecies association of common predators (*Nabis* spp., *Geocoris* spp., *Orius* spp., *Hippodamia* sp., *Chrysoperla* sp.) and herbivores found predators were associated with various primary consumers 163 times and with various other predators 191 times, suggesting that predators may feed extensively on other predators in natural ecosystems (Ellington *et al.*, 1997). Therefore, one critical ecological consideration in open-field releases is that the augmented biological control agents may become prey themselves.

It has long been established that other predators may kill lacewings in natural situations. In Texas cotton fields, the green lynx spider, *Peucetia viridans* (Hentz), was observed feeding on prey species in eight different insect orders; however, predaceous arthropods (e.g., *Hippodamia convergens* Guérin-Méneville, *C. rufilabris*) constituted more than half of the spiders' diet (Nyffeler *et al.*, 1987). In soya beans, releases of 20, 60, or 120 *C. carnea* eggs in cages where *Helicoverpa zea* (Boddie) and *Trichoplusia ni* (Hübner) eggs were placed did not reduce defoliative damage; Barry *et al.* (1974) suggested that predation of lacewing eggs by *Orius insidiosus* (Say) that penetrated the cages explained the ineffectiveness of *C. carnea* in their tests. Ants have often been cited as predators of lacewing eggs (e.g. Vinson & Scarborough, 1989; Morris *et al.*, 1998). In Georgian pecan orchards, the red imported fire ant, *Solenopsis invicta* Buren, was a major predator of eggs, larvae, and pupae of *C. rufilabris*, and of the pupae of *Allograpta obliqua* (Say), but had little effect on the eggs of *Hippodamia convergens* (Tedders *et al.*, 1990). It is not surprising that such predator–predator interactions would be repeated when lacewings are used in augmentative release programmes. Dreistadt *et al.* (1986) tested releases of *C. carnea* eggs on tulip trees to control the aphid *Illinoia liriodendri* (Monell) and found that 98% of the eggs were removed by the Argentine ant, *Linepithema humile* (Mayr).

Recent studies of predator–predator or intraguild predation have stressed ecological relationships among resident predators and lacewing larvae. For example, studies found that releases of *C. carnea* had little effect on densities of cotton aphid, *Aphis gossypii* Glover, because of intraguild predation on lacewing larvae by predator species commonly found in cotton (nabid, reduviid, and geocorid bugs) (Rosenheim & Wilhoit, 1993). Further studies showed reduced effectiveness of lacewings (as aphid predators) when other predators were present (Rosenheim *et al.*, 1993). In most of these ecologically based, intraguild studies, lacewing eggs and larvae fare poorly against other predators, and the suppression of targeted herbivores is reduced. In many

studies, lacewing larvae were not observed killing other predaceous insects or spiders, but were often prey. For example, Dinter (1998) found that erigonid spiders switched between herbivores and predators and caused nearly 100% mortality of lacewing larvae in the absence of alternative prey while lacewing larvae did not kill spiders. In fact, in one study, *C. carnea* larvae were considered, along with the cotton aphid, as prey for the generalist predator *Zelus renardii* Kolenati (Cisneros & Rosenheim, 1997). While both the *Z. renardii* and *C. carnea* fed on aphids, even the youngest instars of *Z. renardii* were shown to cause substantial lacewing mortality and reduce effective aphid suppression by released lacewings. Phoofolo & Obrycki (1998) studied intraguild relationships between larvae of two coccinellids [*Coleomegilla maculata* (DeGeer) and *Harmonia axyridis* Pallas] and *Chrysoperla carnea*. They showed that while lacewings could develop on coccinellid eggs and, conversely, coccinellids could develop on lacewing eggs, the resulting adults were smaller compared with those reared on phytophagous insects [e.g. pea aphids, *Acyrthosiphon pisum* (Harris)]. In some combinations of intraguild food supply (e.g. *C. carnea* larvae fed *H. axyridis* eggs) preimaginal development was not completed. And, of course, it has long been known that lacewing larvae can be cannibalistic (Canard & Duelli, 1984; Phoofolo & Obrycki, 1998), which can interfere with placement of viable lacewings in the field (Daane & Yokota, 1997).

Lacewings do not, however, always lose in intraguild competition. Paired with different larval stages of either the lady beetle *Coleomegilla maculata lengi* Timberlake or the predaceous midge *Aphidoletes aphidimyza* (Rondani), *Chrysoperla rufilabris* was generally able to capture and kill the other aphid predators, although development stages of the predators played a key role in these confrontations (Lucas *et al.*, 1997, 1998; Lucas, 1998). Similarly, in laboratory studies, third-instar *C. carnea* killed and preyed on fourth-instar *Coleomegilla maculata* when confined with or without prey (pea aphids) (Phoofolo & Obrycki, 1998). Lacewings have also been observed feeding on parasitised stages of the obscure mealybug, *Pseudococcus viburni* (Signoret), navel orangeworm, *Amyelois transitella* (Walker), and cotton aphid (K.M. Daane, personal observations) and Stark & Hopper (1988) quantified the amount of *Chrysoperla carnea* predation on tobacco budworm, *Helicoverpa virescens* (Fabricius), larvae parasitised by *Microplitis croceipes* (Cresson). Therefore, lacewings can disrupt activity of parasitoids attacking the targeted pest species. Rosenheim *et al.* (1995) provide a more complete discussion of intraguild predation.

Other predator–predator interactions can include chemical cues, such as oviposition-deterring allomones. Ružička (1998) found that lacewing adults (*C. carnea* and *Chrysopa commata* Kis & Ujhelyi, *C. oculata* Say, and *C. perla* L.) laid fewer eggs on substrates that had been exposed previously to conspecific or heterospecific first-instar larvae. Similarly, Ružička & Havelka (1998) found that *Aphidoletes aphidimyza* (Rondani) females also laid fewer eggs on aphid-infested plants that were previously exposed to unfed first-instar larvae of *Chrysopa* and *Chrysoperla* species, or second-instar larvae of *Coccinella septempunctata*. These results suggest that responses to oviposition-deterring allomones, which are often considered important in producing a more even egg distribution, might also produce predator–predator interference and, therefore, create an ecological disturbance at the release site.

Parasitoids
Another biotic component of the release environment is species abundance and composition of lacewing parasitoids. *Chrysoperla* are attacked by a number of parasitoids (Lyon, 1979; Gerling & Bar, 1985; Ruberson *et al.*, 1989, 1995). In some locations and years, the effect of parasitism can be quite significant, with highest parasitism levels typically resulting from pupal parasitoids. The effect of resident parasitoids at the release site has not often been considered a component of the release success, and yet there is some evidence that lacewing releases can actually increase lacewing parasitism levels by providing hosts at critical periods during the season. For example, in one of two trials, Ehler *et al.* (1997) found that scelionids (*Telenomus chrysopae* Ashmead and *T. tridentatus* Johnson & Bin) parasitised considerably more eggs in release plots (~30%) than in no-release plots (~2%). Similarly, in pecan orchards that had repeated releases of *C. carnea* eggs throughout the season, there was a considerable increase in *Chrysoperla* parasitism rates, such that overall lacewing densities (released and resident populations) were lower in release than non-release fields (W.L. Tedders, personal communication).

14.4 LACEWINGS AND PESTICIDE USE IN THE RELEASE ENVIRONMENT

Pesticides
Pesticides constitute an important component of many agroecosystems, and just as other natural enemies can influence lacewing survivorship, some pesticides can alter the release environment to affect natural-enemy survivorship. Here, *Chrysoperla carnea* may have a clear advantage over other released or resident natural enemies. Numerous researchers have shown *C. carnea* has a relatively broad tolerance to many insecticides, particularly during the larval stage (Grafton-Cardwell & Hoy, 1985; Singh & Varma, 1986; Pree *et al.*, 1989; Mizell & Schiffhauer, 1990). For example, comparisons of insecticides used to suppress the Colorado potato beetle, *Leptinotarsa decemlineata* (Say), showed considerable differences in mortality of two common predators *Coleomegilla maculata* DeGeer and *Chrysoperla carnea* (Hamilton & Lashomb, 1997). Therefore, selection of pesticide use at the release site can influence release effectiveness by reducing performance or mortality, and/or by reducing intraguild predation at the release site.

While *C. carnea* has been shown to have greater tolerance to many pesticides, a blanket statement of lacewing tolerance to insecticide residues may not be appropriate. First, insecticide tolerance varies geographically among *C. carnea* populations; e.g., *C. carnea* from areas in California with a history of heavy pesticide usage are more tolerant to insecticides than are populations from areas with less insecticide usage (Grafton-Cardwell & Hoy, 1985). Second, there are differences between closely related chrysopid genera (e.g. Rumpf *et al.*, 1997) and even within genera. For example, *C. rufilabris* generally displays lower insecticide tolerance than does *C. carnea* (Lawrence *et al.*, 1973; Lawrence, 1974) and *C. externa* (Banks) has a limited response to insecticides (Beije, 1993; Albuquerque *et al.*, this volume). Insectary managers should take these differences into consideration in developing stock material.

In the past decade, there has been increased use of 'softer' insecticides that have lower toxicity to non-target animals. Nevertheless, these soft insecticides also have the potential to cause mortality or reduce effectiveness of beneficial insects. Sewify & El Arnaouty (1998) tested the fungus *Verticillium lecanii* (Zimm.) against *C. carnea* larvae under high relative humidities and found that the fungus was highly pathogenic to third-instar larvae, impaired their feeding and searching capacity, and decreased emergence of adults. Feeding of the larvae with infected aphids had similar effects, decreasing lacewing fecundity. The results show the necessity to determine the timing of field releases of both entomopathogenic fungi and entomophagous predators and the need for host-selective isolates of *V. lecanii*. Another product, *Bacillus thuringiensis* (Berliner), has been shown to work well in combination with beneficial arthropods, although recent studies have demonstrated that *B. thuringiensis* Cry1Ab toxin is toxic to *C. carnea* when delivered in an artificial diet (Hilbeck *et al.*, 1998*b*).

Transgenic *Bacillus thuringiensis*

The effect of transgenic plant material on beneficial insects has become an important and timely concern. In one laboratory study, the effects of feeding corn pollen expressing a *Bacillus thuringiensis* subsp. *kurstaki* Cry1Ab protein were determined on three predatory species, including *Chrysoperla carnea*. In this work, no acute detrimental effects of the transgenic *B. thuringiensis* pollen on preimaginal development and survival were found (Pilcher *et al.*, 1997). Further, in two years of field study, Pilcher *et al.* (1997) observed no detrimental effects on lacewing abundance, suggesting that *B. thuringiensis* corn pollen will not affect natural-enemy movement in corn. In contrast, a recent study used transgenic *B. thuringiensis* var. *kurstaki* in corn plants to study the effect of *B. thuringiensis*-fed herbivores [*Ostrinia nubilalis* (Hübner) and *Spodoptera littoralis* (Boisduval)] on *C. carnea* (Hilbeck *et al.*, 1998*a*). Results found mean total immature mortality for chrysopid larvae raised on *B. thuringiensis*-fed prey was 62% compared with 37% when raised on *B. thuringiensis*-free prey. The results suggest that the reduced fitness of chrysopid larvae was associated with *B. thuringiensis*. For example, there was a prolonged development time of chrysopid larvae raised on *B. thuringiensis*-fed *O. nubilalis*, which was probably caused by a combined effect of *B. thuringiensis* exposure and nutritional deficiency caused by sick prey.

14.5 PREDATOR–PREY RELATIONSHIPS AT THE RELEASE SITE

The common prey species (e.g. Balduf, 1939; Elkarmi *et al.*, 1987) and larval and adult nutritional requirements (e.g. Tauber *et al.*, 1974; Canard *et al.*, 1984; Zheng *et al.*, 1993a, b) have been described for *Chrysoperla carnea* and *C. rufilabris*. However, while the pest species targeted in a release programme may be acceptable prey and meet minimum nutritional requirements under laboratory conditions, differences among prey species can have a profound effect on the ecological interactions in the field. For example, lacewing larvae must consume many prey items over an extended period of time to develop to the third instar, where most prey consumption occurs (Principi & Canard, 1984). In this respect, the lacewing's preference for prey species, prey nutritive quality, prey density, or the lacewing's ability to capture prey in all development stages (of both the predators and prey) may be important for inundative release and crucial for inoculative release.

Prey acceptance

Some chrysopid larvae are highly prey specific, e.g. *Chrysopa slossonae* Banks larvae have only been found associated with the woolly alder aphid, *Prociphilus tesselatus* (Fitch) (Tauber & Tauber, 1987). However, the lacewing species commonly used in augmentation (*Chrysoperla carnea* and *C. rufilabris*) are considered generalist predators, feeding on whatever prey are found in the habitat where the lacewing eggs are deposited. Nevertheless, the biological characteristics and nutritional quality of individual prey species are different and can influence *C. carnea* and *C. rufilabris* feeding preference, which, in turn, can determine the success or failure of an augmentation programme.

In corn, where lacewing releases might be targeted against multiple pests, there were considerable differences in lacewing (*Chrysoperla carnea* and *Chrysopa oculata*) survival among different corn insect pest species tested. All *Chrysoperla carnea* larvae died when fed European corn-borer, *Ostrinia nubilalis*, neonates because the predator would get entangled in the silk produced by these larvae (Obrycki *et al.*, 1989). In this case, a nutritionally acceptable prey was inappropriate in the field because of its silk production. In another example, augmentative releases of *C. carnea* and *C. rufilabris* failed to significantly reduce black bean aphid, *Aphis fabae* Scopoli, populations because, in part, there was an incompatibility between black bean aphids and lacewing larvae (Ehler *et al.*, 1997). Because aphids have long been considered a preferred prey for both *C. carnea* and *C. rufilabris*, it is surprising that Ehler *et al.* (1997) reported this incompatibility. However, prey location, developmental stage, density, and nutritional quality may become important considerations for prey selection in the field, even for an otherwise acceptable prey species in the laboratory.

There has been a great deal of discussion of the importance of alternative prey species for resident generalist predators. However, with augmentative releases of *Chrysoperla* spp., the presence of prey species other than the targeted species may disrupt release effectiveness as released lacewings may preferentially feed on secondary pests (or predators) rather than on the targeted herbivore. For example, Bergeson & Messina (1997, 1998) showed that releases of *C. plorabunda* to control the Russian wheat aphid, *Diuraphis noxia* (Mordvilko), were less effective when the bird cherry-oat aphid, *Rhopalosiphum padi* L., was present. Released lacewings captured *R. padi* more often than *D. noxia*, and spent more time consuming aphids in the mixed-species treatment than in the pure *D. noxia* treatment. Because of their 'generalist' predator label, prey preferences may not be given appropriate consideration when releasing *C. carnea* and *C. rufilabris*. Similar results were reported by Chang (1998): when multiple prey species were present, *C. plorabunda* fed disproportionately on thrips in the field, thereby reducing its effectiveness against other prey when thrips were present.

Prey quality

Nutritional quality of prey can affect the predator's overall effectiveness, because maximal prey consumption is based on predator development and survival throughout its larval period. There are numerous examples of prey species or stage affecting *Chrysoperla carnea* development or survival. In cotton in Israel, *C. carnea* was found to be an ineffective predator of sweet-potato whitefly and this was attributed to the predator's nutritional demands that were met only marginally by feeding on whiteflies, as well as behavioural preferences (Gerling *et al.*, 1997). In tests with another whitefly species, the greenhouse whitefly, *Trialeurodes vaporari-*

orum (Westwood), *C. carnea* larvae fed on all stages, but none of the lacewings survived to pupation (Senior & McEwen, 1998). The authors concluded that inundative releases of lacewing larvae might be effective as a supplementary control method; however, inoculative releases, which are generally better favoured, would not be successful. Prey development stage can also be important. Klingen *et al.* (1996) showed that *C. carnea* had slightly different development times and survival when fed eggs versus larvae of the moth *Mamestra brassicae* L. Osman & Selman (1996) showed that while *C. carnea* accepted all development stages of *Drosophila melanogaster* Meigen, fly pupae and adults proved to be relatively inferior as a prey source. Thus, the nutritional aspects of the prey are important, especially for inoculative release programmes where a numerical response of the predator is desired.

Prey host material also affects prey quality, which in turn, can affect predator nutrition. Legaspi *et al.* (1996) found that *C. rufilabris* fed on silverleaf whitefly, *Bemisia argentifolii* Bellows & Perring, reared on cucumbers and cantaloups developed more rapidly, showed increased survival, and weighed more, compared to those reared on poinsettia and lima bean, which did not survive to the pupal stage. The authors concluded that silverleaf whitefly reared on poinsettia or lima bean may have been nutritionally inadequate for *C. rufilabris* development.

Prey capture efficiency

The relative importance of prey capture efficiency has not been fully investigated for released lacewings, especially considering the many different species of targeted pests. Prey capture efficiency (and often prey density) must be considered in the development of an augmentation programme as release rates should be a function of the predator–prey numerical ratio, predator feeding capacity, and prey capture success. This is another area of lacewing field ecology that needs closer attention. Laboratory feeding studies are often conducted with a single prey developmental stage and an overabundance of individual prey. While this provides information on maximum prey consumption, it does not take into consideration potential predator–prey interactions encountered in the field. Commercial release rates based on laboratory feeding studies do not take into consideration field capture efficiencies, which may be dramatically different in the total number of prey killed. For example, during its larval development period, *C. carnea* can kill >250 fifth-instar variegated grape leafhoppers, *Erythroneura variabilis* Beamer (Daane *et al.*, 1994), while in field-cage trials with this same leafhopper species, lacewing feeding is dramatically lower (Daane *et al.*, 1996). Observations of lacewing and leafhopper field interactions showed third- to fifth-instar leafhoppers often escape predation from first-instar lacewings. Further, the variegated leafhoppers are relatively uniform in their distribution on the grape leaf and *C. carnea* larvae often missed nearby leafhopper nymphs (lacewings have a random search pattern until prey contact is made, after which the lacewing often remains in the vicinity of where the first prey was consumed [Bansch, 1966]).

14.6 CONCLUSIONS

Ridgway *et al.* (1977) outlined the biological requirements for effective augmentation: (1) an ability to rear predictable quantities of insects of known quality, (2) an ability to store, transport, and release the natural enemy in such a manner that it can compete effectively at the release site, and (3) an understanding of the ecological parameters governing the principal interactions between the biological control agent released and the target pest. There has been recent progress in many of these areas of research, particularly in development of rearing methods, release methodology, lacewing biology, and artificial attractants (Tauber *et al.*, 2000). Nevertheless, once lacewings are released into the field, there remain many important and as yet unanswered questions about their effectiveness under varying ecological parameters.

Summarised here are some of the ecological studies of lacewings naturally occurring or augmented in agroecosystems. A review of these studies indicates that problems are most often encountered when there is a mismatch between the lacewing's biological attributes and other ecological factors in the environment, which includes both abiotic and biotic factors. Of note, while there have been many different target pests and wide variations of ecological conditions at the release site, most studies are conducted with two *Chrysoperla* species (*C. carnea*-complex and *C. rufilabris*). The importance of these species in natural ecosystems or in release programmes is not questioned (see Wang & Nordlund, 1994; Daane *et al.*, 1998; Tauber *et al.*,

2000). However, as outlined, there are situations in which the use of these species may be inappropriate due to the ecological interactions at the release environment. The likelihood of changing the biology of *C. carnea* or *C. rufilabris* to fit all predator or habitat characteristics is small. There are, however, numerous chrysopid species that are underutilised in biological control and that have great potential for use in augmentation programmes.

Chrysoperla externa is an example of one of the alternative *Chrysoperla* species investigated. *Chrysoperla externa* ranges over all Latin America and north into Florida and South Carolina; it has also been inadvertently introduced into the Hawaiian Islands (Nuñez, 1989; Wang & Nordlund, 1994). Since it is commonly found in agricultural habitats and preys upon a wide range of insect pest species, its biology has been investigated and found by Barclay (1990) to be similar to that of *C. carnea*. Tauber *et al.* (1997) showed that *C. externa* has great potential for mass-rearing and use in augmentative biological control. Adult *C. externa*, held at 10 °C, had a 'shelf life' survival of at least four months and, after storage, reproduction began quickly, predictably, and with little variation among individuals. Carvalho *et al.* (1998*b*) investigated the reproductive potential and predation capacity of *C. externa* feeding on *Alabama argillacea* (Hübner). The results indicate that the prey species is suitable for mass-rearing, and that *C. externa* larvae have significant potential for pest control in cotton fields. Other *Chrysoperla* species investigated include *Chrysoperla nipponensis* (Okamoto) [= *C. sinica* (Tjeder)], *C. congrua* (Walker) (Kabissa, 1996), and *C. mediterranea* (Holzel) (Carvalho *et al.*, 1998*a*). These are just a few examples of other *Chrysoperla* species that show potential as biological control agents but which have not yet been thoroughly studied to understand or utilise their commercial potential.

Outside of the *Chrysoperla* genus, there is a large number of chrysopids that have potential use in biological control and may have the needed biological attributes to overcome ecological conditions that are problematic for *C. carnea* or *C. rufilabris*. The genus *Mallada*, the trash-carriers, is the largest of the Chrysopidae with at least 122 described species but absent from much of the Neotropics (Brooks & Barnard, 1990: but note that many of these species are now included in *Dichochrysa*). Of this genus, only *M. boninensis* Okamoto has been studied in any detail as a natural enemy in an agricultural crop. For example, in an unsprayed Rhodesian cotton field, 81% of the fertile chrysopid eggs sampled by Brettell (1979) were *M. boninensis*, one of three chrysopid species present. The adults are not predaceous but a larva can destroy a total of 297 *Helicoverpa armigera* eggs. In Taiwan, Lee & Shih (1982) developed a life table for this chrysopid, and using field cages, found that released *M. boninensis* larvae suppressed psyllid populations within four to five days. In Australia, among the 11 species of *Mallada* and *Dichochrysa*, *M. signatus* (Schneider) was found to be an excellent predator of *Heliothis* spp. in cotton, with larval searching success and consumption of *Heliothis* greater than that reported for some other chrysopids (e.g. *C. carnea*) (Samson & Blood, 1980). Since *Mallada s. lat.* spp. are similar to *Chrysoperla* spp. (not predaceous in the adult stage and attracted to protein hydrolysates) and are often common in wild refuges as well as in crops, they may prove to be effective predators in crops if artificial honeydew is applied in anticipation of arthropod pest problems.

There have been numerous field and laboratory studies on the potential use of *Chrysopa* spp. in biological control. *Chrysopa* species to consider include *Chrysopa formosa* Brauer, *C. perla*, and *C. pallens* (Rambur), which are Palaearctic species commonly found in Europe and/or Asia. While there have been field release trials with some *Chrysopa* spp. (e.g. Kawalska, 1976; Lyon, 1986), there is still a great deal of information needed to be gathered. Common Nearctic species include *C. nigricornis* Burmeister and *C. oculata*. Both species have been found to be important predators of economically important pests, with their natural densities often higher than resident *Chrysoperla* spp. (e.g. Jubb & Masteller, 1977; Daane *et al.*, 1994). However, it appears that many of the *Chrysopa* spp. are not the generalist predators that *Chrysoperla carnea* larvae are. For example, Obrycki *et al.*, (1989) found the most suitable prey for *Chrysopa oculata* in corn fields was *Rhopalosiphum maidis* (Fitch); if switched from aphids to stalk-borer, *Papaipema nebris* (Guenee), eggs while in third instar, more than 80% of the lacewing larvae died.

Recent advances in artificial diets and field attractants may increase the spectrum of species used in biological control. Because it is quite probable that some lacewing species have biological attributes that make them efficient predators, even in ecological conditions that are less favourable for *Chrysoperla carnea* or *C. rufilabris*, a call is made here for greater efforts to determine the potential use of a wider array of lacewing species for biological control. Potential species should be compared with *C. carnea* and *C. rufilabris* for their ability to be mass-produced, stored and transported, and released, and to control pests under varying ecological conditions and with different pest species.

REFERENCES

Altieri, M.A. & Letourneau, D.K. (1982) Vegetation management and biological control in agroecosystems. *Crop Protection* 1, 405–430.

Altieri, M.A. & Schmidt, L.L. (1985) Cover crop manipulation in Northern California orchards and vineyards: effects on arthropod communities. *Biological Agriculture and Horticulture* 3, 1–24.

Andow, D.A. (1991) Vegetational diversity and arthropod population response. *Annual Review of Entomology* 36, 561–586.

Atakan, E., Coll, M. & Rosen, D. (1996) Within-plant distribution of thrips and their predators: Effects of cotton variety and developmental stage. *Bulletin of Entomological Research* 86, 641–646.

Balduf, W.V. (1939) *The Bionomics of Entomophagous Insects*, part 2. John S. Swift, New York.

Bansch, R. (1966) On prey-seeking behavior of aphidophagous insects. In *Ecology of Aphidophagous Insects* ed. Hodek, I., pp. 123–128. Academia, Prague.

Barclay, W.W. (1990) Role of *Chrysoperla externa* as a biological control agent of insect pests in Mesoamerican agricultural habitats. MSc thesis, University of California, Berkeley.

Barry, R.M., Hatchett, J.H. & Jackson, R.D. (1974) Cage studies with predators of the cabbage looper, *Trichoplusia ni*, and corn earworm, *Heliothis zea*, in soybeans. *Journal of the Georgian Entomological Society* 9, 71–78.

Beije, C.M. (1993) Comparative susceptibility to two insecticides of the predator *Chrysoperla externa* Hagen and the boll weevil *Anthonomus grandis* Boh. in cotton in Nicaragua. *Proceedings, Experimental and Applied Entomology* 4, 243–249.

Bergeson, E. & Messina, F.J. (1997) Resource- versus enemy-mediated interactions between cereal aphids (Homoptera: Aphididae) on a common host plant. *Annals of the Entomological Society of America* 90, 425–432.

Bergeson, E. & Messina, F.J. (1998) Effect of a co-occurring aphid on the susceptibility of the Russian wheat aphid to lacewing predators. *Entomologia Experimentalis et Applicata* 87, 103–108.

Breene, R.G., Meagher, R.L. Jr, Nordlund, D.A. & Yin-Tung Wang (1992) Biological control of *Bemisia tabaci* (Homoptera: Aleyrodidae) in a greenhouse using *Chrysoperla rufilabris* (Neuroptera: Chrysopidae). *Biological Control* 2, 9–14.

Brettell, J.H. (1979) Green lacewings (Neuroptera: Chrysopidae) of cotton fields in central Rhodesia 1. Biology of *Chrysopa boninensis* Okomoto and toxicity of certain insecticides to the larva. *Rhodesian Journal of Agricultural Research* 17, 141–150.

Brooks, S.J. & Barnard, P.C. (1990) The green lacewings of the world: a generic review (Neuroptera: Chrysopidae). *Bulletin of the British Museum (Natural History) Entomology* 59, 117–286.

Bugg, R.L. & Waddington, C. (1994) Using cover crops to manage arthropod pests of orchards: a review. *Agriculture, Ecosystems and Environment* 50, 11–28.

Carvalho, C.F., Canard, M. & Alauzet, C. (1998a) Influence of temperature on the reactivation of diapausing adults of *Chrysoperla mediterranea* (Holzel) (Neuroptera, Chrysopidae). *Acta Zoologica Fennica* 209, 79–82.

Carvalho, C.F., Souza, B. & Santos, T.M. (1998b) Predation capacity and reproduction potential of *Chrysoperla externa* (Hagen) (Neuroptera, Chrysopidae) fed on *Alabama argillacea* (Hübner) eggs. *Acta Zoologica Fennica*, 209, 83–86.

Canard, M. & Duelli, P. (1984) Predatory behavior of larvae and cannibalism. In *Biology of Chrysopidae*, ed. Canard, M., Séméria, Y. & New, T.R., pp. 92–100. Dr W. Junk, The Hague.

Canard, M., Séméria, Y. & New, T.R. (eds.) (1984) *Biology of Chrysopidae*. Dr W. Junk, The Hague.

Chang, G.C. (1996) Comparison of single versus multiple species of generalist predators for biological control. *Environmental Entomology* 25, 207–212.

Chang, G.C. (1998) *Chrysoperla plorabunda* (Neuroptera: Chrysopidae) larvae feed disproportionately on thrips (Thysanoptera: Thripidae) in the field. *Canadian Entomologist* 130, 549–550.

Cisneros, J.J. & Rosenheim, J.A. (1997) Ontogenetic change of prey preference in the generalist predator *Zelus renardii* and its influence on predator–predator interactions. *Ecological Entomology* 22, 399–407.

Clark, T.L. & Messina, F.J. (1998) Foraging behavior of lacewing larvae (Neuroptera: Chrysopidae) on plants with divergent architectures. *Journal of Insect Behavior* 11, 303–317.

Costello, M.J. & Daane, K.M. (1998a) Effects of cover cropping on pest management: arthropods. In *Cover Cropping in Vineyards: A Grower's Handbook*, ed.

Ingels, C.A., Bugg, R.L., McGourty, G.T. & Christensen, L.P., pp. 93–106. University of California, Division of Agriculture and Natural Resource, Oakland CA.

Costello, M.J. & Daane, K.M. (1998b) Influence of ground covers on spider (Araneae) populations in a table grape vineyard. *Ecological Entomology* 23, 33–40.

Daane, K.M. & Yokota, G.Y. (1997) Release methods affect egg survival and distribution of augmented green lacewings (Chrysopidae: Neuroptera). *Environmental Entomology* 26, 455–464.

Daane, K.M., Yokota, G.Y., Hagen, K.S. & Zheng, Y. (1994) Field evaluation of *Chrysoperla* spp. in augmentative release programs for the variegated grape leafhopper, *Erythroneura variabilis*. In *Proceedings of the 7th Workshop of the IOBC Working Group: Quality Control of Mass Reared Arthropods*, ed. Nicoli, G., Benuzzi, M. & Leppla, N.C., pp. 105–120.

Daane, K.M., Yokota, G.Y., Zheng, Y. & Hagen, K.S. (1996) Inundative release of the green lacewing to control *Erythroneura variabilis* and *E. elegantula* (Homoptera: Cicadellidae) in grape vineyards. *Environmental Entomology* 25, 1224–1234.

Daane, K.M., Hagen, K.S. & Mills, N.J. (1998) Predaceous insects for insect and mite control. In *Mass-Reared Natural Enemies: Application, Regulation, and Needs*, ed. Ridgway, R.L., Hoffmann, M.P., Inscoe, M.N. & Glenister, C.S., pp. 61–115. Thomas Say Publications in Entomology, Entomological Society of America, Lanham MD.

Dinter, A. (1998) Intraguild predation between erigonid spiders, lacewing larvae and carabids. *Journal of Applied Entomology* 122, 163–167.

Dreistadt, S.H., Hagen, K.S. & Dahlsten, D.L. (1986) Predation by *Iridomyrmex humilis* (Hym.: Formicidae) on eggs of *Chrysoperla carnea* (Neu.: Chrysopidae) released for inundative control of *Illinoia liriodendri* (Hom.: Aphididae) infesting *Liriodendron tulipifera*. *Entomophaga* 31, 397–400.

Ehler, L.E., Long, R.F., Kinsey, M.G. & Kelley, S.K. (1997) Potential for augmentative biological control of black bean aphid in California sugarbeet. *Entomophaga* 42, 241–256.

Eigenbrode, S.D., Moodie, S. & Castagnola, T. (1995) Predators mediate host plant resistance to a phytophagous pest in cabbage with glossy leaf wax. *Entomologia Experimentalis et Applicata* 77, 335–342.

Eigenbrode, S.D., Castagnola, T., Roux, M.B. & Steljes, L. (1996) Mobility of three generalist predators is greater on cabbage with glossy leaf wax than on cabbage with a wax bloom. *Entomologia Experimentalis et Applicata* 81, 335–343.

Elkarmi, L.A., Harris, M.K. & Morrison, R.K. (1987) Laboratory rearing of *Chrysoperla rufilabris* (Burmeister), a predator of insect pests of pecans. *Southwestern Entomologist* 12, 73–78.

Ellington, J., Southward, M. & Carrillo, T. (1997) Association among cotton arthropods. *Environmental Entomology* 26, 1004–1008.

Gerling, D. & Bar, D. (1985) Parasitization of *Chrysoperla carnea* (Neuroptera, Chrysopidae) in cotton fields of Israel. *Entomophaga* 30, 409–414.

Gerling, D., Kravchenko, V. & Lazare, M. (1997) Dynamics of common green lacewing Neuroptera: Chrysopidae in Israeli cotton fields in relation to whitefly (Homoptera: Aleyrodidae) populations. *Environmental Entomology* 26, 815–827.

Grafton-Cardwell, E.E. & Hoy, M.A. (1985) Short-term effects of permethrin and fenvalerate on oviposition by *Chrysoperla carnea* (Neuroptera: Chrysopidae). *Journal of Economic Entomology* 78, 955–959.

Hamilton, G.C. & Lashomb, J.H. (1997) Effect on insecticides on two predators of the Colorado potato beetle (Coleoptera: Chrysomelidae). *Florida Entomologist* 80, 10–23.

Heinz, K.M. (1998) Dispersal and dispersion of aphids (Homoptera: Aphididae) and selected natural enemies in spatially subdivided greenhouse environments. *Environmental Entomology* 27, 1029–1038.

Henry, C.S., Wells, M.M. & Pupedis, R.J. (1993) Hidden taxonomic diversity within *Chrysoperla plorabunda* (Neuroptera: Chrysopidae): two new species based on courtship songs. *Annals of the Entomological Society of America* 86, 1–13.

Hilbeck, A., Baumgartner, M., Fried, P.M. & Bigler, F. (1998a) Effects of transgenic *Bacillus thuringiensis* corn-fed prey on mortality and development time of immature *Chrysoperla carnea* (Neuroptera: Chrysopidae). *Environmental Entomology* 27, 4 80–487.

Hilbeck, A., Moar, W.J., Pusztai-Carey, M., Filippini, A. & Bilger, F. (1998b) Toxicity of *Bacillus thuringiensis* Cry1Ab toxin to the predator *Chrysoperla carnea* (Neuroptera: Chrysopidae). *Environmental Entomology* 27, 1255–1263.

Hodek, I. (1993) Habitat and food specificity in aphidophagous predators. *Biological Control Science and Technology* 3, 91–100.

Jubb, G.L. & Masteller, E.C. (1977) Survey of arthropods in grape vineyards of Erie County, Pennsylvania: Neuroptera. *Environmental Entomology* 6, 419–428.

Kabissa, J.C.B., Yarro, J.G., Kayumbo, H.Y. & Juliano, S.A. (1996) Functional responses of two chrysopid predators feeding on *Helicoverpa armigera* (Lep.:

Noctuidae) and *Aphis gossypii* (Hom.: Aphididae). *Entomophaga* 41, 141–151.

Kawalska, T. (1976) Mass rearing and possible uses of Chrysopidae against aphids in glasshouses. *Bulletin OILB/SROP* 1976(4), 80–85.

King, E.G., Hopper, K.R. & Powell, J.E. (1985) Analysis of systems for biological control of crop arthropod pests in the U.S. by augmentation of predators and parasites. In *Biological Control in Agricultural IPM Systems*, ed. Hoy, M. & Herzog, D.C., pp. 201–227. Academic Press, Orlando FL.

Klingen, I., Johansen, N.S. & Hofsvang, T. (1996) The predation of *Chrysoperla carnea* (Neurop., Chrysopidae) on eggs and larvae of *Mamestra brassicae* (Lep., Noctuidae). *Journal of Applied Entomology* 120, 363–367.

Lawrence, P.O. (1974) Susceptibility of *Chrysopa rufilabris* to selected insecticides and miticides. *Environmental Entomology* 3, 146–150.

Lawrence, P.O., Kerr, S.H. & Whitcomb, W.H. (1973) *Chrysopa rufilabris*: effect of selected pesticides on duration of third larval stadium, pupal stage, and adult survival. *Environmental Entomology* 2, 477–480.

Lee, S. J. & Shih, C.I.T. (1982) Biology, predation and field cage release of *Chrysopa boninensis* Okamoto on *Pouraephala psylloptera* Crawford and *Corcyra cephalonica* Stanton. *Journal of Agriculture and Forestry* 31, 129–144.

Legaspi, J.C., Nordlund, D.A. & Legaspi, B.C. Jr (1996) Tri-trophic interactions and predation rates in *Chrysoperla* spp. attacking the silverleaf whitefly. *Southwestern Entomologist* 21, 33–42.

Lingren, P.D., Ridgway, R.L. & Jones, S.L. (1968) Consumption by several common arthropod predators of eggs and larvae of two *Heliothis* species that attack cotton. *Annals of the Entomological Society of America* 61, 613–618.

Lucas, E. (1998) How do ladybirds (*Coleomegilla maculata lengi*) [Coleoptera: Coccinellidae]) feed on green lacewing eggs (*Chrysoperla rufilabris*) [Neuroptera: Chrysopidae]). *Canadian Entomologist* 130, 547–548.

Lucas, E., Coderre, D., & Brodeur, J. (1997) Instar-specific defense of *Coleomegilla maculata lengi* (Col.: Coccinellidae): influence on attack success of the intraguild predator *Chrysoperla rufilabris* (Neur: Chrysopidae). *Entomophaga* 42, 3–12.

Lucas, E., Coderre, D. & Brodeur, J. (1998) Intraguild predation among aphid predators: characterization and influence of extraguild prey density. *Ecology* 79, 1084–1092.

Lyon, J.P. (1979) Lâchers expérimentaux de chrysopes et d'Hyménoptères parasites sur les pucerons en serres d'aubergines. *Annales de Zoologie–Écologie Animale* 11, 51–65.

Lyon, J.P. (1986) Use of aphidophagous and polyphagous beneficial insects in greenhouse. In *Ecology of Aphidophaga*, ed. Hodek, I., pp. 471–474. Dr W. Junk, The Hague.

Messina, F.J., Jones, T.A. & Nielson, D.C. (1995) Host plant affects the interaction between the Russian wheat aphid and a generalist predator, *Chrysoperla carnea. Journal of the Kansas Entomological Society* 68, 313–319.

Messina, F.J., Jones, T.A. & Nielson, D.C. (1997) Host-plant effects on the efficacy of two predators attacking Russian wheat aphids (Homoptera: Aphididae). *Environmental Entomology* 26, 1398–1404.

Mizell, R.F. III & Schiffhauer, D.E. (1990) Effects of pesticides on pecan aphid predators *Chrysoperla rufilabris* (Neuroptera: Chrysopidae), *Hippodamia convergens, Cycloneda sanguinea* (L.), *Olla v-nigrum* (Coleoptera: Coccinellidae), and *Aphelinus perpallidus* (Hymenoptera: Encyrtidae). *Journal of Economic Entomology* 83, 1806–1812.

Morris, T.I., Campos, M., Jervis, M.A., McEwen, P.K. & Kidd, N.A.C. (1998) Potential effects of various ant species on green lacewing, *Chrysoperla carnea* (Stephens) (Neuropt., Chrysopidae) egg numbers. *Journal of Applied Entomology* 122, 401–403.

New, T.R. (1975) The biology of Chrysopidae and Hemerobiidae (Neuroptera) with reference to their usage as biocontrol agents: a review. *Transactions of the Royal Entomological Society of London* 127, 115–140.

Nuñez, E.Z. (1989) Chrysopidae (Neuroptera) del Perú y sus especies más comunes. *Revista de Peruana de Entomologia* 31, 69–75.

Nyffeler, M., Dean, D.A. & Sterling, W.L. (1987). Predation by green lynx spider, *Peucetia viridans* (Araneae: Oxyopidae), inhabiting cotton and woolly croton plants in East Texas. *Environmental Entomology* 16, 355–359.

Obrycki, J.J. (1986) The influence of foliar pubescence on entomophagous species. In *Interactions of Plant Resistance and Parasitoids and Predators of Insects*, ed. Boethel, D.J. & Eikenbary, R.D., pp. 61–83. Horwood, Chichester UK.

Obrycki, J.J., Hamid, M.N., Sajap, A.S. & Lewis, L.C. (1989) Suitability of corn insect pests for development and survival of *Chrysoperla carnea* and *Chrysopa oculata* (Neuroptera: Chrysopidae). *Environmental Entomology* 18, 1126–1130.

Osman, M.Z. & Selman, B.J. (1996) Effect of larval diet on the performance of the predator *Chrysoperla carnea* Stephens (Neuropt., Chrysopidae). *Journal of Applied Entomology* 120, 115–117.

Phoofolo, M.W. & Obrycki, J.J. (1998) Potential for intraguild predation and competition among predatory Coccinellidae and Chrysopidae. *Entomologia Experimentalis et Applicata* 89, 45–55.

Pilcher, C.D, Obrycki, J.J., Rice, M.E. & Lewis, L.C. (1997) Preimaginal development, survival, field abundance of insect predators on transgenic *Bacillus thuringiensis* corn. *Environmental Entomology* 26, 446–454.

Pree, D.J., Archibald, D.E. & Morrison, R.K. (1989) Resistance to insecticides in the common green lacewing *Chrysoperla carnea* (Neuroptera: Chrysopidae) in southern Ontario. *Journal of Economic Entomology* 82, 29–34.

Principi, M.M. & Canard, M. (1984) Feeding habits. In *Biology of Chrysopidae*, ed. Canard, M., Séméria, Y. & New, T.R., pp. 76–92. Dr W. Junk, The Hague.

Ridgway, R.L. & Murphy, W.L. (1984) Biological control in the field. In *Biology of Chrysopidae*, ed. Canard, M., Séméria, Y. & New, T.R., pp. 220–238. Dr W. Junk, The Hague.

Ridgway, R.L., King, E.G. & Carrillo, J.L. (1977) Augmentation of natural enemies for control of plant pests in the western Hemisphere. In *Biological Control by Augmentation of Natural Enemies*, ed. Ridgway, R.L. & Vinson, S.B., pp. 379–416. Plenum Press, New York.

Rosenheim, J.A. & Wilhoit, L.R. (1993) Predators that eat other predators disrupt cotton aphid control. *California Agriculture* 47, 7–9.

Rosenheim, J.A., Wilhoit, L.R. & Armer, C.A. (1993) Influence of intraguild predation among generalist predators on the suppression of an herbivore population. *Oecologia* 96, 439–449.

Rosenheim, J.A., Kaya, H., Ehler, L.E., Morois, J.J. & Jaffee, B.A. (1995) Intraguild predation among biological-control agents: theory and evidence. *Biological Control* 5, 303–335.

Ruberson, J.R., Tauber, C.A. & Tauber, M.J. (1989) Development and survival of *Telenomus lobatus*, a parasitoid of chrysopid eggs: effect of host species. *Entomologia Experimentalis et Applicata* 51, 101–106.

Ruberson, J.R., Tauber, C.A. & Tauber, M.J. (1995) Developmental effects of host and temperature on *Telenomus* spp. (Hymenoptera: Scelionidae) parasitizing chrysopid eggs. *Biological Control* 5, 245–250.

Rumpf, S., Frampton, C. & Chapman, B. (1997) Acute toxicity of insecticides to *Micromus tasmaniae* (Neuroptera: Hemerobiidae) and *Chrysoperla carnea* (Neuroptera: Chrysopidae): LC-50 and LC-90 estimates for various test durations. *Journal of Economic Entomology* 90, 1493–1499.

Ružička, Z. (1998) Further evidence of oviposition–deterring allomone in chrysopids (Neuroptera: Chrysopidae). *European Journal of Entomology* 95, 35–39.

Ružička, Z. & Havelka, J. (1998) Effects of oviposition-deterring pheromone and allomones on *Aphidoletes aphidimyza* (Diptera: Cecidomyiidae). *European Journal of Entomology* 95, 211–216.

Samson, P.R. & Blood, P.R.B. (1980) Voracity and searching ability of *Chrysopa signata* (Neuroptera: Chrysopidae), *Micromus tasmaniae* (Neuroptera: Hemerobiidae) and *Tropiconabis capsiformis* (Hemiptera: Nabidae). *Australian Journal of Zoology* 28, 515–580.

Schultz, B.B. (1988) Reduced oviposition by green lacewings (Neuroptera: Chrysopidae) on cotton intercropped with corn, beans, or weeds in Nicaragua. *Environmental Entomology* 17, 229–232.

Sengonca, Ç. & Frings, B. (1985) Interference and competitive behaviour of the aphid predators, *Chrysoperla carnea* and *Coccinella septempunctata* in the laboratory. *Entomophaga* 30, 245–251.

Senior, L.J. & McEwen, P.K. (1998) Laboratory study of *Chrysoperla carnea* (Stephens) (Neuropt., Chrysopidae) predation on *Trialeurodes vaporariorum* (Westwood) (Hom., Aleyrodidae). *Journal of Applied Entomology* 122, 99–101.

Sewify, G.H. & El Arnaouty, S.A. (1998) The effect of the entomopathogenic fungus *Verticillium lecanii* (Zimm.) on mature larvae of *Chrysoperla carnea* Stephens (Neuroptera, Chrysopidae) in the laboratory. *Acta Zoologica Fennica* 209, 233–237.

Singh, P.P. & Varma, G.C. (1986) Comparative toxicities of some insecticides to *Chrysoperla carnea* (Chrysopidae: Neuroptera) and *Trichogramma brasiliensis* (Trichogrammatidae: Hymenoptera), two arthropod natural enemies of cotton pests. *Agriculture, Ecosystems and the Environment* 15, 23–30.

Smith, M.W., Arnold, D.C., Eikenbary, R.D., Rice, N.R., Shiferaw, A. Cheary, B.S. & Carroll, B.L. (1996) Influence of ground cover on beneficial arthropods in pecan. *Biological Control* 6, 164–176.

Stark, S.B. & Hopper, K.R. (1988) *Chrysoperla carnea* predation on *Heliothis virescens* larvae parasitized by *Microplitis croceipes. Entomologia Experimentalis et Applicata* 48, 69–72.

Tauber, C.A. & Tauber, M.J. (1986) Ecophysiological responses in life-history evolution: evidence for their importance in a geographically widespead insect species complex. *Canadian Journal of Zoology* 64, 875–884.

Tauber, C.A. & Tauber, M.J. (1987) Food specificity in predaceous insects: a comparative ecophysiological and genetic study. *Evolutionary Ecology* 1, 175–186.

Tauber, M.J. & Tauber, C.A. (1974) Dietary influence on reproduction in both sexes of five predacious species (Neuroptera). *Canadian Entomologist* 106, 921–925.

Tauber, M.J. & Tauber, C.A. (1983) Life history traits of *Chrysopa carnea* and *Chrysopa rufilabris* (Neuroptera: Chyrsopidae): influence of humidity. *Annals of the Entomological Society of America* 76, 282–285.

Tauber, M.J., Albuquerque, G.S. & Tauber, C.A. (1997) Storage of nondiapausing *Chrysoperla externa* adults: influence on survival and reproduction. *Biological Control* 10, 69–72.

Tauber, M.J., Tauber, C.A., Daane, K.M. & Hagen, K.S. (2000) New tricks for old predators: implementing biological control with *Chrysoperla. American Entomologist* 46, 26–38.

Tedders, W.L., Reilly, C.C., Wood, B.W., Morrison, R.K. & Lofgren, C.S. (1990) Behavior of *Solenopsis invicta* (Hymenoptera: Formicidae) in pecan orchards. *Environmental Entomology* 19, 44–53.

Treacy, M.F., Benedict, J.H., Lopez, J.D. & Morrison, R.K. (1987) Functional response of a predator (Neuroptera: Chrysopidae) to bollworm (Lepidoptera: Noctuidae) eggs on smoothleaf, hirsute, and pilose cottons. *Journal of Economic Entomology* 80, 376–379.

Vinson, S.B. & Scarborough, T.A. (1989) Impact of the imported fire ant on laboratory populations of cotton aphid (*Aphis gossypii*) predators. *Florida Entomologist* 72, 107–111.

Wang, R. & Nordlund, D.A. (1994) Use of *Chrysoperla* spp. (Neuroptera: Chrysopidae) in augmentative release programmes for control of arthropod pests. *Biocontrol News and Information* 15, 51N–57N.

Zheng, Y., Daane, K.M., Hagen, K.S. & Mittler, T.E. (1993a) Influence of larval food consumption on the fecundity of the lacewing *Chrysoperla carnea. Entomologia Experimentalis et Applicata* 67, 9–14.

Zheng, Y., Hagen, K.S., Daane, K.M. & Mittler, T.E. (1993b) Influence of larval dietary supply on the food consumption, food utilization efficiency, growth and development of the lacewing *Chrysoperla carnea. Entomologia Experimentalis et Applicata* 67, 1–7.

CHAPTER 15

Sampling and studying lacewings in crops

T.R. New and A.E. Whittington

15.1 INTRODUCTION

Neuenschwander (1984) emphasised that much sampling and monitoring of lacewing populations in the past has not been strictly quantitative, but has involved collections to determine presence/absence and gross inferences on the relative abundance of various species on a crop or in an assemblage in an area or defined environment.

Nevertheless, for many purposes there is need for field assessments based on quantitative appraisal based on adequate, replicable sampling protocols, and continued monitoring to determine population sizes, changes in species abundance, phenological traits, and responses to change. Neuenschwander's account of such approaches and the techniques employed remains valid and applicable, and forms the basis for the discussion which follows. However, in the intervening years, the need for improved sampling to establish predictive capability has become more widespread in IPM, and the wisdom of monitoring the establishment, influences, and spread of biological control agents consolidated.

As in any sampling programme, the scale of operation and the precise questions being asked must determine the sampling regime and the intensity of sampling needed. 'Sampling effort' reflects both the duration and intensity of sampling. At the extremes, a single brief collection period may be all that is needed to confirm the presence of a given lacewing species on a crop whilst, at the other, continuous or interval sampling using a variety of techniques over an extended period (of one or more growing seasons or years) may be necessary to reveal population trends, phenology, and the relative abundance of the members of a complex of natural enemies, or the establishment of a newly introduced biological control agent. It is implicit, also, that 'presence and abundance' may not invariably equate to 'desired effect' and, whereas there are considerable problems over extrapolating uncritically from laboratory studies to the field environment (New, 1991) because of restricted prey choice and reduced mobility

in an alien environment, integration of laboratory and field data can indeed often provide useful leads. Measures of 'predatory activity' by adults and larvae may necessitate complementary studies to relate field abundance to voracity and selectivity, and also the use of serological techniques to determine the prey from the gut contents of field-collected predators.

15.2 FIELD APPRAISAL

Because all stages of lacewings are passed within the crop, and data on the presence and abundance of these are used to provide estimates of their likely effects on pests, any of these (eggs, larvae, pupae, adults) may be used as an index for quantification or comparison, either singly or in combinations. For some species, particular stages may be relatively inaccessible; not all arboreal species in orchards form their cocoons in trap bands around tree trunks, for example, so that any available biological knowledge of each particular species may be critical in formulating optimal sampling protocols.

Much of the rationale of taking field samples of lacewings has not advanced substantially since Neuenschwander's (1984) summary. The major approaches to this are exemplified in Table 15.1. The comments below reflect more recent examples of some of these approaches, to demonstrate the contexts of their use and some refinements and caveats. Many of the studies reported in later chapters of this book contain reference to a variety of methods and exemplify the diversity of individualistic approaches which may be needed.

The twin approaches to sampling, in common with those used widely for other animals, are therefore (1) absolute – aiming to define the density per unit of habitat or (2) relative – for comparisons and surveys based on samples not related directly to habitat. Many of the methods used conventionally for lacewings, such as light traps and suction traps, fall into the latter category. The various techniques each have limitations

Table 15.1. *Examples of methods used in sampling various stages of lacewings, and for population estimation*

Sample method	Stage				References
	Egg	Larva	Pupa	Adult	
Beating				•	Viollier & Fauvel, 1984; Monserrat & Marín, 1992
Chemical/insecticide knockdown		•			Pantaleoni & Ticchiati, 1990; Pantaleoni, 1996
Chemical lures				•	McEwen *et al.*, 1994
Direct counts on plant samples	•	•	•	•	El Arnaouty & Sewify, 1998; Gerling *et al.*, 1997; McEwen & Ruiz, 1994; Paulian, 1992, 1998*a, b*
Light traps				•	Monserrat & Marín, 1992; Szentkirályi, 1992; Paulian, 1996
McPhail traps				•	Campos & Ramos, 1983; Santas, 1987; Alrouechdi, 1984
Overwintering chambers				•	Sengonca & Frings, 1987, 1989; McEwen, 1998
Pan traps				•	Ridland, 1988
Pitfall traps			•	•	Mészáros, 1984; Ridland, 1988; Home & Edward, 1995, 1997, 1998
Sticky traps				•	Raspi & Malfatti, 1985; McEwen, 1995; Sziráki, 1996; Udayagiri *et al.*, 1997; Paulian, 1998*a, b*
Suction traps				•	New, 1967*a*
Sweeping				•	Mészáros, 1984; Pantaleoni & Curto, 1990; Paulian, 1998*a*
Trap banding			•		New, 1967*b*; Tamaki & Halfhill, 1968; Fye, 1985; Mizell & Schiffhauer, 1987; Curkovic *et al.*, 1995
Windvane nets				•	Ridland, 1988

and biases and, although extended sampling effort may help to overcome some of these, as may a combination of different sampling techniques, the progressive demands for rapid and effective assessment of lacewings in crops also dictate the limitations of sampling. Many of the techniques employed are discussed by Southwood (1978) or New (1998). In general, approaches toward quantitative sampling and analysis are needed, and these underlie most recent assessment studies of lacewings.

15.3 EGGS

Neuenschwander (1984) noted that, because chrysopid eggs are reasonably conspicuous (because of their stalks) and unlikely to be confused with those of other insects, absolute counts are possible, based on the numbers of

eggs/plant or other vegetation units that can be examined thoroughly and directly. Such counts can be laborious (and, hence, relatively expensive) if undertaken on any large scale, but are valuable in helping to determine the dispersion of chrysopids in crops. Thus, weekly counts of eggs on randomly selected cotton plants provided useful information on population trends (Schultz, 1988). Hemerobiid and coniopterygid eggs are generally much less conspicuous, as they are laid flat on vegetation, and estimates of other, active stages are usually preferred as being easier to perform.

15.4 LARVAE

Larvae of many lacewings are dislodged from vegetation by beating or sweeping, and these methods (sometimes augmented by suction sampling or insecticidal knockdown) are the most frequently used in sampling this stage. Some species (such as *Micromus tasmaniae* (Walker) in Australia: Horne *et al.*, this volume) have larvae which descend from low vegetation and are active on the ground surface. These can be appraised effectively by pitfall traps (with a liquid preservative), but such species are rather exceptional. Direct searches for larvae on vegetation can be made but, as for eggs, can become laborious.

15.5 PUPAE

Little work has been undertaken on quantitative sampling of lacewing cocoons (in essence, pupae), because other stages are generally more accessible. However, for some arboreal species of all three families, counts of cocoons in hessian or corrugated cardboard trap bands can be useful in indicating the relative abundance of various species, not least because this stage is often the overwintering one and counts at times when other stages are not present may provide augmentation of estimates made from other stages, and fit more easily into annual work schedules. Lacewings frequenting orchards and forestry plantations should be considered for this mode of estimation and appraisal.

15.6 ADULTS

Adult lacewings may be captured either directly from their habitats, or by various traps which capture airborne or flying individuals within the general vicinity of their presumed habitats or breeding sites. Both approaches are difficult to quantify, the first because of 'unevenness' of capture and the second also because of difficulties of calibration and uncertainties over the origins of captured material, which may have travelled for considerable distances.

Despite numerous attempts to quantify the 'effectiveness' of all the methods used many such uncertainties remain, due largely to the vagaries of insect behaviour and its influences on amenability to capture. Many species will thus not be captured in direct relationship to their natural abundance, but valuable data on seasonality and incidence can accrue. Fixed numbers of sweeps using a standard net in a uniform crop can indicate relative abundance of different species on different occasions and represents, perhaps, the 'most quantitative' simple estimation –

provided that sampling effort is sufficient to counter local variations in density within the crop. Direct searches are attractive, where feasible, because all stages of lacewings can be appraised simultaneously. Thus, Gerling *et al.* (1997) employed 'top-to-bottom' searches of cotton plants in Israel to help determine the population dynamics of (putative) *Chrysoperla carnea* (Stephens). Importantly in many practical contexts, other species of natural enemies can be appraised at the same time. Only rarely are lacewings treated in isolation from other predators and parasitoids, all of which may need evaluation over the same period. Many of the above techniques can be used to do this, but any selectivity is important to evaluate in order to validate comparative studies. In Delaware (USA) cornfields, coloured sticky cards (red, green, white, or yellow) were used to sample several natural enemies (Udayagiri *et al.*, 1997); trap colour has substantial effects on some species of predators but, apparently had little effect on catches of *C. carnea*.

Chemical lures attractive to chrysopids may be more uniform in effect, simply because many or most individuals present may be susceptible. However, the radius of attraction may depend on weather conditions and any competing odours from the crop or prey organisms and, whereas such influences are very difficult to confirm, they are equally hard to discount! Lures can be used in conjunction with liquid traps which preserve the incoming insects, or with sticky traps to intercept them; such traps have been used in monitoring chrysopid behaviour (Duelli, 1980). Protein hydrolysates have proved useful as attractants for glycophagous chrysopids, but do not generally attract predatory adults.

15.7 LABORATORY APPRAISAL

Laboratory studies on voracity and prey selection/consumption/utilisation inevitably suffer from the confines of a simplified artificial environment and, whereas they may provide useful 'leads' to the real world, their results can often only be extrapolated to the field with considerable caution. Quantifying invertebrate predation in the field is a complex exercise, and there are many approaches to this (Sunderland, 1987, 1988). Very broadly, the options available for Neuroptera include direct observation, field cage trials, use of labelled prey and measuring the recovery of these, and *post mortem* analysis of field-collected material. A further approach noted by Sunderland for some other predators, faecal analysis, can not be used for larval lacewings (as they produce no faeces) but is of limited use for adults by examining faeces microscopically for characteristic exoskeletal fragments of prey. Many of the above suite of methods have been developed and diversified predominantly for other groups of arthropod predators, and have not been validated critically for lacewings.

Direct observation is likely to be of only limited value, because extensive observations are laborious and difficult to undertake. They are of most use in limited arenas (such as laboratory enclosures or field cages) in which video-photography may be able to replace human observers, especially in validating inferred predator–prey associations through detecting capture and consumption, in elucidating behaviour and search patterns and possible evasive responses by the prey species. Much of the predatory process has depended on direct observation for its elucidation (New, 1991). Although estimates of the rate of disappearance of immobile prey placed outside can be made from observation, it is usually not possible to relate this to the activities of a specific predator.

Field cages can be used to help measure predation rates, by caging predator and prey together and comparing the prey population trajectory with that in cages with prey alone. Such trends facilitate measurements of predation rate/predator/day, as Hance (1986) showed for Carabidae feeding on Aphis fabae Scopoli. Complications arise when several possible prey species are thereby exposed to polyphagous predators, and where the caged predator populations markedly exceed normal field levels and lead to interference effects. The method often involves disturbance of the natural system under study, and artefacts may occur. However, most agricultural systems in which lacewing appraisals are likely to be needed are already highly manipulated and artificial, and the short-term practical needs for augmented predator effects may override such considerations in practice.

Recovery of 'labelled prey' (where labels may be dyes, radioactive isotopes, or rare elements in various studies) depends on ability to detect and, preferably, quantify the eaten prey. Autoradiography, for example, can detect predation from presence of isotope-labelled prey; if such predators are exposed to X-ray film, labelled prey remains in their guts are revealed as contrasting spots. Sunderland (1987) noted a number of studies in which approaches to quantification were made by counting the levels of radioactivity in each predator, but the method has numerous problems if exact quantification is needed (Southwood, 1978).

A wide range of *post mortem* methods are based on examination of suspected predators collected from the system under study, but all suffer (as for labelled prey studies) from failure to distinguish between predation and scavenging. Lacewing larvae often take dead or immobile prey, and some will feed from artificial diets, so such ambiguities may lead to false conclusions if the simple food chain expected is more complicated. Errors may arise also from cannibalism and secondary predation, again providing 'positives' of uncertain origins. The range of options includes an array of serological methods, some of which are under rapid refinement to increase their sensitivities. These approaches are most useful for semiquantitative investigation of the fates of key prey species once these have been detected as pests, or targets for enhanced predation.

REFERENCES

Alrouechdi, K. (1984) Les chrysopides (Neuroptera) en Oliveraie. In *Progress in World's Neuropterology, Proceedings of the 1st International Symposium on Neuropterology*, ed. Gepp, J., Aspöck, H. & Hölzel, H. pp. 147–152. Thalerhof, Graz.

Campos, M. & Ramos, P. (1983). Chrysopidos (Neuroptera) capturados en un olivar del sur de Espana. *Neuroptera International* 2, 219–227.

Curkovic, S.T., Barria, P.G. & Gonzalez, R.R. (1995) Observaciones preliminares sobre insectos y acaros presentes en vides, perales, ciruelos y kakis detectados con trampas de agregación. *Acta Entomologica Chilena*, 19, 143–154.

Duelli, P. (1980) Adaptive dispersal and appetitive flight in the green lacewing, *Chrysopa carnea*. *Ecological Entomology* 5, 213–20.

El Arnaouty, S.A. & Sewify, G.H. (1998) A pilot experiment for using eggs and larvae of *Chrysoperla carnea* (Stephens) against *Aphis gossypii* (Glover) on cotton in Egypt. *Acta Zoologica Fennica* 209, 103–106.

Fye, R.E. (1985) Corrugated fiberboard traps for predators overwintering in pear orchards. *Journal of Economic Entomology* 78, 1511–1514.

Gerling, D., Kravchenko, V. & Lazare, M. (1997) Dynamics of common green lacewing (Neuroptera: Chrysopidae) in Israeli cotton fields in relation to whitefly

(Homoptera: Aleyrodidae) populations. *Environmental Entomology* 26, 815–827.

Hance, T. (1986) Expériences de limitation de populations d'*Aphis fabae* par des Carabidae à différentes densités. *Annales de la Société Royale Zoologique de Belgique* 116, 115–124.

Horne, P.A. & Edward, C.L. (1995) The phenology and food preferences of *Labidura truncata* (Dermaptera, Labiduridae) in western Victoria. *Journal of the Australian Entomological Society* 34, 101–104.

Horne, P.A. & Edward, C.L. (1997) Preliminary observations on awareness, management and impact of biodiversity in agricultural systems. *Memoirs of the Museum of Victoria* 56, 21–25.

Horne, P.A. & Edward, C.L. (1998) The effect of tillage on pest and beneficial beetles. *Australian Journal of Entomology* 37, 60–63.

McEwen, P.K. (1995) Attractiveness of yellow sticky traps to green lacewings (Neuropt., Chrysopidae). *Entomologist's Monthly Magazine* 131, 163–166.

McEwen, P.K. (1998) Overwintering chambers for the common green lacewing (*Chrysoperla carnea*): influence of chemical attractant, material and size. *Journal of Neuropterology* 1, 17–21.

McEwen, P.K. & Ruiz, J. (1994). Relationship between non-olive vegetation and green lacewing eggs in a Spanish olive orchard. *Antenna* 18, 148–149.

McEwen, P.K., Jervis, M.A. & Kidd, N.A.C. (1994). Use of sprayed L-tryptophan solution to concentrate numbers of green lacewing *Chrysoperla carnea* in olive tree canopy. *Entomologia Experimentalist et Applicata* 70, 97–99.

Mészáros, Z. (1984) Results of faunistical and floristical studies in Hungarian apple orchards (Apple Ecosystem Research no. 26). *Acta Phytopathologica Academiae Scientiarum Hungaricae* 19, 91–176.

Mizell, R.F. & Schiffhauer, D.E. (1987) Trunk traps and overwintering predators in pecan orchards: survey of species and emergence times. *Florida Entomologist* 70, 238–244.

Monserrat, V.J. & Marín, F. (1992) Substrate specificity of Iberian Coniopterygidae (Insecta: Neuroptera). In *Current Research in Neuropterology, Proceedings of the 4th International Symposium on Neuropterology*, ed. Canard, M. Aspöck, H. & Mansell, M.W., pp. 279–290. Sacco, Toulouse.

Neuenschwander, P. (1984) Sampling procedures for chrysopid populations. In *Biology of Chrysopidae*, ed. Canard, M., Séméria, Y. & New, T.R., pp. 205–212. Dr W. Junk. The Hague.

New, T.R. (1967*a*) The flight activity of some British Hemerobiidae and Chrysopidae (Insecta: Neuroptera) as indicated by suction trap catches. *Proceedings of the Royal Entomological Society of London A* 42, 93–100.

New, T.R. (1967*b*) Trap-banding as a collecting method for Neuroptera and their parasites and some results obtained. *Entomologist's Gazette* 18, 37–44.

New, T.R. (1991) *Insects as Predators*. New South Wales University Press, Sydney.

New, T.R. (1998) *Invertebrate Surveys for Conservation*. Oxford University Press, Oxford.

Paulian, M. (1992) Eco-éthologie des pontes de chrysopes sur maïs en Roumanie (Insecta: Neuroptera: Chrysopidae). In *Current Research in Neuropterology, Proceedings of the 4th International Symposium on Neuropterology*, ed. Canard, M., Aspöck, H. & Mansell, M.W., pp 303–310. Sacco, Toulouse.

Paulian, M. (1996) Green lacewings from the southeast of the Romanian plain, as recorded by light-trapping (Insecta: Neuroptera: Chrysopidae). In *Pure and Applied Research in Neuropterolgy, Proceedings of the 5th International Symposium on Neuropterology*, ed. Canard, M. Aspöck, H. & Mansell, M.W., pp 197–202. Sacco, Toulouse.

Paulian, M. (1998*a*) Occurrence of chrysopids (Neuroptera, Chrysopidae) and activity of their populations within a sugar-beet agro-ecosystem in Romania. *Acta Zoologica Fennica* 209, 203–205.

Paulian, M. (1998*b*) Occurrence of chrysopids (Neuroptera, Chrysopidae) and moving activity of their populations within a peach orchard agro-ecosystem in Romania. *Acta Zoologica Fennica* 209, 207–210.

Pantaleoni, R.A. (1996) Distribuzione spaziale di alcuni Neurotteri Planipenni su piante arboree. *Bollettino dell'Istituto di Entomologia dell'Università di Bologna* 50, 133–141.

Pantaleoni, R.A. & Curto, G.M. (1990) I Neurotteri delle colture agrarie: Crisopidi in oliveti del Salento (Italia meridionale). *Bollettino dell'Istituto di Entomologia dell'Università di Bologna* 45, 167–179.

Pantaleoni, R.A. & Ticchiati, V. (1990) I Neurotteri delle colture agrarie: esperienze sul metodo di campionamento per abbattimento chimico. *Bollettino dell'Istituto di Entomologia dell'Università di Bologna* 45, 143–55.

Raspi, A. & Malfatti, P. (1985) The use of yellow chromotropic traps for monitoring *Dacus oleae* (Gmel.) adults. In *Integrated Pest Control in Olive Groves, Proceedings of the CEC/FAO/IOBC International Joint Meeting*, ed. Cavalloro, R. & Crovetti, A., pp. 428–440. Balkema, Rotterdam.

Ridland, P.M. (1988) Aspects of the ecology of the rice root aphid, *Rhopalosiphum rufiabdominalis* (Sasaki) and the apple-grass aphid, *Rhopalosiphum insertum* (Walker)

(Homoptera: Aphididae) in southeastern Australia. PhD thesis, La Trobe University, Melbourne.

Santas, L.A. (1987) The predators' complex of pear-feeding psyllids in unsprayed wild pear trees in Greece. *Entomophaga* 32, 291–297.

Schultz, B.B. (1988) Reduced oviposition by green lacewings (Neuroptera: Chrysopidae) on cotton intercropped with corn, beans, or weeds in Nicaragua. *Environmental Entomology* 17, 229–232.

Sengonca, Ç. & Frings, B. (1987) Ein künstliches Überwinterungsquartier für die rauberische Florfliege. *Deutsche Landwirtschafts Gesellschaft Mitteilungen* 102, 656–657.

Sengonca, Ç. & Frings, B. (1989) Enhancement of the green lacewing *Chrysoperla carnea* (Stephens) by providing artificial facilities for hibernation. *Turkiye Entomoloji Dergisi* 13, 245–250.

Southwood, T.R.E. (1978) *Ecological Methods with Particular Reference to the Study of Insect Populations.* Chapman & Hall, London.

Sunderland, K.D. (1987) A review of methods of quantifying invertebrate predation occurring in the field. *Acta Phytopathologica et Entomologica Hungarica* 22, 13–34.

Sunderland, K.D. (1988) Quantitative methods for detecting invertebrate predation occurring in the field. *Annals of Applied Biology* 112, 201–224.

Szentkirályi, F. (1992) Spatio-temporal patterns of brown lacewings based on the Hungarian light trap network (Insecta: Neuroptera: Hemerobiidae). In *Current Research in Neuropterolgy, Proceedings of the 4th International Symposium on Neuropterology,* ed. Canard, M. Aspöck, H. & Mansell, M.W., pp. 349–357. Sacco, Toulouse.

Sziráki, G. (1996) Ecological investigations of the Neuropteroidea of oak forests in Hungary (Insecta: Raphidioptera, Neuroptera). In *Pure and Applied Research in Neuropterology, Proceedings of the 5th International Symposium on Neuropterology,* ed. Canard, M., Aspöck, H. & Mansell, M.W., pp. 229–232. Sacco, Toulouse.

Tamaki, G. & Halfhill, J.E. (1968) Bands on peach trees as shelters for predators of the green peach aphid. *Journal of Economic Entomology* 61, 707–711.

Udayagiri, S., Mason, C.E. & Pesek, J.D. (1997) *Coleomegilla maculata, Coccinella septempunctata* (Coleoptera: Coccinelidae), *Chrysoperla carnea* (Neuroptera: Chrysopidae) and *Macrocentrus grandii* (Hymenoptera: Braconidae) trapped on coloured sticky traps in corn habitat. *Environmental Entomology* 26, 983–985.

Viollier, B. & Fauvel, G. (1984) Comparaison de la faune vivant sur 2 espèces de poiriers, *Pyrus amygdaliformis* Vill. et *P. communis* L., en garrigue et dans un verger abandonné de la région de Montpellier. *Agronomie* 4, 11–18.

CHAPTER 16
Interactions with plant management strategies
H. Vogt, E. Viñuela, A. Bozsik, A. Hilbeck, and F. Bigler

16.1 EFFECTS OF PESTICIDES
H. Vogt and E. Viñuela

16.1.1 Introduction
In the last 30 years investigations into side-effects of pesticides on beneficial organisms have gained more and more importance. The increase in knowledge is the basis for reducing the undesirable effects of pesticide applications, which among others, is an important principle of integrated production (Cross & Dickler, 1994). Whereas in former years companies presented data on side-effects of a pesticide on a voluntary basis to the authorities for the registration procedure, in 1986 this became obligatory in Germany (Anon., 1986; Brasse, 1990) and recently in all European countries, as regulated by the Council Directive 91/414/EEC. Factors to be considered when studying side-effects were published by Franz (1974). The IOBC/WPRS[1] working group 'Pesticides and Beneficial Organisms', founded in 1974, has contributed considerably to research of side-effects caused by pesticides. The group is working (in close co-operation with registration authorities, EPPO[2]) on the development, improvement, and validation of test methods (laboratory, semi-field, and field) and on rearing methods, and it carries out joint testing programmes to investigate the impact of pesticides on about 30 beneficial organisms (Hassan *et al.*, 1983, 1985, 1987, 1988, 1991, 1994; Hassan, 1988, 1989, 1992, 1994; Samsøe-Petersen, 1989; Vogt, 1994; Sterk *et al.*, 1999). Comprehensive data on research findings focusing on the interaction of pesticides with entomophagous arthropods have been published by Croft (1990). Our chapter will focus on the following subjects: (1) the main objectives and achievements of the test system for chrysopids as established by the IOBC

working group 'Pesticides and Beneficial Organisms' with outlook on further needs; (2) presentation of results from standard tests, comparing laboratory, semi-field, and field data as well as from special investigations, including electron microscopy.

16.1.2 Lacewings in side-effect testing
Most side-effect tests with lacewings are carried out with the common green lacewing, *Chrysoperla carnea* (Stephens) *s. lat.* as a representative of the Chrysopidae. This is because of its cosmopolitan distribution, its importance as an entomophagous predator in many crops, its wide use in biological control in glasshouse crops, and its ability to be easily reared in large numbers (Hassan, 1975; Greve, 1984; Morrison; 1985). As a foliage-dwelling predator, it is one of the relevant beneficial test species selected for regulatory requirements (Barrett *et al.*, 1994). Quite recently *Micromus tasmaniae* (Walker), a hemerobiid, which is of importance in Australia and New Zealand, has been included in side-effect studies (Rumpf *et al.*, 1997a, b, 1998).

16.1.3 The sequential testing scheme of the IOBC working group 'Pesticides and Beneficial Organisms' and future needs
A sequential scheme for testing side-effects of pesticides on larvae of *Chrysoperla carnea* with a laboratory, a semi-field, and a field method has been established by members of the IOBC working group. The methods have been published by Suter (1978), Bigler (1988), Bigler & Waldburger (1988), Vogt *et al.* (1992) and Vogt (1994). They are in agreement with the test characteristics worked out by the working group (Hassan, 1994), which guarantee adequate exposure of the organisms and allow assessment not only of initial mortality but also of delayed effects during metamorphosis as well as sublethal effects, e.g. reproductive performance, change in behaviour and beneficial capacity. The intention of the sequential testing scheme is that harmless

---

[1] International Organisation for Biological and Integrated Control of Noxious Animals and Plants/Western Palaearctic Regional Section.

[2] European and Mediterranean Plant Protection Organisation.

pesticides are screened out by the first testing tier, i.e. the laboratory test, or the other way round, that harmful effects are detected at this test level with a high probability. Therefore the exposure of the beneficial insect in the laboratory test to a fresh pesticide deposit follows a worst-case scenario (Hassan, 1989; Barrett *et al.*, 1994). The laboratory tests of the IOBC working group mainly consider the contact activity of the pesticides. Of course there are other routes of uptake of a pesticide which can influence its effect, like the direct exposure to spray droplets or via the food chain, but according to several authors as well as to the results of the IOBC group (see below) direct and residual contacts play the most important role (Franz, 1974; Croft, 1990). Depending on the pesticide, its effect via ingestion or other routes of uptake can be more severe. Therefore for each pesticide, in consideration of its mode of action and its mode of use, it has to be carefully decided if the standard residual toxicity test alone is sufficient. Moreover, for new technologies such as genetically modified plants, new test methods have to be established (cf. Hilbeck & Bigler, section 16.3 below). Further factors affecting the susceptibility are the life stage, the age of the organism, and environmental attributes (Franz, 1974; Croft, 1990). It is most important to test different life stages in order to know which one(s) are more or less susceptible. This knowledge will help in risk assessment and in recommendations for more selective use of a pesticide.

Nevertheless, by following the sequential testing scheme it has been proven in most cases that harmful effects were not overlooked and that pesticides that were revealed to be harmless under laboratory conditions did not need further testing with the more complicated test methods in the higher testing tiers. They could be considered harmless under field conditions (Hassan, 1989; Bigler & Waldburger, 1994).

In the case of harmfulness, higher testing tiers are necessary to check if the harmful effect remains under more realistic conditions. Several factors can reduce the impact of the pesticides under semi-field and especially under field conditions, e.g. UV light and rain or the plant surface can influence the degradation of the pesticides. Besides, on the plant there is a higher variability with respect to the residue and the insects can look for hiding places or avoid high-residue areas. On the other hand, the impact of a pesticide might increase under field conditions due to uptake in the food chain of the pesticide via treated prey or to the more natural behaviour of the insect, e.g. a higher mobility when searching for food.

Normally, a single treatment is applied in the standard tests as it is representative for insecticides and acaricides. But it is also possible to treat repeatedly, e.g. in the case of fungicides. Due to the short developmental time of the larvae (e.g. 14 days at 21 °C), however, the intervals between treatments have to be reduced, thus imitating again a worst-case scenario.

Dose-response studies represent another aspect of side-effect investigations and they are needed for a risk assessment with regard to different exposure scenarios, e.g. in off-crop areas contaminated by different drift rates (Forster *et al.*, 1997; Vogt & Just, 1997). All these examples underline the complexity of side-effect research. In order to allow a good judgement of possible impacts of pesticides on non-target organisms a differentiated test system is needed. Further research in developing more practice-oriented, field-approaching test designs, e.g. microcosm experiments, and careful data analysis will help to decide which test design/s is/are most suited to provide realistic estimations on side-effects.

16.1.4 Laboratory, semi-field, and field test characteristics for assessing side-effects on larvae of *Chrysoperla carnea* in comparison

In the laboratory and semi-field test the residual toxicity of pesticides is investigated by exposing first-instar larvae (2–3 days old) on a fresh dried pesticide film on glass or on plants, respectively. The highest rate of the pesticide as recommended for field use is applied, further rates are optional (e.g. dose-response investigations to describe in more detail the activity of the pesticide). In the laboratory test, mortality during larval development until adult emergence is monitored continuously, while in the semi-field test the criterion for assessing the pesticide effect is the recapture rate of adults. With regard to sublethal effects, the reproduction of survivors is always analysed in the laboratory test, while it is optional in the semi-field study. The laboratory method is actually being validated, optimised, and ring-tested by an expert group within a joint initiative of IOBC, BART[1], and EPPO. This was urgently needed to standardise the specific procedure used by various testing laboratories. Besides, test

---

[1] Beneficial Arthropod Regulatory Testing Group.

Table 16.1. *Comparison of the test design for* Chrysoperla carnea *in the semi-field and field*

	Semi-field test	Field (type b) test
Testing unit/replicate	Plastic tray with broad-bean plants	Apple tree (on root stock M9)
Number and age of test larvae per variant	3 × 20 larvae (L1, 2–3 days old)	4 × 300 larvae (L1–L2)
Application of the pesticide	Before releasing the larvae	After releasing the larvae
Location of the test	Outdoors under a transparent PVC-roof	In the orchard or on potted apple trees outdoors
Food supply	Aphids and *Ephestia* or *Sitotroga* eggs	*Sitotroga* or *Ephestia* eggs
Assessment of survivors	Collecting of adults (units are covered with a cotton cloth shortly after the appearance of pupae)	Continuously with bait cards; recaptured larvae are put back on the tree; collecting of cocoons and further observation in the laboratory until emergence of adults
Sublethal effects: fecundity, fertility	Optional with adult survivors (e.g. if effects have been observed under laboratory conditions)	

characteristics have been revised at an international workshop bringing together 35 experts in order to find consensus in the requirements for regulatory testing (Barrett *et al.*, 1994) and these revised characteristics have to be taken into consideration. The expert group has worked out an actual protocol and besides ring-testing is carrying out experiments to investigate the influence of parameters which might need standardisation (e.g. number of test organisms, food supply for larvae and adults: type of food and feeding intervals, reproduction test: number of individuals and sex ratio). The actual status of the laboratory test has been published by Vogt *et al.* (1998).

The field test is carried out with mass-reared larvae (type b test, IOBC), first to second instars. The pesticide is sprayed after the release of larvae. Survival is followed up by recapturing the larvae with bait cards at regular intervals until pupation. Cocoons found in the bait cards are collected and further observed in the laboratory until emergence of adults. A reproduction test can follow.

Although the semi-field and field test have some similarities, there are important differences, which are listed in Table 16.1.

In the field test the larvae are exposed to pesticide droplets during spraying (as this is the case under practical conditions) as well as to the residue. The larvae, according to their natural behaviour, are free to move on the whole tree, within the canopy and on the tree trunk

as well as on the ground underneath the tree. The pesticide is exposed to weathering without restriction. Due to the continuous monitoring of surviving larvae the course of the pesticide effect can be observed in detail, e.g. knockdown effects are not missed, and for insect growth regulators (IGR) it can be determined exactly which developmental stage is affected. Both methods, semi-field and field, are suited for investigating the total effect of a pesticide. Differences with regard to the size of the effect were small when comparing the outcome of both methods (Vogt, 1994). However, the field test is the more realistic approach and depending on the pesticide methodical differences such as the presence of the larvae during spraying can be of great importance for the outcome of the test.

16.1.5 Effects of pesticides on lacewings

Many publications deal with side-effects of pesticides on lacewings, mainly *Chrysoperla carnea*, but it is not the authors' intention to consider them all here. We will focus on recent findings, modern pesticides, our own results, and comparisons between laboratory and field data. Reviews were published by Bigler (1984), Croft (1990), and Bay *et al.* (1993) and information concerning modern pesticides is given in Darvas & Polgár (1998). Bigler & Waldburger (1994) present comprehensive data on laboratory and semi-field investigations and Vogt (1994) provides comparisons between semi-field and field results.

Table 16.2. *Effects of insecticides on larvae of C. carnea in the field (type b test, IOBC) in comparison to laboratory results*

Type of Insecticide	Active ingredient	Common name	Type of formulation	Active ingredient content	Tested dose %	Effect size %[a,b,c,d] Laboratory	Effect size %[a,b,c,d] Field	IOBC category[e] Laboratory[f]	IOBC category[e] Field[g]	References Laboratory	References Field
Neurotoxins											
Organophosphates	Thiocyclam	Evisect	WP	50%	0.03	58.5	28.1[a]	3	2	Bigler & Waldburger, 1994	Vogt, 1994
	Parathion	E 605 forte	EC	507.5 g/l	0.035	100	>95[a]	4	4	Grafton-Cardwell & Hoy, 1985	Wetzel et al., 1991
Synthetic pyrethroid	Phosmet	Imidan	WP	50%	0.25	100	10.5[a]	4	1	Bigler & Waldburger, 1994	Vogt, 1994
	Cyfluthrin	Baythroid 50	EC	50 g/l	0.05	100	95.9[a]	4	4	Vogt, 1994	Vogt, 1994
Natural pyrethroid	Pyrethrine, piperonyl butoxide	Spruzit flüssig	EC	36 g/l 144 g/l	0.1	84	20.2[a]	3	1	Vogt & Händel, 1995; Händel, 1996	Vogt & Händel, 1995; Händel, 1996
Insect growth regulators											
Chitin synthesis inhibitors	Diflubenzuron	Dimilin	WP	25%	0.05	100	59.4[a] + 86.1[c]	4	3-4	Vogt, 1994	Vogt, 1994
	Flufenoxuron	Cascade	DC	10%	0.05	100	63.5[a] + 54.4[c]	4	3	Vogt, 1994	Vogt, 1994
	Teflubenzuron	Nomolt	SC	15%	0.1	100	89.0[a] + 86.5[c]	4	4	Bigler & Waldburger (7th JPTP[e])	Vogt, 1994
	Triflumuron	Alsystin	WP	230 g/kg	0.05	—	100[b]	—	4		Vogt, 1994
	Lufenuron	Match	EC	50 g/l	0.08	100	88.5[a] + 100[c]	4	4	Bigler & Waldburger (8th JPTP[e])	Vogt et al., 1996
Juvenoid	Fenoxycarb	Insegar	WP	25%	0.04	100	52.4[b]	4	3	Rumpf, 1990; Vogt, 1994	Rumpf, 1990; Vogt, 1994
Ecdysone-agonist	nn	CM-001	SC	5%	0.4 lab 0.2 field	29.6	11.2	2	1	Veith, 1997	Veith, 1997
Insectistatica	Azadirachtin	NeemAzal-T/S	EC (51% plant oils)	1%	0.3	100	5.8[b] <1[c,d]	4	1	Vogt & Händel, 1995; Händel, 1996; Vogt et al., 1997	Jakob & Vogt, 1993; Jakob, 1996
	Azadirachtin	Neem-Azal-F	EC	5%	0.2	51.6	19.4[4] 4.4[c,d]	—	1-2	Vogt & Händel, 1995; Händel, 1996; Vogt et al., 1997	Jakob & Vogt, 1993; Jakob, 1996
	Azadirachtin	Neem-Azal-T	EC	5%	0.2		26.3[3] 8[c,d]				
Pyrethroid and pheromone	Cyfluthrin Codlemone	Appeal (Attract & Kill)	Gel	4% 0.065%	5 drops	—	16.7[a] 40.6[b]	—	1-2	Vogt et al., 1996; Wirth, 1997	Vogt et al., 1996; Wirth, 1997
Nitro-guanidin	Imidacloprid	Confidor	WG	665 g/kg	0.1	—	77.2[a]	—	4	Wetzel et al., 1996; Weiss, 1997	Wetzel et al., 1996; Weiss, 1997

		Formulation used in citrus							
Imidacloprid	Confidor		100 g/l	trunk treatment 0.25 g/tree	—	27.1[a]	—	2	Wetzel et al., 1996; Weiss, 1997
	Sulfan, 2 treatments	WG	800 g/kg	0.4	87.3	27.1[a] No reduction[a,b,c,d]	—	3	Vogt & Händel, 1995; Händel, 1996
Fungicide	Sulfan, 4 treatments	WG	800 g/kg	0.4	88.6	6.1[b], no reduction[a,c,d]	—	3	Vogt & Händel, 1995; Händel, 1996

Notes:

[a,b,c,d] Effect size %: [a] larvae, [b] cocoons, [c] adults, [d] total effect = preimaginal mortality and reduction in reproduction (fecundity and fertility).

[e] IOBC categories: 1 = harmless, 2 = slightly harmful, 3 = moderately harmful, 4 = harmful.

[f] Laboratory: 1 = <30%, 2 = 30%–79%, 3 = 80%–99%, 4 = >99% effect.

[g] Field: 1 = <25%, 2 = 25%–50%, 3 = 51%–75%, 4 = >75% effect.

[h] JPTP = Joint Pesticide Testing Programme of the IOBC 'Pesticides and Beneficial Organisms'.

Type of formulation: WP = wettable powder, EC = emulsified concentrate, DC = dispersible concentrate, SC = suspension concentrate, WG = wettable granule.

16.1.6 Effects of pesticides in laboratory and semi-field tests with larvae of *Chrysoperla carnea*

Bigler & Waldburger (1994) tested the toxicity of 152 pesticides in the laboratory and 55 of them in the semi-field by exposing larvae to fresh dried residues. The tests were performed according to the methods described by Bigler (1988) and Bigler & Waldburger (1988). As predicted, insecticides were more harmful than fungicides and herbicides. In laboratory tests the authors found 71.1% of insecticides, 26.2% of fungicides, 20.7% of herbicides, and one out of five plant growth regulators with an effect over 50%. In the semi-field the values above 50% dropped to 50% for insecticides and 6.3% for fungicides. The authors emphasise that 'the data show a relatively high consistency of the results, proving the reliability of the test methods. The hypothesis that products shown harmless in the laboratory do not need further testing in the semi-field and/or field is confirmed in general. However, a few exceptions are observed and we conclude that a careful case-by-case evaluation is necessary.' The authors also ask for additional test methods to study the effects of pesticides on eggs and adults.

16.1.7 Effects of pesticides on larvae of *Chrysoperla carnea* in field tests

Several insecticides, screened out as harmful in the laboratory, have been tested in the field by the author and co-workers at the Biologische Bundesanstalt Institute at Dossenheim (Germany). Table 16.2 summarises the results, in comparison with laboratory data. It is obvious that in no case were lower effects observed in the higher testing tier. Thus again, the sequential testing scheme has been confirmed. Comparing the field and laboratory data, significantly lower effects were found in the field with phosmet, natural pyrethrum, thiocyclam, the azadirachtin products, and sulfur. Even repeated treatments with sulfur up to four times within 3–4 days did not result in harmful effects. Similar results were obtained by Wetzel & Dickler (1994) for *Trichogramma*: in field tests sulfur and natural pyrethrum induced much lower effects than in the laboratory. The most important reason for the lower impact of these pesticides in the field is their quicker degradation. For example, azadirachtin is known to be degraded on the leaf surface due to UV light within a short period, but due to its translaminar action as well

as some systemic characteristics it is still effective on phytophagous insects for up to 6–8 weeks (Kienzle *et al.*, 1997; Vogt *et al.*, 1997). Such characteristics of an insecticide render it rather selective for foliage-dwelling predators.

In the case of thiocyclam a knockdown effect was observed. Whereas recapture of larvae on the treated trees was significantly reduced at the first checking date by more than 60%, recovery occurred during the following days, the recapture rates increased again and differences between treated and untreated trees were small.

In contrast to the above-mentioned pesticides the synthetic pyrethroid cyfluthrin was highly toxic in the field test. Only very few survivors were recaptured over six days of observation (mean number per tree: 4.8 versus 118.2 in the control). The harmful effect also persisted in the case of the chitin-synthesis inhibitors and the juvenoid Insegar. Effects of the chitin-synthesis inhibitors were visible only a few days after the treatments. The recapture rates of surviving larvae dropped continuously as the insecticides caused mortalities during the moulting of the larvae into the next larval stage. Malformed larvae were found often. Some larvae survived and were able to spin a cocoon, but the mortality in this stage was high and only few adults hatched. The most severe effects were observed with Alsystin, Match, and Nomolt (Table 16.3). With fenoxycarb, corresponding to its mode of action, the recapture rates of larvae did not differ between untreated and treated groups. But pupation was delayed on the Insegar-treated trees for one week and the number of cocoons was significantly lower than on the control trees. The proportion/percentage of permanent larvae, intermediate forms between larvae and pupae, pupae without cocoons dying later, and larvae dying before pupation was much higher for Insegar than for the control.

The field tests revealed that the IGRs have a high contact activity. This has been proved by the author in a special laboratory test with diflubenzuron, where the food was not offered on the contaminated glass arena, but the larvae were transferred to uncontaminated petri dishes to feed there for 1 hour during the first days of the test. After each feeding period the larvae were again put back into the testing arenas. The typical IGR effects were visible after a few days and finally all larvae died (H. Vogt, unpublished data). Moreover, these IGRs are very persistent, lasting up to several months (Austin & Hall, 1981; Austin & Carter, 1986; Marshall *et al.*, 1988;

Table 16.3. *Mean number of cocoons per tree during the whole period of the test and percentage reduction*

Insecticide	Active ingredient	Untreated	Treated	Reduction[a]
Alsystin	triflumuron	13.3 ± 5.0	0	100%
Cascade	flufenoxuron	13.7 ± 3.4	5.0 ± 2.0	63.5%
Dimilin	diflubenzuron	10.6 ± 3.0	4.3 ± 2.3	59.4%
Insegar	fenoxycarb	12.6 ± 4.4	6.0 ± 1.5	52.4%
Match	lufenuron	6.5 ± 2.5	0.7 ± 0.9	88.5%
Nomolt	teflubenzuron	24.5 ± 1.3	2.7 ± 2.8	89.0%

Note:
[a] According to Abbott (1925).

Weiss & Vogt, 1994). Thus the harmful effect lasts for a long period. In the case of diflubenzuron, when *C. carnea* larvae were exposed to treated leaves collected in the field at different intervals after the treatment, the size of the effect (mortality according to Abbott, 1925) amounted to 61.5% 40 days after the treatment and to 32.7% after 103 days (Schaaf & Vogt, 1995). In addition, chitin-synthesis inhibitors can impair reproduction of adults. For example topical application of diflubenzuron and flufenoxuron to 1-day-old *C. carnea* adults reduced fecundity slightly (diflubenzuron: 23%, flufenoxuron 17%) and fertility significantly (diflubenzuron: 51%, flufenoxuron: 89%) (Vogt, 1992).

The investigation with imidacloprid is an example of how the mode of application influences the effect size. When spraying, the contact activity of this insecticide reached its full extent resulting in high mortalities of the larvae (effect size: 77%), whereas localised trunk treatment reduced the effects significantly (effect size: 27%).

With regard to sublethal effects, the analysis of reproduction was feasible with adults collected from the field tests with phosmet, natural pyrethrum, the neem preparations, and sulfur. No or negligible reductions in fecundity and fertility were observed.

16.1.8 Effects of pesticides on developmental stages other than larvae
Besides the larvae, of course all developmental stages can be exposed to pesticides. While eggs and pupae are rather well protected and most often less sensitive than larvae (Bartlett, 1964; Grafton-Cardwell & Hoy, 1985), this is not the case for the soft-bodied adults (Croft & Brown, 1975; Bigler, 1984; Croft, 1990; Bozsik, 1991;

1995). With topical application of E 605 (parathion), Roxion (dimethoate), and Karate (λ-cyhalothrin) for example, the adults are revealed to be more sensitive (Table 16.4, cf. Vogt & Just, 1997). When exposing adults and larvae to residues of Spinosad (a naturally derived insecticide from an actinomycete) with Spinosyn A and D as active components, both developmental stages survived the actually recommended field rate for orchards (200–300 ml Spinosad/ha). By increasing the rate, however, adults were more susceptible than the larvae (H. Vogt & E. Viñuela, unpublished data). Considering the different ways of pesticide contamination in adults, the side-effect studies may include topical application, direct spraying, and residual contact as well as ingestion via food and drinking water.

As mentioned above, the eggs are rather resistant. There are only a few cases known, in which they are harmed. F. Bigler & M. Waldburger (unpublished data) sprayed *Chrysoperla* eggs with 13 different pesticides in aqueous solution at the highest recommended field rate. Only rape-seed oil (Telmion) reduced hatching up to 41.5%. All the other pesticides (insecticides: phosalone, lufenuron, abamectine, difenthiuron, pyriproxifen; fungicides: thiram, carbendazim, sulfur; herbicides: mecoprop, chloridazon, metazachlor, cycloxdim) had no or a very low impact on the hatching rate (0%–14%). Senior *et al.* (1997) applying the IGR triflumuron in acetone topically to eggs reported a decrease in survival of successive developmental stages: at 0.01 mg active ingredient 20%, at 0.1 and 1.0 mg active ingredient only 10% of the eggs were able to develop to the adult stages opposed to 55% for the acetone controls. However, it has to be considered that using acetone probably increased the effect of the pesticide and the

Table 16.4. LD_{50} (µg active ingredient/g) 7d of Chrysoperla carnea, L2 and adults

Insecticide	Active ingredient	C. carnea females	C. carnea males	C. carnea larvae (L2)
E 605	Parathion	2.52	2.2	14.83
Roxion	Dimethoate	9.84	6.98	15.48
Karate	λ–cyhalothrin	3.28	1.17	6.39

rates applied were higher than compared to droplets that could reach an egg by a field application.

16.1.9 Pesticide induced effects at the cellular level

Investigations with the electron microscope allow the detection of target sites of a pesticide at the cellular level and help to explain pesticide–induced effects observed macroscopically. Rumpf *et al.* (1992) investigated the fat body of *Chrysoperla carnea* larvae after exposure to fenoxycarb (fresh residue), the fat body being the most active organ in the metabolism of insect larvae and a target for juvenile hormones. Differences between treated and untreated insects appeared with pupation. Whereas shortly after the moult into pupae the fat body of the control animals developed into a mere storage organ for lipid, protein, and glycogen, which provides the insect with energy needed for the further metamorphosis, this storage function was prevented in the fenoxycarb-treated animals. The lack of energy seems to be an important reason why the fenoxycarb-treated insects failed to pupate. A detailed description is given in Rumpf (1990).

Within a joint co-operative project the authors recently investigated the ultrastructure of *C. carnea* larval muscles and changes induced by the insect growth regulator azadirachtin. The methods used were as follows. First-instar larvae were exposed to fresh dry residue of NeemAzal-T/S and Align (see Vogt *et al.*, 1998, for details about exposure). Three days afterwards when typical IGR malformations were observed, larvae were pre-fixed with 2.5% glutaraldehyde in Palade buffer 1 M (pH = 6.9) for 2–4 min at −9 °C, following the technique of Hayat (1986). Insects were then cut into two to three sections to facilitate the fixative penetration and left in glutaraldehyde overnight at 4 °C. After several rinses in Palade buffer at room temperature under a fume hood, sections were again placed in buffer overnight. Post-fixation was carried out in 2% osmium tetroxide in 1 M Palade buffer under the same

conditions, for 1 h. Tissues were rinsed in three 15-min changes in distilled water before being dehydrated in eight 30-min acetone series (20%–100%). Tissues were embedded in three 2-h changes in acetone/Epon-Araldite (2:1; 1:1; 1:2) and then left in 100% Epon-Araldite overnight, before being polymerised at 70 °C for 3 days. Thin sections were prepared on an Ultracut E microtome (Reicher/JWG) and stained with uranyl acetate and lead citrate before being examined in a Zeiss-902 electron microscope.

The following results were obtained. Young *C. carnea* larvae treated with azadirachtin by residual contact showed, among other symptoms, great inhibition of locomotion and uncoordinated movements. Similar symptomology has also been observed in *Ceratitis capitata* (Wiedemann) (Diptera) larvae but after the ingestion of the product (Adán *et al.*, 1998). In *Chrysoperla carnea*, alterations of mobility could be partly attributed to the fact that some larvae showed intestine extrusion or were unable to shed off completely the old cuticle during the moulting process (Vogt *et al.*, 1998). The former factor caused larvae to stick to the floor of the cages preventing them from moving. The latter, being a movement impediment, caused remnants to remain attached to the end of the body of those larvae that underwent abnormal moults.

Under the electron microscope, other abnormalities were noticeable in muscles of treated *C. carnea* larvae, which might also have contributed to the lack of normal movements. Control larvae exhibited typical longitudinal muscles with a dense fibre packing and Z-bands (Z) clearly visible (Fig. 16.1). At higher magnification (Fig. 16.2), functional mitochondria (arrowheads) near the Z-bands (Z) and abundant glycogen (G) could be distinguished among the muscle fibres (F) of a control larva. On the other hand, longitudinal muscles of treated larvae showed distorted areas of fibre arrangement and a much less dense fibre packing than in controls (Fig. 16.3). The observation of another section (Fig. 16.4) revealed the presence of many

Fig. 16.1. Longitudinal muscles from a control *Chrysoperla carnea* larva with a more dense fibre packing than the treated ones. G = glycogen; Z = Z-band (scale bar 1.7 μm).

Fig. 16.2. Functional mitochondria (arrowheads) in the muscles of a control *Chrysoperla carnea* larva. F = fibrils; G = glycogen; Z = Z-bands (scale bar 0.6 μm).

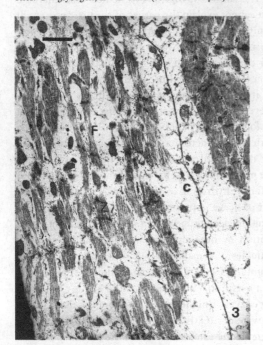

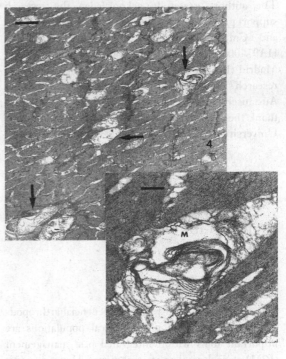

Fig. 16.3 Longitudinal muscles from a *Chrysoperla carnea* second-instar larva treated with azadirachtin (Align 0.15%, 6 μg/cm²) via residual contact showing a very low fibre packing. C = cell union; F = fibrils; Z = Z-band (scale bar 1.7 μm).

Fig. 16.4. Altered mitochondria (arrows) in the muscles of a *Chrysoperla carnea* second-instar larva treated with azadirachtin (NeemAzal-T/S 0.3%, 12 μg/cm²) via residual contact. (scale bar 2.5 μm). Inset: Detail of a degenerated mitochondrion with a myelin figure (M) (scale bar 0.6 μm).

altered mitochondria (arrows), some of them having myelin (M) figures inside (Fig. 16.4 inset), which is a clear sign of degeneration.

The ultrastructural observations noted in skeletal muscles of *C. carnea* suggest a decrease in the muscular mass of treated insects and in the available energy for movement. The inhibition of larval movement detected in *C. carnea* larvae treated with azadirachtin via residual contact, and the lesions observed in muscles under the microscope, seem to indicate a protein involvement in the action of azadirachtin.

The alterations detected in muscles of *C. carnea* larvae are not typically caused by azadirachtin because they are rather similar to those described by Budia & Viñuela (1994) in *Ceratitis capitata* larvae, when insects were reared in the presence of the IGR cyromazine. The similarity of some lesions caused by different IGRs might give rise to the notion that these insecticides, even though they can have very different primary modes of action, may have a similar secondary one.

Acknowledgements

The authors gratefully acknowledge the research support provided by the Spanish Ministry of Education and Culture (Concerted Action Spain–Germany HA97–0005) and the Autonomous Community of Madrid (Project 06M/022/96) to E. Viñuela and the research support provided by Deutscher Akademischer Austauschdienst (DAAD), Bonn, to H. Vogt. We also thank the Electron Microscopy Centre (Complutense University of Madrid) for the micrographic facilities.

16.2 DETERMINATION OF ACETYLCHOLINESTERASE ACTIVITY AS A HELPFUL TOOL FOR ASSESSING PESTICIDE SIDE-EFFECTS IN LACEWINGS

A. Bozsik

16.2.1 Introduction

The reintroduction or release of beneficial arthropods and the preservation of their local populations are important tools within integrated pest management (IPM) as ecologically safe techniques. However pesticides are also used in modern agriculture, as biological control alone most often does not solve all pest and disease problems. The most crucial requirement for

pesticides is that they must be compatible with biological control. Therefore only those pesticides should be used that are most selective and which have no unacceptable adverse effects on beneficial organisms (Hassan, 1989; Cross & Dickler, 1994). It may also be helpful to find and select tolerant or resistant strains of natural enemies (Hoy, 1979, 1985; Grafton-Cardwell & Hoy, 1985). However, we have to be aware that in this case only the selected strain of a species will tolerate the pesticide application and that other populations of that species still can be harmed. A prerequisite for an environmentally safe plant protection is the extensive study of the effects of pesticides on beneficial species.

16.2.2 Methodical background

In general, the assessment of the toxicity of a pesticide to a natural enemy is made empirically, i.e. that standard tests (laboratory, semi-field, field tests) are conducted using recommended rates of commercially available pesticide formulations in order to estimate according to accepted categories their toxic effect (Hassan *et al.*, 1985; Hassan, 1988; Vogt, 1994). In most cases this kind of test cannot explain the nature of the toxic process. It can only give an answer whether the tested rate of a pesticide is toxic or non-toxic to a beneficial organism under consideration. This is of course valuable information for the practical use of the pesticide as it indicates its compatibility or incompatibility with IPM. However, there is another fascinating opportunity but only for the acetylcholinesterase (AChE)-inhibiting insecticides: the determination of the interaction of the AChE of a natural enemy with insecticides. AChE is one of the most important enzymes in insects. It is responsible for the hydrolysis of acetylcholine (chemical transmitter of nerve impulses at synaptic nerve endings) to choline and acetate and thus it must be the target site of inhibition by organophosphates (OP) and carbamate insecticides (Smallman & Mansingh, 1969; Matsumura, 1985). The determination of AChE activity is a useful tool because enhanced activity of detoxifying enzymes (mono-oxidases, esterases) also reduced sensitivity of AChE to the inhibition induced by an insecticide and can be the cause of the tolerance or resistance of an organism. Analysis and characterisation of the insecticidal AChE and its response to inhibitors provides a good method for measuring and comparing natural or changed tolerance of different species or different field populations of a species, for monitoring field applications with regard to potential adverse effects, and for detecting pesticide pressure in a habitat. The aforementioned different populations should be emphasised since in the course of the empirical testing mostly reared individuals are used and great effort is made to select highly uniform individuals as test subjects. Here I have to anticipate that the use of reared organisms in the routine pesticide side-effect testing is essential mainly when sublethal effects (impact on the beneficial capacity, i.e. parasitism, number of fed prey, or reproductive performance, i.e., egg-laying) are studied. However, during rearing the naturally occurring tolerance or resistance of the original sample (Pree *et al.*, 1989) and the fitness of individuals derived from the sample (Jones *et al.*, 1978) may decrease with time; but also from the point of view of population toxicology (in nature a population is always composed of individuals varying in age, developmental stage, weight, and state of health), the use of uniform individuals may be a cause of failure of prediction. Therefore for truly predictive results, individuals of field populations should be caught and tested (Robertson & Worner, 1990). Nevertheless, with careful quality control of insect rearing the danger of loss of tolerance can be avoided or significantly reduced.

The Ellman method (Ellman *et al.*, 1961) is a widely used, simple and flexible *in vitro* method which is amenable to automated techniques able to process a large number of samples in a short period of time with a reasonable degree of precision. Its special advantage – compared with the empirical testing – is that fewer individuals and less time are needed and it gives more precise information about the physiological background of the insects susceptibility. The method itself was originally developed for measuring the inhibition of cholinesterase activity by OPs and carbamates but insecticides of other classes (such as pyrethroids and insect growth regulators) can be tested (Rumpf *et al.*, 1997*b*), if their metabolites are able to react as expressed below. In principle, substrate (in most cases acetylthiocholine) and enzyme [tissue (brain/head) homogenizate] interact, which results in the thiocholine. This reacts with 5,5-dithiobis-2-nitrobenzoate (DTNB) to give the yellow anion of 5-thio-2-nitro-benzoic acid. The rate of the production of the coloured ion is recorded at 412 nm on a spectrophotometer (Silver, 1974). When inhibitor (only effective active ingredient) is added the quantity of coloured ion reduces and this

change, the change in enzyme activity, can be easily measured. Although the method itself is really comprehensive, for characterisation of the tolerance the I_{50} (molar concentration of an anticholinesterase which causes 50% inhibition of AChE; M) and k_i (bimolecular rate constant; M^{-1} min^{-1}) values are used. The smaller I_{50} and the higher k_i values are, the more potent the inhibitor or the more sensitive the AChE is. However, it should be stressed that results gained this way cannot substitute for empirical data because they cannot reflect differential abilities of compounds to penetrate cuticle and gut barriers or their differential solubilities in insect haemolymph (Corbett *et al.*, 1984). Nevertheless, they can exactly characterise inhibitory properties of a compound and resistance mechanisms. In most cases LC_{50} values as good indicators of direct insecticidal potencies should be added.

Experiments were performed as previously described (Bozsik *et al.*, 1996). Biochemical measurements giving the results to be commented as follows here have been carried out by A. Bozsik, E. Haubruge, and C. Gaspar in 1992, 1993, 1995, and 1997 in Belgium (Zoologie générale et appliquée, Faculté des Sciences agronomiques de Gembloux).

16.2.3 Results and discussion

In this section some preliminary results will be presented to demonstrate the use of the Ellman procedure on the example of common green lacewing, *Chrysoperla carnea s. lat.* It should be mentioned that we do not know to which taxon of this complex the individuals used for testing belonged. The only exception is the 1997 Belgian sample containing probably exclusively *Chrysoperla kolthoffi* (Navás) specimens. Five Belgian and one Hungarian populations were tested. Diazinon proved to be the weakest inhibitor. Carbofuran, malaoxon, and carbaryl were the most active pesticides and paraoxon, dicrotophos, chlorfenvinphos, and schradran were moderately harmful to lacewing AChE. Although because of the lack of higher number of repetitions (for calculating the I_{50}, k_i values the heads of 2 or 3 times 40 individuals have been used), it was not possible to calculate significant differences between the values of the various chrysopid populations. It seems that the population collected at the Hungarian uncultivated area was more susceptible to malaoxon compared with the Belgian ones, and similarly the ParkGe92 (Belgium) population showed considerably greater sensitivity to

Table 16.5. *Susceptibility of AChE in* Chrysoperla carnea *adults originating from various habitats to different pesticide active ingredients*

Lacewing populations and pesticides	I_{50} (M)	k_i (M^{-1} min^{-1})	r^b
ParkGe92			
Paraoxon	1.0×10^{-5}	6.79×10^4	0.963
Malaoxon	3.8×10^{-5}	1.82×10^4	0.975
WheatGe93			
Paraoxon	1.9×10^{-4}	3.53×10^3	0.940
Malaoxon	4.8×10^{-5}	1.44×10^4	0.957
ReserveGe93			
Paraoxon	1.3×10^{-4}	5.30×10^3	0.995
Malaoxon	6.1×10^{-5}	1.14×10^4	0.983
UncultGö95			
Malaoxon	4.3×10^{-6}	1.62×10^5	0.980
ExpGe95			
Diazinon	5.4×10^{-3}	1.30×10^2	0.982
Carbaryl	4.8×10^{-5}	1.46×10^4	0.997
Schradran	$>1.0 \times 10^{-3}$	—	—
ExpGe97			
Ethoprophos	$>4.0 \times 10^{-3}$	—	—
Carbofuran	2.5×10^{-7}	2.80×10^6	0.962
Dicrotophos	1.7×10^{-4}	3.97×10^3	0.984
Chlorfenvinphos	4.9×10^{-4}	1.41×10^3	0.911

Notes:

Abbreviations: Ge, Gembloux, Belgium; Gö, Gödöllö, Hungary; Habitats: Park, park; Wheat, wheat field; Reserve, nature reserve; Uncult, uncultivated area; Exp, experimental area; Dates: 92, 1992; 93, 1993; 95, 1995; 97, 1997.
[a] The order of sensitivity of adult lacewing AChE to pesticides was the following: carbofuran > malaoxon > carbaryl > paraoxon > dicrotophos > chlorfenvinphos > schradran > ethoprophos > diazinon.
[b] r, coefficient of correlation of the dose-inhibition relationship (Bozsik *et al.*, 1996 and unpublished data).

paraoxon than the others (Table 16.5). In case of schradran and ethoprophos the accurate I_{50} value could not be computed because not even the highest dilutable concentration caused measurable inhibition. The AChE was practically insensitive against these compounds. The variation of data between the Belgian values was relatively unimportant seeing the malaoxon sensitivity. The serious difference of the Hungarian population from these values may be due to the limited use of malaoxon in Hungary (at present only one malathion product is registered, and it is practically impos-

sible to purchase it). Concerning paraoxon, the variation of the toxicological parameters was much more explicit, and logically the population of the wheat field seems to be the most tolerant. This can be due to the presumably higher selection pressure of the AChE-inhibiting insecticides used here. The similarly increased tolerance of the natural reserve's population can be explained by the immediate neighbourhood of these areas which could render possible the contact of populations and mutual change of their genetic material. The park is quite isolated by a distance of

Table 16.6. *Susceptibility of AChE in* Chrysoperla carnea *third-instar larvae in a wheat field, Gembloux, Belgium, 1993*

Lacewing populations and pesticides	I_{50} (M)	k_i $(M^{-1} min^{-1})$	r^a
Paraoxon	1.9×10^{-6}	3.74×10^{5b}	1.000^b
Malaoxon	1.1×10^{-5}	6.26×10^4	0.884

Notes:
[a] Coefficient of correlation of the dose-inhibition relationship.
[b] Rough estimation because $n = 2$.
Source: A. Bozsik, E. Haubruge & C. Gaspar, unpublished data, 1993.

several kilometres and also by high buildings and extensive hedges from the field and the reserve. The strong correlation of dose-inhibition relationship supported by the coefficient of correlation might prove this evaluation.

A few measurements have also been carried out on lacewing larvae which proved to be slightly more sensitive than adults (Table 16.6). The order of sensitivity seems to be reversed as compared with the adults: paraoxon > malaoxon. Considering that only the malaoxon data may be reliable the difference between the adult and larva values cannot be significant.

In the future pesticides should be developed to be more efficient against pests than against beneficials, and we should not find only by chance some of them that satisfy the need for harmlessness towards natural enemies. The solution may be found by studying the interaction of natural enemies with pesticides (e.g. biochemical characterisation of their metabolising enzymes, mechanisms of detoxification). Central points are differences of defence mechanisms between pests and their enemies, for example, plant-feeding insects are efficient oxidisers but predators and parasitoids are good hydrolytic (using esterases) metabolisers thus pesticides metabolised by hydrolytic processes can be safer for natural enemies than for phytophagous pests (Plapp & Bull, 1985); and also the susceptibility of target sites of pesticides in pest and beneficials may be of importance. The Ellman procedure can be an efficient tool for the last point of this challenging task.

Acknowledgements
The author thanks Dr Heidrun Vogt for helpful comments and suggestions on the manuscript.

16.3 EFFECTS OF *BACILLUS THURINGIENSIS* VIA INGESTION OF TRANSGENIC CORN-FED PREY AND PURIFIED PROTEINS
A. Hilbeck and F. Bigler

16.3.1 Introduction
Engineered insecticidal crop plants containing the gene from *Bacillus thuringiensis* (Berliner) (*Bt*) var. *kurstaki* (Berliner) that encodes for the expression of an insecticidal δ-endotoxin have been commercially produced in North America and other countries in the world since the mid-1990s (corn, cotton, and potatoes). The large-scale commercial cultivation of more and other transgenic Bt crop plants is imminent (Hoyle, 1995; Niebling, 1995; Mellon & Rissler, 1998). Therefore, and because of the simultaneous increase in use of Bt insecticides, Bt proteins are becoming very widespread, highly bioactive substances in the agroecosystems of many countries throughout the world.

Products containing Bt proteins have been used in agriculture for several decades and are commonly considered to have little or no effect on natural enemies of pest insects (Flexner *et al.*, 1986; Melin & Cozzi, 1989; Croft, 1990). This commonly accepted knowledge is primarily based upon previous studies that were designed to test for undesired side-effects of Bt compounds on beneficial insects when used as a foliar insecticide (Croft, 1990). However, Bt insecticides and transgenic plants expressing Bt toxins differ in a number of aspects that do not allow the simple deduction of the long-term selectivity of Bt crop plants from the long record of safe use of Bt insecticides.

The drastically extended temporal and spatial

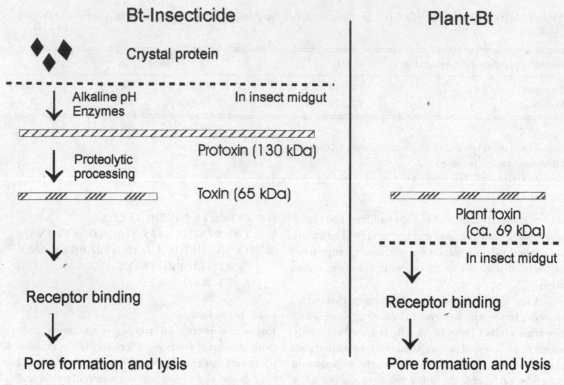

Bt-Insecticide

Crystal protein

Alkaline pH   In insect midgut
Enzymes

Protoxin (130 kDa)

Proteolytic
processing

Toxin (65 kDa)

Receptor binding

Pore formation and lysis

Plant-Bt

Plant toxin
(ca. 69 kDa)

In insect midgut

Receptor binding

Pore formation and lysis

Fig. 16.5. Differences between Bt proteins present in most insecticidal formulations and expressed in transgenic corn plants.

availability of Bt toxins in agroecosystems is ecologically most important. Bt toxins are expressed in various major crop plants in essentially all plant parts throughout the entire field season from germination to senescence of the plants (Perlak, *et al.*, 1990; Koziel *et al.*, 1993). Further, Bt insecticides consist of a mixture of spores and crystalline inclusion bodies that need to undergo a complex activation process to induce the lethal effects in a susceptible insect. In transgenic plants, the Bt protein is present as activated toxin (Fig. 16.5).

With the use of insecticide applications, only chewing herbivores could ingest the Bt proteins, whereas Bt toxins in transgenic plants can also be ingested by sucking arthropods. Consequently, most, if not all, non-target herbivores colonising transgenic Bt plants in the field during the season will ingest plant tissue containing Bt proteins which they may pass on to their natural enemies in a more or less processed form. This ubiquitous and almost permanent availability of Bt proteins in agroecosystems in addition to its modified form of release makes it necessary to verify and

monitor the compatibility of this new pest management strategy with natural enemies (Jepson *et al.*, 1994). Enhancing and preserving naturally occurring biological control is one of the most important components of modern pest management. As suggested by Jepson *et al.* (1994), extended dietary exposure bioassays of natural enemies seem to be one logical approach toward this goal.

In the past, a major obstacle in conducting bioassays with *Chrysoperla carnea* was that the larvae typically suck their food from within a substrate. Mass-rearing techniques of *C. carnea* typically involve either live prey, such as aphids, or insect eggs such as *Ephestia kuehniella* (Hübner) or *Sitotroga cerealella* (Olivier) (Ridgway *et al.*, 1970; Hassan, 1975; Morrison *et al.*, 1975; Morrison & Ridgway, 1976). In previous studies investigating direct effects of Bt proteins on *C. carnea*, the surface of *S. cerealella* eggs was coated with the Bt proteins and subsequently fed to *C. carnea* larvae (Croft, 1990; Sims, 1995). However, because *C. carnea* suck out the egg contents without ingesting the shells, they most likely ingested little to no Bt protein. In

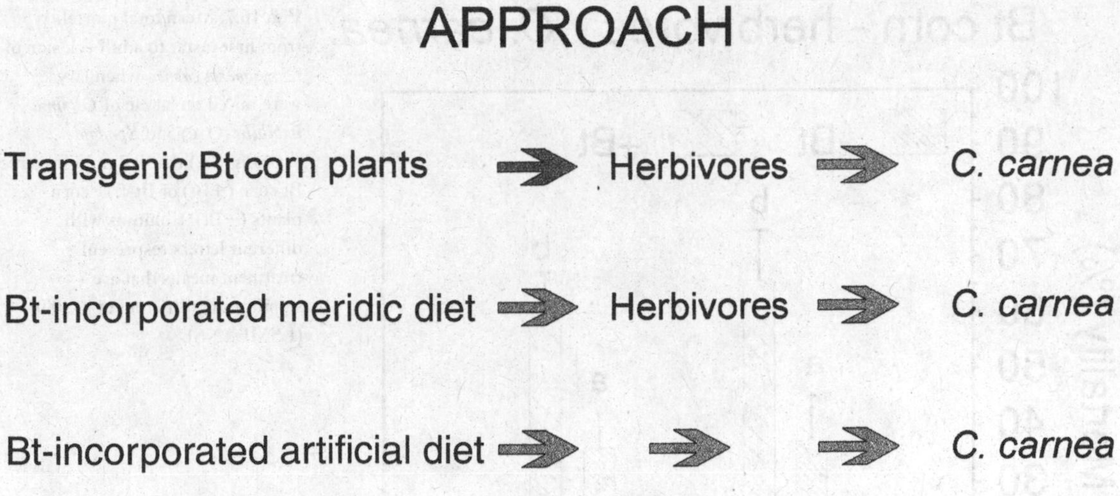

APPROACH

Transgenic Bt corn plants ➡ Herbivores ➡ *C. carnea*

Bt-incorporated meridic diet ➡ Herbivores ➡ *C. carnea*

Bt-incorporated artificial diet ➡ ➡ ➡ *C. carnea*

Fig. 16.6. Overview of conducted experiments.

another study, Bt-containing pollen of transgenic plants was provided to test for side-effects of Cry1Ab toxin on *C. carnea* larvae, expressed for example in transgenic corn (Pilcher *et al.*, 1997). However, concentrations of Cry1Ab toxin in pollen is low (2.57–2.94 μg/g dry weight) compared to that of leaves (Fearing *et al.*, 1997) and predaceous *C. carnea* larvae feed only to a very limited extent, if at all, on pollen.

In an effort to improve efficient mass-production techniques for biological control purposes, attempts have been made to develop an artificial diet for chrysopid larvae. The larvae were either supplied with an artificial diet soaked into a sponge, droplets of an artificial diet, or an artificial diet encapsulated into paraffin spheres (Hagen & Tassan, 1965; Vanderzant, 1969; Bigler *et al.*, 1976; Martin *et al.*, 1978). The encapsulation technique of a liquid artificial diet into tiny paraffin spheres has been markedly improved and mechanised by a German company (STB Control, Aarbergen, Germany) and is being successfully used by this company for mass-rearing of *C. carnea* as a biological control agent.

16.3.2 Methods and results

Prey-mediated and direct effects of Bt proteins including those expressed in transgenic corn (Cry1Ab toxin) on *Chrysoperla carnea* larvae were investigated in three different series of experiments (Fig. 16.6). Concomitantly to the experiments, insecticidal activity and

effects of the various Bt-protein-containing diets on the prey species were monitored by conducting bioassays with *Ostrinia nubilalis* (Hübner) and *Spodoptera littoralis* (Boisduval) (for details see Hilbeck *et al.*, 1998*a,b*, 1999).

16.3.3 Prey-mediated effects of Bt-protein-containing diets on immature *Chrysoperla carnea*

Transgenic Bt-corn

In a first series of experiments, individual *Chrysoperla carnea* larvae were raised on two different lepidopterous herbivore species that had fed either on transgenic Bt corn or Bt-free corn for 24 h prior to being provided as prey to *C. carnea* larvae. The two lepidopterous species used were the European corn-borer, *Ostrinia nubilalis* (target pest), and *Spodoptera littoralis* (non-target pest for Bt) (for technical details see Hilbeck *et al.*, 1998*a*).

Mean total immature mortality of 66% and 59% was detected when chrysopid larvae were raised on Bt-corn-fed *O. nubilalis* or *S. littoralis* larvae, respectively, compared with 37% when raised on Bt-free prey of both species (Fig. 16.7). No significant differences in prey-mediated mortality of *C. carnea* between prey species were detected regardless of the plant variety used. These results suggested that the reduced fitness of chrysopid larvae was associated with Bt.

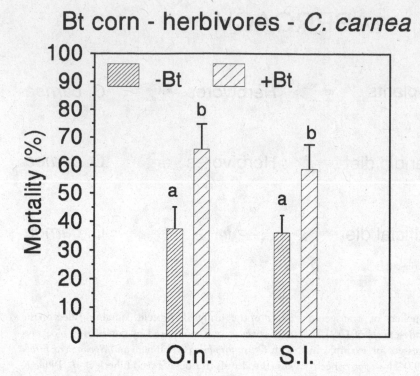

Bt corn - herbivores - *C. carnea*

Fig. 16.7. Mean total mortality from first instar to adult eclosion of *Chrysoperla carnea* when larvae were raised on larvae of *Ostrinia nubilalis* (O.n.) and *Spodoptera littoralis* (S.l.) that fed on transgenic Bt corn (+Bt) or Bt-free corn plants (−Bt). Columns with different letters respresent treatment means that are significiantly different at $p = 0.05$ (LSMEANS).

Bt-incorporated meridic diet

In another series of laboratory experiments, Bt-incorporated diets were used as prey food source instead of transgenic Bt corn. Activated Cry1Ab toxin and the protoxins of Cry1Ab and Cry2A were incorporated into standard meridic diet for *S. littoralis* larvae at various concentrations (Cry1Ab toxin: 100 µg/g diet, 50 µg/g diet, and 25 µg/g diet; Cry1Ab protoxin: 200 µg/g diet, 100 µg/g diet, and 50 µg/g diet, i.e. equivalent concentration of activated toxin; Cry2A protoxin: discriminating concentration 100µg/g diet). Individual *C. carnea* larvae were raised on *S. littoralis* larvae fed one of the described treated meridic diets (for further technical details see Hilbeck *et al.*, 1999).

Mean total immature mortality for chrysopid larvae reared on Bt-fed prey was again always significantly higher than in the control (26%) (Fig. 16.8). Total immature mortality of *C. carnea* reared on prey fed on Cry1Ab toxin 100 was highest (78%) and declined with decreasing toxin concentration. Prey-mediated total mortality of Cry1Ab protoxin exposed chrysopid larvae was intermediate to Cry1Ab toxin and Cry2A protoxin exposed *C. carnea* larvae and did not exhibit a dose response (Fig. 16.8).

16.3.4 Direct effects of activated Cry1Ab toxin

To investigate a direct Bt-induced effect, a novel bioassay technique was used that allowed for incorporation of the activated Cry1Ab toxin into a liquid diet specifically developed for optimal nutrition of *Chrysoperla carnea* (see section 16.3.1, above). This medium was then encapsulated within small paraffin spheres and fed to *C. carnea* larvae. Because only second and third instars can penetrate the skins of the paraffin spheres, two different methods were used to rear chrysopid larvae through first instar. The first method used small foam cubes soaked with non-encapsulated, liquid diet. Activated Cry1Ab toxin (100 µg/ml diet) was mixed into the non-encapsulated diet whereas only an equivalent amount of double-distilled water was added to the diet for the corresponding control. The second method used *Ephestia kuehniella* (Hübner) eggs as prey during first instar. After reaching second instar, all larvae received an encapsulated, artificial diet with or without Cry1Ab, respectively. In a fifth treatment, chrysopid larvae were raised on *E. kuehniella* eggs only (for technical details see Hilbeck *et al.*, 1998*b*).

When reared only on an artificial diet containing Cry1Ab toxin, total immature mortality was signifi-

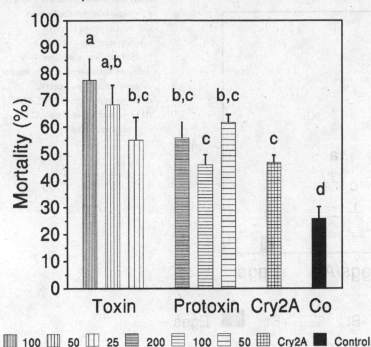

Fig. 16.8. Mean total mortality from first instar to adult eclosion of *Chrysoperla carnea* when raised on *Spodoptera littoralis* that were fed meridic diet containing various Bt proteins (Cry1Ab toxin, Cry1Ab protoxin and Cry2A protoxin) at different concentrations. Columns with different letters respresent treatment means that are significantly different at $p = 0.05$ (LSMEANS).

cantly higher (57%) than in the respective untreated control (30%) (Fig. 16.9). Also, significantly more chrysopid larvae died (29%) having received Cry1Ab later during their larval development compared to the respective control (17%). Only 8% of the larvae died when reared exclusively on *E. kuehniella* eggs. These results demonstrated that activated Cry1Ab was toxic to *C. carnea* at 100 µg/ml diet.

16.3.5 Discussion and conclusions

The results of all three series of experiments consistently demonstrated the susceptibility of immature *Chrysoperla carnea* to Bt proteins (Cry1Ab toxin and protoxin, Cry2A protoxin). The degree of mortality varied depending on the Bt delivery system. Considering the lowest concentration of the Bt toxin Cry1Ab present in transgenic Bt corn plants (<4 µg/g fresh weight; Fearing *et al.*, 1997) compared with the concentrations of the other diets, prey-mediated mortality of immature *C. carnea* was surprisingly high (59%–66%) when using transgenic plants. When feeding the more than 25-fold higher Bt toxin concentration (100 µg Cry1Ab/ml artificial diet) directly to

chrysopid larvae, the induced total immature mortality of *C. carnea* was still significantly higher than in the control but 'only' similar (57%) to the prey-mediated mortality using transgenic plants. Similarly, when incorporating the same high concentration into meridic diet (100 µg Cry1Ab toxin/g meridic diet), prey-mediated total immature mortality of *C. carnea* was 78%, i.e. 21% higher than when feeding this concentration directly to *C. carnea* larvae. Further, also Bt-protoxin-incorporated diets (Cry1Ab and Cry2A) caused significantly higher prey-mediated mortality in immature *C. carnea* than the untreated control, although to a lower degree than the Cry1Ab-toxin-incorporated diet. Reasons for this are unknown but one conceivable explanation is that *Spodoptera littoralis* may process the protoxin/toxin to a product that lethally affects *C. carnea*. Regardless of the mechanism involved, these findings strongly suggest that interactions between Bt-protein and herbivorous prey (*S. littoralis*) increase the toxicity of the prey to *C. carnea* while affecting the prey to a much lesser degree.

When comparing the total immature control mortality rates of all studies occurring over periods ranging

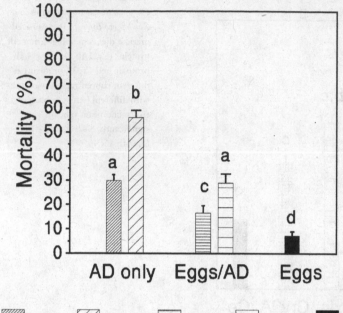

Bt toxin (Cry1Ab) - *C. carnea*

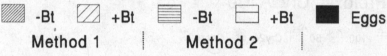

Fig. 16.9. Mean total mortality from first instar to adult eclosion of *Chrysoperla carnea* feeding on different types of Cry1Ab-toxin–containing or Cry1Ab toxin–free diets. Columns with different letters represent treatment means that are significantly different at $p = 0.05$ (LSMEANS). AD = artificial diet, Eggs/AD = eggs and artificial diet.

from 23 to up to 37 days, mortality of *C. carnea* in the trials using transgenic Bt corn plants was approximately 10% higher (37%) than when using meridic diet (26%). Although a separately conducted statistical analysis on control mortalities revealed that these differences were not significant, it indicates that plant-fed *S. littoralis* larvae seem to be less suitable for optimal nutrition of immature *C. carnea* than meridic-diet-fed *S. littoralis* which may have contributed to the overall higher toxicity of Bt-corn-fed prey to immature *C. carnea*. Previous non-target effect trials with *C. carnea* did not exceed 7 to 9 days and a control mortality of 23% was reported (Sims, 1995; Pilcher *et al.*, 1997). However, during the first 7 days of our studies (end of first or middle of second instar depending on treatment) total mean control mortality rates were around 10% or lower which is typical of many lepidopteran bioassays (MacIntosh *et al.*, 1990; Moar *et al.*, 1995).

These studies clearly demonstrate the importance of tritrophic level studies for the assessment of the long-term compatibility of insecticidal plants with important natural enemies. Direct feeding studies using paraffin-encapsulated diet are suitable for a first

screening of compounds like Bt proteins, that exert their lethal effect in the intestinal tract of *C. carnea*. Hence, they should also work for lectins or proteinase inhibitors, that are also already being expressed in transgenic crop plants. But if no additional tritrophic level experiments are conducted, processing of the insecticidal compound in the herbivore's gut is then ignored entirely and, thereby, important ecological information may be missed.

In cropping systems that use Bt crop plants expressing Cry1Ab and where *C. carnea* is an important biological control component, an adverse impact on long-term control efficacy might be expected. But field studies are necessary to evaluate the ecological implications of these findings for IPM and biological control programs.

REFERENCES

Abbott, W.S. (1925) A method of computing the effectiveness of an insecticide. *Journal of Economic Entomology* 18, 265–267.

Adán, A., Soria J., Del Estal, P., Sánchez-Brunete, C. & Viñuela, E. (1998) Differential action of two azadirachtin formulations on the developmental stages of *Ceratitis capitata. Boletín de Sanidad Vegetal Plagas* 24, 1009–1018. (in Spanish)

Anonymous (1986) Gesetz zum Schutz der Kulturpflanzen vom 15. September 1986 (Pflanzenschutzgesetz – PflSchG), Bundesgesetzblatt I, 1505.

Austin, D.J. & Carter, K.J. (1986) Further studies of the deposition and persistence of binapacryl, bupirimate and diflubenzuron on apple foliage and fruit. *Pesticide Science* 17, 74–78.

Austin, D.J. & Hall, K.J. (1981) A method of analysis for the determination of binapacryl, bupirimate and diflubenzuron on apple foliage and fruit, and its application to persistence studies. *Pesticide Science* 12, 459–502.

Barrett, K.L., Grandy, N., Harrison, E.G., Hassan, S.A. & Oomen, P.A. (1994) *Guidance Document on Testing Procedures for Testing Pesticides and Non-Target Arthropods*. SETAC-Europe.

Bartlett, B.B. (1964) Toxicity of some pesticides to eggs, larvae and adults of the green lacewing, *Chrysopa carnea. Journal of Economic Entomology* 57, 366–369.

Bay, T., Hommes, M. & Plate, H.P. (1993) Die Florfliege *Chrysoperla carnea* (Stephens). *Mitteilungen aus der biologischen Bundesanstalt für Land- und Forstwirtschaft, Berlin–Dahlem* 288, 4–175.

Bigler, F. (1984) Biological control by chrysopids: integration with pesticides. In *Biology of Chrysopidae*, ed. Canard, M., Séméria, Y. & New, T.R., pp. 233–245. Dr. W.Junk, The Hague.

Bigler, F. (1988) A laboratory method for testing side-effects of pesticides on larvae of the green lacewing, *Chrysoperla carnea* Steph. (Neuroptera, Chrysopidae). *IOBC/WPRS Bulletin* 11, 71–77.

Bigler, F. & Waldburger, M. (1988) A semi-field method for testing the initial toxicity of pesticides on larvae of the green lacewing, *Chrysoperla carnea* Steph. (Neuroptera, Chrysopidae). *IOBC/WPRS Bulletin* 11, 127–134.

Bigler, F. & Waldburger, M. (1994) Effects of pesticides on *Chrysoperla carnea* Steph. (Neuroptera, Chrysopidae) in the laboratory and semi-field. *IOBC/WPRS Bulletin* 17, 55–69.

Bigler, F., Ferran, A. & Lyon, J.-P. (1976) L'élevage larvaire de deux prédateurs aphidiphages (*Chrysopa carnea* Steph., *Chrysopa perla* L.) à l'aide de différents milieux artificiels. *Annales de Zoologie–Écologie Animale* 8, 551–558.

Bozsik, A. (1991) Effects of chemicals on aphidophagous insects – response of adults of common green lacewing *Chrysoperla carnea* to pesticides. In *Behaviour and Impact of Aphidophaga*, vol. 1, ed. Polgár, L.,

Chambers, R.J., Dixon, A.F.G. & Hodek, I., pp. 297–304. SPB Academic Publishing, The Hague.

Bozsik, A. (1995) Effect of some zoocides on *Chrysoperla carnea* adults (Planipennia, Chrysopidae) in the laboratory. *Anzeiger für Schädlingskunde, Pflanzenschutz, Umweltschutz* 68, 58–59.

Bozsik A., Haubruge, E. & Gaspar, C. (1996) Effect of some organophosphate insecticides on acetylcholinesterase of adult *Coccinella septempunctata* (Coccinellidae). *Journal of Environmental Science and Health, Part B: Pesticides, Food Contaminants and Agricultural Wastes* 31, 577–584.

Brasse, D. (1990) Einführung der obligatorischen Prüfung der Auswirkung von Pflanzenschutzmitteln auf Nutzorganismen in das Zulassungsverfahren. *Nachrichtenblatt für den deutschen Pflanzenschutzdienst* 42, 81–86.

Budia, F. & Viñuela, E. (1994) Ultrastructure of *Ceratitis capitata* larval integument and changes induced by the IGI cyromazine. *Pesticide, Biochemestry and Physiology* 48, 191–201.

Corbett, J.R., Wright, R. & Baillie, A.C. (1984) *The Biochemical Mode of Action of Pesticides*, 2nd edn. Academic Press, New York.

Council Directive 91/414/EEC concerning the placing of plant protection products on the market (1991). *Official Journal of the European Communities* L230, 1–32, 19 August 1991 (OJ L230 of 19–8–91).

Croft, B.A. (1990) *Arthropod Biological Control Agents and Pesticides*. Wiley, New York.

Croft, B.A. & Brown, A.W.A. (1975) Response of arthropod natural enemies to insecticides. *Annual Review of Entomology* 20, 285–335.

Cross, J. & Dickler, E. (1994) *Guidelines for Integrated Production of Pome Fruits in Europe: Technical Guideline III*, 2nd edn. IOBC/WPRS, Montfavet.

Darvas, B. & Polgár, L.A. (1998) Novel-Type Insecticides. In *Specificity and Effects on Non-Target Organisms*, vol. 1, ed. Ishaaya, I. & Degheele, D., pp. 189–259. Springer-Verlag, Berlin.

Ellman, G.L., Courtney, K.D., Andres, V. & Featherstone, R.M. (1961) A new and rapid colorimetric determination of acetylcholinesterase activity. *Biochemical Pharmacology* 7, 88–95.

Fearing, P.L., Brown, D., Vlachos, D., Meghji, M. & Privalle, L. (1997) Quantitative analysis of Cry1A(b) expression in Bt maize plants, tissues, and silage and stability of expression over successive generations. *Molecular Breeding* 3, 169–176.

Flexner, J.L., Lighthart, B. & Croft, B.A. (1986) The effects of microbial pesticides on non-target, beneficial arthropods. *Agriculture, Ecosystems and Environment* 16, 203–254.

Forster, R., Baier, B., Berendes, K.-H., Heimbach, U., Rautmann, D., Süss, A. & Vogt, H. (1997) Vergleichende Laboruntersuchungen zur Sensitivität von Nichtzielorganismen gegenüber Pflanzenschutzmitteln und Möglichkeiten der expositionsabhängigen Risikoabschätzung. *Mitteilungen aus der biologischen Bundesanstalt für Land- und Forstwirtschaft, Berlin–Dahlem* 333.

Franz, J.M. (1974) Die Prüfung von Nebenwirkungen der Pflanzenschutzmittel auf Nutzarthropoden im Laboratorium – ein Sammelbericht. *Zeitschrift für Pflanzenkrankheiten und Pflanzenschutz* 81, 141–174.

Grafton-Cardwell, E.E. & Hoy, M.A. (1985) Intraspecific variability in response to pesticides in the common green lacewing *Chrysoperla carnea* (Stephens) (Neuroptera: Chrysopidae). *Hilgardia* 53, 1–32.

Greve, L (1984) Chrysopid distribution in northern latitudes. In *Biology of Chrysopidae*, ed. Canard, M., Séméria, Y. & New, T.R., pp. 180–186. Dr W. Junk, The Hague.

Hagen, D.S. & Tassan, R.L. (1965) A method of providing artificial diets to *Chrysopa* larvae. *Journal of Economic Entomology* 58, 999–1000.

Händel, U. (1996) Evaluierung und Erweiterung von Methoden zur Prüfung der Auswirkungen von Pflanzenschutzmitteln auf die Florfliege, *Chrysoperla carnea* (Stephens). Diploma thesis, Universität Heidelberg.

Hassan, S.A. (1975) Über die Massenzucht von *Chrysopa carnea* Steph. (Neuroptera, Chrysopidae). *Zeitschrift für angewandte Entomologie* 79, 310–315.

Hassan, S.A. (ed.) (1988) Guidelines for testing the effects of pesticides on beneficials: short description of test methods. *IOBC/WPRS Bulletin* 11, 143.

Hassan, S.A. (1989) Testing methodology and the concept of the IOBC/WPRS Working Group. In *Pesticides and Non-Target Invertebrates*, ed. Jepson, P.C., pp. 1–18. Intercept, Wimborne, UK.

Hassan, S.A. (ed.) (1992) Guidelines for testing the effects of pesticides on beneficial organisms. *IOBC/WPRS Bulletin* 15.

Hassan, S.A. (1994). Activities of the IOBC/WPRS Working group 'Pesticides and Beneficial Organisms'. *IOBC/WPRS Bulletin* 17, 1–5.

Hassan, S.A., Bigler, F., Bogenschütz, H., Brown, J.U., Firth, S.I., Huang, P., Ledieu, M.S., Naton, E., Oomen, P.A., Overmeer, W.P.J., Rieckmann, W., Samsøe-Petersen, L., Viggiani, G. & Zon, A.Q. van (1983) Results of the second joint pesticide testing programme by the IOBC/WPRS Working Group 'Pesticides and Beneficial Organisms'. *Zeitschrift für angewandte Entomologie* 95, 151–158.

Hassan, S.A, Bigler, F., Blaisinger, P., Bogenschütz, H., Brun, J., Chiverton, P., Dickler, E., Easterbrook, M.A., Edwards, P.J., Englert, W.D., Firth, S.I., Huang, P., Inglesfield, C., Klingauf, F., Kühner, C., Ledieu, M.S., Naton, E., Oomen, P.A., Overmeer, W.P.J., Plevoets, P., Reboulet, J.N., Rieckmann, W., Samsøe-Petersen, L., Shires, S.W., Stäubli, A., Stevenson, J., Tuset, J.J., Vanwetswinkel, G. & Zon, A.Q. van (1985) Standard methods to test the side-effects of pesticides on natural enemies of insects and mites developed by the IOBC/WPRS Working Group 'Pesticides and Beneficial Organisms'. *OEPP/EPPO* 15, 214–255.

Hassan, S. A., Albert, R., Bigler, F., Blaisinger, P., Bogenschütz, H., Boller, E., Brun, J., Chiverton, P., Edwards, P, Englert, W.D., Huang, P., Inglesfield, C., Naton E., Oomen, P.A., Overmeer, W.P.J., Rieckmann, W., Samsøe-Petersen, L., Stäubli, A., Tuset, J.J., Viggiani, G. & Vanwetswinkel, G. (1987) Results of the third joint pesticide testing programme by the IOBC/WPRS Working Group 'Pesticides and Beneficial Organisms'. *Zeitschrift für angewandte Entomologie* 103, 92–107.

Hassan, S.A, Bigler, F., Bogenschütz, H., Boller, E., Brun, J., Chiverton, P., Edwards, P., Mansour, E., Oomen, P.A., Overmeer, W.P.J., Polgár, L., Rieckmann, W., Samsøe-Petersen, L., Stäubli, A., Sterk, G., Tavares, K. & Tuset, J.J (1988) Results of the fourth joint pesticide testing programme carried out by the IOBC/WPRS Working group 'Pesticides and Beneficial Organisms'. *Zeitschrift für angewandte Entomologie* 105, 321–329.

Hassan, S.A., Bigler, F., Bogenschütz, H., Boller, E., Brun, J., Calis, J.N.M., Chiverton, P., Coremans-Pelseneer, J., Duso, C., Lewis, G.B., Mansour, F., Moreth, L., Oomen, P.A., Overmeer, W.P.J., Polgár, L., Rieckmann, W., Samsøe-Petersen, L., Stäubli, A., Sterk, G., Tavares, K., Tuset, J.J. & Viggiani, G (1991) Results of the fifth joint pesticide testing programme carried out by the IOBC/WPRS Working Group 'Pesticides and Beneficial Organisms'. *Entomophaga* 36, 55–67.

Hassan, S.A., Bigler, F., Bogenschütz, H., Boller, E., Brun, J., Calis, M., Coremans-Pelseneer, J., Duso, C., Grove, A., Heimbach, U., Helyer, N., Hokkanen, H., Lewis,G.B., Mansour, F., Moreth, L., Polgár, L., Samsøe-Petersen, L., Sauphanor, B., Stäubli, A., Sterk, G., Vainio, A., Veire, M. van de, Viggiani, G. & Vogt, H. (1994) Results of the sixth joint pesticide testing programme of the IOBC/WPRS working Group 'Pesticides and Beneficial Organisms'. *Entomophaga* 39, 107–119.

Hayat, M.A. (1986) *Basic techniques for Transmisssion Electron microscopy.* Academic Press, New York.

Hilbeck, A., Baumgartner, M., Fried, P.M. & Bigler, F.

(1998*a*) Effects of transgenic *Bacillus thuringiensis* corn-fed prey on mortality and development time of immature *Chrysoperla carnea* (Neuroptera: Chrysopidae). *Environmental Entomology* 27, 480–487.

Hilbeck, A., Moar, W.J., Pusztai-Carey, M., Filipini, A. & Bigler, F. (1998*b*) Toxicity of the *Bacillus thuringiensis* Cry1Ab toxin on the predator *Chrysoperla carnea* (Neuroptera: Chrysopidae) using diet incorporated bioassays. *Environmental Entomology* 27, 1255–1263.

Hilbeck, A., Moar, W.J., Pusztai-Carey, M., Filipini, A. & Bigler, F. (1999) Prey-mediated effects of Cry1Ab toxin and protoxin and Cry2A protoxin on the predator *Chrysoperla carnea. Entomologia Experimentalis et Applicata* 91, 305–316

Hoy, M.A. (1979) The potential for genetic improvement of predators for pest management programs. In *Genetics in Relation to Insect Management*, ed. Hoy, M.A. & McKelvey, Jr. J.J., pp. 106–115. Rockefeller Foundation Press, New York.

Hoy, M.A. (1985) Recent advances in genetics and genetic improvement of the phytoseiidae. *Annual Review of Entomology* 30, 345–370.

Hoyle, R. (1995). EPA okays first pesticidal transgenic plants. *Biotechnology* 13, 434–435.

Jakob, G. (1996) Zur Bekämpfung des Apfelschalenwicklers *Adoxophyes orana* F.v.R. (Lepidoptera, Tortricidae) und anderer Apfelschädlinge mit Inhaltsstoffen des Niembaumes *Azadirachta indica* A. Juss. (Meliaceae) unter Berücksichtigung von Nebenwirkungen auf natürliche Feinde. Dissertation, Universität Giessen.

Jakob, G. & Vogt, H. (1993) Einsatz von Niempräparaten gegen *Adoxophyes orana* F.v.R. und Untersuchungen zu Nebenwirkungen. In 6. *Internationaler Erfahrungsautausch über Forschungsergebnisse zum ökologischen Obstbau*, ed. Kienzle, J & Straub, M., pp. 51–61. Staatliche Lehr- und Forschungsanstalt, Weinsberg.

Jepson, P.C., Croft, B.A. & Pratt, G.E. (1994) Test systems to determine the ecological risks posed by toxin release from *Bacillus thuringiensis* genes in crop plants. *Molecular Ecololgy* 3, 81–89.

Jones, S.L., Kinzer, R.E. & Bull, D. (1978) Deterioration of *Chrysopa carnea* in mass culture. *Annals of the Entomological Society of America* 71, 160–162.

Kienzle, J., Schulz, C. & Zebitz, C.P.W. (1997) Two years of experience with the use of NeemAzal in organic fruit orchards. In *Practice-Oriented Results on Use and Production of Neem-Ingredients and Pheromones, Proceedings of the 5th Workshop*, ed. Kleeberg, H. & Zebitz, C.P.W., pp. 27–31, Wetzlar.

Koziel, M.G., Beland, G.L., Bowman, C., Carozzi, N.B., Crenshaw, R., Crossland, J., Dawson, J., Desai, N.,

Hill, M., Kadwell, S., Launis, K., Lewis, K., Maddox, D., McPherson, K., Meghji, M.R., Merlin, E., Rhodes, R., Warren, G.W., Wright, M. & Evola, S.V. (1993) Field performance of elite transgenic maize plants expressing an insecticidal protein derived from *Bacillus thuringiensis. Biotechnology* 11, 194–200.

MacIntosh S.C., Stone, T.B. & Sims, S.R. (1990) Specificity and efficacy of purified *Bacillus thuringiensis* proteins against agronomically important insects. *Journal of Invertebrate Pathology* 56, 258–266.

Marshall, D.B., Pree, D.J. & McGarvey, B.D. (1988) Effects of benzoylphenylurea insect growth regulators on eggs and larvae of the spotted tentiform leafminer *Phyllonorycter blancardella* (Fabr.) (Lepidoptera: Gracilariidae). *Canadian Entomologist* 120, 49–62.

Martin, P.B., Ridgway, R.L. & Schüetze, C.E. (1978) Physical and biological evaluation of an encapsulated diet for rearing *Chrysopa carnea. Florida Entomologist* 61, 145–152.

Matsumura, F. (1985) *Toxicology of Insecticides*, 2nd edn. Plenum Press, New York.

Melin, B.E. & Cozzi, E.M. (1989) Safety to nontarget invertebrates of lepidopteran strains of *Bacillus thuringiensis* and their β-exotoxins. In *Safety of Microbial Insecticides*, ed. Laird, M., Lacey, L.A. & Davidson, E.W., pp. 149–168. CRC Press, Boca Raton, FL.

Mellon, M. & Rissler, J. (1998) *Now or Never: Serious New Plans to Save a Natural Pest Control.* Union of Concerned Scientists, Cambridge MA.

Moar, W.J., Pusztai-Carey, M. & Mack, T.P. (1995) Toxicity of purified proteins and the HD-1 strain from *Bacillus thuringiensis* against lesser cornstalk borer (Lepidoptera: Pyralidae*). Journal of Economic Entomology* 88, 606–609.

Morrison, R.K. (1985) *Chrysopa carnea.* In *Handbook of Insect Rearing*, vol. 1, ed. Singh, P. & Moore, R.F., pp. 419–426. Elsevier, Amsterdam.

Morrison, R.K. & Ridgway, R.L. (1976). Improvements in techniques and equipment for production of a common green lacewing, *Chrysopa carnea. United States Department of Agriculture, Agricultural Research Service* S-143.

Morrison, R.K., House, V.S. & Ridgway, R.L. (1975) Improved rearing unit for larvae of a common green lacewing. *Journal of Economic Entomology* 68, 821–822.

Niebling, K., (1995) Agricultural biotechnology companies set their sights on multi-billion $$ markets. *Genetic Engineering News* 15, 20–21.

Perlak, F.J., Deaton, R.W., Armstrong, T.A., Fuchs, R.L., Sims, S.R., Grenplate, J.T. & Fischhoff, D.A. (1990) Insect resistant cotton plants. *Biotechnology* 8, 939–943.

Pilcher, C.D., Obrycki, J.J., Rice, M.E. & Lewis, L.C. (1997). Preimaginal development, survival, and field abundance of insect predators on transgenic *Bacillus thuringiensis* corn. *Environmental Entomology* 26, 446–454.

Plapp, F.W. Jr & Bull, D.L. (1985) Modifying chemical control practices to preserve natural enemies. In *Increasing in the Effectiveness of Natural Enemies*, ed. King, E.G. & Jackson, R.D., pp. 537–546. Aspect, London.

Pree, D.J., Archibald, D.E. & Morrison, R.K. (1989) Resistance to insecticides in the common green lacewing, *Chrysoperla carnea* (Neuroptera: Chrysopidae) in southern Ontario. *Journal of Economic Entomology* 82, 29–34.

Ridgway, R.L., Morrison, R.K. & Badgley, M. (1970) Mass rearing of green lacewing. *Journal of Economic Entomology* 63, 834–836.

Robertson, J.L. & Worner, S.P. (1990) Population toxicology: suggestions for laboratory bioassays to predict pesticide efficacy. *Journal of Economic Entomology* 83, 8–12.

Rumpf, S. (1990) Wirkungen des Juvenoids Fenoxycarb auf die Larven des Nutzinsektes *Chrysoperla carnea* Steph. Diploma thesis, Universität Heidelberg.

Rumpf, S., Storch, V., Vogt, H. & Hassan, S.A. (1992) Effects of juvenoids on larvae of *Chrysoperla carnea* Steph. (Chrysopidae). *Acta Phytopathologica et Entomologica Hungarica* 27, 557–563.

Rumpf, S., Frampton, C. & Chapman, B. (1997a) Acute toxicity of insecticides to *Micromus tasmaniae* (Neuroptera: Hemerobiidae) and *Chrysoperla carnea* (Neuroptera: Chrysopidae) LC_{50} and LC_{90} estimates for various test durations. *Journal of Economic Entomology* 90, 1493–1499.

Rumpf, S., Hetzel, F. & Frampton, C. (1997b) Lacewings (Neuroptera: Hemerobiidae and Chrysopidae) and integrated pest management: enzyme activity as biomarker of sublethal insecticide exposure. *Journal of Economic Entomology* 90, 102–108.

Rumpf, S., Frampton, C. & Dietrich, D.R. (1998) Effects of conventional insecticides and insect growth regulators on fecundity and other life-table parameters of *Micromus tasmaniae* (Neuroptera: Hemerobiidae). *Journal of Economic Entomology* 91, 34–40.

Samsøe-Petersen, L., Bigler, F., Bogenschütz, H., Brun, J., Hassan, S.A., Helyer, N.L., Kühner, C., Mansour, F., Naton, E., Oomen, P.A., Overmeer, W.P.J., Polgár, L., Rieckmann, W. & Stäubli, A. (1989) Laboratory rearing techniques for 16 beneficial arthropod species and their prey/hosts. *Zeitschrift für Pflanzenkrankheiten und Pflanzenschutz* 96, 289–316.

Schaaf, C. & Vogt, H. (1995) Untersuchungen in Gradationsgebieten des Schwammspinners *Lymantria dispar* L. unter Berücksichtigung verschiedener Bekämpfungsmassnahmen. *Mitteilungen der deutschen Gesellschaft für allgemeine und angewandte Entomologie* 10, 123–128.

Senior, L.J., McEwen, P.K. & Langley, P.A. (1997) Effects of the chitin synthesis inhibitor Triflumuron on eggs and larvae of the green lacewing, *Chrysoperla carnea* (Neuroptera: Chrysopidae). In *New Studies in Ecotoxicology* (Papers resulting from posters given at the Welsh Pest Management Forum Conference: Ecotoxicology, Pesticides and Beneficial Organisms, Cardiff, UK, 14–16 October 1996), ed. Haskell, P.T. & McEwen, P.K., pp. 80–83. Welsh Pest Management Forum, Cardiff.

Silver, A. (1974) *Frontiers of Biology*, Vol. 36, *The Biology of Cholinesterases*. North Holland, Amsterdam.

Sims, S.R. (1995) *Bacillus thuringiensis* var. *kurstaki* (Cry1A(c)) protein expressed in transgenic cotton: effects on beneficial and other non-target insects. *Southwestern Entomologist* 20, 493–500.

Smallman, B.N. & Mansingh, A. (1969) The cholinergic system in insect development. *Annual Review of Entomology* 14, 347–408.

Sterk, G., Hassan, S.A., Baillod, M., Bakker, F., Bigler, F., Blümel, S., Bogenschütz, H., Boller, E., Bromans, B., Brun, J., Calis, J.N.M., Coremans-Pelseneer, J., Duso, C., Garrido, A., Grove, A., Heimbach, U., Hokkanen, H., Jacas, J., Lewis, G., Moreth, L., Polgár, L., Rovesti, L., Samsøe-Petersen, L., Schaub, L, Stäubli, A., Tuset, J.J., Vainio, A., Veire, M. van de, Viggiani, G., Viñuela, E. & Vogt, H. (1999) Results of the seventh joint pesticide testing programme carried out by the IOBC/WPRS Working Group 'Pesticides and Beneficial Organisms'. *BioControl* (previously *Entomophaga*) 44, 99–117.

Suter, H. (1978) Prüfung der Einwirkung von Pflanzenschutzmitteln auf die Nutzarthropodenart *Chrysopa carnea* Steph. (Neuroptera, Chrysopidae) – Methodik und Ergebnisse. *Zeitschrift für landwirtschaftliche Forschung* 17, 37–44.

Vanderzant, E.S. (1969) An artificial diet for larvae and adults of *Chrysopa carnea*, an insect predator of crop pests. *Journal of Economic Entomology* 62, 256–257.

Veith, S. (1997) Untersuchungen zur selektiven Bekämpfung von Tortriciden im Apfelanbau mit dem Ecdyis-Induktor CM 001. Examination thesis, Universität Heidelberg.

Vogt, H. (1992) Untersuchungen zu Nebenwirkungen von Insektiziden und Akariziden auf *Chrysoperla carnea*

Steph. (Neuroptera, Chrysopidae). *Mededelingen van de Faculteit landbouwwetenschapen, Rijksuniversiteit Gent* 57/2b, 559–567.

Vogt, H. (ed.) (1994) Side-effects of pesticides on beneficial organisms: comparison of laboratory, semi-field and field results. *IOBC/WPRS Bulletin* 17, 178.

Vogt, H. & Händel, U. (1995) Beurteilung der Nebenwirkungen von Pflanzenschutzmitteln auf Nichtzielorganismen. *Jahresbericht der biologischen Bundesanstalt für Land- und Forstwirtschaft, Berlin und Braunschweig* 1995, 88–89.

Vogt, H. & Just, J. (1997) Ermittlung der letalen Dosis (LD$_{50}$) von fünf Insektiziden für *Adoxophyes orana* (Lepidoptera, Tortricidae) und *Chrysoperla carnea* (Neuroptera, Chrysopidae). *Mitteilungen aus der biologischen Bundesanstalt für Land- und Forstwirtschaft, Berlin-Dahlem* 333, 75–90.

Vogt, H., Rumpf, S., Wetzel, C. & Hassan, S.A. (1992) A field method for testing effects of pesticides on the green lacewing *Chrysoperla carnea* Steph. (Neuroptera, Chrysopidae) *IOBC/WPRS Bulletin* 15, 176–182.

Vogt, H., Veith, S. & Wirth, J. (1996) Untersuchungen zur Beurteilung der Nebenwirkungen von Pflanzenschutzmitteln auf Nichtzielorganismen. *Jahresbericht der Biologischen Bundesanstalt für Land- und Forstwirtschaft, Berlin und Braunschweig*. 1996, 79–80.

Vogt, H., Händel, U. & Viñuela, E. (1997) Field investigations on the efficacy of NeemAzal-T/S against *Dysaphis plantaginea* (Passerini)(Homoptera:Aphididae) and its effects on larvae of *Chrysoperla carnea* Stephens (Neuroptera: Chrysopidae). In *Practice-Oriented Results on Use and Production of Neem-Ingredients and Pheromones, Proceedings of the 5th Workshop*, Wetzlar, pp. 105–114.

Vogt, H., Degrande, P., Just, J., Klepka, S., Kühner, C., Nickless, A., Ufer, A., Waldburger, M., Waltersdorfer, A. & Bigler, F. (1998) Side effects of pesticides on larvae of *Chrysoperla carnea* (Neuroptera, Chrysopidae): actual state of the laboratory method. In *Ecotoxicology: Pesticides and Beneficial Organisms*, ed. Haskell, P.T. & McEwen, P., pp. 123–136. Kluwer, Dordrecht.

Vogt, H., González, M., Adán, A., Smagghe, G. & Viñuela, E. (1998). Side effects of azadirachtin via residual contact in young larvae of the predator *Chrysoperla carnea*. *Boletin de Sanidad Vegetal Plagas* 24, 67–78. (in Spanish)

Weiss, S. (1997) Untersuchungen zur Wirksamkeit von Stammbehandlungen mit Imidacloprod – einem Insektizid mit neuer Wirkungsweise – auf Schad- und Nutzarthropoden im Obstbau. Examination thesis, Universität Heidelberg.

Weiss, A. & Vogt, H (1994) Populationsdynamik und Parasitierung von Miniermotten am Apfel in Abhängigkeit von biologischen und chemischen Pflanzenschutzmassnahmen. *Mitteilungen der deutschen Gesellschaft für allgemeine und angewandte Entomologie* 9, 379–387.

Wetzel, C. & Dickler, E. (1994) Side effects of sulfur and a natural pyrethroid on *Trichogramma dendrolimi* Matsura (Hym., Trichogrammatidae). *IOBC/WPRS Bulletin* 17, 123–131.

Wetzel, C., Krczal, H. & Hassan, S.A. (1991) Investigations to evaluate the side-effects of pesticides on the green lacewing *Chrysoperla carnea* Steph. (Neuroptera, Chrysopidae) in the field. *Journal of Applied Entomology* 111, 217–224.

Wetzel, C., Weiss, S. & Dickler, E. (1996) Untersuchungen zum Einsatz von Imidacloprid durch Stammapplikation und zu den Nebenwirkungen diesen Verfahrens auf Nutzarthropoden im Apfelanbau. *Jahresbericht der biologischen Bundesanstalt für Land- und Forstwirtschaft, Berlin und Braunschweig*. 1996, 78.

Wirth, J. (1997) Untersuchungen zur Auswirkung der Attract & Kill- Methode auf Ziel- und Nichtzielorganismen. Examination thesis, Universität Heidelberg.

CHAPTER 17

Lacewings, biological control, and conservation

T.R. New

17.1 INTRODUCTION

Release of non-specific predatory insects into a new area implies responsibilities associated with a 'duty of care' for the receiving environment and, from the conservationist's viewpoint, is not always an exercise to be undertaken lightly, despite the seeming urgency of biological control measures against pest arthropods. The twin contexts that are relevant to introductions of lacewings are (1) release of species which are exotic to the areas where they are released (that is, as classical biological control agents, species not present previously in the region of introduction, for which the above precautions may be particularly important) and (2) native species whose populations are enhanced by releases from commercial stocks, or by use of attractants to concentrate field populations in more limited areas. Both approaches are usually part of IPM programmes undertaken with the overall aim of seeking more cost-efficient pest control with the economic and environmental benefits of reduced pesticide use.

Incorporation of predators into such strategies is increasingly undertaken with the aims of short-term gain and longer-term environmental safety. Nevertheless, it is pertinent here to recapitulate some of the responsibilities and concerns that such operations raise, and to emphasise the need to monitor introductions effectively to help improve further the application technologies and quality of the results obtained.

Many such releases are indeed undertaken with due regard to 'acceptable risk', but clear definition of those risks is itself a complex matter. Increasingly, such risks are being dealt with by a combination of regulations, protocols, and guidelines involving some degree of consensus of scientific and regulatory participants. Conflicts of interest in use of biological control may need to be resolved politically, as those decisions have wide societal relevance (Harris, 1990). However, acquiring the primary information needed for optimal decisions must devolve on biologists.

17.2 TOPICS OF CONCERN

Much has been made in the controversies over wise use of classical biological control agents of 'the mistakes of the past', and the effects of non-specific natural enemies on non-target organisms. Once such agents have been introduced and become established, it is usually impossible to eliminate them. The tempo of debate was accelerated markedly by Howarth's (1983, 1991) accounts of the losses of Hawaii's remarkable endemic insect fauna by the spread of exotic natural enemies into natural habitats (see also Nafus, 1994, for similar inferences on Guam). Gagné & Howarth (1984) attributed the extinction of a number of Hawaiian endemic Lepidoptera specifically to errant exotic agents. The major practical outcome from such cases has been the greater realisation that biological control and the conservation of native biota must be considered together in attempts to improve IPM, and that the benefits anticipated from introductions of predators or parasitoids may not always outweigh the possible losses in sensitive or remote environments. More broadly, invasive species are viewed with concern as a major collective threat to native organisms in many parts of the world, and consideration of possible adverse effects of any deliberate introduction now constitutes an integral part of many screening protocols for exotic biological control agents (Gillespie & New, 1998). Any introduction of an alien species may involve some degree of risk, but 'doing nothing' or continuing to use pesticides intensively may cause far more harm than a well-considered biological control strategy.

In short, the era of simply introducing likely control agents without due care for the consequences to the broader environment is largely past, but many practitioners justifiably fear that the imposition of overly restrictive regulations and demands will render the science difficult to pursue, and development costs excessive. The cost–benefit ratios for classical biological control are difficult to derive, but some published estimates run into many millions of dollars annually for particular cases (examples for the United States cited by Coulson & Soper, 1989).

17.3 LACEWINGS AND RISKS

The spectrum of lacewings used, or likely to be used in the near future, as classical biological control agents in this way is small but does not obviate the need for careful and objective assessment of any risks they may incur. One reason for the small number of species employed in IPM is simply that most of these are sufficiently polyphagous to be useful in a variety of contexts and against a broad array of pests – the very qualities which might engender risk to non-target species. There are thus many parallels between the features that render such natural enemies successful and those which cause concerns over their wider effects in natural environments. Yet their effects on non-target species are rarely documented, generally inferred, and often entirely disregarded. Thus 'all Neuroptera' were proposed for inclusion on a list of biological control agents to be exempted from the requirements of a formal environmental impact assessment in the USA (Coulson et al., 1991), using the following proposed criteria: (1) no known plant pest in the taxon listed; (2) historical use of other biological control agents in the taxon listed are without adverse environmental impacts; (3) systematics are known and therefore there is significant understanding of the taxon's basic biology, behaviour, and ecology; (4) no significant detrimental effect of any species in the listed taxon reported in the literature; (5) not harmful to humans and livestock; and (6) limited host range: do not attack, or have only insignificant attack on beneficial natural enemies, other beneficial organisms, or further endanger listed endangered or threatened species.

Several of these draft criteria are difficult to justify so blandly for lacewings, but demonstrate clearly the dilemma of trying to opt for either a collective screening agenda for natural enemies or a more costly case-by-case testing process. It would indeed be premature to claim, for example, that no lacewing is capable of attacking other beneficial organisms, even though some of these have limited host ranges.

Despite the attractions of the collective screening approach, especially for arthropod biological control agents which are rarely monophagous and may be accorded some more generalised 'collective risk ranking', the species-level path is clearly more desirable if an adequate protocol can indeed be devised. Laboratory screening tests for prey specificity, for example, can be notoriously unreliable and the advanced kinds of centrifugal host screening employed widely for weed control agents are usually not as feasible (Sands, 1997), because prey selection is commonly not taxonomically based but governed, rather, by availability and physical vulnerability. By definition, polyphagous predators are likely to feed on non-target taxa in an area of introduction, and may disperse from a release site if given the opportunity to do so. For naturally dispersive species, such as members of the *Chrysoperla carnea* (Stephens) group (Duelli, 1984), provision of overwintering retreats (McEwen & Sengonca, this volume) may help to counter this trend and, importantly, to help stabilise the predator population in the area where its beneficial effects may be important in the future.

However, information on the levels of natural dispersal for most species of lacewings is very limited. Broad trapping surveys, such as by suction traps, show that a high proportion of resident species may be amenable to capture in Britain (New, 1967), Australia (New, 1984) and, surely, elsewhere. The kinds of broad habitat specificity demonstrated for many northern temperate region species, in particular (Monserrat & Marín, this volume), may help in predicting the broader outcomes of individual releases, though many workers would still advocate caution because of unpredictable spreads in new environments. Species characteristic of low vegetation and released on to low-growing field crops, for example, may pose only minimal risk to sensitive arboreal taxa. Whereas it is usually not clear whether habitat specificity in lacewings is prey-mediated, or the converse, any such consistent trends may be valuable in helping to design a pertinent species-level screening protocol before taking a decision to release.

17.4 DISCUSSION

A variety of different approaches is available to help in predictive evaluation of the effects of arthropod natural enemies on their targets (Luck et al., 1988). Problems of adequate screening for undesirable side-effects and quarantine for introduced agents will remain pervasive, with strong cases from conservationists that polyphagous exotic natural enemies should be eliminated from the biological control arena. Such considerations remain one of the strongest broad incentives to foster the use of native beneficial insects wherever possible. Massively increased numbers of native predators may, of course, foster local changes in communities due to increased influences of natural enemies, but are usually likely to be more transient than the 'permanent' addition of further species to the complex, and have until now caused little concern.

Good taxonomy is at the core of our abilities to predict the results of transfers and augmentation. The confusion between sibling taxa of the *C. carnea* group in Europe, for example, has fostered a recent history of international transfers of unrecognised taxa across the continent, and similar translocations have occurred also in North America. Insectary stocks in southern Europe, used particularly to enhance 'natural' populations in more northerly countries, may well constitute different taxa, some of which are likely exotics in the areas of release. Such unwitting range expansions have not been documented adequately, and may account for local variations in 'performance' of presumed identical taxa. Wherever feasible, voucher material of all stocks transported across international boundaries should be archived responsibly and subject to critical taxonomic scrutiny in case they are unexpectedly exotic. As an additional aspect of 'quality control' in long-established laboratory colonies, assessment of the degree of fluctuating asymmetry at intervals to monitor 'stress' may be useful (Clarke, 1993).

In addition to field releases, some chrysopids are among the array of predators that can be used each year for seasonal inoculations into greenhouses, resulting in massive reductions in pesticide use (Hokkanen, 1997). Periodic releases of this nature do not assume permanent establishment of the predator, and conditions usually ensure that this can not occur, because the crops are of short duration and, often, highly seasonal. Despite the possibility that exotic species used in this way may escape, there are few reported incidences of this. Nevertheless, vigilance may be needed in some cases to prevent escape of predators during seasons when they could establish outside.

Augmentation of the use of native beneficial agents in management of both exotic and native pest arthropods is one of the major hopes for the future of biological control (Waage, 1992). The association of exotic pest with native enemy is a 'new association' just as much as that of exotic agent with native pest, and such new associations have contributed about 40% of the successful cases of biological control (Pimentel, 1997). They are of increasing interest. Their use involves manipulating both the insects and their local environment, and often involves the incorporation of cultural controls (Pimentel, 1993, 1997), such as conservation tillage systems, intercropping, and provision of reservoir habitats. in some cases, conservation of native predators can allow new pest species to be eliminated before their effects are felt (Reickert & Lockley, 1984). Hokkanen (1997) noted the desirability of establishing and maintaining guilds of generalist predators to constitute a 'predator buffer' which can respond rapidly to invasive species in the local environment, and otherwise depend on local reservoir habitats and innocuous prey species which are not conservation-sensitive. Several of the cases discussed in this book contribute to the development of this approach. In particular, *Micromus tasmaniae* in Australia (Horne *et al.*, this volume) is an abundant native generalist predator which can build up numbers rapidly in response to aphid populations, and is proving amenable to mass-rearing. Many of the Chrysopidae used in the northern hemisphere, especially North America, also fall into this manipulable category and the early development of attractant food sprays (Hagen & Tassan, 1965) attests to the methods available for predator enhancement in crop environments, and which have continued to diversify. In many situations, strip-harvesting or some equivalent rotation may be feasible, so that there may always be a standing crop to serve as a refuge for such beneficial insects. More direct interference with crops, such as the recent incorporation of *Bacillus thuringiensis* (Berliner) δ-endotoxin genes into cotton, can lead to massive reductions in external pesticide use. The multiple pest management strategies now usual against major insect pests commonly seek to foster well-being of native natural enemies, including relative generalist taxa. Studies on their tolerance to commonly used pesticides (including sublethal effects) become significant in appraising such values (Rumpf *et al.*, 1998).

Use of native species in this way removes the greatest concerns of conservationists – the uncertainties of possible spread of exotic species to become invasive and to feed on native biota of unknown vulnerability. In addition, their use may result in (or, at least, help to foster) retention or restoration of natural habitats with reservoirs of natural prey within heavily degraded landscapes. They are thus a powerful argument for improved agricultural practices ranging from conservation tillage activities to retention and restoration of natural vegetation, and other methods which provide for increased biotic diversity close to croplands. These can have massive benefits for broader conservation, and in enhancing the perceived values of remnant habitat patches. Conservation in agricultural landscapes is recommended widely as a necessary part of future sustainability, and of greater dependence on conservation outside the limited system of reserves such as National Parks (Gillespie & New, 1998). In general, around 36% of global land area is devoted to some form of agriculture (Gerard, 1995), and agriculture is thereby the single most important activity affecting conservation of biological diversity. The recent trends to changes to high-intensity systems tend to decrease markedly the integrity of the natural communities inhabiting such areas. Ensuring the maximum compatibility between pest management and conservation is a major focus for activities related to biological control. Careful manipulation of released biological control agents, and monitoring the effects of enhanced natural populations are integral facets of this.

REFERENCES

Clarke, G.M. (1993) Patterns of developmental stability of *Chrysopa perla* L. (Neuroptera: Chrysopidae) in response to environmental pollution. *Environmental Entomology* 22, 1362–1366.

Coulson, J.R. & Soper, R.S. (1989). Protocols for the introduction of biological control agents in the U.S. In *Plant Protection and Quarantine, vol 3, Special Topics*, ed. Kahn, R.P., pp. 1–35. CRC Press, Boca Raton FL.

Coulson, J.R., Soper, R.S. & Williams, D.W. (1991) *Biological Control Quarantine: Needs and Procedures*. (Workshop report). United States Department of Agriculture, Agriculture Research Service, Beltsville MD.

Duelli, P. (1984) Flight, dispersal, migration. In *Biology of Chrysopidae*, ed. Canard, M., Séméria, Y. & New, T.R., pp. 110–116. Dr W. Junk, The Hague.

Gagné, W.C. & Howarth, F.G (1984) Conservation status of endemic Hawaiian Lepidoptera. *Proceedings of the 3rd Congress of European Lepidopterology*, Cambridge: pp. 74–84.

Gerard, P.W. (1995) *Agricultural Practices, Farm Policy and the Conservation of Biological Diversity*. USDI Biological Sciences Report No. 4, Washington DC.

Gillespie, R.G. & New, T.R. (1998) Compatibility of conservation and pest management strategies. In *Pest Management: Future Challenges*, vol. 2. ed. Zalucki, M.P., Drew, R.A.I. & White, G.G., pp. 195–208. University of Queensland, Brisbane.

Hagen, K.S. & Tassan, R.L. (1965) A method of providing artificial diets to *Chrysopa* larvae. *Journal of Economic Entomology* 58, 999–1000.

Harris, P. (1990) Environmental impact of introduced biological control agents. In *Critical Issues in Biological Control*, ed. Mackauer, M., Ehler, L.E. & Roland, J., pp. 289–300. Intercept, Andover UK.

Hokkanen, H.M.T. (1997) Role of biological control and transgenic crops in reducing use of chemical pesticides for crop protection. In *Techniques for Reducing Pesticide Use*, ed. Pimentel, D., pp. 103–127. Wiley, New York.

Howarth, F.G. (1983) Classical biological control: panacea or Pandora's box? *Proceedings of the Hawaiian Entomological Society* 24, 239–244.

Howarth, F.G. (1991) Environmental impacts of classical biological control. *Annual Review of Entomology* 36, 485–509.

Luck, R.F., Shepard, B.M. & Kenmore, P.E. (1988) Experimental methods for evaluating arthropod natural enemies. *Annual Review of Entomology* 33, 367–391.

Nafus, D.R. (1994) Extinction, biological control and insect conservation on islands. In *Perspectives on Insect Conservation*, ed. Gaston, K.J., New, T.R. & Samways, M.J., pp. 139–154. Intercept, Andover UK.

New, T.R. (1967) The flight activity of some British Hemerobiidae and Chrysopidae as indicated by suction trap catches. *Proceedings of the Royal Entomological Society of London (A)* 42, 93–100.

New, T.R. (1984) Comparative biology of some Australian Hemerobiidae. In *Progress in World's Neuropterology, Proceedings of the 1st International Symposium on Neuropterology*, ed. Gepp, J., Aspöck, H. & Hölzel, H., pp. 153–166. Thalerhof, Graz.

Pimentel, D. (1993) Cultural controls for insect pest management. In *Pest Control and Sustainable Agriculture* ed. Corey, S.A., Dall, D.J. & Milne, W.M., pp. 35–38. CSIRO, Melbourne.

Pimentel, D. (1997) *Techniques for Reducing Pesticide Use*. Wiley, New York.

Reickert, S.E. & Lockley, T. (1984) Spiders as biological

control agents. *Annual Review of Entomology* 29, 299–320.

Rumpf, S., Frampton, C. & Dietrich, D.R. (1998) Effects of conventional insecticides and insect growth regulators on fecundity and other life-table parameters of *Micromus tasmaniae* (Neuroptera: Hemerobiidae). *Journal of Economic Entomology* 91, 34–40.

Sands, D.P.A. (1997) The 'safety' of biological control agents: assessing the impact on beneficial and other non-target hosts. *Memoirs of the Museum of Victoria* 56, 281–285.

Waage, J.K. (1992) Biological control in the year 2000. In *Pest Management and the Environment in 2000*, ed. Kadir, A.A.S.A. & Barlow, H.S., pp. 329–340. CAB International, Wallingford UK.

Case studies

Introduction to Part 4

The Editors

We have deliberately devoted a large section of this book to descriptions of lacewing assemblages and practical examples of the use and manipulation of lacewings in pest control programmes. This allows the reader to evaluate lacewings in action and we hope that it will promote the greater use of lacewings in pest control in the future. We have attempted to provide some geographical balance to this section by including examples from around the world but despite this the section is heavily Eurocentric with little input from Australasia and the Americas and none at all from Asia and Africa. We know that work on the use of lacewings does occur in these areas particularly in Egypt in North Africa and in China and would hope that future editions of this book would include contributions from these areas.

The section has an inevitable bias towards chrysopids which have received the most attention in pest control programmes and we are therefore particularly pleased to see contributions dealing with *Micromus tasmaniae* from Australia, and some reporting on other

Hemerobiidae and on Coniopterygidae from the Iberian peninsula.

In terms of cropping systems covered we have been somewhat more successful in providing a good variety of case studies. Thus, whilst two chapters deal with lacewings in olive groves we also have contributions on strawberry pest control, vineyards, a range of Italian ecosystems, and artichokes in France.

The contributions vary greatly in their approach with some chapters giving specific release rates of lacewings in different cropping systems, and others dealing with more specific potential ways of improving lacewing performance by, for example, conservation of overwintering populations. One chapter tackles the subject of plant substrate specificity, an area of potential importance in the management of field populations of lacewings.

Overall this section demonstrates lacewings in use as effective pest control agents and will hopefully enthuse other workers by acting as a model for their own future pest control efforts.

CHAPTER 18

Micromus tasmaniae: a key predator on aphids on field crops in Australasia?

P.A. Horne, P.M. Ridland, and T.R. New

18.1 INTRODUCTION

The Australian Hemerobiidae are largely endemic, with many of the 34 recorded species (New, 1988) not known elsewhere. *Micromus tasmaniae* (Walker) is by far the most abundant and widespread hemerobiid in Australia and New Zealand (where it is presumed indigenous, though possibly originating from Australia: Wise, 1992), and has attracted attention as the most frequent lacewing on low vegetation, including common occurrence on a variety of field crops. Its predominance is evident in several surveys noted below, and as New Zealand lacks native Chrysopidae on crops, it has attracted attention there as *the* lacewing occurring in agricultural systems. It is, likewise, often abundant in Australia. For example, in suction trap catches near Melbourne, Victoria, over three years it comprised 3847/4080 (94.3%) of all Hemerobiidae (New, 1984). Most other Australian hemerobiids are relatively scarce in such collections, and only *Drepanacra binocula* (Newman) is otherwise likely to be familiar, even to many entomologists. That species also occurs in New Zealand, and like the other New Zealand *Micromus* species, *M. bifasciatus* Tillyard, is relatively scarce and not regarded as a significant component of the crop fauna.

18.2 BIOLOGY

The two common Australasian hemerobiid species differ considerably in biology. *Drepanacra binocula* is predominantly arboreal and is a relatively specialised feeder on Psylloidea, whereas *Micromus tasmaniae* is a more generalised polyphagous feeder on arthropods of low vegetation, though it occurs also on shrubs and trees. The early stages were described by New & Boros (1984). Aspects of its biology have been studied both in Australia (Samson & Blood, 1979, 1980; New, 1984)

and New Zealand (Syrett & Penman, 1981; Leathwick & Winterbourn, 1984), and more recent unpublished studies reported here confirm that it is an important predator of aphids on field agricultural crops in southern Australia, in addition to lucerne (Milne & Bishop, 1987) and horticultural crops such as roses (Maelzer, 1977), and is thereby of considerable interest for biological control appraisal. Serological and laboratory feeding trials in New Zealand (Leathwick & Winterbourn, 1984; Rohitha & Penman, 1986) clearly demonstrated the potential of *M. tasmaniae* as an effective predator on aphids (*Acyrthosiphon* spp.) on lucerne. Recent studies of *M. tasmaniae* have been stimulated by its abundance, especially on low vegetation, and mark part of a determined transition from heavy pesticide use on field crops to IPM, in part forced by withdrawal from use of many of the pesticides upon which growers previously depended.

Overall sex ratio of catches of adults by various methods, such as suction traps (New, 1984) and others noted below, is close to unity, so that females are dispersive: testing of the variance–mean relationships for weekly catches in green pan traps showed no significant differences in the regression lines for males and females (Ridland, 1988).

Larvae are very active. They readily descend from plants such as cereals and move across the ground before climbing other stems. They are thus captured readily in pitfall traps, with regular trapping reflecting periods of incidence and abundance in relation to preceding adult activity measured by other techniques (Ridland, 1988). As Ridland noted, pitfall trap catches are difficult to interpret quantitatively because of confounding activity with density – but only to the same extent as for other groups for which the technique is employed as a regular sampling or monitoring tool. However, it is somewhat unusual to be able to

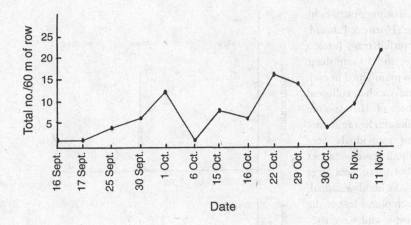

Fig. 18.1. Numbers of *Micromus tasmaniae* captured by suction–trap sampling in potatoes at Robinvale, Victoria, 1997. Each graph point is the number captured on 60 m of potato row.

monitor any lacewing species, albeit tentatively, by this method. By comparison, in New Zealand Leathwick & Penman (1990) showed that suction (D-vac) sampling in lucerne underestimated numbers of first- and second-instar larvae but accurately estimated the third instar. *Micromus tasmaniae* was consistently the most abundant predator in that system, sometimes exceeding 100 individuals per m² of crop. Adult *Micromus* are also amenable to pitfall trap capture. *Micromus tasmaniae* has proved suitable for laboratory culture, where it can be reared easily on aphid prey (Rumpf *et al.*, 1998).

The abundance of *M. tasmaniae* and the limited information available on its biology has fostered interest in its status as a possible key predator in several field crop environments. Most studies until now have been exploratory, but encompass quantitative appraisal in several different field crops in southeastern Australia. Whereas *M. tasmaniae* may not always be the most abundant predator there or in New Zealand (where, for example, spiders, opilionids, Staphylinidae, and Carabidae comprised more than 94% of the more than 12 500 predators captured in pitfall traps in carrot fields: Sivasubramanian *et al.*, 1997), its consistency, apparent responsiveness to aphid presence, and demonstrated voracity in both larval and adult stages suggest that it is indeed a key predator species for use against aphid pests. In contrast, it did not feed on larvae of light-brown apple moth (*Epiphyas postvittana* (Walker), Tortricidae) in laboratory trials (MacLellan, 1973) and was not considered important against pest Noctuidae on cotton (Samson & Blood, 1980).

18.3 SURVEYS ON FIELD CROPS

The following examples from recent studies in Victoria, involving different cropping systems and localities, illustrate its potential.

18.3.1 Intensive horticulture: potatoes

A 100-ha potato crop near Robinvale (34.37 °S, 142.50 °E) was sampled for insects for 13 weeks in 1997, from crop emergence to a week before harvesting, using a suction apparatus (Electrolux 'Weed eater'® blower vac with a mesh bag fitted to retain insects). Three replicates, each of 20 m of potato row, were sampled each week, and aphid numbers on the crop were estimated by direct counts on 100 true leaves. Numbers of *Micromus tasmaniae* adults showed a gradual increase (Fig. 18.1), despite the low numbers of aphids present, which reached a density of 2 aphids/100 leaves in only one week. Even in such relatively short-duration crops, *M. tasmaniae* is present soon after crop emergence, and Fig. 18.1 indicates that it is capable of building up rapidly in numbers, apparently with generational or immigration peaks over the sampling period. In the relative absence of aphids, it is likely that the lacewings fed on larvae of Lepidoptera such as *Chrysodeixis* and *Phthorimaea operculella*.

18.3.2 Broad-acre field crops: cereals and legumes

Populations of insects were monitored by pitfall traps in three paddocks in the Wimmera (36.39 °S, 142.50 °E) used to demonstrate conservation farming practices by Longerenong Agricultural College (Horne & Edward, 1995, 1997, 1998). A grid of 20 pitfall traps (each a plastic container 110 mm diameter and 50 mm deep, half filled with ethylene glycol) was maintained in each 4.6-ha paddock for 2.5 years, and catches collected weekly over this time. Total numbers of *Micromus* are recorded in Fig. 18.2, in which adults and larvae are not distinguished; more than 90% of individuals were larvae. Populations were present in all three paddocks in this highly altered landscape, but there were large differences in abundance. Catches were highly seasonal, with annual peaks concurrent in each plot. Most of the lacewings were associated with crops, and were presumably feeding on aphids. Somewhat unexpectedly, a canola crop in one block (in 1994) yielded large numbers of aphids but no *M. tasmaniae*.

18.3.3 Broad-acre field crops: barley near Melbourne

Ridland's (1988) study summarised here is the most comprehensive field appraisal of *Micromus tasmaniae* yet made in the region. From 1982 to 1985, the relative abundance and phenology of cereal aphids and their natural enemies was studied in barley crops at Mt Derrimut, Deer Park (37.47 °S, 44.47°E), some 22 km west of Melbourne. Landing rates of aphids and aphidophagous predators were estimated from catches with green pan traps containing ethylene glycol. Population levels of aphids were assessed periodically within each crop and the occurrence and activity of surface-dwelling arthropods monitored with pitfall traps. Additional information on the flight activity of aphids and their natural enemies was obtained from four windvane interception nets from April 1983 to April 1986.

The 1-ha experimental plot (100-m side) was subdivided into 25 equal subplots of 400 m^2 (20-m sides) with 50 green pan traps (2 per subplot). In 1984, only 10 pan traps were used, along the western side of the study plot. The traps were serviced twice a week, when aphids and other arthropods of interest were removed and stored in vials of 70% alcohol.

The green pan trap (after lrwin, 1980) consisted of a circular green polyvinyl chloride bowl, 138 mm inside diameter rim and 115 mm diameter base, 50 mm deep. Each pan was loaded with 300 ml of ethylene glycol to give a catching surface of diameter 131 mm, and thus a nominal area of 0.0134 m^2. Each pan was supported by

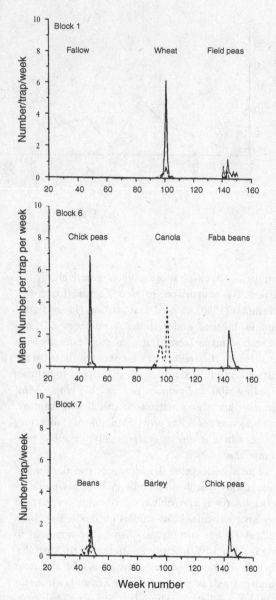

Fig. 18.2. Numbers of *Micromus tasmaniae* captured in pitfall traps at Longerenong, Victoria, 1995–7. The highly seasonal catches occurred in all three blocks, and on a variety of crops. Solid line, lacewings (adults and larvae); dashed line, aphids.

a steel ring attached by a clamp to a vertical steel pole, so that it could be raised to canopy height as the crop developed.

Four windvane nets (Gaynor & Atkinson, 1980) were erected at the field site on 6 April 1983 and oper-

Table 18.1. *Mean total number and sex ratio (given as % males) of aphidophagous predators caught per 10 green pan traps in barley crops at Mt Derrimut, Victoria, in three successive years*

	1982		1983		1984	
	N	% males	N	% males	N	% males
Neuroptera						
Micromus tasmaniae	59.6	58	721.8	42	35	54
Drepanacra binocula	0		0		2	
Mallada signatus	0.2		0		0	
Syrphidae						
Melangyna viridiceps	244	48	2972.4	85	220	65
Simosyrphus grandicornis	43	37	21.4	51	29	38
Sphaerophoria ?macrogaster	0		6	27	2	0
Coccinellidae						
Coccinella transversalis	1.4		0		0	
Diomus notescens	0.4		1.2		1	

ated until 6 April 1986. Two nets, 20 m apart, were located on the northern and southern boundaries. Twenty pitfall traps (after Price & Shepard, 1980) were placed around the perimeter of the plot, and changed weekly during the cropping period. Each trap was a polystyrene container, diameter 120 mm and depth 97 mm, containing 200 ml of ethylene glycol.

The most commonly caught aphidophagous predators in the pan traps (Table 18.1) were *M. tasmaniae* and the hoverfly *Melangyna viridiceps* (Macquart), and both were most abundant in 1983. Two other syrphid species were caught much less frequently in pan traps, but two coccinellids, *Drepanacra binocula*, and the chrysopid *Mallada signatus* (Schneider) were caught only rarely. The data summarised in Table 18.1 reflect the predominance of *Micromus tasmaniae* in field crops in the region. The phenology and relative abundance of larvae and adults caught in all three trap types in 1983 is shown in Fig. 18.3. The pattern of adult catches in the windvane traps indicated that the peak flight activity was in late spring and early summer. Very few specimens were caught at other times, and flight activity was generally not detected until August. Significant activity in 1983 first occurred in the green pans in mid-September (week 38).

A DARABUG phenological model (McDonald, 1990) based on the New Zealand data of Syrett &

Penman (1981), namely a lower threshold temperature of 5.8 °C, supplemented by data from Rumpf *et al.* (1998) was used to relate the seasonal trapping data from all of the trap types used. Degree–day accumulations were calculated with the modified DEGDAYS programme (Watson & Beattie 1996), with the arbitrary starting date for the temperature accumulations set at 1 April. Larvae were first caught in pitfall traps in week 37, and the first adults in week 33. Over that period degree–day (DD) accumulation above the base threshold reached 190 DD. Eggs are predicted to develop to second instar after 170 DD, and the peaks of adult activity recorded in weeks 43 and 44 in all trap types occurred 151 DD and 200 DD respectively from the first peak of larvae (second and third instars) in week 40. The predicted development time from newly moulted third instar to newly emerged adult is 198 DD. The second peak of larvae (in week 45) was 260 DD after the first peak of activity in all trap types (week 40). The phenological model therefore suggests that third instar larvae trapped in week 45 would have been derived from eggs laid in week 41.

The period of reproduction of *M. tasmaniae* coincided with the main period of (admittedly low) aphid build-up in the crop. The major ambiguity in the above interpretation rests on the significance of larval catches in the pitfall traps. The sizes of the catches are

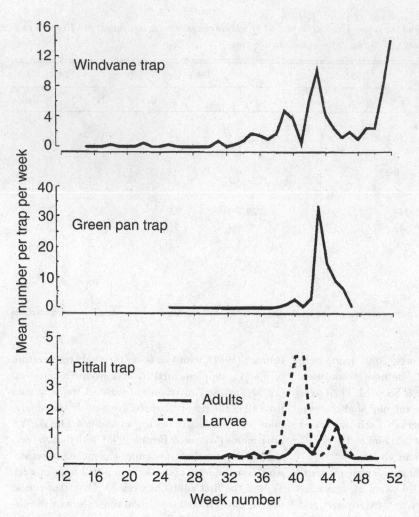

Fig. 18.3. Numbers of *Micomus tasmaniae* captured at Mt Derrimut, Victoria by three different capture regimes in 1983. Adult and larval catches are separated for pitfall trap catches. (After Ridland, 1988.)

influenced by many external factors, and it is not known whether different instars vary in their dispersal tendencies, so that the age structure in trap catches may well differ from that in the actual population.

18.4 DISCUSSION

Further investigation of *Micromus tasmaniae*, in both field and laboratory, is needed to explore further its role in regulating aphid populations and, possibly, a variety of other agricultural pests, in Australia and New Zealand. It is clearly one of the most promising native polyphagous predators in such systems, and recent studies of pesticide tolerances (Rumpf *et al.*, 1997, 1998) confirm its likely important role in IPM. Rumpf

et al. (1998) considered that it has high potential in IPM operations in the region.

The initial studies reported here solidly endorse that appraisal. One very important factor yet to be evaluated clearly is the integration of insect growth regulators (IGRs) and lacewings in IPM strategies. IGRs are often assumed to be totally IPM–compatible, and promoted as such. However, some evidence demonstrates that IGRs can have severe detrimental effects on populations of beneficial insects (Hattingh & Tate, 1995; Hattingh, 1996). Rumpf *et al.* (1998) demonstrated that some IGRs can have more severe effects than organophosphate and pyrethroid insecticides on *M. tasmaniae*, so that the relevance of studying their effects in more detail is increasingly apparent.

ACKNOWLEDGEMENTS

PAH thanks Cindy Edward for her considerable help in his surveys reported here. PMR thanks Bruce Tomkins for his assistance in field surveys.

REFERENCES

Gaynor, D.L.& Atkinson, D.S. (1980) A catching chamber for a wind net aphid trap. *New Zealand Entomologist* 7, 193–195.

Hattingh, V. (1996) The use of insect growth regulators – implications for IPM with citrus in southern Africa as an example. *Entomophaga* 41, 513–518.

Hattingh, V. & Tate, B. (1995) Effects of field-weathered residues of insect growth regulators on some Coccinellidae (Coleoptera) of economic importance as biological control agents. *Bulletin of Entomological Research* 85, 489–493.

Horne, P.A. & Edward, C.L. (1995) The phenology and food preferences of *Labidura truncata* (Dermaptera, Labiduridae) in western Victoria. *Journal of the Australian Entomological Society* 34, 101–104.

Horne, P.A. & Edward, C.L. (1997) Preliminary observations on awareness, management and impact of biodiversity in agricultural systems. *Memoirs of the Museum of Victoria* 56, 21–25.

Horne, P.A. & Edward, C.L. (1998) The effect of tillage on pest and beneficial beetles. *Australian Journal of Entomology* 37, 60–63.

Irwin, M.E. (1980) Sampling aphids in soybean fields. In *Sampling Methods in Soybean Entomology*, ed. Kogan, M. & Herzog, D.C., pp. 239–259. Springer-Verlag, New York.

Leathwick, D.M. & Penman, D.R. (1990) The efficiency of sampling for aphid predators from lucerne. *Proceedings of the 42nd New Zealand Weed and Pest Control Conference*, pp. 81–85.

Leathwick, D.M. & Winterbourn, M.J. (1984) Arthropod predation on aphids in a lucerne crop. *New Zealand Entomologist* 8, 75–80.

McDonald, G. (1990) DARABUG: a computer program for simulating development rates of insect pests in Victorian agriculture. Technical report, no. 185, Department of Agriculture and Rural Affairs, Melbourne.

MacLellan, C.R. (1973) Natural enemies of the light-brown apple moth, *Epiphyas postvittana*, in the Australian Capital Territory. *Canadian Entomologist* 105, 681–700.

Maelzer, D.A. (1977) The biology and main causes of changes in numbers of the rose aphid, *Macrosiphum rosae* (L.) on cultivated roses in South Australia. *Australian Journal of Zoology* 25, 269–84.

Milne, W.M. & Bishop, A.L. (1987) The role of predators and parasites in the natural regulation of lucerne aphids in eastern Australia. *Journal of Applied Ecology* 24, 893–905.

New, T.R. (1984) Comparative biology of some Australian Hemerobiidae. In *Progress in World's Neuropterology, Proceedings of the 1st International Symposium on Neuropterology*, ed. Gepp, J., Aspöck, H. & Holzel, H., pp. 153–166. Thalerhof, Graz.

New, T.R. (1988) A revision of the Australian Hemerobiidae (Insecta: Neuroptera). *Invertebrate Taxonomy* 2, 339–411.

New, T.R. & Boros, C. (1984) The early stages of *Micromus tasmaniae* (Neuroptera: Hemerobiidae). *Neuroptera International* 2, 213–217.

Price, J.F. & Shepard, M. (1980) Sampling ground predators in soybean fields. In *Sampling Methods in Soybean Entomology*, ed. Kogan, M. & Herzog, D.C., pp. 532–543. Springer-Verlag, New York.

Ridland, P.M. (1988) Aspects of the ecology of the rice root aphid, *Rhopalosiphum rufiabdominalis* (Sasaki) and the apple-grass aphid, *Rhopalosiphum insertum* (Walker) (Homoptera: Aphididae) in southeastern Australia. PhD thesis, La Trobe University, Melbourne.

Rohitha, B.H. & Penman, D.R. (1986) Flight of the bluegreen aphid, *Acyrthosiphon kondoi* Shinji (Homoptera: Aphididae). III. Comparison of trapping methods for *A. kondoi* and natural enemies. *New Zealand Journal of Zoology* 13, 215–220.

Rumpf, S., Frampton, C. & Chapman, B. (1997) Acute toxicity of insecticides to *Micromus tasmaniae* (Neuroptera: Hemerobiidae) and *Chrysoperla carnea* (Neuroptera: Chrysopidae): LC_{50} and LC_{90} estimates for various test durations. *Journal of Economic Entomology* 90, 1493–1499.

Rumpf, S., Frampton, C. & Dietrich, D.R. (1998) Effects of conventional insecticides and insect growth regulators on fecundity and other life-table parameters of *Micromus tasmaniae* (Neuroptera: Hemerobiidae). *Journal of Economic Entomology* 91, 34–40.

Samson, P.R. & Blood, P.R.B. (1979) Biology and temperature relations of *Chrysopa* sp., *Micromus tasmaniae* and *Nabis capsiformis*. *Entomologia Experimentalis et Applicata* 25, 253–259.

Samson, P.R. & Blood, P.R.B. (1980) Voracity and searching ability of *Chrysopa signata* (Neuroptera: Chrysopidae), *Micromus tasmaniae* (Neuroptera: Hemerobiidae) and *Tropiconabis capsiformis* (Hemiptera: Nabidae). *Australian Journal of Zoology* 28, 575–80.

Sivasubramanian, W., Wratten, S.D. & Klimaszewski, J.

(1997) Species composition, abundance and activity of predatory arthropods in carrot fields, Canterbury, New Zealand. *New Zealand Journal of Zoology* 24, 205–212.

Syrett, P. & Penman, D.R. (1981) Developmental threshold temperatures for the brown lacewing, *Micromus tasmaniae* (Neuroptera: Hemerobiidae). *New Zealand Journal of Zoology* 8, 281–283.

Watson, D.M. & Beattie, G.A.C. (1996) Degree–day models in New South Wales: climatic variation in the accuracy of different algorithms and geographical bias correction procedures. *Australian Journal of Experimental Agriculture* 36, 717–729.

Wise, K.A.J. (1992) Distribution and zoogeography of New Zealand Megaloptera and Neuroptera (Insecta). In *Current Researches in Neuropterology, Proceedings of the 4th International Symposium on Neuropterology*, ed. Canard, M., Aspöck, H. & Mansell, M.W., pp. 390–395. Sacco, Toulouse.

CHAPTER 19

Preliminary notes on *Mallada signatus* (Chrysopidae) as a predator in field crops in Australia

P.A. Horne, T.R. New, and D. Papacek

19.1 INTRODUCTION

The manipulative use of Chrysopidae as biological control agents in Australia is in its infancy. The family is well represented, with slightly more than 50 species described (New, 1996), but many of these are recorded only infrequently in Australia and occur almost entirely in forests and other naturally vegetated habitats. Many are also limited in their distribution within Australia. The species encountered most frequently in southern Australia are *Apertochrysa edwardsi* (Banks), *Mallada innotatus* (Walker), *M. signatus* (Schneider), and *Plesiochrysa ramburi* (Schneider). Indeed, a survey of Chrysopidae on native *Acacia* trees over three years in Victoria yielded only three species (*A. edwardsi, M. signatus, P. ramburi*), with *A. edwardsi* by far the most abundant (New, 1983). However, this species is predominantly arboreal, and has not been reported commonly on field crops. *Apertochrysa edwardsi* is Bassian, the two *Mallada* species are widespread in Australia, and *P. ramburi* occurs over much of the western Pacific region, where its larvae have been reported from many crops. The larvae of all these species were described by Boros (1984), but many details of their biology have not been clarified, and there are few records of their prey or feeding activity. However, they are clearly the most likely candidates among Australian Chrysopidae for investigation for their potential in biological control.

This note considers the potential for one of these common species, *M. signatus*, as a species currently attracting considerable attention as a polyphagous predator on pests of field crops.

19.2 ROLE IN BIOLOGICAL CONTROL

In eastern Australia, *Mallada signatus* is commonly found only in small numbers on field crops. Thus, Ridland's (1988) surveys of cereal aphid predators near Melbourne yielded very few individuals of this – or of any other chrysopid. It is usually much less abundant than the hemerobiid *Micromus tasmaniae* (Walker). Occasional greater abundance, and its association with a considerable variety of trees and low vegetation has stimulated efforts to investigate its predatory role, as an apparently vigorous and voracious polyphagous native predator. As for other species of *Mallada*, adults feed on nectar and pollen, and the debris-carrying larvae are predatory.

Abundance in cotton fields in Queensland led to initial studies of its influence on *Helicoverpa* (Noctuidae; Samson & Blood, 1979, 1980), and its voracity appeared to be greater than that of some analogous chrysopids on cotton in North America (New, 1984). *Mallada signatus* was 'more voracious' than *Micromus tasmaniae* on *H. punctigera* (Wallengren), and was clearly the more suitable predator for attempting to manipulate against this important pest. Larvae also searched more effectively for prey than the nabid *Tropiconabis capsiformis* (Germar) (Samson & Blood, 1980).

More recently, *Mallada signatus* has been considered suitable for pest control operations on a variety of field and greenhouse crops, against groups such as mites, aphids, whitefly, and various other Homoptera, as well as immature Lepidoptera.

The development of this interest is noted here, and is of more general interest in marking increased appreciation of the first native chrysopid investigated for use as a biological control agent in Australia. In particular, it seems likely to have a role in suppression of pest Lepidoptera in parts of eastern Australia; in Queensland, for example, it has been reported to feed on eggs of the macadamia nut-borer (*Cryptophlebia ombrodelta* (Lower), Tortricidae). In Victoria, PAH has found *M. signatus* in large numbers on grapevines in the

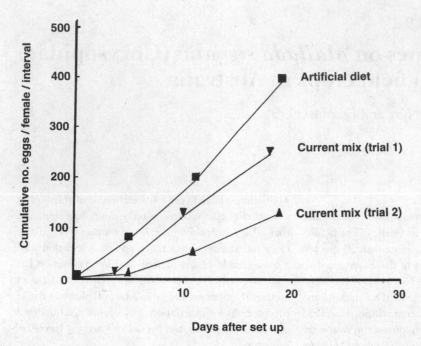

Fig. 19.1. Egg production by *Mallada signatus* over three weeks when fed solely on the currently used artificial diet (two replicate trials shown) and on another formulation at present under trial.

Yarra Valley, and considers it one of the most important predators of pests of wine grapes in the state. At one vineyard where no chemical pesticides had been applied, PAH recorded 126 *Mallada* per 100 leaves inspected, and it apparently feeds on eggs and caterpillars of the polyphagous light-brown apple moth (*Epiphyas postvittana* (Walker), Tortricidae) in that crop. In such environments it has a role complementary to that of *Micromus tasmaniae*, which is not an efficient predator of Lepidoptera (MacLellan, 1973; Samson & Blood, 1980). The large size of *Mallada signatus* may render it even more versatile than the hemerobiid, but it may prove possible to use (and manipulate) these two species together on some crops.

19.3 PRODUCTION OF STOCKS

Mallada signatus has proved to be amenable to laboratory rearing, and is now raised commercially by Bugs for Bugs P/L in Queensland. Current production is approximately 50000 individuals per week, on a diet based on *Sitotroga* (Lepidoptera) eggs. Each female is capable of producing up to 600 eggs over a four-week period, following a preoviposition period of around a week. The optimal egg-laying period is about three weeks. Preliminary trials using a dietary supplement

have produced the kind of result exemplified in Fig. 19.1, so that the new diet appears very promising in increasing production levels in mass-rearing operations.

Eggs are harvested three times per week, and eclosion is around 95%. The rearing regime involves setting up 120 adults in each production chamber, with the proportion of females in the range of 50%–60%.

The lacewings are distributed to customers as eggs near larval emergence. Because larvae are cannibalistic, distribution at this stage helps to reduce losses, but delays in transmission can lead to serious losses. Eggs are despatched packed in ground rice hulls and are distributed over the crop by hand, shaking the container like a salt dispenser.

The effectiveness of releases depends on monitoring the crops, as in any similar operation. Recommended release levels vary with the context, but field releases two or three times at intervals of 10–14 days are recommended to establish continuing populations (Papacek *et al.*, 1995). This interval corresponds to the period when the previous release has reached the cocoon stage and when field populations of larvae will be low. For general releases in large areas, a minimum of 500–1000 larvae/ha is recommended. Releases of 1–5 larvae per plant in nurseries is often ample.

Further study of this promising predator species is currently under way. We are confident that it will assume a greater role among the augmentable native predators in a variety of field crops in eastern Australia.

ACKNOWLEDGEMENTS

PAH thanks Cindy Edward for help in accumulating the field and laboratory data noted here.

REFERENCES

Boros, C.B. (1984) Descriptions of the larvae of six Australian species of *Chrysopa* Leach, s. lat. (Neuroptera: Chrysopidae). *Australian Journal of Zoology* 32, 833–849.

MacLellan, C.R. (1973) Natural enemies of the light-brown apple moth, *Epiphyas postvittana*, in the Australian Capital Territory. *Canadian Entomologist* 105, 681–700.

New, T.R. (1983) Aspects of the biology of *Chrysopa edwardsi* Banks (Neuroptera, Chrysopidae) near Melbourne, Australia. *Neuroptera International* 1, 165–174.

New, T.R. (1984) Chrysopidae: ecology on field crops. In *Biology of Chrysopidae*, ed. Canard, M., Séméria, Y. & New, T.R., pp. 160–167. Dr W. Junk, The Hague.

New, T.R. (1996) Neuroptera. In *Zoological Catalogue of Australia*, vol. 28, ed. Wells, A., pp. 1–104. CSIRO, Australia.

Papacek, D., Llewellyn, R., Altmann, J., Ryland, A. & Seymour, J. (1995) *The Good Bug Book*. Australasian Biological Control Inc., Department of Primary Industry Queensland, Rural Industries Research and Development Corporation, Richmond, New South Wales.

Ridland, P.M. (1988) Aspects of the ecology of the rice root aphid, *Rhopalosiphum rufiabdominalis* (Sasaki) and the apple-grass aphid, *Rhopalosiphum insertum* (Walker) (Homoptera: Aphididae) in southeastern Australia. PhD thesis, La Trobe University, Melbourne.

Samson, P.R. & Blood, P.R.B. (1979) Biology and temperature relations of *Chrysopa* sp., *Micromus tasmaniae* and *Nabis capsiformis*. *Entomologia Experimentalis et Applicata* 25, 253–259.

Samson, P.R. & Blood, P.R.B. (1980) Voracity and searching ability of *Chrysopa signata* (Neuroptera: Chrysopidae), *Micromus tasmaniae* (Neuroptera: Hemerobiidae) and *Tropiconabis capsiformis* (Hemiptera: Nabidae). *Australian Journal of Zoology* 28, 575–580.

CHAPTER 20

An evaluation of lacewing releases in North America

K.M. Daane and the late K.S. Hagen

20.1 INTRODUCTION

The study and use of green lacewings in North America has a long history that includes some of the initial research (1940s–70s) on lacewing augmentation, biology, field ecology, and mass-production techniques. Much of this early work helped develop guidelines for the commercial programmes that emerged in the 1980s. Nevertheless, research on the proper use and efficacy of augmentation programmes with lacewings lagged behind concurrent improvements in mass-production methods for parasitoids and increases in their commercial use, especially in glasshouse systems in Europe. During this past decade, research has once again focused on lacewing field ecology and augmentation and, as a result, there have been substantial advances in the use of *Chrysoperla* in North American agriculture. The progress includes (1) systematic revisions of green lacewings that make correct identification and evolutionarily based biological comparisons a reality, (2) improvements in the methodology for mass-production, (3) applying information from chemical ecology and seasonality to conserve and manipulate natural populations, and (4) rigorous experimental evaluation of release methodology and lacewing efficacy (see Tauber *et al.*, 2000). In this chapter, we describe the advance from experimental to commercial use of lacewings and highlight three recent studies that improved consumer acceptance of the augmentation or conservation of lacewings by providing efficacy studies and setting commercial release guidelines.

20.2 EARLY USE OF LACEWINGS IN NORTH AMERICA

The first field releases of *Chrysoperla carnea* (Stephens) were made against the obscure mealybug, *Pseudococcus viburni* (Signoret), infesting California pears treated with DDT to control the codling moth, *Cydia pomonella* (L.). In the unsprayed blocks, it became clear that the mealybug was being controlled by *Chrysoperla carnea* (Doutt, 1948). These observations led to tests of the mass-culture and release of *C. carnea*. Three releases of 250 eggs per tree reduced the mealybug infestation rate of the fruit down to 12%, while 58%–68% of the pears were infested in the DDT treated blocks, where no lacewing releases were made (Doutt & Hagen, 1949, 1950).

It was the 1960s before *C. carnea* was again manipulated in the field, with experiment releases focused on lepidopteran pests in row crops, particularly bollworms in cotton (Table 20.1) Initial trials, conducted in field cages, demonstrated that *C. carnea* larvae could reduce bollworm densities. Ridgway & Jones (1968, 1969) obtained 76%–99% reduction of *Heliocoverpa virescens* (Fabricius) and a 74%–90% reduction of *H. zea* (Boddie) with releases of *C. carnea* eggs or larvae. Lingren *et al.* (1968) obtained similar results, with a 96% reduction of *H. virescens*. However, in this study *C. carnea* were released at rates near 1 000 000 larvae per ha! Other cage trials followed, with van den Bosch *et al.* (1969), Lopez *et al.* (1976), and Stark & Whitford (1987) releasing *C. carnea* larvae to reduce bollworms and their resulting damage to the cotton bolls. Similarly, another lepidopteran pest of cotton, the pink bollworm, *Pectinophora gossypiella* (Saunders), was reduced by 33%–80% by releasing third-instar *C. carnea* (Irwin *et al.*, 1974). While research in the southwestern USA investigated lacewing releases on cotton against bollworms, Shands *et al.* (1972*a*, *b*, *c*, *d*, *e*) conducted a series of experiments on potatoes against aphids in the northeastern USA.

Each study added information that was useful for the commercialisation of lacewing releases. For example, Shands *et al.* (1972*c*) described prey preference differences of *C. carnea* against the aphids *Myzus persicae* (Sulzer) and *Macrosiphum euphorbiae* (Thomas), indicating that even within similar taxonomic prey groups there can be differences in release

Table 20.1. *Lacewings tested in augmentation programmes in North America*

Targeted pest	Crop/site condition[a]	Release stage	Release rate(s)	Release method[b]	Pest reduction[c]	Reference[d]
Chrysoperla carnea (Stephens)						
mealybug	pears/field	eggs	750/tree	exp	52%	Doutt & Hagen, 1949, 1950
bollworms	cotton/cage	larvae	1 000 000/ha	exp	96%	Lingren et al., 1968
bollworms	cotton/cage	eggs	494 000–1 975 000/ha	exp	82%–95%	Ridgway & Jones, 1968, 1969
bollworms	cotton/cage	larvae	61 000–988 000/ha	exp	74%–99%	Ridgway & Jones, 1968
bollworms	cotton/field	larvae	720 000/ha	exp	96%	Ridgway & Jones, 1969
bollworms	cotton/cage	larvae	37 000–150 000/ha	exp	41%–83%	van den Bosch et al., 1969
aphids	cotton/cage	larvae	218 000/ha	exp	48%	Shands et al., 1972c
aphids	potato/field	larvae	361 000/ha	exp	33%	Tamaki & Weeks, 1973
bollworms	sugar beets/field	eggs	358 600/ha	exp	none	Barry et al., 1974
aphids	soya bean/cage	larvae	1–12/plant	exp	26%–80%	Irwin et al., 1974
gelechiid (eggs)	cotton/cage	larvae	247 000/ha	exp	24%–81%	Lopez et al., 1976
bollworms	cotton/cage	eggs	8000/tree	com	none	Dreistadt et al., 1986
aphids	tulip tree/field	eggs	8000–12 000/tree	exp	>80%	Hagley & Miles, 1987
spider mites	peach/gh	larvae	1.5/plant	exp	43%–75%	Stark & Whitford, 1987
bollworm (eggs)	cotton/cage	eggs	335 000/ha	exp	0%–47%	Hagley, 1989
aphids	apple/field	eggs	100 000–400 000/gh	com	potential	Heinz & Parrella, 1990
aphids	marigolds/gh	eggs	100 000–416 700/ha	exp	none	Rosenheim & Wilhoit, 1993
aphids	cotton/field	eggs	123 500/ha	com	none	Edelson et al., 1993
aphids	broccoli/field	eggs	7400–37 100/ha	com	0%–20%	Daane et al., 1996
cicadellid	grape/field	eggs	9800–89 900/ha	exp	29%–42%	Daane et al., 1996
cicadellid	grape/cage	larvae	83 300–416 700/ha	com	none	Daane et al., 1996
aphids	sugar beet	eggs				Ehler et al., 1997
Chrysoperla rufilabris (Burmeister)						
chrysomelid	potato/cage	larvae	40 000–80 900/ha	exp	38%–84%	Nordlund et al., 1991
whitefly	ornamentals/gh	larvae	50–200/plant	com	>80%	Breene et al., 1992
aphids	ornamentals/field	larvae	200–300/plant	com	none	Raupp et al., 1994
aphids	white fir/field	larvae	100/plant	com	>90%	Ehler & Kinsey, 1995[e]
cicadellid	grape/field	larvae	19 800/ha	exp	31%	Daane et al., 1996
Chrysoperla externa (Hagen)						
noctuid	soya bean/field	eggs	110 000–350 000/ha	com	none	Barclay, 1990

Notes:

[a] Release conditions are grouped as 'field' if in an open agricultural or landscape ecosystem, 'cage' if in closed cages, or 'gh' if in a glasshouse.

[b] Release methodology rated as 'com' if commercial methodologies were used, 'exp' if rate or methodology would be prohibitive in a commercial operation.

[c] Reduction is the percentage change between the release and control plots.

[d] Information was compiled through a literature search using AGRICOLA, BIOS, and summary articles; only replicated trials are included. Experimental trials using cages to study predator biology, rather than augmentative release, are not included.

[e] Large-scale trials were combined with applications of an insecticidal soap.

effectiveness. Shands *et al.* (1972*d*) investigated the spatial distribution of released lacewings, which aided the concurrent development of release methods (e.g. Shands *et al.*, 1972*e*; Jones & Ridgway, 1976; Ables *et al.*, 1979). Stark & Whitford (1987) conducted one of the more elaborate studies, using a series of third-instar *C. carnea* releases against varying densities of *H. virescens* eggs on cotton plants. Fifty-five per cent of the *H. virescens* eggs were attacked at the lowest bollworm density (predator:prey ratio of 2.8:1), while 42% were attacked at the highest density (predator: prey ratio of 0.8:1), producing a type II functional response. From these studies, they calculated that *C. carnea* larvae had an average search rate of 1.08×10^{-5} ha per predator–day. Such information on lacewing prey preferences, foraging abilities, and field ecology derived from these cage trials was useful in determining the theoretical release rates needed for pest control.

These early field studies provided valuable information that is closely linked to concurrent studies on *Chrysoperla* biology, ecology, and systematics; mass-production; release methodologies; adult chrysopid diet and chemical attraction (for reviews see Canard & Duelli, 1984; Tauber *et al.*, 2000). Nevertheless, only a few of the studies tested *C. carnea* outside cages, and development of effective release programmes in the open field was not yet realised. In one open-field study, Ridgway & Jones (1969) did show a 96% reduction of bollworms after releasing 721 200 *C. carnea* larvae per ha. However, no reduction of bollworms occurred when eggs were released and the experimental design may not have accounted for the effects of intraguild predation, which became a key issue of lacewing releases in cotton in the 1990s (this volume, Chapter 14). Further, the best results were obtained with larval releases and prey:predator release rates were high: ranging from 2.5:1 to 30:1 (these release rates are equivalent to 62 000–741 000 lacewings per ha). At that time, the use of larvae and release rates over 100 000 *C. carnea* per ha was not economically competitive with concurrent insecticide programmes. The recent advances in mass-rearing and release methods (e.g., Nordlund, 1993; Nordlund & Correa, 1995; Gardner & Giles, 1996*a, b*; Tauber *et al.*, 1997), our understanding of predator–prey interactions (this volume, Chapter 6) and chemical ecology (Chapters 5 and 17) has helped develop commercially viable release programmes in North America.

To illustrate recent advances in the use of lacewings, we highlight studies on lacewing releases conducted in forest nurseries against aphids, in glasshouses against whiteflies, and in vineyards against leafhoppers. Each study helped advance commercialisation of lacewing releases.

20.3 CASE STUDIES IN NORTH AMERICA

20.3.1 Lacewings in forest nurseries

Mindarus kinseyi Voegtlin, is a newly described pest of white fir seedlings in California nurseries. This aphid was first detected in 1987 and, shortly after, Ehler & Kinsey (1995) conducted a series of studies to determine its pest status, the presence and effectiveness of resident natural enemies, and the effectiveness of different 'soft' IPM programmes, which included the release of green lacewings.

An initial survey found aphid populations infested up to 50% of the first- and second-year seedlings. The most severe damage is to first-year seedlings, which can be killed or severely stunted in both the present and following seasons if aphid populations are left unchecked. The survey also revealed a number of resident natural enemies on the seedlings, including four species of syrphid larvae and the convergent lady beetle, *Hippodamia convergens* Guérin-Menéville. No parasitoids were found. During this period, researchers gathered needed information to describe *M. kinseyi* life history and develop sampling methods. A control programme was tested that relied on releases of *Chrysoperla carnea* and *C. rufilabris* (Burmeister) and applications of M-Pede™ insecticidal soap.

Two laboratory studies tested the effectiveness of lacewings as *M. kinseyi* predators. In the first, sets of 20 first-year seedlings were arranged to simulate field conditions. Each seedling bore approximately 100 aphids. Either 1 or 5 *C. rufilabris* were released per seedling and the number of aphids on each determined at 7 and 14 days after release. Results showed that aphid densities were not affected at release rates of 1 lacewing larva per seedling, but aphids were eliminated from seedlings receiving 5 larvae per seedling. In the second study, the number of aphids eaten in the predator's larval stage was determined. Aphids (third instar to adult stages) were caged on field-collected seedlings, with small sections of infested foliage isolated for the feeding study. Newly emerged first-instar *C. carnea* and *C. rufilabris* were provided daily with an excess number of aphid prey and each cage was inspected daily to determine lacewing development stage and the number of aphids killed. Results show *C. carnea* and *C. rufilabris* consumed >100 aphids during the 10–12 days required from hatch to pupation. Researchers concluded that *M. kinseyi* is a suitable host for these predators and, because seedlings are often infested with >100 aphids per seedling, more than 1 lacewing per seedling may be required for control.

Ehler & Kinsey (1995) also investigated the effectiveness of insecticidal soap. Their work showed a 96% reduction of aphids with a single application within label rates. While lacewings are generally considered tolerant to a number of conventional insecticides (Grafton-Cardwell & Hoy, 1985), the soap also killed *C. rufilabris* larvae. However, there was no residual effect, indicating that insecticidal soap applications could be followed by lacewing releases as soon as the soap had dried (often within hours).

With two effective and compatible control programmes, the final research phase was a demonstration of the management programme for *M. kinseyi*. First-year seedlings infested with aphids were treated with either insecticidal soap (label rate) or a release of 2 *C. rufilabris* larvae per seedling (early trials showed release of *C. rufilabris* eggs were ineffective), or both. Results showed this combination of 'soft' control measures adequately suppressed aphid populations and could be adopted immediately by commercial nurseries. However, the treatments did not provide season-long control because of the short residual effect of the insecticidal soap and lack of a second generation from the released lacewings. Therefore, a monitoring programme and repeated control measures were suggested. Note that both insecticidal soap and *C. rufilabris* treatments are applied at commercial rates and are cost-competitive with other chemical treatments.

20.3.2 Lacewings in vineyards

Leafhoppers, *Erythroneura variabilis* Beamer and *E. elegantula* Osborn, are the primary insect pests in California's table and raisin grape-producing regions (Wilson *et al.*, 1992; Daane & Costello, 2000). As an alternative to carbamate and organophosphate insecticides, some growers were releasing lacewings to control leafhoppers, but no guidelines for their use or scientific evaluation of their effectiveness were available. The standard commercial programme released *C. carnea* eggs mixed with corn grit at peak nymph densities of the first and second (of three) leafhopper generations. Daane *et al.* (1994, 1996) and Daane & Yokota (1997) evaluated release effectiveness.

Before release trials began, vineyards were surveyed for leafhopper predators. Spiders were by far the most common predator, comprising >85% of the collected specimens (Costello & Daane, 2000). Five lacewing species were found: *C. carnea*, *C. comanche* Banks, *Chrysopa coloradensis* Banks, *C. nigricornis* Burmeister, and *C. oculata* Say. The great majority of these lacewings were collected on the ground cover, where aphids were a preferred prey. On the vines, *Chrysoperla carnea* and *C. comanche* were the most common species collected; however, their densities were low, <1 larvae per 1000 leaves sampled. Concurrent laboratory studies showed that *C. carnea*, *C. rufilabris,* and *C. comanche* preyed on leafhoppers. Provided with a continual supply of fourth- and fifth-instar leafhoppers, *C. comanche* larva killed >250 leafhoppers during a 10-day development period. Thus, lacewings are effective predators of leafhoppers under laboratory conditions.

Before commercial programmes were evaluated, a series of experiments were conducted in field cages and small (three-vines) plots to determine lacewing success under field conditions. In each case, first-instar *C. carnea* were hand-placed on leaves. Release rates tested were equivalent to 1–4× the commercial rates used. Results from cage studies showed a significant decrease of leafhopper in *C. carnea*-release cages, with the greatest difference (42% reduction) at the highest release rate tested (88956 *C. carnea* per ha). Similarly, results from small-plot studies showed a significant 31%–34% reduction in leafhoppers at release rates 2× the average commercial rate (19768 *C. carnea* per ha), although there was no effect at the commercial release rate (9884 *C. carnea* per ha). After these cage and small-plot studies, the effectiveness of commercial release programmes was then tested in on-farm trials using large, replicated *C. carnea*-release and no-release plots. *C. carnea* was released at rates of 7413–37065 eggs per ha per leafhopper generation. Results showed a weak trend of lower leafhopper numbers in *C. carnea*-release plots as compared with no-release plots; the amount of reduction varied from 0%–37% and was significant in 9 of 20 trials.

In summary, the average reduction of leafhoppers in *C. carnea*-release plots, as compared with no-release plots, was 29.5% in cages, 15.5% in three-vine plots, and 9.6% in commercial vineyards. Of many possibilities, the release method used was the most plausible explanation for differences in lacewing effectiveness among trials (e.g. commercial release of eggs vs. hand-release of larvae). The commercial operation of lacewing delivery can be subdivided into production of insectary material, release of viable lacewings, release stage (egg vs. larvae), and proper timing of the release. Typical egg hatch of insectary material stored at 25–27 °C and 20%–40% RH was >90%; therefore, shipment of viable material from the insectary to the grower was not considered a problem and we focused attention on release methodology used in the vineyards.

Two commercial release methods were compared. In the first delivery system, a mixture of lacewing eggs and corn grit was placed in paper cups, which were distributed to every fifth vine in every other row. Egg hatch was low (~60%) in the paper cups, compared with egg hatch when lacewings were reared in individual cells (~91%). The poor egg hatch was attributed primarily to cannibalism (see Bar & Gerling, 1985). Larval dispersal from paper cups with corn grit was 25% lower than that from cups without corn grit, resulting in incomplete distribution of lacewings throughout the vineyard. In a second delivery system, lacewing eggs, combined with corn grit, were dropped onto the vines from above, sifted through a funnel that was mounted on the back of a flat-bed trailer. Egg hatch was ~62%, similar to the paper-cup method. However, egg distribution throughout the vineyard was much improved, with eggs delivered to each vine and vine row. Delivery of eggs was slightly uneven: more eggs were dropped at the beginning (~11 eggs per vine) of each release batch than at the end (~5 eggs per vine).

In a second experiment, using a non-commercial delivery system, release rates between 6175 and 1235000 eggs per ha were tested. Releases timed to ~50%–70% leafhopper egg hatch had a greater effect on leafhopper densities than releases timed to peak leafhopper nymph densities. However, no correlation between release rate and prey density was found and, as in other experiments, cannibalism was observed and suspected of reducing lacewing effectiveness at the higher release rates. In the third experiment, we tested the effectiveness of egg vs. larval releases. In egg release plots, there was ~70% egg mortality and leafhopper densities were not significantly different from no-release plots. In larval release plots, ~50% of the larvae survived until the third instar and there was a significant reduction in leafhopper densities.

The accumulated evidence suggests that lacewing releases are not yet an effective tool for leafhopper control in vineyards. Most importantly, in all trials (cage, small-plot, and on-farm) when leafhopper densities were above the suggested economic injury level (15–20 nymphs per leaf), the reduction in leafhopper number was frequently not sufficient to lower the leafhopper density below the economic injury threshold. Reductions of leafhopper numbers in cage, small-plot, and on-farm studies were far lower than the estimated reduction based on laboratory consumption experiments. One problem was that growers released eggs, rather than larvae, because of the lowered cost and easier handling. The results suggest that egg-delivery systems used in vineyards were not effective and current improvements in delivery systems and the use of food sprays should be tested to improve lacewing releases in North American vineyards.

20.3.3 Lacewings in glasshouses

Historically, predator releases have fared better in glasshouse systems as compared with open-field crops (Hussey & Scopes, 1985; van Lenteren & Woets, 1988; Daane *et al.*, 1998). In the 1990s, researchers in North America developed commercial release programmes utilising lacewings for control of glasshouse pests of ornamentals. Here, we summarise work by Breene *et al.* (1992) on their trials with the sweet-potato whitefly, *Bemisia tabaci* (Gennadius), on *Hibiscus*. Unlike research described in previous case studies, Breene *et al.* had considerable background information on which to draw. Researchers had previously described natural enemies of the sweet-potato whitefly (e.g. Gerling, 1986; Dean & Schuster, 1995) and tested the augmentative release of parasitoids (Parrella *et al.*, 1992). Against the greenhouse whitefly, *Trialeurodes vaporariorum* (Westwood), releases of a parasitoid were combined with supplementary releases of *Chrysoperla carnea* to control whitefly densities below economically damaging levels (Heinz & Parrella, 1990; Parrella & Heinz, 1991) Therefore, there was considerable evidence that lacewings could be used to suppress white-fly populations.

Breene *et al.* (1992) infested *Hibiscus* with sweet-potato whitefly until pest densities were at the 'threshold' level used for insecticide treatment. In a series of experiments, treatments consisted of 2, 5, 25, 50, or 100 *C. rufilabris* larvae per plant released twice (two weeks apart). Additional treatment factors included either *Hibiscus* plants touching, to increase predator or prey movement between plants, or plants not contiguous. Whitefly densities were measured with counts of adults on sticky cards and eggs and nymphs on sub-samples of leaves. Results clearly demonstrated that whitefly density was significantly higher in no-release controls. In most trials the 0 and 5 *C. rufilabris* per plant treatments had significantly higher whitefly numbers (as compared with the higher *C. rufilabris* release rates) and in some trials whitefly densities were reduced even at 5 *C. rufilabris* per plant. To provide a commercial evaluation, all tested plants were ranked from 1 to 5, with 1 representing unmarketable plants (resulting from whitefly, honeydew, and sooty mould) and 5 representing clean plants. All plants that received *C. rufilabris* were marketable, untreated plants were most often unmarketable and the differences between treated and untreated caged were described as 'visibly striking'.

Researchers conclude that *C. rufilabris* can be used commercially for whitefly control. Release rates ⩾25 *C. rufilabris* per plant produced marketable *Hibiscus* in all trials, with very little difference in whitefly densities among the higher release rates. Researchers released lacewings by hand either on each plant or in the centre of 12 plants with leaves either touching or not touching. In each case, lacewings dispersed from the release point and whitefly populations were suppressed, although data suggest that there was poorer dispersal or increased cannibalism when lacewings were released at a centre location. Additionally, results from a companion study showed that *C. rufilabris* were not affected by residual activity of insecticidal soaps.

20.4 CONCLUSIONS

Early augmentation studies on lacewings in North America highlighted the potential of lacewing field releases. The biological requirements for effective augmentation include (1) an ability to rear predictable quantities of natural enemies of known quality, (2) store, transport, and release the natural enemies such that they can compete effectively at the release site, and (3) an understanding of the ecological parameters governing the principal interactions between the biological control agent released and the target pest. The three more recent studies highlighted describe research needed to move potential lacewing release programmes to commercial applications and to improve existing commercial programmes. Each case study had common traits, or base information: description of existing natural enemies and lacewing abundance, feeding studies with lacewings against the targeted pests, field-cage or small-plot studies, and then field trials with commercial release methods and rates.

Information should be gathered either from the literature or new data collected on the feeding preference of lacewing to targeted pests. Breene *et al.* (1992) could rely on the literature and on previous trials with *Chrysoperla* releases against whiteflies (Heinz & Parrella, 1990). Both Ehler & Kinsey (1995) and Daane *et al.* (1994) completed feeding studies, which showed that in *C. carnea* and *C. rufilabris* the number of aphids and leafhoppers consumed was similar to other laboratory predation studies (see Principi & Canard, 1984). Their work also indicated that >75% of total leafhopper consumption by lacewing larvae occurred during the third instar. The results suggest that release timing or release stage may be critical for field success. Comparing consumption to developmental time indicates that after field release there will be an ~7-day period during which few leafhoppers will be killed while the lacewings develop to the third instar. Yet, for most commercial practices, lacewings are released in the egg stage at peak pest densities.

To understand the predator–prey relationship and ecological interactions in the field, information on pest biology is also needed. In the vineyard and glasshouse studies highlighted, there was considerable information available about the targeted pests (variegated leafhopper and sweet-potato whitefly). Ehler & Kinsey (1995) worked with a newly described pest and properly studied pest biology before conducting field-release experiments.

It is also essential to survey the targeted release area to determine abundance and species composition of resident natural enemies. This work is essential for two reasons. First, resident lacewing density can be higher than release densities. Rosenheim & Wilhoit (1993) found the number of naturally deposited lacewing eggs in cotton fields to be far greater than the number of eggs that could be economically released. Second, surveys can show which lacewing species are the most common at the target site. For example, Daane *et al.* (1994) found that one of the more important species in the vines (*C. comanche*) was not commercially available. In response to that survey, some insectaries began production of *C. comanche*, in addition to *C. carnea* and *C. rufilabris*, for release in vineyards. Third, released lacewings are subject to predator–predator interactions at the release site (this volume, Chapter 14) and information on other predator species may help release decisions. In the Ehler & Kinsey (1995) and Daane *et al.* (1994), research completed this work before proceeding. Breene *et al.* (1992) had the advantage of working in a glasshouse system, which often provides better isolation of predators and a better controlled environment.

Before testing commercial operations, small-plot or cage studies should be conducted at the target site. There are obvious advantages to releasing insects in cages for study and, while the results cannot always be related to field performance, small-plot studies help to control predator and pest numbers to better determine the functional response of released lacewings in an environment closely resembling the open-field site. Prior to open-field releases, cage studies also have the advantage of measuring maximum lacewing effectiveness in the absence of other predators and reducing the amount of predator and prey dispersal.

Large-scale trials should be conducted using commercial release methods. In most North American trials, commercial release methodology was not used (Table 20.1) We believe, however, that the release methodology weighs heavily in the evaluation of an augmentation programme. In the case studies, both Breene *et al.* (1992) and Ehler & Kinsey (1995) released pre-fed larvae, while Daane *et al.* (1996) tested egg releases. In each case, release methods and rates were within the economic range of commercial programmes. Two primary factors control costs: use of eggs or larvae and the release rate. In large-scale open-field releases, such as vineyards, orchards, or row crops, the expense of pre-fed lacewing releases may not be economically compet-

itive with insecticide programmes. It was also common for researchers to hand-place test organisms onto test plants and in many of the trials the lacewings were released as well-developed larvae (e.g., Lingren *et al.*, 1968; Lopez *et al.*, 1976; Beglyarov & Smetnik, 1977; Rao & Chandra, 1984), while in commercial operations the egg is the most commonly released stage. Such manipulations of tested predators will lead to efficacy data that indicate a greater impact on pest densities than might be achieved in a commercial release programme. Nevertheless, for the more valuable nursery crops or with many glasshouse ornamentals or vegetables, the smaller size of the crop, and therefore lower release rate needed allows for larval releases. Also, note that in most studies, release rates were far higher than commercial rates (Table 20.1). For example, the tested *C. carnea* release rates in field crops varied from ~7500 eggs per ha to ~2000000 eggs per ha (Ridgway & Jones, 1968). Commercial release rates for field crops are more commonly between 5000–30000 eggs per ha.

Finally, Breene *et al.* (1992) and Ehler & Kinsey (1995) included in their work the compatibility of lacewing releases with insecticidal soap sprays. This information provides growers with an important additional control tool by describing methods to use both augmentation and insecticide programmes together. The overall importance of each case study is the evaluation and/or development of commercial programmes. The recent advances in lacewing biology and field ecology, combined with improved rearing methods and a growing public concern for less-toxic control, presents the opportunity for expanded use of lacewings in field crops.

REFERENCES

Ables, J.R., Reeves, B.G. & Morrison, R.K. (1979) Methods for the field release of insect parasites and predators. *Transactions of the American Society of Agricultural Engineers* 22, 59–62.

Bar, D. & Gerling, D. (1985) Cannibalism in *Chrysoperla carnea* (Stephens) (Neuroptera, Chrysopidae). *Israel Journal of Entomology* 19, 13–22.

Barclay, W.W. (1990) Role of *Chrysoperla externa* as a biological control agent of insect pests in Mesoamerican agricultural habitats. MSc thesis, University of California, Berkeley.

Barry, R.M., Hatchett, J.H. & Jackson, R.D. (1974) Cage

studies with predators of the cabbage looper, *Trichoplusia ni*, and corn earworm, *Heliothis zea*, in soybeans. *Journal of the Georgian Entomological Society* 9, 71–78.

Beglyarov, G.A. & Smetnik, A.I. (1977) Seasonal colonization of entomophages in the U.S.S.R. In *Biological Control by Augmentation of Natural Enemies*, ed. Ridgway, R.L. & Vinson, S.B., pp. 283–328. Plenum Press, New York.

van den Bosch, R., Leigh, T.F., Gonzalez, D. & Stinner. R.E. (1969) Cage studies on predators of the bollworm in cotton. *Journal of Economic Entomology* 62, 1486–1489.

Breene, R.G., Meagher, R.L. Jr, Nordlund, D.A. & Yin-Tung Wang (1992) Biological control of *Bemisia tabaci* (Homoptera: Aleyrodidae) in a greenhouse using *Chrysopa rufilabris* (Neuroptera: Chrysopidae). *Biological Control* 2, 9–14.

Canard, M. & Duelli, P. (1984) Predatory behavior of larvae and cannibalism. In *Biology of Chrysopidae*, ed. Canard, M., Séméria, Y. & New, T.R., pp. 92–100. Dr W. Junk, The Hague.

Costello, M.J. & Daane, K.M. (2000) Abundance of spiders and insect predators on grapes in central California. *Journal of Arachnology*, 27, 531–538.

Daane, K.M. & Costello, M.J. (2000) Variegated and Western grape leafhopper. In *Raisin Production Manual*, ed. Christensen, L.P., pp. 173–181. University of California, Division of Agriculture and Natural Resources, Berkeley, CA.

Daane, K.M. & Yokota, G.Y. (1997) Release methods affect egg survival and distribution of augmentated green lacewings (Chrysopidae: Neuroptera). *Environmental Entomology* 26, 455–464.

Daane, K.M., Yokota, G.Y., Hagen, K.S. & Zheng, Y. (1994) Field evaluation of *Chrysoperla* spp. in augmentative release programs for the variegated grape leafhopper, *Erythroneura variabilis*. In *Proceedings of the 7th Workshop of the IOBC Working Group, Quality Control of Mass Reared Arthropods*, ed. Nicoli, G., Benuzzi, M. & Leppla, N.C., pp. 105–120.

Daane, K.M., Yokota, G.Y., Zheng, Y. & Hagen, K.S. (1996) Inundative release of the green lacewing to control *Erythroneura variabilis* and *E. elegantula* (Homoptera: Cicadellidae) in grape vineyards. *Environmental Entomology* 25, 1224–1234.

Daane, K.M., Hagen, K.S. & Mills, N.J. (1998) Predaceous insects for insect and mite control. In *Mass-Reared Natural Enemies: Application, Regulation, and Needs*, ed. Ridgway, R.L., Hoffmann, M.P., Inscoe, M.N. & Glenister, C.S, pp. 61–115. Thomas Say Publications in Entomology, Entomological Society of America, Lanham MD.

Dean, D.E. & Schuster, D.J. (1995) *Bemisia argentifolii* (Homoptera: Aleyrodidae) and *Macrosiphum euphorbiae* (Homoptera: Aphididae) as prey for two species of Chrysopidae. *Environmental Entomology* 24, 1562–1568.

Doutt, R.L. (1948) Effect of codling moth sprays on natural control of the Baker mealybug. *Journal of Economic Entomology* 41, 116.

Doutt, R.L. & Hagen, K.S. (1949) Periodic colonization of *Chrysopa californica* as a possible control of mealybugs. *Journal of Economic Entomology* 42, 560.

Doutt, R.L. & Hagen, K.S. (1950) Biological control measures applied against *Pseudococcus maritimus* on pears. *Journal of Economic Entomology* 43, 94–96.

Dreistadt, S.H., Hagen, K.S. & Dahlsten, D.L. (1986) Predation by *Iridomyrmex humilis* (Hym.: Formicidae) on eggs of *Chrysoperla carnea* (Neu.: Chrysopidae) released for inundative control of *Illinoia liriodendri* (Hom.: Aphididae) infesting *Liriodendron tulipifera*. *Entomophaga* 31, 397–400.

Edelson, J.V., Magaro, J.J. & Browning, H. (1993) Control of insect pests on broccoli in southern Texas: a comparison between synthetic organic insecticides and biorational treatments. *Journal of Entomological Science* 28, 191–196.

Ehler, L.E. & Kinsey, M.G. (1995) Ecology and management of *Mindarus kinseyi* Voegtlin (Ahidoidea: Mindaridae) on white-fir seedlings at a California nursery. *Hilgardia* 62, 1–62.

Ehler, L.E., Long, R.F., Kinsey, M.G. & Kelley, S.K. (1997) Potential for augmentative biological control of black bean aphid in California sugarbeet. *Entomophaga* 42, 241–256.

Gerling, D. (1986) Natural enemies of *Bemisia tabaci*, biological characteristics and potential as biological control agents: a review. *Agriculture, Ecosystems and the Environment* 17, 99–110.

Gardner, J. & Giles, K. (1996a) Handling and environmental effects on viability of mechanically dispersed green lacewing eggs. *Biological Control* 7, 245–250.

Gardner, J. & Giles, K. (1996b) Mechanical distribution of *Chrysoperla rufilabris* and *Trichogramma pretiosum*: survival and uniform discharge after spray dispersal in aqueous suspension. *Biological Control* 8, 1–5.

Grafton-Cardwell, E.E., & Hoy, M.A. (1985) Short-term effects of permethrin and fenvalerate on oviposition by *Chrysoperla carnea* (Neuroptera: Chrysopidae). *Journal of Economic Entomology* 78, 955–959.

Hagley, E.A.C. (1989) Release of *Chrysoperla carnea* Stephens (Neuroptera: Chrysopidae) for control of the green apple aphid, *Aphis pomi* DeGeer (Homoptera: Aphididae). *Canadian Entomologist* 121, 309–314.

Hagley, E.A.C. & Miles, N. (1987) Release of *Chrysoperla carnea* Stephens (Neuroptera: Chrysopidae) for control of *Tetranychus urticae* Koch (Acarina: Tetranychidae) on peach grown in a protected environment structure. *Canadian Entomologist* 119, 205–206.

Heinz, K.M. & Parrella, M.P. (1990) Biological control of insect pests on greenhouse marigolds. *Environmental Entomology* 19, 825–835.

Hussey, N.W. & Scopes, N. (1985) *Biological Pest Control, the Glasshouse Experience.* Cornell University Press, Ithaca, New York.

Irwin, M.E., Gill, R.W. & Gonzalez, D. (1974) Field-cage studies of native egg predators of the pink bollworm in southern California cotton. *Journal of Economic Entomology* 67, 193–196.

Jones, S. L. & Ridgway, R.L. (1976) Development of methods for field distribution of eggs of the insect predator *Chrysopa carnea* Stephens. *United States Department of Agriculture, Agricultural Research Service, S-124.*

van Lenteren, J. & Woets, J. (1988) Biological and integrated control in greenhouses. *Annual Review of Entomology* 33, 239–269.

Lingren, P.D., Ridgway, R.L. & Jones, S.L. (1968) Consumption by several common arthropod predators of eggs and larvae of two *Heliothis* species that attack cotton. *Annals of the Entomological Society of America* 61, 613–618.

Lopez, J.D. Jr., Ridgway, R.L. & Pinnell, R.E. (1976) Comparative efficacy of four insect predators of the bollworm and tobacco budworm. *Environmental Entomology* 5, 1160–1164.

Nordlund, D.A. (1993) Improvements in the production system for green lacewings: a hot melt glue system for preparation of larval rearing units. *Journal of the Georgian Entomological Society* 28, 338–342.

Nordlund, D.A. & Correa, J.A. (1995) Improvements in the production systems for green lacewings: an adult feeding and oviposition unit and hot wire egg harvesting system. *Biological Control* 5, 179–188.

Nordlund, D.A., Vacek, D.C. & Ferro, D.N. (1991) Predation of Colorado potato beetle (Coleoptera: Chrysomelidae) eggs and larvae by *Chrysoperla rufilabris* (Neuroptera: Chrysopidae) larvae in the laboratory and field cages. *Journal of Entomological Science* 26, 443–449.

Parrella, M.P. & Heinz, K.M. (1991) Biological control of insect pests on greenhouse marigolds. *Environmental Entomology* 19, 825–835.

Parrella, M.P., Bellows, T.S. & Gill, R.J (1992) Sweetpotato whitefly: prospects for biological control. *California Agriculture* 46, 25–26.

Principi, M.M. & Canard, M. (1984) Feeding habits. In *Biology of Chrysopidae*, ed. Canard, M., Séméria, Y. & New, T.R., pp. 76–92. Dr W. Junk, The Hague.

Rao, R.S.N. & Chandra, I.J. (1984) *Brinchochrysa scelestes* (Neur.: Chrysopidae) as a predator of *Myzus persicae* (Hom.: Aphididae) on tobacco. *Entomophaga* 29, 283–285.

Raupp, M.J., Hardin, M.R., Baxton, S.M. & Bull, B.B. (1994) Augmentative release for aphid control on landscape plants. *Journal of Arboriculture* 20, 241–249.

Ridgway, R.L. & Jones, S.L. (1968) Field-cage releases of *Chrysopa carnea* for suppression of populations of the bollworm and the tobacco budworm on cotton. *Journal of Economic Entomology* 61, 892–898.

Ridgway, R.L. & Jones, S.L. (1969) Inundative releases of *Chrysopa carnea* for control of *Heliothis* on cotton. *Journal of Economic Entomology* 62, 177–180.

Rosenheim, J.A. & Wilhoit, L.R. (1993) Predators that eat other predators disrupt cotton aphid control. *California Agriculture* 47, 7–9.

Shands, W.A., Simpson, G.W. & Brunson, M.H. (1972a) Insect predators for controlling aphids on potatoes. 1. In small plots. *Journal of Economic Entomology* 65, 511–514.

Shands, W.A., Simpson, G.W. & Storch, R.H. (1972b) Insect predators for controlling aphids on potatoes. 3. In small plots separated by aluminium flashing strip-coated with a chemical. *Journal of Economic Entomology* 65, 799–805.

Shands, W.A., Simpson, G.W. & Gordon, C.C. (1972c) Insect predators for controlling aphids on potatoes. 4. Spatial distribution of introduced eggs of two species of predators in small fields. *Journal of Economic Entomology* 65, 805–809.

Shands, W.A., Simpson, G.W. & Gordon, C.C. (1972d) Insect predators for controlling aphids on potatoes. 5. Numbers of eggs and schedules for introducing them in large field cages. *Journal of Economic Entomology* 65, 810–816.

Shands, W.A., Gordon, C.C. & Simpson, G.W. (1972e) Insect predators for controlling aphids on potatoes. 6. Development of a spray technique for applying eggs in the field. *Journal of Economic Entomology* 65, 1099–1103.

Stark, S.B. & Whitford, F. (1987) Functional response of *Chrysopa carnea* (Neur.: Chrysopidae) larvae feeding on *Heliothis virescens* (Lep.: Noctuidae) eggs on cotton in field cages. *Entomophaga* 32, 521–527.

Tamaki, G. & Weeks, R.E. (1973) The impact of predators on populations of green peach aphids on field-grown sugarbeets. *Environmental Entomologist* 2, 345–349.

Tauber, M.J., Albuquerque, G.S. & Tauber, C.A. (1997) Storage of nondiapausing *Chrysoperla externa* adults: influence on survival and reproduction. *Biological Control* 10, 69–72.

Tauber, M.J., Tauber, C.A., Daane, K.M. & Hagen, K.S. (2000) New tricks for old predators: implementing

biological control with *Chrysoperla*. *American Entomologist* 46, 26–38.

Wilson, L.T., Barnes, M.M., Flaherty, D.L., Andris, H.L. & Leavitt, G.M. (1992) Variegated grape leafhopper. In *Grape Pest Management*, ed. Flaherty, D.L., Christensen, L.P., Lanini, W.T., Marois, J.J., Phillips, P.A. & Wilson, L.T., pp. 202–213. University of California, Division of Agriculture and Natural Resources Publications, Berkeley CA.

CHAPTER 21

Chrysoperla externa and *Ceraeochrysa* spp.: potential for biological control in the New World tropics and subtropics

G.S. Albuquerque, C.A. Tauber, and M.J. Tauber

The Neotropical chrysopid fauna is one of the richest in the world; 21 genera and more than 300 species have been described (Brooks & Barnard, 1990) and numerous taxa await description. Despite this diversity, very little is known about the life histories of most species and the systematics of the group is poorly resolved (Penny, 1984; Adams & Penny, 1985, 1986; Tauber & Adams, 1990). Nevertheless, during the last decade interest in the chrysopids has increased, especially with regard to their use in biological control.

21.1 *CERAEOCHRYSA* AND *CHRYSOPERLA*: GENERA OF PRIMARY IMPORTANCE TO BIOLOGICAL CONTROL

Of the eight chrysopine genera that occur in the Neotropics, we propose that *Chrysoperla* and *Ceraeochrysa* have the greatest potential for use in biological control. We base this conclusion on a number of life history and behavioural traits that these two taxa express. Species in both genera occur in a variety of habitats (e.g. dry and moist forests, grasslands, horticultural settings) and are commonly associated with agricultural crops (Muma, 1959a; Adams, 1982; Olazo, 1987; Brooks & Barnard, 1990). Their larval prey includes a large number of economically important insect pests (see below). Moreover, both genera are well adapted for mass-rearing and use in augmentative biological control (Núñez, 1988a, b; Nordlund & Morrison, 1992; Albuquerque et al., 1994; Wang & Nordlund, 1994; López-Arroyo et al., 1999a, c, 2000).

Chrysoperla externa (Hagen) shares many features with the Holarctic *Chrysoperla carnea* (Stephens) s. lat., which is produced commercially and marketed for release throughout North America, Europe, and parts of Asia, and it is an excellent candidate for mass-production and use in biological control programs in Latin America (Albuquerque et al., 1994). Furthermore, recent investigations by E. Núñez in Peru, C. F. Carvalho and his collaborators in Brazil, and J. I. López-Arroyo, C. A. Tauber, and M. J. Tauber in the USA indicate that a number of *Ceraeochrysa* species, too, are amenable to mass-rearing and use in commercial agriculture. Their larval habit of carrying packets of trash, their chemically defended eggs, and the ability of adults to escape from spider webs may offer considerable defence against ants, other predators, and parasitoids that sometimes interfere with biological control.

Although we recognise that other Neotropical chrysopid groups (especially *Nodita* and *Chrysopodes*) may have potential for use in biological control, there is virtually no literature on their biology. Thus, we restrict this chapter to *Ceraeochrysa* spp. and *Chrysoperla externa*. Our review assesses the current knowledge of these two groups in relation to a series of crucial steps in the practice of biological control, beginning with systematics and comparative biology, and concluding with the evaluation of biological control projects in the field.

21.2 SYSTEMATICS

Virtually all aspects of biological control depend on having a sound systematics base (e.g. Knutson, 1981; DeBach & Rosen, 1991). Systematics provides stable names that enable communication and access to the scientific literature; it also offers a comparative phylogenetic perspective that is essential for understanding the biological traits of pests and their natural enemies. The logical conclusion is that the systematics of Neotropical chrysopids should receive significantly more attention and support. ·

Ceraeochrysa is the most commonly collected Neotropical genus of lacewings; it is very diverse, with 44 species described to date (Brooks & Barnard, 1990; Penny, 1997) and numerous undescribed species. Many species are widely distributed over a variety of ecosystems (Adams, 1982; Adams & Penny, 1985, 1986), and a number, e.g. *C. cubana* (Hagen), *C. cincta* (Schneider), *C. everes* (Banks), *C. claveri* (Navás), *C. caligata* (Banks), *C. scapularis* (Navás), *C. sanchezi* (Navás), *C. valida* (Banks), and *C. smithi* (Navás) have been reported from agroecosystems (Adams & Penny, 1986; Serrano *et al.*, 1987; Núñez, 1988a; Freitas & Adams, 1994; M. J. & C. A. Tauber, unpublished data).

Currently, the systematics of *Ceraeochrysa* is under study. Dr N. Penny in the USA has described several new species and is preparing a revision based on the adults (e.g. Penny, 1997). Meanwhile, some identifications can be made with a key to Amazonian species (Adams & Penny, 1985). C.A. Tauber, with colleagues (T. De Leon, G.S. Albuquerque), is preparing keys and descriptions of the larval stages (e.g. Tauber *et al.*, 1998). It is noteworthy that economically important *Ceraeochrysa* present significant systematics problems. The *C. cincta* group contains numerous undescribed species; moreover, the species-status of many geographically differentiated populations is unknown. Similarly, several cryptic species resemble *C. cubana*, the Neotropical species most commonly featured in biological control reports. Resolution of the systematics of these taxa is essential for understanding their biology and for their effective and reliable utilisation in biological control.

Relative to *Ceraeochrysa*, the genus *Chrysoperla* is considerably less diverse; only seven species are recorded from the New World tropics and subtropics (Brooks, 1994). Of these, *C. externa* is the most abundant and widespread; it occurs in open grasslands from South Carolina in the United States to southern South America (Tauber, 1974; Adams & Penny, 1985; Brooks, 1994). Frequently, its populations reach high densities in agricultural crops. However, it is noteworthy that other *Chrysoperla* species can co-occur with *C. externa* and thus accurate identification is essential.

Unfortunately, determination of *Chrysoperla* spp. is very difficult at this time. Although a world-wide key for adults exists (Brooks, 1994), it is useful for only certain taxa within the genus. The larvae of only two Neotropical *Chrysoperla* [*C. externa* and *C. rufilabris* (Burmeister)] have been described (Tauber, 1974).

21.2.1 Recommendations (systematics)

The poor state of *Ceraeochrysa* and *Chrysoperla* systematics imposes handicaps for the use and commercialisation of lacewings. We make the following recommendations to help alleviate some of the problems and to ensure that data gained from current studies are associated with accurate identifications and specimens for future verification.

1. Foremost, we recommend that voucher specimens (adults and larvae, if possible) from all biological studies be preserved in an appropriate university collection or publicly accessible museum and that the deposition site be indicated in publications. Given the status of museums in many Latin American countries, this recommendation may sometimes require the use of collections outside the country of the study; in some cases, deposition of specimens in more than one collection may be advantageous or necessary. All specimens should bear appropriate data (locality of collection, name of collector, lot number, and publication reference, if available).

2. We recommend that species identifications be verified by a systematist who works with the group. In many instances, this systematist can arrange for deposition of voucher specimens in a museum. Similarly, involvement of a systematist early in the planning and implementing of biological control projects can prevent unfortunate problems and advance the progress of biological control projects.

21.3 FIELD DATA: PREY AND CROP ASSOCIATIONS, SEASONAL CYCLES

An early step in any biological control project is to assess the species composition of natural enemies associated with the crop and the targeted pest. The primary chrysopid species for which there are substantial published collection records associated with prey and host-plant data are *Ceraeochrysa cubana* and *Chrysoperla externa* (Table 21.1); few records exist for *Ceraeochrysa cincta* (Núñez, 1988 *a*, *b*; Mason *et al.*, 1991; Murata *et al.*, 1998). All three species occur in a wide variety of crops,

Table 21.1. *Potential prey (pests) of Ceraeochrysa cubana and Chrysoperla externa and their respective host plants (crops)*

Predator	Prey[a]	Host plant	References
Ceraeochrysa cubana	Anagasta kuehniella (Zeller) (Lep.: Pyralidae) (eggs)	Stored grains	Santa-Cecilia et al., 1997
	Ascia monuste orseis (Latreille) (Lep.: Pieridae) (eggs and larvae)	Cabbage	Murata et al., 1996
	Bemisia argentifolii Bellows & Perring (Hom.: Aleyrodidae)	Hibiscus, tomato	Dean & Schuster, 1995
	Brevicoryne brassicae (Linnaeus) (Hom.: Aphididae)	Cabbage	Fernandes et al., 1996
	Chrysomphalus aonidum (Linnaeus) (Hom.: Diaspididae)	Citrus	Muma, 1957
	Dialeurodes citrifolii (Morgan) (Hom.: Aleyrodidae)[b]	Citrus	Muma, 1957
	Eotetranychus sexmaculatus (Riley) (Acari: Tetranychidae)	Citrus	Muma, 1957
	Lepidosaphes beckii (Newman) (Hom.: Diaspididae)[b]	Citrus	Muma, 1957
	Macrosiphum euphorbiae (Thomas) (Hom.: Aphididae)	Eggplant, tomato	Dean & Schuster, 1995
	Panonychus citri (McGregor) (Acari: Tetranychidae)[b]	Citrus	Muma, 1957
	Pinnaspis aspidistrae (Signoret) (Hom.: Diaspididae)[b]	Citrus	Vitorino & Carvalho, 1991; Santa-Cecilia et al., 1997
	Sitotroga cerealella (Olivier) (Lep.: Gelechiidae) (eggs)	Stored grains	Serrano et al., 1988
	Toxoptera citricida (Kirkaldy) (Hom.: Aphididae)[b]	Citrus	Venzon et al., 1996; Santa-Cecilia et al., 1997
Chrysoperla externa	Alabama argillacea (Hübner) (Lep.: Noctuidae) (eggs and larvae)	Cotton	Ribeiro et al., 1991; Silva et al., 1998
	Anagasta kuehniella (Zeller) (Lep.: Pyralidae) (eggs)	Stored grains	Aun, 1986; Ribeiro et al., 1991
	Aphis gossypii Glover (Hom.: Aphididae)	Cotton	Figueira et al., 1997
	Ascia monuste orseis (Latreille) (Lep.: Pieridae) (eggs and larvae)	Cabbage	Murata et al., 1996; Pessoa et al., 1996
	Brevicoryne brassicae (Linnaeus) (Hom.: Aphididae)	Cabbage	Fernandes et al., 1996
	Cydia pomonella (Linnaeus) (Lep.: Tortricidae) (eggs and larvae)	Apple	Núñez, 1988a, b
	Diatraea saccharalis (Fabricius) (Lep.: Pyralidae) (eggs and larvae)	Sugarcane	Fernandes et al., 1995; Caetano et al., 1997
	Helicoverpa virescens (Fabricius) (Lep.: Noctuidae) (eggs and larvae)	Cotton, tobacco	Núñez, 1988a
	Helicoverpa zea (Boddie) (Lep.: Noctuidae) (eggs and larvae)	Maize, tomato	Núñez, 1988a, b
	Myzus persicae (Sulzer) (Hom.: Aphididae)	Cabbage, cotton, potato, tomato	Albuquerque et al., 1994
	Palpita quadristigmalis (Guenée) (Lep.: Pyralidae) (eggs and larvae)	Olive	Núñez, 1988a, b
	Parlatoria cinerea Doane & Hadden (Hom.: Diaspididae)	Citrus	Gravena et al., 1993
	Phthorimaea operculella Zeller (Lep.: Gelechiidae) (eggs and larvae)	Potato	Ru et al., 1975; Crouzel & Botto, 1977
	Pseudoplusia includens (Walker) (Lep.: Noctuidae) (eggs)	Soyabean	Ru et al., 1975
	Schizaphis graminum (Rondani) (Hom.: Aphididae)	Sorghum, wheat	Fonseca et al., 1998
	Selenaspidus articulatus (Morgan) (Hom.: Diaspididae)	Citrus	Xavier et al., 1997

Sitotroga cerealella (Olivier) (Lep.: Gelechiidae) (eggs and larvae)	Stored grains	Núñez, 1988b; Albuquerque et al., 1994
Spodoptera eridania (Cramer) (Lep.: Noctuidae) (eggs and larvae)	Potato	Núñez, 1988a
Spodoptera frugiperda (J.E. Smith) (Lep.: Noctuidae) (eggs)	Alfalfa, maize	Núñez, 1988a, b
Trichoplusia ni (Hübner) (Lep.: Noctuidae) (eggs)	Cabbage	Ru et al., 1975
Tuta absoluta (Meyrick) (Lep.: Gelechiidae) (eggs)	Tomato	Carneiro & Medeiros, 1997

Notes:

[a] Prey on which larvae have been reared successfully in the laboratory, despite differences in developmental and/or survival rates.

[b] Prey that are probably suitable only in association with other types of prey.

including fruit trees, vegetables, and grains, and their larvae feed on a broad range of pests such as mites, aphids, mealybugs, and whiteflies (Serrano *et al.*, 1987; Núñez, 1988*a*; Mattioli *et al.*, 1992; Venzon & Carvalho, 1992; Silva *et al.*, 1994; Dean & Schuster, 1995; Scomparin *et al.*, 1996; Venzon *et al.*, 1996). Analysis of the gut contents of field-collected adults indicate that *Chrysoperla* and *Ceraeochrysa* adults are honeydew, nectar, and pollen feeders (Brooks & Barnard, 1990). On maize, laboratory experiments indicate that *Chrysoperla externa* larvae prefer to spin their cocoons on the plant (leaves, mostly) and less frequently on the soil surface or in litter (Miwa *et al.*, 1996).

Data on the seasonal cycles and behaviour of Neotropical chrysopids are very scanty. Collection records indicate that adults of *C. externa* and many *Ceraeochrysa* species occur throughout the year. Field samples from the state of Minas Gerais, Brazil, showed that *C. cubana* adults were more abundant during the dry season (May to September) than in the rainy season (Souza *et al.*, 1995). However, there are no data on the reproductive status of the adults and/or the presence of eggs or larvae in the field. In a small study out-of-doors at Campos dos Goytacazes, Rio de Janeiro State (latitude 22°S), *Chrysoperla externa* reproduced during all seasons and had nine generations in one year (G.S. Albuquerque, unpublished data). No diapause was evident. Other laboratory studies with *C. externa* populations from Arica, Chile indicate that at least a proportion of some populations may enter reproductive diapause (Albuquerque *et al.*, 1994). Information on the occurrence and seasonal timing of this diapause in nature is lacking.

21.3.1 Recommendations (field data)

1. Field surveys and observations of the habitat- and prey-associations of predaceous insects in the field should be conducted early in the development of biological control projects. Periodic, long-term sampling of all life stages and potential prey would provide an especially valuable seasonal component that would help in manipulating the predator–prey interactions. It is crucial that the results of these efforts be included in publications so as to make them available to the scientific community.

2. Outdoor rearings (e.g. in field cages) under natural conditions are needed to elucidate the seasonal cycles of the chrysopids.

21.4 MASS-PRODUCTION

The commercialisation of augmentative biological control depends on the ability of insectaries to mass-produce and supply natural enemies economically. Achieving this objective requires efficient and standardised rearing procedures that are based on reliable quantitative biological data. Below we review relevant biological characteristics of *Chrysoperla externa* and *Ceraeochrysa* that pertain to mass-rearing.

21.4.1 Development and reproduction

Studies such as those mentioned below indicate that *Chrysoperla externa*'s developmental biology is similar to that of the commercially mass-produced *C. carnea*, and that the rate and efficiency of mass-producing it could approach that of *C. carnea*. For example, *C. externa* immatures have excellent survival and fast, uninterrupted, well-synchronised development under long-day conditions and a broad range of temperatures (16–27°C) (Albuquerque *et al.*, 1994). Development from egg to adult emergence requires $\sim$ 22 days at 27°C (lower thermal threshold, $t = 11.8$°C; thermal requirement for total development, $K = 320$ HDD).

Chrysoperla externa larvae develop well on prey [*Anagasta kuehniella* (Zeller) and *Sitotroga cerealella* (Olivier) eggs] that are currently used in commercially producing *C. carnea* (Aun, 1986; Albuquerque *et al.*, 1994), and they may resemble other congeners (*C. carnea* and *C. rufilabris*) in their favourable response to artificial larval diets. In this regard, convenient and economical diets are currently under development in the United States (e.g. Cohen & Smith, 1998). Moreover, like *C. carnea* adults, *C. externa* adults are not predaceous; rather, they feed on honeydew, nectar, and pollen. Under laboratory conditions, they reproduce well when fed an artificial diet of protein and carbohydrate (Botto & Crouzel, 1979; Cañedo & Lizárraga, 1988; Ribeiro *et al.*, 1993; Albuquerque *et al.*, 1994; Carvalho *et al.*, 1996).

Data on the development of three species of *Ceraeochrysa* (*C. cubana*, *C. cincta*, and *C. smithi*) under a range of temperatures and dietary regimes indicate that these species also do well when compared with the commercially mass-produced *Chrysoperla carnea*. Developmental rates are generally linearly related to temperature (between 15°C and 30°C), and development from oviposition to emergence of adults requires from ~28 to ~35 days at 25°C (Venzon & Carvalho, 1993; Venzon et al., 1996; López-Arroyo et al., 1999a; G.S. Albuquerque, unpublished data). Lower thermal thresholds for preimaginal development range from 9.2°C to 11.2°C, and completion of development requires 423–467 HDD (Venzon et al., 1996; López-Arroyo et al., 1999a). These values are slightly higher than those for C. carnea and C. externa, but it is unlikely that they would hinder insectary rearing.

Given currently available data, direct comparisons of developmental rates among Ceraeochrysa species are often not possible because the temperatures and developmental stages tested, as well as the prey provided to the larvae, differed among the tests. However, in one study in which C. smithi, C. cubana, and C. cincta from Florida (USA), were examined concurrently, all three developed at very similar rates (López-Arroyo et al., 1999a). Average preimaginal development (egg to adult emergence) for each species required 26.4–27.3 days at 24°C [larval prey: a mixture of S. cerealella eggs and Myzus persicae (Sulzer), the green peach aphid].

Other studies indicate that developmental times may vary among geographic populations and/or sibling species of C. cubana and C. cincta. For example, larval development in a population of C. cubana from the Atlantic coast of Brazil required 10.9 days (G.S. Albuquerque, unpublished data), whereas the same species required 15.5 days under the same temperature and prey regime, in a population from Lavras (southeastern Brazil) (Venzon et al., 1996). Differences in developmental times may also occur in 'C. cincta' (cf. Núñez, 1988b; López-Arroyo et al., 1999a); however, in this case the differences may reflect species-specific variation.

Larvae of Ceraeochrysa spp. appear to have a relatively broad range of acceptable prey (e.g. Table 21.1). Three species showed very high rates of survival when larvae were reared on Sitotroga or Anagasta eggs, the diets used in commercial production of Chrysoperla carnea (Santa-Cecília et al., 1997; López-Arroyo et al., 1999c). In general, a combination of aphids and moth eggs accelerated larval development in comparison with that on aphids alone, but the differences were not large. Artificial diets that were developed for Chrysoperla also may be appropriate for rearing Ceraeochrysa larvae, but further testing is necessary. Although several diets for rearing Chrysoperla larvae had significant negative effects on the survival, pupal weight, and reproduction of Ceraeochrysa (Serrano et al., 1988), at least one proved to be a good supplement to prey (Vasquez et al., 1998). These findings indicate that with some additional research, artificial diets may drastically reduce the cost of mass-rearing; it would be of particular value to examine the responses of Ceraeochrysa to the solid and semi-solid diets that the US Department of Agriculture (e.g. Cohen & Smith, 1998) currently is developing for Chrysoperla larvae.

Adults of all Ceraeochrysa species studied to date (C. cubana, C. cincta, and C. smithi) express the same dietary requirements for reproduction as Chrysoperla adults. They are not predaceous and they exhibit good reproduction (high fecundity, long oviposition periods, high rates of egg fertility) when provided an artificial diet of brewer's yeast or protein hydrolysate of yeast and carbohydrates (Moraes & Carvalho, 1991; Venzon & Carvalho, 1992; López-Arroyo et al., 1999a). This characteristic of the adults greatly facilitates mass-rearing.

21.4.2 Automated mass-production

Recently there have been several significant advances in developing labour-saving and space-efficient systems for mass-producing Chrysoperla in the United States. These include compact holding units for adults, mechanical devices for feeding adults and harvesting eggs, mechanised methods for presenting larval diet, and automated systems for packaging larval-rearing units (Nordlund & Greenberg, 1994; Nordlund & Correa, 1995; Nordlund et al., this volume). When fully developed, these mechanised systems could enhance production greatly and reduce costs drastically. We see no reason why these new techniques cannot be adapted to the mass-production of Neotropical Chrysoperla and Ceraeochrysa spp.

21.4.3 Storage

Storage of entomophagous insects is a key element often missing in the cost-effective production of natural enemies (Tauber & Helgesen, 1978; Ravensberg, 1992; Ruberson et al., 1999). Storage capabilities offer insectaries the opportunity to synchronise the supply of natural enemies with peak periods of demand, and they can greatly facilitate the efficient distribution of natural enemies.

Techniques for long-term storage of natural enemies frequently centre around the diapausing stage (e.g. for Chrysoperla carnea, see Chang et al., 1995). In the case of *Chrysoperla* spp., the adult is the diapausing stage, and the adults of some geographic populations of *C. externa* enter diapause under conditions of low temperatures and short daylengths (Albuquerque *et al.*, 1994). Nevertheless, either diapausing or non-diapausing adults can be stored under low temperature (~10 °C) for at least 4 months (Tauber *et al.*, 1997). Moreover, after storage, reproduction begins quickly, predictably, and with little variation among individuals, and the rates of oviposition are high.

In contrast to *Chrysoperla*, Neotropical *Ceraeochrysa* spp. are not known to enter diapause (see López-Arroyo *et al.*, 1999a), and long-term storage capabilities are more problematic. However, many species exhibit relatively long preoviposition periods and one Nearctic species enters diapause in the larval stage (Tauber *et al.*, 1998). It is possible that these traits could be of value in developing long-term storage for Neotropical *Ceraeochrysa*.

Both *Chrysoperla externa* and *Ceraeochrysa* spp. have capabilities for short-term storage of eggs (Saini, 1997; López-Arroyo *et al.*, 2000). *Chrysoperla externa* eggs can be held for ~3 weeks at 10 °C and *Ceraeochrysa cubana* and *C. smithi* can survive for 2 weeks at 15 °C. Both storage regimes result in some reduction in post-storage larval and adult quality, but this reduction does not seem to be large.

21.4.4 Quality control

Commercial cultures of lacewings can become contaminated with unwanted species, and sometimes shipments are mailed without the contamination being detected (O'Neil *et al.*, 1998). For biological and commercial reasons, it is crucial that cultures be monitored periodically for correct species identity. Given the inadequate systematic status of Neotropical chrysopids, this requirement is not easily met (see 'Systematics' section above). The involvement of a systematist in mass-rearing efforts could help alleviate the problem of undetected contamination.

Cultures of *Chrysoperla carnea* sometimes show significant deterioration during continuous mass-rearing (Jones *et al.*, 1978). This deterioration can be prevented or reversed by altering physical conditions appropriately (Chang *et al.*, 1996). It is not known whether cultures of *C. externa* experience similar deterioration. In one case, *C. externa* was reared for 45 generations with no introduction of new individuals from nature and with no apparent deterioration (Crouzel &

Botto, 1977). Similarly, studies of *Ceraeochrysa cubana* indicate only small negative effects (slight decreases in developmental rates but no changes in viability) after cultures were maintained for four generations (Silva *et al.*, 1994). In general, monitoring of the developmental and reproductive performance of stock during mass-rearing is necessary to maintain quality.

21.4.5 Recommendations (mass-production)

Given the significant advantages that *Chrysoperla externa* and *Ceraeochrysa* spp. show for mass-rearing, marketing, and use, we recommend that priority be placed on several key issues that will make the mass-production of these species economical and commercially feasible. The mass-rearing of these species by both commercial insectaries and cottage industries should be considered.

1. Low-cost, convenient artificial diets for rearing larvae are under development in the United States. These diets should be evaluated for rearing *Chrysoperla externa* and *Ceraeochrysa* spp. The effects of these diets over multiple generations should also be examined.

2. The seasonality of *Ceraeochrysa* should be evaluated under laboratory and field conditions. Stages that are resistant to temperature extremes or drought should be identified. Results from such studies may indicate avenues to pursue for developing techniques for long-term storage. Analysis of geographical variation in seasonal responses to physical factors should be included in these studies.

3. The application of automated methods to mass-producing *Chrysoperla externa* and *Ceraeochrysa* spp. should be examined.

4. Protocols for ensuring the quality of commercially mass-reared lacewings should be developed. These protocols should include efficient procedures for confirming the species identity, as well as monitoring the developmental and reproductive performance of commercial cultures.

21.5 USE OF LACEWINGS IN PEST MANAGEMENT

Lacewings are generally considered to have important roles in pest management in two ways: augmentative biological control (large releases of commercially produced eggs or larvae) and habitat manipulation (use of food sprays to attract or arrest adults and stimulate oviposition). Both tactics require significant knowledge of the developmental, reproductive, and seasonal biology of lacewings.

Tauber *et al.* (2000) summarised the factors influencing the efficient and economical use of *Chrysoperla carnea* and *C. rufilabris* in pest management programmes in temperate North America. The data and the approaches used in collecting these data may be directly applicable to *C. externa* and *Ceraeochrysa* spp. in Neotropical America. Below we discuss some of the issues that are currently under study.

21.5.1 Augmentation

Considerable advances have been made in developing methods and establishing effective rates for releasing *Chrysoperla*. For example, lacewings are generally sold and dispensed as eggs; however, field tests indicate that, in some cases, releases of young larvae that have been fed may be more effective (Daane *et al.*, 1998). Therefore, it is important to evaluate the biological and economic advantages of releasing either one or the other developmental stage and to devise efficient methods for dispensing the larval stage into the field.

Significant advances have been made in developing delivery systems that apply lacewings (eggs and larvae) at uniform, desired rates. Mechanical systems that use either liquid or solid carriers have been tested and show substantial potential for commercial adoption (Tauber *et al.*, 2000). For example, in Argentina, a mechanical device attached to a tractor gave uniform distribution of *C. externa* larvae on cotton (20 000 to 250 000 first instars per hectare), with rice husks as a carrier (Polak *et al.*, 1996). Similar results were obtained with aerial applications of 10 000 to 50 000 *Chrysoperla* larvae mixed with an equal weight of *Sitotroga* eggs and 4 kg of rice husks per hectare of cotton (Polak *et al.*, 1998). Also, tests have begun to determine commercially feasible rates for release of *C. carnea* eggs and larvae against specific pests (e.g. Daane *et al.*, 1996; Daane & Yokota, 1997); although neither *C. externa* nor *Ceraeochrysa* have been tested, these procedures can be adapted to them.

Studies of the reproductive biology of *Chrysoperla externa* and *Ceraeochrysa cincta* indicate that females of these species require repeated or multiple matings during their lifetimes to sustain high rates of fertile oviposition (Ribeiro & Carvalho, 1991; López-Arroyo *et al.*, 1999b). Such information has direct implications for how lacewing releases are made in the field; i.e. to increase the probabilities of multiple mating, and thus to sustain oviposition, a few, large releases are preferable to many, small releases. Studies of the reproductive biology of other agriculturally important *Ceraeochrysa* species are necessary to determine appropriate release methods.

21.5.2 Habitat manipulation

The behavioural responses of *Chrysoperla* adults to chemical stimuli associated with habitats and food can be used to augment populations in targeted areas. Food sprays that contain these chemicals can attract or arrest adults and stimulate oviposition (Hagen, 1987). It is likely that *Ceraeochrysa*, whose adult diets appear similar to those of *Chrysoperla*, can also be manipulated with food sprays. In olfactometer tests, *Ceraeochrysa cubana* adults were attracted to artificial honeydew, and in field tests, its oviposition increased after application of the diet to tomato plants (Dean, 1994, in Schuster *et al.*, 1996).

Both the chemical composition of the food sprays and the timing of their application are crucial in determining how successful they will be in pest management. *Chrysoperla* adults perceive the attractant (a kairomone contained in honeydew) only in the presence of a synomone (e.g. a volatile emitted from the leaves of certain plants) (van Emden & Hagen, 1976; Flint *et al.*, 1979). Also, to attract lacewings and stimulate oviposition, food sprays must contain both protein hydrolysates and sugar or honey (Hagen & Tassan, 1966). Finally, to be effective, the sprays must be applied early in the season, before natural honeydew becomes abundant.

Two studies tested the use of food sprays on *Chrysoperla externa*. In one case, the food sprays were combined with augmentative releases of *C. externa* adults in soya bean and maize (Barclay, 1990). Augmentative releases alone did not increase the number of lacewing eggs or cause a reduction in the three targeted noctuid pest species. However, when food sprays accompanied the augmentative releases, the densities of lacewing adults and eggs in maize fields increased significantly. In another case, a significantly larger number of *C. externa* eggs was observed on maize plants sprayed with molasses solution when compared to unsprayed plants (Freitas & Scaloppi, 1996). These results indicate that with some well-focused research, food sprays have considerable potential in pest management.

Table 21.2. *Tolerance to pesticides by different stages of Chrysoperla externa and Ceraeochrysa cubana*

Product	Chrysoperla externa			Ceraeochrysa cubana			References
	Egg	Larva	Adult	Egg	Larva	Adult	
Abamectin	Y	Y		Y	Y	Y	Ribeiro et al., 1988; Ferreira et al., 1993; Moraes & Carvalho, 1993
Alpha-cypermethrin				Y	N	N	Mattioli et al., 1992
Bacillus thuringiensis				Y	Y	Y	Mattioli et al., 1992
Bifenthrin					N		Ferreira et al., 1993
Bromopropylate				Y	Y	Y	Ferreira et al., 1993; Moraes & Carvalho, 1993
Buprofezin	Y	Y	Y	Y	Y	Y	Ferreira et al., 1993; Carvalho et al., 1994a, b; Velloso et al., 1995a, b
Carbosulfan				Y	N		Ferreira et al., 1993
Chlorfuazuron		N		Y	N	Y	Carvalho et al., 1994a, b; Velloso et al., 1995a
Clofentezine				Y	Y		Ferreira et al., 1993
Cyfluthrin				Y	N	N	Mattioli et al., 1992
Cyhexatin				Y	Y		Ferreira et al., 1993
Cyromazine	Y	Y	Y	Y	Y	Y	Carvalho et al., 1994a, b; Velloso et al., 1995a, b
Deltamethrin				Y	N	N	Mattioli et al., 1992
Dicofol				Y	Y		Ferreira et al., 1993
Diethion	N	N		Y			Ribeiro et al., 1988
Diflubenzuron		N		Y	N	Y	Mattioli et al., 1992; Carvalho et al., 1994a, b; Velloso et al., 1995a
Fenbutatin-oxide			N	Y	Y	Y	Mattioli et al., 1992; Ferreira et al., 1993; Pássaro et al., 1993; Santa-Cecília et al., 1993
Fenitrothion				Y	N	N	Santa-Cecília et al., 1993
Fenpropathrin				Y	N	N	Ferreira et al., 1993; Moraes & Carvalho, 1993; Santa-Cecília et al., 1993
Fenthion		N					Ribeiro et al., 1988
Fenvalerate				Y	Y	N	Santa-Cecília et al., 1993
Flufenoxuron		N	N	Y	N	Y	Mattioli et al., 1992; Ferreira et al., 1993; Carvalho et al., 1994a, b; Velloso et al., 1995a
Fosetyl-Al			N				Pássaro et al., 1993
Hexythiazox		Y			Y		Ferreira et al., 1993
Malathion		N					Ribeiro et al., 1988
Mineral oil			N				Pássaro et al., 1993
Multimethil alquenol			N				Pássaro et al., 1993
Pirimicarb				Y	Y		Mattioli et al., 1992
Piriproxifen	Y	Y	Y	Y	N	N	Velloso et al., 1995a, b

Quinomethionate	Y	Y	Y	Mattioli et al., 1992
Sulfur	Y	Y	Y	Moraes & Carvalho, 1993
Teflubenzuron	Y	N	Y	Carvalho et al., 1994a, b; Velloso et al., 1995a
Tetradifon	Y	Y	Y	Ferreira et al., 1993; Moraes & Carvalho, 1993
Triazophos		N	N	Ferreira et al., 1993
Triflumuron	Y	N	Y	Carvalho et al., 1994a, b; Velloso et al., 1995a
Vamidothion		N		Pássaro et al., 1993

21.5.3 Resistance to potentially disruptive factors

A number of factors in the agroecosystem can interfere with the action of natural enemies. In some cases, these factors can be altered to increase the effectiveness of biological control. For example, the types of pesticides can be modified or the timing of their application can be synchronised with the occurrence of less susceptible stages of biological control agents. Similarly, releases of natural enemies can be timed or applied in ways so as to avoid interference by ants and other unwanted natural enemies.

Insecticides

Both *Chrysoperla externa* and *Ceraeochrysa* spp. appear to be particularly useful in integrated pest management because they express relatively broad tolerance to pesticides that are commonly used in a variety of crops (see Table 21.2). However, susceptibility varies among the various developmental stages. For example, *C. cubana* eggs are highly tolerant to all insecticides and acaricides that have been tested, whereas larvae and adults are tolerant to only some of these pesticides (e.g. abamectin, bromopropylate, buprofezin, cyromazine, fenbutatin-oxide, quinomethionate, and tetradifon), but not to others (e.g. alpha-cypermethrin, cyfluthrin, deltamethrin, fenitrothion, and fenpropathrin). Similarly, *Chrysoperla externa* eggs appear to tolerate many insecticides and acaricides, whereas the larval and adult stages are relatively susceptible to them. This stage-specific variation should be considered in timing pesticide applications.

Unwanted natural enemies

Recent reports from the United States suggest that intraguild predation can sometimes reduce the effectiveness of *Chrysoperla* in biological control (Rosenheim *et al.*, 1995; James *et al.*, 1999). In this regard, both *Ceraeochrysa* and *Chrysoperla* spp. are susceptible to attack by a broad range of natural enemies: predators, parasitoids, and perhaps pathogens (Muma, 1959*b*; Dreistadt *et al.*, 1986; Ruberson *et al.*, 1995; Eisner *et al.*, 1996; Medeiros & França, 1996; Ehler *et al.*, 1997). This susceptibility should be considered in developing pest management procedures. For example, in some circumstances the release of *Chrysoperla* larvae (rather than eggs) may be necessary, or it may be preferable to release *Ceraeochrysa* (rather than *Chrysoperla*) because they have a number of traits (trash-carrying larvae, chemically defended eggs, and adult defensive behaviour) that may protect them, at least to some degree, from interference (Eisner & Silberglied, 1988; Masters & Eisner, 1990; Mason *et al.*, 1991; Eisner *et al.*, 1996).

The impact of pathogens on lacewings is poorly known. *Ceraeochrysa cubana* showed high tolerance to *Bacillus thuringiensis* (Berliner) (Dipel 25 PM formulation) sprayed over eggs, first-instar larvae, and adults (Mattioli *et al.*, 1992). However, recent studies demonstrate that *Chrysoperla carnea* larvae are vulnerable to *B. thuringiensis* toxins that are being incorporated into maize, potato, and other crops (Hilbeck *et al.*, 1998). The large-scale use of these transgenic plants could have significant negative effects on lacewings and other predators, and we urge caution in their use.

21.5.4 Recommendations (lacewings in pest management)

Focused research on a number of key issues would increase the effectiveness and hasten the commercialisation of lacewings in pest management systems. We feel that the primary areas of concentration should include studies that:

1. Develop methodology and rates for augmentative releases of lacewings against specific pests. Tests should be conducted under field conditions in open and closed cages so that the effect of intraguild predation and parasitism can be included in the assessment.
2. Explore in detail the reproductive biology of *Ceraeochrysa* spp. Results of such studies may help improve methods for mass-rearing and for making augmentative releases.
3. Determine the effectiveness of food sprays in enhancing lacewing (*Chrysoperla* and *Ceraeochrysa*) populations and in reducing pest populations. Investigations should include tests that evaluate when to apply food sprays or chemical attractants in relation to the phenology of the pest and the lacewings.

21.6 EVALUATION

As in all biological control projects, the process of evaluation constitutes a crucial final step. Approaches for evaluating the action of lacewings within a pest management context are under development, and recent advances have been reviewed (Daane *et al.*, 1998; Tauber *et al.*, 2000). The issues raised in these publications should be considered carefully in order to promote the development of effective, economical, and environmentally sound pest management tactics in Neotropical America.

ACKNOWLEDGEMENTS

We thank Athayde Tonhasca Jr (Universidade Estadual do Norte Fluminense), for critique of an earlier draft of this manuscript. This work was supported, in part, by the Brazilian Council for Scientific and Technological Development Grant 300504/96–9, Foundation for Scientific Support in the State of Rio de Janeiro Grant E-26/171.152/96 (GSA), National Science Foundation Grant INT-9817231, NRI/CSREES Competitive Grant 98–35316–6775, and Regional Project W-185 (MJT and CAT).

REFERENCES

Adams, P.A. (1982) *Ceraeochrysa*, a new genus of Chrysopinae (Neuroptera) (Studies in New World Chrysopidae, part II). *Neuroptera International* 2, 69–75.

Adams, P.A. & Penny, N.D. (1985) Neuroptera of the Amazon Basin: Part 11a. Introduction and Chrysopini. *Acta Amazonica* 15, 413–479.

Adams, P.A. & Penny, N.D. (1986) Faunal relations of Amazonian Chrysopidae. In *Recent Research in Neuropterology, Proceedings of the 2nd International Symposium on Neuropterology*, ed. Gepp, J., Aspöck, H. & Hölzel, H., pp. 119–124. Thalerhof, Graz.

Albuquerque, G.S., Tauber, C.A. & Tauber, M.J. (1994) *Chrysoperla externa* (Neuroptera: Chrysopidae): life history and potential for biological control in Central and South America. *Biological Control* 4, 8–13.

Aun, V. (1986) Aspectos da biologia de *Chrysoperla externa* (Hagen, 1861) (Neuroptera, Chrysopidae). MS thesis, Escola Superior de Agricultura 'Luiz de Queiroz', Universidade de São Paulo.

Barclay, W.W. (1990) Role of *Chrysoperla externa* as a biological control agent of insect pests in Mesoamerican agricultural habitats. MSc thesis, University of California, Berkeley.

Botto, E.N. & Crouzel, I.S. (1979) Dietas artificiales y capacidad de postura de *Chrysopa lanata lanata* (Banks) en condiciones de laboratorio. *Acta Zoologica Lilloana* 35, 745– 758.

Brooks, S.J. (1994) A taxonomic review of the common green lacewing genus *Chrysoperla* (Neuroptera: Chrysopidae). *Bulletin of the British Museum (Natural History) Entomology* 63, 137–210.

Brooks, S.J. & Barnard, P.C. (1990) The green lacewings of the world: a generic review (Neuroptera: Chrysopidae). *Bulletin of the British Museum (Natural History) Entomology* 59, 117–286.

Caetano, A.C., Bortoli, S.A., Murata, A.T. & Narciso, R.S. (1997) Capacidade de consumo de *Chrysoperla externa* (Neuroptera: Chrysopidae) em diferentes presas, sob condições de laboratório. In *Abstracts, 16 Congresso Brasileiro de Entomologia*, Salvador, Brazil, p. 109.

Cañedo, D.V. & Lizárraga, A. (1988) Dietas artificiales para la crianza en laboratorio de *Chrysoperla externa* (Hagen) (Neuroptera, Chrysopidae). *Revista Peruana de Entomologia* 31, 83–85.

Carneiro, J.R. & Medeiros, M.A. (1997) Potencial de consumo de *Chrysoperla externa* (Neuroptera: Chrysopidae) utilizando ovos de *Tuta absoluta* (Lepidoptera: Gelechiidae). In *Abstracts, 16 Congresso Brasileiro de Entomologia*, Salvador, Brazil, p. 117.

Carvalho, C.F., Canard, M. & Alauzet, C. (1996) Comparison of the fecundities of the Neotropical green lacewing *Chrysoperla externa* (Hagen) and the West-Palaearctic *Chrysoperla mediterranea* (Hölzel) (Insecta: Neuroptera: Chrysopidae). In *Pure and Applied Research in Neuropterology, Proceedings of the 5th International Symposium on Neuropterology*, ed. Canard, M., Aspöck, H. & Mansell, M.W., pp. 103–107. Sacco, Toulouse.

Carvalho, G.A., Salgado, L.O., Rigitano, R.L.O. & Velloso, A.H.P.P. (1994*a*) Efeitos de reguladores de crescimento de insetos sobre ovos e larvas de *Ceraeochrysa cubana* (Hagen, 1861) (Neuroptera, Chrysopidae). *Ciência e Prática* 18, 49–55.

Carvalho, G.A., Salgado, L.O., Rigitano, R.L.O. & Velloso, A.H.P.P. (1994*b*) Efeitos de compostos reguladores de crescimento de insetos sobre adultos de *Ceraeochrysa cubana* (Hagen) (Neuroptera: Chrysopidae). *Anais da Sociedade Entomológica do Brasil* 23, 335–339.

Chang, Y.F., Tauber, M.J. & Tauber, C.A. (1995) Storage of the mass-produced predator *Chrysoperla carnea* (Neuroptera: Chrysopidae): influence of photoperiod, temperature, and diet. *Environmental Entomology* 24, 1365–1374.

Chang, Y.F., Tauber, M.J. & Tauber, C.A. (1996) Reproduction and quality of F_1 offspring in *Chrysoperla carnea*: differential influence of quiescence, artificially-induced diapause, and natural diapause. *Journal of Insect Physiology* 42, 521–528.

Cohen, A.C. & Smith, L.K. (1998) A new concept in

artificial diets for *Chrysoperla rufilabris*: the efficacy of solid diets. *Biological Control* 13, 49–54.

Crouzel, I.S. & Botto, E.N. (1977) Ciclo de vida de *Chrysopa lanata lanata* (Banks) y algunas observaciones biológicas en condiciones de laboratorio. *Revista de Investigaciones Agropecuarias, Serie 5*, 13, 1–14.

Daane, K.M. & Yokota, G.Y. (1997) Release strategies affect survival and distribution of green lacewings (Neuroptera: Chrysopidae) in augmentation programs. *Environmental Entomology* 26, 455–464.

Daane, K.M., Yokota, G.Y., Zheng, Y. & Hagen, K.S. (1996) Inundative release of common green lacewings (Neuroptera: Chrysopidae) to suppress *Erythroneura variabilis* and *E. elegantula* (Homoptera: Cicadellidae) in vineyards. *Environmental Entomology* 25, 1224–1234.

Daane, K.M., Hagen, K.S. & Mills, N.J. (1998) Predaceous insects for insect and mite control. In *Mass-Reared Natural Enemies: Application, Regulation, and Needs*, ed. Ridgway, R.L., Hoffmann, M.P., Inscoe, M.N. & Glenister, C.S., pp. 61–115. Thomas Say Publications in Entomology, Entomological Society of America, Lanham MD.

Dean, D.E. & Schuster, D.J. (1995) *Bemisia argentifolii* (Homoptera: Aleyrodidae) and *Macrosiphum euphorbiae* (Homoptera: Aphididae) as prey for two species of Chrysopidae. *Environmental Entomology* 24, 1562–1568.

DeBach, P. & Rosen, D. (1991) *Biological Control by Natural Enemies*. Cambridge University Press, Cambridge.

Dreistadt, S.H., Hagen, K.S. & Dahlsten, D.L. (1986) Predation by *Iridomyrmex humilis* (Hym.: Formicidae) on eggs of *Chrysoperla carnea* (Neu.: Chrysopidae) released for inundative control of *Illinoia liriodendri* (Hom.: Aphididae) infesting *Liriodendron tulipifera*. *Entomophaga* 31, 397–400.

Ehler, L.E., Long, R.F., Kinsey, M.G. & Kelley, S.K. (1997) Potential for augmentative biological control of the black bean aphid in California sugarbeet. *Entomophaga* 42, 241–256.

Eisner, T. & Silberglied, R.E. (1988) A chrysopid larva that cloaks itself in mealybug wax. *Psyche* 95, 15–19.

Eisner, T., Attygalle, A.B., Conner, W.E., Eisner, M., MacLeod, E. & Meinwald, J. (1996) Chemical egg defense in a green lacewing (*Ceraeochrysa smithi*). *Proceedings of the National Academy of Sciences, USA* 93, 3280–3283.

van Emden, H.F. & Hagen, K.S. (1976) Olfactory reactions of the green lacewing *Chrysopa carnea* to tryptophan and certain breakdown products. *Environmental Entomology* 5, 469–473.

Fernandes, O.A., Bortoli, S.A., Freitas, S., Gravena, S. & Madlum, S. (1995) Parâmetros biológicos na criação massal de *Chrysoperla externa* (Neuroptera –

Chrysopidae) alimentada com larvas de *Diatraea saccharalis* (Lep. – Pyralidae). In *Abstracts, 15 Congresso Brasileiro de Entomologia*, Caxambú, Brazil, p. 88.

Fernandes, M.C., Murata, A.T., Pessoa, R. & Bortoli, S.A. (1996) Study of consumption capacity of the cabbage aphid *Brevicoryne brassicae* (Hemiptera – Homoptera: Aphididae) by *Chrysoperla externa* and *Ceraeochrysa cubana* (Neuroptera: Chrysopidae). In *Annals of the 5th Symposium of Biological Control*, Iguassu Falls, Brazil, p. 279.

Ferreira, M.N., Carvalho, C.F., Salgado, L.O. & Rigitano, R.L.O. (1993) Seletividade de acaricidas para larvas de *Ceraeochrysa cubana* (Hagen, 1861) (Neuroptera, Chrysopidae) em laboratório. *Ciência e Prática* 17, 71–77.

Figueira, L.K., Souza, B., Santos, T.M., Moura, N.A. & Carvalho, C.F. (1997) Biologia da fase larval de *Chrysoperla externa* (Hagen, 1861) (Neuroptera: Chrysopidae) alimentadas com *Aphis gossypii* Glover, 1877 (Homoptera: Aphididae). In *Abstracts 16 Congresso Brasileiro de Entomologia*, Salvador, Brazil, p. 69.

Flint, H.M., Salter, S.S. & Walters, S. (1979) Caryophyllene: an attractant for the green lacewing. *Environmental Entomology* 8, 1123–1125.

Fonseca, A.R., Carvalho, C.F. & Souza, B. (1998) Potencial de predação de *Chrysoperla externa* (Hagen, 1861) (Neuroptera: Chrysopidae) sobre *Schizaphis graminum* (Rondani, 1852) (Homoptera, Aphididae), sob condições de laboratório. In *Abstracts, 17 Congresso Brasileiro de Entomologia*, Rio de Janeiro, Brazil, p. 818.

Freitas, S. & Adams, P. (1994) Taxonomia de predadores Chrysopidae (Neuroptera) de ocorrência no Brasil. In *Annals of the 4th Symposium of Biological Control*, Gramado, Brazil, p. 260.

Freitas, S. & Scaloppi, E.A.G. (1996) Efeito da pulverização de melaço em plantio de milho sobre a população de *Chrysoperla externa* (Hagen) e distribuição de ovos na planta. *Revista de Agricultura* 71, 251–258.

Gravena, S., Yamamoto, P.T. & Fernandes, O.D. (1993) Biologia de *Parlatoria cinerea* (Hemiptera: Diaspididae) e predação por *Chrysoperla externa* (Neuroptera: Chrysopidae). *Científica* 21, 149–156.

Hagen, K.S. (1987) Nutritional ecology of terrestrial insect predators. In *Nutritional Ecology of Insects, Mites, Spiders, and Related Invertebrates*, ed. Slansky, F. Jr. & Rodriguez, J.G., pp. 533–577. Wiley, New York.

Hagen, K.S. & Tassan, R.L. (1966) The influence of protein hydrolysates of yeasts and chemically defined diets upon the fecundity of *Chrysopa carnea* Stephens (Neuroptera). *Vestník Ceskoslovenské Spolecnosti Zoologické* 30, 219–227.

Hilbeck, A., Moar, W.J., Pusztai-Carey, M., Filippini, A. &

Bigler, F. (1998) Toxicity of *Bacillus thuringiensis* Cry1Ab toxin to the predator *Chrysoperla carnea* (Neuroptera: Chrysopidae). *Environmental Entomology* 27, 1255–1263.

James, D.G., Stevens, M.M., O'Malley, K.J. & Faulder, R.J. (1999) Ant foraging reduces the abundance of beneficial and incidental arthropods in citrus canopies. *Biological Control* 14, 121–126.

Jones, S.L., Kinzer, R.E., Bull, D.L., Ables, J.R. & Ridgway, R.L. (1978) Deterioration of *Chrysopa carnea* in mass culture. *Annals of the Entomological Society of America* 71, 160–162.

Knutson, L. (1981) Symbiosis of biosystematics and biological control. In *Biological Control in Crop Production, BARC Symposium no. 5*, ed. Papavizas, G.C., pp. 61–78. Allanheld, Osmun, Totowa NJ.

López-Arroyo, J.I., Tauber, C.A. & Tauber, M.J. (1999a) Comparative life histories of the predators *Ceraeochrysa cincta, C. cubana*, and *C. smithi* (Neuroptera: Chrysopidae). *Annals of the Entomological Society of America* 92, 208–217.

López-Arroyo, J.I., Tauber, C.A. & Tauber, M.J. (1999b) Intermittent oviposition and remating in *Ceraeochrysa cincta* (Neuroptera: Chrysopidae). *Annals of the Entomological Society of America* 92: 587–593.

López-Arroyo, J.I., Tauber, C.A. & Tauber, M.J. (1999c) Effects of prey on survival, development, and reproduction of trash-carrying chrysopids (Neuroptera: *Ceraeochrysa*). *Environmental Entomology* 28, 1183–1188.

López-Arroyo, J.I., Tauber, C.A. & Tauber, M.J. (2000) Survival of stored lacewing eggs and quality of subsequent stages (Neuroptera: *Ceraeochrysa*). *Biological Control* 18, 165–171.

Mason, R.T., Fales, H.M., Eisner, M. & Eisner, T. (1991) Wax of a whitefly and its utilization by a chrysopid larva. *Naturwissenschaften* 78, 28–30.

Masters, W.M. & Eisner, T. (1990). The escape strategy of green lacewings from orb webs. *Journal of Insect Behavior* 3, 143–157.

Mattioli, E., Carvalho, C.F. & Salgado, L.O. (1992) Efeitos de inseticidas e acaricidas sobre ovos, larvas e adultos do predador *Ceraeochrysa cubana* (Hagen, 1861) (Neuroptera, Chrysopidae) em laboratório. *Ciência e Prática* 16, 491–497.

Medeiros, M.A. & França, F.H. (1996) Parasitismo de ovos de *Chrysoperla* sp. (Neuroptera: Chrysopidae) em milho-doce. In *Annals of the 5th Symposium of Biological Control*, Iguassu Falls, Brazil, p. 264.

Miwa, F., Freitas, S. & Ferreira, R.J. (1996) Estratégia de empupação de *Chrysoperla externa* (Neuroptera; Chrysopidae) em plantas de milho. In *Annals of the 5th*

Symposium of Biological Control, Iguassu Falls, Brazil, p. 133.

Moraes, J.C. & Carvalho, C.F. (1991) Influência da fonte de carboidratos sobre a fecundidade e longevidade de *Ceraeochrysa cubana* (Hagen, 1861) (Neuroptera: Chrysopidae). *Ciência e Prática* 15, 137–144.

Moraes, J.C. & Carvalho, C.F. (1993) Seletividade de acaricidas a ovos, larvas e adultos de *Ceraeochrysa cubana* (Hagen, 1861) (Neuroptera, Chrysopidae). *Ciência e Prática* 17, 388–392.

Muma, M.H. (1957) Effects of larval nutrition on the life cycle, size, coloration, and longevity of *Chrysopa lateralis* Guer. *Florida Entomologist* 40, 5–9.

Muma, M.H. (1959a) Chrysopidae associated with citrus in Florida. *Florida Entomologist* 42, 21–29.

Muma, M.H. (1959b) Hymenopterous parasites of Chrysopidae on Florida citrus. *Florida Entomologist* 42, 149–153.

Murata, A.T., Pessoa, R., Fernandes, M.C., Bortoli, S.A. & Freitas, S. (1996) Consumption capacity of *Chrysoperla externa* and *Ceraeochrysa cubana* (Neuroptera: Chrysopidae) using eggs of *Ascia monuste orseis* (Latr., 1819) (Lep. – Pieridae), under laboratory conditions. In *Annals of the 5th Symposium of Biological Control*, Iguassu Falls, Brazil, p. 41.

Murata, A.T., Narciso, R.S. & Bortoli, S.A. (1998) Aspectos biológicos de *Ceraeochrysa cincta* (Neuroptera, Chrysopidae) em três substratos alimentares. In *Abstracts 17 Congresso Brasileiro de Entomologia*, Rio de Janeiro, Brazil, p. 625.

Nordlund, D.A. & Correa, J.A. (1995) Improvements in the production systems for green lacewings: an adult feeding and oviposition unit and hot wire egg harvesting system. *Biological Control* 5, 179–188.

Nordlund, D.A. & Greenberg, S.M. (1994) Facilities and automation for the mass production of arthropod predators and parasitoids. *Biocontrol News and Information* 4, 45N–50N.

Nordlund, D.A. & Morrison, R.K. (1992) Mass rearing of *Chrysoperla* species. In *Advances in Insect Rearing for Research and Pest Management*, ed. Anderson, T.E. & Leppla, N.C., pp. 427–439. Westview Press, Boulder, CO.

Núñez, E. (1988a) Chrysopidae (Neuroptera) del Perú y sus especies más comunes. *Revista Peruana de Entomologia* 31, 69–75.

Núñez, E. (1988b) Ciclo biológico y crianza de *Chrysoperla externa* y *Ceraeochrysa cincta* (Neuroptera, Chrysopidae). *Revista Peruana de Entomologia* 31, 76–82.

Olazo, E.V.G. (1987) Los neurópteros asociados con los cultivos citricos de la provincia de Tucumán y descripción de una nueva especie de *Nomerobius* (Hemerobiidae). *CIRPON, Revista de Investigación* 5, 37–54.

O'Neil, R.J., Giles, K.L., Obrycki, J.J., Mahr, D.L., Legaspi, J.C. & Katovich, K. (1998) Evaluation of the quality of four commercially available natural enemies. *Biological Control* 11, 1–8.

Pássaro, P.C.M., Fernandes, O.D., Ramos, M.A., Fernandes, O.A. & Freitas, S. (1993) Avaliação da toxicidade de alguns agrotóxicos sobre adultos de *Chrysoperla externa* (Neuroptera, Chrysopidae). In *Abstracts 14 Congresso Brasileiro de Entomologia*, Piracicaba, Brazil, p. 567.

Penny, N.D. (1984) Recent progress in the taxonomy of South American Neuroptera. In *Progress in World's Neuropterology, Proceedings of the 1st International Symposium on Neuropterology*, ed. Gepp, J., Aspöck, H. & Hölzel, H., pp. 75–77. Thalerhof, Graz.

Penny, N.D. (1997) Four new species of Costa Rican *Ceraeochrysa* (Neuroptera: Chrysopidae). *Pan-Pacific Entomologist* 73, 61–69.

Pessoa, R., Fernandes, M.C., Murata, A.T. & Bortoli, S.A. (1996) Potencial de consumo de *Chrysoperla externa* e *Ceraeochrysa cubana* (Neuroptera: Chrysopidae) sobre larvas de curuquerê da couve *Ascia monuste orseis* (Latreille, 1819) (Lepidoptera: Pieridae). In *Annals of the 5th Symposium of Biological Control*, Iguassu Falls, Brazil, p. 377.

Polak, M.G.A., Ruiz, A.J., Videla, G.W., Olazo, E.G. & Contreras, G.B. (1996) Aplicación de larvas de *Chrysoperla externa* sobre cultivo de algodón utilizando un sistema mecanizado. In *Annals of the 5th Symposium of Biological Control*, Iguassu Falls, Brazil, p. 293.

Polak, M.G.A., Contreras, G.B., Olazo, E.G. & Mazza, S. (1998) Aplicación aérea de larvas de *Chrysoperla* (Neuroptera: Chrysopidae). In *Annals of the 6th Symposium of Biological Control*, Rio de Janeiro, Brazil, p. 64.

Ravensberg, W.J. (1992) Production and utilization of natural enemies in western European glasshouse crops. In *Advances in Insect Rearing for Research and Pest Management*, ed. Anderson, T.E. & Leppla, N.C., pp. 465–487. Westview Press, Boulder, CO.

Ribeiro, M.J. & Carvalho, C.F. (1991) Aspectos biológicos de *Chrysoperla externa* (Hagen, 1861) (Neuroptera, Chrysopidae) em diferentes condições de acasalamento. *Revista Brasileira de Entomologia* 35, 423–427.

Ribeiro, M.J., Matioli, J.C. & Carvalho, C.F. (1988) Efeito da avermectina-B1 (MK-936) sobre o desenvolvimento larval de *Chrysoperla externa* (Hagen) (Neuroptera; Chrysopidae). *Pesquisa Agropecuária Brasileira* 23, 1189–1196.

Ribeiro, M.J., Carvalho, C.F. & Matioli, J.C. (1991) Influência da alimentação larval sobre a biologia de adultos de *Chrysoperla externa* (Hagen, 1861) (Neuroptera: Chrysopidae). *Ciência e Prática* 15, 349–354.

Ribeiro, M.J., Carvalho, C.F. & Matioli, J.C. (1993) Biologia de adultos de *Chrysoperla externa* (Hagen, 1861) (Neuroptera, Chrysopidae) em diferentes dietas artificiais. *Ciência e Prática* 17, 120–130.

Rosenheim, J.A., Kaya, H.K., Ehler, L.E., Marois, J.J. & Jafee, B.A. (1995) Intraguild predation among biological-control agents: theory and evidence. *Biological Control* 5, 303–335.

Ru, N., Whitcomb, W.H., Murphey, M. & Carlysle, T.C. (1975) Biology of *Chrysopa lanata* (Neuroptera: Chrysopidae). *Annals of the Entomological Society of America* 68, 187–190.

Ruberson, J.R., Tauber, C.A. & Tauber, M.J. (1995) Developmental effects of host and temperature on *Telenomus* spp. (Hymenoptera: Scelionidae) parasitizing chrysopid eggs. *Biological Control* 5, 245–250.

Ruberson, J.R., Nechols, J.R. & Tauber, M.J. (1999) Biological control of arthropod pests. In *Handbook of Pest Management*, ed. Ruberson, J.R., pp. 417–448. Marcel Dekker, New York.

Saini, E.D. (1997) Almacenaje de huevos de *Chrysoperla externa* (Hagen) (Neuroptera: Chrysopidae). In *Abstracts 16 Congresso Brasileiro de Entomologia*, Salvador, Brazil, p. 115.

Santa-Cecília, L.V.C., Souza, B., Carvalho, C.F., Marínho, C.S. & Souto, R.F. (1993) Seletividade de alguns inseticidas e acaricidas a *Ceraeochrysa cubana* (Hagen, 1861) (Neuroptera: Chrysopidae) em condições de laboratório. In *Abstracts 14 Congresso Brasileiro de Entomologia*, Piracicaba, Brazil, p. 600.

Santa-Cecília, L.V.C., Souza, B. & Carvalho, C.F. (1997) Influência de diferentes dietas em fases imaturas de *Ceraeochrysa cubana* (Hagen) (Neuroptera, Chrysopidae*). Anais da Sociedade Entomológica do Brasil* 26, 309–314.

Schuster, D.J., Funderburk, J.E. & Stansly, P.A. (1996) IPM in tomatoes. In *Pest Management in the Subtropics, Integrated Pest Management – A Florida Perspective,* ed. Rosen, D., Bennett, F.D. & Capinera, J.L., pp. 387–411. Intercept, Andover UK.

Scomparin, C.H.J., Freitas, S. & Xavier, A.L.Q. (1996) Espécies de crisopídeos (Neuroptera, Chrysopidae) associadas às plantas de citros e às plantas da cobertura vegetal do solo. In *Annals of the 5th Symposium of Biological Control*, Iguassu Falls, Brazil, p. 74.

Serrano, C.V., Luque, J.E. & Villanueva, A. (1987)

Reconocimiento e identificación de las especies de chrysopas verdes (Neuroptera: Chrysopidae) en la zona de Puerto Wilches. *Agronomía Colombiana* 4, 16–18.

Serrano, C.V., Luque, J.E. & Villanueva, A. (1988) Uso de dietas artificiales para la cría de larvas y adultos de *Ceraeochrysa cubana* (Hagen) (Neuroptera: Chrysopidae). *Agronomía Colombiana* 5, 60–68.

Silva, G.A., Carvalho, C.F. & Souza, B. (1998) Aspectos biológicos de *Chrysoperla externa* (Hagen, 1861) (Neuroptera: Chrysopidae) alimentada com lagartas de *Alabama argillacea* (Hübner, 1818) (Lepidoptera, Noctuidae). In *Abstracts 17 Congresso Brasileiro de Entomologia*, Rio de Janeiro, p. 820.

Silva, R.L.X., Carvalho, C.F. & Venzon, M. (1994) Aspectos biológicos das fases imaturas de *Ceraeochrysa cubana* (Hagen, 1861) (Neuroptera, Chrysopidae) em quatro gerações sucessivas em laboratório. *Ciência e Prática* 18, 13–17.

Souza, B., Carvalho, C.F. & Santa-Cecília, L.V.C. (1995) Flutuação populacional de *Ceraeochrysa cubana* (Hagen) (Neuroptera: Chrysopidae) em citros. In *Abstracts 15 Congresso Brasileiro de Entomologia*, Caxambú, Brazil, p. 83.

Tauber, C.A. (1974) Systematics of North American chrysopid larvae: *Chrysopa carnea* group (Neuroptera). *Canadian Entomologist* 106, 1133–1153.

Tauber, C.A. & Adams, P.A. 1990. Systematics of the Neuropteroidea: present status and future needs. In *Systematics of the North American Insects and Arachnids: Status and Needs*, ed. Kosztarab, M. & Schaefer, C.W., pp. 151–164. Virginia Polytechnic Institute and State University, Blacksburg VA.

Tauber, M.J. & Helgesen, R.G. (1978) Implementing biological control systems in commercial greenhouse crops. *Bulletin of the Entomological Society of America* 24, 424–426.

Tauber, C.A., De León, T., López-Arroyo, J.I. & Tauber, M.J. (1998) *Ceraeochrysa placita* (Neuroptera: Chrysopidae): generic characteristics of larvae, larval descriptions, and life cycle. *Annals of the Entomological Society of America* 91, 608–618.

Tauber, M.J., Albuquerque, G.S. & Tauber, C.A. (1997) Storage of nondiapausing *Chrysoperla externa* adults: influence on survival and reproduction. *Biological Control* 10, 69–72.

Tauber, M.J., Tauber, C.A., Daane, K.M. & Hagen, K.S. (2000) Commercialization of predators: recent lessons from green lacewings (Neuroptera: Chrysopidae: *Chrysoperla*). *American Entomologist* 46, 26–38.

Vasquez, E.E., Schuster, D.J. & Hunter, W.B. (1998) Alternative method for encapsulation of artificial diet used in rearing *Ceraeochrysa cubana* (Hagen) larvae (Neuroptera: Chrysopidae). *Journal of Entomological Science* 33, 316–318.

Velloso, A.H.P.P., Rigitano, R.L.O. & Carvalho, G.A.(1995a) Efeitos de compostos reguladores de crescimento de insetos sobre ovos e larvas de *Chrysoperla externa* (Hagen, 1861) (Neuroptera, Chrysopidae). In *Abstracts 15 Congresso Brasileiro de Entomologia*, Caxambú, Brazil, p. 648.

Velloso, A.H.P.P., Rigitano, R.L.O. & Carvalho, G.A. (1995b) Efeitos de compostos reguladores de crescimento de insetos à fase adulta de *Chrysoperla externa* (Hagen, 1861) (Neuroptera, Chrysopidae). In *Abstracts 15 Congresso Brasileiro de Entomologia*, Caxambú, Brazil, p. 649.

Venzon, M. & Carvalho, C.F. (1992) Biologia da fase adulta de *Ceraeochrysa cubana* (Hagen, 1861) (Neuroptera, Chrysopidae) em diferentes dietas e temperaturas. *Ciência e Prática* 16, 315–320.

Venzon, M. & Carvalho, C.F. (1993) Desenvolvimento larval, pré-pupal e pupal de *Ceraeochrysa cubana* (Hagen) (Neuroptera: Chrysopidae) em diferentes dietas e temperaturas. *Anais da Sociedade Entomológica do Brasil* 22, 477–483.

Venzon, M., Carvalho, C.F. & Silva, R.L.X. (1996) Effects of various diets and temperature on larval development in the Neotropical green lacewing *Ceraeochrysa cubana* (Hagen) (Insecta: Neuroptera: Chrysopidae). In *Pure and Applied Research in Neuropterology, Proceedings of the 5th International Symposium on Neuropterology*, ed. Canard, M., Aspöck, H. & Mansell, M.W., pp. 251–257. Sacco, Toulouse.

Vitorino, N.A.M. & Carvalho, C. F. (1991) Biologia de larvas de *Ceraeochrysa cubana* (Hagen, 1861) (Neuroptera – Chrysopidae) com algumas presas. In *Abstracts 13 Congresso Brasileiro de Entomologia*, Recife, Brazil, p. 79.

Wang, R. & Nordlund, D.A. (1994) Use of *Chrysoperla* spp. (Neuroptera: Chrysopidae) in augmentative release programmes for control of arthropod pests. *Biocontrol News and Information* 15, 51N–57N.

Xavier, A.L.Q., Freitas, S. & Scomparin, C.H.J. (1997) Avaliação da capacidade de predação de *Chrysoperla externa* (Hagen, 1861) (Neuroptera, Chrysopidae) sobre a cochonilha *Selenaspidus articulatus* (Morgan, 1889) (Hemiptera, Diaspididae). In *Abstracts 16 Congresso Brasileiro de Entomologia*, Salvador, Brazil, p. 135.

CHAPTER 22

Comparative plant substrate specificity of Iberian Hemerobiidae, Coniopterygidae and Chrysopidae

V.J. Monserrat and F. Marín

22.1 INTRODUCTION

The biology of the species in the order Neuroptera is generally poorly known. In the species of Coniopterygidae and Hemerobiidae, data on the plant substrate specificity are very limited or completely unknown in many taxa, with the majority of existing records being limited to Holarctic species. The best-known fauna is the European one, where there are many data on the capture of numerous species on different plants. However, there is little published information on the autoecology of Coniopterygidae (Meinander, 1972, 1990; Aspöck *et al.*, 1980) and Hemerobiidae (Hinke, 1975; Laffranque & Canard, 1975; Miermont & Canard, 1975; Neuenschwander, 1975, 1976; Samson & Blood, 1979; Garland, 1981; New, 1984), and ecological studies on both families are very scarce (see references in Aspöck *et al.*, 1980 and New, 1989).

In contrast, the biology of the species in the family Chrysopidae is relatively well known. Data on the preferences of green lacewings for plant substrates are very abundant in many taxa, with the majority of existing records likewise being limited to Holarctic species. The best-known fauna is also the European one, where there are many data on the capture of numerous species on different plants, and also there are many articles on the autoecology of chrysopids (see references in Aspöck *et al.*, 1980; Canard *et al.*, 1984; New, 1989).

With reference to the Iberian fauna of these three families, there is apparently much information about the faunistics in the numerous papers published by L. Navás during the early part of this century. However, these references are very unreliable and should not be considered due to the many misidentifications detected, chiefly in Coniopterygidae and Hemerobiidae (Monserrat, 1984, 1986*a*, 1986*b*, 1990). In the second half of this century, several authors have provided some data on the faunistics and biology of certain Iberian species (see references in: Aspöck *et al.*, 1980; Aspöck & Aspöck, 1989).

To compensate for this lack of knowledge, we have carried out systematic samplings in continental Spain over the last 22 years (see Monserrat & Marín, 1992, 1994, 1996; Marín, 1994; Marín & Monserrat, 1995*a*, *b*). These results give us much information on the biology and distribution of Iberian Coniopterygidae, Hemerobiidae, and Chrysopidae, which represent one of the richest faunas in Europe, with 120 of the 164 European species (Aspöck *et al.*, 1980; Aspöck, 1992).

As is well known, Coniopterygidae, Hemerobiidae, and Chrysopidae are renowned as predaceous insects in adult and especially in larval stages, often being used as agents in biological control (New, 1986, 1989). However, in certain chrysopid and hemerobiid species it has been shown that some exhibit a glyco-phytophagous or omnivorous diet (Stelzl & Gepp, 1987; New, 1989; Stelzl, 1990, 1991, 1992). Thus their association with certain plant substrates is evident: indirect when they are preying on small phytophagous arthropods, and direct when they feed on honeydews, pollen, and other plant substances. These factors, without taking account of other climatic or ecological requirements, can associate particular species with certain kinds of plant substrates, sometimes resulting in a very close relationship among them (Killington, 1936, 1937; Meinander, 1972; Zelený, 1978, 1984; Aspöck *et al.*, 1980; New, 1984, 1986). To determine this relationship, we have studied the plant substrate specificity that the Coniopterygidae, Hemerobiidae, and Chrysopidae species show in the Iberian Peninsula (Monserrat & Marín, 1992, 1994, 1996).

22.2 MATERIALS AND METHODS

Between 1976 and 1994, we carried out systematic sampling in peninsular Spain, generally using the constituent provinces as area units for the data. Twenty-three provinces have been sampled, representing 55.1% of the Spanish peninsular area and 46.4 % of the Iberian surface area (Monserrat & Marín, 1992, 1994, 1996; Marín, 1994; Marín & Monserrat, 1995*a*, *b*). To study each province, a sufficiently representative number of sampling localities was selected, and periodically material was collected using the method noted in Monserrat & Marín (1992, 1994, 1996).

On the basis of these published data, as well as other unpublished material recently collected by us, the plant substrate preferences of Iberian hemerobiid, coniopterygid, and chrysopid species were analysed and discussed, in order to establish the degree of association among these families and Iberian plants, mainly shrubs and trees. The available data was considered adequately significant in terms of specimens studied (20 676) and the coverage and diversity of the Iberian area sampled.

22.3 RESULTS AND DISCUSSION

We have obtained data for 104 neuropteran species (36 hemerobiids, 38 coniopterygids, and 40 chrysopids) collected on 109 different plant species, with another 10 species collected only by light and Malaise trap. Table 22.1 provides a list of the collected species. This table shows all known Iberian hemerobiid, coniopterygid, and chrysopid species, except the following species that are dubiously or very scarcely recorded in the Iberian fauna, without any real data about their presence on any plant substrates: *Wesmaelius concinnus* (Stephens 1836), *Wesmaelius fassnidgei* (Killington 1933), *Sympherobius maculipennis* (Kimmins 1929), *Nothochrysa capitata* (Fabricius 1793), *Nineta vittata* (Wesmael 1841) and *Chrysopa dubitans* (McLachlan 1887).

In Table 22.2, we summarised all data recorded by the authors (Monserrat & Marín, 1992, 1994, 1996; Marín, 1994; Marín & Monserrat, 1995*a*, *b*). In this table, the total abundance (A = total number of specimens collected on any plant substrate of each Neuroptera species), the number of different plant substrates where hemerobiid, coniopterygid, and chrysopid species were collected (N), and the habitat amplitude (HA) (Simpson, 1949) have been noted.

22.4 HABITAT AMPLITUDE

The degree of preference of the neuropteran species for the plant substrates constitutes one of the most interesting aspects of their applied biology. Without taking into account other environmental factors, this preference is related to the presence of certain prey species on a specific vegetable substrate or to the honeydews, pollens, and other specific substances of the plants on which Neuroptera species feed.

Figure 22.1 represents the qualitative values of the number of different plant substrates where different Iberian hemerobiid, coniopterygid, and chrysopid species have been found that would permit us to know the range of the plant substrate preferences (Monserrat & Marín, 1992, 1994, 1996). These values indicate that most of the species occur on a few plant substrates. Only 36 species have been found on 10 or more plant substrates, some of them being the most eurytopic ones (*Wesmaelius subnebulosus*, *Sympherobius pygmaeus*, *S. elegans*, *Coniopteryx borealis*, *Semidalis aleyrodiformis*, *C. tjederi*, *Mallada flavifrons*, *Chrysopa formosa*, *M. picteti*, *Chrysoperla carnea s. lat.*, etc.).

However, this parameter (*N*) is only qualitative and its use can greatly modify the real preferences for a certain plant substrate. Captures of many of the hemerobiid, coniopterygid, and chrysopid species have been very scarce, and could be considered as accidental (Monserrat & Marín, 1992, 1994, 1996).

With the object of assessing the qualitative data, we have used the expression 'habitat amplitude' (HA), according to the formula of Simpson (1949), which reflects the relative abundance of each Neuroptera species on the different plant substrates. In the cases which we studied, the limits of this index varied between 1 and 109 (maximum number of plant substrates investigated). According to this, the higher values of HA represent more eurytopic species and, conversely, the lower values indicate more stenotopic species.

Figure 22.2 shows the hemerobiid, coniopterygid, and chrysopid species arranged according to their HA values. The most eurytopic Iberian species by far are, in decreasing order, *M. flavifrons*, *Coniopteryx borealis*,

Table 22.1. *Relevant species of Hemerobiidae, Coniopterygidae, and Chrysopidae collected in the Iberian Peninsula*

Hemerobiidae Latreille 1803	Coniopterygidae Burmeister 1839	Chrysopidae Schneider 1851
1. *Hemerobius humulinus* Linnaeus, 1758	37. *Aleuropteryx loewii* Klapálek, 1894	75. *Nothochrysa fulviceps* (Stephens, 1836)
2. *Hemerobius nitidulus* Fabricius, 1777	38. *Aleuropteryx minuta* Meinander, 1965	76. *Hypochrysa elegans* (Burmeister, 1839)
3. *Hemerobius micans* Olivier, 1792	39. *Aleuropteryx juniperi* Ohm, 1968	77. *Italochrysa stigmatica* (Rambur, 1842)
4. *Hemerobius lutescens* Fabricius, 1793	40. *Aleuropteryx iberica* Monserrat, 1977	78. *Italochrysa italica* (Rossi, 1790)
5. *Hemerobius stigma* Stephens, 1836	41. *Aleuropteryx remane* Rausch, H. Aspöck & Ohm, 1978	79. *Nineta flava* (Scopoli, 1763)
6. *Hemerobius perelegans* Stephens, 1836	42. *Aleuropteryx maculata* Meinander, 1963	80. *Nineta guadarramensis* (Pictet, 1865)
7. *Hemerobius pini* Stephens, 1836	43. *Helicoconis hirtinervis* Tjeder, 1960	81. *Nineta pallida* (Schneider, 1851)
8. *Hemerobius simulans* Walker, 1853	44. *Helicoconis pseudolutea* Ohm, 1965	82. *Chrysopidia ciliata* (Wesmael, 1841)
9. *Hemerobius gilvus* Stein, 1863	45. *Helicoconis iberica* Ohm, 1965	83. *Chrysopa perla* (Linnaeus, 1758)
10. *Hemerobius contumax* Tjeder, 1932	46. *Helicoconis parvicosa* Ohm, 1965	84. *Chrysopa phyllochroma* Wesmael, 1841
11. *Hemerobius handschini* Tjeder, 1957	47. *Helicoconis hispanica* Ohm, 1965	85. *Chrysopa pallens* (Rambur, 1838)
12. *Wesmaelius nervosus* (Fabricius, 1793)	48. *Nimboa adelae* Monserrat, 1985	86. *Chrysopa regalis* Navás, 1915
13. *Wesmaelius subnebulosus* (Stephens, 1836)	49. *Nimboa espanoli* Ohm, 1973	87. *Chrysopa viridana* Schneider, 1845
14. *Wesmaelius quadrifasciatus* (Reuter, 1894)	50. *Coniopteryx atlasensis* Meinander, 1963	88. *Chrysopa formosa* Brauer, 1850
15. *Wesmaelius navasi* (Andreu, 1911)	51. *Coniopteryx lopetsederi* H. Aspöck, 1963	89. *Chrysopa nierembergi* Navás, 1908
16. *Wesmaelius malladai* (Navás, 1925)	52. *Coniopteryx atlantica* Ohm, 1963	90. *Chrysopa nigricostata* Brauer, 1850
17. *Wesmaelius ravus* (Withycombe, 1923)	53. *Coniopteryx kerzhneri* Meinander, 1971	91. *Rexa lordina* Navás, 1919
18. *Wesmaelius helveticus* (Aspöck & Aspöck, 1964)	54. *Coniopteryx perisi* Monserrat, 1976	92. *Cunctochrysa baetica* (Hölzel, 1972)
19. *Wesmaelius reisseri* Aspöck & Aspöck, 1982	55. *Coniopteryx ketiae* Monserrat, 1985	93. *Cunctochrysa albolineata* (Killington, 1935)
20. *Sympherobius elegans* (Stephens, 1836)	56. *Coniopteryx tineiformis* Curtis, 1834	94. *Mallada alarconi* (Navás, 1915)
21. *Sympherobius pygmaeus* (Rambur, 1842)	57. *Coniopteryx parthenia* (Navás & Marcet, 1910)	95. *Mallada flavifrons* (Brauer, 1850)
22. *Sympherobius pellucidus* (Walker, 1853)	58. *Coniopteryx borealis* Tjeder, 1930	96. *Mallada clathratus* (Schneider, 1845)
23. *Sympherobius fuscescens* (Wallengren, 1863)	59. *Coniopteryx ezequi* Monserrat, 1984	97. *Mallada granadensis* (Pictet, 1865)
24. *Sympherobius fallax* Navás, 1908	60. *Coniopteryx haematica* McLachlan, 1868	98. *Mallada inornatus* (Navás, 1901)
25. *Sympherobius gratiosus* Navás, 1908	61. *Coniopteryx drammonti* Rousset, 1964	99. *Mallada ventralis* (Curtis, 1834)
26. *Sympherobius riudori* Navás, 1915	62. *Coniopteryx esbenpeterseni* Tjeder, 1930	100. *Mallada picteti* (McLachlan, 1880)
27. *Sympherobius klapaleki* Zelený, 1963	63. *Coniopteryx tjederi* Kimmins, 1934	101. *Mallada venosus* (Rambur, 1842)
28. *Megalomus hirtus* (Linnaeus, 1761)	64. *Coniopteryx lentiae* H. Aspöck & U. Aspöck, 1964	102. *Mallada prasinus* (Burmeister, 1839)
29. *Megalomus tortricoides* Rambur, 1842	65. *Coniopteryx arcuata* Kis, 1965	103. *Mallada subcubitalis* (Navás, 1901)
30. *Megalomus pyraloides* Rambur, 1842	66. *Parasemidalis fuscipennis* (Reuter, 1894)	104. *Mallada ibericus* (Navás, 1903)
31. *Megalomus tineoides* Rambur, 1842	67. *Parasemidalis triton* Meinander, 1976	105. *Mallada genei* (Rambur, 1842)
32. *Drepanepteryx phalaenoides* (Linnaeus, 1758)	68. *Hemisemidalis pallida* (Withycombe, 1924)	106. *Peyerimhoffina gracilis* (Schneider, 1851)
33. *Micromus paganus* (Linnaeus, 1767)	69. *Conventzia psociformis* (Curtis, 1834)	107. *Chrysoperla carnea s. lat.* (Stephens, 1836)
34. *Micromus variegatus* (Fabricius, 1793)	70. *Conventzia pineticola* Enderlein, 1905	108. *Chrysoperla mediterranea* (Hölzel, 1972)
35. *Micromus angulatus* (Stephens, 1836)	71. *Semidalis aleyrodiformis* (Stephens, 1836)	109. *Chrysoperla ankylopteryformis* Monserrat & Diaz-Aranda, 1989
36. *Micromus lanosus* (Zelený, 1962)	72. *Semidalis vicina* (Hagen, 1861)	110. *Chrysoperla mutata* (McLachlan, 1898)
	73. *Semidalis pluriramosa* (Karny, 1924)	111. *Brinckochrysa nachoi* Monserrat, 1977
	74. *Semidalis pseudouncinata* Meinander, 1963	112. *Suarius iberiensis* Hölzel, 1974
		113. *Suarius walsinghami* Navás, 1914
		114. *Suarius tigridis* (Morton, 1921)

Table 22.2. *Summary of data on relevant species of Hemerobiidae, Coniopterygidae, and Chrysopidae collected in the Iberian Peninsula*

Species[a]	Total number of specimens collected	Number of different plant substrates	Habitat amplitude[b]	Species[a]	Total number of specimens collected	Number of different plant substrates	Habitat amplitudes[b]
1	108	12	2.6	43	1	1	1
2	342	6	1.1	44	267	20	2.8
3	265	8	1.3	45	0	0	0
4	59	6	1.5	46	13	7	4.3
5	617	26	2.1	47	11	4	1.8
6	6	2	1.4	48	0	0	0
7	7	2	1.3	49	10	6	4.2
8	5	3	2.3	50	160	13	4.9
9	66	7	2.2	51	43	13	6.8
10	14	1	1	52	0	0	0
11	21	3	1.8	53	42	4	1.5
12	1	1	1	54	12	4	2.1
13	106	41	6.3	55	1	1	1
14	1	1	1	56	76	8	2.7
15	11	6	2.9	57	1876	22	1.9
16	1	1	1	58	399	24	12.8
17	2	1	1	59	438	5	1.4
18	0	0	0	60	409	14	2
19	1	1	1	61	86	9	3
20	103	14	4.1	62	161	15	5.2
21	588	24	4.1	63	191	26	8.2
22	2	2	2	64	147	14	1.7
23	163	10	1.2	65	231	9	2.5
24	3	2	1.8	66	4	4	4
25	16	3	2.6	67	4	2	1.6
26	9	4	3.8	68	365	10	2.6
27	47	7	4	69	549	29	4.4
28	17	4	2.4	70	59	5	2.6
29	1	1	1	71	1495	44	9.3
30	2	1	1	72	105	10	4.1
31	1	1	1	73	0	0	0
32	1	1	1	74	485	8	1.9
33	2	2	2	75	19	5	2
34	3	3	3	76	3	2	1.8
35	6	3	2.6	77	8	2	1.8
36	17	3	1.4	78	40	8	4.2
37	872	18	3.7	79	27	7	4.3
38	18	2	1.1	80	131	6	1.8
39	76	12	6.1	81	26	7	1
40	332	6	1.1	82	40	4	2.4
41	11	2	1.2	83	53	9	4.4
42	71	3	1.1	84	0	0	0

Table 22.2. (*cont.*)

Species[a]	Total number of specimens collected	Number of different plant substrates	Habitat amplitudes[b]	Species[a]	Total number of specimens collected	Number of different plant substrates	Habitat amplitudes[b]
85	120	21	9.8	100	387	31	10.1
86	24	8	3.8	101	18	9	6
87	508	28	3.8	102	2954	52	4.5
88	89	21	11.9	103	201	23	4.4
89	2	2	2	104	314	24	6.1
90	41	6	2.1	105	146	15	6
91	3	1	1	106	26	3	1.7
92	76	9	2.2	107	8234	91	9.8
93	33	8	4.8	108	328	14	1.3
94	21	3	2.8	109	0	0	0
95	906	51	13.3	110	2	2	2
96	30	10	4.2	111	106	3	1.4
97	142	16	2	112	0	0	0
98	50	10	4.8	113	0	0	0
99	81	7	3.6	114	0	0	0

Notes:

[a] Numbers refer to species listed in Table 22.1.

[b] These parameters were calculated without considering captures from light, Malaise and Moericke traps.

Chrysopa formosa, M. picteti, C. pallens, Chrysoperla carnea s. lat. and *S. aleyrodiformis*, with HA values above 9. Evidently their higher ecological versatility facilitates colonisation of more diverse habitats (Fig. 22.1), and so the species increases its dispersal potential resulting in a more diverse ecological and geographical distribution.

The remaining Iberian species are relatively stenotopic species, with HA values of generally less than 8. The most stenotopic Iberian species, associated mainly with single substrates, are found in this group. Monserrat & Marín (1992, 1994, 1996) show the close relationship of, for example, *Aleuopteryx iberica* with *Quercus rotundifolia; Nineta guadarramensis, N. fulviceps,* and *Hemerobius gilvus* with *Q. pyrenaica; Sympherobius gratiosus* with *Populus* spp.; *H. lutescens* with *Q. robur; Micromus lanosus* with *Corylus avellana; H. micans* with *Fagus sylvatica; Brinckochrysa nachoi* with *Eucalyptus globulus; C. mediterranea* with *Pinus halepensis; Coniopteryx parthenia, S. fuscescens,* and *H. nitidulus* with *P. sylvestris; Peyerimhoffina gracilis* and *H. contumax* with *Abies alba; Aleuopteryx maculata* and *C. ezequi* with *Juniperus thurifera; A. minuta* with *Tamarix*

gallica; etc. This group of species has, in general, very precise ecological requirements which limit their potential to colonise, and so their geographical distribution is usually restricted to that of the specific plant substrates (Aspöck *et al.*, 1980) and therefore their use as specific agents in biological control seems very appropriate.

Generally, Iberian Hemerobiidae have very low HA values and consequently must be very specific in their selection of food, phytophagous arthropods or plant substances. Some authors have indicated this fact for certain species (Nakahara, 1954; Yang, 1980; New, 1984).

The habitat amplitude of the Iberian Coniopterygidae and Chrysopidae shows similar results but with higher HA values than in the Hemerobiidae (Monserrat & Marín, 1996). Therefore, the predaceous hemerobiids can act as much more specific biological control agents than other families. There are no previous studies that quantify the plant substrate specificity in Hemerobiidae, Coniopterygidae, and Chrysopidae, with the exception of Monserrat & Marín (1992, 1994, 1996), but some authors have given valuable data about

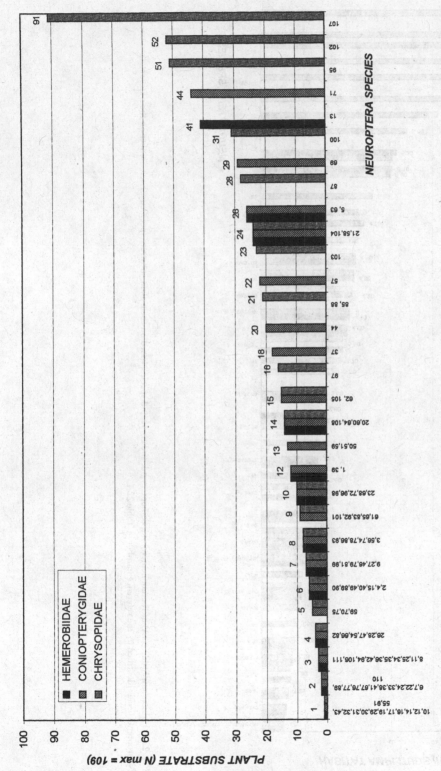

Fig. 22.1. Number of different plant substrates (N) on which Iberian hemerobiid, coniopterygid, and chrysopid species have been collected. Each number on the x axis refers to one Neuroptera species identified by the reference number in Table 22.1.

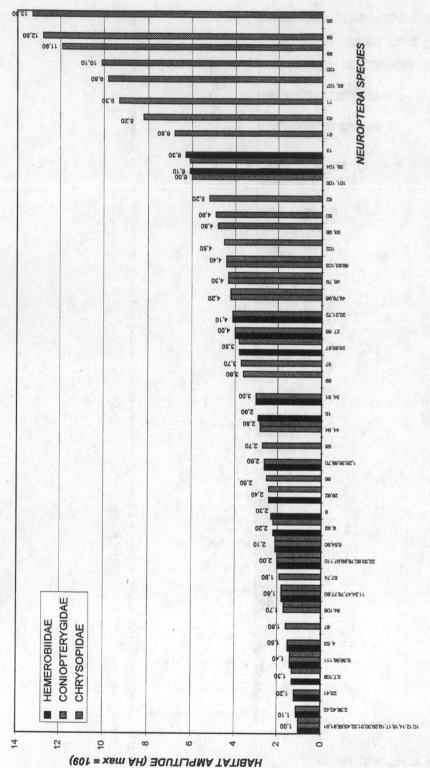

Fig. 22.2. Habitat amplitude (HA) of Iberian hemerobiid, coniopterygid, and chrysopid species, based on their plant substrate specificity. Each number on the x axis refers to one Neuroptera species identified by the reference number in Table 22.1.

the preferences for plant substrates in certain geographic areas, showing also in many species a great specificity (Aspöck *et al.*, 1980; also Eglin-Dederding, 1980, 1984, 1986; Pantaleoni, 1982; Zelený, 1984; Marín & Monserrat, 1987; Marín, 1994).

22.5 RELATIONSHIP BETWEEN HEMEROBIID, CONIOPTERYGID, AND CHRYSOPID SPECIES AND THE PLANT SUBSTRATES

In attempting to characterise definite groups of Iberian Coniopterygidae, Chrysopidae, and Hemerobiidae which could be more or less linked to one or more plant species with ecological or taxonomic affinities, Monserrat & Marín (1992, 1994, 1996) have undertaken in each family studied a cluster analysis and compiled a quantitative similarity dendrogram. Taking into account the results obtained, and new data obtained by the authors (Marín, 1994; Marín & Monserrat, 1995*a*, *b*), we can define the groups of Neuroptera species listed in Table 22.3.

In conclusion, there is generally a marked association between the Iberian hemerobiid, coniopterygid, and chrysopid species and plant substrates. In certain species the preferences are not clearly defined due to the low number of specimens collected, or because most of them were collected in light and Malaise traps, which fact excludes a clear association with the considered plants.

Future investigations will provide more data on the substrate specificity of the hemerobiid, coniopterygid, and chrysopid species studied here, and more importantly in those species not clearly associated with the Iberian plant substrates considered. Likewise, studies in other countries could provide interesting data for comparing other possible plant substrate preferences.

REFERENCES

Aspöck, H. (1992) The Neuropteroidea of Europe: a review of present knowledge (*Insecta*). In *Current Research in Neuropterology, Proceedings of the 4th International Symposium on Neuropterology*, ed. Canard, M., Aspöck, H. & Mansell, M.W., pp. 43–56. Sacco, Toulouse.

Aspöck, H., Aspöck, U. & Hölzel, H. (1980) *Die Neuropteren Europas*, 2 vols. Göcke & Evers, Krefeld.

Aspöck, U. & Aspöck, H. (1982) Eine neue Species der Genus *Wesmaelius* Krüger aus Spanien (Neuropteroidea: Planipennia: Hemerobiidae). *Entomologische Zeitschrift* 92, 289–292.

Canard, M., Séméria, Y. & New, T.R. (eds.) (1984) *Biology of Chrysopidae*. Dr W. Junk Publishers, The Hague.

Eglin-Dederding, W. (1980) Die Netzflüger des Schweizerischen Nationalparks und seiner Umgebung (Insecta: Neuropteroidea). *Ergebnisse der Wissenschaftliche Untersuchungen Schweizerischen Nationalpark* 15, 281–315.

Eglin-Dederding, W. (1984) Probleme beim Netzflüglerfang in den Gebirgswäldern des Schweizerischen Nationalparks 1938–1978 (Insecta, Neuropteroidea). *Entomologische Gesellschaft Basel* 34, 54–57.

Eglin-Dederding, W. (1986) Netzflügler und Schnabelfliegen (Neuropteroidea, Mecoptera). *Ergebnisse der Wissenschaftliche Untersuchungen Schweizerischen Nationalpark* 12, 169–198.

Garland, J.A. (1981) Effect of low-temperature storage on oviposition in *Hemerobius stigma* Steph. (Neuroptera, Hemerobiidae). *Entomologist's Monthly Magazine* 116, 149–150.

Hinke, F. (1975) Autökologische Untersuchungen an mitteleuropäischen Neuropteren. *Zoologische Jahrbücher (Systematik, Geographic und Biologie)* 102, 303–330.

Killington, F.J. (1936) *A Monograph of the British Neuroptera*, vol 1. Ray Society, London.

Killington, F.J. (1937) *A Monograph of the British Neuroptera*, vol. 2. Ray Society, London.

Laffranque, J.P. & Canard, M. (1975) Biologie du prédateur aphidiphage *Boriomyia subnebulosa* (Stephens) (Neuroptera, Hemerobiidae); études au laboratoire et dans les conditions hivernales du sud-ouest de la France. *Annales Zoologie– Écologie Animale* 7, 331–343.

Marín, F. (1994) Las comunidades de neurópteros de la provincia de Albacete (Insecta: Neuropteroidea). *Al-Basit* 34, 247–304.

Marín, F. & Monserrat, V.J. (1987) Los neurópteros del encinar ibérico (Insecta, Neuropteroidea). *Boletin de Sanidad Vegetal Plagas* 13, 347–359.

Marín, F. & Monserrat, V.J. (1995*a*) Contribución al conocimiento de los neurópteros de Valencia (Insecta, Neuroptera). *Bolitin de la Asociación española de Entomología* 19, 35–49.

Marín, F. & Monserrat, V.J. (1995*b*) Contribución al conocimiento de los neurópteros de Zaragoza (Insecta, Neuropteroidea). *Zapateri, Revista aragone de entomología* 5, 109–126.

Meinander, M. (1972) A revision of the family Coniopterygidae (Planipennia). *Acta Zoologica Fennica* 136, 1–357.

Meinander, M. (1990). The Coniopterygidae (Neuroptera: Planipennia). A check-list of the species of the world,

Table 22.3. *Groups of Neuroptera species linked to particular plant substrates*

Where collected	Species	Comments
Abies alba	*Sympherobius pellucidus, Hemerobius pini, H. contumax, Wesmaelius nervosus, W. quadrifasciatus, Nineta pallida, Peyerimhoffina gracilis*	These species are relatively stenotopic ones with biology in Iberian Peninsula almost unknown
Pinus, especially *P. halepensis*	*Aleuopteryx loewii, Mallada venosus, M. ibericus, M. genei, Chrysoperla mediterranea*	
Pinus, especially *P. pinaster*	*Chrysopa regalis*	
Pinus, especially *P. sylvestris*	*H. handschini, W. subnebulosus, H. stigma, S. fuscescens, H. nitidulus, W. ravus, W. reisseri, Coniopteryx parthenia, Conwentzia pineticola, Chrysopa nigricostata, M. clathratus*	*Wesmaelius subnebulosus* has been collected mainly on *Pinus* substrate, but it has been also abundantly collected on other plants, especially of genus *Quercus*, revealing its high eurytopism (Aspöck et al., 1980; Pantaleoni, 1982).
Juniperus species, especially *J. thurifera*	*A. maculata, Coniopteryx ezequi, Semidalis pseudouncinata, Hemisemidalis pallida, M. alarconi*	
Corylus avellana	*Micromus lanosus, Chrysopidia ciliata*	
Species very stenotopic; Eurosiberian broad-leaved trees and shrubs, with a marked association with *Fagus sylvatica*	*H. simulans, H. micans*	
Wet Mediterranean substrates of genus *Quercus*, mostly associated with *Q. faginea*	*Sympherobius pygmaeus*	
Broad-leaved trees and shrubs of the genus *Quercus* in the Eurosiberian or Mediterranean Region, associated mostly with *Q. pyrenaica*	*S. klapaleki, H. gilvus, Coniopteryx arcuata, C. esbenpeterseni, C. drummonti, C. lentiae*	
Eurosiberian broad-leaved trees and shrubs, particularly *Quercus robur*	*H. humulinus, H. perelegans, W. malladai, H. lutescens, C. timeiformis, Conwentzia psociformis, Semidalis aleyrodiformis, Nothochrysa fulviceps, Nineta guadarramensis, N. flava, Cunctochrysa albolineata, Mallada ventralis, Chrysopidia ciliata, Chrysopa perla*	*S. aleyrodiformis*
Dry Mediterranean substrates of genus *Quercus*, associated mostly with *Q. rotundifolia*	*A. iberica, Coniopteryx haematica, Helicoconis pseudolutea, Italochrysa italica, M. picteti, Chrysopa pallens, C. viridana, M. prasinus, Cunctochrysa baetica, M. granadensis, M. flavifrons, M. inornatus, Chrysoperla carnea s. lat., Chrysoperla formosa, Chrysoperla mutata.*	*C. Carnea s.lat.* presents a high eurytopism, having also been collected on many different plant substrates in Iberian Peninsula

Riverside or Mediterranean phanerophytes in wet microclimates, like *Ulmus* sp. and *Fraxinus angustifolia*	*Conioptteryx tjederi*
Mostly on phanerophytes of the Mediterranean region, like *Pistacia lentiscus* or *Quercus coccifera*	*S. vicina, C. borealis*
Mostly on *Tamarix gallica*	*A. minuta, M. subcubitalis*
Mostly on *Eucalyptus globulus*	*Brinckochrysa nachoi*
Only on *Olea europaea*	*Rexa lordina*
Only on herbaceous vegetation, like Gramineae and *Stipa tenacissima*	*I. stigmatica*
Mainly or exclusively by light	*Sympherobius elegans, W. navasi, S. fallax, Megalomus tineoides, Micromus variegatus, Megalomus pyraloides, Micromus angulatus, M. paganus, S. riudori, Drepanepteryx phalaenoides, Megalomus torricoides, S. gratiosus, Nimboa adelae, A. remane, C. perisi, H. hispanica, C. loipetsederi, C. kershneri, C. atlasensis, Semidalis pluriramosa, C. atlantica, H. iberica, Chrysopa phyllochroma, Chrysoperla ankylopteryiformis, Suarius iberiensis, S. walsinghami, S. tigridis* — These species were collected mostly with very low captures on plant substrates, therefore excluding their close association with the plants studied. The biology of some of them is almost completely unknown, and some of them are recently described taxa or recorded in Europe for the first time (Monserrat & Díaz-Aranda, 1989a, b)
Mainly by Malaise traps that do not permit a clear association with the studied plants	*W. helveticus, M. hirtus*
Without clear preference for any plant substrate	*A. juniperi, N. espanoli, Parasemidalis fuscipennis, Coniopteryx ketiae, H. panticosa, P. trion, Hypochrysa elegans, Chrysopa nierembergi*
Without data about their substrate specificity	*W. concinnus, W. fassnidgei, S. maculipennis, N. capitata, N. vittata, C. dubitans* — For these data are not confirmed about their presence on any plant substrates in the Iberian Peninsula (see Monserrat & Marín, 1994, 1996)

descriptions of new species and other new data. *Acta Zoologica Fennica* 189, 1–95.

Miermont, Y. & Canard, M. (1975). Biologie du prédateur aphidiphage *Eumicromus angulatus* (Neur:Hemerobiidae): études au laboratoire et observations dans le sud-ouest de la France. *Entomophaga* 20, 179–191.

Monserrat, V.J. (1984). Correcciones a las citas de Coniopterígidos (Insecta, Planipennia: Coniopterygidae) dadas por L. Navás. *Miscellánia Zoológica* 8, 145–151.

Monserrat, V.J. (1986a) Longinos Navás, his neuropterological work and collection. In *Recent Research in Neuropterology, Proceedings of the 2nd International Symposium on Neuropterology*, ed. Gepp, J., Aspöck, H. & Hölzel, H., pp. 173–176. Thalerhof, Graz.

Monserrat, V.J. (1986b) Sipnosis de los hemeróbidos de la Península Ibérica (Neuroptera, Planipennia, Hemerobiidae). *Actas VIII Jornadas Asociación española de Entomología*1200–1223. Sevilla.

Monserrat, V.J. (1990) Revisión de la obra de L.Navás, II: El género *Micromus* Rambur, 1842 (Neuropteroidea, Planipennia: Hemerobiidae). *Graellsia* 46, 175–190.

Monserrat, V.J. & Díaz-Aranda, L.M. (1989a) Nuevos datos sobre los crisópidos ibéricos (Neuroptera, Planipennia: Chrysopidae). *Boletin de la Asociación española de Entomología* 13, 251–267.

Monserrat, V.J. & Díaz-Aranda, L.M. (1989b) *Suarius walsinghami* (Navás, 1914) nuevo crisópido para la fauna europea (Neuroptera, Chrysopidae). *Nouvelle Revue d' Entomologie* (N.S.) 6, 407–411.

Monserrat, V.J. & Marín, F. (1992) Substrate specificity of Iberian Coniopterygidae (*Insecta: Neuroptera*). In *Current Research in Neuropterology, Proceedings of the 4th International Symposium on Neuropterology*, ed. Canard, M., Aspöck, H. & Mansell, M.W., pp. 279–290. Sacco, Toulouse.

Monserrat, V.J. & Marín, F. (1994) Substrate specificity of Iberian Chrysopidae (Insecta: Neuroptera). *Acta Oecologica* 15, 119–131.

Monserrat, V.J. & Marín, F. (1996) Plant substrate specificity of Iberian Hemerobiidae (Insecta: Neuroptera). *Journal of Natural History* 30, 775–787.

Nakahara, W. (1954) Early stages of some Japanese Hemerobiidae, including two new species. *Kontyu* 22, 41–46.

Neuenschwander, P. (1975) Influence of temperature on the immature stages of *Hemerobius pacificus*. *Environmental Entomology* 4, 215–220.

Neuenschwander, P. (1976) Biology of the adult *Hemerobius pacificus*. *Environmental Entomology* 5, 96–100.

New, T.R. (1984) Comparative biology of some Australian Hemerobiidae. In *Progress in World's Neuropterology, Proceedings of the 1st International Symposium on Neuropterology*, ed. Gepp, J., Aspöck, H. & Hölzel, H., pp. 153–166. Thalerhof, Graz.

New, T.R. (1986) A review of the biology of Neuroptera Planipennia. *Neuroptera International* Supplementary Series 1, 57 pp.

New, T.R. (1989) Planipennia (Lacewings). In *Handbook of Zoology*, Vol. 4, *Arthropoda: Insecta*, Part 30. Walter de Gruyter, Berlin.

Pantaleoni, R.A. (1982) Neuroptera Planipennia del Comprensorio delle Valli di Comacchio: Indagine Ecologica. *Bollettino dell'Istituto di Entomologia dell'Università di Bologna* 37, 1–73.

Samson, P.R. & Blood, P.R. (1979) Biology and temperature relationships of *Chrysopa* sp., *Micromus tasmaniae* and *Nabis capsiformis*. *Entomologia Experimentalis et Applicata* 25, 253–259.

Simpson, E.M. (1949) Measurement of diversity. *Nature* 163, 688.

Stelzl, M. (1990) Nahrungsanalytische Untersuchungen an Hemerobiiden-Imagines (Insecta, Planipennia). *Mitteilungen der deutsche Gessellschaft für allgemeine und angewandte Entomologie* 7, 670–676.

Stelzl, M. (1991) Untersuchungen zu Nahrungsspektren mitteleuropäischer Neuropteren-Imagines (Neuropteroidea, Insecta). Mit einer Diskussion über deren Nützlichkeit als Opponenten von Pflanzenschädlingen. *Journal of Applied Entomology* 111, 469–477.

Stelzl, M. (1992) Comparative studies on mouthparts and feeding habits of adult Raphidioptera and Neuroptera (Insecta: Neuropteroidea). In *Current Research in Neuropterology, Proceedings of the 4th International Symposium on Neuropterology*, ed. Canard, M., Aspöck, H. & Mansell, M.W., pp. 341–347. Sacco, Toulouse

Stelzl, M. & Gepp, J. (1987) Nahrungsanalytische Untersuchungen an Imagines von *Hemerobius micans* (Olivier)(Planipennia, Hemerobiidae). *Mitteilungen der Naturwissenschaftlichen Veriens für Steiermark* 17, 185–188.

Yang, J.K. (1980) Three new species of *Sympherobius* from China (Neuroptera: Hemerobiidae). *Acta Agriculture Universitatis Pekinensis* 6, 87–90.

Zelený, J. (1978) Les fluctuations spatio-temporelles des populations de Névroptères aphidiphages (Planipennia) (comme élément indicateur de leur spécificité). *Annales Zoologie–Écologie Animale* 10, 359–366.

Zelený, J. (1984) Chrysopid ocurrence in west paleartic temperate forests and derived biotopes. In *Biology of Chrysopidae*, ed. Canard, M., Séméria, Y. & New, T.R., pp. 151–160. Dr W. Junk, The Hague.

CHAPTER 23

Lacewings in Sardinian olive groves

R.A. Pantaleoni, A. Lentini, and G. Delrio

23.1 INTRODUCTION

Chrysopid taxocoenoses of the olive agroecosystem are well known and have been studied in nearly all Mediterranean countries (Canard & Laudého, 1977, 1980; Canard, 1979; Canard *et al.*, 1979; Alrouechdi *et al.*, 1980*a*, *b*; Neuenschwander & Michelakis, 1980; Neuenschwander *et al.*, 1981; Neuenschwander, 1982; Campos & Ramos, 1983; Yayla, 1983; Alrouechdi, 1984; Liber & Niccoli, 1988). Pantaleoni & Curto (1990) suggested that they were influenced by environmental conditions such as the spatial pattern of trees, chemical treatments, and the surrounding vegetation, besides zoogeography.

The results of adult captures, where the species of the genus *Dichochrysa* often dominate, are, however, in contrast with the larval collection results, where *Chrysoperla carnea* (Stephens) *s. lat.* is by far the most abundant species, to the point that this species is practically considered the only chrysopid species developing on olive trees (Alrouechdi *et al.*, 1980*a*; Neuenschwander & Michelakis, 1980). Some doubts on this conclusion were raised by Pantaleoni & Curto (1990) and Pantaleoni *et al.* (1993) who suggested that the efficiency of different sampling methods can be influenced by the behaviour of the different chrysopid genera and for these reasons the presence of some species can easily be underestimated.

The key pests in Sardinian olive agricultural systems are the olive fly, *Bactrocera oleae* (Rossi), the olive moth, *Prays oleae* (Bernard), and the black olive scale, *Saissetia oleae* (Olivier). Chrysopids can prey on the last two and are considered important predators of the olive moth (Ramos & Ramos, 1990).

The purpose of this study was to observe the species composition and the seasonal trends of abundance of chrysopids and other lacewings in Sardinian olive groves and to obtain preliminary data on their role in the control of olive pests.

23.2 MATERIAL AND METHODS

The observations were carried out in two plots of an olive-growing area near Sassari (northern Sardinia, Italy) during 1993. The first one was a 1-ha square with vase-trained centuries-old olive trees (cultivar 'Bosana'), in 12×12-m tree spacing. The second one was a series of hedgerows of 20-year-old olive trees (cultivar 'Sivigliana'). The plots were surrounded mainly by olive groves but also, especially around the hedgerow plot, by vineyards and other orchards.

Neither plot was treated with pesticides but the surrounding olive groves were treated with Methidation against an outbreak of olive scale on 14 July and 21 September and with protein bait sprays poisoned with Deltamethrin against olive fly on 22 July and 6 August. Weeds were controlled in late spring by mowing.

In order to monitor adult lacewings, yellow sticky traps (20×20 cm) were suspended at head height on the outside of the canopy of olive trees. Five traps were used in each plot with at least 20 m between each trap.

Direct counting of all lacewing life stages was done on a random sample of 250 twigs, 20 cm long, in each plot. Out of these, 50 twigs per plot were cut and transferred to the laboratory to determine the number of phytophagous insects.

In each plot, 25 randomly selected branches were beaten and predators, namely spiders, bugs, ladybirds and lacewings, were collected onto an 80-cm white square cloth tray.

There were 32 sampling dates in total. Between mid-March and mid-September samples were collected weekly, while at the beginning and at the end of the year sampling intervals were longer.

All preimaginal stages of lacewings were reared in the laboratory until adults were obtained for species recognition. Nevertheless, most specimens could not be determined at a species level because a large number of

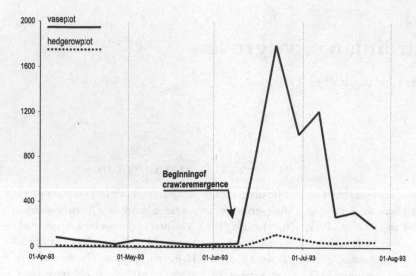

Fig. 23.1. *Saissetia oleae*: seasonal trends of counts (number per twig) in the two experimental plots.

preimaginal stages were parasitized or dead and a large number of adults from yellow traps had deteriorated. In addition the current problems in the taxonomic definition of *Chrysoperla*, and to some extent of *Dichochrysa*, hinder the correct determination of these species.

In order to observe the attack of chrysopid larvae, nymphs and adults of Heteroptera, from an olive twig infested with a known number of healthy *Prays oleae* eggs was isolated for 12 hours together with one predator in a plastic vial, in the laboratory.

23.3 RESULTS

23.3.1 Prey
The most abundant phytophagous insects that make up the potential prey for chrysopids were the olive moth *Prays oleae*, the black scale *Saissetia oleae*, the olive psylla *Euphyllura* sp., and psocids. Attacks on leaves by the mites *Aceria oleae* (Nalepa) and *Oxycenus maxwelli* (Keifer) were also recorded in summer.

The trends of black scale counts in both plots are shown in Fig. 23.1. In the vase plot the scale number was far higher than in the hedgerow plot and scale density increased abruptly in mid-June, when the crawler emergence started. The maximum occurred at the end of June.

The seasonal trends of abundance of olive moth, olive psylla, and psocids are shown in Fig. 23.2.

Larvae of the olive moth phyllophagous generation were found at a low density from January to mid-

April. Larvae and pupae of the anthophagous generation were found in the flowering period between the end of May and early June. The carpophagous-generation eggs occurred in June and early July with a peak of oviposition in the first half of June. In the three samplings from the end of June to early July the predation on eggs reached a mean of 60%.

Few adults of olive psylla were found before early April and after June. The peaks of nymph population occurred between mid-April and mid-May.

In the olive grove the most common psocids were *Ectopsocus briggsi* McLachlan and *Trichopsocus dalii* (McLachlan). The populations were abundant in the early spring and reached a peak in mid-June. A great reduction of density was recorded after mid-July.

23.3.2 Coniopterygidae and Hemerobiidae
Seven species of Coniopterygidae were found: *Coniopteryx borealis* Tjeder, the most abundant species, *C. drammonti* Rousset, *C. haematica* McLachlan, *C. esbenpeterseni* Tjeder, *Parasemidalis fuscipennis* (Reuter) *Conwentzia psociformis* (Curtis), and *Semidalis aleyrodiformis* (Stephens). The captures of adults and those of larvae showed their peaks from July on (Fig. 23.3). By using the yellow traps, adults were far more abundant in the hedgerow plot than in the vase plot, while by beating the opposite was found (Table 23.1).

Finally, only two adults, of an indeterminate *Sympherobius* (Hemerobiidae), were caught by yellow traps (Table 23.1).

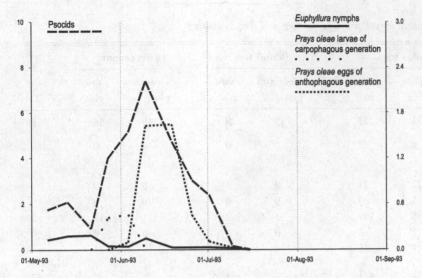

Fig. 23.2. Seasonal trends of counts (number per twig) of some chrysopid potential prey (mean of samples in both plots).

23.3.3 Chrysopidae

All the stages of four different chrysopid taxa belonging to *Chrysoperla*, *Cunctochrysa*, and *Dichochrysa* were found on olive trees. The first two genera were represented by one species (or group of species) each, *Chrysoperla carnea s. lat.* and *Cunctochrysa baetica* (Hölzel), respectively, while the last one was represented by two species: *Dichochrysa prasina* (Burmeister) *s. lat.*, laying eggs singly and *D. flavifrons* (Brauer) *s. lat.*, laying eggs in clusters. Furthermore, only one adult of *Chrysopa formosa* Brauer was captured in the yellow traps (Table 23.1).

Of 855 chrysopid eggs sampled all the year round 42.6% were parasitised by *Telenomus chrysopae* (Ashmead) (Hymenoptera: Scelionidae) and 15.1% died in the laboratory. The larvae obtained from the remaining eggs belonged to *Chrysoperla* (20.0%), *Dichochrysa* (19.4%), and *Cunctochrysa* (2.9%). The eggs were more abundant in the vase plot than in the hedgerow plot (7.89% vs. 21.1%). Only 27.6% of all the eggs were in clusters (Table 23.2). Unexplainable egg mortality was registered in the three samples obtained between the end of June and the beginning of July in the vase plot. Eighty out of the 115 indeterminate single eggs reported in Table 23.2 came from those samples.

Sampled larvae belonged to *Chrysoperla* (87.3%), *Cunctochrysa* (8.1%), and *Dichochrysa* (4.6%) (Table 23.3). *Cunctochrysa* and *Dichochrysa* were more abundant in the vase plot than in the hedgerow plot.

Sampling methods influenced the capture proportions of the three larval instars of all species. By beating, the third-instar larvae reached 61.2% and the first instar reached only 7.2%, while by direct counting the third instar remained under 35% and the first instar exceeded 31%. The sampling method influenced also the survey of *Dichochrysa* larvae; in fact by beating only 3.2% of larvae belonging to this genus were collected while by direct counting 10.3%.

As regards the adults, *Chrysoperla* was by far the most abundant taxon. Beating was an ineffective collecting method for chrysopid adults which as a result were caught in lower numbers. All chrysopids were more abundant in the vase plot. *Chrysoperla* exceeded 80% of total chrysopid captures in yellow-trap collections but reached about 30%–40% by other sampling methods. *Dichochrysa* were relatively very abundant in samples obtained by direct counting (Table 23.1). The sex was determined of only 24% of chrysopids captured by yellow traps: 80.6% of these individuals were male.

In *Chrysoperla*, overwintering as adults, egg presence began early in the year, but stayed low until the end of May. Egg numbers reached a maximum at the end of July and after that fell gradually, remaining low from September to November. This trend was probably affected by high egg mortality at the end of June and certainly by parasitism on the eggs. Larvae appeared at the beginning of April and showed three peaks, in May, late June, and late July, probably corresponding to many

Table 23.1. *Lacewing adults collected by yellow traps, beating, and direct counting*

Taxa	Yellow traps		Beating tray		Direct counting		Total
	Hedgerow plot	Vase plot	Hedgerow plot	Vase plot	Hedgerow plot	Vase plot	
Coniopterygidae	584	232	15	29	19	18	897
Hemerobiidae	0	2	0	0	0	0	2
Chrysopidae							
Unidentified	22	23	0	0	0	0	45
Chrysopa	1	0	0	0	0	0	1
Chrysoperla	88	190	2	2	1	11	294
Cunctochrysa	1	2	0	0	0	0	3
Dichochrysa	18	35	4	1	6	24	88
Total	714	484	21	32	26	53	1330

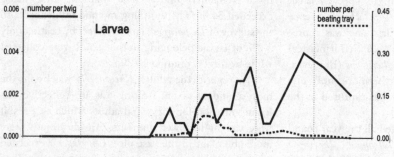

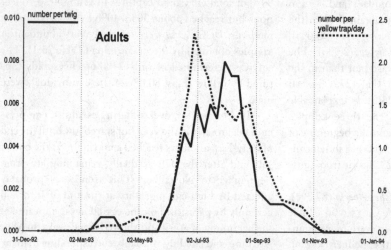

Fig. 23.3. Coniopterygidae: seasonal trends of larva and adult captures (mean of samples in both plots).

Table 23.2. *Chrysopid eggs collected by direct counting*

	Single		Cluster		
Taxa	Hedgerow plot	Vase plot	Hedgerow plot	Vase plot	Total
Parasitised	30	244	0	90	364
Unidentified	14	115	—	—	129
Chrysoperla	72	99	—	—	171
Cunctochrysa	7	18	—	—	25
Dichochrysa	6	14	51	95	166
Total	129	490	51	185	855

generations. From September on, larvae were found in small numbers only. The adults of the winter generation were found only in the few captures of late March. The adults born during the year appeared quite late in June. Their number increased quickly in July, reaching a maximum in early August, followed by a consistent decrease until the end of October (Fig. 23.4).

In *Dichochrysa*, overwintering as second- or third-instar larvae, the first eggs were found around mid-May. The successive trend was clearly bimodal with two peaks, corresponding roughly to June and August. The last ovipositions were in late autumn. The collections of the few larvae present were distributed throughout the year. By direct counting the adults showed a bimodal trend also with maxima in May and September (Fig. 23.5).

The eggs of *Cunctochrysa*, overwintering as prepupae, were found from mid-May to mid-September, in a bimodal trend with maxima in June and August. The seasonal trend of larvae captures was very similar to that of the eggs and showed a first generation in June that was numerically more consistent than the second generation of August (Fig. 23.6). The few captured adults occurred in August.

23.3.4 Other predators
Spiders were the most abundant predators and showed the highest number of taxa. The collected individuals belonged to 18 families and more than 40 genera (C. Arnò, personal communication). The most abundant families were Theridiidae, among the web spiders, and Salticidae, Clubionidae, Philodromidae, and Thomisidae, among the hunting spiders.

In our study, among the heteropteran bugs, only

the flower bugs and the zoophagous or phytozoophagous mirids were considered. At the end of June this group reached the highest density of all predators, according to beating-method counts. In fact, the three most abundant taxa, Anthocoridae, *Campyloneura virgula* (Herrich-Schäffer) and *Pilophorus* spp., showed a maximum abundance in the same period.

Only two species of ladybirds were abundant, *Chilocorus bipustulatus* (L.) and *Exochomus quadripustulatus* (L.). Both of them were also far more abundant in the vase plot than in the hedgerow plot.

The seasonal trends of abundance of the main predator groups, including lacewings, are shown in Fig. 23.7.

23.3.5 Laboratory observations on predation of *Prays oleae* eggs
In laboratory the eggs of olive moth were actively attacked by the *Chrysoperla carnea* larvae and by nymphs and adults of the mirids *Campyloneura virgula* and *Pilophorus* sp. and anthocorids.

Nearly all the eggs attacked by Heteroptera were completely emptied with the upper surface directly in contact with the inferior surface, while about 70% of those attacked by lacewings were less collapsed with slightly crumpled sides. The remaining 30% of attacked eggs were detached from the support or had a similar aspect to those predated by Heteroptera.

23.4 DISCUSSION
Reports of coniopterygids on olive trees are extremely scarce and indefinite (Neuenschwander, 1982; Yayla, 1983). During our research, however, they were numerically abundant and belonged to a high number of species. The presence of mites in both plots favoured such a colonisation. Adult and larva captures were almost equal in the two plots by direct counting and higher in the vase plot by beating. Adult captures with the yellow traps were, on the other hand, much more numerous in the hedgerow plot. This is very probably due to individuals coming from the fruit orchards near this plot (see for comparison the case reported by Pantaleoni & Alma, Chapter 25, this volume). Unfortunately the inability to determine the preimaginal stages, the complex procedure needed for the identification of females – for which individuals from sticky traps are poorly suited – and the need for particular

Table 23.3. *Lacewing larvae collected by beating and direct counting*

		Beating tray		Direct counting		
Taxa	Instar	Hedgerow plot	Vase plot	Hedgerow plot	Vase plot	Total
Coniopterygidae	First	0	0	1	0	1
	Second	4	5	3	3	15
	Third	2	9	4	7	22
	All	6	14	8	10	38
Chrysopidae						
Chrysoperla	First	4	11	6	6	27
	Second	32	37	7	10	86
	Third	69	68	6	13	156
	All	105	116	19	29	269
Cunctochrysa	First	0	3	0	2	5
	Second	1	8	0	1	10
	Third	2	7	0	1	10
	All	3	18	0	4	25
Dichochrysa	First	0	0	0	4	4
	Second	0	1	1	1	3
	Third	2	5	0	0	7
	All	2	6	1	5	14
Total		116	154	28	48	346

sampling methods for the eggs prevented further detailed study.

In contrast to the preceding family, we already have a fair amount of information on the chrysopids, which, however, sometimes seems contradictory and which therefore necessitates interpretation also in the light of our findings.

In an olive grove in Provence (southern France), Alrouechdi *et al.* (1980*a*, *b*) sampled chrysopid eggs by direct counting for three years. In all years more than 90% of the eggs reared were *Chrysoperla carnea*. The remaining percentage was subdivided between *Dichochrysa flavifrons*, *D. prasina*, and *Chrysopa pallens* (Rambur). To this latter species, aphidophagous with predator adults, were attributed all the grouped eggs not reared. The eggs in these groups rose above 35% of the total in the second year of the test.

Our data differ from those of the above authors in some essential aspects. Above all the eggs laid in clusters, about 30% of the total, belonged indisputably to *D. flavifrons s. lat.* Furthermore only about 80% of the eggs laid singly were *Chrysoperla carnea* and of the inclusive total the eggs attributable to this species were only slightly more numerous than those of *Dichochrysa*. Lastly also *Cunctochrysa baetica* was present in a small, but not negligible, percentage.

Up to this time the only data available on larvae was that of Neuenschwander & Michelakis (1980) showing that by means of sondage of olive trees with a pyrethroid fog 439 naked and 115 debris-carrying chrysopid larvae of second and third instars were captured. Of these 93 *Chrysoperla carnea*, 27 *D. flavifrons*, and 10 *D. zelleri* (Schneider) were reared, the remaining being dead or parasitised. Furthermore Alrouechdi *et al.* (1980*a*) using trap bands collected nearly 200 chrysopid cocoons; of the about 30 not parasitised they obtained almost exclusively *C. carnea*. In Sardinia this species is also the most abundant chrysopid in the larvae collections.

To partly justify the wide discrepancy between the results of the adult captures [see Neuenschwander *et al.* (1981), reported below] and those of the larvae

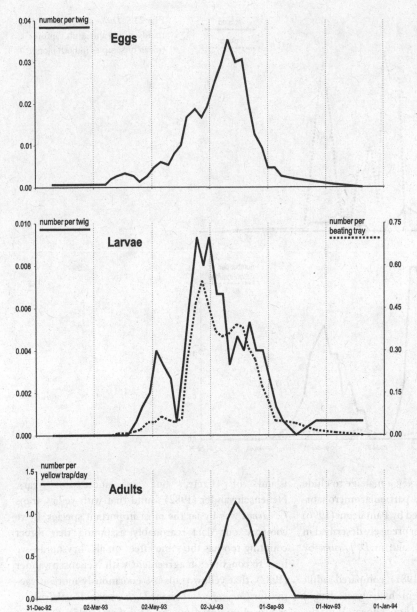

Fig. 23.4. *Chrysoperla*: seasonal trends of egg, larva and adult captures (mean of samples in both plots).

Neuenschwander & Michelakis (1980) suggested that with some sampling methods the fast-moving naked larvae of *Chrysoperla* are more easily collected than the less-active debris-carriers (*Dichochrysa* and others), a point taken up again by Neuenschwander (1986). Confirmation of this supposition could very probably also be found in our data. The percentage of larvae of *Dichochrysa* collected with the beating, in fact, was only one third of that obtained with direct counting. The

debris-carrying larvae would therefore have fewer tendencies to let themselves fall when the plant is shaken or treated with insecticide. According to our results however this tendency is greater in first-instar larvae of all the chrysopids in respect to those of third instar, as already suggested by Pantaleoni & Ticchiati (1990) after experiences of chemical knockdown in apple orchards and in accordance with data of Neuenschwander & Michelakis (1980) reported above.

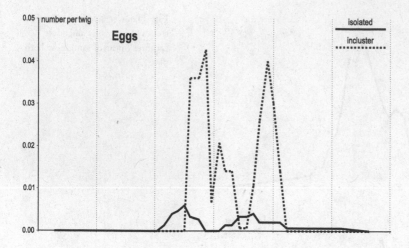

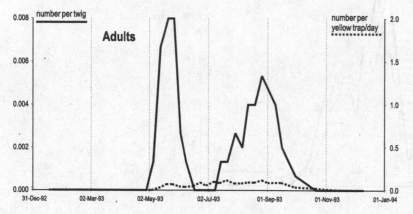

Fig. 23.5. *Dichochrysa*: seasonal trends of egg and adult captures (mean of samples in both plots).

In addition, the larvae of *Dichochrysa* manage to elude capture because they frequent a particular microhabitat. This behaviour, demonstrated by Pantaleoni (1996) on garden trees, had already been precisely described in *D. flavifrons* by Principi (1956), and in *D. prasina* by Withycombe (1923).

Neuenschwander *et al.* (1981) compared adult chrysopid captures with protein hydrolysate–baited McPhail traps and with chemical knockdown. With the first method *D. flavifrons* resulted as the most abundant species with 50.5% of the individuals identified, followed by *D. zelleri* with 33.1% and by *C. carnea* with 13.2%. With the second method, *D. flavifrons* was still the most abundant with 37.2% followed by *C. carnea* with 31.1% and by *D. zelleri* with 15.2%. Under the assumption that sondage with pyrethroid has the same effect on all species, the authors showed that *D. flavifrons* and *D. zelleri* are more attracted to the McPhail traps than *C. carnea*. Using the same olive groves on the

island of Crete, but in successive years, Neuenschwander (1982) found that with yellow traps *C. carnea* was by far the most important species. If we compare our data reasonably assuming that direct counting too has the same effect on all chrysopids, we have to conclude, in agreement with Neuenschwander (1982), that yellow traps are enormously more attractive for *Chrysoperla* than for *Dichochrysa*. Furthermore, as already pointed out by McEwen (1995), the same yellow traps capture a high percentage of males and their efficiency seems therefore to be influenced both by the species and by the sex of the chrysopids.

This situation can seem confused with *C. carnea* as easily the most abundant species in the larvae collections and in the adult captures with yellow traps. On the other hand the species of *Dichochrysa* are predominant in the adult captures with net, chemical knock down, McPhail traps, and direct counting, and the eggs of these can also reach high densities. The interpretation

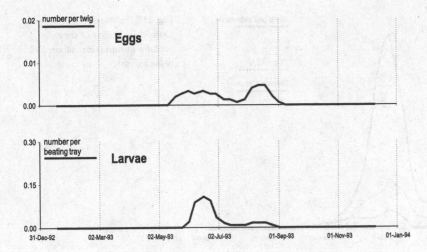

given by Neuenschwander & Michelakis (1980), Alrouechdi *et al.* (1980*a*), and partly by Duelli (1984), according to which *C. carnea* is the only species which develops on olive groves while the *Dichochrysa* utilises them only as a refuge plant, does not appear satisfactory. *Dichochrysa* probably also firmly colonises the olive grove until it is, in certain cases, more abundant than *Chrysoperla*. This is demonstrated by the abundance of the eggs and of the adults while the scarcity of larvae is attributable to the inefficiency of some sampling methods influenced both by the behavioural characteristics and by the microhabitat of these species. *Chrysoperla carnea* is the only chrysopid the larvae of which continually frequent the canopy of the olive tree. It also presents behavioural characteristics of flight, dispersal, and migration that favour its capture with yellow traps [see Duelli (1984) for a review on the subject].

The seasonal trends of all chrysopid taxa were preliminarily discussed by Pantaleoni *et al.* (1993). The data regarding *C. carnea* are in accordance with a possible development of three generations, as found by Alrouechdi *et al.* (1981) in Provence (France) and by Pantaleoni & Curto (1990) in Salento (Italy). The bimodal trend of *Dichochrysa* was previously found in arid environments (Pantaleoni, 1990) and was attributed to a dispersal phenomenon of the second-generation adults (July–August) towards more favourable environments, while the first (May–June) and the third (September–October) generations showed a complete development *in situ*. However, aestivation cannot be excluded. The results concerning *Cunctochrysa* are in

accordance with the findings of Pantaleoni (1984), who studied *C. baetica* in the Pianura Padana (northern Italy) and in the laboratory, in which almost two-thirds of the individuals of the first larvae generation enter diapause after cocoon-spinning. The remaining one-third emerge from the cocoon in the same year, originating a second flight of lower numbers than the first one.

The numbers of eggs, larvae, and adults of chrysopids collected during this research were clearly less abundant in the hedgerow plot than in the vase plot where there was a high density of black scale and consequently high production of honeydew. In agreement with this result, Alrouechdi *et al.* (1980*b*) found that adult chrysopid captures with McPhail traps were influenced by presence of high numbers of *Saissetia oleae* and abundance of its honeydew. In turn, this latter influenced the number of eggs laid on the infested plants, which was one and a half times higher than those on non-infested plants. On the other hand even though predation on black scale by lacewings is considered not very effective, it has been widely documented in literature (see e.g. Argyriou, 1967, 1985; Viggiani *et al.*, 1973, 1975; Viggiani & Bianco, 1974; Lal & Naji, 1980; Bagnoli, 1983; Delrio, 1983; Paparatti, 1984; Orphanides, 1986, 1988).

Chrysopids are considered the major oophagous predators of the *Prays* carpophagous generation, being of economic importance in this role (Montiel Bueno, 1981; Ramos *et al.*, 1983, 1987, 1988*a*, *b*; Ramos & Ramos, 1990). Although the efficiency of chrysopid larvae for the control of *Prays* eggs in relation to the pest density (Alrouechdi *et al.*, 1981; Niccoli & Boni,

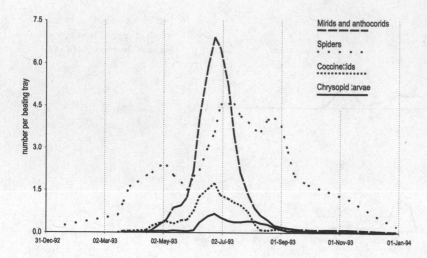

Fig. 23.7. Seasonal trends of captures by beating of the main predator groups (mean of samples in both plots).

1984; Campos & Ramos, 1985) and the chronological relationship between number of chrysopid larvae and moth eggs (Alrouechdi *et al.*, 1981) have been studied, this predatory activity of lacewings has never been verified with direct observations in natural conditions. In fact the predation of the eggs has generally been attributed to the action of the chrysopids without taking into account the presence of other predators.

In the Sardinian olive groves used in this study the mirids and anthocorids were numerically dominant. The presence of predator mirids and anthocorids inside the olive grove has been sufficiently documented by Yayla (1983, 1984). Indications about their role are provided again by Yayla (1984) who indicates them as predators of various Homoptera and by Sacchetti (1990) who lists the anthocorids as predators of olive moth eggs. In the laboratory we have shown that the main species of these two families can prey on the eggs of *Prays oleae* and that the appearance of the eggs attacked by them corresponds with that of most of the eggs preyed in the field.

The coccinellids, particularly the two most common species, *Chilocorus bipustulatus* and *Exochomus quadripustulatus*, even though decidedly coccoidophagous (see e.g. Argyriou & Katsoyannos, 1977) and dependent on the strong infestations of *Saissetia oleae*, are polyphagous predators able to feed on other prey (Yinon, 1969; Mendel *et al.*, 1985) and could attack the eggs of *Prays*.

Spiders are very abundant in olive groves and are continuously present throughout the year but, at present, it is not possible to accurately define their role

as predators of *P. oleae* eggs. Certainly they have already been indicated as predators of olive psylla (Abdul Baki & Ahemed, 1985) and of moth larvae (Sacchetti, 1990; Triggiani, 1971) and it is known that in general they can feed not only on moving prey but also on eggs (Nyffeler *et al.*, 1990).

The role of the lacewings as a factor in the control of the eggs of the *Prays* carpophagous generation should be re-examined and quantified more precisely. In fact the high abundance of other predators bring into doubt that the predation of *P. oleae* eggs can be attributed entirely to the chrysopids. On the other hand, the estimation of the predation in the field is very complex due to the difficulty in attributing it to a particular group of organisms. However, the efficiency of the chrysopids can be underestimated in the routine surveys because in some cases they detach the egg from the substrate leaving no trace of their action.

REFERENCES

Abdul Baki, M.H. Ali & Ahemed, M.S. (1985) Ecological studies on olive psyllid *Euphyllura straminea* Log at Mosul Region with special reference to its natural enemies. *Iraqi Journal of Agricultural Sciences 'Zanco'* 3, 14 pp.

Alrouechdi, K. (1984) Les Chrysopides (Neuroptera) en oliveraie. In *Progress in World's Neuropterology, Proceedings of the 1st International Symposium on Neuropterology*, ed. Gepp, J, Aspöck, H. & Hölzel, H., pp. 147–152. Thalerhof, Graz.

Alrouechdi, K., Lyon, J.P., Canard, M. & Fournier, D. (1980*a*) Les chrysopides (Neuroptera) récoltés dans un oliveraie du sud-est de la France. *Acta oecologica/oecologia Applicata* 1, 173–180.

Alrouechdi, K., Pralavorio, R., Canard, M. & Arambourg, Y. (1980*b*) Repartition des adultes et des pontes de Chrysopides (Neuroptera) récoltés dans une oliveraie de Provence. *Neuroptera International* 1, 65–74.

Alrouechdi, K., Pralavorio, R., Canard, M. & Arambourg, Y. (1981) Coïncidence et relations prédatrices entre *Chrysopa carnea* (Stephens) (Neur., Chrysopidae) et quelques ravageurs de l'olivier dans le sud-est de la France. *Mitteilungen der schweizerischen entomologischen Gesellschaft* 54, 281–290.

Argyriou, L.C. (1967) The scales of olive trees occuring in Greece and their entomophagous insects. *Annales de l'Institut Phytopathologique Benaki* (N.S.) 8, 66–73.

Argyriou, L.C. (1985) The soft scale of olive trees in Greece. In *Integrated Pest Control in Olive Groves, Proceedings of the International Joint Meeting*, ed. Cavalloro, R. & Crovetti, A., pp. 147–151. A.A. Balkema, Rotterdam.

Argyriou, L.C. & Katsoyannos, P. (1977) Coccinellidae species found in the olive-groves of Greece. *Annales de l'Institut Phytopathologique Benaki* (N.S.) 11, 331–345.

Bagnoli, B. (1983) Entomophagous insects of *Saissetia oleae* (Oliv.) in olive groves in Tuscany. In *Entomophagous Insects and Biotechnologies against Olive Pests, Proceedings of the E.C. Experts' Meeting on Entomophages and Biological Methods in Integrated Control in Olive Orchards* ed. Cavalloro, R. & Piavaux, A., pp. 7–14. A.A. Balkema, Rotterdam.

Campos, M. & Ramos, P. (1983) Chrisopidos (Neuroptera) capturados en un olivar del sur de Espana. *Neuroptera International* 2, 219–27.

Campos, M. & Ramos, P. (1985) Some relationships between the number of *Prays oleae* eggs laid on olive fruits and their predation by *Chrysoperla carnea*. In *Integrated Pest Control in Olive-Groves, Proceedings of the International Joint Meeting*, ed. Cavalloro, R. & Crovetti, A., pp. 237–241. A.A. Balkema, Rotterdam.

Canard, M. (1979) Chrysopides (Neuroptera) récoltés dans les oliveraies en Grèce. *Biologia Gallo-Hellenica* 9, 237–240.

Canard, M. & Laudého, Y. (1977) Les Névroptères capturés au piège de McPhail dans les oliviers en Grèce. 1: l'Île d'Aguistri. *Biologia Gallo-Hellenica* 7, 65–75.

Canard, M. & Laudého, Y. (1980) Les Névroptères capturés au piège de McPhail dans les oliviers en Grèce. 2: La Région d'Akrefnion. *Biologia Gallo-Hellenica* 9, 139–146.

Canard, M., Neuenschwander, P. & Michelakis, S. (1979) Les Névroptères capturés au piège de McPhail dans les oliviers en Grèce. 3: la Crète occidentale. *Annales de la Société Entomologique de France* (N.S.) 15, 607–615.

Delrio, G. (1983) The entomophagous insects of *Saissetia*

oleae (Oliv.) in Sardinia. In *Entomophagous Insects and Biotechnologies against Olive Pests, Proceedings of the E.C. Experts' Meeting on Entomophages and Biological Methods in Integrated Control in Olive Orchards* ed. Cavalloro, R. & Piavaux, A., pp. 15–23. A.A. Balkema, Rotterdam.

Duelli, P. (1984) Flight, dispersal, migration. In *Biology of Chrysopidae*, ed. Canard, M., Séméria, Y. & New, T.R., pp. 110–116. Dr W. Junk, The Hague.

Lal, O.P. & Naji, A.H. (1980) Observations on the predators of the black olive scale, *Saissetia oleae* Bern. (Homoptera: Coccidae) in the Socialist Peoples Libyan Arab Jamahiriya. *Zeitschrift fur Pflanzenkrankheiten und Pflanzenschutz* 87, 27–31.

Liber, H. & Niccoli, A. (1988) Observations on the effectiveness of an attractant food spray in increasing chrysopid predation on *Prays oleae* (Bern.) eggs. *Redia* 71, 465–482.

McEwen, P.K. (1995) Attractiveness of yellow sticky traps to green lacewings (Neuropt., Chrysopidae). *Entomologist's Monthly Magazine* 131, 163–166.

Mendel, Z., Podoler, H. & Rosen, D. (1985) A study of the diet of *Chilocorus bipustulatus* (Coleoptera: Coccinellidae) as evident from its midgut content. *Israel Journal of Entomology* 19, 141–146.

Montiel Bueno, A. (1981) Factores de regulación de las poblaciones de *Prays oleae* (Bern.). *Boletin del Servicio de Defensa contra Plagas e Inspeccion Fitopatologica* 7, 133–40.

Neuenschwander, P. (1982) Beneficial insects caught by yellow traps used in mass-trapping of the olive fly, *Dacus oleae*. *Entomologia Experimentalis et Applicata* 32, 286–296.

Neuenschwander, P. (1986) Sampling procedures for chrysopid populations. In *Biology of Chrysopidae*, ed. Canard, M, Séméria, Y. & New, T.R., pp. 205–212. Dr W. Junk, The Hague.

Neuenschwander, P. & Michelakis, S. (1980) The seasonal and spatial distribution of adult and larval Chrysopids in olive-trees in Crete. *Acta Oecologica/Oecologia Applicata* 1, 93–102.

Neuenschwander, P., Canard, M. & Michelakis, S. (1981) The attractivity of protein hydrolysate baited McPhail traps to different chrysopid and hemerobiid species (Neuroptera) in a Cretan olive orchard. *Annales de la Société Entomologique de France* (N.S.) 17, 213–220.

Niccoli, A. & Boni, F. (1984) Osservazioni sulla distribuzione delle uova di *Prays oleae* Bern. sui frutti e sull'incidenza dei fattori di mortalita. *Redia* 47, 515–525.

Nyffeler, M., Breene, R.G., Dean, D.A. & Sterling, W.L. (1990) Spiders as predators of arthropod eggs. *Journal of Applied Entomology* 109, 490–501.

Orphanides, G.M. (1986) Situation actuelle de la lutte biologique contre la cochenille noire de l'olivier, *Saissetia oleae*, à Chypre. In *Réunion sur la Protection Phytosanitaire de l'Olivier*. pp. 99–113. Sfax, Tunisie.

Orphanides, G.M. (1988) *Current Status of Biological Control of the Black Scale*, Saissetia oleae *(Olivier), in Cyprus.* Technical Bulletin of the Agricutural Research Institute, Cyprus no. 100.

Pantaleoni, R.A. (1984) Neuroptera Planipennia del comprensorio delle Valli di Comacchio: le neurotterocenosi del *Quercetus ilicis* e del *Populus nigra pyramidalis*. *Bollettino dell'Istituto di Entomologia 'Guido Grandi' dell'Università di Bologna* 39, 61–74.

Pantaleoni, R.A. (1990) I Neurotteri (Neuropteroidea) della Valle del Bidente-Ronco (Appennino Romagnolo). *Bollettino dell'Istituto di Entomologia 'Guido Grandi' dell'Università di Bologna* 44, 69–122.

Pantaleoni, R.A. (1996) Distribuzione spaziale di alcuni Neurotteri Planipenni su piante arboree. *Bollettino dell'Istituto di Entomologia 'Guido Grandi' dell'Università di Bologna* 50, 133–141.

Pantaleoni, R.A. & Curto, M.G. (1990) I Neurotteri delle colture agrarie: Crisopidi in oliveti del Salento (Italia meridionale). *Bollettino dell'Istituto di Entomologia 'Guido Grandi' dell'Università di Bologna* 45, 167–180.

Pantaleoni, R.A. & Ticchiati, V. (1990) I Neurotteri delle colture agrarie: esperienze sul metodo di campionamento per abbattimento chimico. *Bollettino dell'Istituto di Entomologia 'Guido Grandi' dell'Università di Bologna* 45, 143–155.

Pantaleoni, R.A., Lentini, A. & Delrio, G. (1993) Crisopidi in oliveti della Sardegna. Risultati preliminari. In *Atti del Convegno: Tecniche, Norme e Qualità in Olivicoltura*, (*Potenza 15–17 December 1993*), pp. 879–890. Università degli Studi della Basilicata, Potenza.

Paparatti, B. (1984) Bibliografia della *Saissetia oleae* (Olivier) e cenni bibliografici sui suoi principali parassitoidi e predatori. *Agricoltura Italiana* 1984, 77–115.

Principi, M.M. (1956) Contributi allo studio dei Neurotteri italiani. XIII. Studio morfologico, etologico e sistematico di un gruppo omogeneo di specie del Gen. *Chrysopa* Leach (*C. flavifrons* Brauer, *prasina* Burm. e *clathrata* Schn.). *Bollettino dell'Istituto di Entomologia dell'Università di Bologna* 21, 319–410.

Ramos, P., Campos, M. & Ramos, J.M. (1983) Present status of research on biological control of the olive moth in Spain. In *Entomophagous Insects and Biotechnologies against Olive Pests, Proceedings of the E.C. Experts' Meeting Entomophages and Biological Methods in Integrated Control in Olive Orchards* ed. Cavalloro, R. & Piavaux, A., pp. 127–144. A.A. Balkema, Rotterdam.

Ramos, P., Campos, M. & Ramos, J.M. (1987) Evolución del ataque de *Prays oleae* Bern. al fruto del olivo: I. Estudio de parámetros y sus relaciones. *Boletín de Sanidad Vegetal Plagas* 13, 129–142.

Ramos, P., Campos, M. & Ramos, J.M. (1988*a*) Evolutión del ataque de *Prays oleae* Bern. al fruto del olivo: II. Evolución de puestas, estabilización de parámetros y ecuaciones predictivas. *Boletín de Sanidad Vegetal Plagas* 14, 265–279.

Ramos, P., Campos, M. & Ramos, J.M. (1988*b*) Evolutión del ataque de *Prays oleae* Bern. al fruto del olivo: III. – Distributión y agregatión de puestas. *Boletín de Sanidad Vegetal Plagas* 14, 343–55.

Ramos, P. & Ramos, J.M. (1990) Veinte años de observaciones sobre la depredación oófaga en *Prays oleae* Bern. Granada (España), 1970, 1989. *Boletín de Sanidad Vegetal Plagas* 16, 119–127.

Sacchetti, P. (1990) Osservazioni sull'attività e sulla bio-etologia degli entomofagi di *Prays oleae* (Bern.) in Toscana. I. I predatori. *Redia* 73, 243–259.

Triggiani, O. (1971) La *Margaronia unionalis* (piralide dell'olivo). *Entomologica* 7, 29–47.

Viggiani, G. & Bianco, M. (1974) Ripercussioni dei trattamenti chimici contro *Saissetia oleae* (Oliv.), sull'entomofauna utile dell'olivo. *Bollettino del Laboratorio di Entomologia Agraria 'Filippo Silvestri' Portici* 31, 99–104.

Viggiani, G., Fimiani, P. & Bianco, M. (1973) Ricerca di un metodo di lotta integrata per il controllo della *Saissetia olea* (Oliv.). In *Atti Giornate Fitopatopatologiche*. pp. 251–259. Cooperativa Libraria Universitaria Editric, Bologna.

Viggiani, G., Pappas, S. & Tzoras, A. (1975) Osservazioni su *Saissetia oleae* (Oliv.) e i suoi entomofagi nell'isola di Corfù. *Bollettino del Laboratorio di Entomologia Agraria 'Filippo Silvestri' Portici* 32, 156–167.

Withycombe, C.L. (1923) Note on the economic value of the Neuroptera, with special reference to the Coniopterygidae. *Annals of Applied Biology* 9, 112–125.

Yayla, A. (1983) Antalya ili zeytin zararlilari ile dogal düsmanlarinin tesbiti üzerinde ön çalismalar. *Bitki Koruma Bülteni* 23, 188–206.

Yayla, A. (1984) Antalya ve çevresi zeytin agaçlarinda rastlanan faydali hepteropter'lerin taninmalari, konukçulari ve etkinlikleri üzerinde arastirmalar. *Zirai Mücadele Ve Zirai Karantina Genel Müdürlügü Arastirma Eserleri Serisi* 3, 34 pp.

Yinon, U. (1969) Food consumption of the armored scale lady-beetle *Chilocorus bipustulatus* (Coccinellidae). *Entomologia Experimentalis et Applicata* 12, 139–146.

CHAPTER 24

Lacewing occurrence in the agricultural landscape of Pianura Padana

R.A. Pantaleoni

24.1 INTRODUCTION

Techniques of manipulation of agroecosystems have an important role among the strategies of integrated pest management (IPM) (van den Bosch & Telford, 1964) and its successive developments, i.e. vegetational management (Altieri & Letourneau, 1982) or ecosystem management (Speight, 1983). This technique is a complex agronomic practice leading to diversification of the agroecosystem which creates a less suitable environment for the development of the phytophagous populations (Delucchi, 1997). In this context the natural enemies are augmented either by removing and mitigating adverse elements or by providing lacking requisites (DeBach, 1974).

It appears evident that to reach these aims more must be known about the ecology of the various guilds of useful, harmful, and innocuous insects or at least about the main species. Such knowledge must necessarily concern not only crops, but also all the surrounding territory in relation to it. Numerous species, in fact, spend only a part of their life in cultivated fields needing other habitats for activities such as nesting, reproduction, overwintering, or for simple refuge (Maini, 1995).

In relation to chrysopids and other lacewings, attention has been directed for years more towards the field release of artificially reared individuals (augmentation method) than towards the above aspects. Ridgway & Kinzer (1974) and Ridgway & Murphy (1984), reviewing the use of lacewings in biological control, only once mention environmental manipulations speaking almost exclusively about the food attractants. These techniques are now well developed but their results depend on the natural populations in neighbouring areas, and on which more information is also needed.

At the Istituto di Entomologia 'Guido Grandi',

University of Bologna, under the direction of Maria Matilde Principi, a research group worked for a long time on the use of chrysopids in biological control. This topic was linked to the implementation of the regional IPM programme in Emilia Romagna. The group operated along two main lines. First it was dedicated to testing and improving artificial rearing systems and to studying environmental factors, in particular the photoperiod, that regulate the development of the pre-imaginal stages and adult egg-laying rhythms (for a review see Principi, 1992, 1993). This research helped towards the setting-up of an insectary (Biolab, Cesena) to trade chrysopids (Celli *et al.*, 1991). The second research line regarded the chrysopid occurrence in ecosystems and agroecosystems of the southeastern Pianura Padana. This information had to provide the basic knowledge for future development of environmental manipulation techniques for crop protection. A review of this research carried out between 1979 and 1990 and possible practical implications are discussed in this work.

24.2 THE LANDSCAPE

24.2.1 Geography

The Pianura Padana (northern Italy) is approximately 46000 km^2 with a maximum length of 396 km and an average width of about 120 km. It is confined by the Alps and the Apennines and crossed in a west–east direction by the course of the River Po (the antique 'Padus', from which the plain's name is derived) (Gribaudi, 1956). The research was carried out in the southeast quadrant, about 8500 km^2, south of the watershed of the river Adige and east of the Panaro, the last tributary to the south of the Po (Fig. 24.1). This concerned the provinces of Bologna, Rovigo, Ferrara, Ravenna and Forlì. Approximately 1700000 people live

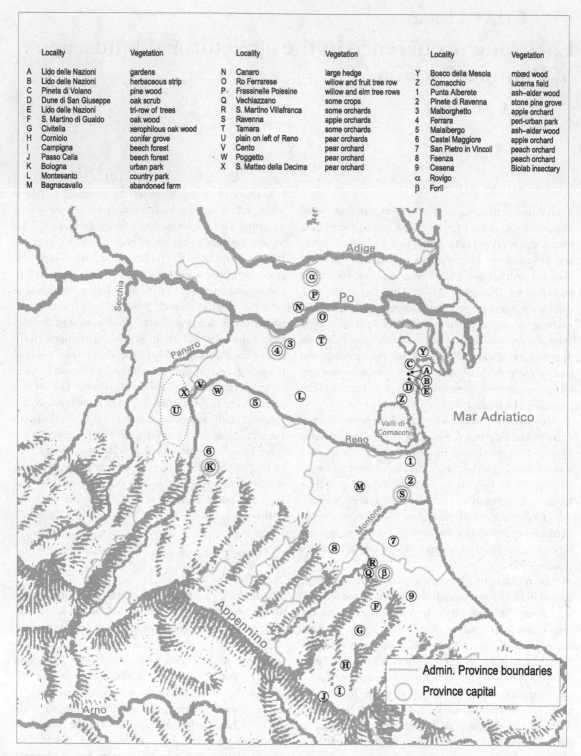

	Locality	Vegetation
A	Lido delle Nazioni	gardens
B	Lido delle Nazioni	herbaceous strip
C	Pineta di Volano	pine wood
D	Dune di San Giuseppe	oak scrub
E	Lido delle Nazioni	tri-row of trees
F	S. Martino di Gualdo	oak wood
G	Civitella	xerophilous oak wood
H	Corniolo	conifer grove
I	Campigna	beech forest
J	Passo Calla	beech forest
K	Bologna	urban park
L	Montesanto	country park
M	Bagnacavallo	abandoned farm

	Locality	Vegetation
N	Canaro	large hedge
O	Ro Ferrarese	willow and fruit tree row
P	Frassinelle Polesine	willow and elm tree rows
Q	Vechiazzano	some crops
R	S. Martino Villafranca	some orchards
S	Ravenna	apple orchards
T	Tamara	some orchards
U	plain on left of Reno	pear orchards
V	Cento	pear orchard
W	Poggetto	pear orchard
X	S. Matteo della Decima	pear orchard

	Locality	Vegetation
Y	Bosco della Mesola	mixed wood
Z	Comacchio	lucerna field
1	Punta Alberete	ash–alder wood
2	Pinete di Ravenna	stone pine grove
3	Malborghetto	apple orchard
4	Ferrara	peri-urban park
5	Malalbergo	ash–alder wood
6	Castel Maggiore	apple orchard
7	San Pietro in Vincoli	peach orchard
8	Faenza	peach orchard
9	Cesena	Biolab insectary
α	Rovigo	
β	Forlì	

Fig. 24.1. Map of southeastern Pianura Padana, with location of cited places.

in this area (ISTAT, 1990). The region is characterised by average annual temperatures of between 12°C and 14°C, January averages between 1°C and 4°C, July averages from 22°C to 24°C. Annual rainfall is on average between 600 mm and 900 mm and presents an autumnal maximum and a summer or winter minimum (Mori, 1956).

24.2.2 Potential vegetation

The whole of the Pianura Padana falls in the Middle-European bioclimatic zone. The vegetation belts in the southeastern plain and in the neighbouring Apennine sector are described by Pignatti (1980).

The three main forest formations of the area are herein briefly described (Pignatti, 1998). The Illyrian holm-oak woodland, *Orno-Quercetum ilicis*, is a wood of holm oak (always prevalent) mixed with flowering ash. Of the holm-oak formations it is the one that develops in the coolest and dampest climatic situations forming the point of contact between the evergreen and deciduous wood in the Italian peninsula. In the southeastern Pianura Padana it is limited to the well-stabilised coastal dune ridges. The oak–hornbeam forest, *Ornithogalo-Carpinetum*, is a mixed wood of English–oak and hornbeam with the presence of elm, hedge maple, ash, and other tree species. It is the climax of the Pianura Padana, where it is found in areas rich in water, but with well-drained soil, and also extends over the surrounding hills. Today it is reduced in the inner plain to a few examples of relict fragments. The ash–alder wood, *Carici remotae-Fraxinetum oxycarpae*, is a mixed wood of ash, elm, and poplar which grows along the rivers or in the depressions that have high groundwater level in spring. Once certainly widespread, they have now almost disappeared due to land reclamation and drainage.

Along the coast, in well-conserved local situations such as in the Bosco della Mesola, it is possible to record the serial and chain relations between the different forest formations. The consolidated dune is occupied by Illyrian holm-oak woodland, which is then substituted by oak–hornbeam forest; in the damp lowlands the ash–alder wood becomes established (Piccoli *et al.*, 1983).

The age of oak woods and the other broad-leaved tree formations, in the present form, goes back to about 5000 BC or slightly further, when in the Pianura Padana the wild-pine palaeoclimax was definitively substituted by the actual climax (Bertolani Marchetti, 1966/7, 1969/70).

The confining hilly areas and lower mountain slopes of the Apennines present a more complex mosaic of vegetation in which mixed oak woods predominate but with important floristic differences in respect to areas in the plain. In fact, here are found north Italian *Quercus cerris* woods, north Italian *Quercus pubescens* woods, hop–hornbeam woods and others. In the mountainous areas, above 800–1000 metres, beech forests and silver-fir forests grow (Ferrari, 1987; Pignatti, 1998).

24.2.3 Landscape evolution

The native ancient forest probably began being reduced by human activity in the Bronze Age between 2000 and 1000 BC, but with the Roman conquest vast wooded areas were transformed into cultivated fields. This deforestation was not interrupted until the Middle Ages between the 4th and 10th centuries when the population abandoned the countryside. The consequences on the land were damage and floods, contraction of cultivated land and the return of woods, waste land, and damp areas (Tomaselli & Tomaselli, 1973). Palynologic and stratigraphic analysis confirm a wood crisis in Roman times and a recovery in the Middle Ages (Bertolani Marchetti & Cupisti, 1970).

With the disastrous flood of the 'rotta di Ficarolo' (AD 1152) the geography of the southeastern Pianura Padana underwent catastrophic evolution with the changes in course to the north of the main branch of the Po, the transformation to swamp of the southern delta branches and most of its Apennine tributaries. Hence an enormous increase of wet areas, named 'valli', occurred in all of the low plain (Corbetta *et al.*, 1981). This event made land reclamation necessary and this was begun by the monastic communities in mediaeval times and continued up to the 1970s (Bondesan, 1990). Contemporarily a type of agricultural system, named 'piantata', was adopted, to enable an efficient drainage in territories with a very superficial groundwater level.

The 'piantata' system, known since antique times, consists of planting lines of trees holding up vines along the main side of the fields, usually maples, elms, and poplars. Often, two small ditches flanked the row of trees. The dimensions of the fields and the distance between the 'piantate' have increased over the centuries in relation to the work capacity of the ploughing systems. It is thought that in the Middle Ages the

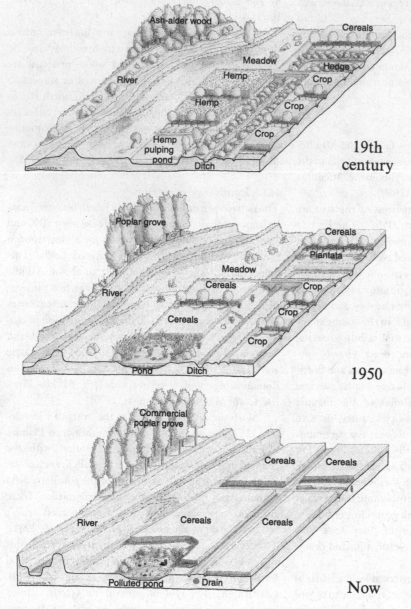

Fig. 24.2. Landscape evolution in the southeastern Pianura Padana. (Drawing by Massimo Saretta, slightly modified; from Agostini, 1993, by permission of '*Il divulgatore*' Bologna).

distance between rows was 6–7 m as against 38 m at the beginning of the 20th century. During this later period it reached its maximum extent and was present on more than 4500 km² (Agostini, 1993).

The advent of mechanisation almost completely eliminated the 'piantata'. Ditches, small fields, and rows of trees also represent obstacles for agricultural machinery (Giardini, 1977). This coincided with the rapid expansion of the maize monoculture or moreso

with the cultivation of cereals, sugar beet, and, more recently, soya-bean (Paoletti, 1985; Paoletti & Lorenzoni, 1989) (Fig. 24.2).

The southeastern Pianura Padana is now intensely cultivated and hedges, woods, and even isolated trees are rare if not entirely absent. The only patches of uncultivated vegetation are those that surround cities and towns (parks, gardens, avenues with trees, etc.) or along streets and canals. In some areas fruit groves, more or less alternated with annual crops, are extremely widespread. The natural areas, woods and wet zones ('valli') are concentrated along the coast. This, intensely urbanised for the sake of tourism, appears in some tracts as an uninterrupted stretch of gardens tens of kilometres in length and some hundreds of metres deep (Pirola & Chiusoli, 1976).

The old naturalised plantations of stone pine, on dune ridges along the coast, should also be mentioned amongst the non-agricultural environments (Pignatti, 1998).

24.3 LACEWING OCCURRENCE IN LANDSCAPE UNITS

24.3.1 Coastal areas

The occurrence of the lacewings present in five coastal habitats near the Valli di Comacchio was studied in the period 1979–83 using insect net samples for periods varying from one to three years (Table 24.1) (Pantaleoni, 1982, 1984).

Site A was made up of a group of gardens bordered by many ornamental shrub hedges with numerous trees, among which occur conifers (*Pinus*, Cupressaceae), and a lot of flower beds. Only in this site a New Jersey light trap was used contemporaneous with sweeping. Site B was made up of a strip of uncultivated land some metres deep along an irrigation canal with exclusively herbaceous vegetation. Site C was represented by a tract of pine wood aged about 60 years partly of marine pine (*P. pinaster*) (strip neighbouring the sea) and partly of stone pine (*P. pineae*) (strip lying behind), planted on dune ridges and laid upon a pre-existing holm-oak woodland and perhaps, in some places, hornbeam–oak forest. Site D was a modest strip of holm oak and pubescent-oak scrub on internal dunes. Lastly, Site E was composed of a triple row of Lombardy poplar (*Populus nigra pyramidalis*) placed along the edge of a lucerne field.

Sites A, B and E are found within urbanised areas and could be considered unnatural. Throughout the research they were to constitute a useful comparison with physiognomically similar agricultural environments. At the time, however, ecological conditions of the whole area were quite good with the absence of pesticide use and with horticultural intervention being limited to mowing and pruning.

The most abundant species were chrysopids.

Chrysopa formosa in particular predominately on grasses (site B) or shrubs (site A). *Dichochrysa prasina* was dominant, or at least very abundant, in the presence of sparse shrubs and trees (sites A, D, E). *Chrysoperla carnea* was confirmed to be a widely euryoecious species representing in each site about one-fifth of the total captures. *Chrysopa dorsalis* and *C. viridana* were, on the other hand, strongly influenced by the presence of their plant hosts: pine for the first, deciduous oak for the latter. *Chrysopa pallens* occurred rather irregularly. Various species of hemerobiids, *Wesmaelius subnebulosus*, *Hemerobius humulinus*, *Sympherobius pygmaeus*, were well spread but not very abundant anywhere. Among the coniopterygids *Coniopteryx esbenpeterseni* was the most abundant species in the gardens of site A, *Semidalis aleyrodiformis* on English oaks of site C and *C. borealis* on poplars of site E.

The seasonal trends of captures in general assumed a characteristic aspect, with a single spring maximum due mainly to the abundance of the species of the genus *Chrysopa*. *Chrysoperla carnea* showed a capture peak always later, sometimes, as in site A, even in autumn (Fig. 24.3).

Altogether in a territory of a few square kilometres 30 species were found belonging to the three preceding families: the 28 species listed in Table 24.1 plus *Conwentzia psociformis* and *Wesmaelius nervosus* found later (Pantaleoni, 1986).

Unfortunately, it has not been possible to study some coastal forest areas of great interest. Only fragmentary data are known from Bosco della Mesola: *Micromus angulatus*, *Chrysopa formosa*, *Chrysoperla carnea*, and *Dichochrysa* sp. by samples with Malaise trap in 1994/5 (Civico Museo di Storia naturale di Ferrara, unpublished data). In the Punta Alberete ash–alder wood *Chrysopa perla*, *C. pallens*, *Chrysoperla carnea*, and *D. prasina* were collected in a once-only net sample. The latter two species are present too in the Pinete di Ravenna and in the bordering areas (Pantaleoni, 1990b).

24.3.2 Hills and mountains

In order to obtain an overall picture, the Neuroptera taxocoenoses of the hilly and mountainous areas near to the southeastern Pianura Padana were also studied. After a year (1985) of preliminary faunistic research in just over 40 localities of the Apennines of Romagna, in 1986/7 five sites located along the altitudinal gradient

Table 24.1. *Relative abundance of species collected in coastal areas*

Relative abundance[a]	Site A[1] Gardens Lido delle Nazioni 1979–81 insect net	Site A[2] Gardens Lido delle Nazioni 1979–81 light trap	Site B Herbaceous grasses Lido delle Nazioni 1979–80 insect net	Site C Pine wood on scrub Pineta di Volano 1979–80 insect net	Site D Holm-oak formation Dune di S. Giuseppe 1982–3 insect net	Site E Poplar filare Lido delle Nazioni 1983 insect net
66%–100%	Chrysopa formosa	Chrysopa formosa	Chrysopa formosa			Dichochrysa prasina
33%–66%	Chrysoperla carnea	Chrysoperla carnea	Chrysoperla carnea	Chrysopa pallens Chrysopa formosa Chrysopa dorsalis Chrysoperla carnea	Chrysopa viridana Dichochrysa prasina	Chrysoperla carnea
16%–33%	Dichochrysa prasina				Chrysopa formosa Chrysoperla carnea Chrysopa pallens	
8%–16%	Dichochrysa flaviifrons	Dichochrysa flaviifrons	Chrysopa abbreviata	Dichochrysa flaviifrons		
4%–8%	Dichochrysa flaviifrons Coniopteryx esbenpeterseni	Wesmaelius subnebulosus Dichochrysa prasina		Semidalis aleyrodiformis	Sympherobius pygmaeus	Coniopteryx borealis Sympherobius pygmaeus Dichochrysa flaviifrons Chrysopa pallens
1%–4%	Sympherobius pygmaeus Chrysopa abbreviata Chrysopa dorsalis Wesmaelius subnebulosus Hemerobius micans Chrysopa pallens Coniopteryx haematica	Chrysopa abbreviata Chrysopa pallens Sympherobius pygmaeus Chrysopa dorsalis Micromus angulatus Hemerobius humulinus Hemerobius stigma		Hemerobius humulinus Hemerobius micans	Cunctochrysa baetica Dichochrysa flaviifrons Chrysopa abbreviata Coniopteryx arcuata	Chrysopa abbreviata Chrysopa formosa Chrysopa viridana Hemerobius humulinus Chrysopa dorsalis
max 1%	Hemerobius humulinus Semidalis pseudouncinata Micromus angulatus Semidalis aleyrodiformis Chrysopa viridana Cunctochrysa baetica Chrysopa nigricostata Coniopteryx borealis Coniopteryx tineiformis Coniopteryx arcuata			Chrysopa abbreviata Dichochrysa prasina Semidalis pseudouncinata Nineta flava	Hemerobius humulinus Hemerobius micans Hemerobius gilvus	Wesmaelius submebulosus Hemerobius handschini

Note:
[a] The species are in order of decreasing abundance.

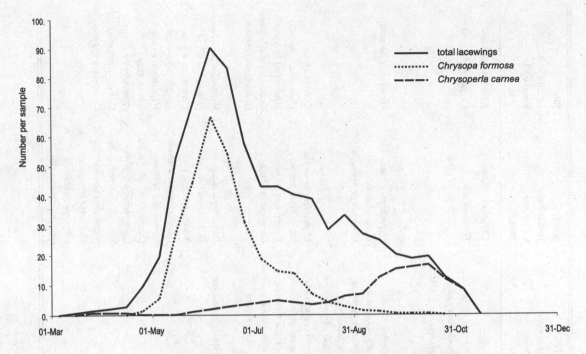

Fig. 24.3. Seasonal trends of captures by insect net in gardens of site A, 1979–80. (From Pantaleoni, 1986.)

of the Bidente–Ronco valley were studied using samples with insect net (Table 24.2) (Pantaleoni, 1988, 1990*a*).

Site F (135m a.s.l.) was one of the remaining strips of pre-Apennine oak wood with pubescent-oak predominance, but a few durmast oak. Site G (250–300 m a.s.l.) was a little valley with xerophilous pubescent–oak formations. Site H (500–550 m a.s.l.) was represented by a reforestation of conifers, mainly pine, but also fir and in lesser measure silver fir. Site I (1,000–1,075 m a.s.l.) was composed of a glade and of the surrounding beech forest. Lastly, site J (1,300–1,350 a.s.l.) was made up of the beech forest and glades of a brief tract of the Apennine ridge.

The data recorded can be summarised briefly as follows. The genus *Chrysopa* was always only rather uncommon. *Dichochrysa prasina* on the contrary reached high abundance in the oak woods at the lower altitudes (sites F, G) confirming its preference for sunny and breezy habitats with open vegetation of deciduous trees. *Chrysoperla carnea* was particularly common in the beech woods probably following seasonal phenomena of dispersion (see Pantaleoni, 1990*a*). Hemerobiids, in contrast to the coniopterygids, were

rather abundant, in particular *Hemerobius gilvus* in the oak woods, *H. handschini* in the conifer groves, and *H. micans* in the beech forests.

Altogether in the Apennines of Romagna, 51 species were found belonging to the coniopterygids, hemerobiids, and chrysopids (Pantaleoni, 1988).

24.3.3 Parks and low plain woods
Within the low plain areas the only scraps of uncultivated vegetation that have a sufficiently complex structure and a two-dimensional development are represented by parks and the wood fragments that have survived by chance.

The lacewings present in three parks were studied in 1989 by means of insect nets. The three sites taken into consideration were: K, the experimental garden of the Istituto di Entomologia 'Guido Grandi' of Bologna with a surface of about 3000 m^2, in an urban area, with vegetation kept in a natural state; L, a country park of 3–4 ha with ancient trees and thick undergrowth; M, a farm of about 10 ha, not cultivated for roughly 30 years and still presenting the traditional agricultural system with hedges and 'piantate' (Table 24.3) (Pantaleoni, 1995).

Table 24.2. Relative abundance of species collected by insect net in Apennine areas

Relative abundance[a]	Site F — Pre-Apennine oak wood, S. Martino di Gualdo, 1986/7	Site G — Xerophilous oak wood, Civitella, 1986/7	Site H — Conifer wood, Corniolo, 1986/7	Site I — Beech forest, Campigna, 1986/7	Site J — Beech forest, Passo Calla, 1986
33%–66%	Dichochrysa prasina	Dichochrysa prasina	Dichochrysa flavifrons Peyerimhoffina gracilis	Chrysoperla carnea Hemerobius micans	Hemerobius micans
16%–33%	Chrysoperla carnea		Hemerobius handschini		Chrysoperla carnea
8%–16%	Hemerobius gilvus	Hemerobius gilvus Chrysoperla carnea Dichochrysa zelleri			
4%–8%	Dichochrysa zelleri Semidalis aleyrodiformis	Chrysopa viridana Hemerobius micans Sympherobius pygmaeus	Hemerobius stigma Chrysoperla carnea Coniopteryx pygmaea Nineta pallida Semidalis aleyrodiformis	Hypochrysa elegans Hemerobius lutescens Coniopteryx pygmaea	Hemerobius lutescens
1%–4%	Chrysopa viridana Dichochrysa flavifrons Sympherobius pygmaeus Italochrysa italica Hemerobius micans Dichochrysa clathrata	Hemerobius humulinus Helicoconis pseudolutea Dichochrysa flavifrons Coniopteryx arcuata Hypochrysa elegans Semidalis aleyrodiformis	Conwentzia pineticola Helicoconis pseudolutea Dichochrysa prasina Chrysopa dorsalis Hemerobius micans	Hemerobius humulinus Dichochrysa prasina Hemerobius gilvus Chrysopa viridana	Chrysopa perla Hemerobius contumax Hemerobius humulinus Cunctochrysa albolineata Dichochrysa ventralis
max 1%	Chrysopa pallens Coniopteryx esbenpeterseni Hemerobius humulinus Chrysopa walkeri	Hemerobius handschini Coniopteryx borealis Cunctochrysa baetica Conwentzia pineticola Chrysopa dorsalis Chrysopa pallens Coniopteryx esbenpeterseni	Hemerobius humulinus	Dichochrysa flavifrons Dichochrysa ventralis Nineta pallida Wesmaelius subnebulosus Hemerobius contumax Cunctochrysa albolineata Coniopteryx tineiformis Sympherobius pellucidus Wesmaelius nervosus Micromus variegatus Chrysopa pallens Coniopteryx borealis	Wesmaelius subnebulosus Hemerobius gilvus Hemerobius stigma Sympherobius elegans Hypochrysa elegans Nineta flava Peyerimhoffina gracilis Dichochrysa prasina Coniopteryx pygmaea

Note:

[a] The species are in order of decreasing abundance.

Table 24.3. *Relative abundance of species collected by insect net in parks s.lat.*

Relative abundance[a]	Site K Urban park Bologna 1989	Site L Country park Montesanto 1989	Site M Abandoned farm Bagnacavallo 1989
33%–66%	*Conwentzia psociformis*	*Chrysopa formosa*	
16%–33%	*Semidalis aleyrodiformis*	*Coniopteryx esbenpeterseni*	*Coniopteryx esbenpeterseni* *Dichochrysa prasina* *Dichochrysa flavifrons*
8%–16%	*Chrysopa viridana* *Chrysoperla carnea*	*Hemerobius humulinus*	*Semidalis aleyrodiformis*
4%–8%	*Dichochrysa prasina* *Dichochrysa flavifrons*	*Chrysopa pallens* *Dichochrysa prasina*	*Chrysopa pallens* *Coniopteryx borealis*
1%–4%	*Hemerobius humulinus*	*Coniopteryx borealis* *Dichochrysa flavifrons* *Micromus angulatus* *Chrysoperla carnea* *Hemerobius micans* *Chrysopa perla*	*Chrysoperla carnea* *Coniopteryx haematica* *Hemerobius humulinus* *Hemerobius micans* *Chrysopa formosa*
max 1%	*Sympherobius pygmaeus* *Chrysopa pallens* *Coniopteryx esbenpeterseni* *Coniopteryx lentiae* *Hemerobius micans* *Wesmaelius subnebulosus* *Sympherobius luqueti* *Cunctochrysa baetica* *Nineta guadarramensis*	*Micromus variegatus*	

Note:

[a] The species are in order of decreasing abundance.

In the urban park in Bologna (a city near to the Apennine range) the lacewing species were particularly numerous: the 16 listed in Table 24.3 plus *Chrysopa formosa* already recorded for the same site (Principi, 1958; Pantaleoni & Letardi, 1998). The most abundant species were those associated with the deciduous oaks, such as *Conwentzia psociformis*, *Semidalis aleyrodiformis*, and *Chrysopa viridana*.

In the two rural parks (true 'islands' immersed in very ample, intensely cultivated areas) the number of species falls to just over ten. In site L the most abundant species was *C. formosa*, followed by *Coniopteryx esbenpeterseni* and *Hemerobius humulinus*. Species typically attached to oaks were not found, even though these trees were very numerous. On the other hand *S. aleyrodiformis* was abundant on the English oaks of site M, but the dominant species was *C. esbenpeterseni* together with *Dichochrysa prasina* and *D. flavifrons*.

The constant finding of preimaginal stages proved that, except in a few cases, all the found species occurred permanently in each site.

The odd sporadic captures of lacewings in the

large suburban parks of the city of Ferrara (R.A. Pantaleoni, unpublished data) provided, other than some of the more common species already named, the hemerobiid *H. perelegans*.

A series of samples on the riparian vegetation of the Po near Canaro (Rovigo) gave disappointing results with only two species found, *Chrysoperla carnea* and *D. flavifrons*, in very low numbers of individuals (Pantaleoni & Sproccati, 1987).

Unfortunately, for the few remaining plain woodlands, very little data is known. Results of research carried out in these environments by the Civico Museo di Storia naturale of Ferrara are still not completely available. In the 'Tenuta la Comune', near Malalbergo, *Wesmaelius subnebulosus*, *Micromus angulatus*, *H. humulinus*, and *Chrysopa perla* were found using a Malaise trap in 1993/4. Paoletti *et al.* (1989) studied an environment of this type in the northeastern Pianura Padana, the Lison wood, also giving data on lacewings, unfortunately identified only at genus level. On the marginal trees of the wood coniopterygids, *Hemerobius*, *Chrysopa*, *Chrysoperla*, and *Dichochrysa* were found.

24.3.4 Hedges and tree rows

The hedges or, in the territory studied, the rows of the 'piantata' practically develop linearly but their structure can be more or less complex in relation to the length, to the vegetation pattern, and to the management practices.

In 1982 in a belt of land on either side of the Po mainly growing maize, some hedges were studied using the usual samples by nets to find out the occurrence of lacewings (Pantaleoni & Sproccati, 1987). The sites studied were: N, large hedges about 100 m in length made up of willows, poplars, elms (some of large dimensions), and fruit trees grown wild again, flanking a small abandoned ditch overgrown with herbaceous vegetation; O, a row of willows along the access road and garden of a farm; P, rows of elms, willows, and fruit trees of two closely neighbouring farms.

The dimension and vegetation structure of the sites, as expected, strongly influenced the Neuroptera taxocoenoses. The number of species found in sites N, without doubt the most ample and varied, total altogether 12 belonging to three families, while those collected in site O and P are only 6 belonging to the chrysopids (Table 24.4). In all the sites *Chrysoperla*

carnea was easily the most abundant species always followed by *Chrysopa formosa*. Another three chrysopids were always present: *C. pallens*, *Dichochrysa flavifrons*, and *D. prasina*.

Further samples in a very small garden of a farmhouse in the same zone (Canaro) gave only three species, *C. formosa*, *Chrysoperla carnea*, and *D. flavifrons*.

In 1983, research was concentrated to within site N to determine the seasonal variations of chrysopid abundance, by means of samples with net and by beating, and of the aphid colonies, by means of direct inspection (Fig. 24.4) (Pantaleoni & Sproccati, 1987).

The chrysopids showed a secondary peak between the end of May and the beginning of June in which *Chrysopa formosa* was prevalent and a main peak in August, due particularly to *Chrysoperla carnea*. The data of net and beating captures were very similar. With this second method however adults were collected belonging only to the genera *Chrysoperla* and *Dichochrysa*; this brought about the almost complete absence of spring captures mainly made up of *Chrysopa*. There was no quantitative correlation between aphids and adult chrysopids: their curves appear completely unconnected. There was a relation between aphid abundance and chrysopid larvae captures, belonging almost exclusively to *C. formosa*.

24.3.5 Orchards

Fruit production represents one of the principal economic activities of Emilia Romagna. The first Italian IPM programme carried out on orchards was developed in this region so that available data for these environments is relatively abundant.

Excluding some vague citations by Golfari (1937, 1946, 1947), the first published data were those by Principi (1958) for an apple orchard at Castel Maggiore (Bologna) and then by Principi *et al.* (1967) and Castellari *et al.* (1967) for apple orchards at Malborghetto (Ferrara). Only four lacewings, in very low numbers, were recorded by inspection on aphid colonies: *Hemerobius humulinus*, *Chrysopa formosa*, *C. pallens*, and *Chrysoperla carnea*. It should be remembered that in those years organophosphate insecticides and DDT were heavily used to control pests.

Again Castellari (1980), with five years of research (1975–9) in peach orchards of Romagna (San Pietro in

Table 24.4. *Relative abundance of species collected by insect net on hedges*

Relative abundance[a]	Site N great hedge Canaro 1982/3	Site O willow and fruit-tree row Ro Ferrarese 1982	Site P willow and elm rows Frassinelle Polesine 1982
66%–100%		*Chrysoperla carnea*	
33%–66%	*Chrysoperla carnea*		*Chrysoperla carnea*
16%–33%	*Chrysopa formosa*	*Chrysopa formosa*	*Chrysopa formosa* *Dichochrysa prasina*
8%–16%	*Dichochrysa prasina* *Dichochrysa flavifrons*		*Dichochrysa flavifrons*
4%–8%	*Chrysopa pallens*		
1%–4%	*Chrysopa perla* *Hemerobius humulinus* *Coniopteryx borealis*	*Chrysopa pallens* *Dichochrysa prasina*	*Chrysopa pallens*
max 1%	*Coniopteryx haematica* *Semidalis aleyrodiformis* *Micromus angulatus*	*Dichochrysa flavifrons*	

Note:

[a] The species are in order of decreasing abundance.

Vincoli and Faenza, Ravenna), published a fundamental contribution to ecological knowledge of *Coniopteryx esbenpeterseni* occurring in large numbers as an efficient mite predator.

In 1979/80 Monari studied in apple orchards (site S) the occurrence of chrysopids and other predators in relation to the aphid infestations (Table 24.6) (data published partially in Pasqualini *et al.*, 1982). Sampling started at the end of April and finished in mid-July in 1979 and mid-August in 1980. Inspection of the aphid colonies found only a few larvae of *Chrysopa pallens*. *Chrysopa formosa*, *C. pallens*, *D. prasina*, and *Chrysoperla carnea* were collected with insect net. The latter was present only in the July–August period of 1981.

In the same period 1979/80 in the province of Forlì the chrysopid taxocoenoses of vineyards, peach, and apple orchards were studied (Table 24.5) (Pantaleoni & Tisselli, 1985). The vineyard data (site Q[3]) are the only ones relating to the southeastern Pianura Padana. In September there are contemporaneously the highest number of individuals and the highest number of species. In this month practically all the *Chrysopa viridana* were captured. In the peach orchard (site R[1]) four species were found, the same as those found by Monari (see above), while in the apple orchard (site R[2]) there were only two: *C. pallens* and *Chrysoperla carnea*. The latter represented about 90% of the capture in both orchards.

In 1984 in site T chemical knockdown samples were used in three orchards, apple, pear and peach, and contemporaneously the presence of chrysopid egg-laying and aphid colonies were examined by visual surveys (Table 24.6) (Pantaleoni & Ticchiati, 1988). The total species collected were only four, two hemerobiids and two chrysopids. The overlap between the curves of aphid infestation and the lacewing presence was almost nil. The Neuroptera were however strongly influenced by the pesticide treatments, still based on organophosphates, carried out on the crops (Fig. 24.5).

The following year chemical knockdown samples were used in two plots (apple and peach) to determine the spatial pattern of *C. carnea* in orchards (Pantaleoni

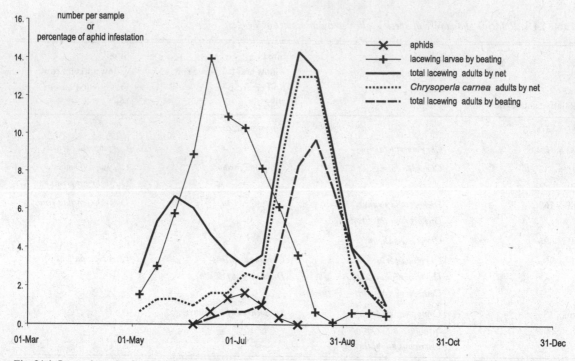

Fig. 24.4. Seasonal trends of aphid infestation and lacewing captures in hedge of site N, 1983. (From Pantaleoni & Sproccati, 1987.)

& Ticchiati, 1990). A clumped pattern was obtained only in the apple orchard and was probably due to the internal variability of the plot and not to the bio-ecological characteristics of the species. In agreement with Taylor's power law (Taylor, 1961, 1984) the aggregation index *b* for this chrysopid has in fact been calculated obtaining a particularly low value (1.11) near random (Pantaleoni & Ticchiati, 1990).

In 1987 in the area U, with samples collected by beating, research was carried out on predatory insects present in three groups of four pear orchards each. Group 'a' was mainly treated with several organophosphates and pyrethroids, group 'b' with Anziphosmethyl, and group 'c' with *Bacillus thuringiensis* and Diflubenzuron. Lacewings captured are listed in Table 24.7 (Nicoli *et al.*, 1988). Apparently the species found were much more numerous than in previous surveys, but, in fact, for each plot their number varied only from two to five. There was clearly a reduction in the total number of species, and also individuals, passing from group 'a' to group 'c'. *Chrysoperla carnea* was the most common species only in the pear orchards of group 'a'; in the other two groups, in relation to greater mite population densities, *Coniopteryx esbenpeterseni* repre-

sented 50% of the captures. The samples collected by beating however make it possible to record the presence of coniopterygids more efficiently than by net.

The preceding surveys were repeated in three other plots during 1990, in the plains area on the border of the provinces of Ferrara and Bologna. The principal differences with the 1987 studies were higher frequency of samples and choice of three orchards in which only low quantities of chemical pesticides were used, decreasing in the following order: site V, site W, site X (Table 24.7) (Marzocchi & Pantaleoni, 1995). In this case the number of species and individuals collected also increased with the reduction in pesticide use. Especially the hemerobiids reacted positively reaching in site X unusual high abundance for the plain. The seasonal trend of abundance of the coniopterygids presented a peak in August, that of the hemerobiids in July and October, and that of the chrysopids in October (Fig. 24.6).

Lastly, it is of interest to report for comparison the unpublished data coming from an untreated plot of apple orchard located to the extreme north of Pianura Padana (San Martino Buonalbergo, Verona) and regarding only the chrysopids collected by net in 1985

Table 24.5. *Relative abundance of species collected by insect net on some crops in the Forlì province*

Relative abundance[a]	Site Q¹ *Vicia faba minor* fodder crop Vecchiazzano 1979/80	Site Q² potato crop Vecchiazzano 1979/80	Site Q³ vineyard Vecchiazzano 1979/80	Site R¹ peach orchard S. Martino Villafranca 1979/80	Site R² apple orchard S. Martino Villafranca 1979/80
66%–100%		*Chrysoperla carnea*		*Chrysoperla carnea*	*Chrysoperla carnea*
33%–66%	*Dichochrysa prasina*		*Chrysoperla carnea*		
16%–33%	*Chrysopa formosa* *Chrysoperla carnea*		*Dichochrysa prasina*		
8%–16%		*Chrysopa formosa*	*Chrysopa formosa* *Chrysopa viridana*		*Chrysopa pallens*
4%–8%			*Chrysopa pallens*	*Dichochrysa prasina*	
1%–4%	*Dichochrysa clathrata* *Chrysopa pallens*		*Dichochrysa flavifrons* *Dichochrysa clathrata*	*Chrysopa formosa* *Chrysopa pallens*	

Note:

[a] The species are in order of decreasing abundance.

Table 24.6. *Relative abundance of species collected in some orchards*

Relative abundance[a]	Site S Apple orchards Ravenna 1979–80 insect net	Site T[1] Apple orchard Tamara 1984 chemical knockdown	Site T[2] Pear orchard Tamara 1984 chemical knockdown	Site T[3] peach orchard Tamara 1984 chemical knockdown	Outside Site Apple orchard S. Martino Buonalbergo 1985 insect net
66%–100%		*Chrysoperla carnea*	*Chrysoperla carnea*	*Chrysoperla carnea*	*Chrysopa perla*
33%–66%	*Chrysopa pallens*				
16%–33%	*Chrysopa formosa* *Chrysoperla carnea*			*Micromus angulatus*	*Chrysoperla carnea*
8%–16%	*Dichochrysa prasina*				*Chrysopa formosa*
4%–8%			*Micromus angulatus*	*Chrysopa pallens*	
1%–4%				*Hemerobius humulinus*	*Chrysopa pallens* *Dichochrysa prasina* *Chrysopa phyllochroma* *Dichochrysa flavifrons*
max 1%	*Micromus angulatus*	*Micromus angulatus*			

Note:

[a] The species are in order of decreasing abundance.

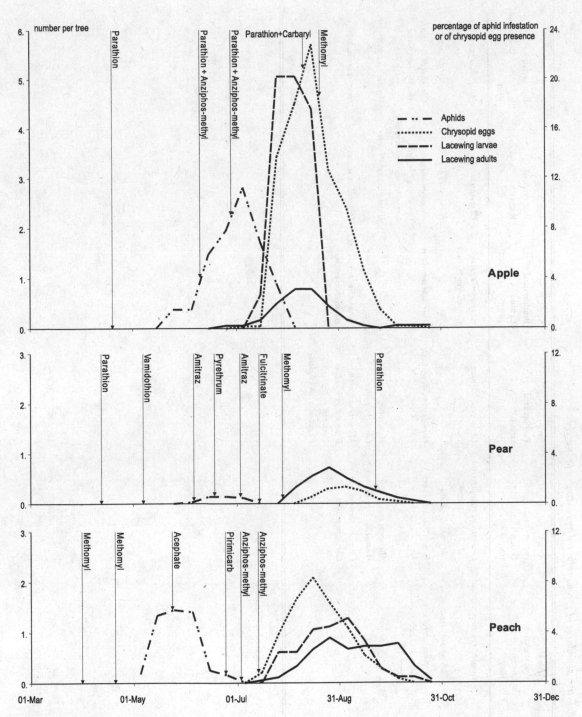

Fig. 24.5. Seasonal trends of aphid infestation, egg presence on twigs and lacewing captures by chemical knockdown in the orchards of site T, with indication of insecticide treatments, 1984. (From Pantaleoni & Ticchiati, 1988.)

Table 24.7. *Relative abundance of species collected by beating in pear orchards*

Relative abundance[a]	Area U Pear orchards Group a 1987	Area U Pear orchards Group b 1987	Area U Pear orchard Group c 1987	Site V Pear orchard Cento 1990	Site W Pear orchard Poggetto 1990	Site X Pear orchard S. Matteo della Decima 1990
33%–66%	*Coniopteryx esbenpeterseni*	*Coniopteryx esbenpeterseni*	*Chrysoperla carnea*	*Chrysoperla carnea* *Coniopteryx borealis*	*Chrysoperla carnea*	*Hemerobius humulinus*
16%–33%	*Chrysoperla carnea*	*Chrysoperla carnea*			*Hemerobius humulinus*	*Chrysoperla carnea*
8%–16%	*Micromus angulatus* *Coniopteryx haematica*	*Hemerobius humulinus*	*Coniopteryx esbenpeterseni*	*Micromus angulatus*	*Micromus angulatus* *Coniopteryx borealis* *Coniopteryx esbenpeterseni*	*Coniopteryx borealis* *Micromus angulatus*
4%–8%	*Chrysopa pallens*	*Micromus angulatus* *Wesmaelius subnebulosus*	*Hemerobius humulinus* *Chrysopa pallens*		*Wesmaelius subnebulosus*	*Wesmaelius subnebulosus*
1%–4%		*Chrysopa formosa* *Chrysopa pallens*	*Micromus angulatus* *Coniopteryx borealis* *Coniopteryx haematica* *Wesmaelius subnebulosus*	*Hemerobius humulinus* *Hemerobius micans*	*Hemerobius micans*	*Hemerobius micans*
max 1%						*Dichochrysa flavifrons*

Note:

[a] The species are in order of decreasing abundance.

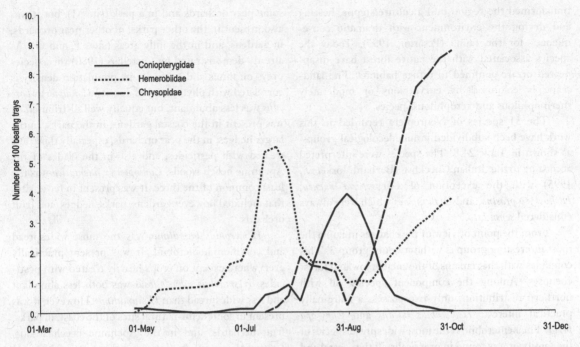

Fig. 24.6. Seasonal trends of lacewing captures (average of three plots) in pear orchards of sites V, W, and X, 1990. (From Marzocchi & Pantaleoni, 1995.)

(Table 24.6) (A. Biondani, unpublished data). Genus *Chrysopa* was present with four species out of seven. Among these *C. perla* represented almost two-thirds of all captures.

24.3.6 Crop fields

Data regarding lacewing occurrence on crop fields are limited to a very few specific studies, but fortunately various indirect information is also available.

Within the sphere of the research on crops in the province of Forlì (site Q), Pantaleoni & Tisselli (1985) studied the chrysopid taxocoenoses of *Vicia faba* var. *minor* (grown as a fodder crop) and of the potato (Table 24.5). A rather high number of species were found on the fodder crop, among others, those ecologically associated with tree and shrub *Chrysopa pallens*, *Dichochrysa clathrata*, and *D. prasina*. On the potato *Chrysoperla carnea* represented almost 90% of the captures.

Apart from a very high density of *Chrysopa formosa* recorded for a lucerne crop near Comacchio (Ferrara) (Pantaleoni, 1986), it is always and only *Chrysoperla carnea* that is cited as present on the crop fields, so too

in the northeastern Pianura Padana. Ragusa & Paoletti (1985) and Paoletti *et al.* (1989) found it on maize and soya-bean, Zandigiacomo (1985) on broad beans, and again Vidano & Arzone (1987) on maize in the western Pianura Padana. Unfortunately other records provide only general indications 'chrysopids', see for example Ciampolini *et al.* (1987). Lastly, in some cases such as in weekly inspection of 400 stems in three wheat plots near to the pear orchards of sites V, W, and X no lacewings were found (Marzocchi & Pantaleoni, 1995).

24.3.7 Faunistic analysis

It is very probable that in Neuroptera, as within other animal groups (Brandmayr, 1982), the fauna complex of lowland forest, especially oak–hornbeam forest including clearings and ash–alder wood, was composed of hygrocolous, mesophilous, and sciophilous species. In the southeastern Pianura Padana this outline is complicated by a rising from south towards north of Mediterranean floro-faunistic elements in two corridors placed along the coast and at the base of the Apennines (Zangheri, 1950; Contarini, 1995). The activities of people over the centuries have, however, transformed the region into a culture steppe, heating and drying the environment with dramatic consequences for the fauna (Pesarini, 1995). Today the species associated with the native forest have disappeared or are confined to refuge habitats. The landscape is colonised by euryoecious or moderately thermophilous and xerophilous species.

The 51 species of Neuroptera recorded in this work have been subdivided in nine 'ecological groups' as shown in Table 24.8. The species were interpreted according to the 'Italian Checklist' (Bernardi Iori *et al.*, 1995) with the exception of *Chrysoperla carnea*, *Dichocrysa prasina*, and *D. flavifrons* which were always considered *sensu lato*.

From the point of view of the changes in fauna the most interesting group is without doubt group 3. This coincides with the remains of the ancient lowland forest coenoses. Among the component species, all with northern distribution, only two possess a marginally practical interest: *Hemerobius micans* and *Chrysopa perla*. The hemerobiid is the most widespread species of the family; it was found in practically all the coastal and Apennine sites, in the parks, and in some pear orchards subjected to low chemical insecticide treatments. The chrysopid on the other hand was present not only in the remnant plain woods, in some parks, in larger hedges but also in the clearings of the Apennine ridge. It was probably the most abundant species on the 'piantate' as it is again the most abundant species in damp and fresh habitats of other sectors of Pianura Padana (Arzone, 1979; A. Biondani, personal communication).

Groups 5 and 6 are composed of strongly xerothermophilous Mediterranean species that occupy the coastal belt and the pre-Apennine hills. This latter area hosts a much greater number of species than the first.

The species attached to deciduous oaks (group 2) are mainly present along the coast and in the hilly oak woods. They penetrate to the inner plain following their own host plants confined in the parks and in the hedges. Near the hilly areas they can also sometimes move on to crops for brief periods (vineyard of site Q). However further away from the Apennine oak woods these species become more sporadic (see Table 24.3).

Eleven species quite regularly occur on the crops: three coniopterygids, three hemerobiids, and five chrysopids (group 1).

The most common coniopterygid was *Coniopteryx esbenpeterseni*. It was the most abundant lacewing in

some pear orchards and in a park (site M), but it was also present in the other parks, in other pear orchards, in gardens, and in the hilly areas (sites F, and G). As already demonstrated by Castellari (1980) this species preys on mites and normally its population density is correlated with phytophagous density. *Coniopteryx borealis* was less abundant, but equally well distributed. It was present in the coastal gardens, in the parks, in the larger hedges, in the pear orchards, especially those less treated with pesticides, and also in the glades of the Apennine beech woods. *Coniopteryx haematica* was the least common of the three. It was present in fewer sites that included however gardens, parks, hedges, and fruit orchards.

Hemerobius humulinus was the most widespread and common hemerobiid. It was present practically everywhere except on crops heavily treated with pesticides. *Wesmaelius subnebulosus* was both less abundant and less widespread than *H. humulinus*. However, it was present in some semi-natural sites of the coast, in parks, fruit orchards, and in the Apennine beech woods. Lastly, *Micromus angulatus* was the most widespread hemerobiid on the crops. Unlike *H. humulinus* this species can in fact exploit some aspects of its ethology, such as in particular its ability to develop on low vegetation and to overwinter as adult, which enables an easier and more widespread colonisation of the culture steppe.

Chrysopa formosa was clearly the most abundant species in the coastal sites and in a park (site L), and was still very abundant on the crops bordering the hilly areas (site Q), in the parks, in the hedges, and it was also present in some fruit orchards. This aphidophagous chrysopid is potentially able to develop on low vegetation and could be the most abundant lacewing in the southeastern Pianura Padana. However the spinning of the cocoon a few centimetres below the soil surface (Principi, 1947) means it does not survive ploughing and this limits its range enormously.

Chrysopa septempunctata, although never very abundant, was widespread and present almost everywhere. Unlike the previous species it is associated with trees and shrubs (Principi, 1940). Strictly aphidophagous and with a good flight capacity it is able to colonise even isolated fruit orchards quite easily.

Chrysoperla carnea is an *r*-strategy species, able to colonise every habitat even if only temporarily suitable for its development [refer to Duelli (1984*a, b*) on the

Table 24.8. *Ecological groups of lacewings recorded in this research*

Group	Species	Notes
1	*Coniopteryx borealis* Tjeder, 1930 *Coniopteryx haematica* McLachlan, 1868 *Coniopteryx esbenpeterseni* Tjeder, 1930 *Hemerobius humulinus* Linnaeus, 1758 *Wesmaelius subnebulosus* (Stephens, 1836) *Micromus angulatus* (Stephens, 1836) *Chrysopa formosa* Brauer, 1850 *Chrysopa pallens* (Rambur, 1838) *Chrysoperla carnea* (Stephens, 1836) *s. lat.* *Dichochrysa flavifrons* (Brauer, 1850) *s. lat.* *Dichochrysa prasina* (Burmeister, 1839) *s. lat.*	Euryoecious species, present with a certain regularity on the crops of the plain
2	*Conwentzia psociformis* (Curtis, 1834) *Semidalis aleyrodiformis* (Stephens, 1836) *Sympherobius pygmaeus* (Rambur, 1842) *Chrysopa viridana* Schneider, 1845	Associated with deciduous oaks
3	*Coniopteryx tineiformis* Curtis, 1834 *Hemerobius micans* Olivier, 1792 *Wesmaelius nervosus* (Fabricius, 1793) *Micromus variegatus* (Fabricius, 1793) *Chrysopa perla* (Linnaeus, 1758) (*sensu* Schneider, 1851) *Nineta flava* (Scopoli, 1763)	Hygrocolous and sciophilous species present in a few refuge habitats of the plain and in beech woods of the Apennine ridge
4	*Hemerobius lutescens* Fabricius, 1793 *Sympherobius elegans* (Stephens, 1836) *Cunctochrysa albolineata* (Killington, 1935) *Dichochrysa ventralis* (Curtis, 1834) *Hypochrysa elegans* (Burmeister, 1839)	Species found exclusively in beech woods of the Apennine ridge
5	*Coniopteryx arcuata* Kis, 1965 *Hemerobius gilvus* Stein, 1863 *Cunctochrysa baetica* (Hölzel, 1972)	Xerothermophilous Mediterranean species present in the coastal belt and in the pre-Apennine hills
6	*Helicoconis pseudolutea* Ohm, 1965 *Italochrysa italica* (Rossi, 1790) *Chrysopa walkeri* McLachlan, 1893 *Dichochrysa clathrata* (Schneider, 1845) *Dichochrysa zelleri* (Schneider, 1851)	Xerothermophilous Mediterranean species present only in the pre-Apennine hills
7	*Conwentzia pineticola* Enderlein, 1905 *Semidalis pseudouncinata* Meinander, 1963 *Hemerobius handschini* Tjeder, 1957 *Hemerobius stigma* Stephens, 1836 *Chrysopa dorsalis* Burmeister, 1839	Species confined to conifers (mainly *Pinus* and Cupressaceae) present in the coastal belt and in the Apennine areas
8	*Coniopteryx pygmaea* Enderlein, 1906 *Hemerobius contumax* Tjeder, 1932 *Sympherobius pellucidus* (Walker, 1853)	Species confined to conifers (mainly *Abies* and *Picea*) present only in the Apennine areas

Table 24.8 (*cont.*)

Group	Species	Notes
	Nineta pallida (Schneider, 1845)	
	Peyerimhoffina gracilis (Schneider, 1851)	
9	*Coniopteryx lentiae* Aspöck & Aspöck, 1964	Species with little-known ecology or linked to particular
	Hemerobius perelegans Stephens, 1836	biogeographical distributions
	Sympherobius luqueti Leraut, 1991	
	Chrysopa abbreviata Curtis, 1834	
	Chrysopa nigricostata Brauer, 1850	
	Chrysopa phyllochroma Wesmael, 1841	
	Nineta guadarramensis principiae Monserrat, 1980	

eco-ethology of this species]. This chrysopid was present in all the sites studied. Its relative abundance reached very high levels in the more degraded environments, such as crops heavily treated with pesticides, but also in hedges, and stayed much lower in natural or semi-natural areas and in the parks.

The two *Dichochrysa* species have, as far as we know, similar ecological requirements. *Dichochrysa prasina* was the most abundant lacewing in open natural habitats (sites E, F, G). It was also common in gardens, parks, hedges, and some crops. *Dichochrysa flavifrons* on the other hand was slightly less widespread and abundant. For both of these the critical vital period for survival in agroecosystems is overwintering as free larvae (Principi, 1940, 1956). Winter insecticide treatments, which are carried out on the trunks of fruit trees and frequently adopted in Emilia Romagna, in fact greatly affect the larvae sheltering in the cracks of the bark.

Neuroptera taxocoenoses decreased in richness from coastal and Apennine territories to the semi-natural areas of the inner plain (parks, hedges, wood remnants) and from these to the crops. Only some pear orchards in which a very advanced IPM was applied (site W, and X) showed a slightly contrary trend. Human activities are the cause of alteration of habitats suitable for many species, especially the use of pesticides. The effects of the heavy insecticide treatments are clearly demonstrated by the results obtained in the fruit orchards in site T (Fig. 24.5).

The habitat characteristics, such as dimensions and distance from the Apennine oak-wood area, would appear to have a great influence on the distribution of the species associated with deciduous oaks (group 3).

Nevertheless there are insufficient data to enter the debate on the equilibrium theory of insular biogeography (MacArthur & Wilson, 1963) and on its applicability to pest management schemes (Price, 1976; Rey & McCoy, 1979; Liss *et al.*, 1986; Duelli, 1988).

In the coastal gardens (site A) the seasonal trend of lacewing captures showed a spring maximum prevalently due to *Chrysopa formosa* (Fig. 24.3), in the large hedge (site N) this maximum becomes a secondary peak (Fig. 24.4), while it disappears altogether in the pear orchards (sites V, W, and X) (Fig. 24.6). The tendency of this chrysopid, or of *C. perla* which substitutes it in some situations, to disappear is of high practical importance. These species have a life cycle correlated in some way with the density of aphids, which are their preferred prey. Almost 70% of annual captures are concentrated in fact in May and June. About two-thirds of the first larval brood go into diapause and overwinter as prepupae while the remaining one-third emerge giving life in the same year to a second generation which can be followed by a third (Principi, 1947; Pantaleoni, 1982).

At the same time as the decrease in *C. formosa* there is a relative increase of *Chrysoperla carnea* which shows a seasonal maximum in late summer or in autumn. However this trend is characteristic of habitats with trees, such as gardens, hedges, and orchards. This chrysopid in fact colonises the crop fields more quickly, when environmental conditions are still favourable for development. It then moves on after the harvest towards hedges or fruit orchards. The arrival from the crop fields anticipates the search for and the reaching of overwintering sites that almost certainly are chosen within the same orchards, which represent in many areas of southeastern Pianura Padana the only woody vegetation.

The migration phenomena of entomophages from and towards hedges, wood remnants, spontaneous vegetation, and orchards of the Pianura Padana were pointed out by Paoletti & Lorenzoni (1989), Nicoli & Marzocchi (1993), Marzocchi & Pantaleoni (1995), and others. They always concern species overwintering at the adult stage such as ladybirds and flower bugs. Of the lacewings, other than *C. carnea*, *Micromus angulatus* also appears to have this behaviour (L. Marzocchi, personal communication).

24.4 CONCLUSIONS

How many lacewings lived on crops of the Pianura Padana at the time of maximum diffusion of the 'piantata'? Of course, there is no way of knowing for sure, but it is possible that there were at least 20 corresponding to the first three groups of Table 24.8. This number is not very different from that recorded in the coastal gardens (site A) where the vegetation has a similar physiognomy to that of the 'piantata' systems.

According to Pimentel *et al.* (1989) one of the principles that underlie a low-input sustainable agriculture is adapting the agricultural system to the environment of the region, including soil, water, climate, and biota present at the site. In this context traditional agricultural practices can provide a solid starting-point for modern pest control techniques of agroecosystem manipulation. In the southeastern Pianura Padana it might not be possible to go back to the 'piantata' but the testing of techniques such as minimum-tillage, no-tillage, intercropping of perennial and annual crop with wild plants, or intercropping of two or several crops has now become necessary.

In the agroecosystems of Pianura Padana the absolute priority should be the restoration of minimal survival conditions for *Chrysopa formosa*, or *C. perla*. Even without allocating a part of territory to non-productive functions creating uncultivated strips, a good result could be obtained by adopting minimum-tillage or no-tillage techniques in some annual crops (Paoletti, 1983). The two chrysopids could permanently colonise the fields especially if there are neighbouring reservoir areas. Their cocoon-spinning under the surface of the ground might also enable integrated management with some herbicides without damaging lacewing populations.

The restoration of the hedges and other woody areas would seem to be another necessary practice. These habitats would act as winter refugia for species overwintering as adults and would maintain a greater species diversity in the landscape.

ACKNOWLEDGEMENTS

The author wishes to thank A. Biondani and L. Marzocchi for provision of data.

REFERENCES

Agostini, N. (1993) Alberi, siepi e maceri. *Il divulgatore*, Bologna 16(6).

Altieri, M.A. & Letourneau, D.K. (1982) Vegetation management and biological control in agroecosystems. *Crop Protection* 1, 405–430.

Arzone, A. (1979) Indagini sui limitatori naturali di *Psylla pyri* (L.) in Piemonte. *Bollettino del Laboratorio di Entomologia Agraria 'Filippo Silvestri' Portici* 36, 131–149.

Bernardi Iori, A., Kathirithamby, J., Letardi, A., Pantaleoni, R.A. & Principi, M.M. (1995) Neuropteroidea (Megaloptera, Raphidioptera, Planipennia), Mecoptera, Siphonaptera, Strepsiptera. In *Checklist delle specie animali italiane*, vol. 62, ed. Minelli, A., Ruffo, S. & La Posta, S., pp. 1–20. Calderini Editore, Bologna.

Bertolani Marchetti, D. (1966/7) Vicende climatiche e floristiche dell'ultimo glaciale e del postglaciale in sedimenti della laguna Veneta. *Memorie di Biogeografia Adriatica* 7, 193–225.

Bertolani Marchetti, D. (1969/70) Climax e paleoclimax della pianura padano-veneta. *Memorie di Biogeografia Adriatica* 8, 69–77.

Bertolani Marchetti, D. & Cupisti, M. (1970) Aspetti della vegetazione postglaciale nel Modenese. Analisi polliniche in una trivellazione al Collegio Universitario di Modena. *Emilia Preromana* 6, 185–194.

Bondesan, M. (1990) L'area deltizia padana: caratteri geografici e geomorfologici. In *Il Parco del Delta del Po. Studi ed immagini. L'ambiente come risorsa*, pp. 9–48. Spazio Libri Editore, Ferrara.

van den Bosch, R. & Telford, A.D. (1964) Environmental modification and biological control. In *Biological Control of Insect Pests and Weeds*, ed. DeBach, P., pp. 459–488. Chapman & Hall, London.

Brandmayr, P. (1982) Lineamenti principali del paesaggio zoocenotico della pianura padano-veneta: passato e presente. In: *Quaderni sulla 'Struttura delle zoocenosi terrestri'*, vol. 4 *I boschi primari della pianura padano-veneta*, pp. 137–149. Consiglio Nazionale delle Ricerche, Rome.

Castellari, P.L. (1980). Indagini biologiche su *Coniopteryx (Metaconiopteryx) esbenpeterseni* Tjeder (Neur. Coniopterygidae), predatore di Acari Tetranichidi sul Pesco. *Bollettino dell'Istituto di Entomologia dell'Università di Bologna* 35, 157–180.

Castellari, P.L., Giunchi, P. & Principi, M.M. (1967) Problemi riguardanti la difesa del Melo dalle infestazioni di alcune specie di Afidi. In *Atti Giornate Fitopatopatologiche*, pp. 309–320. Cooperativa, Libraria Universitaria Editrice. Bologna.

Celli, G., Maini, S. & Nicoli, G. (1991) *La fabbrica degli insetti. Più insetti e meno pesticidi per una nuova agricoltura*. Franco Muzzio Editore, Padua.

Ciampolini, M., Grossi, A. & Zottarelli, G. (1987) Danni alla soia per attacchi di *Metcalfa pruinosa*. *L'Informatore agrario* 43, 101–103.

Contarini, E. (1995) L'influsso climatico mediterraneo sui popolamenti a Coleotteri della Padania (s. lat.) orientale. *Quaderni della Stazione Ecologica del Civico Museo di Storia Naturale di Ferrara* 9, 229–242.

Corbetta, F., Zanutti Censoni, A.L., & Zarrelli, R. (1981) Antropizzazione e depauperamento floristico-vegetazionale nella 'Bassa' bolognese. *Archivio Botanico e Biogeografico Italiano* 57, 113–132.

DeBach, P. (1974). *Biological Control by Natural Enemies*. Cambridge University Press, Cambridge.

Delucchi, V. (1997) Una nuova frontiera: la gestione ambientale come prevenzione. In *Atti della Giornata sulle strategie bio-ecologiche di lotta contro gli organismi nocivi*, pp. 35–57. Istituto di Ricerca per il Controllo Biologico dell'Ambiente (Consiglio Nazionale delle Ricerche) & Istituto di Entomologia Agraria dell'Università di Sassari, Sassari, Italy.

Duelli, P. (1984a) Dispersal and oviposition strategy in *Chrysoperla carnea* (Steph.). In: *Progress in World's Neuropterology, Proceedings of the 1st International Symposium on Neuropterology*, ed. Gepp, J., Aspöck, H. & Hölzel, H., pp. 133–146. Thalerhof, Graz.

Duelli, P. (1984b) Flight, dispersal, migration. In *Biology of Chrysopidae*, ed. Canard, M, Séméria, Y. & New, T.R., pp. 110–116. Dr. W. Junk, The Hague.

Duelli, P. (1988) Aphidophaga and the concepts of island biogeography in agricultural areas. In *Ecology and Effectiveness of Aphidophaga*, ed. Niemczyk, E. & Dixon, A.F.G., pp. 89–93. SPB Academic Publishing, The Hague.

Ferrari, C. (1987) La vegetazione forestale dell'Emilia-Romagna. In *I boschi dell'Emilia-Romagna*, ed. Bagnaresi, U. & Ferrari, C., pp. 61–120. Regione Emilia Romagna, Bologna.

Giardini, L. (1977) *Agronomia generale*. Patron Editore, Bologna.

Golfari, L. (1937) Contributi alla conoscenza dell'Entomofauna del Pero (*Pirus communis* L.) I. *Bollettino dell'Istituto di Entomologia dell'Università di Bologna* 9, 206–249.

Golfari, L. (1946) Etologia ed ecologia dell'Afide farinoso del Pesco, *Hyalopterus arundinis* Fabr. *Bollettino dell'Istituto di Entomologia della R. Università di Bologna* 15, 129–170.

Golfari, L. (1947) Appunti sulla etologia di alcuni Afidi del Pesco (*Anuraphis schwartzi* Börn., *Anuraphis persicae-niger* Smith, *Myzus persicae* Sulz.). *Bollettino dell'Istituto di Entomologia dell'Università di Bologna* 16, 115–128.

Gribaudi, D. (1956) Monti e pianure. Le Alpi e la Padánia. In *L'Italia fisica*, ed. Sestini, A., pp. 169–206. Touring Club Italiano, Milano.

ISTAT Istituto centrale di statistica (1990) *Comuni, comunità montane, regioni agrarie al 31 dicembre 1988. Codici e dati strutturali*. Istituto Centrale di Statistica, Rome.

Liss, W.J., Gut, L.J., Westigard, P.H. & Warren. C.E. (1986) Perspectives on arthropod community structure, organization, and development in agricultural crops. *Annual Review of Entomology* 31, 455–478.

MacArthur, R.H. & Wilson, E.O. (1963) An equilibrium theory of insular zoogeography. *Evolution* 17, 373–387.

Maini, S. (1995) Rimboschimenti e siepi nelle aree agricole: positiva influenza sull'entomofauna utile. *L'Informatore Fitopatologico* 45, 13–17.

Marzocchi, L. & Pantaleoni, R.A. (1995) Indagine sui principali entomofagi predatori (Insecta Heteroptera, Neuroptera et Coleoptera) in pereti della Pianura Padana. *Bollettino dell'Istituto di Entomologia 'Guido Grandi' dell'Università di Bologna* 49, 21–40.

Mori, A. (1956) Il clima. In *L'Italia fisica*, ed. Sestini, A., pp. 21–63. Touring Club Italiano, Milano.

Nicoli, G. & Marzocchi, L. (1993) Valorizzazione di insetti predatori ai fini della lotta naturale. In *Biodiversità negli agroecosistemi*, ed. Paoletti, M.G., Favretto, M.R., Nasolini, T., Scaravelli, D. & Zecchi, G., pp. 81–95. Osservatorio Agroambientale, Cesena.

Nicoli, G., Corazza, L., Cornale, R. & Marzocchi, L. (1988) Indagine sugli insetti predatori in pereti a diversa gestione fitoiatrica. In *Atti XV Congresso Nazionale Italiano Entomologia*, pp. 489–496. Accademia Nazionale Italiana di Entomologia, L'Aquila.

Pantaleoni, R.A. (1982) Neuroptera Planipennia del comprensorio delle Valli di Comacchio: indagine ecologica. *Bollettino dell'Istituto di Entomologia dell'Università di Bologna*, 37, 1–73.

Pantaleoni, R.A. (1984) Neuroptera Planipennia del comprensorio delle Valli di Comacchio: le neurotterocenosi del *Quercetum ilicis* e del *Populus nigra pyramidalis. Bollettino dell'Istituto di Entomologia 'Guido Grandi' dell'Università di Bologna* 39, 61–74.

Pantaleoni, R.A. (1986) I Neurotteri delle colture agrarie. Aspetti generali e stato delle ricerche nella Pianura Padana sud-orientale. *Dimensione Ambiente* (Ferrara), 19–20, 17–33.

Pantaleoni, R.A. (1988) La Neurotterofauna dell'Appennino Romagnolo. In *Atti XV Congresso Nazionale Italiano Entomologia*. pp. 633–640. Accademia Nazionale Italiana di Entomologia, L'Aquila.

Pantaleoni, R.A. (1990*a*) I Neurotteri (Neuropteroidea) della Valle del Bidente-Ronco (Appennino Romagnolo). *Bollettino dell'Istituto di Entomologia 'Guido Grandi' dell'Università di Bologna* 44, 69–122.

Pantaleoni, R.A. (1990*b*) I Neurotteri (Insecta Neuropteroidea) delle collezioni 'Zangheri' (Museo di Storia Naturale della Romagna) e 'Malmerendi' (Museo Civico di Scienze Naturali di Faenza). *Bollettino del Museo Civico di Storia Naturale di Verona* 17, 277–291.

Pantaleoni, R.A. (1995) Neurotteri (Insecta Neuropteroidea) della Pianura Padana: i parchi urbani e rurali come zone di 'rifugio faunistico'. *Quaderni della Stazione Ecologica del Civico Museo di Storia Naturale di Ferrara* 9, 393–397.

Pantaleoni, R.A. & Letardi, A. (1998) I Neuropterida della collezione dell'Istituto di Entomologia 'Guido Grandi' di Bologna. *Bollettino dell'Istituto di Entomologia 'Guido Grandi' dell'Università di Bologna* 52, 15–45.

Pantaleoni, R.A. & Sproccati, M. (1987) I Neurotteri delle colture agrarie: studi preliminari circa l'influenza di siepi ed altre aree non coltivate sulle popolazioni di Crisopidi. *Bollettino dell'Istituto di Entomologia 'Guido Grandi' dell'Università di Bologna* 42, 193–203.

Pantaleoni, R.A. & Ticchiati, V. (1988) I Neurotteri delle colture agrarie: osservazioni sulle fluttuazioni stagionali di popolazione in frutteti. *Bollettino dell'Istituto di Entomologia 'Guido Grandi' dell'Università di Bologna* 43, 43–57.

Pantaleoni, R.A. & Ticchiati, V. (1990) I Neurotteri delle colture agrarie: esperienze sul metodo di campionamento per abbattimento chimico. *Bollettino dell'Istituto di Entomologia 'Guido Grandi' dell'Università di Bologna* 45, 143–155.

Pantaleoni, R.A. & Tisselli, V. (1985) I Neurotteri delle colture agrarie: rilievi sui Crisopidi in alcune coltivazioni del forlivese. *Bollettino dell'Istituto di Entomologia 'Guido Grandi' dell'Università di Bologna* 40, 51–65.

Paoletti, M.G. (1983) La pedofauna dell'agroecosistema a mais. Prospettive di ricerca e sviluppo con le tecniche di minima lavorazione e non lavorazione. In *Atti XII Congresso Nazionale Italiano Entomologia*, pp. 481–489. Accademia Nazionale Italiana di Entomologia, Turin.

Paoletti, M.G. (1985) La pedofauna al passaggio dal bosco alla monocoltura nella bassa pianura veneta. In *Atti XIV Congresso Nazionale Italiano Entomologia*, pp. 459–461. Accademia Nazionale Italiana di Entomologia, Palermo.

Paoletti, M.G., & Lorenzoni, G.G. (1989) Agroecology patterns in northeastern Italy. *Agriculture, Ecosystems and Environment* 27, 139–154.

Paoletti, M.G., Favretto, M.R., Ragusa, S. & zur Strassen, R. (1989) Animal and plant interactions in the agroecosystems: the case of woodland remnants in northeastern Italy. *Ecology International Bulletin* 17, 79–91.

Pasqualini, E., Briolini, G., Memmi, M. & Monari, S. (1982) Prove di lotta guidata contro gli Afidi del Melo. *Bollettino dell'Istituto di Entomologia dell'Università di Bologna* 36, 159–171.

Pesarini, F. (1995) Il popolamento animale della Padania. Stato delle conoscenze e problemi irrisolti. *Quaderni della Stazione Ecologica del Civico Museo di Storia Naturale di Ferrara* 9, 21–33.

Piccoli, F., Gerdol, R. & Ferrari, C. (1983) Carta della vegetazione del Bosco della Mesola (Ferrara). *Atti dell'Istituto di Botanica e del Laboratorio Crittogamico dell'Università di Pavia*, series 7 (2), 3–23.

Pignatti, S. (1980) I piani di vegetazione in Italia. *Giornale Botanico italiano* 113, 411–428.

Pignatti, S. (1998) *I boschi d'Italia. Sinecologia e biodiversità.* Unione Tipografico-Editrice Torinese, Turin.

Pimentel, D., Culliney, T.W., Buttler, I.W., Reinemann, D.J. & Beckman, K.B. (1989) Low-input sustainable agriculture using ecological management practices. *Agriculture, Ecosystems and Environment* 27, 3–24.

Pirola, A. & Chiusoli, A. (1976) La carta fisionomico-strutturale della vegetazione. In *Carta delle vocazioni faunistiche della regione Emilia-Romagna*, pp. 65–105. Regione Emilia Romagna, Bologna.

Price, P.W. (1976) Colonization of crops by arthropods: non-equilibrium communities in soybean fields. *Environmental Entomology* 5, 605–611.

Principi, M.M. (1940) Contributi allo studio dei Neurotteri

Italiani. I. *Chrysopa septempunctata* Wesm. e *Chrysopa flavifrons* Brauer. *Bollettino dell'Istituto di Entomologia della R. Università di Bologna* 12, 63–144.

Principi, M.M. (1947) Contributi allo studio dei Neurotteri Italiani. V. Ricerche su *Chrysopa formosa* Brauer e su alcuni suoi parassiti. *Bollettino dell'Istituto di Entomologia dell'Università di Bologna* 16, 134–175.

Principi, M.M. (1956) Contributi allo studio dei Neurotteri italiani. XIII. Studio morfologico, etologico e sistematico di un gruppo omogeneo di specie del Gen. *Chrysopa* Leach (*C. flavifrons* Brauer, *prasina* Burm. e *clathrata* Schn.). *Bollettino dell'Istituto di Entomologia dell'Università di Bologna* 21, 319–410.

Principi, M.M. (1958) Le selettività degli insetticidi sistemici. Ripercussioni dei trattamenti con 'Systox' e 'Metasystox' sugli Insetti utili. *Frutticoltura* 20, 385–390.

Principi, M.M. (1992) Lo stato di diapausa negli Insetti ed il suo manifestarsi in alcune specie di Crisopidi (Insecta Neuroptera) in dipendenza dell'azione fotoperiodica. *Bollettino dell'Istituto di Entomologia 'Guido Grandi' dell'Università di Bologna* 46, 1–30.

Principi, M.M. (1993) Protezione integrata e produzione integrata delle colture agrarie: realizzazioni e prospettive. *Bollettino dell'Istituto di Entomologia 'Guido Grandi' dell'Università di Bologna* 47, 79–100.

Principi, M.M., Castellari, P.L. & Giunchi, P. (1967) Observations sur les infestations de pucerons et leurs prédateurs et parasites dans des parcelles traitées avec des produits phytiatriques polyvalents ou sélectifs. *Entomophaga* Mémoire hors Série 3, 103–107.

Ragusa, R. & Paoletti, M.G. (1985) Phytoseiid mites of corn and soybean agroecosystems in the low-laying plain of Veneto (N-E Italy). *Redia* 68, 69–89.

Rey, J.R. & McCoy, E.D. (1979) Application of island biogeographic theory to pest of cultivated crops. *Environmental Entomology* 8, 577–582.

Ridgway, R.L. & Kinzer, R.E. (1974) Chrysopids as predators of crop pests. *Entomophaga* Mémoire hors Série 7, 45–51.

Ridgway, R.L. & Murphy, W.L. (1984) Biological control in the field. In *Biology of Chrysopidae*, ed. Canard, M, Séméria, Y. & New, T.R., pp. 220–228. Dr. W. Junk, The Hague.

Speight, M.R. (1983) The potential of ecosystem management for pest control. *Agriculture, Ecosystems and Environment* 10, 183–199.

Taylor, L.R. (1961) Aggregation, variance and the mean. *Nature* 189, 732–735.

Taylor, L.R. (1984) Assessing and interpreting the spatial distributions of insect populations. *Annual Review of Entomology* 29, 321–357.

Tomaselli, C. & Tomaselli, E. (1973) Appunti sulle vicende delle foreste padane dall'epoca romana ad oggi. *Archivio Botanico e Biogeografico Italiano* 49, 85–101.

Vidano, C. & Arzone, A. (1987) Natural enemies of *Zyginidia pullula* (Rhynchota Auchenorrhyncha). In *Proceedings of the 6th Auchenorrhyncha Meeting*, ed. Vidano, C. & Arzone, A., pp. 581–590. Consiglio Nazionale delle Ricerche progetto finalizzato 'Incremento della Produttività delle Risorse Agricole', Turin.

Zandigiacomo, P. (1985) Osservazioni sulla entomofauna della fava nel Veneto e nel Friuli. *L'Informatore Agrario* 41, 79–82.

Zangheri, P. (1950) Il posto della Romagna nel quadro della biogeografia dell'Italia. *Studi Romagnoli* 1, 335–361.

CHAPTER 25

Lacewings and snake-flies in Piedmont vineyards (northwestern Italy)

R.A. Pantaleoni and A. Alma

25.1 INTRODUCTION

Vineyards are an ancient Mediterranean agroecosystem, which spread all over the world in geographic areas with adequate climatic conditions. Wherever *Vitis vinifera* L. was introduced, it came into contact with new pests, such as the Nearctic phylloxerid *Viteus vitifolii* (Fitch) which also produced dramatic effects in the area of origin of this plant (Goidanich, 1960), and new biocoenoses were established (Vidano, 1988).

Grapevine pests are well known and have been thoroughly listed by Englert & Maixner (1992) for Europe, and by Delrio *et al.* (1989) and Pollini (1998) for Italy. Further studies on the main pests and their biocoenoses were made by Vidano (1988). On the other hand, only some guilds of useful insects have been studied in depth. This is the case, for example, of the parasitoids of *Lobesia botrana* (Denis & Schiffermüller) and *Empoasca vitis* (Göthe), about which there is a vast literature including, for the geographic area of our interest and for the many references cited, the papers by Vidano *et al.* (1988), Cerutti *et al.* (1989), and Marchesini & Dalla Montà (1994). Information on predators is on the other hand very scarce, and, with few exceptions, concerns mainly Orthoptera, anthocorids, mirids, and phytoseid mites (Duso & Girolami, 1983; Arzone *et al.*, 1988; Camporese & Duso, 1996).

In most different geographic areas, lacewings are present on grapevines as predators of all the main groups of pests. Thus according to some recent reports, they are predators of mites in Australia (James & Whitney, 1991) and South Africa (Schwartz, 1993), of thrips in Switzerland (Remund & Boller, 1989), of scale insects in the USA (Grimes & Cone, 1985) and Chile (Ripa & Rojas, 1990), of caterpillars in Germany (Buchholz & Schruft, 1995) and Italy (Addante & Moleas, 1996), and of leafhoppers in Italy (Arzone *et al.*, 1988) and California (Daane *et al.*, 1996) where some grape-growers use inundative releases of green lacewings, *Chrysoperla* spp., for leafhopper control. Snakeflies, in contrast, have been reported only in a few cases and only Pantaleoni (1990a) and Letardi (1994) pointed out that the inocelliid *Parainocellia bicolor* (A. Costa) is rather abundant in Italian vineyards.

In spite of this constant presence there are practically no papers on the Neuropterida taxocoenoses of vineyards and on the role they, or single species, have inside these agroecosystems. The only exceptions are the studies by Jubb & Mastellar (1977) on lacewings of commercial Concord grape vineyards in Pennsylvania, by Schruft *et al.* (1983) on lacewings and snake-flies in Germany, and by Pantaleoni & Tisselli (1985) and Caffarelli & Letardi (1991) only on chrysopids in Italy.

The main aim of this work was to determine the species composition and seasonal abundance trends of Neuropterida in three Piedmontese (northwestern Italy) commercial vineyards and the different environmental features influencing them.

25.2 MATERIALS AND METHODS

The observations were carried out in three vineyards of grape-growing areas in the province of Turin (Piedmont, Italy) in 1995 and 1996. The three vineyards (A, B, and C), with managed ground cover and vertical trellis and pergola systems, had the following environmental features: A, vineyard surrounded by vineyards; B, vineyard surrounded by orchards of kiwi, apple, peach, and other stone-fruit trees; C, vineyard next to woods of broad-leaf trees and shrubs, such as chestnut, oak, black locust, hazel, bramble, hawthorn, etc.

All the investigated vineyards were IPM managed and only fungicide treatments (on average six to eight applications each session) were made and no insecticide was sprayed.

Table 25.1. *Semiquantitative report of the main pests found in the investigated Piedmontese vineyards*

| Plant pest | Vineyard | | |
	A	B	C
Panonychus ulmi (C.L. Koch, 1836)	+	+	+
Drepanothrips reuteri Uzel, 1895	+ +	−	−
Metcalfa pruinosa (Say, 1830)	+ +	−	+
Scaphoideus titanus Ball, 1932	+	+	+
Zygina rhamni Ferrari, 1882	+	+	+
Empoasca vitis (Göthe, 1875)	+ +	+ +	+ +
Planococcus spp.	+	+	+
Pulvinaria vitis (Linnaeus, 1758)	+	+	+
Lobesia botrana (Denis & Schiffermüller, 1775)	+ +	+	+ +
Eupoecilia ambiguella (Hübner, 1796)	+ +	+	+ +

Notes:

− = absence

+ = presence

+ + = considerable presence

These data remained constant in the biennium 1995–6.

There were 38 sampling dates in total: 16 in 1995, from mid-May to end October; 22 in 1996, from the beginning of April to end October. Samples were collected about every ten days.

In order to monitor adult lacewings, ten yellow sticky traps (25 × 40 cm) were set up in each plot at about 1.2 m from the ground. The traps were placed with an equilateral triangle design (four rows of alternatively three and two traps; the side being of 40–50 m) so to be regularly spread all over the surface of each plot.

Lacewing adults were also collected using in every plot a sweep-net along five transects of 40–50 m, two of which were inside the plot and three were on the borders. While checking the IPM programme, it was possible to find ten main pests of the Piedmontese vineyards (Table 25.1). When necessary, investigations on the pests of neighbouring fields and observations on the presence of preimaginal stages of some predators were carried out.

25.3 RESULTS

Altogether 23 Neuropterida species were found: 1 Inocelliidae, 6 Coniopterygidae, 7 Hemerobiidae, and 9 Chrysopidae (Table 25.2). The species collected in each vineyard were respectively 15 in A, 13 in B, and 16 in C.

Using the sweep-net, 361 specimens were collected, while 2557 were caught by means of yellow sticky traps. The percentage of each species with the two sampling methods changed sometimes considerably (enormously for *Parainocellia bicolor*). In three cases, excluding single specimens, captures were made using only one of the two sampling methods. Differences were found also in sex ratios, which generally appeared strongly male-biased in the yellow sticky traps.

The position of the yellow traps in each plot greatly influenced the number of captured Neuropterida (Table 25.3). But in general no correlation could be found between the position and environmental variables, except for traps 1–B and 10–C which had the highest captures. The two traps were placed near to corners bordering on two sides, respectively with orchards in vineyard B and woods in vineyard C. The traps of vineyard A showed a variation range clearly much smaller than that of the two other groups of traps.

Also the position of the sampled sides using a sweep-net influenced the captures (Table 25.4). But in

Table 25.2. *List of species and number of collected specimens in the biennium 1995/6*

Taxa	A		B		C		Total
	Net	Yellow traps	Net	Yellow traps	Net	Yellow traps	
Inocelliidae							
Parainocellia bicolor (A. Costa, 1855)	2	552	0	133	4	740	1431
Coniopterygidae							
Coniopteryx borealis Tjeder, 1930	0	0	0	21	0	0	21
Coniopteryx tineiformis Curtis, 1834	0	0	1	2	0	0	3
Coniopteryx esbenpeterseni Tjeder, 1930	0	0	2	197	0	0	199
Coniopteryx lentiae Aspöck & Aspöck, 1964	9	1	0	0	1	2	13
Coniopteryx unidentified	7	1	8	50	7	6	79
Conwentzia psociformis (Curtis, 1834)	0	0	1	0	0	0	1
Semidalis aleyrodiformis (Stephens, 1836)	1	0	0	0	8	0	9
unidentified	0	0	0	57	0	3	60
Hemerobiidae							
Hemerobius humulinus Linnaeus, 1758	3	14	6	13	13	58	107
Hemerobius micans Olivier, 1792	0	7	9	5	0	8	29
Wesmaelius subnebulosus (Stephens, 1836)	0	1	0	0	0	0	1
Sympherobius pygmaeus (Rambur, 1842)	0	1	0	1	2	0	4
Drepanepteryx phalaenoides (Linnaeus, 1758)	0	0	0	0	0	1	1
Micromus angulatus (Stephens, 1836)	0	6	0	17	0	7	30
Micromus variegatus (Fabricius, 1793)	0	11	1	15	8	36	71
unidentified	0	2	0	8	0	23	33
Chrysopidae							
Chrysopa formosa Brauer, 1850	1	0	0	0	1	0	2
Chrysopa perla (Linnaeus, 1758)	0	0	2	2	1	4	9
Chrysopa viridana Schneider, 1845	0	1	0	0	0	0	1
Chrysoperla carnea s. lat. (Stephens, 1836)	65	301	148	45	40	167	766
Chrysopidia ciliata (Wesmael, 1841)	0	1	0	0	4	4	9
Cunctochrysa albolineata (Killington, 1935)	0	0	1	0	0	0	1
Dichochrysa flavifrons (Brauer, 1850)	3	2	0	1	0	3	9
Dichochrysa inornata (Navás, 1901)	0	0	0	0	1	0	1
Dichochrysa prasina (Burmeister, 1839)	1	4	0	0	1	0	6
unidentified	0	7	0	12	0	3	22
Total	92	913	178	579	91	1065	2918

this case the effect of some environmental factors seems clearer. In vineyard A there was a substantial homogeneity representing the actual uniformity of the area. In vineyard B 68.5% of the total of specimens were collected on side 5, next to an apple orchard. Finally, in vineyard C twice the number of specimens was captured on the sides bordering on the wood than within the vineyard.

25.3.1 Inocelliidae

While collections using the sweep-net were sporadic, *Parainocellia bicolor* represented more than half of all Neuropterida collected by means of yellow sticky traps (60.46% in A, 22.97% in B, and 69.48% in C), and captures were almost exclusively composed of males: 93.60% in A, 93.75% in B, and 95.96% in C. Such particular data made us verify the actual presence in the

Table 25.3. *Total number of Neuroptera collected by each yellow trap in the three vineyards in 1995 and 1996*

Plot	Trap									
	1	2	3	4	5	6	7	8	9	10
A	68	99	101	93	127	63	65	108	134	55
B	221	18	37	102	34	23	54	36	37	17
C	78	109	52	106	103	120	174	42	54	227

field of larvae which were always found in good numbers by means of direct observations on the bark of the plants. Preliminary experiments in the laboratory showed that this species was an efficient predator of numerous pests in all life stages.

The seasonal trend of the crop is shown in Fig. 25.1. *Parainocellia bicolor* had only one swarming period in all three sites, between mid-April and mid-July with a peak at the end of May to the beginning of June.

25.3.2 Coniopterygidae
Vineyards A and C gave similar results both for abundance and species presence. In both sites very few specimens were captured by means of yellow traps, while fewer than 20 were caught using the sweep-net. The species found were *Coniopteryx lentiae* and *Semidalis aleyrodiformis*. The latter was collected in C only by means of the sweep-net on the sides next to the wood.

In vineyard B the results were completely different: the Coniopterygidae comprised more than half of the lacewings collected by means of the yellow traps, with a percentage of males of 86.59%. The species found, besides a single specimen of *Conwentzia psociformis*, were three *Coniopteryx: borealis, tineiformis,* and *esbenpeterseni*. The last was by far the most abundant one. Of this species 137 out of 197 specimens (87.26%) were captured in trap 1–B placed on the side bordering on the peach orchard.

The seasonal trend of captures of the genus *Coniopteryx* using yellow traps is shown in Figure 25.2. It shows a single peak corresponding, in both years, to the beginning of August, a period in which on the neighbouring peach stands there were strong infestations of tetranychid mites.

25.3.3 Hemerobiidae
While the number of species found in each vineyard was almost the same (6 in A and C, 5 in B), the number of specimens collected was much higher in C with regard to B and above all with regard to A. The most common Hemerobiidae was *Hemerobius humulinus*, followed by *Micromus variegatus*, *M. angulatus*, and *H. micans*, while the three remaining species, *Wesmaelius subnebulosus*, *Sympherobius pygmaeus*, and *Drepanepteryx phalenoides*, were captured in small numbers. The *Hemerobius* were captured by means of sweep-net especially in C on the sides bordering on the wood (12 specimens of 13) and in B on the side next to the apple stand (11 out of 15). This happened also for *M. variegatus* (7 out of 8 in C and 1 of 1 in B); *M. angulatus* was caught only by means of yellow traps.

The seasonal trend of the presence of *H. humulinus* using yellow traps is shown in Fig. 25.3. It shows a considerably different trend in the two years. During 1995 we had a single rather late peak placed, depending on the plot, between the end of August and the middle of September and probably due to the overlapping of two generations in vineyard C. In 1996 there were three successive peaks in vineyards B and C, corresponding to three generations, at end April, end June, and beginning of September, while in vineyard A only the first swarming was recorded.

25.3.4 Chrysopidae
Chrysoperla carnea was the most common species in all three vineyards, using both sampling methods, with a percentage on the total of chrysopids varying from 83.33% in C to 98.67% in B using a sweep-net and from 93.75% in B to 97.10% in A using yellow traps. Fewer than ten specimens were captured for each of the remaining species.

Table 25.4. *Total number of specimens collected by sweep-net samplings along each side in 1995 and 1996*

Vineyard		Side	Number collected
A	1	boundary with vineyard	18
	2	boundary with vineyard	9
	3	internal	17
	4	internal	19
	5	boundary with uncultivated field	29
B	1	boundary with kiwi orchard	18
	2	boundary with kiwi and peach orchard	21
	3	internal	11
	4	internal	6
	5	boundary with apple orchard	122
C	1	boundary with wood	32
	2	boundary with wood	27
	3	internal	13
	4	internal	14
	5	boundary with road	5

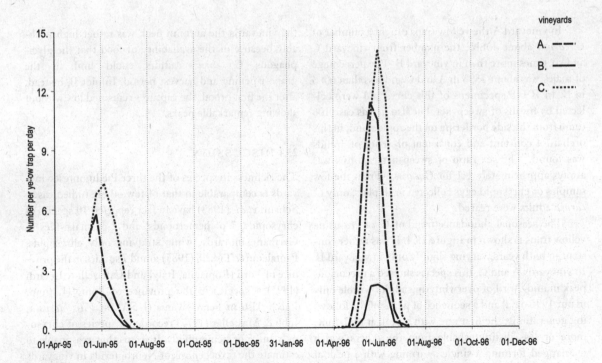

Fig. 25.1. The seasonal abundance trend of *Parainocellia bicolor* in the three vineyards in the years 1995–6.

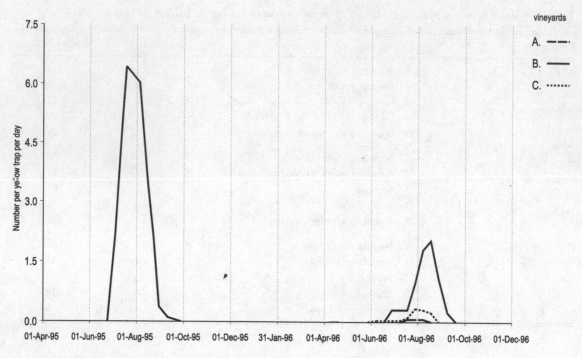

Fig. 25.2. The seasonal abundance trend of *Coniopteryx* spp. in the three vineyards in the years 1995–6.

In vineyard A the yellow traps caught a number of *C. carnea* about double the number from vineyard C and six times more than in vineyard B. The percentage of males was about 75% in A and C and less than 60% in B. In B 148 specimens of this chrysopid were collected by means of sweep–net, but also in this case 106 came from the side bordering on the apple stand; in this orchard a constant and consistent presence of aphids was found. The sex ratio of sweep–net captures was always approximately 1:1 for *C. carnea*. From the few samples of chrysopid eggs collected in all plots, only *C. carnea* adults were reared.

The seasonal abundance trend of *C. carnea* using yellow traps is shown in Figure 25.4. It was rather constant in both years, varying slightly only in vineyard B. In vineyards A and C, this species showed a precocious peak in mid–April of overwintering adults (visible only in our 1996 data) and a sequence of peaks of the following generations: the first one with a peak at mid–June, more or less distinguishable from the others which overlapped forming a single swarming with a peak at end July. Finally, there was a peak in autumn at the beginning of October made of overwintering adults. In

the vineyards the autumn peak was rather high probably because of the availability of food that the glyciphagous *C. carnea* adults could find in the grape-ripening and harvest period. In plot B, instead, after the first brood, the captures remained low without showing remarkable peaks.

25.4 DISCUSSION

The richness in species of the three Piedmontese vineyards is comparable to that of few other studied cases. Schruft *et al.* (1983) have in fact reported 10 species of chrysopids, 7 of hemerobiids, and 2 of snake-flies in German vineyards; while studying only chrysopids, Pantaleoni & Tisselli (1985) found 7 species in the province of Forlì (Romagna, Italy) and Caffarelli & Letardi (1991) 4 species in the province of Rome (Latium, Italy). Also in Pennsylvania (USA) on *Vitis labrusca*, Jubb & Mastellar (1977) recorded 5 species of chrysopids and 5 of hemerobiids. It seems thus reasonable to estimate the taxocoenosis of Neuropterida in vineyards of 10–20 species. Only some of these species permanently dwell in this agroecosystem and breed there.

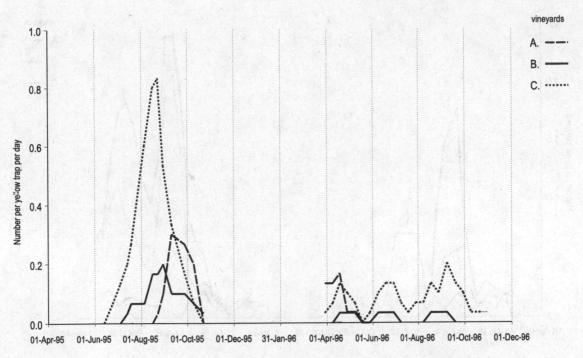

Fig. 25.3. The seasonal abundance trend of *Hemerobius humulinus* in the three vineyards in the years 1995–6.

The most considerable novelty which emerged from our data is the strong presence of *Parainocellia bicolor*. This snake-fly certainly carries out its whole life cycle in vineyards, it is extremely abundant, and represents a capital element of the guild of predators feeding on pests hidden in the grapevine bark. The seasonal abundance trend found in Piedmont agrees with the already known biological data, with a short and synchronous swarming period which is remarkably influenced by local climatic features (Pantaleoni, 1990*b*). It is surprising how such a big and abundant predator has always been unnoticed by applied entomologists, besides the first reports by Pantaleoni (1990*a*) and Letardi (1994). Maybe this is due to the short and precocious swarming period of the adults and to the exclusively bark-dwelling habits.

We do not have any previous reference on coniopterygids to compare with. Surely the results obtained in vineyard B were influenced by the nearby orchards. The abundance of *Coniopteryx esbenpeterseni* and *C. borealis* seems to be determined, besides by prey availability, by the microclimatic features caused by the position and exposition of the host plants (Pantaleoni,

1984), conditions which become optimal really in palmette or vase orchards where they are both common (Nicoli *et al.*, 1988; Marzocchi & Pantaleoni, 1995). *Coniopteryx esbenpeterseni*, in particular, seems strongly linked to peach trees as a predator of mites (Castellari, 1980). Likewise it is probable that *C. lentiae*, the only abundant species in the exclusively viticolous area of vineyard A, has ecological needs corresponding to the situation of the studied vineyards with luxuriant grapevines forming thick rows with a fresh and relatively humid microclimate. It is, however, feasible that *Coniopteryx* species may carry out all their life cycle in vineyards, above all on the occasion of infestations of mites, their preferred prey.

The presence of hemerobiids in the vineyard biocoenoses where these predators are collected with a certain continuity seems linked more to the whole agroecosystem than to the grapevine plant. The *Micromus* species are, in fact, almost certainly linked to herbaceous plants hosting numerous aphid colonies (Pantaleoni, 1982), as the data of Jubb & Mastellar (1977) also show for Nearctic species. The *Hemerobius* species, which are aphid-eaters too and are fundamentally linked to the tree

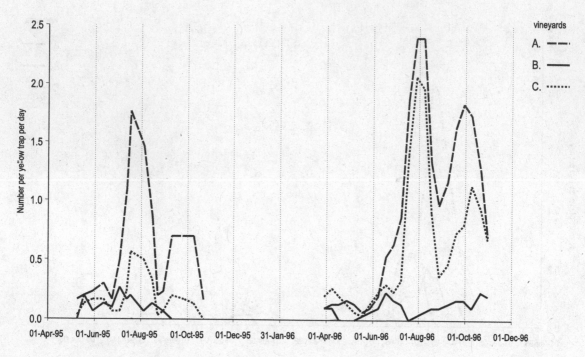

Fig. 25.4. The seasonal abundance trend of *Chrysoperla carnea* in the three vineyards in the years 1995–6.

and shrub layers (Pantaleoni, 1982), in spite of a greater euryoecy, seem rather to find a temporary shelter in the vineyards; in the plots B and C, in fact, we collected them by means of a sweep-net especially along the border with environments rich in aphids. The seasonal abundance trend corresponds to that reported by Pantaleoni (1982, 1990a).

Among chrysopids, *Chrysoperla carnea* is surely the most abundant species in the vineyard and this agrees with Schruft *et al.* (1983), Pantaleoni & Tisselli (1985), Caffarelli & Letardi (1991), and our data. The usual finding of preimaginal stages in vineyards and the biological features of this species, above all the dispersal and oviposition strategies (Duelli, 1984), show that it manages to carry out its whole life cycle on this plant preying on all kinds of small arthropods as a result of its strong polyphagy. Another group of chrysopids living in vineyards as adults are the *Dichochrysa* species, which are constantly reported by all above-mentioned authors, even if in low numbers in Piedmont. According to some observations (Marchesini & Dalla Montà, 1994; R.A. Pantaleoni & C. Duso, unpublished data), their larvae live on grapevines and can prey on

several pests showing, however, a certain preference for scale insects. Finally, the same considerations concerning hemerobiids may be made for *Chrysopa* species, which are also commonly found on this crop.

The interactions existing between vineyards, weeds, neighbouring crops and other plants (hedges, borders, isolated trees, etc.) are extremely complex as far as insects are concerned (Vidano, 1988). Predators can move, much more than plant pests, even only temporarily, from one environment to another under the influence of various stimuli, such as the need for prey or shelter, or of particular microclimatic features.

Besides the ecology of single species, more or less strictly linked to particular vegetation layers or plant species, it is interesting to understand what influence the characteristics of the landscape structure has in the light of the differences we found between the three vineyards here considered.

The collections in vineyard A, in an exclusively viticolous area, surrounded by other vineyards, showed the greatest space homogeneity both using yellow traps (Table 25.3) and the sweep-net (Table 25.4). The species found were practically only those linked somehow to grapevines – *Parainocellio bicolor, Coniopteryx lentiae, Chrysoperla carnea, Dichochrysa* spp., *Hemerobius* spp. – to which one must add the low-density species linked to herbaceous plants – both *Micromus* and *Chrysopa formosa* – and sporadic individuals of lacewings linked to the few isolated trees present – *Semidalis aleyrodiformis, Wesmaelius subnebulosus, Sympherobius pygmaeus, Chrysopa viridana, Chrysopidia ciliata*, and *Cunctochrysa albolineata*.

The captures in vineyard B, surrounded by orchards, were strongly influenced by the presence of neighbouring crops. *Parainocellia bicolor* and *Chrysoperla carnea* were recorded in this plot in only low numbers with yellow traps. On the contrary, using a sweep-net, the chrysopid was shown to be extremely abundant, but only on the side next to the apple stand. Also the seasonal trend of collections showed some anomalies. *Coniopteryx borealis* and *C. esbenpeterseni* were caught in high numbers by means of yellow traps almost exclusively on the side of the peach stand, while *Hemerobius humulinus* and *H. micans* captures by means of the sweep-net took place mostly on the apple stand side.

The collections in vineyard C, surrounded by woods, were also influenced by the neighbouring vegetation. But here the abundance of specimens and the variety of species were the highest of all. Particularly abundant were the aphidophagous hemerobiids, *Semidalis aleyrodiformis* linked to oak, and *Chrysopa perla* and *Chrysopidia ciliata* belonging to fundamentally humid environments.

On the whole the knowledge of vineyard Neuropterida is still extremely fragmentary. Through further investigations, the real role of the different species in the control of the main grapevine pests will be important to ascertain. It will also be necessary to evaluate the influence on the lacewing fauna of crop, agronomic, and other features of the very vineyards and of the neighbouring environments.

REFERENCES

Addante, R. & Moleas, T. (1996) Effectiveness of mating disruption method against *Lobesia botrana* (Den. et Schiff.) (Lepidoptera – Tortricidae) in Apulian vineyards. *Bulletin OILB-SROP* 19, 247–251.

Arzone, A., Vidano, C. & Arnò, C. (1988) Predators and parasitoids of *Empoasca vitis* and *Zygina rhamni* (Rhynchota Auchenorrhyncha). In *Proceedings of the 6th Auchenorrhyncha Meeting*, ed. Vidano, C. & Arzone, A., pp 623–629. Consiglio Nazionale delle Ricerche progetto finalizzato 'Incremento della Produttività delle Risorse Agricole, Turin.

Buchholz, von U. & Schruft, G. (1995) Räuberische Arthropoden auf Blüten und Früchten der Weinrebe (*Vitis vinifera* L.) als Antagonisten des Einbindigen Traubenwicklers (*Eupoecilia ambiguella* Hbn.) (Lep., Cochylidae). *Journal of Applied Entomology* 118, 31–37.

Caffarelli, V. & Letardi, A. (1991) *I Neurotteri delle colture agrarie: rilievi sui Crisopidi nell'area viticola di Cerveteri.* RT/INN/91/04. Ente per le nuove Tecnologie, l'Energia e l'Ambiente, Rome.

Camporese, P. & Duso, C. (1996) Different colonization patterns of phytophagous and predatory mites (Acari: Tetranychidae, Phytoseiidae) on three grape varieties: a case study. *Experimental and Applied Acarology* 20, 1–22.

Castellari, P.L. (1980) Indagini biologiche su *Coniopteryx (Metaconiopteryx) esbenpeterseni* Tjeder (Neur. Coniopterygidae), predatore di Acari Tetranichidi sul Pesco. *Bollettino dell'Istituto di Entomologia dell'Università di Bologna* 35, 157–80.

Cerutti, F., Delucchi, V., Baumgärtner, J. & Rubli, D. (1989) Ricerche sull'ecosistema 'vigneto' nel Ticino: II. La colonizzazione dei vigneti da parte della cicalina *Empoasca vitis* Goethe (Hom., Cicadellidae, Typhlocybinae) e del suo parassitoide *Anagrus atomus* Haliday (Hym., Mymaridae), e importanza della flora circostante. *Mitteilungen schweizerischen entomologischen Gesellschaft* 62, 253–267.

Daane, K.M., Yokota, G.Y., Zheng, Y. & Hagen, K.S. (1996) Inundative release of common green lacewings (Neuroptera: Chrysopidae) to suppress *Erythroneura variabilis* and *E. elegantula* (Homoptera: Cicadellidae) in vineyards. *Environmental Entomology* 25, 1224–1234.

Delrio, G., Floris, I., Luciano, P., Ortu, S. & Prota, R. (1989) Lotta integrata contro i fitofagi delle principali colture arboree della Sardegna. *S.IT.E./Atti* 8, 71–81.

Duelli, P. (1984) Dispersal and oviposition strategy in *Chrysoperla carnea* (Steph.). In *Progress in World's Neuropterology, Proceedings of the 1st International Symposium on Neuropterology*, ed. Gepp, J., Aspöck, H. & Hölzel, H., pp. 133–146. Thalerhof, Graz.

Duso, C. & Girolami, V. (1983) Ruolo degli Antocoridi nel controllo del *Panonychus ulmi* Koch nei vigneti. *Bollettino dell'Istituto di Entomologia dell'Università di Bologna* 37, 157–169.

Englert, W.D. & Maixner, M. (1992) Biological and integrated control in European vineyards. In *Biological Control and Integrated Crop Protection: Towards*

Environmentally Safer Agricolture ed. van Lenteren, J.C. et al., Minks, A.K. & de Ponti, O.M.B. pp. 39–48. Pudoc Scientific Publisher, Wageningen.

Goidanich, A. (1960) Fillossera della vite. In *Enciclopedia agraria italiana*, vol. 4, pp. 682–698. Ramo Editoriale degli Agricoltori, Rome.

Grimes, E.W. & Cone, W.W. (1985) Life history, sex attraction, mating, and natural enemies of the grape mealybug, *Pseudococcus maritimus* (Homoptera: Pseudococcidae). *Annals of the Entomological Society of America* 78, 554–558.

James, D.G. & Whitney, J. (1991) Biological control of grapevine mites in inland south-eastern Australia. *Australian New Zealand Wine Industry Journal* 6, 210–214.

Jubb, G.L. Jr & Mastellar, C. (1977) Survey of arthropods in grape vineyards of Erie County, Pennsylvania: Neuroptera. *Environmental Entomology* 6, 419–428.

Letardi, A. (1994) Dati sulla distribuzione italiana di Megaloptera Sialidae, Raphidioptera Inocelliidae e Planipennia Mantispidae, con particolare riferimento all'Italia centrale (Neuropteroidea). *Bollettino della Società Entomologica Italiana* 125, 199–210.

Marchesini, E. & Dalla Montà, L. (1994) Observations on natural enemies of *Lobesia botrana* (Den. & Schiff.) (Lepidoptera, Tortricidae) in Venetian vineyards. *Bollettino di Zoologia Agraria e di Bachicoltura* 26, 201–30.

Marzocchi, L. & Pantaleoni, R.A. (1995) Indagine sui principali entomofagi predatori (Insecta Heteroptera, Neuroptera et Coleoptera) in pereti della Pianura Padana. *Bollettino dell'Istituto di Entomologia 'Guido Grandi' dell'Università di Bologna* 49, 21–40.

Nicoli, G., Corazza, L., Cornale, R. & Marzocchi, L. (1988) Indagine sugli insetti predatori in pereti a diversa gestione fitoiatrica. In: *Atti XV Congresso Nazionale Italiano Entomologia*, pp. 489–496. Accademia Nazionale Italiana di Entomologia, L'Aquila.

Pantaleoni, R.A. (1982) Neuroptera Planipennia del comprensorio delle Valli di Comacchio: indagine ecologica. *Bollettino dell'Istituto di Entomologia dell'Università di Bologna* 37, 1–73.

Pantaleoni, R.A. (1984) Neuroptera Planipennia del comprensorio delle Valli di Comacchio: le neurotterocenosi del *Quercetum ilicis* e del *Populus nigra pyramidalis*. *Bollettino dell'Istituto di Entomologia 'Guido Grandi' dell'Università di Bologna* 39, 61–74.

Pantaleoni, R.A. (1990*a*) Un nuovo ausiliario nel vigneto: *Parainocellia bicolor* (Costa). *L'Informatore Fitopatologico* 40, 39–43.

Pantaleoni, R.A. (1990*b*) I Neurotteri (Neuropteroidea) della Valle del Bidente–Ronco (Appennino Romagnolo). *Bollettino dell'Istituto di Entomologia 'Guido Grandi' dell'Università di Bologna* 44, 69–122.

Pantaleoni, R.A. & Tisselli, V. (1985) I Neurotteri delle colture agrarie: rilievi sui Crisopidi in alcune coltivazioni del forlivese. *Bollettino dell'Istituto di Entomologia 'Guido Grandi' dell'Università di Bologna* 40, 51–65.

Pollini, A. (1998). *Manuale di Entomologia applicata*. Edagricole, Bologna.

Remund, U. & Boller, E. (1989) Thripse im Ostschweizer Rebbau: 1. Problemstellung, Artenspektrum und Lebensweise. *Schweizerische Zeitschrift Obst Weinbau* 125, 173–188.

Ripa, S.R. & Rojas, P.S. (1990) Manejo y control biologico del chanchito blanco de la vid. *Revista Fruticola* 11, 82–87.

Schruft, G., Wegner, G., Muller, R.-D. & Samples, J. (1983) Das Auftreten von Florfliegen (Chrysopidae) und anderen Netzflügern (Neuroptera) in Rebanlagen. *WeinWissenschaft* 38, 186–194.

Schwartz, A. (1993) Occurrence of natural enemies of phytophagous mites on grapevine leaves following application of fungicides for disease control. *South Africa Journal of Enology and Viticulture* 14, 16–17.

Vidano, C. (1988) Entomofauna di ecosistemi naturali e incolti in agroecosistemi con particolare riguardo al vigneto. In *Atti XV Congresso Nazionale Italiano Entomologia*, pp. 451–470. Accademia Nazionale Italiana di Entomologia, L'Aquila.

Vidano, C., Arnò, C. & Alma, A. (1988). On the *Empoasca vitis* intervention threshold on vine (Rhynchota Auchenorrhyncha). In *Proceedings of the 6th Auchenorrhyncha Meeting*, ed. Vidano, C. & Arzone, A., pp. 525–537. Consiglio Nazionale delle Ricerche progetto finalizzato 'Incremento della Produttività delle Risorse Agricole', Turin.

CHAPTER 26

Control of aphids by *Chrysoperla carnea* on strawberry in Italy

M.G. Tommasini and M. Mosti

26.1 THE STRAWBERRY CROP IN ITALY

From an FAO estimate made in 1996, about 26% of the world production of strawberries, *Fragaria* × *ananassa* (Duchesne), is grown in Europe and mostly in Spain and Italy, with about 380 000 tonnes of product per annum (Sbrighi *et al.*, 1998). In the Mediterranean area, strawberries are cultivated both in the open field and in greenhouses (mainly plastic tunnels) with a spring and summer production period.

In Italy, the phytosanitary condition of the strawberry crop varies greatly with latitude and in the warmer areas a wider range of problems commonly occur, for example a hot and dry climate is favourable to the development of pests such as the red spider mite, *Tetranychus urticae* Koch. For instance outbreaks of western flower thrips (WFT), *Frankliniella occidentalis* (Perg.), are usually higher and cause more severe outbreaks in southern regions than in northern ones. WFT can overwinter in quiescence mainly when plastic covers are set up in the autumn or early winter (Tommasini & Maini, 1995). On the other hand, aphids can be particularly harmful in temperate regions such as in northern Italy. Towards central and southern Italy aphids become less of a problem (Nicoli, 1992). The more frequent aphid species on strawberry in Italy are *Macrosiphum euphorbiae* (Thom.) and *Chaetosiphon fragaefolii* (Cock.) (Benuzzi *et al.*, 1991*b*), with occasional problems with *Aphis gossypii* Glov. and *Myzus persicae* (Sulz.) (Galazzi & Nicoli, 1992). Also growing techniques can influence plant development and the damage caused by some arthropod pests or diseases. The main pests occurring on strawberry in the Mediterranean area, especially in Italy, are summarised in Table 26.1.

The use of plants originating from disease-free material is common practice nowadays and reduces the risks of severe diseases after transplanting. Other culti-vation practices help growers to reduce problems and the need for sprays on strawberry cultivation; for example plastic mulch decreases decay problems and weed control and pruning reduces the incidence of some diseases as well as the overwintering population of some pests (e.g. noctuids, aphids, mites). On the other hand, the tendency on the world market towards the production of strawberries all year round has led to the development of some new cultivation techniques, such as the 'soil-less crop' and 'reflowering varieties', which can increase the problem of some pests (i.e. spider mites and thrips).

On protected strawberries the use of bumblebees [*Bombus terrestris* (L.)] for pollination, common in northern Europe and in southern Italy on tomato, is starting to increase, positively influencing the use of biological control and integrated pest management (IPM) due to the necessity of avoiding the use of wide-spectrum pesticides.

26.2 MAIN INCENTIVE FOR BIOLOGICAL CONTROL AGAINST APHIDS ON STRAWBERRY IN ITALY

Aphids are important pests on strawberries either in the open field or in greenhouses, but in the protected crop the biological cycle of the pest is faster so infestations are more rapid and more intensive; resistance symptoms are more frequent and early cultivation makes the action of naturally occurring beneficials less effective. In addition to worker safety concerns, soil pollution and phytotoxicity are also important reasons to develop a biological control strategy against pests (Celli *et al.*, 1984; Principi, 1984). Aphids sometimes appear in Italian strawberry-growing greenhouses 7–14 days before harvesting. Furthermore strawberries must be harvested two or three times a week so, in order to avoid

Table 26.1. *General guidelines for IPM in protected strawberry crops in Italy*

Target pest	Control agent	Release rate
Tetranychus urticae	*Phytoseiulus persimilis*	4–6 predators/m^2 (cold areas) to 10–20 predators/m^2 (warm areas): multiple releases
Frankliniella occidentalis	*Orius laevigatus*	2–4 predators/m^2 in total: multiple releases
Macrosiphum euphorbiae	*Chrysoperla carnea*	20 second-instar larvae/m of row.
Chaetosiphon fragaefolii	*Harmonia axyridis*	5–10 young larvae/m^2 (on hot-spots)
Aphis gossypii	Aphid parasitoids	0.25 parasitoids/m^2 (weekly releases)
Myzus persicae	Natural pyrethrum or other selective insecticides	Hot-spot treatments or general treatments
Lepidopteran larvae	*Bacillus thuringiensis* ssp. *kurstaki* Neurotoxic insecticides	General treatments (sprays) or baits
Otiorrhynchus spp.	*Heterorhabditis* spp.	50 000–150 000 entomopathogenic nematodes/m of row
Plant pathogens	Copper, sulfur, and occasionally other selective fungicides	

chemical residuals, it is difficult to use pesticides in a 'traditional' way (Bonomo *et al.*, 1991). Direct experience demonstrates that an incautious use of chemicals can aggravate the pest situation instead of improving it: synthetic pyrethroids (e.g. Delthamethrin) are highly toxic to beneficials and this action has a long persistence; Pyrazophos kills eggs of *Phytoseiulus persimilis* Athias-Henriot, a predator of *T. urticae*; the use of both pesticides is linked with *T. urticae* infestations (Benuzzi *et al.*, 1991*b*). Therefore research on biological control in strawberries started and developed in greenhouses and only later was this technology applied in the open field. Strawberries are one of the first crops on which biological control has been applied commercially in the northeast of Italy (Emilia-Romagna region).

26.3 *CHRYSOPERLA CARNEA*: DEVELOPMENT OF THE RELEASE TECHNIQUE

The common green lacewing, *Chrysoperla carnea* (Stephens), occurs more often than other chrysopids in the agroecosystem (Pantaleoni, 1986) and this has been one of the reasons why this species was preferred to carry out the first experiments of biological control

against aphids in 1983 (Celli *et al.*, 1986). The lacewing is suitable for most environments and it is resistant to climatic adversity. Furthermore, it is polyphagous and suitable for mass-rearing in artificial conditions (Principi, 1984).

Some authors have used the polyphagous attribute of chrysopids to apply them to different pests. Chang & Huang (1995) have used with success *Mallada basalis* (Walker) against tetranychid mites on strawberry; but on strawberry in Europe the attention of researchers has focused on aphid control. There are many studies on this predator as a biological control agent, and in strawberries most of them have been carried out in Italy (Principi, 1984; Celli *et al.*, 1986; Benuzzi & Nicoli, 1988; Manzaroli & Mosti, 1990; Benuzzi *et al.*, 1991*a*, 1992; Bonomo *et al.*, 1991). Moreover *C. carnea* is entomophagous only at the larval stages (adults feed on honeydew, nectar, and pollen) so it is possible to rear adults grouped in big mass-rearing units without cannibalistic problems.

However, *C. carnea* adults undergo an obligate post-emergence migration (Duelli, 1980). After emergence adults can fly for several kilometres meaning that lacewings are not suitable for inoculating large infested areas. If this was done then aphid control would occur

Table 26.2. *Relative amounts of prey consumed by* Chrysoperla carnea *larvae instars on three different prey species*

Prey species	First instar		Second instar		Third instar		Total
	N	%	N	%	N	%	N
Macrosiphum euphorbiae	17.7	8.0	32.1	14.4	172.6	77.6	222.4
Chaetosiphon fragaefolii	21.7	5.6	51.1	13.2	313.5	81.2	386.3
Ephestia kuehniella (eggs)	24.3	2.2	91.7	8.5	967.8	89.3	1083.8

Source: Nicoli *et al.* (1991).

far from the desired areas. Instead it is advisable to release a large and concentrated number of eggs or larvae in the infested area; these individuals will then control aphids directly (Pantaleoni, 1986).

In 1982 in the Entomology Institute 'G. Grandi' of Bologna mass-rearing of *C. carnea* was started for experimental purposes. The following year, in the northeast of Italy, 90 *C. carnea* eggs per metre of row arranged in pairs (i.e. 80 eggs per m²) were released on a strawberry crop in plastic tunnels, as soon as aphids appeared, obtaining a good biological control (Celli *et al.*, 1984). In 1984 and 1985 the experiment was repeated in seven commercial greenhouses. The predator:prey ratio varied from 1:1 up to 1:4.

C. carnea eggs were distributed on strawberries in two ways: (1) on cardboard pieces where adults laid their eggs; (2) on cardboard pieces where eggs, detached from original egg-laying site, were glued with *Ephestia kuehniella* Zell. eggs (Celli *et al.*, 1986).

The latter method of release was used to reduce cannibalistic attacks among the first larval stage and to provide some food before the larvae started climbing strawberry leaves. Egg release started as soon as the presence of aphids was noticed (usually at the end of April in the northeast of Italy). Good aphid control was obtained with both methods until the end of the crop cultivation. Only larvae can prey upon aphids and under Italian climatic conditions newly hatched larvae could reach the adult stage in 14–21 days during the spring. *Chrysoperla carnea* could only prey during the period when aphids were most active and most damaging on strawberries. In May, climatic conditions in greenhouses became less favourable for *Macrosiphum euphorbiae* and *Chaetosiphon fragaefolii*; strawberry plants changed their metabolism and became less suitable for aphids. Moreover, in May several species of beneficials (predators and parasitoids: Coleoptera,

Diptera, Hymenoptera, and mites) are active in the open field and they can take over in greenhouses, particularly where wide-spectrum pesticides are absent (Trumble *et al.*, 1983; Manzaroli & Mosti, 1990).

However, according to Celli *et al.* (1986), newly hatched *C. carnea* larvae have some difficulty moving on hairy strawberry sprouts and leaves, and first instars are small and not very voracious. Ants can be a serious problem because they can prey on both eggs and younger larvae. To avoid this problem every piece of cardboard had to be sprayed with an ant repellent. Moreover, the action of *C. carnea* is delayed because eggs have to hatch and larvae have to reach the more voracious instars. In spite of the good results reached with eggs and fewer problems of production, preservation, and shipment, it soon became apparent that the use of the larvae instead of eggs would be more efficient (Celli *et al.*, 1986; Principi, 1984). Lacewings are commonly sold and released at the egg stage. Larval release appeared more effective than egg release as observed also by Daane & Yokota (1997) due to an increase of the survival of lacewings to the third instar which is the stage in which they consume the most prey. A similar result was previously demonstrated by Nicoli *et al.* (1991) in a laboratory experiment, where different prey were supplied to *C. carnea* larvae (Table 26.2). Another example of the advantages of larval over egg lacewing release was demonstrated by Adashkevich & Kuzina (1976) working on the control of Colorado potato beetle, *Leptinotarsa decemlineata* (Say).

In 1987 the first greenhouse trials using second instar larvae of *C. carnea* produced by the Italian biofactory 'Biolab' (Benuzzi & Nicoli, 1988) started and larvae releases gave better results compared to egg releases. The release of the second instar appeared to be the most reliable probably due to the better mobility of the larvae on the strawberry leaves. Furthermore, a lower number of beneficials was necessary per surface unit to control the aphid infestation and control was faster.

26.4 *CHRYSOPERLA CARNEA*: RELEASE TECHNIQUE

Research carried out in Italy, particularly in the Pianura Padana, resulted in an efficient integrated pest management of aphids on strawberry in greenhouses. Good aphid control was achieved by releasing 18 second-instar larvae per m² (20 larvae per metre of strawberry row arranged in pairs). The release of *C. carnea* larvae is advised when 20% of young leaves are infested; a good sampling programme is to check 100 young leaves per 500 m² of greenhouse (Benuzzi *et al.*, 1991b). If it is not possible to put into practice such an accurate sampling programme, then it is advisable to release lacewing larvae as soon as aphids appear. We can obtain a better result by concentrating the release in the most infested plants. According to circumstances it is possible to reduce the release amount at 10 larvae per m² or to increase it as high as 30 larvae per m².

Chrysoperla carnea can be used, in Italy, early in the season in unheated plastic tunnels and even in the open field remaining active over a large range of temperatures (±0°C to >15°C) (Nicoli *et al.*, 1991).

The release of *C. carnea* on protected strawberries has greatly influenced the choice of a compatible control strategy against other strawberry pests in the Pianura Padana (Emilia Romagna region, Italy) such as *T. urticae* and noctuids (e.g. *Mamestra* spp. L., *Agrotis* spp. Ochs., *Agrochola lychnidis* (D. et S.), *Phlogophora meticulosa* (L.)] (Antoniacci & Nasolini, 1998).

The use of lacewings on strawberries must be part of a complete integrated pest management. Red spider mite can be controlled by *Phytoseiulus persimilis*; *Orius laevigatus* (Fieber) has been demonstrated to be effective against thrips, and *Bacillus thuringiensis* Berliner ssp. *kurstaki* controls Lepidoptera. Also, to control pathogens, it has been necessary to choose selective fungicides, as for example sulfur against mildew (*Oidium* sp. LK.EM.Sacc); *Botrytis cinerea* Pers.:F.r. can be controlled by frequently airing the greenhouses, whereas in the open field fungicides are commonly used (e.g. procimidone, vinclozolin, iprodione).

Advanced aphid infestations are best knocked down with pyrethrum. Products extracted by natural pyrethrum are harmful if they are used directly on chrysopid larvae, but they can be used some days before releasing *C. carnea* (Manzaroli & Mosti, 1990).

26.5 MARKETING OF *CHRYSOPERLA CARNEA*

In Italy, prior to 1996, *C. carnea* larvae were supplied inside cardboard tubes (c. 550 cm³). The tube contained about 500 second-instar larvae spread in cardboard-shavings; a small quantity of *E. kuehniella* eggs were also placed inside the tube as food. Cardboard-shavings and *E. kuehniella* eggs were adopted to prevent cannibalistic problems inside the tubes. Nowadays *C. carnea* larvae are delivered in plastic bottles of 450 cm³. The bottle contains c. 1000 second-instar larvae; instead of cardboard-shavings chaff buckwheat is used. Thanks to the dispersal method and to *E. kuehniella* eggs it is possible to store the bottles for 7 days at 8°C without damage to the beneficials. Growers use the beneficials directly, spreading the bottle's contents on strawberries. The buckwheat helps to provide a uniform distribution of the larvae on the crops. For special purposes even detached eggs can be delivered.

This technique has developed mainly in Emilia Romagna, in the northeast of Italy. In this region the areas of strawberry under integrated pest management were: 22 ha of greenhouses in 1987; 108 ha of greenhouses and 94 ha of open field in 1988. During 1988 and 1989, 400 growers were involved, and 40.5 tonnes of strawberries were produced with IPM (Mosti & Barducci, 1988). The growers were grouped in co-operative societies and *C. carnea* was released on infested crops: during 1987 aphids appeared only on 6% of the total area while in 1988 aphid frequency was much higher. The cost of the beneficials was shared among all growers involved.

The success reached in greenhouses with the release of *C. carnea* against aphids was for some marketing companies the first stimulus towards a quality certification on the product, which could reach a higher price on the market. Growers soon found a remarkable interest in IPM mainly because of the possibility of developing a 'green label' certifying environmentally safe production techniques and absence of pesticide residue on fruits which provided new commercial opportunities (Celli, 1987).

In the last ten years the development of IPM on strawberry has been very fast and biological control has given an important impetus to implement such a strategy. Nowadays about 1230 ha of strawberries are cultivated in Emilia Romagna region; on about 20% IPM is applied and 1% is organic production. In 1997 and 1998, *C. carnea* larvae were released against aphids on about 32% of strawberries in protected crop (*c.* 50 ha) in the Emilia Romagna region.

26.6 PERSPECTIVES

Notwithstanding the considerable interest for strawberry IPM by growers and marketing companies, in the past years there has not been an increase in the application of lacewings for the control of aphids, which have found a rather stable market substantially in the temperate Italian regions (mostly in Emilia Romagna) where aphids remain key strawberry pests.

Strawberry production in the temperate Italian regions is becoming less and less competitive with that of the southern areas of the Mediterranean basin, and the area of strawberry cultivation may decline. The reasons for this situation in the Italian market are twofold: (1) the higher production costs in the northeast regions compared to those of warmer areas: the farms are smaller so the cost of technical assistance is higher per surface unit.(2) the production is late and generally coincides with the peak of the product on the market.

In conclusion, although the technical application of lacewings on strawberries is available and effective, there are few opportunities to develop it further in Italy. However, the techniques developed for strawberries could be applied to vegetable crops (some trials are in progress on sweet pepper), and on ornamentals.

ACKNOWLEDGEMENTS

Dr L. Antoniacci (Regional Crop Advisory Service of Emilia Romagna, Italy) is thanked for providing data on the Emilia Romagna growing area; Dr P. McEwen (Cardiff University, UK) and Dr G. Nicoli (University of Bologna, Italy) are thanked for their critical reading of the manuscript.

REFERENCES

Adashkevich, B.P. & Kuzina, N.P. (1976) Chrysopids in vegetable crops. *Review of Applied Entomology* A64, 6355.

Antoniacci, L. & Nasolini, T. (1998) Produzione biologica di fragola nel Cesenate. *Frutticoltura* 5, 41–44.

Benuzzi, M. & Nicoli, G. (1988) *Lotta Biologica e Integrata Nelle Colture Protette: Strategie e Tecniche Disponibili.* Editore Centrale Ortofrutticola, Cesena, 167 pp.

Benuzzi, M., Manzaroli, G. & Mosti, M. (1991*a*) Prove di lotta biologica con larve di *Chrysoperla carnea* (Steph.) contro gli afidi della fragola. In *Abstracts of the National Meeting of Strawberry*, Verona, 8 November 1991 pp. 255–260.

Benuzzi, M., Manzaroli, G. & Nicoli, G. (1991*b*) Lotta biologica e integrata sulla fragola. *Frutticoltura* 9, 63–67.

Benuzzi, M., Manzaroli, G. & Nicoli, G. (1992) Biological control in protected strawberry in northern Italy. *OEPP/EPPO Bulletin* 22, 445–448.

Bonomo, G., Catalano, V. & Sparta, S. (1991) Esperienze di lotta biologica e integrata nella fragolicoltura marsalese. *L'Informatore Agrario* 47, 97–100.

Celli, G. (1987) The so-called 'organic strawberries'. *L'Informatore Agrario* 43, 66–68 (in Italian).

Celli, G., Nicoli, G. & Corazza, L. (1984) Primi risultati di lotta biologica contro gli afidi della fragola in coltura protetta con il neurottero predatore *Chrysoperla carnea* Steph. In *Proceedings of Meeting 'Una giornata sulla Lotta Biologica'*, Forlì, 5 April 1984, pp. 51–61.

Celli, G., Corazza, L., Nicoli, G., Burchi, C., Cornale, R. & Benuzzi, M. (1986) Lotta biologica con *Chrysoperla carnea* (Steph.) (Neuroptera: Chrysopidae) agli afidi della fragola in serra. Due anni di esperienze. *Abstract Giornate Fitopatologiche* 1, 93–102.

Chang, C.P. & Huang, S.C. (1995) Evaluation of the effectiveness of releasing green lacewing, *Mallada basalis* (Walker) for the control of tetranychid mites on strawberry. *Plant Protection Bulletin Taipei* 37, 41–58.

Daane, K.M. & Yokota, G.Y. (1997) Release strategies affect survival and distribution of green lacewings (Neuroptera: Chrysopidae) in augmentation programs. *Environmental Entomology* 26, 455–464.

Duelli, P. (1980) Adaptative dispersal and appetitive flight in the green lacewing, *Chrysopa carnea*. *Ecological Entomology* 5, 213–220.

Galazzi, D. & Nicoli, G. (1992) *Chrysoperla carnea*. *Informatore Fitopatologico* 3, 25–30.

Manzaroli, G. & Mosti, M. (1990) Fragola in coltura protetta: lotta biologica e integrata. *Demetra* 20, 73–76.

Mosti, M. & Barducci, S. (1988) Lotta biologica ed integrata. *Terra e Vita* 18, 74–75.

Nicoli, G. (1992) Lotta biologica ed integrata. *Terra e Vita* 19, 58–59.

Nicoli, G., Galazzi, D., Mosti, M. & Burgio, G. (1991) Embryonic and larval development of *Chrysoperla carnea* (Steph.) (Neur. Chrysopidae) at different temperature regimes. *OILB/SROP Bulletin* 14, 43–49.

Pantaleoni, R. A. (1986) I Neurotteri delle colture agrarie. *Dimensione Ambiente* 8, 17–34

Principi, M.M. (1984) I Neurotteri Crisopidi e le possibilità della loro utilizzazione in lotta biologica e in lotta integrata. In *Proceedings of Meeting 'Una Giornata sulla Lotta Biologica'*, Forlì, 5 Aprile 1984, pp. 1–50.

Sbrighi, M., Faedi, W., Baruzzi, G., Lieten, P., López-Aranda, J.M., Morkocic, M., Roudeillac, P. & Nourisseau, J.C. (1998) La fragolicoltura europea sta cambiando. *Frutticoltura* 5, 11–19.

Tommasini, M.G. & Maini, S. (1995) *Frankliniella occidentalis* and other thrips harmful to vegetable and ornamental crops in Europe. *Wageningen Agricultural University Papers* 95, 1–42.

Trumble, J.T., Oatman, E.R. & Voth, V. (1983) Thresholds and samplings for aphids in strawberries. *California Agriculture* Nov.–Dec. 1983, 20–21.

CHAPTER 27

Artificial overwintering chambers for *Chrysoperla carnea* and their application in pest control

P.K. McEwen and Ç. Sengonca

27.1 INTRODUCTION

Leather *et al.* (1993) asked the pertinent question 'What is overwintering?' They answered themselves by quoting Mansingh (1971) who defines insect hibernation as 'a physiological condition of growth retardation or arrest, primarily designed to overcome lower than optimum temperatures during winter or summer'. Mansingh (1971) subclassified insect hibernation into different categories depending on the insect's response to winter conditions. The three different categories are quiescence, oligopause and diapause representing, in sequence, increasingly highly evolved systems of dormancy. The common green lacewing (*Chrysoperla carnea*) *s. lat.* can be considered to enter diapause. After Leather *et al.* (1993) this is characterised by (1) a definite preparatory phase usually initiated by a temperature independent factor such as photoperiod; (2) absence of feeding by the insect during winter; and (3) the return of favourable conditions does not result in immediate termination of diapause – a complex series of events such as accumulation of heat units is required first.

Common green lacewings diapause in large clusters of adult insects. Observed overwintering sites include unheated parts of buildings, barns, the underside of tree bark, and leaf litter (Canard & Principi, 1984), with the choice of site partly determined by sibling species (Thierry *et al.*, 1994). Normally, due to a combination of fluctuating temperature and predation, winter mortality is high (Sengonca & Frings, 1987). However lacewings will colonise artificial overwintering chambers especially designed for the purpose and in these overwinter survival can approach 100% (Sengonca & Henze, 1992).

The lacewing overwintering chamber (Fig. 27.1) was developed by Sengonca & Frings (1987, 1989) and has been further investigated by McEwen (1998). This chapter provides a simple review of what is known about these overwintering devices and how to use them, and explores the possibility of using the devices as practical portable pest-control units.

27.2 ACCEPTANCE OF THE CHAMBERS

27.2.1 Effect of construction material

Both Sengonca & Frings (1989) and McEwen (1998) compared lacewing catches in plastic and wooden boxes. Sengonca & Frings (1989) found that plywood was a superior material to plastic and this was confirmed by McEwen (1998). In both studies approximately four times as many lacewings colonised wooden chambers as plastic chambers. It is presumed that this is due to the superior thermal properties of the wood which is likely to protect insects better from changes in external temperature.

27.2.2 Effect of orientation and position of the chamber

Sengonca & Frings (1989) examined the effect of chamber orientation on lacewing catches and recorded that strong air movements and circulation inside the chambers leads insects to leave the chambers. For this reason chambers pointing downwind from the main wind direction are preferred. It is also noteworthy that if chambers are not firmly secured (if for example they are hung in trees) then lacewings will also disperse (P.K. McEwen, unpublished data). Few lacewings will colonise chambers placed in trees or near to buildings (P.K. McEwen, unpublished data). The best position to maximise lacewing colonisation is 150 cm from the ground, well away from trees, buildings and other large objects (Sengonca & Frings, 1987).

Fig. 27.1. Artificial overwintering chamber for lacewings.

27.2.3 Relationship between chamber size and rate of colonisation

The chamber developed by Sengonca & Frings (1987, 1989) measured 25–35 cm cubed. Some of the chambers tested by McEwen (1998) were of comparable size (30×30×30 cm) but others were half this size (15×15 ×15 cm). With only one-eighth of the volume of the larger chambers, the smaller chambers attracted approximately one quarter of the number of lacewings. The smaller chambers had an inside surface area one quarter that of the larger chambers (1350 cm^2 versus 5400 cm^2), indicating that surface area is the determining factor as to how many lacewings colonise. However, subsequent work by McEwen (unpublished) showed far greater variation in lacewing numbers taken from chambers of the same size (from 0–80 lacewings per small 15×15×15 cm chamber) suggesting that this relationship may not always hold true.

27.2.4 Specificity of the chambers

Where chambers remain dry inside they are colonised almost exclusively by lacewings. In UK trials in 1996 these were identified as true *Chrysoperla carnea* (P. Duelli, personal communication). Sengonca & Henze (1992) also recorded low numbers of non-lacewing organisms (spiders and Diptera) and reported that this was unaffected by location. When the boxes are damp lacewing numbers drop dramatically and housefly (*Musca domestica*) and spider numbers increase (P.K. McEwen, unpublished data). The absence of large numbers of lacewing predators when chambers are kept dry is important given the reported high numbers of spiders at natural lacewing overwintering sites (Honek, 1977; Frings, 1988) which probably partly accounts for the high overwinter mortality reported above.

27.2.5 Effect of filling material

Sengonca & Frings (1989) experimented with different filling materials by partitioning boxes into three sections, each connected by a 1-cm gap in the dividing wall to allow lacewings to move from one area to another. The three filling materials tested were leaves, hay, and straw. Straw was the preferred material with more than 20 times as many lacewings colonising straw-filled as hay- or leaf-filled chambers.

27.2.6 Effect of colour

Sengonca & Frings (1989) examined the effect of colour (white, brown, red, green) and found no significant effect on lacewing colonisation rate. There is some evidence for an attraction by common green lacewings towards the colour yellow in the field (McEwen, 1995) and further work in this area might be useful. However there is no guarantee that lacewings will respond to the same visual cues in the autumn when searching for overwintering sites as they do when searching for habitats or food in the summer.

27.2.7 Effect of crop

Sengonca & Henze (1992) placed overwintering chambers at three different locations, namely a fallow field, a field sown with winter barley, and a field of sugarbeet ready for harvest. Peak movement of lacewings into all the chambers was in late October, but first flights were 2 weeks later in the sugarbeet fields than the other sites, possibly due to the greater cover and protection offered by the sugarbeet. In terms of total lacewings colonising the chambers at the different sites there were significantly fewer insects at the barley field site, indicating an effect of crop.

27.2.8 Effect of chemical attractant

A number of potential lacewing attractants have previously been identified, including acid hydrolysed L-tryptophan (Hagen *et al.*, 1976; Van Emden & Hagen, 1976; McEwen *et al.*, 1994), terpenyl acetate (Caltagirone, 1969), and caryophyllene (Flint *et al.*, 1979). In the laboratory only caryophyllene, out of this list, was found to evoke a response in green lacewing adults in a wind tunnel (unpublished results). McEwen

(1998) incorporated caryophyllene into some of his overwintering chambers but failed to detect any effect of the attractant on lacewing numbers. It is interesting to note that in another wind tunnel study, the American species *Chrysoperla plorabunda* (formerly also *C. carnea*) reacted positively to L-tryptophan, whereas *C. carnea* in Switzerland in exactly the same set-up did not react (P. Duelli, personal communication). This might suggest that some populations of lacewings could be attracted into overwintering chambers. However, even where lacewings are responding strongly to chemical cues they may respond to different cues when searching for overwintering sites rather than for oviposition sites or food. This would mean that what little is known about responses to chemical attractants would not apply to the overwintering situation.

27.3 EFFECT OF LACEWINGS ON PEST POPULATIONS

It is easy to carry out simplistic calculations on the possible effects of increasing lacewing overwintering survival on pest populations the following spring. For example Principi & Canard (1984) indicated that on average each *Chrysoperla carnea* larva will consume 511 *Bemisia tabaci* whitefly during the course of its development. Assuming that a female surviving the winter as a result of the presence of a lacewing chamber lays 200 eggs the following spring and that 100 of these eggs hatch and successfully reach adulthood, each despatching 500 whitefly on their way, then one female lacewing will result in the death of 50000 whitefly. Clearly other factors intervene and control at this level is not likely. Nevertheless the presence of extra lacewings in the field should demonstrate increased pest control. In fact, the evidence for increased pest control is not good. Sengonca & Henze (1992) placed overwintering chambers containing lacewings in two wheat fields in spring 1990 when the first aphids appeared and subsequently monitored aphid population development and the mean numbers of lacewing eggs, larvae, pupae, and adults in the release and in a control site. Whilst it is clear from their results that lacewings leaving diapause stayed on the site with adults and eggs recorded earlier and in higher numbers than in the control site, increased pest control was not demonstrated. The authors concluded that this was because too few lacewings were released but there must also be a possibility that the lacewings leaving diapause were unable to find sufficient proteinaceous food to mature large quantities of eggs. Such food is required by lacewings in order to maximise egg production (McEwen & Kidd, 1995). McEwen (unpublished data) attempted to address this problem by placing overwintering chambers containing lacewings in glasshouses planted with beans in the spring of 1997. He sprayed some plants with an artificial food comprising yeast autolysate, sucrose, and sugar (4:7:10) and also placed the food on yellow feeding targets. The aim of this study was to demonstrate (a) that lacewings would feed on this material and (b) that this would result in large-scale oviposition on the beans. Although lacewings were observed feeding on both the plants and the targets, suggesting that this method should be explored further, the lacewings were subsequently eaten by spiders before they could oviposit. The experiment was repeated in spring 1998 but glasshouses were first treated with a smoke bomb to kill any resident spiders. Despite lacewings being present in the glasshouse for several weeks following the release of the insects no lacewing eggs were found on the plants and the experiment can be considered to be a failure. The reasons for this are unclear. Resumption of oviposition following diapause is believed to be nearly immediate (C.A. Tauber, personal communication) and significant oviposition from the overwintered lacewings had been expected. Clearly there are other factors at work here and further research is required to clarify the situation.

27.4 CONCLUSION

The conclusion from the foregoing work is that chambers will only function properly (maximum lacewing colonisation and survival) if various guidelines are followed. These are:

- Construct the chambers from wood not plastic
- Follow the design outlined by Sengonca & Frings (1987, 1989)
- Place chambers away from trees and buildings in areas where there has been no or minimal insecticide spraying
- Face chamber pointing away from the prevailing wind direction
- Chambers should be placed about 150–180 cm above the ground
- Tightly pack the chambers with straw

- Place chambers in the field in late summer in order to maximise colonisation

It should also be noted that colonisation by lacewings is poor or non-existent in the event that the contents of the chamber become damp, and design has been modified for the UK market to include an overhanging lid to reduce entry by water. McEwen (unpublished data) also indicates that the louvred bottom section of the box can be replaced with a solid panel, thus reducing manufacturing costs and allowing the box to be placed on, for example, bird tables. In general our information on the performance of these chambers is incomplete. More work is required to explore the effects of conserving overwintering populations of lacewings on pests the following spring. In addition the influence of sibling species on chamber colonisation is unclear despite the evidence that sibling species show differential overwintering behaviour (Thierry et al., 1994). The chambers appear to work better in some countries or regions than in others and the reasons for this require further investigation.

REFERENCES

Caltagirone, L.E. (1969) Terpenyl acetate bait attracts Chrysopa adults. Journal of Economic Entomology 62, 1237

Canard, M. & Principi, M.M. (1984) Development of Chrysopidae. In Biology of Chrysopidae, ed. Canard, M., Séméria, Y. & New, T. R., pp. 57–75. Dr W. Junk, The Hague.

Flint, H.M., Salter, S.S. & Walters, S. (1979) Caryophyllene: an attractant for the green lacewing. Environmental Entomology 8, 1123–1125.

Frings, B. (1988) Unterssuchungen uber die Moglichkeiten der Erhaltung und Forderung von Nutzlingen im Zuckerrubenanbau. Dissertation, Universität Bonn.

Hagen, K.S., Greany, P., Sawall, E.F. Jr & Tassan, R.L. (1976) Tryptophan in artificial honeydews as a source of an attractant for adult Chrysopa carnea. Environmental Entomology 5, 458–468.

Honek, A. (1977) Annual variation in the complex of aphid predators; investigation by light trap. Acta Entomologica Bohemoslovaca 74, 345–348.

Leather, S.R., Walters, K.F.A. & Bale, J.S. (1993) The Ecology of Insect Overwintering. Cambridge University Press, Cambridge.

Mansingh, A. (1971) Physiological classification of dormancies in insects. Canadian Entomologist 103, 983–1009.

McEwen, P.K. (1995) Attractiveness of yellow sticky traps to green lacewings (Neuropt., Chrysopidae). *Entomologist's Monthly Magazine* 131, 163–166.

McEwen, P.K. (1998) Overwintering chambers for green lacewings (*Chrysoperla carnea*): effect of chemical attractant, material and size. *Journal of Neuropterology*, 1, 17–21.

McEwen, P.K. & Kidd, N.A.C. (1995) The effects of different components of an artificial honeydew on fecundity and longevity in the green lacewing, *Chrysoperla carnea. Entomologia Experimentalis et Applicata* 77, 343–346.

McEwen, P.K., Jervis, M.A. & Kidd, N.A.C. (1994) Use of sprayed L-tryptophan solution to concentrate numbers of the green lacewing *Chrysoperla carnea* in olive tree canopy, *Entomologia Experimentalis et Applicata* 70, 97–99.

Principi, M.M. & Canard, M. (1984) Feeding habits. In *Biology of Chrysopidae*, ed. Canard, M., Séméria, Y. & New, T. R., pp. 76–92. Dr W. Junk, The Hague.

Sengonca, Ç. & Frings, B. (1987) Ein künstliches Überwinterungsquartier für die rauberische Florfliege. *DLG Mitteilungen* 102, 656–657.

Sengonca, Ç. & Frings, B. (1989) Enhancement of the green lacewing *Chrysoperla carnea* (Stephens) by providing artificial facilities for hibernation. *Turkiye Entomoloji Dergisi* 13, 245–250.

Sengonca, Ç. & Henze, M. (1992) Conservation and enhancement of *Chrysoperla carnea* (Stephens) (Neuroptera: Chrysopidae) in the field by providing hibernation shelters. *Journal of Applied Entomology* 114, 497–501.

Thierry, D., Cloupeau, R. & Jarry, M. (1994) Variation in the overwintering ecophysiological traits in the common green lacewing West-Palaearctic complex (Neuroptera: Chrysopidae). *Acta Oecologica* 15, 1–17.

Van Emden, H.F. & Hagen, K.S. (1976) Olfactory reactions of the green lacewing *C. carnea* to tryptophan and certain breakdown products. *Environmental Entomology* 5, 469–473.

CHAPTER 28

Lacewings in Andalusian olive orchards

M. Campos

28.1 INTRODUCTION

Olive cultivation is of great importance in the Mediterranean Basin, representing 97% of world production. Spain, with 27% of world production, is the foremost olive-growing country in the world in terms of surface area under cultivation, and 60% of the national production is concentrated in Andalusia (Porras Piedra et al., 1995; Barranco et al., 1997).

Factors that have a negative impact on this crop include pests, diseases, and weeds, for which losses in yield can reach 30% (De Andrés Cantero, 1991). Among the 18 recognised insect pests, the three most serious are: *Bactrocera oleae* (Gmelin), the olive fruit fly; *Prays oleae* (Bernard), the olive moth; and *Saissetia oleae* (Olivier), the olive scale (Bellido, 1975; Campos, 1976; Briales, 1984; Civantos, 1995). Currently, protection against these pests relies on concepts of integrated control based on knowledge of the environment and of the population dynamics within the agricultural systems and, in this sense, both chemical as well as biological methods are used. The aim is to reduce the insect populations to levels at which the damage caused does not exceed certain economic thresholds (Civantos & Sanchez, 1994; Civantos, 1995).

The utilisation of entomophagous insects plays a key role within the widest concept of integrated olive-pest control. Most of the work performed to date on natural enemies in southern Spain has focused on parasitoids (Jimenez et al., 1969; Campos & Ramos, 1981; Briales & Campos, 1985; Montiel Bueno & Santaella, 1995) rather than on the predators (Morris, 1997). However, numerous studies have been conducted on chrysopids, given that lacewing larvae are major oophagous predators of the olive moth (Montiel Bueno, 1981; Ramos et al., 1983a, b). These predators are especially beneficial during the oviposition period of the pest on the fruits, the period of maximum economic impact on the olive. In addition, they are also known predators of the olive scale *S. oleae* as well as of less harmful pests such as the olive psyllid *Euphyllura olivina* Costa and *Aspidiotus nerii* Bouche (Alrouechdi et al., 1981b; Arambourg, 1986; De Andrés, 1991).

28.2 CHRYSOPIDS WITHIN THE OLIVE ORCHARD

Chrysopids form part of the permanent complex of insect predators within the ecosystem of the olive orchard (Fimiani, 1965; Viggiani et al., 1973; Canard & Laudeho, 1977, 1980; Canard et al., 1979; Alrouechdi, 1980; Alrouechdi et al., 1980; Neuenschwander & Michelakis, 1980; Bagnoli, 1983; Campos & Ramos, 1983; Delrio, 1983; Sacchetti, 1990; Pantaleoni, 1991; Bento, 1994). In Andalusian orchards, 13 species have been recorded in McPhail and yellow traps: *Chrysoperla carnea, Mallada flavifrons, M. genei, M. prasina, Rexa lordina, M. picteti, M. venosa, Cunctochrysa baetica, Chrysopa viridana, M. subcubitalis, C. formosa, C. septempunctata* and *Italochrysa italica*. The most abundant species are *Chrysoperla carnea* and *M. flavifrons*, which account for 80% of the total captures (Campos & Ramos, 1983). According to these authors, the overall population of adult chrysopids present over almost the entire year begins to increase in March, reaching a single well-defined maximum in July. Afterwards the populations decline, most sharply in November. During the three first months of the year, the only species present in the orchard is *C. carnea*. In April, species increase in diversity until June, when 10 of the 13 species are found. In the second half of the year, species richness surpasses that of the first half, with seven species recorded in October. With respect to the sexes captured, in general, females outnumber males, especially in autumn, a trend also reported by Alrouechdi (1980), who indicated greater longevity among females and the early hatching of the males.

With regard to the presence of the different species of chrysopids, Campos & Ramos (1983) indicated that

C. carnea occupies a dominant position in Andalusian olive orchards, as in olive orchards in France (Alrouechdi *et al.*, 1980), Greece (Canard, 1978), Italy (Sacchetti, 1990; Pantaleoni, 1991), and Portugal (Bento, 1994). This species, which overwinters as an imago, is the only chrysopid present over the entire year, this also being the case in olive orchards in other areas (Canard *et al.*, 1979; Delrio, 1983), with a small peak in numbers in March and a subsequent larger peak in June to July (1.06 adults/day/trap). *Mallada flavifrons* appears around mid-April and reaches a maximum in August, maintaining similar levels during June and July (0.58 adults/day/trap) and even the first part of September, after which its numbers diminish until it disappears in December. *Mallada genei* is captured from the end of June until the middle of October, presenting a maximum in August (0.18 adults/day/trap). This species, essentially Mediterranean, is also present in Greek olive orchards (Canard & Laudeho, 1977). *Mallada prasina* makes its appearance in mid-April and disappears in October. Its numbers in Andalusian olive orchards are low (0.07 adults/day/trap), while in France it is the most frequent chrysopid (Alrouechdi *et al.*, 1980). *Rexa lordina* is present only during the spring, whereas *M. picteti*, also captured in low numbers, is found from April to October.

28.3 BIOECOLOGY OF *CHRYSOPERLA CARNEA* (STEPHENS)

This insect has three generations per year in Spain (Campos, 1989), the first beginning approximately in March, the second at the end of May and beginning of June, and the third in August to September. The second generation produces the greatest population density. The adult populations in olive orchards vary from year to year, although the fluctuations are not as sharp as those reported by Honek (1977*a*). Reproductive activity normally takes place from the first of March to the end of November and, in certain years, even extends through December and January (Campos, 1989), as observed by Alrouechdi & Canard (1979), given that this species presents a facultative reproductive diapause (Principi, 1991). In other zones, the reproductive period of this chrysopid is shorter (Honek, 1977*b*; Alrouechdi *et al.*, 1981*b*). The percentage of gravid females varies according to the year, and eggs are laid principally on the petioles of the leaves.

Among the limiting factors of the *C. carnea* populations is parasitism which has been the subject of numerous studies (Séméria, 1976, 1981; Alrouechdi & Panis, 1980; Alrouechdi *et al.*, 1981*a*; Pantaleoni, 1987; Ruberson *et al.*, 1989). Nevertheless, in Andalusian olive orchards, the insect fauna of this chrysopid and its activity are not as crucial (Campos, 1986) as in the olive orchards of southeastern France (Alrouechdi & Panis, 1980; Alrouechdi *et al.*, 1981*a*) or Greece (Neuenschwander & Michelakis, 1980). The eggs are attacked exclusively by *Telenomus acrobates*, with an average parasitisation rate of 6.6%. Parasitoids of larvae and cocoons are more harmful, this parasitisation reaching 20%; the parasitoid species recorded are: *Isodromus punticeps*, *Helorus* sp., and *Pachyneuron concolor* (secondary species). With respect to the adults, the only parasitoid is *Chrysopophthorus chrysopimaginis*, which affects 3.3% of the *Chrysoperla carnea* adults, of which 86.4% are females. Other chrysopids attacked by this parasitoid include *Mallada flavifrons*, *M. picteti*, and *M. prasina*. Ants, prominent among the insect fauna of Andalusian olive orchards (Morris, 1997), also attack chrysopid eggs. *Crematogaster scutelaris* Ol. and *Tapinoma nigerimum* Nyl. are the most effective predators of chrysopid eggs, with the latter being probably the most damaging due to both its large nest size and its activity. *Plagiolepis pygmaea* (Latr.), *Camponotus micans* Nyl., and *C. foreli* Em. are the least efficient predators. *Formica subrufa* Roger is intermediate in terms of its ability to prey on lacewing eggs, but a small sample size prevents any firm conclusion to be drawn (Morris *et al.*, 1998).

28.4 CONSERVATION AND AUGMENTATION OF *CHRYSOPERLA CARNEA* POPULATIONS

The high level of olive moth egg predation evident in the olive in some years (see below) suggests that control could be enhanced in other years through the manipulation of the lacewing population. Thus, as part of a wider study of the biological control of olive pests, numerous authors have investigated the role of the lacewing in olive-pest management to improve pest-control efficiency.

McEwen *et al.* (1994) have shown that it is possible to concentrate adults of the green lacewing *C. carnea* in areas of olive orchards that have been sprayed with solutions containing acid hydrolysed L-tryptophan, but these researchers observed that without the provision of proteinaceous material for maintenance and egg production, the adults thus concentrated did not increase the lacewing egg population (McEwen et al., 1993*b*), thereby failing to improve pest control.

The attempt to enhance predation of pests by *C. carnea* by spraying of artificial honeydew in olive orchards in order to attract adult lacewings to the crop and provide food for their larval progeny did not appreciably increase predation on the pest eggs (Liber & Niccoli, 1988; McEwen et al., 1993*a*).

One aspect of green lacewing ecology that is at present poorly understood is the role of non-olive vegetation within the olive orchard. McEwen & Ruiz (1994) have demonstrated a clear association between non-olive vegetation and green lacewing eggs, a finding with important olive-management implications. In commercial olive orchards in Andalusia, the non-olive vegetation is usually removed by ploughing or by the application of herbicides early in the season to reduce water competition between the olive trees and weeds. This destroys a significant proportion of the green lacewing population in the olive orchard.

One of the problems in the use of pesticides in the olive orchard is the negative impact on the rich auxiliary insect fauna (Heim, 1985). It has been determined that organophosphate treatments against attacks by *Prays oleae* reduce the activity of *C. carnea*, and thus in sprayed olive trees the percentage of eggs eaten decreases significantly, augmenting the damage in these trees (Fimiani, 1965; Ramos et al., 1978).

28.5 RELATIONSHIP BETWEEN *PRAYS OLEAE* AND *CHRYSOPERLA CARNEA*

Prays oleae produces three generations per year, attacking olive leaves, flowers, and fruits, and causing significant damage and crop loss (Campos, 1976; Civantos, 1995). The larvae of *C. carnea* feed principally on eggs and larvae of the olive moth, an observation which has been confirmed in the serological test, enzyme-linked immunosorbent assay, ELISA (Morris, 1997). Although their action can be detected in all three generations of *P. oleae*, the predatory activity is most important on the moth eggs laid on olive fruits, in some years resulting in a dramatic decrease in the attack of the olive moth (Ramos et al., 1984, 1988*b*).

Among the factors that affect the relationship between these two insects, apart from their own population levels, is the temporal and distributional coincidence with respect to the egg-laying by the moth, this occurring roughly at 36-day intervals on average. In relation to the second factor, it is known that *P. oleae* is unable to distinguish between olive fruits with and without eggs of its own species (Fournier et al., 1980), so that there is normally a certain degree of egg accumulation on the fruits. During 17 years of study in the Andalusian olive orchard, the most frequent category of egg number per fruit was one egg per fruit, followed by two and three. The category of four and more eggs per fruit was significantly less frequent (Campos & Ramos, 1984; Ramos et al., 1988*a*), and no relationship was found between egg density and the response of the predator. This contradicts the results of Alrouechdi et al. (1981*b*), who found that the proportion of eggs eaten increased when eggs were in groups of three or more per fruit.

The predatory activity of *C. carnea* is undoubtedly highly important, since in 75% of the 17 years studied, predation of *P. oleae* eggs exceeded 50% (Ramos et al., 1987, 1988*b*). However, this predation is of value only when all the eggs laid on a particular fruit are destroyed, since only one surviving egg can produce a larva which in turn can destroy an olive fruit. Therefore, true predatory efficiency should be expressed as the percentage of protected fruits (Alrouechdi, 1980), i.e. fruits for which all the moth eggs were eaten, and this value decreased when the number of eggs per fruit increased. In Andalusian olive orchards, in 50% of the years studied, the number of protected fruits surpassed 60%.

In summary, the true action of the predator *C. carnea* is substantial in the Andalusian olive orchard, the larvae of this species being a factor in the natural reduction of *P. oleae*. Therefore, for the improvement of pest protection in olive orchards, natural chrysopid populations should be conserved, and their biotic potential should be increased.

REFERENCES

Alrouechdi, K. (1980). Les chrysopides en verger d'oliviers. Bioécologie de *Chrysoperla carnea* (Stephens) (Neuroptera:Chrysopidae), relations comportamentales et trophiques avec certaines espèces phytophages. PhD thesis, Université de Paris.

Alrouechdi, K. & Canard, M. (1979). Mise en évidence d'un biotype sans diapause photopériodique dans une population méditerranéenne de *Chrysoperla carnea* (Stephens) (Insectes: Neuroptera). *Séances l'Académie des Sciences Paris* 289, 553–555.

Alrouechdi, K. & Panis, A. (1980) Les parasites de *Chrysoperla carnea* (Steph.) (Neuroptera, Chrysopidae) sur olivier en Provence. *Agronomie* 1, 139–141.

Alrouechdi, K., Lyon, J.-P., Canard, M. & Fournier, D. (1980) Les chrysopides (Neuroptera) récoltés dans une oliveraie de sud-est de la France. *Acta Oecologica* 1, 173–180.

Alrouechdi, K., Canard, M., Pralavorio, R. & Arambourg, Y. (1981*a*) Influence du complexe parasitaire sur les populations de chrysopides (Neuroptera) dans un verger d'oliviers du sud-est de la France. *Zeitschrift für angewandte Entomologie* 91, 411–417.

Alrouechdi, K., Pralavorio, R., Canard, M. & Arambourg, Y. (1981*b*) Coincidence et relations prédatrices entre *Chrysopa carnea* Stephens (Neuroptera: Chrysopidae) et quelques ravageurs de l'olivier dans le sud-est de la France. *Bulletin de la Société Entomologique Suisse* 54, 281–290.

Arambourg, Y. (1986) *Traite d'Entomologie Oleicole*. Consejo de Oleicola Internacional, Madrid.

Bagnoli, B. (1983) Entomophagous insects of *Saissetia oleae* (Oliv.) in olive groves in Tuscany. In *Entomophagous Insects and Biotechnologies against Olive Pests, Proceedings of the E.C. Experts' Meeting on Entomophages and Biological Methods in Integrated Control in Olive Orchards*. Greece, ed. Cavalloro, R. & Piavaux A., A.A. Balkema, Rotterdam.

Barranco, D., Fernandez-Escobar, D. & Rallo, L. (1997) *El Cultivo del Olivo*. Mundi-Prensa, Madrid.

Bellido, L. (1975). Contribución al estudio de *Prays oleae* Bern (Lep. Hyponomeutidae) en Córdoba. PhD thesis, Universidad Córdoba.

Bento, A. (1994) Studo sobre a traça da oliveira (*Prays oleae* Bern.) na terra quente transmontana na óptica da protecçao integrada. MSc thesis, Universidade Lisboa.

Briales, M.J. (1984) Contribución al estudio bioecológico de *Saissetia oleae* (Oliv.) (Homoptera, Coccoidea, Coccidae) en Granada. PhD thesis, Universidad Granada.

Briales, M.J. & Campos, M. (1985) Contribución al estudio de la entomofauna parasitaria de *Saissetia oleae* (Olivier, 1971) (Hom., Coccidae) en la zona de Iznalloz (Granada). *Asociación española de Entomología* 9, 5–62.

Campos, M. (1976) Contribución al estudio de la entomofauna del olivo en España. Observaciones bioecológicas sobre *Prays oleae* Bern. (Lepidoptera, Hyponomeutidae). PhD thesis, Universidad Granada.

Campos, M. (1986) Influencia del complejo parasitario sobre las poblaciones de *Chrysoperla carnea* (Neuroptera, Chrysopidae) en olivares del sur de España. *Neuroptera International* 4, 97–105.

Campos, M. (1989) Observaciones sobre la bioecología de *Chrysoperla carnea* (Stephens) (Neuroptera: Chrysopidae) en el sur de España. *Neuroptera International* 5, 159–164.

Campos, M. & Ramos, P. (1981) Contribución al estudio de la entomocenosis de *Prays oleae* Bern. en Granada. *Acta Oecologica, Oecologica Applicata* 2, 27–35.

Campos, M. & Ramos, P. (1983) Chrisópidos (Neuroptera) capturados en un olivar del sur de España. *Neuroptera International* 2, 219–227.

Campos, M. & Ramos, P. (1984) Some relationships between the number of *Prays oleae* eggs laid on olive fruits and their predation by *Chrysoperla carnea*. In *Proceedings of International Joint Meeting CEC/FAO/IOBC*. pp. 237–241.

Canard, M. (1978) Chrysopides (Neuroptera) récoltés dans les oliveraies en Grèce. *Biologia Gallo-Hellenica* 8, 237–242.

Canard, M. & Laudeho, Y. (1977) Les névroptères capturés au piège de McPhail dans les oliviers en Grèce. 1: L'île d'Aguistri. *Biologia Gallo-Hellenica* 7, 65–75.

Canard, M. & Laudeho, Y. (1980) Les névroptères capturés au piège de McPhail dans les oliviers en Grèce. 2: La region d'Akrefnion. *Biologia Gallo-Hellenica* 9, 139–146.

Canard, M., Neuenschwander, P. & Michelakis, S. (1979) Les névroptères capturés au piège de McPhail dans les oliviers en Grèce. 3: La Crète occidentale. *Annales de la Société Entomologique Française* 15, 607–615.

Civantos, M. (1995) Desarrollo del control integrado en el olivar español. *Olivae* 59, 75–81.

Civantos, M. & Sanchez, M. (1994) Nuevos métodos de lucha contra plagas y enfermedades en olivar. *Fruticultura* 62, 69–76.

De Andrés Cantero, F. (1991) *Enfermedades y plagas del olivo*. Riquelme y Vargas, Jaén.

Derlio, G. (1983) The entomophagous insects of *Saissetia oleae* (Oliv.) in Sardinia. In *Entomophagous Insects and Biotechnologies against Olive Pests, Proceedings of the E.C. Experts' Meeting on Entomophages and biological methods in integrated control in olive orchards*, Greece, pp. 15–23.

Fimiani, P. (1965) Effetti del 'Sevin' sull'entomofauna dell'olivo e degli agrumi. *Istituto di Entomologia Agraria Portici* 30, 3–8.

Fournier, D., Pralavorio, R. & Arambourg, Y. (1980) La compétition larvaire chez *Prays oleae* Bern. (Lep., Hyponomeutidae) et ses relations avec quelques paramètres démographiques. *Acta Oecologica, Oecologica Applicata* 1, 233–246.

Heim, G. (1985). Effect of insecticial sprays on predators and indifferent arthropods found on olive trees in the north of Lebanon. In *Proceeding of the International Joint Meeting* CEC/FAO/IOLB, Pisa, pp. 456–465.

Honek, A. (1977a). The life cycle of *Chrysopa carnea* Stephens (Neuroptera) in central Europe. *Acta Entomologica Bohemoslovacia* 78, 76–86.

Honek, A. (1977b). Annual variation in the complex of aphid predators: investigation by light trap. *Acta Entomologica Bohemoslovacia* 74, 345–348.

Jimenez, A., Arroyo, M. & Mellado, L. (1969). Ensayos preliminares de la introducción en España de *Opius concolor* Szepl. parasito endófago de *Dacus oleae* Rossi. *Boletín Patología Vegetal Agraria* 31, 29–32.

Liber, H. & Niccoli, A. (1988). Observations on the effectiveness of an attractant food sprays in increasing chrysopid predation on *Prays oleae* (Bern.) egg. *Redia* 71, 467–482.

McEwen, P. K. & Ruiz, J. (1994). Relationship between non-olive vegetation and green lacewing eggs in Spanish olive orchard. *Antenna* 18, 148–150.

McEwen, P. K., Jervis, M.A. & Kidd, N.A.C. (1993a). The effect on olive moth (*Prays oleae*) populationn levels, of applying artificial food to olive trees. In *A.N.P.P. 3rd International Conference Pests Agriculture*, Montpellier, pp. 361–368.

McEwen, P.K., Jervis, M.A. & Kidd, N.A.C. (1993b). Manipulation of naturally occurring populations of the green lacewing *Chrysoperla carnea* (Stephen) as a means to control the olive moth *Prays oleae* (Bern.). In *The Development of Environmentally Safe Pest Control Systems for European Olive*. AGRE:0013: Eclair 209 Tech. Annex, 1992/93.

McEwen, P.K., Jervis, M.A. & Kidd, N.A.C. (1994). Use of a sprayed L-tryptophan solution to concentrate numbers of the green lacewing *Chrysoperla carnea* in olive tree canopy. *Entomologia Experimentalis et Applicata* 70, 97–99.

Montiel Bueno, A. (1981) Factores de regulación de las poblaciones de *Prays oleae* (Bern.). *Boletín Servicio de Plagas* 7, 133–140.

Montiel Bueno, A. & Santaella, S. (1995). Evolución de la población de *Saissetia oleae* OLIV en condiciones naturales. Periodos susceptibles de control biológico. *Boletín Sanidad Vegetal Plagas* 21, 445–455.

Morris, I.T. (1997). Interrelaciones entre olivos, plagas y depredadores. PhD thesis, Universidad Granada.

Morris, I.T., Campos, M., Kidd, N.A.C., Jervis, M.A. & McEwen, P.K. (1998) Potential effects of various ant

species on green lacewing (*Chrysoperla carnea*) egg number. *Journal of Applied Entomology*, 122, 401–403.

Neuenschwander, P. & Michelakis, S. (1980). The seasonal and spatial distribution of adults and larval chrysopids on olive-trees in Crete. *Acta Oecologica, Oecologica Applicata*, 1, 93–102.

Pantaleoni, R.A. (1987) Studi sui parassitoidi (Hym. Ichneumonidae et Pteromalidae) di *Chrysoperla carnea* (St.) (Neur. Chrysopidae) che ricercano ed aggrediscono gli stadi racchiusi nel bozzolo. *Bollettino dell'Istituto di Entomologia 'Guido Grandi' dell'Università di Bolognà* 41, 241–255.

Pantaleoni, R.A. (1991) I neurotteri delle colture agrarie: Crisopidi in oliveti del Salento (Italia meridionale). *Bollettino dell'Istituto di Entomologia 'Guido Grandi' dell'Università di Bologa* 45, 167–179.

Porras Piedra, A., Cabrera de la Colina, J. & Soriano Martín, M. L. (1995) *Olivicultura y Elaiotecnia*. Universidad de Castilla–La Mancha

Principi, M.M. (1991) Lo stato di diapausa negli insetti e suo manifestarsi in alcune specie di Crisopidi (Insecta, Neuroptera) in dipendenza dell'azione fotoperiodica. *Bollettino dell'Istituto di Entomologia 'Guido Grandi' dell'Università di Bologna* 46, 1–30.

Ramos, P. Campos, M. & Ramos, J.M. (1978) Osservazioni biologiche sui trattamenti contro la tignola dell'olivo (*Prays oleae* Bern., Lep. Plutellidae) *Bollettino del Laboratorio di Entomologia Agraria 'Fillippo Silvestri' di Portici*, 35, 16–24.

Ramos, P., Campos, M. & Ramos, J.M. (1983a) The economic importance of predators in reducing infestation of olive fruits by *Prays oleae* Bern. in Granada (Spain). *Olea* 30–31.

Ramos, P., Campos, M. & Ramos, J.M. (1983b). État actuel des recherches sur la lutte biologique contre la teigne de l'olivier en Espagne. In *Entomophagous Insects and Biotechnologies against Olive Pests, Proceedings of the E.C. Experts' Meeting on Entomophages and Biological Methods in Integrated Control on Olive Orchards. Entomophages and Biological Methods in Integrated Control in Olive Orchards*, ed. Cavalloro, R. & Piavaux, A., pp. 128–144. A.A. Balkema, Rotterdam.

Ramos, P., Campos, M. & Ramos, J.M. (1984) Estabilización del ataque de *Prays oleae* Bern. y de la actividad de los depredadores oófagos sobre el fruto del olivo. *Boletín Servicio de Plagas* 10, 239–243.

Ramos, P., Campos, M. & Ramos, J.M. (1987) Evolución del ataque de *Prays oleae* Bern. al fruto del olivo. I. Estudio de parametros y sus relaciones. *Boletín Sanidad Vegetal Plagas* 13, 129–12.

Ramos, P., Campos, M. & Ramos, J.M. (1988a) Evolución

del ataque de *Prays oleae* Bern. al fruto del olivo. II. Evolución de puestas, estabilización de parametros y ecuaciones predictivas. *Boletín Sanidad Vegetal Plagas* 14, 265–278.

Ramos, P., Campos, M. & Ramos, J.M. (1988*b*) Evolución del ataque de *Prays oleae* Bern. al fruto del olivo. III. Distribución y agregación de puestas. *Boletín Sanidad Vegetal Plagas* 14, 343–355.

Ruberson, J.R., Tauber, C.A. & Tauber, M.J. (1989) Development and survival of *Telenomus lobatus*, a parasitoid of chrysopid eggs: effect of host species. *Entomologia Experimentalis et Applicata* 51, 101–106.

Sacchetti, P. (1990) Osservazioni sull'attività e sulla bio-etologia degli entomofagi di *Prays oleae* (Bern.) in Toscana. I. I predatori. *Redia* 73, 243–259.

Séméria, Y. (1976) Un braconide, *Chrysopophthorus chrysopimaginis* Goidanich (Hymenoptera, Braconidae) parasite des imagos de Chrysopinae (Planipennia, Chrysopidae). *Annales de la Société Linnéenne de Lyon* 3, 102–107.

Séméria, Y. (1981) Signification de deux formes de sur-parasitisme, dont une inédite, chez les adultes de chrysopine (Planipennia, Chrysopidae), produits par le braconide (Hymenoptera): *Chrysopophthorus chrysopimaginis* Goid. *Neuroptera International* 1, 207–209.

Viggiani, G., Fimiani, P. & Bianco, M. (1973) Ricerda di un metodo di lotta integrata per el controlo della *Saissetia oleae* (Oliv.). *Atti Giornate Fitopatologiche* 251–259.

CHAPTER 29

The green lacewings of Romania, their ecological patterns and occurrence in some agricultural crops

M. Paulian

29.1 INTRODUCTION

The impact of Chrysopidae in the limiting of arthropod populations, especially those with a soft integument such as aphids, lepidopterous larvae, leafhoppers and mites has been frequently noted (Ridgway & Murphy, 1984; Baudry, 1996). The generally negative pressure of pesticides upon the environment compared to cases of agricultural crops managed by biological and integrated protections emphasises this point.

In Romania the following study on Neuroptera species provides some evidence of aspects such as:

- cataloguing the correct identity of lacewing species from this geographical area, the biotopes where these occurred, and their ecological patterns;
- assessment of the species that occur in the agricultural crops of the region and their ecological and behavioural aspects related with these agroecosystems;
- assessment of the factors that influence their occurrence and the effects generated by their presence;
- examination of the factors that allow the conservation of lacewing populations.

29.2 CHRYSOPIDAE OF ROMANIA

In Romania the natural environment is characterised by high diversity. In the framework of a temperate continental climate, topography and vegetation is heterogeneous (varied), providing numerous different habitats for a rich fauna.

In an earlier list and classification for Romania (Kis et al., 1970) the Chrysopidae were represented by 6 genera and 24 species. Following our study using the revised classification of Chrysopidae (Brooks & Barnard, 1990), the list for Romania was increased to 29 species belonging to 9 genera (Table 29.1). This represents 47% of the chrysopid species recorded from Europe (Aspöck, 1992). This number is lower than that from neighbouring regions. In Yugoslavia, the number of species recorded is 35, but additional species occur in the warmer climatic regions such as Slovenia, Croatia, and Macedonia, and are associated with particular vegetation types (Devetak, 1992). By comparison, in Bulgaria 28 species are known (Popov, 1990). In Bulgaria, *Chrysopa dasyptera* is not reported, although this species is recorded in Hungary (Szabo & Szentkirály, 1981) and in Romania (Paulian, 1996), and *Chrysopa hungarica* (Kis et al., 1970), is poorly recorded, although it is found in south and southeast Romania (in the Romanian Plain and Dobrogea) which are nearest to Bulgaria.

29.3 THE COMMON GREEN LACEWING COMPLEX IN ROMANIA

The green lacewing *Chrysoperla carnea*, has been long considered a single cosmopolitan species, morphologically slightly variable, although having local biotypes differing in seasonal and life history characters (Sheldon & MacLeod, 1974; Alrouechdi & Canard, 1979). According to recent research *C. carnea* includes a number of sibling species and a diversity of faunal elements of this complex occurs in the northern hemisphere (Canard, 1998). The nomenclature used to characterise these insects remains uncertain (Thierry & Adams, 1992).

Seven species for the genus *Chrysoperla* Steinmann have been proposed for Europe (Leraut, 1991) Three of these belong to the *Chrysoperla carnea* complex (Brooks, 1994): *Chrysoperla carnea* s. str., *Chrysoperla lucasina* (Lacroix), and *Chrysoperla kolthoffi* (*sensu*

Table 29.1. *Check-list of Chrysopidae recorded from Romania*

1. *Italochrysa italica* (Rossi 1790)	16. *Chrysoperla renoni* (Lacroix 1933)
2. *Chrysopa abbreviata* Curtis 1834	17.*Chrysotropia ciliata* (Wesmael 1841)
3. *Chrysopa dasyptera* McLachlan 1872	18.*Cunctochrysa albolineata* (Killington 1935)
4. *Chrysopa dorsalis* Burmeister 1839	19. *Dichochrysa flavifrons* (Brauer 1850)
5. *Chrysopa formosa* Brauer 1850	20. *Dichochrysa prasina* (Burmeister 1839)
6. *Chrysopa hungarica* Klapalek 1899	21. *Dichochrysa ventralis* (Curtis 1834)
7. *Chrysopa nigricostata* Brauer 1850	22. *Dichochrysa zelleri* (Schneider 1851)
8. *Chrysopa pallens* (Rambur 1838)	23. *Nineta flava* (Scopoli 1763)
9. *Chrysopa perla* Linnaeus 1758	24. *Nineta pallida* (Schneider 1846)
10. *Chrysopa phyllochroma* Wesmael 1841	25. *Nineta vittata* (Wesmael 1841)
11. *Chrysopa viridana* Schneider 1845	26. *Peyerimhoffina gracilis* (Schneider 1851)
12. *Chrysopa walkeri* McLachlan 1893	27. *Hypochrysa elegans* (Burmeister 1839)
13. *Chrysoperla carnea carnea* Stephens 1836	28. *Nothochrysa capitata* (Fabricius 1793)
14. *Chrysoperla lucasina* Lacroix 1912	29. *Nothochrysa fulviceps* (Stephens 1836)
15.*Chrysoperla kolthoffi* (Navás 1927)	

Leraut, 1991). Further discussion of this complex is found in Henry *et al.* (Chapter 3, this volume). These reproductively isolated species are clearly distinguished by courtship song, seasonal differences in mating period, and habitat differences (Thierry *et al.*, 1992).

According to this research, we did a preliminary assessment of the real identity of the *Chrysoperla* species in Romania, using data concerning their distribution and ratio. The main geographical regions where collecting was carried out are given in Table 29.2. After initial examination of 'the common green lacewing' from Romania, it was clear that there are in fact four taxa belonging to *Chrysoperla* in the Romania fauna (Paulian *et al.*, 1996). *C. kolthoffi* (*sensu* Leraut 1991) is the dominant species (61%), constituting more than half of the total specimens of *Chrysoperla*, being found in all representative sites, with a frequency of 91% in some places (e.g. Rodna).

Chrysoperla lucasina represented more than one-third of the total (38%). It is not really a minority species, other than in one place (Rodna); on the contrary, it is dominant in one biotope, viz. the mountains of Furnica (Sinaia). This species occurs in equal numbers with *C. kolthoffi* in two places, respectively in Oltenita, where the collecting has been undertaken very near the Danube, and in Uzlina, from a swamp, where reeds are abundant. *Chrysoperla carnea s. str.*, is globally rare, constituting less than 1% of the total, occurring only in three of the intensively surveyed locations.

Besides the three species of the *C. carnea* complex, a male specimen was found, which was identified as *Chrysoperla renoni*. The occurrence of this significant novelty indicated the existence of more species belonging to *Chrysoperla* in Romania. This specimen is interesting because of its ecological characteristics, originating from the Uzlina area, which contains natural ecosystems and the typical vegetation of the Danube Delta (Paulian & Andriescu, 1996). This is a very interesting, if not unique biome in Europe, but there has been no special study on the faunistic characters of this location. The Neuroptera of this biome are thus only known through the general work of Kis *et al.* (1970), and later by the European survey of Aspöck *et al.* (1980).

Initially, *C. renoni* was only known and described from some specimens taken in the region Deux–Sèvres, France. It is possible that it has often been confused with morphologically similar species, such as *Chrysoperla mediterranea* (Hölzel) and *C. carnea*. *Chrysoperla renoni* prefers moist and swampy lowland habitats (Leraut, 1992), and those caught in the Danube Delta confirm this suspected preference. It is a new species (Paulian *et al.*, 1996) and its occurrence here

Table 29.2. *Number and frequency of species of the Chrysoperla carnea complex in Romania*

Number and place	carnea N	carnea Frequency %	kolthoffi N	kolthoffi Frequency %	lucasina N	lucasina Frequency %	renoni N	renoni Frequency %	Total by place	Method of collection
1. Arad			1	100					1	Sweeping hedges
2. Beclean			3	100					3	Sweeping hedges
3. Beius	1	100							1	McPhail trap
4. Bucuresti	2	1	148	61	92	38			242	Light trap
5. Crasna			7	100					7	Sweeping hedges
6. Dabuleni	1	1	58	57	38	39	1	1	98	Sweeping hedges
7. Furnica			40	37	65	61	1	1	107	Light trap
8. Huedin			7	87	1	13			8	Light trap
9. Mitreni			29	78	8	22			37	Light trap
10. Oltenita			10	50	10	50			20	Light trap
11. Oradea			1	100					1	Sweeping hedges
12. Rodna			43	91	4	9			47	Light + sweeping grass
13. Scobilteni			6	100					6	Sweeping hedges
14. Sinaia	2	6	20	63	8	25	2	6	32	Light trap
15. Tirnaveni			1	100					1	Sweeping trees
16. Troianu			18	72	7	28			25	Light trap
17. Uzlina			14	45	16	52	1	3	31	Sweeping hedges
Total by species	6	—	406	—	250	—	6	—	668	
Frequency %	—	1	—	61	—	37	—	1		

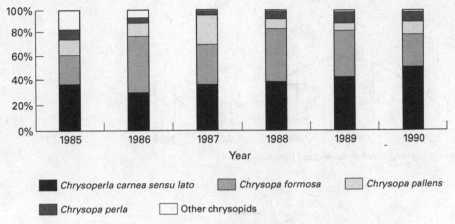

Fig. 29.1. The ratio of major chrysopid species collected by light trapping in Romania.

considerably enlarges the distribution eastwards from the previous known range of the region Deux–Sevres and the Parisian Basin.

29.4 ECOLOGICAL PATTERNS OF CHRYSOPIDS IN ROMANIA

The great diversity of environmental conditions induces a very large and specific distribution of the green lacewings, each of which has a particular seasonal strategy, already described from some regions and/or biotypes, to best adjust their life cycles to local conditions (Tauber & Tauber, 1981; Canard, 1998; Volkovich, 1998).

Chrysopidae are uniform in their morphological structures, but not in their biological and ecophysiological characters (Canard, 1998). A comprehensive study of seasonal adaptations of closely related species and especially of those dominant in each geographical region is very interesting for understanding which factors influence the seasonal cycles, the relationship among them, and their capability to suppress injurious insects to agriculture.

We carried out a study to determine the structure and the level of the green lacewing populations, the ratio between species, the geographical distribution in various areas and biotopes of Romania, and some ecological traits, during investigations in recent years. The aim was to clarify the structure of lacewing populations in agricultural biotopes in Romania, focusing on eco-

logical traits of related Chrysopidae, on the possibility of biological control and their protection as natural predators in crop pest management.

29.4.1 The green lacewing species and their flight activity

Some ecological features were clarified by interpreting catches of adults from light traps. During six consecutive years (1985–90) a long-term light-trap programme, working with mercury vapour bulbs, was situated at the experimental field station in Bucuresti-Băneasa, situated on the Romanian Plain, in the southeast of the country. The environment adjacent to the traps varied annually, but always included perennial crops on one side (orchard and vineyard) and annual crops on the other, in addition to various weeds and spontaneous vegetation growing in field borders and uncultivated places.

The data records are given and indicate the dominant species, their flight activity and seasonal occurrence. Sixteen species were collected (Paulian, 1996), of which only four Chrysopidae were regularly present and abundant, constituting 95% of the total catch. These species were: *Chrysoperla carnea s. lat*, *Chrysopa formosa* (Brauer), *C. pallens* (Rambur) and *C. perla* (Linnaeus). The population structure varied slightly, throughout the year, but the relative frequency of the components did not fluctuate (Fig. 29.1). We consider:

Chrysoperla carnea and *Chrysopa formosa* as dominant, as they constituted $40.1 + 36.4 = 76.5\%$ of the

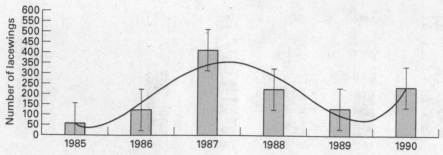

Fig. 29.2. Annual distribution of lacewing populations observed from flight activity.

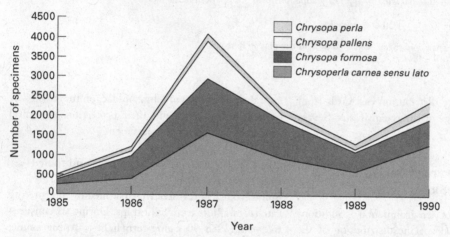

Fig. 29.3. Annual distribution and ratio between the four main lacewings in Romania.

total, *Chrysopa pallens* and *C. perla* frequent as they constituted 13.2 + 6.9 = 20.1%,

Cunctochrysa albolineata (Killington), *Chrysopa phyllochroma* Wesmael, *Dichochrysa ventralis* (Curtis), and *Chrysopa nigricostata* Brauer rare because they were regularly present, but each constituted less than 1% – and finally the seven other species are considered casual because they were recorded only occasionally and collectively did not reach 1% of the total.

29.4.2 Annual flight activity

The pattern of lacewings caught by light trapping shows that population levels fluctuated yearly, in a cycle apparently repeated after five years (Fig. 29.2). Annual distribution of the relative abundance between the four main sympatric chrysopid species remains constant between those species throughout this cycle (Fig. 29.3).

29.4.3 The seasonal cycle of flight activity

Seasonal fluctuation of chrysopid flight activity and the relative abundance between species followed a normal distribution pattern (Fig. 29.4) with the maximum flight activity in July. Limits of abundance fluctuated each year in relation to climatic and environmental conditions.

Chrysoperla carnea s. lat. represented about 40% of the total captures. Adults flew from March up to October, sometimes November, resulting in about seven months of activity each year. The earlier specimens were captured in the traps in spring, when the minimum temperatures reached about 7 °C (Fig. 29.5), whereas the latest ones in autumn were obtained under conspicuously higher minimum temperatures, of around 12 °C. In fact we consider that the start of flight activity in spring depends mainly on temperature, while the termination of flight activity in autumn depends on other factors, but mainly photoperiod.

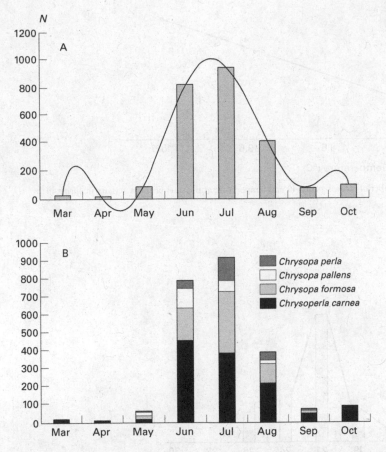

Fig. 29.4. The seasonal fluctuation (A) and mean ratio (B) of flight activity of green lacewings in Romania.

There is a strong correlation between temperature and the number of specimens collected in light traps (Figs. 29.6, 29.7).

Chrysoperla carnea adults collected in 1990 were grouped in 5-day intervals and the general distribution was then analysed (Fig. 29.8A), providing evidence of two reproductive generations. The early spring flight performed by overwintering adults gave rise to the first larval brood, which in turn produced a conspicuous peak of newly emerged adults at the end of June. The second larval brood then developed and resulted in the second flight peak at the end of July and beginning of August. Surprisingly, the number of adults collected was relatively greater from the overwintering generation than from the ensuing generation. This was probably a consequence both of a good survival of overwintering adults in the studied biotopes and in neighbouring wintering places, and of better synchrony

of activity in post-diapausing individuals. In addition, further fluctuation in capture numbers during the whole active flight season (and more extensively during the period of high activity in June to August) can be explained by climatic conditions. Duelli (1986) clearly showed that the effect of temperature, rainfall and the power of flight activity at the end of September and beginning of October represents the adult pre-overwintering activity.

Chrysopa formosa is the second dominant lacewing, both by its frequency of capture and the high numbers of specimens. This species, which is found under approximately the same crop conditions as *Chrysoperla carnea*, showed good adaptive and invasive capability from May to September. Flight activity during the growing season indicated at least two broods (Fig. 29.8B), with the possibility of a third generation in the warmer regions in the south of Romania, or in warmer

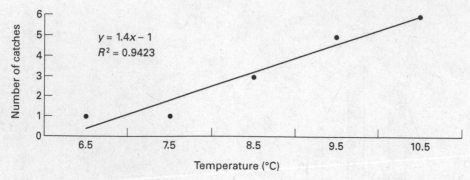

Fig. 29.5. The relationship between the number of lacewings caught and temperature at the start of flight activity in spring in Romania.

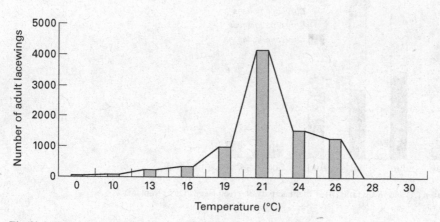

Fig. 29.6. The number of lacewings caught in light traps in relation to temperature.

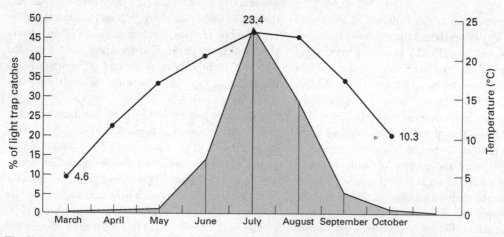

Fig. 29.7. Seasonal lacewing fluctuations in relation to mean monthly temperature.

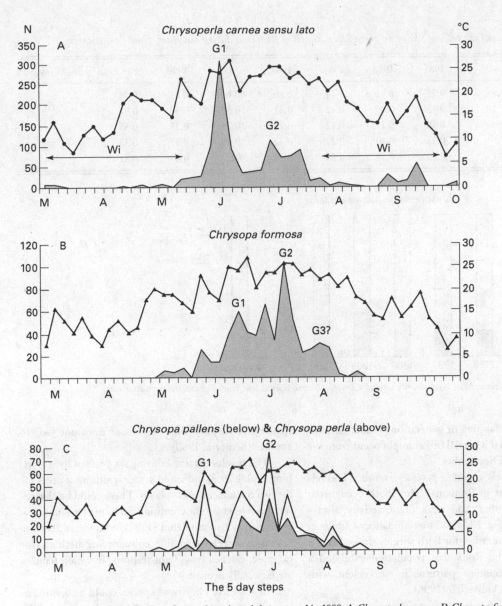

Figures 29.8. Annual distribution of green lacewing adults captured in 1990. A, *Chrysoperla carnea;* B, *Chrysopa formosa;* C, *Chrysopa perla* (white area) and *Chrysopa pallens* (grey area). G1, G2, G3 represent the first, second, and third generations. Wi, diapausing winter generation.

years, where complete larval development occurs. Similar results were obtained by Volkovich (1998) in Russia, who also promoted the idea of a third generation for *Chrysopa formosa* in the forest–steppe zone, but not on a permanent basis, because it seems to be related to warmer autumns. After overwintering of prepupae within the cocoons, the new adults emerged in spring, sometimes very early (end of March in 1990) if weather was mild (minimum and maximum temperatures 9 °C and 21 °C, respectively) and dry (relative humidity 85%, monthly rainfall 3 mm). The flight of the second generation occurred at the end of June, and another larval brood developed in summer, followed by a new adult frequency peak largely spread out from the end of

Table 29.3. *Annual sex ratio of the main green lacewings from Romania during six years, from light trapping*

Species	1985	1986	1987	1988	1989	1990	Sex ratio (mean of 6 years)
Chrysoperla carnea	0.73	0.82	0.74	0.80	0.66	0.81	0.78
Chrysopa formosa	0.48	0.49	0.42	0.52	0.50	0.58	0.47
Chrysopa phyllochroma	0.58	0.25	0.72	0.25	0.75	0.25	0.55
Chrysopa pallens	0.23	0.40	0.27	0.42	0.49	0.37	0.35
Chrysopa perla	0.50	0.59	0.50	0.55	0.52	0.49	0.51

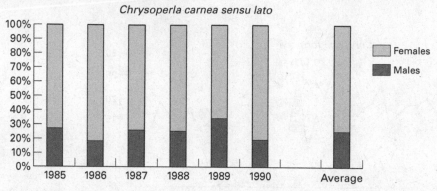

Fig. 29.9. Annual frequency (%) of the sex ratio of *Chrysoperla carnea sensu lato*, from light trapping in Romania.

July. Later, overlapping of generations renders it difficult to ascertain if a partial brood might occur from the adults caught in September.

Chrysopa pallens and *C. perla* are common and relatively abundant in Romania. They both occur from May to September, and peak flight activity always occurs in July (Fig. 29.8C). The abundance of adults in the summer generation for both species showed that a large part of the Romanian population exhibited a bivoltine development pattern, in agreement with observations by Volkovich (1996).

29.4.4 Sex ratio

The data recorded for the chrysopid species in Romania are insufficient to give a conclusive statement regarding the sex ratio (an expression of the number of females/number of males + number of females). The values related by some authors (reported by Canard & Principi, 1984) differ significantly for different species either side of a hypothetical sex ratio of 0.50, fluctuating sometimes in the favour of males and other times in favour of the females. Even within a single species, fluc-

tuations of the sex ratio may occur from one year to another (Séméria, 1980).

The results obtained during six years of light trapping (Table 29.3) indicated some significant aspects for the five main lacewings species. These could be characteristic at least for the conditions in Romania, since they are contrary to Pantaleoni (1982), who found that the sex ratios are related to different sampling methods (for example, catches from light trapping in which females are especially prevalent).

At least five chrysopid species could be divided in three distinct categories, according to the values of the sex ratio and especially by their yearly fluctuation. The first category is represented by the species of *Chrysoperla* 'carnea' complex, in which the mean annual sex ratio in all six years was always female-biased (Fig. 29.9). The second category, for which the mean annual sex ratio was also constant in the six years of observation, but in which there was an equilibrium between both sexes (0.50). This group includes the species *Chrysopa formosa* and *C. perla*. In *C. perla*, however, the sex ratio in May and June was male-

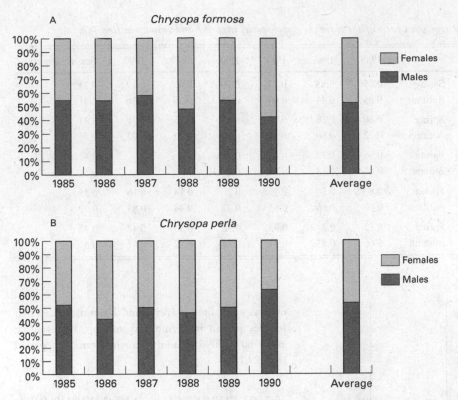

Fig. 29.10. The yearly frequency (%) of the sex ratio of *Chrysopa formosa* (A) and *Chrysopa perla* (B) as seen by light trapping.

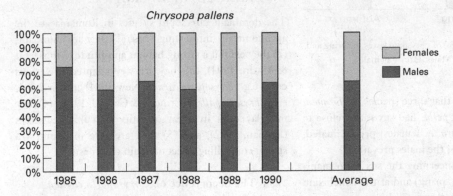

Fig. 29.11. Annual frequency (%) of the sex ratio of *Chrysopa pallens* from light trapping in Romania.

biased, becoming balanced in July, confirming previous observations about the protandry of newly emerged adults by Vannier (1961) and Canard (1973), both in the laboratory and in the field (Fig. 29.10). The third and smallest category (to which *C. pallens* belongs) has a fluctuating annual sex ratio; but overall the results of

yearly captures resulting in many more males than females, the average ratio being 0.35. Such a sex ratio probably did not represent the actual natural balance, but instead resulted from the more intensive activity of male and their stronger attraction to light traps (Fig. 29.11). An overall analysis of the mean sex ratio values,

Table 29.4. *Average and yearly sex ratio of the main chrysopid species in spring and autumn in Romania*

Species		1985	1986	1987	1988	1989	1990	Sex ratio (average)
Chrysoperla carnea	Spring	0.65	0.85	0.77	0.90	0.50	0.79	0.74
	Autumn	0.85	0.44	0.60	0.85	0.39	0.56	0.61
Chrysopa formosa	Spring	0.66	0.28	0.50	0.53	0.25	0.27	0.41
	Autumn	0.12	0.80	0.42	0.50	0.72	0.72	0.54
Chrysopa phyllochroma	Spring	0.36	0.33	0.66	0.36	—	—	0.42
	Autumn	0.63	0.33	0.88	0.2	—	—	0.51
Chrysopa pallens	Spring	0.8	0.5	0.35	0.2	0.14	0.16	0.35
	Autumn	0.66	0.66	0.52	0.37	0.48	0.83	0.50
Chrysopa perla	Spring	0.42	0.2	0.45	0.50	0.33	0.42	0.42
	Autumn	0.66	0.85	0.33	0.54	0.72	0.61	0.61

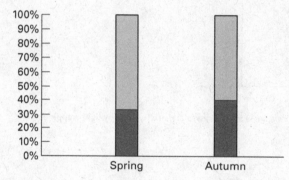

Fig. 29.12. The frequency (%) of chrysopid sexes in spring and autumn in Romania. Black = Males, Grey = Females.

for all six years, showed that three species, *C. formosa*, *C. phyllochroma*, and *C. perla*, had sex ratios close to half; in *Chrysoperla carnea*, females predominated, while in *Chrysopa pallens*, the males prevailed.

Further analysis concerning the sex ratio manifested at the beginning (spring) and at the conclusion (autumn) of activity of the main lacewings species (Table 29.4), revealed that, in all species except *Chrysoperla carnea*, the males appear first and prevail in spring, while in autumn, the females prevail. This situation may be largely determined by climatic conditions, with milder conditions in some years favouring the presence of males. This phenomenon was noted by Canard (1973) for *Chrysopa perla*, and our data confirm this conclusion, at least for this species. Total captures

over six years, both in spring and autumn, showed that females prevail in spring and autumn, representing about 60%–70% of adults caught from each period (Fig. 29.12).

29.5 OCCURRENCE AND BEHAVIOUR OF GREEN LACEWINGS IN AGRICULTURAL CROPS

The dominant chrysopid species in Romania are the agronomically interesting ones: (1) they are abundant, (2) they exhibit a strong habitat amplitude (Monserrat & Marín, 1994), (3) they are well adapted to various crops e.g. *Chrysoperla carnea* (New, 1984) or to orchards e.g. *Chrysopa pallens* (Principi & Canard, 1984), (4) they may lay eggs in great quantity in cultivated areas (Paulian, 1992), and (5) they are able to manifest a strong controlling effect on main crop pests.

29.5.1 The occurrence of *Chrysoperla carnea* species in the (agro)ecosystems

By electrophoretic and genetic methods both in field and laboratory studies, Thierry *et al.* (1992) and Cianchi & Bullini (1992) identified different strains of *C. carnea* species, on the basis of some ecophysiological peculiarity, such as the mode of overwintering, which could characterise the natural populations (Thierry *et al.*, 1996). True *C. carnea* and its sibling species exhibit seasonal differences in mating period and

Table 29.5. *Seasonal occurrence (%) and density (calculated from numbers of specimens per m^2) of species from the* Chrysoperla carnea *species complex in agricultural crops in Romania*

Crops	Apr	May	June	July	Aug	Sept	Oct	Nov	%▲	%●	Density
Wheat		●	▲●						40	60	6.2
Barley		●	▲●						20	80	5.3
Corn			▲●	▲●	▲●	●	●□		53	40	5.3
Sorghum			●	▲●	▲●	●			30	70	8.9
Soya bean				▲●	●				20	80	3.2
Sunflower				●	●				—	100	2.2
Potato		●		●	●				—	100	4.4
Alfalfa		●	▲●	▲●	▲●				33	66	3
Tobacco				▲●		▲●	▲●		50	50	2
Apple			▲●	▲●	▲●	▲●	▲●		79	21	6
Pear			▲●	▲●	▲●	●			70	30	10
Peach			▲●	▲●	▲●	●			52	48	1
Apricot			▲●	▲●	▲●	●			52	48	6
Plum			▲	▲	▲				100	—	5
Cherry				▲●	▲●	▲			80	20	3
Vine			▲●	▲●	▲●	▲●	▲●		60	40	4

Notes:

▲ = *Chrysoperla kolthoffi,* ● = *Chrysoperla lucasina,* □ = *Chrysoperla renoni.*

Table 29.6. *Seasonal occurrence (%) and density (calculated from numbers of specimens per m^2) of species from the* Chrysoperla carnea *species complex in natural habitats in Romania*

Natural habitat	Apr	May	June	July	Aug	Sept	Oct	Nov	%▲	%●	Density
Rush		●	▲●	▲●	▲●	●	●	●	20	80	6.7
Swamp vegetation	▲	▲●	▲●	▲●	▲●	▲●			40	60	5.3
Pioneer vegetation			●	▲●	▲●				30	70	1.2
Locust tree		●	●	●	▲●	▲●			40	60	2.2
Willow		▲	▲	▲●	▲●	▲	▲		60	40	2.2
Poplar				▲●	▲●	▲			70	30	2.2
Pine			▲	▲	▲				100	—	7.3

Notes:

▲ = *Chrysoperla kolthoffi,* ● = *Chrysoperla lucasina.*

habitat. Furthermore, the study of temporal changes in overwintering sites has revealed marked differences between the three forms (Thierry *et al.*, 1996).

Similarly, we have tried to show some of the ecological peculiarities, temporal and spatial separation of '*carnea*' species during their active period, both in the main agricultural crops and natural environment and also their movement between and within those habitats.

29.5.2 *Chrysoperla 'carnea'* species and their seasonal occurrence

The structure and the distribution of the common green lacewing populations were investigated in 16 cultivated crops and on seven different plant species which constituted the components of the neighbouring natural ecosystems, to determine their preference, and to allow us to characterise these habitats. The dominant species with respect to its habitat amplitude as with the number of specimens collected is *C. carnea* (Stephens), which is found in all crops and biotopes investigated.

As is known *C. carnea* overwinters in rolled dry leaves and in ivy tufts, whereas *C. kolthoffi* is preferentially found indoors. In spring, the two forms coexist for some weeks in both dry leaves and ivy. At the end of the spring, in general, only *C. kolthoffi* can be found in bushes and trees, while *C. lucasina* is only found in ivy in winter (Thierry *et al.*, 1996).

The preliminary results (Tables 29.5 and 29.6) indicates some of these related species and their seasonally active periods. The main representative species of the '*carnea*'- complex (*C. kolthoffi* and *C. lucasina*) occurred both in crops and in natural habitats in Romania. The ratio between these species and the time of their occurrence in each biotope differed and provided the basis on which we can document some of their characteristics.

Chrysoperla lucasina is more frequent and abundant in field crops, in which it is the first species to appear, although it seems that it prefers crops managed by hoeing, while *C. kolthoffi* seems to prefer trees, where it was sometimes the only one that occurred there.

In natural habitats, both *C. kolthoffi* and *C. lucasina* are present, from April to November, but in different abundances, which mirrored the idea already proposed for agricultural crops that *C. lucasina* is more frequent and abundant on the herbaceous plants, while *C. kolthoffi* preferred trees.

ACKNOWLEDGEMENTS

I wish to express my grateful thanks to Michel Canard and Dominique Thierry for their careful revision of the material. Many thanks also to Timothy New and Andy Whittington for the revision of the English manuscript.

REFERENCES

Alrouechdi, K. & Canard, M. (1979) Mise en évidence d'un biotype sans diapause photopériodique dans une population méditeranéenne de *Chrysoperla carnea* (Stephens) (Insecta: Neuroptera). *Comptes Rendus Hebdomadaires des Séances de l'Académie des Sciences Paris* 289, 553–555.

Aspöck, H. (1992) The Neuropteroidea of Europe: a review of present knowledge (Insecta). In *Current Research in Neuropterology, Proceedings of the 4th International Symposium on Neuropterology*, ed. Canard, M., Aspöck, H. & Mansell, M.W., pp. 43–56. Sacco, Toulouse.

Aspöck, H., Aspöck, U. & Hölzel, H. (1980) *Die Neuropteren Europas*. Göcke & Evers, Krefeld.

Baudry, O. (1996) *Reconnaître les Auxiliaires: Vergers et Vignes/Recognising Natural Enemies: Orchards and Vineyards*. Paris, France.

Brooks, S.J. (1994) A taxonomic review of the common green lacewing genus *Chrysoperla* (Neuroptera: Chrysopidae). *Bulletin of the British Museum (Natural History) Entomology* 63, 137–210.

Brooks, S.J. & Barnard, P.C. (1990) The green lacewings of the world: a generic review (Neuroptera: Chrysopidae). *Bulletin of the British Museum (Natural History) Entomology* 59, 117–286.

Canard, M. (1973) Voltinisme, diapause et sex-ratio de *Chrysopa perla* (L.) (Neuroptera, Chrysopidae) dans le sud-ouest. *Annales de Zoologie–Écologie Animale* 5, 29–37.

Canard, M. (1998) Life history of green lacewings in temperate climates: a review (Neuroptera, Chrysopidae). *Acta Zoologica Fennica* 209, 65–74.

Canard, M. & Principi, M.M. (1984) Development of Chrysopidae. In *Biology of Chrysopidae*, ed. Canard, M., Séméria, Y. & New, T.R., pp. 57–76. Dr W. Junk, The Hague.

Cianchi, R. & Bullini, L. (1992) New data on sibling species in chrysopid lacewings: the *Chrysoperla carnea* (Stephens) and *Mallada prasinus* (Burmeister) complexes (Insecta: Neuroptera: Chrysopidae). In *Current Research in Neuropterology, Proceedings of the 4th International Symposium on Neuropterology*, ed. Canard, M., Aspöck, H. & Mansell, M.W., pp. 99–104. Sacco, Toulouse.

Devetak, D. (1992) Present knowledge of Megaloptera, Raphidioptera and Neuroptera of Yugoslavia (Insecta: Neuropteroidea). In *Current Research in Neuropterology, Proceedings of the 4th International Symposium on Neuropterology*, ed. Canard, M., Aspöck, H. & Mansell, M.W., pp. 105–106. Sacco, Toulouse.

Duelli, P. (1986) Flight activity patterns in lacewings (Planipennia: Chrysopidae). In *Recent Research in Neuropterology, Proceedings of the 2nd International Symposium on Neuropterology*, ed. Gepp, J., Aspöck, H. & Hölzel, H., pp. 165–170. Thalerhof, Graz.

Kis, B., Nagler, C. & Mîndru, C. (1970) *Fauna Republicii Socialiste România. Neuroptera (Planipennia). Insecta* 8(6). Academia Republicii Socialiste România, Bucuresti, România.

Leraut, P. (1991) Les *Chrysoperla* de la faune de France (Neuroptera: Chrysopidae). *Entomologica Gallica* 2, 75–81.

Leraut, P. (1992) Les Névroptéres des Alpes centrales françaises. *Entomologica Gallica* 3, 59–65.

Monserrat, V. J. & Marín, F. (1994) Substrate specificity of Iberian Coniopterygidae (Insecta: Hemerobiidae). In *Current Research in Neuropterology, Proceedings of the 4th International Symposium on Neuropterology*, ed. Canard, M., Aspöck, H. & Mansell, M.W., pp. 279–290. Sacco, Toulouse.

New, T.R. (1984) Chrysopidae: ecology on field crops. In *Biology of Chrysopidae*, ed. Canard, M., Séméria, Y. & New, T.R., pp. 160–167. Dr W. Junk, The Hague.

Pantaleoni, R.A. (1982) Neuroptera Planipennia del comprensorio delle Valli di Comacchio: indagine ecologica. *Bollettino dell'Instituto di Entomologia dell'Università di Bologna* 37, 1–73.

Paulian, M. (1992) Eco-éthologie des pontes de chrysopes sur maïs en Roumanie. In *Current Research in Neuropterology, Proceedings of the 4th International Symposium on Neuropterology*, ed. Canard, M., Aspöck, H. & Mansell, M.W. pp. 303–310. Sacco, Toulouse.

Paulian, M. (1996) Green lacewings from the southeast of the Romanian Plain, as recorded by light-trapping (Insecta: Neuroptera: Chrysopidae). In *Pure and Applied Research on Neuropterology, Proceedings of the 5th International Symposium on Neuropterology*, ed. Canard, M., Aspöck, H. & Mansell, M.W., pp. 197–202. Sacco, Toulouse.

Paulian, M. & Andriescu, I. (1996) Chrysopidae and Hemerobiidae recorded from crops and adjacent natural habitats in the Danube Delta, Romania (Insecta: Neuroptera). In *Pure and Applied Research in Neuropterology, Proceedings of the 5th International Symposium on Neuropterology*, ed. Canard, M., Aspöck, H. & Mansell, M. W., pp. 203–206. Sacco, Toulouse.

Paulian, M., Canard, M., Thierry, D. & Cloupeau, R. (1996) Les *Chrysoperla* Steinmann de Roumanie (Neuroptera: Chrysopidae). *Annales de la Société Entomologique de France* (N.S.) 32, 285–290.

Popov, A.K. (1990) Zur Verbreitung der Chrysopiden (Neuroptera) in Bulgarien. *Acta Zoologica Bulgarica* 39, 47–52.

Principi, M. & Canard, M. (1984) Feeding habits. In *Biology of Chrysopidae*, ed. Canard, M., Séméria, Y. & New, T.R., pp. 76–92. Dr W. Junk, The Hague.

Ridgway, R.L. & Murphy, W.L. (1984) Biological control in the field. In *Biology of Chrysopidae*, ed. Canard, M., Séméria, Y. & New, T.R., pp. 220–228. Dr. W. Junk, The Hague.

Séméria, Y. (1980) Observations sur l'auto-ecologie et la syn-écologie des principales especes de Chrysopinae (Neuroptera, Planipennia) du sud-est de la France, des genres *Anisochrysa* Nakahara et *Chrysoperla* Steinmann. *Neuroptera International* 1, 4–25.

Sengonca, Ç., Griesbach, M. & Löchte, C. (1995) Geeignete Räber-Beute Verhältnisse für den Einsatz von *Chrysoperla carnea* (Stephens) – Eirn zur Bekämpfung von Blattläusen an Zuckerrüben unter Labor- und Freilandbedingungen. *Zeitschrift für Pflantzenkrankheiten und Pflantzenschutz* 102, 113–120.

Sheldon, J.K. & MacLeod, E.G. (1974) Studies on the biology of Chrysopidae. IV. A field and laboratory study of the seasonal cycle of *Chrysopa carnea* Stephens in central Illinois (Neuroptera: Chrysopidae). *Transactions of the American Entomological Society* 100, 437–512.

Szabó, S. & Szentkirály, F. (1981) Comunities of Chrysopidae and Hemerobiidae (Neuroptera) in some apple orchards. *Acta Phytopatologica Academiae Scientarum Hungariae* 16, 157–169.

Tauber, C.A. & Tauber, M.J. (1981) Seasonal responses and their geographic variation in *Chrysopa downesi:* ecophysiological and evolutionary considerations. *Canadian Journal of Zoology* 59, 370–376.

Thierry, D. & Adams, P.A. (1992) Round table discussion on the *Chrysoperla carnea* complex (Insecta: Neuroptera: Chrysopidae). In *Current Research in Neuropterology, Proceedings of the 4th International Symposium on Neuropterology*, ed. Canard, M., Aspöck, H. & Mansell, M.W., pp. 367–377. Sacco, Toulouse.

Thierry, D., Cloupeau, R. & Jarry, M. (1992) La chrysope commune *Chrysoperla carnea* (Stephens) sensu lato dans le centre de la France: mise en évidence d'un complexe d'espèces (Insecta: Neuroptera: Chrysopidae). In *Current Resarch in Neuropterology, Proceedings of the 4th Symposium on Neuropterology*, ed.

Canard, M., Aspöck, H. & Mansell, M.W., pp. 379–392. Sacco, Toulouse.

Thierry, D., Cloupeau, R. & Jarry, M. (1996) Variation in the overwintering ecophysiological traits in the common green lacewing West-Palaeartic complex (Neuroptera: Chrysopidae). *Acta Oecologica* 15, 593–606.

Vannier, G., (1961) Observations sur la biologie de quelques chrysopides (Névroptères, Planipennes). *Bulletin du Muséum National d'Histoire Naturelle Paris* 33, 396–405.

Volkovich, T.A. (1996) Effects of temperature on diapause induction in *Chrysopa perla* (Linnaeus) (Insecta: Neuroptera: Chrysopidae). In *Pure and Applied Research in Neuropterology, Proceedings of the 5th International Symposium on Neuropterology,* ed. Canard, M., Aspöck, H. & Mansell, M. W., pp. 259–267. Sacco, Toulouse.

Volkovich, T.A. (1998) Environmental control of seasonal cycles in green lacewings (Neuroptera, Chrysopidae) from the forest–steppe zone of Russia. *Acta Zoologica Fennica* 209, 263–275.

CHAPTER 30

Biological control with *Chrysoperla lucasina* against *Aphis fabae* on artichoke in Brittany (France)

J.C. Maisonneuve

30.1 INTRODUCTION

In Brittany artichokes, *Cynara scolymus* L., are cultivated widely along the Channel coast, where the
climate is favourable to this crop. Aphids are considered
as the main pests of this vegetable in this area, mainly:
Capitophorus horni (Börner) (the green aphid) and *Aphis
fabae* (Scop.) (the black aphid) (Maisonneuve *et al.*,
1981; Collet, 1997). To control these aphids, the
growers are developing a system of integrated pest
management (IPM), in order to increase the quality of
this important crop of our country: at present 50000
metric tonnes are produced on about 7–9000 hectares.

In many countries *Chrysopa* spp. or *Chrysoperla*
spp. are used to control aphids on field vegetable crops.
As part of the interest in this predator (Lyon, 1976,
1979; Sengonca, 1995) to control aphids, *Chrysoperla
lucasina* has been investigated in 1995 and 1996 in artichokes and shows some promise (Malet *et al.*, 1994) and
is easy to rear.

In order to appreciate the impact of this beneficial
insect on artichoke, in our experiments we have examined the development of aphids and some lady beetles
(Coccinellidae). These experiments are described here,
to indicate the preliminary assessment for using
Chrysopidae to control aphids on artichokes in France.

30.2 MATERIALS AND METHODS

The experiments below are described and discussed
with relevance to the years 1996 and 1997. The work
was carried out on second-year crops of 'Camus de
Bretagne' artichokes. The duration of a plantation is
usually three years.

Chrysoperla lucasina was mass-reared and supplied
by the laboratory of the Plant Protection Service
(S.R.P.V.) in Brest. The larvae were fed with eggs of
Ephestia kuehniella, and the adults were fed with a
mixture of yeast, pollen, and honey. Eggs of *C. lucasina*
were collected after laying and mixed in boxes of 0.5
litre, with buckwheat husk and some eggs of *E. kuehniella*. To overcome cannibalism inside the boxes, four
eggs were required for each first-instar larva.

Chrysoperla lucasina larvae were released, at the
end of the first-instar stage, about five days after collection of the eggs, at a density of 20 larvae per m^2.

30.2.1 Experimental conditions: 1996

The total area for this experiment was 1728 m^2, divided
into 24 elementary plots, of 72 plants each, measuring
about 6 m × 12 m, in which there were four treatments
and six replicates. The work was carried out on a
second-year crop, which had been planted in April 1995
and in which the harvest began on 11 June and finished
on 22 July.

The four treatments were:
1. Control untreated   [CU]
2. Control treated   [CT]
   - 2.1. Croneton (ethiofencarb): 24 April 96
   - 2.2. Pirimor (pirimicarb): 7 June 96
3. Five releases, each 15 days (18 April, 2 May, 15
   May, 29 May, 13 June), of 4 first and second
   instars per m^2 [5*4]
4. Three releases, each 15 days (9 May, 22 May, 5
   June), of 7 first and second instars per m^2 [3*7]

First release was in the middle of April, when the temperature is mild, for treatment [5*4]. On the other
hand, for treatment [3*7] first release began when *Aphis
fabae* were first detected at the beginning of the flight
period. Larvae were mixed with buckwheat husk and
spread on the plants.

Assessment procedure: eight plants (11%) were
marked in each plot of 72 plants, and examined weekly.
In order to assess the developmental pattern and interactions of these insect populations, all the insects (pests

513

and beneficials) were recorded from the whole plant (leaves and flower heads). For the last assessment on each plot, for commercial reasons, the presence or the absence of aphids was recorded on all flower heads.

Wild beneficial insects (Coccinellidae, Syrphidae, Aphididae, etc.), observed during the course of the experimentation were recorded each week, on the assumption that their predatory action is complementary to that of the chrysopids.

30.2.2 Experimental conditions: 1997

Work during this year was in association with a commercial grower, on a total experimental area of 12400 m², divided into two elementary plots, each with two treatments.

The two treatments were:
1. Control treated   [CT]
2. Three releases, each 15 days (15 April, 29 April, 13 April), of 4 first and second instars per m²   [3*4]

The first release was in the middle of April, when the temperature is mild, for treatment [3*4].

Assessment procedure: in this case we produced an aphid-map of the field, showing the presence or the absence of aphids, recorded on *all* flower heads. The most important thing, as before, is to know if the flower heads were free from aphids or not. These experiments were twinned with one other in an experimental station where the population's dynamics were studied.

30.3 RESULTS

30.3.1 Experimental results: 1996

The change in *Aphis fabae* numbers per plant is shown in Figure 30.1. The first aphids are seen at the beginning of May, three weeks after the first release in the treatment [5*4], which event gave the signal for the first release of treatment [3*7]. In the two trials where *Chrysoperla lucasina* were introduced, the beneficial insects delayed the arrival of *A. fabae* on flower heads, in comparison with treatment [CU]. Five weeks separate the observation of the first aphids. Nevertheless [5*4] and [3*7] showed an unequal effectiveness.

The percentage of plants carrying *A. fabae* on 9 July 1996 (Fig. 30.2) shows that the two treatments were homogeneous [C T] and [5*4] (test Newman–Keuls threshold = 5%), even if this strategy shows the number of aphids two weeks before. At that date, half of the

harvest was complete and all the flower heads had been inspected to record the number of aphids.

A statistical test (Newman–Keuls threshold = 5%) of the mean number of larvae of ladybird beetles (*Coccinella septempunctata* L.) (Fig. 30.3) shows two different groups:
A: [CU]
B: [CT], [3*7], [5*4]
It would appear that there is competition between the introduced, *Chrysoperla lucasina* and wild *Coccinella septempunctata*, and that chemical sprays also reduce ladybird numbers.

30.3.2 Experimental results: 1997

30.4 DISCUSSION AND CONCLUSIONS

30.4.1 Pests

Capitophorus horni and *Aphis fabae* are the two major aphids of artichoke in Britanny (Grousson, 1974; Robic, 1975; Robert & Rouzé-Jouan, 1976; Such, 1978; Le Bohec, 1981; Maisonneuve *et al.*, 1981; Collet, 1997). *Capitophorus horni* does not cause any damage to flower heads, because it stays only on leaves. Some authors suggest that this species is able to transmit viruses, but there are no symptoms on the plants.

Aphis fabae is completely different: it arrives in main flight periods each year and is able to hide inside the flower head, at which time insecticides are not effective, and the flower heads are commercially damaged.

The purpose of the phytosanitary protection of artichoke is thus not so much to protect the plant from aphids, but to prevent the infestation of aphids on or inside the head. The head must be free from aphids or mummies.

30.4.2 Beneficial insects

The strain of *Chrysoperla lucasina* used in this study was described under the name *C. carnea* and is in fact synonymous with that species (Canard, 1987). However, there is much debate over the sibling species in this complex, so the name *C. lucasina* has been retained here. This beneficial chrysopid has been reported from other countries to feed on *A. fabae* on sugarbeet (Sengonca *et al.*, 1995).

Each larva is able to eat more than 100 aphids and can feed at temperatures as low as 10°C (Nicoli *et al.*, 1991). Feeding is not restricted to aphids, since other

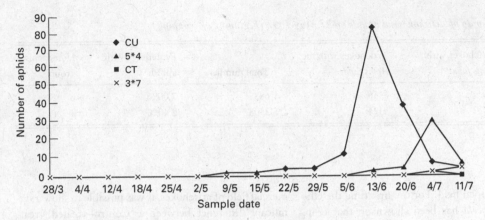

Fig. 30.1. Dynamics of *Aphis fabae* (all stages). For details of experimental treatments, see text.

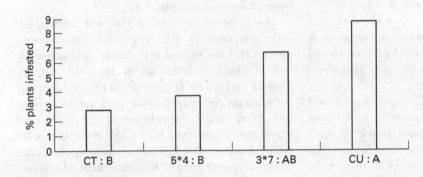

Fig. 30.2. Percentage of plants with *Aphis fabae* on 9 July 1996. For details of experimental treatments, see text.

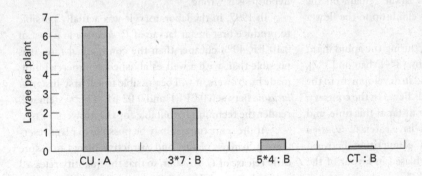

Fig. 30.3. Mean number of larvae of ladybirds: homogenous groups. For details of experimental treatments, see text.

pests such as mites and lepidopteran eggs are also known to form part of the diet of *C. lucasina*.

30.4.3 Release strategy

Lyon (1979) released two different stages of *C. lucasina* (eggs and second-instar larvae) and recommended a release rate of one second-instar larva per ten aphids on peppers in greenhouses. Nicoli *et al.* (1991) reported that, in Italy, 18 larvae per m² were released on strawberry in unheated plastic tunnels. In contrast, 20 first-instar larvae per m² were released in this study, with positive results on artichoke.

30.4.4 Findings:

Figures 30.1 and 30.2 show that it is possible to control *A. fabae*, in Brittany, with the predator *C. lucasina*, on

Table 30.1. *Aphid map of artichoke and aphids on 27 May 1997 (beginning of cropping)*

Treatment	Artichokes with *Aphis fabae*	Artichokes without *Aphis fabae*	Total number	Percentage with aphids	Homogenous group
[CT]	479	6356	6835	7.00%	b
[3*4]	84	3424	3508	2.40%	a

Note: $\chi^2 = 95.9$

artichokes in the open field. For the first time the efficiency of *Chrysoperla* has been shown on that crop, which has a great sensitivity to this pest. This result, which needs further replications, will allow us to build a new IPM strategy based on Nicoli *et al.* (1991).

Although *Capitophorus horni* arrives early and remains on the leaves and does not cause damage to the artichoke crop, it is easily controlled by parasitoids and fungi and also provides food for early releases of *Chrysoperla lucasina*.

In treatment [3*7], where *C. lucasina* was released at the beginning of the first flight of *A. fabae*, fewer aphids were seen, on the whole plant, than in the other treatments. However, there were more flower heads with aphids, reducing commercial yield, than in the [CT] or the [5*4] treatments. We can explain this, by the fact that young *C. lucasina* larvae remain on the leaves of artichoke, before they climb up to the flower heads.

Another hypothesis is that, during the aphid flight period, the larvae are stronger on [5*4] than on [3*7], since their earlier release resulted in development to the next instar by the time the aphids flew. On the contrary, [3*7] larvae were still in the first instar at this time, and the older coccinellid larvae may have eaten *C. lucasina* in preference to aphids. This competition between *Chrysoperla* and *Coccinella* was biased in favour of the aphids in the case of treatment [3*7].

The activity of the entomofauna was important, and must be integrated, because there is competition between *Chrysoperla lucasina* and *Coccinella septempunctata*. In the past, we always believed that the natural entomofauna was complementary to the introduced beneficials and not in competition.

Table 30.1 showed that it was possible to reduce the number of larvae released in the field, as the density of 12 larvae per m^2 was able to control the pests at commercially acceptable levels. For the first time, on a commercial field of artichokes, it was possible to show a significant difference between a 'Control-treated' treatment (b) and a treatment under biological control (a), where the results obtained with *Chrysoperla lucasina* are better than chemical application.

Nevertheless, to develop this new strategy of plant protection, it will be necessary to convince growers that this method is profitable. The estimated cost of chemical protection of artichoke is from 1000 FF to 1500 FF (1 Euro = 6.50 FF), per hectare, including the cost of application. With the commercial cost of first-instar-larvae in France of 0.14 FF, the biological control treatment, used in this example, cost 16 800 FF. Based on these figures it is easy to conclude that biological control using *C. lucasina* on artichokes is not commercially profitable. Other data indicate that this hypothesis is wrong.

In 1997, in the laboratory, it was actually possible to produce first-instar larvae of *C. lucasina* at a cost of 0.10 FF, 40% cheaper than the commercial rate. It is possible that, with a well-established commercial unit, made by growers, it will be possible to sell first-instar *C. lucasina* between 0.01 FF and 0.02 FF. This system will render the technique profitable for the growers.

At the same time, it may be possible to plant reservoir or 'banker plants' and attractive plants to assure early release of *C. lucasina*, so that the quantity released can be decreased. Furthermore, modification of the law in Europe, regarding the uses of phytosanitary products and the obligations of growers, is likely to be a factor that will increase the cost of chemical treatment.

Finally, all the components of integrated control will be studied, point per point, including all the prophylactic measures. The last condition will be to educate growers to these new methods. Under these conditions biological control with *C. lucasina* will be cheaper than chemical control; the grower will abide by the law and the consumer will have food safety.

ACKNOWLEDGEMENTS

My thanks go to European Community (Program 5b) Ministry of Agriculture, Regional Council of Bretagne and P.A.O., for their help; and to the growers for their trust.

REFERENCES

Canard, M., Séméria, Y. & New, T.R. (1984) *Biology of Chrysopidae*. Dr W. Junk, The Hague.

Collet, J.M. (1997) Pucerons de l'artichaut en Bretagne: plus de trente ans d'histoire. *Aujourd'hui et Demain*, 53, 9–13.

Galazzi, D. & Nicoli, G. (1992) *Chrysoperla carnea*. *Informatore Fitopatologico* 3, 25–30.

Grousson, C. (1974) Étude écologique de pucerons dans la zône légumière des Côtes du nord – cas particulier de deux espèces présentes sur artichaut – *Aphis fabae* Scop. – *Capitophorus horni* Borner – Conséquences agronomiques. *Mémoire d'études ENITHP Angers* (49), 89 p.

Le Bohec, J. (1981) Les pucerons de l'artichaut. Étude particulière de *Capitophorus horni* Borner et d'*Aphis fabae* Scop en Bretagne. *Journées ACTA, Les Pucerons des cultures, Paris* 2–4/03/81, 265–274.

Lyon, J.P. (1976) Les populations aphidiennes en serre et leur limitation par utilisation expérimentale de divers entomophages. *Bulletin OILB/SROP* 1976/4, 64–76.

Lyon, J.P. (1979) Lâchers expérimentaux de Chrysopes et d'Hyménoptères parasites sur pucerons en serres d'aubergine. *Annales Zoologie– Écologie Animale* 11, 51–65.

Maisonneuve, J.C. & Collet, J.M. (1995) Un nouveau prédateur contre les pucerons: *Chrysopa lucasina*. *Aujourd'hui et Demain*, 47, 5–6.

Maisonneuve, J.C., Such, A., Paitier, G. & Vannier, M.P. (1981) Lutte contre les pucerons de l'artichaut en Bretagne par le Service de la Protection des Végétaux. In *Journées ACTA, Les Pucerons des cultures*, Paris, 2–4 March 1981, pp. 291–297.

Maisonneuve, J.C., Couture, I., Marrec, C., Courbet, S. & Collet, J.M. (1997) Biological control with *Chrysoperla lucasina* against *Aphis fabae* on artichoke in Brittany (France): first results. *Medelingen van de Faculteit Landbouww Gentetenschappen Rijksuniversitait*, 62, 455–459.

Malet, J.C., Noyer, C., Maisonneuve, J. C. & Canard, M. (1994) *Chrysoperla lucasina* (Lacroix) (Neur., Chrysopidae), prédateur potentiel du complexe méditerranéen des *Chrysoperla* Steinmann: premier essai de lutte biologique contre *Aphis gossypii* Glover (Hom., Aphididae) sur melon en France méridionale. *Journal of Applied Entomology* 118, 429–436.

Nicoli, G., Galazzi, D., Mosti, M. & Burgio, G. (1991) Embryonic and larval development of *Chrysoperla carnea* (Steph.) (Neur., Chrysopidae) at different temperature regimes. *Bulletin OILB/SROP* 14, 43–49.

Quemeneur, Y. (1991) Lutte raisonnée contre puceron vert et puceron noir de l'artichaut. *Aujourd'hui et Demain* 30, 27–30.

Robert, Y. (1981) Pucerons noirs sur artichaut. Mise au point biologique sur le problème lié au groupe *Aphis fabae* Scop. Conséquences agronomiques. In *Journées ACTA, Les Pucerons des cultures*, Paris 2–4 March 1981, 275–276.

Robert, Y. & Rouzé-Jouan J. (1976) Activité saisonnière de vol des pucerons [Hom. Aphididae] dans l'ouest de la France. Résultats de neuf années de piégeage (1967–1975). *Annales de la Société Entomologique de France* (N.S.), 12, 671–690.

Robic, R., (1975) Observations écologiques de pucerons dans la zône légumière des Côtes du nord – contribution à l'étude de la biologie de deux espèces présentes sur artichaut – *Aphis fabae* Scop. – *Capitophorus horni* Borner – Conséquences agronomiques. *Mémoire d'études ENITHP ANGERS* (49).

Sengonca, Ç., Griesbach, M. & Lochte, C. (1995) Suitable predator–prey ratios for the use of *Chrysoperla carnea* (Stephens) eggs against aphids on sugar beet under laboratory and field conditions. *Journal of Plant Diseases and Protection* 102(2), 113–120.

Such, A. (1978) Les problèmes phytosanitaires concernant la culture de l'artichaut en Bretagne. Contribution à l'étude d'une méthode de lutte contre les pucerons. *Capitophorus horni* Borner et *Aphis fabae* Scop. *Mémoire d'études – SRPV Bretagne Brest* – 41 p.

PART 5

Conclusion

CHAPTER 31

Lacewings in crops: towards the future

The Editors

In this book we have brought together some of the numerous studies which address the importance of lacewings in the crop environment; sufficient, we hope, to demonstrate the wide variety of contexts in which their values as predators are acknowledged and exploited, and that sound scientific understanding of the biology of individual taxa is critical in facilitating and extending their use. Lacewings have long been appreciated as predators, and some of the cases presented here exemplify the subtleties and innovations needed to increase their impacts. The variety of themes included constitutes a contemporary 'state-of-the-art' summary of the perceived roles of lacewings in crops. The constraints under which integrated pest control using polyphagous (generalist) natural enemies must proceed are also evident.

Lacewings are not as well understood as some other groups of predatory insects on crops, particularly ladybirds (Coleoptera: Coccinellidae). However, the widely distributed species of *Chrysoperla*, in particular, may be ranked among the most widespread and popular predators of small arthropods in many crop contexts. Yet, in many parts of the world other taxa are coming to play their part. *Chrysoperla* does not occur in Australia or New Zealand, for example, and such generalist predators would be highly unlikely to be approved for introduction there. Emphasis on diversifying the options for pest management must then focus on native species, and such additional studies (in many parts of the world) are expanding our appreciation of lacewings as versatile predators in an increasingly diverse arena.

However, it must be accepted – despite widespread evangelism to the contrary – that the spectrum of Neuroptera likely to be employed and manipulated in crop pest management is likely to remain small. At most, it is unlikely to expand beyond the three families treated here, and even Chrysopidae contains numerous segregates whose deliberate use against crop pests

seems highly improbable. In contrast, perceptive analysis of sibling groups such as *C. carnea s. lat.* will progressively yield an unsuspected multiplicity of combinations for subtle integration. Just as past confusions between entities in this complex have led to overgeneralisations on chrysopid biology, recognition that two or more sibling species may commonly coexist creates additional opportunities for their use in biological control.

The major importance of lacewings is, and will remain, as generalist predators amenable to mass production and precisely controlled releases, and several authors in this book describe recent technological advances to make these processes ever more efficient. The key to fostering commercial interest in this activity, the use of a given species in a variety of crops, counters effectively increasing calls for use of specialist species with fewer effects on other local biota. The natural primary habitat 'preference' of many lacewings appears to be for major vegetation types, rather than for particular plant species, but the roles of prey specificity (or 'preference') in governing habitat limitation are very poorly understood. They are an important component of future study. Thus, the selection of either coniferous or broad-leaved trees by predators strongly suggests some mediating role of prey. As in many herbivore–plant interactions, the roles of allelochemicals in predator behaviour may be of critical importance.

The wealth of natural enemies available for use in crop pest management is only now coming to be appreciated, with recent enumerations of up to several hundred taxa present without introduction. Lacewings, clearly, are a numerically small component of this richness.

As our knowledge of lacewings increases, so also the contexts and scales of pest management needs will change. Despite the wide current debate over use of genetically modified crops for human consumption, and attendant concerns over both ethics and possible adverse side-effects, the major feature of agricultural development has been to replace low-technology systems in many parts of the world by increasingly 'high-input' systems in which assurance of product quality (and, hence, economic return) is paramount. Despite some reversion in parts of the northern hemisphere to lower-technology agriculture and diversification, this trend is likely to continue. Recent trials with cotton and other crops which genetically incorporate pesticides such as *Bacillus thuringiensis* are likely to proliferate, and the influences of such crops on natural enemies of their herbivore pests are largely unknown. Many groups of so-called 'soft' pesticides, likewise, need further study to explore how they can most effectively be used.

Provision of food for burgeoning human populations, and maintaining public health whilst doing this, is perhaps humanity's greatest challenge as we enter the new millennium. A possible global population of around ten thousand million people by the middle of the 21st century century commands massive agricultural intensification, with increasing needs for interdisciplinary and innovative pest management strategies. These will involve integrating the use of natural enemies, including lacewings, with an almost imponderable array of transgenic crops and biopesticides. Fundamental knowledge of the organisms is augmented (even revolutionised) by information technology and molecular biology, areas which have not yet impinged to any major extent on our uses of lacewings in crops. Delivery of sustainable pest management systems must be based eternally on good biological knowledge and understanding; the problems and progress exemplified in this volume for lacewings are a microcosm of those applied to entomology as a whole. They range from establishing a reliable systematic framework, backed firmly by ecological and biological information, to effective implementation of IPM in the most suitable and cost effective ways. Increased focus on native species, rather than introduced exotics, is a key to a more global perspective which will draw less heavily from implied generalities based on studies of very few – and, perhaps, atypical – taxa, and will reveal the full potentialities of lacewings on crops. The need to monitor all attempts of manipulating predators is self-evident in contributing to this greater general knowledge. 'More research' is not simply a clarion call from those seeking funding! It is the key for revealing the many secrets lacewings still hold, and to their increased applied values in the future.

Yet, lacewings are more than tools for human use. Epithets such as 'golden eye' attest to their broader aesthetic appeal, as part of the variety of life forms appreciated by people in many walks of life, who may be concerned over the sustainability of natural ecosystems. Many conservationists acknowledge readily the importance of agricultural systems in assuring landscape integrity. Crop production and pastoralisation constitute the largest components of land alteration by people, and there are increasing pressures to harmonise the changes more effectively with broader biodiversity conservation so that the hospitality of this diverse matrix can be enhanced. Lacewings, both as predators on crops, and components of natural diversity, are symbols of the needs for holistic landscape management for the future, and of conserving future options for the increasing variety of pest management strategies which people will, inevitably, need.

Taxonomic index of generic and specific names

This index covers all the generic and specific names used in this book. Family group names are appended as an abbreviation in each instance, so as to indicate to which family each taxon belongs. These are abbreviated as: Co (Coniopterygidae), Ch (Chrysopidae), H (Hemerobiidae). The name '*carnea*' has been deliberately excluded from this index, because, firstly, the use of the name is undergoing considerable transformation and secondly, it appears so frequently that few pages would be absent from the list.

In some instances in the text, generic placement, endings of specific epithets, authorities and dates vary slightly from author to author. Wherever possible, these have been corrected in this index, but some division of opinion concerning taxonomic placement remains, mainly because of recent taxonomic advances. In some instances, the correct generic placement of species is still under revision (e.g. placement of species in *Mallada* or *Dichochrysa*), while in other cases the type designation is debatable and thus the original combination has been used here.

Page numbers in bold indicate that the taxon has been illustrated.

General index